Dielectrophoresis

Dielectrophoresis

Theory, Methodology and Biological Applications

First Edition

Ronald Pethig
Emeritus, University of Edinburgh,
United Kingdom

WILEY

This edition first published 2017

Registered Offices
John Wiley & Sons, Inc., 111 River Street, Hoboken, NJ 07030, USA
John Wiley & Sons, Ltd., The Atrium, Southern Gate, Chichester, West Sussex, PO19 8SQ, UK

Editorial Office
The Atrium, Southern Gate, Chichester, West Sussex, PO19 8SQ, UK

For details of our global editorial offices, customer services, and more information about Wiley products visit us at www.wiley.com.

Wiley also publishes its books in a variety of electronic formats and by print-on-demand. Some content that appears in standard print versions of this book may not be available in other formats.

Library of Congress Cataloging-in-Publication Data

Names: Pethig, Ronald.
Title: Dielectrophoresis : theory, methodology, and biological applications / Ronald Pethig.
Description: Hoboken, NJ : John Wiley & Sons, Inc., 2017. | Includes bibliographical references and index.
Identifiers: LCCN 2016043341 | ISBN 9781118671450 (cloth) | ISBN 9781118671412 (epub) |
ISBN 9781118671436) (epdf)
Subjects: LCSH: Dielectrophoresis. | Electrophoresis. | Electrophysiology. | Cells–Electric properties. |
Diagnostic imaging.
Classification: LCC QC585.7.D5 P48 2017 | DDC 534/.4–dc23 LC record available at
https://lccn.loc.gov/2016043341

Cover Images: Background image: Photo ephemera/Gettyimages;
Inset images: Courtesy of the author
Cover Design: Wiley

Set in 10/12pt WarnockPro by Aptara Inc., New Delhi, India

10 9 8 7 6 5 4 3 2 1

To my family
Angela, Richard & Jane, Helen & Jez, Nell, Max, Florence and William,
for their enthusiastic and constant loving support.

Contents

Index of Worked Examples

Preface

Herb Pohl's seminal book, *Dielectrophoresis: The Behavior of Neutral Matter in Nonuniform Electric Fields,* was published in 1978. The aim of this present text is to describe the development since then of the theory and practice of this subject. The primary focus is on the biomedical applications of dielectrophoresis (DEP), so many of the chapters are written with a multidisciplinary readership in mind. However, the theories and techniques described here are valid for all types of particles – animate and inanimate. The subject has changed dramatically since 1978. Up to that time only 16 scientific reports on biological applications of DEP had appeared in the scientific literature, with 12 of them describing work performed by Herb and his postgraduate students at Oklahoma State University. One of these papers deserves special mention, namely that written in 1966 with his MSc student, Ira Hawk. They describe, in the journal *Science* (vol. 152, 1966), the first demonstration of a purely physical technique (i.e., DEP) that can be used to distinguish and separate live and dead cells simultaneously. Furthermore, the live cells that had been exposed to the DEP field for several minutes were found to be viable and capable of cell culture. A macroscopic pin-plate electrode arrangement, composed of a rounded 0.66 mm stainless-steel wire facing a flat steel plate, was used in these experiments. Microfabrication and microfluidic techniques, taken for granted now in this subject, were not available to Herb in 1978. Apart from the impact of microtechnologies, this present book has also to take into account the fact that, at its time of completion (July 2016), more than 300 published papers are devoted solely to the DEP behaviour of yeast cells, with more than 3000 other papers of relevance to biomedical applications of DEP. Herb's initial interest in the motion of particles induced by nonuniform AC fields (an effect he was later to term *dielectrophoresis*) was directed towards industrial applications such as the removal of carbon-black filler from polyvinyl chloride samples. However, as I was privileged to witness at first hand, he gained most amusement from observing the DEP behaviour of bioparticles. In this way, Herb was able to describe in his book, in some detail, the DEP characterization of yeast cells and several types of bacteria, as well as preliminary results for blood cells, chloroplasts, green algae and mitochondria. These results act as the springboard for this book.

I suspect that I am not alone in finding more enjoyment in writing and reading about the biomedical applications of DEP than of its use to separate carbon black from PVC, or particulate matter from petroleum, for example. How can other such studies (potentially important as they may be) induce the same 'buzz' as viewing the geometrical distinction between life and death in the form of the Argand plots shown in Figure 11.9 of this book? Can studies of inanimate particles be as amusing as observing viable *Giardia* rotating in the opposite sense to nonviable ones in a rotating electric field? Such entertainment will not occur with particles extracted from oil, for example, unless they are bacteria such as oil-eating *Alcanovorax.* This explains, in part, why this present text is restricted to the DEP behaviour of biological particles. An exception is the inclusion of polymer beads because they are used widely in biomedical and biosensor devices, with DEP able to monitor the extent of attachment to them of target bioparticles. There is also a pragmatic reason for focussing on biomedical applications of DEP. A search in the autumn of 2015, using the *Web of Science Core Collection* and other library data bases, revealed the existence of at least 4000 publications on the theory, technology and application of DEP. Of relevance to the subject matter of this present text are also the many hundreds of scientific papers on the theories of dielectric phenomena, as well as those that describe the dielectric and electrokinetic properties of cells, bacteria, viruses together with bio-macromolecules such as proteins and nucleic acids. By largely excluding conference abstracts for possible citation, as well as papers not addressing a bio-related topic or not readily available through normal library resources, the number of candidates for citation was reduced to around 3000 publications. To avoid the text assuming the character of a list of disjointed citations, an attempt has been made to summarize the development of bio-DEP over the past half-century through only around 800 references to relevant work. This does not completely mirror important contributions to the subject made by my own co-workers and many researchers from other laboratories. I apologize to those who inspect the index of cited authors and are disappointed to find that their innovative work has either not been described adequately or is not cited at all. Among past colleagues not cited at all is John Morgan, who submitted his PhD thesis 'Dielectrophoretic Studies of Biological Materials' in 1978, whilst for Paul Carnochan only one image (Figure 11.2) from his PhD thesis 'Dielectric Properties of Biological Cell Suspensions', submitted in 1982, records his valuable contribution. An omission of work from the citation index does not reflect its perceived lack of novelty or importance – it has simply suffered from the culling exercise performed for reasons explained above (or from an unfortunate oversight on my part).

Another objective of this book is to make large parts of its content agreeably accessible to those trained in the biomedical sciences – not just engineering and physical science graduates. For those engaged in biomedical applications of DEP, the guidance and involvement of those trained in the molecular and life sciences is greatly desired and in most cases can be considered as essential. However, most published works on DEP appear in journals of engineering or the physical sciences and are largely unhelpful in addressing the 'so what, who cares?' questions of interest and relevance to those trained in the life and medical sciences. Chapter 1 addresses a common question about how the technique of DEP can compete against other microfluidic methods for cell manipulation and separation, such as flow cytometry, electrophoresis and magnetophoresis. Electrophoresis is a method well understood by biologists, but its similarity to the term *dielectrophoresis* is not helpful in discouraging the impression that DEP represents no more than an esoteric extension of what they already know. The purpose of Chapter 2 is to describe, in broad terms, how the special features of DEP lend to it the promise of providing important contributions to cell biology, particularly to such areas as drug discovery, medical diagnostics and regenerative medicine. As already stated, bearing in mind that an increasing number of scientists trained in the biomedical fields are entering the subject area, the nontheoretical sections of the text, throughout this book, are written in a style that is hopefully suitable for an interdisciplinary readership. To assist this and to help maintain the narrative, separate boxes and worked examples are used throughout the book to act as pedagogical material and to divert the more formal and quantitative details away from the main text.

In the preface of a special issue of the *Journal of Electrostatics* (Vol. 21, 119–364, 1988) to honour the memory of Herb Pohl, I mentioned that his devotion to science and generous nature had once been revealed to me by his statement that 'senior scientists should act rather as oak trees, to give shelter and provide growing conditions to the younger ones'. I also suggested that he would have gained much satisfaction and pleasure to see how some of his acorns had matured. In this spirit, I wish to take this opportunity to thank and acknowledge the contributions that the following, as young researchers at Bangor, made to my own understanding of DEP and to the content of this book: Talal Al-Ameen, W. Michael Arnold, Julian P. H. Burt, Paul Carnochan, Ka-Lok Chan, Colin Dalton, Peter R. C. Gascoyne, Andrew D. Goater, Clair Hodgson, Michael P. Hughes, Ying Huang, Richard S. Lee, Gary M. Lock, Zu-Hong Lu, Gerard H. Markx, Anoop Menachery, Hywel Morgan, John R. Morgan, Jonathan A. R. Price, Mark S. Talary, Xiao-Bo Wang and Xiao-Feng Zhou. It is with some pride that I know the DEP community will recognize the names of some fine oak trees in this list. We benefited from having the following with us on sabbatical leave or year-long fellowships: Ralph Hölzel, Takashi Inoue, Thomas B. Jones, Juliette Rousselet, Miguel Sancho, Herman P. Schwan and Junya Suehiro. A special mention should be given to John Tame, who operated the photolithography and clean-room facilities at Bangor. In the summer of 1986 he was asked if he could fabricate for us an array of gold, interdigitated, microelectrodes on a microscope slide. After being informed what we intended to do with it, he impishly responded: "We usually keep our electronic devices away from water, but I'll give it a try". For nearly 20 years thereafter (until his untimely death in 2004) he provided various microelectrode arrays for the DEP and electrokinetic studies of the researchers mentioned above. In 2005 a new clean-room facility at Bangor was named and dedicated to his memory.

At the School of Engineering in Edinburgh I have appreciated moral support and helpful interactions with Professors Alan Murray, Ian Underwood, Anthony Walton, as well as Drs Andrew Downes, Stewart Smith, Adam Stokes and Jon Terry. I enjoyed working with Colin Chung, Massimo Muratore and Srinivas Velugotla during their PhD research projects. I also acknowledge very fruitful research interactions at Edinburgh with Dr Paul de Sousa of the Centre for Clinical Brain Sciences, Dr Steve Pells of the MRC Centre for Regenerative Medicine and Professor Christopher D. Gregory of the MRC Centre for Inflammation Research. Chris Gregory deserves special mention – with the objective of making the text accessible to biologists he undertook the herculean task of reading many of the draft chapters, cleaning out much bio-unfriendly material. His research is directed towards understanding how apoptotic cancer cells condition their microenvironment. His input was thus especially valuable for those sections of Chapter 11 dealing with cell death – in so doing he performed what was not even required of Hercules, namely to replace the removed material. During the writing of this book I have appreciated valuable e-mail exchanges with Massimo Camarda, Cesare Cametti, Rodrigo Martinez-Duarte, Nic Green, Mike Hughes, Ralph Hölzel, Hywel Morgan and Joel Voldman. Special thanks also go to Anne Parkinson for translating Mossotti's paper of 1850 into English and so helping to clarify certain aspects of Chapter 6; Professor Peter Sarre of Nottingham University for facilitating my access to that university's original copy of George Green's *Essay*. I have greatly appreciated the efforts of the international team working for John Wiley & Sons, especially the meticulous attention to detail given by David Michael in copyediting the manuscript.

Those with sharp eyes might notice a 'smiley face' in Chapter 6. This is by way of a personal tribute to Herb Pohl, who in 1975 introduced me to this symbol and how to use it. Finally, the last lines on the dust cover of Herb Pohl's book of 1978 read: *A far wider range of potential applications exists than Professor Pohl has been able to include. The book should thus provide stimulating reading for imaginative research workers in the physical, medical and biological sciences*. It is probably pushing an ambition too far – but hopefully this present effort goes part way to achieving the same sentiment.

The electronic supplemental content to support use of this text is available online at http://booksupport.wiley.com.

Ronald Pethig
Edinburgh

Nomenclature

Symbol	Definition	Value/units
a, b, c	The semiaxes of an ellipsoid	m
A_n	Depolarization factor along axis n	
A	Unit area	m^2
α	Polarizability factor	$C\ V^{-1}\ m^{-1}$
α^*	Complex polarizability factor	$C\ V^{-1}\ m^{-1}$
B	Magnetic flux density	$Wb\ m^{-2}$
C	Molar concentration	$10^3\ mol\ m^{-3}$
C_{mem}	Cell membrane capacitance	$C\ V^{-1}\ m^{-2}$
CM	Clausius–Mossotti factor	
D	Electric flux density	$C\ m^{-2}$
D	Diffusion coefficient	$m^2\ s^{-1}$
De	Dean number	
D_H	Hydraulic diameter (wetted perimeter)	m
e	Charge on an electron	$(1.602 \times 10^{-19}\ C)$
E	Electric field	$V\ m^{-1}$
E	Electric field vector	$N/C\ (V\ m^{-1})$
f_{x01}	Low-frequency DEP crossover	Hz
f_{x02}	Second (high-frequency) DEP crossover	Hz
F_{DEP}	Dielectrophoretic force	N
F_e	Force exerted on a test charge in a field E	N
F_{MAG}	Magnetophoretic force	N
F_s	Sedimentation force acting on a particle	N
H	Enthalpy	J
i	Complex unit ($i^2 = -1$)	
$\hat{\imath}, \hat{\jmath}, \hat{k}$	Unit vectors along the x-, y-, z-Cartesian axes.	
Im[. . .]	Imaginary part of complex parameter inside brackets	
J	Current density	$A\ m^{-2}$
k	Boltzmann constant	$(1.3806 \times 10^{-23}\ J\ K^{-1})$
K_n	Knudson number	
m	Electric dipole moment	C m
N	Number density	m^{-3}
N_A	Avogadro's number	$(6.022 \times 10^{23}\ mol^{-1})$
p	Dipole moment	C m
$p^{(n)}$	Linear multipole of nth order	$C\ m^n$
P	Polarization vector	$C\ m^{-2}$
q	Electric charge (typically a test charge)	C
Q	Electric charge	C
r	Radial distance (Eqn 1.2)	m
$\hat{r}$	Unit radial vector	
R	Particle radius	m
R_f	Fluid channel flow resistance	$Pa\ m^{-3}\ s$

Symbol	Definition	Value/units
Re[. . .]	Real part of parameter inside brackets	
Re	Reynolds number	
S	Entropy	$\mathrm{J\,T^{-1}}$
t	Time	s
T	Surface tension	$\mathrm{N\,m^{-1}}$
T	Absolute temperature	K
U	Potential energy	J
v	Velocity (or volume)	$\mathrm{m\,s^{-1}}$ ($\mathrm{m^3}$)
V	Electric potential	$\mathrm{J\,C^{-1}}$ (V)
W	Work done	J
Z	Atomic number	
Z_e	Electrode impedance	Ohm (Ω)
χ	Susceptibility	
∇	Del vector operator	
δ	Dielectric increment (or decrement)	
$\Delta\varepsilon'$	Magnitude of dielectric dispersion	
$\varepsilon_{\mathrm{cyt}}$	Cytoplasm relative permittivity	
ε_{m}	Suspending medium relative permittivity	
$\varepsilon_{\mathrm{mem}}$	Membrane relative permittivity	
ε_{o}	Permittivity of free space	($8.854\times10^{-12}\ \mathrm{F\,m^{-1}}$)
ε_{r}	Relative permittivity of a dielectric	
ε_{w}	Bacterial cell wall permittivity	
ε_m^*	Complex permittivity of suspending medium	
ε_p^*	Complex permittivity of particle	
ε'	Real component [Re] of complex permittivity	
ε''	Imaginary component [Im] of complex permittivity	
ε_{s}	Low-frequency relative permittivity	
ε_{∞}	High-frequency relative permittivity	
ϕ	Angle (or electrical potential)	rad (V)
Φ_D	D-field flux	C
Φ_E	E-field flux	C
η	Dynamic viscosity	Pa s
λ	Charge density (linear)	$\mathrm{C\,m^{-1}}$
μ_{o}	Magnetic permeability of free space	($4\pi\times10^{-7}\ \mathrm{H\,m^{-1}}$)
μ_{e}	Electrophoretic mobility	$\mathrm{m^2\,V^{-1}\,s^{-1}}$
μ_{eo}	Electro-osmotic mobility	$\mathrm{m^2\,V^{-1}\,s^{-1}}$
θ	Angle	Radians
ρ	Charge density (volume)	$\mathrm{C\,m^{-3}}$
ρ	Mass density (Eqn (1.1)	$\mathrm{kg\,m^{-3}}$
σ	Charge density (surface)	$\mathrm{C\,m^{-2}}$
σ_p	Particle conductivity	$\mathrm{S\,m^{-1}}$
σ_m	Particle suspending medium conductivity	$\mathrm{S\,m^{-1}}$
σ_m^*	Complex conductivity of suspending medium	$\mathrm{S\,m^{-1}}$
σ_p^*	Complex conductivity of particle	$\mathrm{S\,m^{-1}}$
σ'	Real component [Re] of complex conductivity	
σ''	Imaginary component [Im] of complex conductivity	
τ	Relaxation time	s
τ_{MW}	Interfacial (Maxwell–Wagner) relaxation time	s
ω	Angular frequency	$\mathrm{rad\,s^{-1}}$
Ω_o	Rate of rotation	$\mathrm{rad\,s^{-1}}$
ξ (T_η)	Frictional torque	N m
ζ	Zeta potential	V

1

Placing Dielectrophoresis into Context as a Particle Manipulator

1.1 Introduction

For those interested in etymology, deciphering the origin and hence probable meaning of compound words ending in *phoresis* is relatively straightforward. Based on Greek translation, such compound words imply something to do with 'carrying things around'. For example, stating that *electrophoresis* describes an object being carried (i.e., moved) by an electrical effect is therefore an acceptable definition. For a particle to be set into motion requires the imposition on it of an external force. An example is the buoyancy force acting on a particle suspended in a fluid – the particle will either sink or rise under the action of gravity, depending on whether its specific density is greater or less than that of the surrounding fluid. If the particle finds itself in a flowing fluid, it will also experience a viscous drag force and be accelerated to the speed of the local fluid flow. The particle can be solid or take the form of a fluid droplet or gas bubble. This book's focus is the use of *dielectrophoresis* as a means to spatially manipulate bioparticles such as cells, bacteria, viruses, proteins and nucleic acids. In May 2013, in the United States, two sessions were devoted to this subject at an international conference on Advances in Microfluidics and Nanofluidics. In the flyer that promoted the conference, it was stated:

> As dielectrophoresis (DEP) is arguably one of the fundamental pillars of microfluidic manipulation and given the continued advances in this mature field, we will be organising special sessions on dielectrophoresis with the aim of promoting interaction between researchers that work on fundamentals and applications of DEP across different communities and disciplines.

Various methods can be used to manipulate particles in a microfluidic device, so what justification is there to state that dielectrophoresis can be singled out as 'one of the fundamental pillars'? Why is DEP considered an important topic for a conference on microfluidics and nanofluidics? Why devote a whole book to the subject? Some answers are provided in this chapter by reviewing those forces that can be used to manipulate bioparticles in microfluidic devices. It is not intended as a comprehensive review, but covers sufficient ground to set dielectrophoresis into context and highlight some of its special features and advantages. Bearing in mind that an increasing number of scientists trained in the biomedical fields are entering the subject area, the text is written in a style intended to be suitable for an interdisciplinary readership. To help maintain the word flow, boxes and worked examples are used in this chapter (and throughout the book) to divert the more formal and quantitative details away from the main text.

Dielectrophoresis is the induced motion of a particle when it is placed in an electric field gradient. In Chapter 2, we find that one advantage of this method is that it scales favourably with a reduction in dimensions of the electrodes used to generate the electric field. It is therefore ideally suited for applications in microfluidic devices designed to perform, for example, as an electronically controllable 'laboratory on a chip' or 'micro-total analysis' system. Although the terms are often used interchangeably, *lab on chip* is used to describe devices that integrate several laboratory processes, whereas *micro-total analysis systems* are considered to integrate *all* laboratory processes required for an analysis. For both cases, fluid flow in one or more channel networks, fabricated into or from a single solid substrate, is an essential element of the analytical or preparative function of the device [1–6]. It is also generally accepted that to qualify as a microfluidic device, at least one of its fluidic dimensions should be in the range 1 μm ~1 mm.

The fundamental features and potential advantages of using microfluidic devices for biomedical assays and processes will now be outlined.

Dielectrophoresis: Theory, Methodology and Biological Applications, First Edition. Ronald Pethig.

1.2 Characteristics of Micro-Scale Physics

A simple form of a microfluidic device would be, for example, a channel etched into a glass substrate of length 1 cm and internal cross section 10 μm × 10 μm, equipped with an inlet and outlet fluid port. One envisaged application of such a simple structure would be to study how thrombocytes (platelets) in a flowing fluid interact with immobilized proteins. The proteins can be immobilized by coating them onto the internal surfaces of the channel. With our specified dimensions the channel has an internal volume of 10^{-9} dm^3 (1 nL). One small droplet of water that leaks from a tap has a volume about 20 000 times larger than this! Physical effects or forces, such as surface tension that controls the size of a water droplet, may play relatively minor roles in our macro-scale world of activity, but can dominate in microfluidic devices. The ability to accommodate such forces, either by minimizing their disruptive effects or using them to advantage, is an important aspect of the design and operation of a microfluidic device.

The following are practical examples of dominant physical phenomena at the micro scale:

- Microfluidic devices tend to have a large ratio of their surface area to volume. Consider a spherical chamber of radius R. This has a surface area of $4\pi R^2$ and a volume of $(4\pi R^3)/3$. The ratio of these two parameters is $3/R$. Therefore, as the radius R decreases the ratio of surface area to volume increases. For example, a 10 dm × 10 dm × 10 dm cube has a surface-to-volume ratio of 40 m^{-1}, whereas for the 1 cm × 10 μm × 10 μm channel considered above, this ratio increases to 4×10^5 m^{-1}. Scaling down the dimensions of a fluidic device thus provides the opportunity for suspended particles to interact with a large surface area. This can represent a desired outcome, as in the study of platelet-protein interactions, or lead to an undesirable result such as the adventitious adherence of cells to the internal walls of narrow-bore tubing.
- In micro devices, capillary action and other surface energy effects can be greater than gravitational forces. This can result in an upward or transverse fluid movement, or even block downward fluid flow in a capillary.
- A small drop of fluid placed in the inlet of a microfluidic device can evaporate very rapidly.
- Fluids that are brought together in a microfluidic circuit do not mix easily. Any mixing that does occur arises mainly from the diffusion of solutes across the boundaries between separate laminar flows of fluid.
- Solute particles that are heavier than the surrounding fluid settle to the channel bottom very quickly.
- Small fluid volumes will almost immediately take on the temperature of the environment and cool down or heat up very quickly.

1.2.1 Exploiting Micro-Scale Physics

Some of the physical effects outlined above may be undesirable in the design and operation of microfluidic devices. However, they can also be exploited as powerful tools. Examples of such physical effects and their advantages include:

- Fluid flow in microchannels is almost always laminar, characterized by the parallel flow of the individual lamellar elements of the fluid (see Chapter 12, section 12.4.5). The flow has a parabolic velocity profile across a channel, with zero velocity at each channel wall and a maximum velocity at the centre of the channel. These characteristics can be turned to advantage in the design of particle separation devices, where an external force drives target particles into different parts of the fluid velocity profile or across the boundary between adjacent fluid streams.
- A large surface-to-volume ratio provides an intrinsic compatibility between the use of a microfluidic system and surface-based assays.
- At micro dimensions, diffusion becomes a viable approach to move particles, mix fluids and control reaction rates. Small drug molecules, for example, can diffuse at rates of ~10 mm/s at 25 °C in aqueous solutions. This allows the establishment of controlled concentration gradients in flowing systems, as well as rapid and complete equilibration of small molecular weight particles across a microchannel. Relatively fast reaction times are therefore possible when molecular diffusion lengths are of the order of the microchannel dimensions.
- Unaided by centrifugation, sedimentation becomes a viable means to separate dispersed particles by density across small channel dimensions. For example, red blood cells will sediment in a 100 μm deep channel in about 1 minute and generate a 50 μm layer of plasma in the process.
- Evaporation of small quantities of fluids can be extremely rapid because of a typically large surface-to-volume ratio. This effect can be used for the concentration of suspended particles.
- The energy associated with surface tension can be used to drive liquids through microfluidic devices. By chemically treating the surfaces of a microchannel to be hydrophilic, water will be driven through the channel without any applied pressure. This flow is driven by

the attractive energy between the water and the channel wall surface. Surface tension effects can be controlled electrically using the technique known as *electrowetting on dielectric* (EWOD) described later in this chapter.
- It is possible to design passive fluidic devices that utilize inherent properties of the fluid and its microenvironment (e.g., capillary force, evaporation, heat transfer, diffusion) for fluid movement, mixing, heating, cooling and catalysing chemical reactions. Thus, disposable stand-alone devices can be designed that require no external power source or instrumentation, yet still perform many, if not all, of the functions typically associated with full-scale automated chemical analysis devices containing pumps, mixers and heating elements.

Other advantageous characteristics of microfluidic devices are associated with economic considerations. For example, polymer-based microfluidic structures can be mass produced at very low unit cost, allowing them to be disposable. Micro devices require only small volumes of sample and reagents (down to picolitres) and produce only small amounts of waste. They are also amenable to high throughput by processing multiple samples and assays in parallel.

1.3 Microfluidic Manipulation and Separation of Particles

A critical action for many biological and medical diagnostic procedures is the selective manipulation and separation of particles. By *manipulation* we mean the relocation of a particle with respect to its position within a fluidic device or to that of neighbouring particles. An extension of this is particle *separation*, which implies the physical isolation within or outside of a device of the target particles from a mixture of different particle types.

1.3.1 Defining the Performance of Cell Manipulators and Separators

The performance of a particle manipulator can be measured in terms of how quickly, precisely, reproducibly and how many target particles can be relocated to a specific site or sites. The concept of a particle separation device is quite straightforward – the input is a heterogeneous mixture of particles and the output consists of target particles that can be collected or totally isolated from the unwanted particles. The performance of such a device is often given in terms of its *throughput*. Throughput can be expressed in terms of either the volumetric flow rate that can be handled by the device (e.g., mL/s) or the number of particles that can be processed in a given time (e.g., 100 000 cells per second). However, we also want to know how well the device performs as a separator. If for every 100 000 cells in a sample there are 100 target cells, can it deliver 100 target cells per second with no contamination from unwanted cells? The language employed to evaluate this is not as straightforward as defining its throughput. For example, suppose we have cells that have been brought back to physiological temperature after a period of cryopreservation. It is common experience to find that many of these cells will have suffered as a result of being frozen for a long time and will be nonviable (dead). We will want our cell separation procedure to provide us with as many viable cells as possible and very few dead ones. An evaluation of this is variously called the recovery rate, the target capture efficiency or the yield, which should be close to 100%. If we define the viable, live, cells as the target cells and the dead ones as the unwanted cells, the yield (or recovery rate, capture efficiency) can be defined as the ratio of the number of viable cells collected at the output to the total number of viable cells contained in the original cryopreserved sample fed into the separator:

$$\text{Yield} = \frac{\text{Target cells}_{output}}{\text{Target cells}_{input}}$$

We can determine the yield parameter by performing viability assays for the input and output samples and use this to compare the use of new buffers and cryopreservation procedures designed to maintain cell viability. However, if we are using a cell separator to isolate cancer cells from peripheral blood (to assess chemotherapy treatment, for example) we have (as yet) no accurate method to determine the number of cancer cells that exist in blood samples taken from patients. In this case, the use of terms such as yield, recovery rate or capture efficiency cannot be used to describe the performance of a separator used to extract cancer cells from blood. However, if a known number of cancer cells are 'spiked' into a blood sample of defined volume, it is possible to determine the yield of a procedure to isolate them.

When using a separator to isolate specific cells for therapy, an important parameter to define is the *purity* of the output. This can be defined as the fractional content of target cells in the output sample:

$$\text{Purity} = \left(\frac{\text{Target cells}}{\text{Target cells} + \text{Unwanted cells}}\right)_{output}$$

Flow cytometry can be used to determine the concentration of the target cells and the total cell count. In some cases a cell separator is used to increase the percentage of target cells in a heterogeneous cell population prior to

PCR analysis, for example. This is referred to as *enrichment* and is quantified using the following relationship:

$$\text{Enrichment} = \frac{(\text{Target cells/Unwanted cells})_{output}}{(\text{Target cells/Unwanted cells})_{input}}$$

The subtle relationship between enrichment, yield (or recovery rate, capture efficiency) and purity can be obtained by combining the relationships given above for yield and enrichment, to give:

$$\text{Enrichment} = \text{Yield}\frac{(\text{Unwanted cells})_{input}}{(\text{Unwanted cells})_{output}}$$

1.4 Candidate Forces for Microfluidic Applications

Some forces scale down well for applications in microfluidic devices and others do not. For example, centrifugation is used to separate particles based on differences in their size and density. A centrifugal force occurs as a reaction to centripetal acceleration ($R\omega^2$) acting on a mass moving with angular velocity ω (radians / s) along a path of radius of curvature R. This does not scale favourably with miniaturization, because with all other factors remaining constant a reduction of R results in a reduction of the centrifugal force. Another factor that does not favour the incorporation of centrifugation into lab-on-chip devices is that it requires a rotating component. To subject a cell suspension to the relatively gentle centrifugal acceleration of 100 times the gravitational acceleration (100 g) requires that the sample is positioned at the end of an 18 cm long arm that is rotating at ~1000 revs/min. This rotation rate has to be increased a hundredfold if we wish to maintain the centrifugal force but reduce the rotor arm length to 18 μm for example.

Examples of forces that do scale well with miniaturization include those that induce electrophoresis and dielectrophoresis. These effects become more dominant as the applied electric field is increased and for a fixed voltage applied across an electrode pair the field increases with reduction of the spacing between them. Indeed, using microelectrodes and modest applied voltages of 10 V or less, electric fields as large as 10^6 V/m can be generated, together with useful associated electrokinetic forces that are difficult to achieve at a macroscale. As the dimensions of a fluidic system, such as channel width and height, decrease the influences of laminar flow increase and can be exploited in various ways. The forces that can usefully be scaled down to operate in microfluidic devices are given in Table 1.1. Examples of how each one of these forces can be used to selectively manipulate and separate cells will now be outlined:

Table 1.1 Forces used to manipulate and separate cells in microfluidic devices.

Type of force	Example	Basis of selectivity
Mechanical	Filtration/sieving	Size and deformability
Hydrodynamic	Field flow fractionation Inertial forces Hydrophoresis	Size and various physico-chemical properties
Acoustic	Acoustophoresis	Size, density and compressibility
Optical	Optophoresis Laser tweezers	Refractive index
Electrical	1. Electrophoresis 2. Dielectrophoresis 3. FACS[a]	1. Surface charge 2. Dielectric polarization 3. Antibody label
Magnetic	i) Magnetophoresis ii) MACS[b]	i) Para- or diamagnetic ii) Antibody/ferritin label
Surface	Cell rolling Cell patterning	Cell adhesion

Notes: [a]FACS: fluorescence activated cell sorting; [b]MACS: magnetic activated cell sorting.

1.4.1 Mechanical

Particles can be selectively separated according to their size by mechanical filtration or sieving. The particles are suspended in a fluid (in which they do not dissolve) and are then flowed through microstructured perforations or constrictions. Membranes composed of pure nylon fibres or polycarbonate containing pores of precise diameter have been widely used to separate blood cells by size and deformability. Such membranes can easily clog, so that the recovery of target cells from them is not straightforward. For such reasons efforts have been directed towards replacing conventional membranes with physical structures such as weirs and arrays of microposts or pillars that are built into fluidic channels. In a weir structure the microchannel takes a sudden reduction of cross section. Particles that are too big to pass through such structures are retained. In some cases, especially at high fluid flow rates, the deformability of a particle determines whether it is captured or not. Particle deformity is an important consideration for the selective retention of blood cells because, as shown in Table 1.2, there are overlaps of the size distributions of the various blood cells.

An informative investigation of the mechanical filtration of erythrocytes (red blood cells) from leukocytes (white blood cells) in whole blood was performed by Wilding *et al.* [7]. Blood was passed between microposts or over weirs etched into a silicon substrate in a chamber capped with a glass top. This is shown schematically in

Table 1.2 The volume (fL = 10^{-15} L), diameter, density, concentration ranges and properties of cells in normal adult human blood.

Cell	Volume diameter	Density (g/cm^3)	Cell count (cells/mL)	Properties
Erythrocyte	95 ± 20 fL 7.5~8.0 μm	1.09–1.10	$4.8 \pm 0.9 \times 10^9$	Biconcave disks, anucleated
Erythroblasts	270–4000 fL 8~20 μm	1.07~1.08	$<1 \times 10^4$	Nucleated, (immature erythrocytes)
Platelet	9 ± 6 fL 1~3 μm	1.04–1.06	$2.7 \pm 0.5 \times 10^8$	Round or oval
Leukocytes Lymphocyte (90% population) (10% population)	 500~2000 fL 10~12 μm 12~16 μm	 1.055–1.065	 $2.7 \pm 1.3 \times 10^6$	 Mononucleated (small cytoplasm content) (larger cytoplasm content, often with large granules)
Granulocytes	500~2500 fL			
Neutrophil	12~15 μm	1.075–1.085	$4.6 \pm 2.9 \times 10^6$	Polynucleated, very granulated cytoplasm
Eosinophil	12~17 μm	1.075–1.085	$2.3 \pm 2 \times 10^5$	Polynucleated, larger granules than neutrophils
Basophil	10~14 μm	1.075–1.085	$7 \pm 6 \times 10^4$	Polynucleated, intermediate granularity
Monocytes	900–4000 fL 12~20 μm	1.055–1.065	$4 \pm 2 \times 10^5$	Mononucleated (nucleus often irregular or kidney shaped)

Sources: Bain, B. J. (1995) *Blood Cells: A Practical Guide.* 2 edn. Blackwell Science, Osney Mead. De Waele, M., Foulton, W. and Renmans, W. (1988) Hematologic values and lymphocyte subsets in fetal blood. *Amer. J. Clin. Prac.* **89**, 742–746. Lynch, D. C., Yates, A. P. and Watts, M. J. (1996) *Haematology*, Churchill Livingstone, New York, NY.
Notes: Cell counts for adults vary due to demographic factors (e.g., sex, age, ethnic origin and geographical location) and biological factors (e.g., diurnal variation, pregnancy, menstruation, menopause, exercise, cigarette smoking, alcohol intake). About 70% of the lymphocytes are T cells (approximately two-thirds CD4 and one-third CD8), 5–10% are B cells and the remainder are non-T, non-B-cells.

Figure 1.1. The objective was to demonstrate that isolation of leukocytes from the erythrocytes, followed by the polymerase chain reaction (PCR) for the DNA released

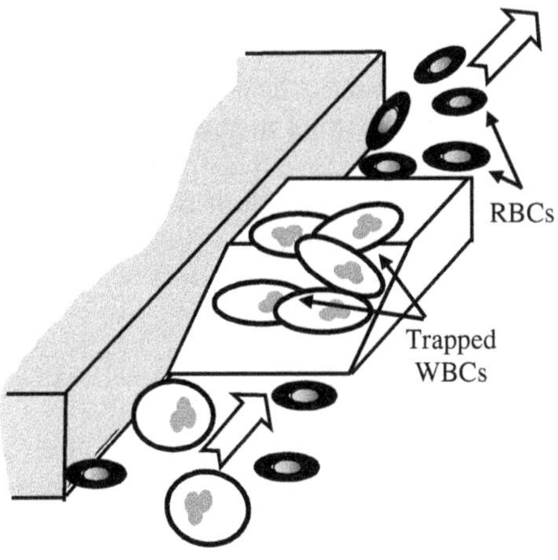

Figure 1.1 Schematic of a weir-type microfilter used to separate blood cells. A small gap between the top of the weir and a glass cover plate provides active filtration of cells based on size and deformability. In this example white blood cells (WBCs) are trapped on the weir whereas red blood cells (RBCs) flow freely over it. (Based on Wilding *et al.* [7].)

directly from the trapped leukocytes, could be performed as sequential processes in a single microfluidic chamber. Removal of the erythrocytes from the whole blood sample was required because haemoglobin protein molecules that can leak from them inhibit the PCR process.

Wilding *et al.* [7] found that sieving of blood cells was influenced by several factors, namely: the deformability of the cells; their concentrations; the pressure applied to produce the fluid flow; the viscosity of the fluid; and the physical gap between microposts and above the silicon weirs. Erythrocytes readily passed through gaps as small as 3 μm, whereas the larger leukocytes (diameters in excess of 15 μm) could only squeeze through gaps larger than 7 μm. The filtration mechanism shown in Figure 1.1 was presumed to rely on trapping the leukocytes in the narrow gap between the top of the silicon weir and the Pyrex glass cover, but cell adhesion may also have played a role.

In other studies, Mohamed *et al.* [8] demonstrated the use of a micromachined silicon device for separating foetal cells from maternal blood, based on differences in cell size and deformability. The device consisted of four sections of successively narrower channels along the flow axis. These channels did not take the form of continuous structures with side walls, but as a series of pillars. In total the device contained more than three million of such 'channels'. This design allowed the cells to deform and recover as they passed between channels and to migrate around regions where the cell flow was locally hindered or clogged. The nucleated foetal erythrocytes, ranging in diameter from 9 to 12 μm, could deform and pass through a channel as small as 2.5 μm wide and 5 μm deep. The larger leukocytes, ranging in diameter from 10 to 20 μm, could not deform to the same extent and were retained by the 2.5 μm wide and 5 μm deep channels. Later studies, using the same filtering device, demonstrated that cultured cancer cells spiked into whole blood could be recovered, based solely on their size and deformability [9].

1.4.2 Hydrodynamic

1.4.2.1 Basic Principles

The separation of particles using hydrodynamics often relies on the principle that macroscopic particles subjected to viscous drag forces in laminar fluid flow will stay within their fluid streamlines. An extreme example of this is the way that the coloured strands of a certain brand of toothpaste remain in place and do not mix together as the paste is squeezed slowly from its tube (see Figure 12.9). We can classify this as a *deterministic* effect – the individual streamlines of paste will flow in a predictable way. If particles within a streamline are small enough to be buffeted about by the thermally induced motions of the fluid's molecules, we have a *stochastic* process because they will diffuse in a random manner across adjacent fluid streamlines. In a nondeterministic, stochastic regime we are unable to separate particles according to the principles described in this section.

All fluid flow, whether in a channel or around an object, can be broadly classified as either laminar or turbulent. As described in Chapter 12 (12.4.5), which of these fluid flow conditions is dominant depends on the relative importance of the inertial forces versus viscous shear forces in the flow. An inertial force is the concept we use to understand the principle of inertia as embodied in Newton's First Law of Motion (an object not subject to any net external force moves at a constant velocity). In fluidics we can relate this to the translational momentum (mass × velocity) of a unit volume of a fluid element. This is given by the product (ρv) of the fluid density ρ and the bulk velocity v of the fluid flow. A fluid element can also have rotational inertia, which refers to the fact that the angular momentum of the fluid element will remain unchanged unless an external torque is applied. The shear forces that act to damp out translational and angular momentum of the fluid are viscous in nature. They occur at the channel walls and between fluid streamlines. The ratio of the inertial forces and viscous shear forces is a dimensionless parameter, known as the channel Reynolds number Re, given by the relationship:

$$Re = \frac{\rho v L}{\eta} \tag{1.1}$$

in which η is the fluid's dynamic viscosity and the parameter L is the effective wetted (hydraulic) diameter of the channel. For low values of Re the viscous damping caused by shear at the channel walls and between fluid streamlines quickly removes translational and rotational kinetic energy from a fluid element and the flow is laminar. Laminar fluid flow is characterized by a parabolic velocity profile (see Figure 1.3). As a rough guide, described in more detail in Chapter 12, for a Reynolds number above ~1000, the shear between streamlines is unable to dampen out the inertia of transverse and rotational fluid motions. As a result, the laminar streamline structure is destroyed and the fluid flow becomes turbulent throughout the channel. For aqueous fluids we have $\rho \sim 10^3$ kg m^{-3} and $\eta \sim 10^{-3}$ Pa s, so that the factor (ρ/η) in Equation (1.1) has a value of $\sim 10^6$ m^{-2} s. In microfluidic devices we typically have flow velocities much less than 1 cm/s and L rarely exceeds 10 cm. In Equation (1.1) we therefore have $vL < 10^{-3}$ m^2/s, giving us a Reynolds number of less than 1000. Unless a very high fluid flow velocity is achieved (driven by high pressure in a channel fabricated to withstand such pressure) it is not possible to induce high Reynolds number conditions. Therefore, as a working rule, we can assume that flow in a microfluidic device is laminar. In such flow the fluid stream velocity is zero at the boundary layer next to a channel wall or a wetted object's surface and increases with distance away from such boundaries.

Equation (1.1) defines the Reynolds number for the fluid flow inside a channel in the absence of suspended particles or fixed objects such as posts. We can define a Reynolds number Rp that includes details of a particle by multiplying Re by the dimensionless parameter (R^2/L^2) where R is the radius of the particle:

$$Rp = \frac{R^2}{L^2} Re = \frac{\rho v R^2}{\eta L} \tag{1.2}$$

For low values of the particle's Reynolds number ($Rp \ll 1$) viscous drag of the fluid will act over its surface and accelerate it to the local velocity of a laminar fluid streamline. For $Rp \sim 1$ inertial forces can lift the particle away from a channel wall and also cause it to cross fluid streamlines. If the walls of the fluid channel are not straight but curved, rotational flow called Dean flow [10], caused by fluid vortices induced by the channel curvature, can also cause particles to cross streamlines and alter their position in the channel. This effect is characterized by the Dean number *De*:

$$De = \delta^{1/2} Re \tag{1.3}$$

In Equation (1.3), δ is the ratio $L/2R_c$, with R_c being the radius of curvature of the channel of hydraulic diameter L. The position of a particle in a curved channel can be influenced by drag associated with Dean flow when $De > 1$. It has also been demonstrated that particles can be separated according to their size using slanted obstacles in a channel [11,12]. Transverse flow streams perpendicular to the direction of the main fluid flow result from an anisotropic fluidic resistance of the slanted obstacles. Their top or bottom areas represent a higher flow resistance than their side areas. Particles subjected to the lateral pressure gradients induced by this flow resistance anisotropy are moved from the one sidewall of a channel to the other sidewall. This effect has been coined 'hydrophoresis' and is defined as the movement of suspended particles under the influence of a microstructure-induced pressure field [11].

Whether or not a particle (defined by its centroid) follows a fluid flow stream depends also on the rate at which it can diffuse in that fluid. The constant thermally induced motions of molecules in fluids ensures that when one fluid stream is placed adjacent to another one, as in laminar flow, its molecules percolate between flow streams in a process called diffusion. It is convenient to separate the actual diffusion process into two conceptual transport mechanisms: a molecular process modelled as a statistical random walk that is proportional to the degree of kinetic energy in the system and an advective process in which molecules are carried along by the average velocity of the flow. The common practice is to restrict the word *diffusion* to describe the first process and label the second process *advection* (*convection* if heat is being transferred). As described in Chapter 12 the relative importance of these two conceptual transport mechanisms is given by the Peclet Number P_e, the ratio of advection and diffusion:

$$P_e = \frac{vL}{D} \tag{1.4}$$

where D is the particle's diffusion coefficient. If $Pe < 1$ the diffusion rate is smaller than the advection rate over the characteristic distance L. The flow is deterministic because a particle is confined to a fluid streamline. Cells suspended in aqueous fluids exhibit D values of the order 10^{-15}~10^{-16} m^2/s (see Table 12.3). From Equation (1.4) such values of the diffusion coefficient ensure that when working with cells Pe will greatly exceed 1.0 in any practical microfluidic device. Thus, in low Reynolds number fluid flow and in the absence of an externally applied force cells will remain within their fluid streamlines.

1.4.2.2 Practical Examples of Applying Hydrodynamic Forces

When a microchannel splits into two channels that carry different fluid flow rates, the suspended particles will follow different streamlines depending on their locations in the channel leading up to the bifurcation. This is governed by the ratio of the fluid flow rates in the bifurcations and the difference in the shear force that acts on the surface of a cell when near to the channel wall compared to that when it is in the middle of the channel. This shear difference contributes to a lift force that pushes a cell away from direct contact with the channel wall. This is known as the Bifurcation Law or Zweifach–Fung effect, depicted in Figure 1.2 and has been explored by Yang

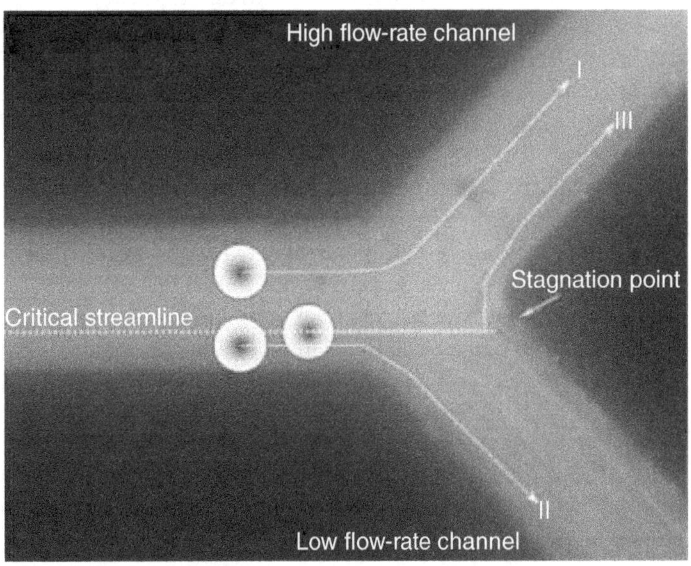

Figure 1.2 Schematic of the Zweifach–Fung effect where a particle with its centroid on the critical streamline is directed to the high flow rate channel in a bifurcation. (Yang *et al.* [13] reproduced with permission of the Chemical Society.)

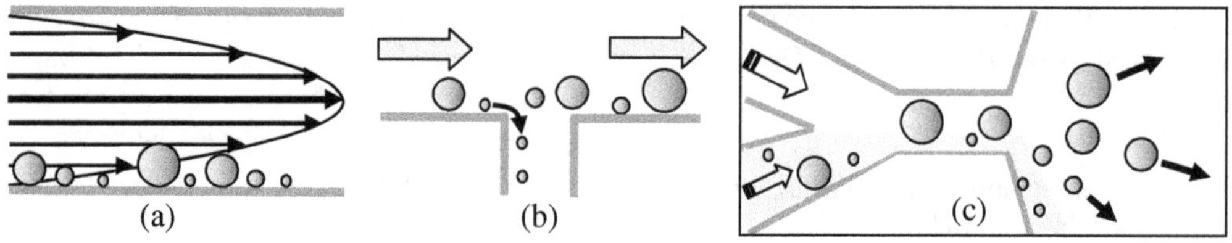

Figure 1.3 (a) Particles of various sizes are shown close to the wall of a channel that sustains a pressure-driven laminar flow of fluid. The parabolic profile of the fluid velocity results in the larger particles moving more rapidly and eluting from the channel more quickly than the smaller particles. This is the basis of steric field flow fractionation [14–16]. (b) If the fluid flow rate into a side channel is sufficiently low, only the smaller particles contained within the fluid flowing near the sidewall will be withdrawn into it from the main channel [17]. (c) Particles can be separated according to their size by a spreading flow profile at the exit of a pinched section of channel joining two flow streams. The extent of particle separation is controlled by adjusting the flow rates in the two inlets to the pinched section. (Based on Yamada *et al.* [19].)

et al. [13] as a method to separate blood plasma from blood cells.

As shown in Figure 1.3(a), in the situation where particles in a laminar fluid flow are forced close to a channel wall, the larger particles will have a larger velocity than the smaller ones. The larger particles will on average be in the faster flowing streamlines because their centroids cannot be closer to the wall than their radii. This is the basis of the various forms of field flow fractionation (FFF) pioneered by Giddings [14, 15] and extensively reviewed by Roda *et al.* [16]. In FFF a force (field) is applied perpendicularly to the parabolic flow to drive the particles (or analytes) to a channel wall or into different laminar flows due to differences in their size, density and other physical features such as shape, rigidity and surface properties. The applied field can be a centrifugal force or electrical field, for example. Because the larger particles will be swept downstream more rapidly they can be selectively collected from the fluid exit of the channel before the smaller particles are eluted. This mode of elution is referred to as steric or hyperlayer FFF, respectively, depending on whether the particles are all brought to a wall of the channel or distributed into different laminar streams, respectively. For submicron particles, where diffusion down their concentration gradient becomes a dominant process, the elution order of analyte size and mass is reversed. This is known as the normal elution mode.

Yamada *et al.* [17, 18] have exploited the properties of laminar flow for the size-selective filtration of leukocytes from blood and size-dependent separation of liver cells. The operating principle is outlined in Figure 1.3(b), which shows a narrow side channel branching off from a broad main channel. The fluid flow rates in the main and side channel will depend on their effective fluidic resistance values and the pressures applied to drive them. When the relative flow rate into the side channel is sufficiently low, only the fluid stream near the main channel wall will be withdrawn into the side channel. As shown in Figure 1.3(b), in this flow state particles whose diameters are larger than a certain value will not enter the side channel, even if they are flowing near the wall of the main channel and have a diameter smaller than the cross sectional size of the side channel. Depending on their size, shape and other factors such as surface roughness, the particles will also experience lift forces that push them away from the channel wall. This will influence the selective filtration of the particles.

Yamada *et al.* [19] also introduced the concept of pinched flow fractionation (PFF) outlined in Figure 1.3(c). A liquid containing the particles to be separated by size is continuously introduced into another channel containing particle-free fluid. By adjusting the flow rates of these two fluids, the particles can be restricted to flow at one sidewall of a 'pinched' section of the fluid channel. At the end of the pinched section, where the flow velocity profile spreads out, the larger particles are directed toward the centre of the channel and the smaller particles are directed within their slower moving stream lines towards the sidewall. Consequently, slight differences of particle elevations in the pinched channel section are significantly amplified in the broadened outlet. Particles are separated according to their size by deterministic lateral displacement in the laminar flow.

Davis *et al.* [20] have described an interesting version of size-dependent particle separation by deterministic lateral displacement in a process they refer to as *bumping*. As depicted in Figure 1.4, the particles flow through an array of microposts. Each row of posts is offset laterally with respect to the preceding row. Particles below a critical diameter follow streamlines cyclically through the gaps, moving in an average forward flow direction. Particles above this critical diameter cannot follow such a streamline and are 'bumped' by hydrodynamic lateral drag into the sequential streamline at each post. Thus, such particles do not move parallel to the fluid flow but at an angle determined by the ratio of post offset to row-row spacing. Davis *et al.* have demonstrated that this procedure can fractionate whole blood by separating the erythrocytes from the leukocytes and allow them to be

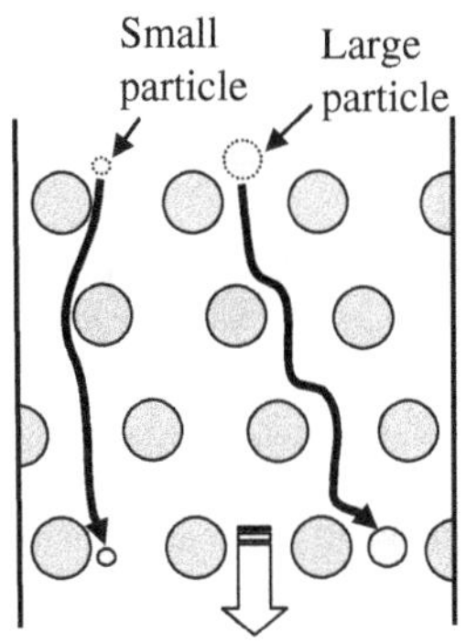

Figure 1.4 The size-dependent separation by deterministic displacement of particles flowing through an array of microposts. Small particles stay within a flow stream that skirts the surface of posts in adjacent rows, whereas large particles are displaced laterally at each post. The extent of lateral separation of the particles is determined by the centre-to-centre post separation, the diameter of the posts, and the relative shift of the post centres in adjacent rows (based on Davis *et al.* [20]).

collected in separate fluid exit ports [20]. By modifying this method of deterministic lateral displacement, Holm *et al.* [21] were able to separate from human blood the living parasites (trypanosomes) that cause sleeping sickness.

As the rate of fluid flow is increased in curved channels and the particle size is increased relative to the channel diameter, the particle Reynolds number given by Equation (1.2) increases so that inertial effects can become significant. Interesting examples of this have been demonstrated by Di Carlo *et al.* [22] for the flow of particles in curved channels. The superposition of lift forces from the channel walls with centrifugal forces arising from the fluid and particles was observed to induce precise ordering of initially scattered particles both longitudinally along the direction of fluid flow and laterally across the channel. This inertial self-ordering effect is shown schematically in Figure 1.5. A noteworthy application of the combined effects of inertial forces and Dean flow is a spiral microfluidic device, shown in Figure 1.5, for separating asynchronous mammalian cell lines according to their cell cycle [23]. This was achieved by exploiting the relationship between cell diameter and cell cycle and provided enriched subpopulations of viable cells in the G0/G1, S and G2/M phases. A comprehensive theoretical study and modelling of inertial focusing dynamics in spiral microchannels has been reported by Martel and Toner [24]. They conclude that the rich variety of inertial focusing dynamics observed in curved channels offers the potential of wide applications and advantages for future generations of microfluidic devices. Further study is also required to elucidate the underlying physical mechanisms and their associated limitations.

1.4.3 Acoustic

An acoustic radiation force [25] can be exerted on a particle in a fluid stream using an ultrasonic transducer located at the wall of the channel. The form of transducer commonly consists of piezoelectric ceramic rings sandwiched and bolted between two metal blocks. When a DC voltage is applied to the piezoceramics, they expand and the pressure applied to the blocks is transmitted into the bulk of the fluid. An applied AC voltage causes the transducer to vibrate at the frequency of the applied voltage and this frequency is higher than that detectable by human ears. The longitudinal pressure wave created in the bulk fluid will travel at the speed of sound, which for water is around 1500 m/s. The wavelength λ is related to the frequency f and speed v by the relationship $\lambda = v/f$. So, for the case where $f = 15$ MHz and $v = 1500$ m/s, we have a wavelength of 100 μm. If the transducer faces a sound reflector, the interactions between the emitted and the reflected acoustic wave create regions of minimum and maximum pressure amplitude changes (nodes and antinodes, respectively). This effect is maximized by

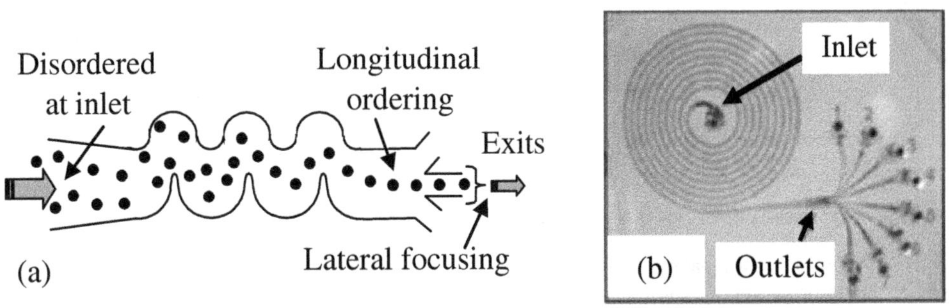

Figure 1.5 (a) The continuous inertial focusing, ordering and separation of particles can be achieved by increasing the particle Reynolds number in a microchannel designed to have asymmetrically curved walls. The combination of lift forces at the channel walls and centrifugal forces acting on the particles and fluid can result in both longitudinal ordering and lateral focussing of particles (based on Di Carlo *et al.* [22]). (b) The spiral microfluidic design developed for cell cycle synchronization by Lee *et al.* [23]. (Reproduced with permission of the Chemical Society.)

placing the sound reflector a multiple of half wavelengths from the transducer, to create a resonant standing wave [26].

The acoustic radiation force consists of two components. One component involves the gradient of the potential energy of the wave interacting with the compressibility difference between the particle and fluid, whilst the other involves the gradient of the kinetic energy interacting with the specific density differences between the particle and fluid [25]. The particles experience a force that is directly proportional to their volume – in other words large particles experience a greater acoustic force than small ones. Whether or not a particle is directed towards or away from a standing wave pressure node depends on its density and compressibility compared to that of the surrounding medium. Cells are of greater density and are less compressible than an aqueous medium and both of these factors result in their being directed to where minimal pressure amplitude changes occur – i.e., towards a pressure node. Particles of the same density as the surrounding medium will move towards a pressure antinode if they are more compressible than the medium. An illustration of the degrees of particle separation created in a half-wavelength acoustic standing wave is shown schematically in Figure 1.6. In this illustration the acoustic force is generated at right angles to the direction of the fluid flow. The suspended particles interact with the acoustic force as they flow along the channel and become spatially separated according to their size, density and compressibility. Different fractions of the particle mixture can then be collected downstream through different fluid exit ports [27,28].

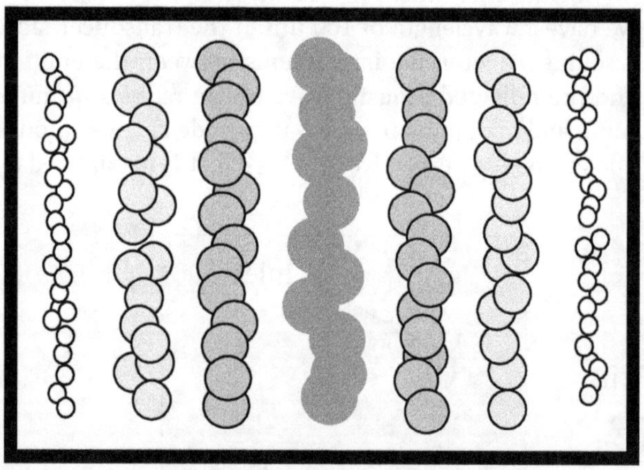

Figure 1.6 A schematic of particles moving at different rates towards the pressure node at the centre of a half-wavelength acoustic standing wave, according to their size, density, compressibility, and acoustic pressure amplitude (based on Petersson *et al.* [27]). The acoustic force is generated across the channel, orthogonal to the fluid flow direction (into the page). The different particle fractions can be collected downstream from several exit ports [27,28].

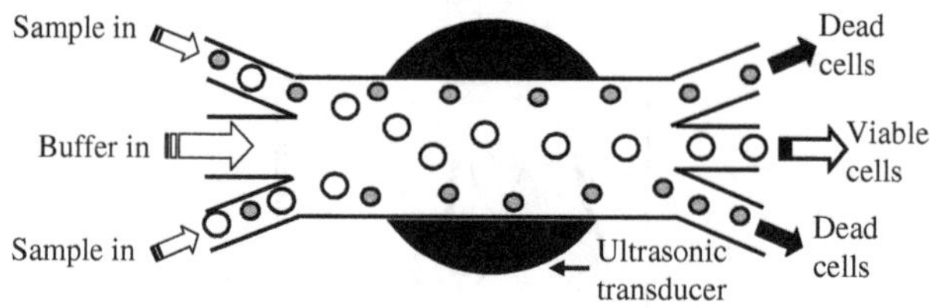

Figure 1.7 Viable (white) and dead (black) cells are separated by injecting them together with a buffer solution at a 1 : 3 flow ratio into a fluidic channel. An acoustic radiation force preferentially focuses the larger viable cells toward the acoustic pressure node within the centre streamline but is insufficient to move the smaller dead cells away from the channel walls. The viable and dead cells are collected from separate exit fluid ports (based on Yang and Soh [29]).

In an aqueous suspension of mixed viable and dead (nonviable) cells the dead ones tend to be more buoyant (less dense) and smaller than the viable ones. In an acoustic standing wave the viable cells should therefore have a greater tendency to move towards a pressure node than the dead ones. This has been demonstrated, as depicted in Figure 1.7, for the case of viable and dead breast cancer cells in a cell separation device constructed by Yang and Soh [29]. Acoustophoresis appears to be particularly well suited for processing fluids of high cell content in microfluidic devices. For example, whole blood is particularly difficult to flow through microchannels because of its high viscosity and clogging tendency. However, Lenshof *et al.* were able to produce plasma from whole blood in a sequential blood cell removal procedure in an acoustic force microdevice having a multiple outlet configuration [30]. The quality of the resulting plasma fulfilled the standard defined by the Council of Europe for plasma transfusion, namely an erythrocyte concentration less than 6×10^6/mL. Furthermore, the plasma was directly linked in the device to a microarray for the detection of a prostate specific antigen via fluorescence readout without any signal amplification at clinically relevant levels of 0.19 to 21.8 ng/mL.

These applications of acoustophoresis have used bulk acoustic standing waves, which require the microchannel walls to exhibit good acoustic reflection properties such as those fabricated from silicon. However, microfluidic devices are commonly made of polymeric materials using soft lithography techniques, which have poor acoustic reflection properties. Attaching a piezoceramic transducer to the interior of a microchannel is also not straightforward. To overcome these limitations, surface acoustic waves (SAWs) can be generated using microelectrodes that have been deposited onto a piezoelectric substrate. SAWs are sound waves that propagate along the surface of an elastic material, with most of its acoustic radiation force confined within a thin layer close to the substrate surface. The energy density of a SAW is therefore high, making it an effective way for to focus, separate and direct particles or fluid droplets in a microchannel. Reviews of such applications of surface acoustic waves have been provided by Yeo and Friend [31] and Wang and Zhe [32].

1.4.4 Optical

A beam of photons, in other words a light beam, carries momentum. If a beam of photons is reflected off a particle's surface, the particle experiences a radiation pressure related to the optical momentum transfer to it. A qualitative assessment of the potential magnitude of this has been given by Ashkin [33] as follows:

> By focusing a laser beam of modest power, about 1 W, to a spot size of about a wavelength λ, one can subject a dielectric sphere 1 μm in diameter to the very high light intensity of about 10^8 W/cm^2. Assuming the light is reflected from the sphere with an average reflectivity of 10%, one achieves an acceleration of approximately 10^6g, where g is the acceleration due to gravity.

Ashkin and his colleagues at the Bell Laboratories went on to demonstrate [34, 35] that particles can be trapped and manipulated using highly focused laser beams, in a technique now generally referred to as *optical tweezers*. Of particular relevance to us is that they were able to demonstrate [35] that infrared laser beams could trap and manipulate individual cells such as yeast, *E. coli* and red blood cells without damaging them. The cells could be manipulated by optophoresis at velocities up to 100 μm/s. Organelles within individual protozoa could also be manipulated. A similar effect has been demonstrated using the DEP tweezer shown in Figure 10.44, which was able to manipulate chloroplasts located inside *Eremosphaera* cells.

An understanding of how particles can be optically trapped is relatively straightforward for particles of diameter significantly larger than the wavelength of the light beam. In this case, as shown in Figure 1.8, simple optics can be used to follow the paths of individual rays of light as they are refracted on entering and leaving a dielectric sphere. This refraction results in a light ray leaving the sphere at a different angle to that of its incidence, which in turn implies that there has been a change of momentum of the light ray. According to Newton's Third Law of Motion there will be an equal and opposite momentum change on the sphere. We say that there has been a transfer of optical momentum from the light beam to the sphere. For most practical examples of an optical tweezer the intensity of light from the laser has a Gaussian profile of intensity, as depicted in Figure 1.8 (a), composed of a transverse combination of an electrical and magnetic field (an electromagnetic wave) in what is termed a TEM mode. If a spherical particle of refractive index greater than the surrounding medium is displaced from the central axis of maximum light intensity, the rays of greater intensity will impart a larger momentum change towards the beam centre than that imparted away from the centre by the rays of weaker intensity. This imparts a lateral force on the sphere towards the maximum beam intensity. Once located at the centre of a focused beam, as shown in Figure 1.8 (b), there is a symmetrical refraction of the individual light rays and hence no net lateral force. This cancels out the scattering force of the laser beam and results in the stable trapping of the sphere just below the point of focus of the laser beam. A spherical particle of refractive index less than that of the surrounding medium will be pushed away from the laser beam.

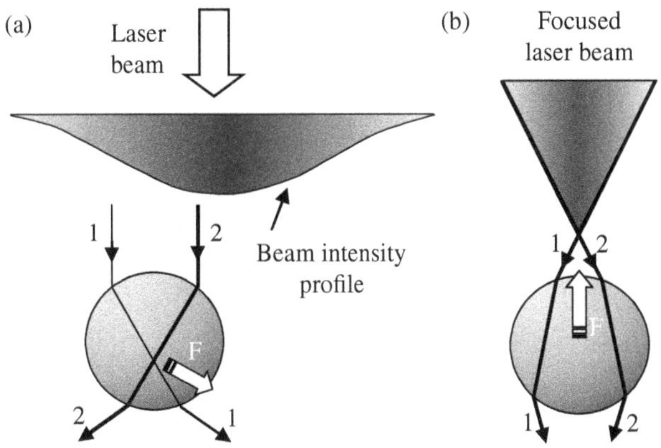

Figure 1.8 (a) A particle of refractive index larger than the surrounding medium is located away from the axis of maximum light intensity of an incident laser beam. Because the optical momentum transfer of a refracted ray such as ray 2 is larger than that of a weaker ray 1, a resultant force F acts on the particle to direct it towards the high-intensity region of the laser beam. (b) The refracted rays of a particle subjected to a focused light beam are symmetrical. The net force F acting against the optical momentum transfer to the particle thus has no lateral component and acts against the force of gravity and light scattering towards the focus point.

MacDonald *et al.* [36] have demonstrated the principle of an optical particle sorter using a diffractive laser beam splitter to create a three-dimensional, dynamically reconfigurable, lattice of optical interference patterns. The sorting by size of drug delivery microcapsules was demonstrated (see Figure 1.9), as well as the sorting by refractive index of polymer and silica particles. This work demonstrates the main advantage of optophoresis, namely that the light beam can be focussed deep within the fluid sample to give three-dimensional control of the

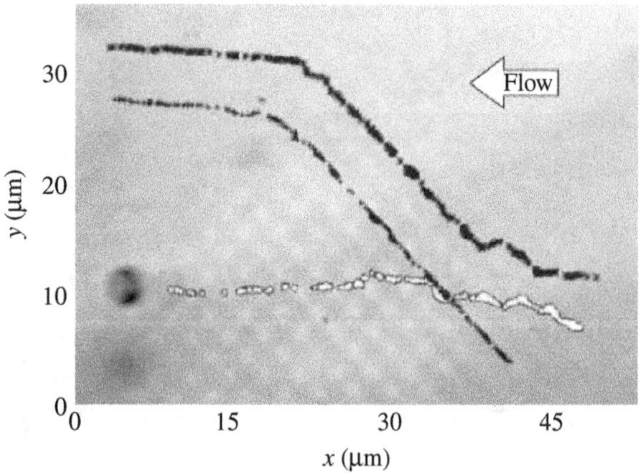

Figure 1.9 The optical fractionation by size of protein microcapsules as they flow from right to left across a three-dimensional optical lattice created by passing a laser beam through a diffractive beam splitter. The video track (white) of a 4 μm diameter capsule shows no deflection, whereas significant optophoretic angular deflection is seen from the black tracks of two 2 μm diameter capsules. (MacDonald *et al.* [36], reproduced with permission of Nature Publishing Group.)

position of a trapped particle. Multiple traps can be produced using separate steering mirrors and beam splitters. The development of holographic optical tweezers [37] makes it possible to trap hundreds of particles in complicated patterns, with simultaneous and easy control of the axial position of the traps.

1.4.5 Electrical

An electric field, E, is created in the space surrounding a distribution of electric charges at *rest*. This field will exert a force $F = qE$ on any other charge q that is present in the field. In other words, the electric charge q does not itself contribute to the electric field E acting on it. The electric force on a positive charge ($+q$) acts along the same direction as the field vector, whilst that on a negative charge ($-q$) acts against the field vector direction. The electric force does not act at an angle to the electric field. If the electrical charges responsible for creating the field E are moving, they create an additional magnetic field B. The resultant electric and magnetic force acting on a particle in such a combination of fields is considered in section 1.4.6.

1.4.5.1 Flow Cytometry

As the name implies, *flow cytometry* is the measurement of cells in a fluid flow system. Cells are delivered one at a time through a microchannel past a point of measurement. The earliest form of flow cytometer uses the Coulter counting principle, as invented by Wallace H. Coulter and disclosed in his US Patent (2 656 508) of 1953. This principle relies on the fact that particles moving in an electric field cause measurable disturbances of the field that are proportional to the volumes of the particles. The practical requirements are that the particles should be suspended in a conducting liquid; the electrical field should be physically constricted so that the movement of particles in the field causes detectable changes in the electric current; and the concentration of the particles should be low enough that they pass one at a time through the physical constriction, preventing an artifact known as coincidence. Direct current (DC) or low frequency alternating (AC) fields are used, so that a viable cell with an intact membrane will appear as an insulating particle and cause a significant transient change in the electrical resistance of the fluid when detected in the constricted channel. By monitoring the pulses in electric current, the number of particles for a given volume of fluid can be counted. The size of the electric current pulse is related to the size of the particle, enabling a particle size distribution to be measured. The most important application of the Coulter counter has been in the characterization of blood cells and is the standard method for obtaining red and white blood cell counts for medical diagnostic purposes. Morgan *et al.* [38, 39] have made significant progress in developing microimpedance cytometry with integrated microfluidics for the detection of bacteria and for performing a full blood count in clinical samples. This demonstrates its potential clinical utility for point-of-care diagnostic purposes.

The flow cytometry method mostly used in research laboratories is fluorescence-activated cell sorting (FACS). An essential element of this method is application of the electric force, given by the relationship $F = qE$, to physically separate cells according to the charge added to a fluid droplet that contains each one. Cells are delivered one at a time in a fast flowing buffer sheath past a point at which a laser beam is focussed. Light scattered by the cells at right angles to the laser beam (side scatter) and light scattered in a forward direction (forward scatter) is measured. Side scatter is mainly influenced by the optical homogeneity (granularity) of the cells, whilst the forward scatter can provide a good estimation of cell size. Fluorescence induced by the laser light is also measured and, depending on the type of instrument, up to ten different fluorescent wavelengths can be monitored. Some algae and bacteria may naturally fluoresce, but otherwise fluorescent chemicals attached to antibodies are used to selectively bind to and so tag specific surface proteins (antigens) on the cells. Fluorescent chemicals and fluorescent nanocrystals (known as quantum dots) can also be used to tag cell components such as nuclei, DNA and chromosomes. Droplets of the buffer solution are then created using a piezoelectric vibrator (as in an inkjet-printer head) so as to encapsulate a single cell in each droplet. The next steps are where an electrical effect and force are introduced. Each droplet passes through a metal ring to which an electric charge is applied. The magnitude and polarity of this charge is controlled by the fluorescence wavelength detected and hence on the antibody label and phenotype of the droplet-encapsulated cell. A charge of polarity opposite to that applied to the ring is induced on the droplet, which then passes through an electric field generated between two charged metal plates. The electric force acting on a charged droplet deflects it into a designated collection tube, whilst uncharged droplets drop directly into a waste container. This employs the so-called *forward* sorting algorithm, where droplet-encapsulated cells are directed to the waste exit port until a threshold fluorescence signal is detected and an appropriate voltage signal is applied to the droplet charging ring. This can slow down the process of seeking rare cells in a sample and multiple measurements of the same cell are not possible. Sample volume throughputs for different FACS instruments range from 12 μL/min to 60 μL/min. For high throughput applications, 1–2 μL volumes are robotically sampled from each 10 μL well volume of a 96-well plate, with cell number densities up to 1000 per μL. A schematic of a basic FACS instrument is shown in Figure 1.10 and represents a simplification of details provided by Shapiro *et al.* [40].

Conventional FACS instruments are costly, technologically complex and require trained personnel to operate them. Clogging of the cell exit nozzle, the sterilization of the fluid chamber between runs and malfunctions of the droplet-encapsulation and charging processes are quite common reasons for instrument down time. Relatively large sample volumes are also required (typically a few μL rather than pL or less) and significant background fluorescence arising from the cell suspension medium and chamber material can be present. Fu *et al.* [41, 42] were the first to report efforts to avoid these problems by exploiting microfluidics to produce an integrated micro-FACS device, bringing with it many of the advantages listed in section 1.2.1. The planar configuration of the device allowed the use of high numerical aperture optics and so increased the sensitivity of fluorescence detection. The small size of the laser interrogation site also reduced the background fluorescence from the media and channel materials. A novel 'reverse' sorting algorithm was implemented by Fu *et al.* [41,42], in which the cells were driven at a high rate by pressure-driven flow from the input to the waste reservoir. Upon fluorescent detection of a target cell, the flow was stopped and the cell sent back to the input. When the cell passed through the detection region a second time, it was directed into the collection channel and fast transport of cells from the input to the waste was resumed. Due to the simple fabrication process and inexpensive materials, these micro-FACS devices can be disposed so as to eliminate any cross-contamination from previous runs. A similar microfluidic cell sorting system has been developed by Dittrich and Schwille [43] who employed fluorescence correlation spectroscopy to detect the tagged cells and pulses of electro-osmotic force, rather than pressure-driven flow, was used to deflect the cell stream into waste and hold reservoirs. Cho *et al.* [44] have further advanced the development of micro-FACS technology by developing a device that operates by integrating microfluidics, optics and acoustics to achieve a throughput better than 1000 cells/s. This was achieved using an integrated piezoelectric lead-zirconate-titanate actuator, with a response time of ~0.1 ms, to hydrodynamically manipulate sub-nanolitre volumes of fluid in which a single cell is suspended.

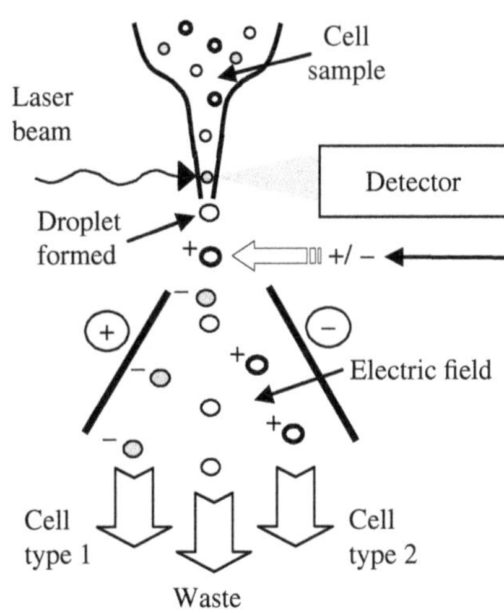

Figure 1.10 A simplified schematic is shown of a FACS instrument. Cells flow one at a time through a fluorescence measurement site before each one is encapsulate in a droplet. The droplet is charged according to the fluorescence signal of its entrapped cell, and is then deflected by an electric field into an appropriate reservoir. This example shows the separation and collection of two cell types, with removal as waste of cells that have not been 'recognized' by the two flourorescent antibody probes.

1.4.5.2 Electrophoresis and Dielectrophoresis

These two electrokinetic effects are introduced in detail in Chapter 2 and will not be dealt with in depth at this point. In brief, electrophoresis is the effect where a particle carrying a net charge Q is induced to move in an electric field. In most practical situations the field is uniform and does not vary with time. The electric force $F = QE$ acting on the particle accelerates it until there is a balance between the frictional viscous force exerted by the surrounding fluid and the electric force. At this point the particle will move at a steady-state velocity v.

The electrophoretic mobility μ_e of a particle is defined as the velocity v that is induced per unit electric field and so is given as $\mu_e = v/E$. All mammalian cells so far studied exhibit a *negative* electrophoretic mobility under normal physiological conditions – they move in the *opposite* direction to that of the applied electric field. This corresponds to the cells carrying a net negative charge. Some bacteria have been observed to exhibit positive electrophoresis and so carry a net positive charge. The μ_e values for different mammalian cells at physiological pH generally lie in the range from -0.5×10^{-8} to -3.5×10^{-8} m^2 V^{-1}s^{-1}. The most studied cell type is the human erythrocyte, with an established μ_e value of -1.1×10^{-8} m^2 V^{-1}s^{-1} and is often used to calibrate electrophoresis equipment as a standard reference. A useful insight of the magnitude of steady-state velocities and cell migration distances typically observed in cell electrophoresis measurements can be obtained by expressing the electrophoretic mobility in units of μm/s per volt/cm. Thus, for a cell with $\mu_e = -1.0 \times 10^{-8}$ m^2 V^{-1}s^{-1}, it will travel 0.6 mm towards the anode after 60 s exposure to a field of 1000 V/m in a stationary fluid. Spatial separation and hence purification of a heterogeneous cell sample by electrophoresis should therefore be possible for μ_e differences between the target and other cells of 10–20%.

Large DC fields are typically used in electrophoresis. This can result in unwanted motion of the fluid medium as a result of heat-induced convection. This can be avoided by using a gel, instead of an aqueous solution, for suspending the particles between the electrodes. For example, an agarose matrix can be used to separate mixed populations of DNA and RNA samples according to their fragment lengths. Nucleic acid molecules are negatively charged and shorter fragments will exhibit larger electrophoretic mobilities because they can migrate more quickly through the pores of the agarose gel than the longer fragments. This is an example of molecular sieving. An added advantage is that the separated entities remain fixed in position after removal of the electric field. Proteins are separated according to their net charge, because the pores of an agarose gel are too large to sieve the proteins according to their size. An immobilized pH gradient can be established in a gel by adding a mixture of ampholytes (molecules that can act as either an acid or a base). Different proteins can then be separated in a method, known as isoelectric focusing or electrofocusing, which takes advantage of the fact that the net charge on a protein depends on the pH of its surrounding medium. A pH value will exist, known as the isoelectric point, pI, where a protein carries no net charge. A protein that is in a pH region below its pI value will be positively charged and migrate towards the cathode. At pH values above the isoelectric point it will be negatively charged and migrate towards the anode. A mixture of proteins will therefore be focused into narrow bands on the gel, each one positioned at a location in the pH gradient that corresponds to the pI point of each protein type. Very high resolution is possible, corresponding to spatial separation of proteins whose net charge differs by just one electronic charge. Gel electrophoresis can also be used to sort nanoparticles.

Dielectrophoresis depends on the fact that a particle, for example a cell, when exposed to an electric field can be electrically polarized and assume the form of an electric dipole moment **p**. Equal and opposite charges ($+Q$ and $-Q$) will appear on opposite sides of the particle, so that for a particle of diameter d the induced dipole moment is given by $\mathbf{p} = Qd$. As explained in Chapter 2, the orientation of the moment, with or against the field direction, will depend on whether the electrical polarizability of the particle is greater or less than that of the surrounding medium. If the electric field is uniform the positive and negatively charged poles of the moment will each experience the electric force $F = QE$. The polarized particle will experience a torque and align its induced electric poles with the field and in order to minimize its electrical potential energy. However, the particle will not be induced to move laterally. If the particle is located in an electric field gradient, the two poles ($+Q$ and $-Q$) of the moment **p** will not experience the same electric force and the particle will move. This induced motion is called dielectrophoresis and the force F_{DEP} responsible for inducing this motion is given by:

$$F_{DEP} = (\mathbf{p} \cdot \nabla)E \tag{1.5}$$

This equation is derived in Chapter 2 (Box 2.4). The symbol ∇ is a mathematical device known as a differential operator. It 'operates' on the electric field E so as to provide a quantitative way to define how the electric field gradient varies in three dimensions. As an analogy we obtain the equivalent of a 3D image of a mountain, so we can judge how the gradient of ascent (or descent) depends on the route we take. Equation (1.5) provides the magnitude and direction of the resultant dielectrophoretic force acting on a small polarized particle. This in turn provides us with the means to determine the direction and induced acceleration of the particle. In Chapter 2 (Boxes 2.4 and 2.5) it is further shown that the dielectrophoretic force depends on the parameter ∇E^2. This square law dependence on the electric field indicates that we can apply either a direct current (DC) voltage, or an alternating current (AC) voltage, to the electrodes in a dielectrophoresis device. The dielectric properties of a cell can vary significantly over the frequency range from 10 kHz to 10 MHz and they can also be sensitive to subtle changes in the physiological state of a cell. The implications of these in terms of the theory, experimentation and biomedical applications of dielectrophoresis are described throughout this book.

Example 1.1 Dielectrophoretic Force Acting on a Cell at a Low Frequency
A 5 V, 1 kHz, voltage applied to an electrode creates an electric field and field gradient of 2.6×10^4 V/m and -2.5×10^8 V/m^2, respectively, at a distance of 20 μm from the electrode edge. At 200 μm from the electrode edge the corresponding field and field gradient values are 1.3×10^4 V/m and -2×10^7 V/m^2, respectively. (The negative field gradient value reflects the fact that the field decreases in magnitude with distance from the electrode.) The dielectric properties of a 10 μm diameter cell and the surrounding aqueous medium are such that the induced dipole moment of the cell is -4×10^{-21} $C.m$ and -2×10^{-21} $C.m$, at a distance of 20 μm and 200 μm, respectively, from the electrode edge. (A negative induced moment implies that the polarizability of the cell is less than that of the suspending medium. An induced moment of absolute magnitude $\sim 10^{-21}$ $C.m$ appears to be very small, but is in fact enormous when compared to that of the permanent dipole moment of a water molecule (6×10^{-30} $C.m$) or a protein molecule ($5 \sim 10 \times 10^{-28}$ $C.m$).)

Calculate the dielectrophoretic force acting on the cell at a distance of 20 μm and 200 μm from the electrode edge.

Solution 1.1 The dielectrophoretic force F_{DEP} is given by Equation (1.5):

$$F_{DEP} = (\mathbf{p} \cdot \nabla)E$$

At 20 μm from the electrode ($\mathbf{p} = -4 \times 10^{-21}$ C.m, $\nabla E = -2.5 \times 10^8$ V/m^2):

$$\begin{aligned} F_{DEP} &= (-4 \times 10^{-21} \text{C m})(-2.5 \times 10^8 \text{ V m}^{-2}) \\ &= 1 \times 10^{-12} \text{ C V m}^{-1} = 1 \text{ pN}. \end{aligned}$$

At 200 μm from the electrode:

$$\begin{aligned} F_{DEP} &= (-2 \times 10^{-21} \text{C m})(-2 \times 10^7 \text{ V m}^{-2}) \\ &= 4 \times 10^{-2} \text{ pN} \end{aligned}$$

(Conversion of C V m^{-1} to newtons makes use of the relationship $F = qE$.)

A newton (1 N) is the force required to accelerate a mass of 1 kg at the rate of 1 m per second per second. A force of 1 pN (10^{-12} N) will thus have a significant effect when acting on a cell typically of mass $\sim 10^{-12}$ kg.

Example 1.2 Dielectrophoretic Force Acting on a Cell at a High Frequency
The frequency of the 5 V signal applied to the electrode described in Example 1.1 is changed from 1 kHz to 10 MHz. At this frequency the induced dipole moment of the cell is 8×10^{-21} C m and 4×10^{-21} C m, at a distance of 20 μm and 200 μm, respectively, from the electrode edge. (The change from a negative to positive moment indicates that the cell is now more polarizable than the surrounding medium. This implies that the electric field now penetrates into the cell interior and that the medium is less conductive than the cytoplasm.)

Calculate the dielectrophoretic force acting on the cell at a distance of 20 μm and 200 μm from the electrode edge.

Solution 1.2 We will assume negligible polarization effects at the electrode, so that the change of electrical frequency will not alter the field and field gradient values. Thus, at 20 μm from the electrode with $\mathbf{p} = 8 \times 10^{-21}$ C m and $\nabla E = -2.5 \times 10^8$ V/m^2:

$$\begin{aligned} F_{DEP} &= (8 \times 10^{-21} \text{ C m})(-2.5 \times 10^8 \text{ V m}^{-2}) \\ &= -2 \text{ pN} \end{aligned}$$

At 200 μm from the electrode:

$$\begin{aligned} F_{DEP} &= (4 \times 10^{-21}\ C.m)(-2 \times 10^7\ V.m^{-2}) \\ &= -8 \times 10^{-2}\ pN \end{aligned}$$

The negative value indicates that the dielectrophoretic force will direct the cell up the field gradient towards the high field at the electrode edge. This is known as *positive* dielectrophoresis. (The positive forces derived in Example 1.1 indicate that the cell is directed by *negative* dielectrophoresis down the field gradient and away from the electrode.)

Examples 1.1 and 1.2 highlight two important features, namely that a dielectrophoretic force acting on a cell will vary significantly as a function of distance away from an electrode edge and that the effective electrical polarizability of a cell can vary greatly in both magnitude and polarity as a function of the applied electric field frequency.

1.4.5.3 Electrowetting on Dielectric (EWOD)

The surface tension force described in Box 1.1 can be modified and controlled electrically. An early demonstration of this, of relevance to lab-on-chip technologies, was given by Pollack *et al.* [45] who described the manipulation of discrete microdroplets along a linear array of electrodes. This could provide the means to integrate microfluidic systems without the need for conventional pumps, valves or channels. Such systems would be flexible, power efficient and capable of performing complex and highly parallel microfluidic processing tasks. To achieve this effect a voltage is applied, as shown in Figure 1.12, between a conducting liquid droplet (e.g., an

Box 1.1 Surface Tension and Young's Equation

Surface tension is a significant and useful force in microfluidic devices. Its origin lies in the difference between the attractive forces acting on a molecule in a bulk liquid and at a free surface between the liquid and air. A molecule in the fluid bulk experiences mutually attractive (van der Waals) forces with neighbouring molecules and, for aqueous solutions, hydrogen-bond forces are also significant. A molecule at the surface is attracted by a reduced number of neighbours and so has a raised potential energy. The creation of a new liquid surface is thus energetically costly. This is why small volumes of fluid assume a spherical shape, and trickles of water break up into spherical droplets, to minimize the total surface area.

If U is the total cohesive energy per molecule in the fluid bulk, then this is halved to a value of $U/2$ for a molecule located at a flat surface. The surface tension created per unit area of surface is related directly to this cohesive energy reduction. For a characteristic molecular dimension R, the effective molecular area is R^2 and the surface tension is $U/(2R^2)$. Surface tension is thus directly proportional to the intermolecular attraction and inversely proportional to the molecular size. Water has a significantly larger surface tension than oils and alcohol, for example, reflecting not only the relatively small size of the water molecule but also the cohesive energy supplied by hydrogen-bonds in bulk water.

Surface tension T is defined as the ratio of the surface force F to the length d along which the force acts ($T = F/d$) and thus has units of force per unit length (equivalent to energy per unit area) and acts tangentially to the free surface. As shown in Figure 1.11, a drop of liquid on a solid surface has three interfaces, namely the solid-liquid, the liquid-air and the solid-air interface. A line on the solid surface (the xy plane) defines the boundary separating these three interfacial areas. The contact angle θ is defined as the angle formed at this three phase boundary between the tangent to the liquid surface and the xy plane. A tension exists in each interface, and different values for these result in different liquids adopting different contact angles relative to different solid surfaces. An equilibrium situation exists when the horizontal components of the surface free energies balance. From Figure 1.11 the equilibrium condition is readily seen to be described by the following relationship, known as Young's equation:

$$T_{S-A} = T_{S-L} + T_{L-A}\cos\theta \tag{1.6}$$

where $T_{S\text{-}A}$, $T_{S\text{-}L}$ and $T_{L\text{-}A}$ are the surface tensions at the solid-air, the solid-liquid and the liquid-air interfaces, respectively and θ is the contact angle defined above. A liquid with low surface tension (low surface energy) resting on a solid surface of higher surface energy will spread out on the surface forming a contact angle θ less than 90°. The liquid is said to wet the surface – if the liquid is water we say the surface is hydrophilic. If the surface energy of the liquid exceeds that of the solid, the liquid will form a bead and θ will have a value between 90° and 180°. In this case we have a nonwetting liquid relative to the surface, corresponding to a hydrophobic surface when considering aqueous liquids.

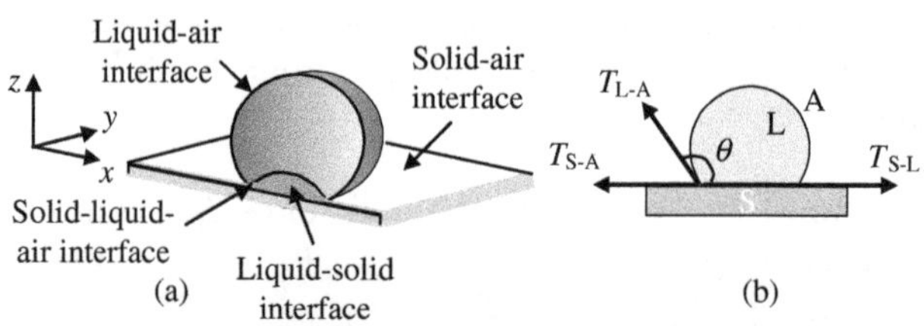

Figure 1.11 (a) A drop of liquid in air on a solid surface has three interfaces (liquid-solid, liquid-air, solid-air). (b) The contact angle θ is defined as the angle formed between the tangent to the liquid surface and the xy-plane at the boundary between the three interfaces. At equilibrium the horizontal components of the surface free energies balance, and this is expressed in the form of Young's equation (Equation (1.6) in Box 1.1).

electrolyte) at rest on a dielectric layer and a counter electrode positioned below this dielectric. The electrode wire shown in Figure 1.12 was replaced by Pollack *et al.* [45] with a linear array of interdigitated electrodes formed of thin chrome layers on glass. For maximum EWOD effect the dielectric surface should be of poor wettability, manifested as a contact angle greater than 90° at the interface between the dielectric and the liquid. For an aqueous droplet this is achieved with a hydrophobic surface. The resulting charge that accumulates at the dielectric-liquid interface leads to a change in contact angle from above 90° to less than 90°, as shown in Figure 1.12. This is equivalent to a transition from a nonwetting to a wetting state. This effect is known as electrowetting on dielectrics (EWOD). A theoretical treatment of this is given in Box 1.2.

Example 1.3 EWOD – Voltage Control of Contact Angle

In an EWOD device of the form shown in Figure 1.12, a water droplet exhibits a contact angle of 110° on a dielectric surface with no voltage applied to the counter electrode. The dielectric is a PTFE layer of thickness 5 μm of relative permittivity 2.0. What voltage should be applied

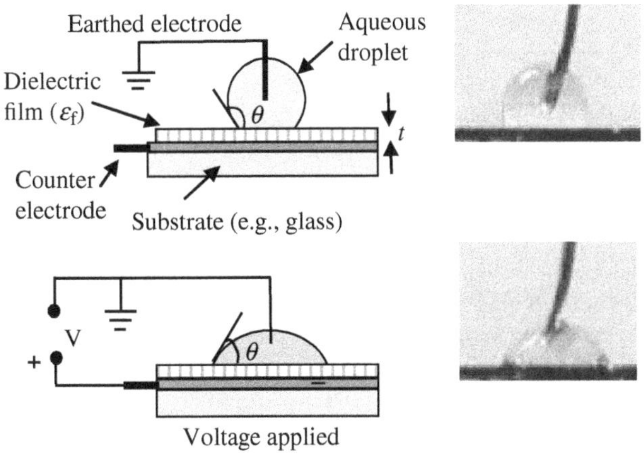

Figure 1.12 Electrowetting on dielectric (EWOD) is achieved by applying a voltage between a conducting liquid droplet at rest on a thin dielectric film of poor wettability, situated above a counter electrode. The resulting charge that accumulates at the film-liquid interface results in the contact angle θ falling from above to below 90°, which is equivalent to a transition from a hydrophobic to a hydrophilic state if the droplet is aqueous.

to the counter electrode to change the contact angle to 70°?

Solution 1.3 The voltage is calculated using a rearrangement of Equation (1.9) in Box 1.2:

$$V = \sqrt{\frac{2tT_{L\text{-}A}\,(Cos\theta(V) - Cos\theta(0))}{\varepsilon_o \varepsilon_r}}$$

Using a value for $T_{L\text{-}A}$ of 70×10^{-3} Nm^{-1} (the surface tension of the water-air interface) we calculate the voltage required to reduce the contact angle from 110° to 70° as follows:

$$V = \sqrt{\frac{2(5 \times 10^{-6}\,m)(70 \times 10^{-3}\,Nm^{-1})\,(Cos\,70° - Cos\,110°)}{(8.854 \times 10^{-12}\,Fm^{-1}) \times 2}}$$
$$= 164\,\text{V}$$

(To derive the unit of Volt requires unravelling $\sqrt{(\text{Nm/F})}$ by equating force (newtons N) in units of coulomb volt/metre and from capacitance (farad F) being defined as coulomb / volt.)

From Solution 1.3 we find that a voltage of 164 V is required to lower θ from 110° to 70° for an aqueous droplet ($T_{L\text{-}A} = 70 \times 10^{-3}$ Nm^{-1}) if PTFE (Teflon) of thickness $t = 5$ μm is used as the dielectric. The field across the dielectric is V/t and so in this case has a value of 32 MV m^{-1}, which is not much lower than the dielectric strength value of 60 MV m^{-1} for PTFE. Increasing the voltage can result in charge injection into the dielectric, followed by electrical breakdown. Equation (1.9) indicates that the EWOD effect is enhanced if the dielectric permittivity ε_r is increased. Some polymers can be vapour deposited as thin hydrophobic dielectric films – a common example being various forms of poly(p-xylene) known as parylene that have values for ε_r of around 3 and a dielectric strength ~7 MV m^{-1}. Other dielectrics being investigated for EWOD applications include the high-κ (high dielectric constant) oxides that have replaced silicon dioxide as the gate material in the latest CMOS devices. Silicon oxynitride, for example, has a value for ε_r of around 8, a dielectric strength greater than 1000 MV m^{-1} and can be formed as a submicron layer on conducting silicon.

Through the suitable physical arrangement and electrical switching of electrodes, EWOD can be used to control the motion and delivery of fluid droplets in microfluidic devices. A droplet situated mainly on an electrode element, but also overlapping an adjacent electrode area, can be induced to relocate onto this neighbouring electrode by bringing the first electrode to earth potential and applying a voltage of sufficient magnitude to the neighbouring electrode (so as to change the leading dielectric-liquid contact angle from above to below 90°). A droplet can also be split into two separate portions by applying voltages to both of its adjacent electrodes. Refinements of EWOD devices remain an active research area [46,47]. Jones [48] has reviewed the fundamental aspects of the effects of electrostatic fields upon the hydrostatic behaviour of liquids and concluded that the term 'electrowetting' should be restricted in its use to denote the effect of the electric field upon the contact angle. The translational forces acting to move droplets are in his view better referred to as examples of the net electromechanical force resulting from electric field nonuniformity and should be recognized as liquid dielectrophoresis. The electric fields applied to droplets in most EWOD devices are certainly nonuniform, but the possible dielectrophoretic effects resulting from this do not yet appear to have been incorporated into the theoretical treatments of EWOD effects.

1.4.6 Magnetic

As described in Box 1.3, magnetic fields are created by electric charges that are in motion. When a particle is placed in a magnetic field an effect known as Faraday's Law of Magnetic Induction occurs, which influences the magnetic dipole moments of the atomic and molecular structure of the particle. The origins of these magnetic moments are the angular momentum (called the spin) of electrons in an atomic orbital and the electrical current loops associated with the orbital paths of electrons.

Box 1.2 Electrowetting on Dielectric (EWOD)

The liquid droplet shown in Figure 1.12 is in contact with a solid dielectric layer of thickness *t* and absolute permittivity ε_{dl}. A voltage *V* is applied between the conducting liquid and a counter electrode situated beneath the dielectric film. Before the voltage is applied we assume that the solid-liquid interface is not electrically charged. When a voltage is applied the conducting liquid drop and the counter electrode form a capacitor *C* of value proportional to the area $A_{S\text{-}L}$ formed by the solid-liquid interface at the base of the droplet. The surface capacitance *C* (per unit area) is given by $C = \varepsilon_{dl}/t$. The wetted dielectric surface will attain a charge of magnitude $Q = VC$, and the water molecules will gain cohesive energy arising from their dipole attraction to these surface charges. This in turn will reduce the surface potential energy of the water molecules at the solid-liquid surface. To a first-order approximation the electrostatic energy ($^1/_2CV^2$) stored in the capacitor can be incorporated into an expression for the voltage-dependent solid-liquid surface tension (interfacial energy) $T_{S-L}(V)$ to give:

$$T_{S-L}(V) = T_{S-L}(0) - \frac{\varepsilon_{dl}}{2t}V^2 \tag{1.7}$$

where $T_{S-L}(0)$ is the surface tension with no voltage applied. The V^2 dependence indicates that either a direct current (DC) voltage, of positive or negative polarity, or an alternating current (AC) voltage can be applied across the dielectric layer. For an AC rather than a DC voltage, V^2 is replaced by V^2_{peak} in Equation (1.7).

The contact angle θ will be modified according to Young's equation (see Box 1.1):

$$T_{S-A} = T_{S-L}(0) + T_{L-A}\cos\theta(0) = T_{S-L}(V) + T_{L-A}\cos\theta(V) \tag{1.8}$$

From Equations (1.7) and (1.8) we obtain:

$$\cos\theta(V) = \cos\theta(0) + \frac{\varepsilon_o\varepsilon_r}{2tT_{L-A}}V^2 \tag{1.9}$$

Box 1.3 Magnetic Field and Force

As described in section 1.4.5, an electric field, E, is created by a distribution of electrical charges at *rest*. This field exerts a force on any other charge that is present in the field. For a *positive* charge the electric force acts in the *same* direction as the electric field vector. Magnetic interactions can be described as follows:

- A *moving* charge or electrical current creates a magnetic field, in addition to an electric field, in the surrounding space.
- The magnetic field exerts a force on any *moving* charge or electrical current that is present in this magnetic field.

The magnitude and direction of the force *F* acting on a charge *q* moving with velocity *v* in a magnetic field B is given by the vector cross product $F = qv \times B$. Thus, the magnetic force does *not* act, as for an electric force, along the field direction but at a direction that is *perpendicular* to both the magnetic field and the direction of travel of the charge. The magnetic force is proportional to the component of the charge's velocity that is perpendicular to the field. When that component is zero, namely when the velocity vector is parallel or antiparallel to the magnetic field vector, the magnetic force is zero. The largest magnetic force occurs when the velocity vector is perpendicular to the magnetic field vector and is given by $F = qvB$. It follows that the units of magnetic field B are the same as the units of F/qv, namely newtons·sec/coulomb·metre (N s/C m), or newtons/ampere.metre (N/A.m).

The direction of a magnetic force is given by the so-called 'right-hand screw rule': wrap the fingers of the right hand around a line that is perpendicular to the plane of *v* and B, so that they curl around in the direction from vector *v* to vector B. The thumb then points in the direction of the force acting on a moving positive charge (or the direction of advancement of a screw if turned in the same direction). This rule also gives the vector direction of the magnetic dipole moment *m* produced by a current-carrying loop. For a negative moving charge, or reversal of the current in the loop, the direction of the force and dipole moment is reversed.

When a charged particle moves in a location where both an electric field and a magnetic field exists, the particle experiences a force that is the vector sum of the electric (qE) and magnetic force ($qv \times B$). This is known as the Lorenz force law. The formal determination of the magnitude and direction of the field B thus involves three steps: (i) Place a particle of known charge in the field, and determine E by measuring the force on the particle when it is stationary; (ii) measure the force when the particle is moving at velocity *v*; (iii) repeat this measurement for a velocity *v* in another direction. The magnetic field B is given by the value that satisfies the Lorenz force law for these three results.

In most molecular structures the electrons occur in pairs with opposite spin, so that their associated magnetic moments cancel and the total magnetic moment of all the atomic current loops is also zero. However, the orbital paths of the electrons are altered by an external magnetic field to produce current loops and induced magnetic moments that do not cancel out. The additional magnetic field generated by these induced current loops is directed against the external field. A simple analogy is to imagine an induced current of electrons circulating around a benzene ring, like current in a single loop of wire. This will produce a magnetic field that reacts against the external one, which is analogous to the back emf generated by an inductor in an AC circuit. Materials that exhibit this property are called *diamagnetic* materials. When the external magnetic field is removed this diamagnetism disappears. Water, cells and micro-organisms are diamagnetic. The magnetic field at any point in a diamagnetic material is slightly *less* than it would be if the material were to be replaced by vacuum. We can quantify this effect by defining the relative permeability μ_r of a material as the ratio of the permeability μ of that material to the permeability of vacuum μ_o (i.e., $\mu_r = \mu/\mu_o$). The value of μ_o is by definition equal to $4\pi \times 10^{-7} \approx 1.257 \times 10^{-6}$ henries per metre (or newtons per ampere squared). Diamagnetic materials thus have a relative permeability value (μ_r) slightly less than unity – typically in the range 0.99990 to 0.99999. The amount by which the relative permeability is less than unity is called the volume magnetic susceptibility χ (i.e., $\chi = \mu_r - 1$) and is a negative and small dimensionless quantity (e.g., for water $\chi \approx -9\times10^{-9}$).

If the atomic structural unit of a material contains unpaired electrons, or the magnetic moments of their atomic current loops do not cancel, each unit will have a small magnetic moment. The magnetic moment of a current loop is defined as the product of the current and area of the loop and so has units of ampere m^2. The magnetic moment is directed at right angles to the loop area and when exposed to a magnetic field a torque is exerted so to align it with the field. The magnetic field within the material is thus slightly larger than the external field. These materials are known as *paramagnetic* materials and the induced magnetism is not retained when the external magnetic field is removed. Paramagnetic materials typically have a relative permeability value in the range 1.000001 to ~1.05 and thus have a small, positive, magnetic susceptibility (e.g., platinum $\chi \approx 1.2 \times10^{-8}$; oxygen gas $\chi \approx 1.3 \times10^{-6}$; brass $\chi \approx 5 \times10^{-2}$ [49]).

A *ferromagnetic* material exhibits a large, positive, magnetic susceptibility that is retained when the external magnetic field is removed. In ferromagnetic materials strong interactions between atomic magnetic moments causes them to align into magnetic domains, even in the absence of an externally applied magnetic field. In the unmagnetized state these domains are randomly oriented, but when an external field is applied a relatively large torque is exerted on the domains to align them all with the field. The magnetic field within a ferromagnetic material can be significantly larger than the external one, with a corresponding relative permeability and magnetic susceptibility in the range from 1000 to 100 000. As the strength of the external field is increased, the alignment of the domains reaches a limiting stage (called saturation) and no further induced magnetization occurs. On removing the external field, some of this saturation magnetization is retained (called remanent magnetism) and the material takes the form of a permanent magnet. The most common ferromagnetic materials are compounds formed from iron, nickel, cobalt or manganese. In some of these compound materials, the atoms have opposing magnetic moments of unequal magnitude and so still exhibit a spontaneous magnetization. They are known as *ferrimagnetic* materials, examples of which are ferrites and magnetic garnets. The oldest known magnetic material, magnetite (Fe_3O_4), is ferromagnetic.

1.4.6.1 Magnetophoresis

A particle placed in a uniform magnetic field will be magnetized. This magnetization takes the form of a net induced magnetic moment, of magnitude and polarity −m if diamagnetic, or +m if either paramagnetic or ferromagnetic. A torque will be exerted on this moment having magnitude and direction given by the vector product of the net moment and the magnetic field. This torque will tend to align a magnetic moment −m *against* the field direction and align a moment +m along the *same* direction as the field. However, the particle will not experience a net force to cause it to move laterally. (A common example in student texts on magnetism is to show that the net force on a current-carrying loop of wire in a uniform magnetic force is zero, because the forces on opposite sides of the loop cancel each other. However, the net torque acting on the loop is usually not zero.) A particle placed in a *nonuniform* magnetic field will, however, experience a net lateral force and if freely suspended will move. (The product of the loop current and the local magnetic field will not sum to zero around the complete loop, because one side of the loop will experience a larger magnetic field than the other.) The lateral movement of a particle induced by a magnetic field is known as *magnetophoresis*.

The magnetophoretic force F_{MAG} acting on a particle in a nonuniform field depends on the strength B and local gradient ∇B of the magnetic field. The magnitude and polarity of this force also depends on the particle's volume (V_p), the difference between the magnetic susceptibilities per unit volume of the particle (χ_p) and

Box 1.4 Magnetophoresis

Magnetization of a material is defined as its magnetic dipole moment per unit volume. When a magnetized particle is placed in a *uniform* magnetic field it will experience a torque but no translational force. It will twist like a compass needle so that its north seeking pole points along the direction of the Earth's magnetic field. To cause the particle to move requires a magnetic field gradient. A particle of total magnetic moment *m* placed in a nonuniform magnetic field ∇B experiences a magnetophoretic force F_{MAG} given by the relationship:

$$F_{MAG} = (m \cdot \nabla)B \tag{1.10}$$

where the symbol ∇ is the grad vector operator used to specify the spatial gradient variation of B. The magnetic dipole moment m is given by:

$$m = (\chi_p - \chi_m)V_pH \tag{1.11}$$

where $(\chi_p - \chi_m)$ is the difference between the magnetic susceptibilities of the particle (χ_p) and the medium it has displaced (χ_m), and V_p is the volume of the particle. In Equation (1.11), H is known as the magnetic field intensity or magnetizing field. The equivalent field B is given by $B = \mu H$, where μ is the *absolute* permeability of the medium, calculated as $\mu = \mu_o(1 + \chi_m)$ with $\mu_o = 4\pi \times 10^{-7}$ N/A^2.

From Equations (1.10) and (1.11)

$$F_{MAG} = \frac{(\chi_p - \chi_m)V_p(B \cdot \nabla)B}{\mu} \tag{1.12}$$

Equations (1.10) and (1.12) are similar in form to Equation (1.5) describing dielectrophoresis, and discussed further in Chapter 2. In fact, for a magnetically linear particle of radius *R*, the effective dipole method and the Clausius–Mossotti function described in Chapter 2 remain the same for magnetophoresis, with the permeabilities of the particle and medium replacing the corresponding permittivity values used in the dielectrophoresis equation [50, p. 65]. Furthermore, because it is usually the case that no time-varying electric fields or currents exist in the medium surrounding the particle, from one of Maxwell's equations we have $\nabla \times B = 0$. We can, in Equation (1.12), therefore express $(B \cdot \nabla)B$ as $^1/_2\nabla B^2$ (see Box 2.5 for the vector transformation). This implies that magnetophoresis can employ an alternating applied magnetic field (although the author is not aware of this being a common act).

the surrounding medium (χ_m). The mathematical form of these relationships is given by Equation (1.10) in Box 1.4.

The magnetic force F_{MAG} may be interpreted as the net result of the combination of the magnetic force acting on the particle itself and the magnetic 'buoyancy' exerted by the surrounding medium. From Equation (1.12) it follows that if the particle is para- or ferro-magnetic ($\chi_p > 0$) and the surrounding medium is aqueous (i.e., diamagnetic with $\chi_m < 0$) the magnetophoretic force is positive. The force will act to direct the particle against the field gradient, towards a region of high magnetic flux density. Iwasaka *et al.* [51] observed that muscle cells, yeast and platelets suspended in an aqueous medium exhibited negative magnetophoresis – they were directed towards lower magnetic field strengths. This was interpreted to mean that the cells exhibited a larger diamagnetic effect than the aqueous medium.

The function of an erythrocyte is to carry oxygen from the lungs to where it is required in the blood's circulatory system. The cell is packed with haemoglobin, which is a protein that contains four haem groups to each of which an oxygen molecule can bind. The specific oxygen binding site is an iron atom in the centre of a haem group. In the absence of oxygen binding, the iron atom has an unpaired electron and so lends a paramagnetic property to the haemoglobin molecule (and hence also to the erythrocyte). In the deoxygenated state, the susceptibility of erythrocytes has been estimated to be $+3.88 \times 10^{-6}$ [52]. However, in fully oxygenated blood this unpaired electron aligns itself in an antiparallel spin to that of the unpaired electron in the bound oxygen. The haemoglobin molecule and the erythrocyte are thus diamagnetic. In fully oxygenated blood erythrocytes exhibit a magnetic susceptibility of $-(2.1{\sim}3.5) \times 10^{-6}$ [52, 53]. Leukocytes do not contain haemoglobin or any other molecular structure containing an unpaired electron and so are diamagnetic in both oxygenated and deoxygenated blood – having an average magnetic susceptibility in both states of -0.13×10^{-6}. These differences in their magnetic properties have been exploited to separate erythrocytes and leukocytes from diluted blood [52, 53]. 89.7% separation of the erythrocytes from diluted whole blood was achieved in a continuous flow magnetophoretic device described by Han and Frazier, using an external magnetic field of 0.2 tesla and an electroplated nickel wire to create a local and highly nonuniform magnetic field [52]. (See Example 1.5 to appreciate the significance of this innovative step.)

Example 1.4 Magnetophoretic Force Acting on Cells
A cylindrical 'Alnico' magnet is fixed into a chamber so that one of its circular end faces lies flush with an internal chamber wall. The chamber contains a mixture of erythrocytes and leukocytes suspended in a deoxygenated aqueous medium.

Estimate the magnetophoretic force acting on each cell type at the following locations along the magnet's central axis: (i) 50 μm and (ii) 500 μm into the chamber. Assume that the cells are of equal size, with a radius of 5 μm.

Solution 1.4 The method to determine the force F_{MAG} is given by Equation (1.12) in Box 1.4:

$$F_{MAG} = \frac{(\chi_p - \chi_m)V_p(B \cdot \nabla)B}{\mu}$$

In a deoxygenated state, the susceptibility (χ_p) of erythrocytes and leukocytes can be taken as $+3.88 \times 10^{-6}$ and -0.13×10^{-6}, respectively [52, 53] and χ_m for an aqueous solution has a value of -9×10^{-8} [49]. The factor $(\chi_p - \chi_m)$ is thus of value 3.79×10^{-6} for an erythrocyte and -4×10^{-8} for a leukocyte. The *absolute* permeability of the medium is given by $\mu_o(1 + \chi_m) \approx \mu_o = 4\pi \times 10^{-7}$ N/A^2. The volume V_p for the cells is equal to $4\pi R^3/3 = 5.2 \times 10^{-16}$ m^3. The factor (B.∇)B in Equation (1.12) can be evaluated using the following relationship:

$$B = \frac{B_r}{2}\left[\frac{L + x}{\sqrt{(L + x)^2 + R^2}} - \frac{x}{\sqrt{R^2 + x^2}}\right] \qquad (1.13)$$

This equation gives the magnetic field B at a distance x from the pole face of a cylindrical magnet, along the central axis of the magnet [54]. B_r is the remanence of the magnet, of length L and radius R. The value of B_r for a permanent 'Alnico' magnet, made of an Al-Ni-Co cast alloy, ranges from 0.6 to 1.4 tesla [49]. We will assume $B_r = 1$ tesla, which can also be expressed as 1 newton per ampere metre (1 N/A.m) and values of 5 mm for both the radius and length of the magnet.

From Equation (1.13) with $L = R = 5$ mm we find that the magnetic field decreases slightly from 0.35 N/A.m at x = 50 μm to 0.32 N/A.m at 500 μm. The value for the magnetic field gradient ∇B remains almost constant at −70 N/A.m^2 over this distance, giving values for $(B.\nabla)B$ of −24.5 N^2/A^2.m^3 at x = 50 μm and −22.4 N^2/A^2.m^3 at 500 μm.

i) Calculation of the magnetophoretic force F_{MAG} acting on a deoxygenated erythrocyte at a distance of 50 μm into the chamber is as follows:

$$F_{MAG} = \frac{(3.79 \times 10^{-6})(5.2 \times 10^{-16}\, m^3)(-24.5\, N^2\, A^{-2}\, m^{-3})}{4\pi \times 10^{-7}\, NA^{-2}}$$
$$= -3.84 \times 10^{-14} \text{ newtons (N)}$$

The negative sign indicates that the force acts *against* the magnetic field and its gradient, so that the erythrocytes are directed up the field gradient *towards* the maximum field at the pole face of the magnet. This is the same effect as commonly observed when a magnet is used to attract iron filings. Deoxygenated erythrocytes and iron filings are both paramagnetic.

The value of F_{MAG} for the leukocytes is calculated by inserting a value of -4×10^{-8} for the parameter $(\chi_p - \chi_m)$ into Equation (1.12) instead of $+3.79 \times 10^{-6}$ as used for an erythrocyte, whilst keeping unchanged all the other parameter values. This leads to a value for F_{MAG} of $+4.05 \times 10^{-16}$ newtons, which is ~100 times smaller than the force acting on an erythrocyte at the same distance from the magnet. However, the positive sign for the force indicates that the leukocytes are directed down the field gradient and away from the magnet – in the opposite direction to that of the erythrocytes.

ii) Calculations of F_{MAG} for the situation where the cells are located 500 μm into the chamber are performed by changing the value of (B.∇)B (from −24.5 to −22.4 N^2/A^2.m^3) whilst using the appropriate values of $(\chi_p - \chi_m)$ for an erythrocyte and leukocyte. This gives the following results:

$F_{MAG} = -3.5 \times 10^{-14}$ N for erythrocytes located 500 μm into the chamber

and $F_{MAG} = +3.7 \times 10^{-16}$ N for leukocytes located 500 μm into the chamber.

These results show that a magnetic field produced by a magnet will penetrate deeply into a microfluidic chamber, with little change in the field gradient. The magnetophoresis effect will therefore be exerted almost uniformly throughout a microchamber. We have used a short 'stubby' magnet in this example. From Equation (1.13) it can be seen that increasing the ratio of magnet length to its radius, whilst maintaining the same volume, will increase the magnetic field close to the pole face. However, the field of a pencil-shaped magnet will decrease more rapidly with distance than a 'stubby' magnet of the same volume. The practical advantages to be gained from altering the magnet's dimensions are relatively small for microfluidic applications.

Example 1.5 Magnetophoretic Velocity
Estimate the magnetophoretic velocities of the erythrocytes and leukocytes at locations of 50 μm and 500 μm away from the magnet described in Example 1.4.

Solution 1.5 The magnetophoretic forces calculated for Example 1.4 will accelerate the cells until a balance with the fluid viscous drag is reached. At this point the cells will reach a steady velocity v_m. As explained more fully in Chapter 2, the pertinent viscous drag force is equal to $6\pi\eta R v_m$, where η is the fluid viscosity and

Box 1.5 Superparamagnetism

A ferromagnetic or ferrimagnetic material normally undergoes a transition to a paramagnetic state above what is known as its Curie temperature. However, if sufficiently small (i.e., below 3~50 nm) ferromagnetic or ferrimagnetic nanoparticles can exhibit a transition to the paramagnetic state below the Curie temperature. They are said to exhibit superparamagnetism. These particles are so small that they take the form of a single magnetic domain, in which all of the magnetic moments of the atoms in the nanoparticle cooperate to form a single, giant, magnetic moment. There is a coupling of all the angular momentums (spins) of the unpaired electrons in all the atomic orbitals, and physicists refer to this as the 'macro-spin approximation'. Such nanoparticles can be magnetized by an external magnetic field and their effective susceptibility is considerably larger than that of a normal paramagnetic material – hence the term 'superparamagnetic'.

R is the radius of the cell. The steady-state magnetophoretic velocity can therefore be evaluated using the relationship:

$$v_m = \frac{F_{MAG}}{6\pi\eta R} \tag{1.14}$$

The viscosity of water is 10^{-3} Pa.s (pascal second) and our cells have a radius of 5 μm. Using the values obtained in Example 1.4 for F_{MAG}, then from Equation (1.14) for an erythrocyte at 50 μm from the pole face:

$$v_m = \frac{-3.84 \times 10^{-14}\,N}{6\pi(10^{-3}\,Pa.s)(5 \times 10^{-6}\,m)}$$
$$= -4.1 \times 10^{-7}\ \text{N/Pa.m.s}$$

A pascal is a measure of stress or pressure, with units of N/m^2. The value obtained for v_m above is thus equivalent to −0.4 μm/s – a very small velocity towards the magnet. The corresponding velocity at 500 μm can be calculated as −0.37 μm/s. It would take a deoxygenated erythrocyte about 20 s to travel a distance equivalent to one diameter.

Based on the values for the magnetophoretic force calculated in Example 1.4 for a leukocyte, we can expect its induced velocity to be even smaller (~a hundredfold less) and practically imperceptible. We conclude from this example that the magnetophoresis of cells, based solely on their own intrinsic magnetic properties, is not a significant effect if we rely on a single external magnet to produce both the field and field gradient. We need to increase the magnitude of the factor (B.∇)B. One way to achieve this is to use two or more magnets with opposing magnetic fields [54]. For example, with two opposing magnets the field strength will be zero at the midpoint between them and increase rapidly towards each pole face. This increases the complexity of a device and may also be difficult to incorporate into a microfluidics design. Inspection of Equation (1.13) reveals that the magnitude of (B.∇)B will increase as the dimensions of the magnetic field source become smaller. Practical examples of this include the use of a current carrying wire, as demonstrated by Hans and Frazier [52] and the use of micro-magnet arrays as described in section 1.5.1.

1.4.6.2 Magnetic Nanoparticles

Magnetic nanoparticles are commonly formed from compounds of nickel, cobalt or iron. Particles above ~150 nm diameter exhibit normal bulklike static and dynamic magnetic properties, but smaller particles deviate from this. For example, 5 nm magnetite (Fe_3O_4) nanoparticles display superparamagnetic behaviour at room temperature [55]. Superparamagnetism is described in Box 1.5. The particles can be coated with an organic shell or molecules that chemically bind preferentially to a biological entity, such as a specific sequence of DNA or a particular protein receptor on a cell membrane. Larger (≥ 50 nm) magnetic particles can be externally tagged to a cell by co-culturing them with the cell sample, whilst those less than 10 nm diameter can be internalized into a cell by endocytosis. The number of magnetic nanoparticles entering the cells by endocytosis depends on the concentration of the co-cultured extracellular magnetic nanoparticles. About 2~5 million magnetic nanoparticles of 10 nm diameter are required to be internalized for a cell to exhibit significant magnetophoresis [56, 57]. An advantage that magnetic tagging has over fluorescence labelling is that at normal temperatures the magnetism of a magnetic nanoparticle is stable over time and is not affected by reagent chemistry or quenched as a result of photobleaching. Detection of a fluorescent tag can also be compromised by background fluorescence produced by biological agents or glass and polymeric substrates. No significant magnetic background occurs from such sources and the magnetic fields of magnetic nanoparticles are also not screened by aqueous reagents or biological cells and tissues.

Example 1.6 Magnetic Labelling of Cells

Magnetized beads of diameter ~3 μm are available commercially, with each bead having a quoted magnetization of ~10^{-13} A.m^2. Estimate the magnetophoretic force and resulting velocity of such a bead when located 50~500 μm from the magnet described in Example 1.4.

Solution 1.6 We need to be careful concerning the term *magnetization*. This term should be used to describe the effective magnetic moment per unit volume of a material, with units of ampere per metre ($A.m^{-1}$). We can view a magnet as the electrical equivalent of a current-carrying solenoid, of length L and cross-sectional area A., with volume LA m^3. By definition, the magnetic dipole moment of a current-carrying loop is given by the product of the current and the loop area and so has dimensions of $A.m^2$. This Example quotes the magnetization of each bead as 10^{-13} $A.m^2$ and so refers to the magnetic moment m given in Equation (1.10) of Box 1.4. (Magnetization should strictly speaking be given units of $A.m^{-1}$.)

The magnetophoretic force F_{MAG} acting on each magnetic bead is thus given by Equation (1.10):

$$F_{MAG} = (\mathrm{m} \cdot \nabla)\,\mathrm{B}$$

Inserting into this equation $m = 10^{-13}$ $A.m^2$ and $\nabla B = -70$ $N/A.m^2$ we have:

$$F_{MAG} = (10^{-13}\,\mathrm{A\,m^2})(-70\,\mathrm{N\,A^{-1}m^{-2}}) = -7 \times 10^{-12}\,\mathrm{N}$$

From Equation (1.14) we calculate the steady state velocity of the bead as:

$$v_m = \frac{F_{MAG}}{6\pi\eta R} = \frac{-7 \times 10^{-12}\,N}{6\pi(10^{-3}\,Pa.s)(1.5 \times 10^{-6}\,m)} = -2.5 \times 10^{-4}\,\mathrm{m/s}.$$

These values obtained for the magnetophoretic force and velocity are considerably larger than those obtained for the blood cells and is the reason why suitably labelled magnetic beads can be used to selectively manipulate target cells.

Example 1.7 Magnetic Labelling of Cells
Estimate the number of 10 nm diameter maghemite (γFe_2O_3) nanoparticle that would need to be internalized by endocytosis into a monocyte of diameter 15 μm for it to exhibit a magnetophoretic velocity of 200 μm/s in the chamber of Example 1.4. Assume the remanent magnetization of a maghemite nanoparticle is 5×10^4 $A.m^{-1}$.

Solution 1.7 To calculate the value of F_{MAG} that produces a velocity of 200 μm/s (towards the magnet) we rearrange Equation (1.14):

$$F_{MAG} = 6\pi\eta\,R\,v_m = 6\pi(10^{-3}\,\mathrm{Pa\,s})(7.5 \times 10^{-6}\,\mathrm{m})(-2 \times 10^{-4}\,\mathrm{ms^{-1}}) = -28\,\mathrm{pN}$$

To achieve this force will require an effective magnetic moment for the cell given by rearranging Equation (1.10):

$$\mathrm{m} = \frac{F_{MAG}}{\nabla B} = \frac{-2.8 \times 10^{-11}\,N}{-70\,N.A^{-1}.m^{-2}} = 4 \times 10^{-13}\,\mathrm{A\,m^2}$$

We are provided with the information that the magnetization value of a maghemite nanoparticle is 5×10^4 $A.m^{-1}$. The magnetic moment of a single 10 nm diameter nanoparticle m_p is thus given by the product of this magnetization value and the volume of the particle:

$$m_p = (5 \times 10^4\,A.m^{-1})\left[\frac{4}{3}\pi(10^{-8})^3\,m^3\right] = 2.1 \times 10^{-19}\,\mathrm{A\,m^2}$$

The number n of such nanoparticles required to be internalized by a 15 μm diameter cell to give it a total magnetic moment of 4×10^{-13} $A.m^2$, is thus:

$$n = \frac{4 \times 10^{-13} A.m^2}{2.1 \times 10^{-19} A.m^2} = 1.9 \times 10^6$$

This number of internalized nanoparticles is typically reported in the literature [56, 57]. The volume taken up by 2×10^6 nanoparticles, each of diameter 10 nm is $\sim 10^{-18}$ m^3 and will occupy only ~0.06% of the internal volume of a 15 μm diameter cell.

Magnetic nanoparticles have been applied in various areas of biomedicine to deliver therapeutic drug, gene and radionuclide agents, destroy tumours by radio frequency hyperthermia, to act as contrast enhancement agents for magnetic resonance imaging and for the magnetic separation of labelled cells and other biological entities [58, 59]. Furthermore, most of the necessary functions required in a lab-on-chip bioassay can also be accomplished using magnetic nanoparticles, such as sample purification, mixing, target labelling and the spatial manipulation, transport and isolation of the biological target [59]. The separation of magnetically labelled cells is known as magnetic activated cell sorting (MACS). The process is shown schematically in Figure 1.13. For positive cell selection or enrichment, magnetic particles coated with antibodies specific for a cell surface protein of interest are bound to the target cells by incubating the

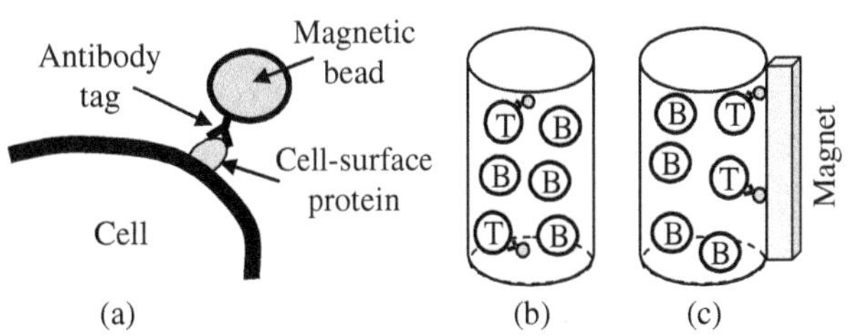

Figure 1.13 A schematic is shown of magnetic activated cell selection (MACS). (a) A magnetic bead coated with an antibody specific to a cell-surface protein is shown bound to a cell. (b) Beads labelled with an antibody specific for the CD3 protein on T cells are added to a mixture of B and T cells. (c) The labelled beads bind to the T cells and are attracted to a magnet. The B cells can be separated from the T cells by pipetting or flushing them out of the tube.

cells and particles in a reagent tube. Magnetically tagged cells can then be isolated by placing a magnetic near the tube and pipetting off the supernatant (or by flushing the supernatant through the tube). With negative selection, the target (magnetically tagged) cells are the unwanted cells and the cells of interest can either be pipetted out or eluted through the tube. The choice of positive or negative cell selection depends on the antibodies available and the subsequent use of the cells. The main disadvantage of MACS is that it is a discontinuous process, requiring additional steps for sample preparation before and after sorting. Having to magnetically label a target cell requires the use of expensive labelled magnetic beads, which may then require removal after the cell separation process. This may be relatively straightforward for externally tagged cells, but is not possible where the nanoparticles have been internalized.

An interesting application of magnetically tagged particles has been described by Xie *et al.* [60], who used carboxyl-functionalized magnetic particles to preferentially adsorb leukocytes and proteins from whole blood. After isolation and lysis of the leukocytes, it was found that the released genomic DNA was readily adsorbed onto the nanoparticles in isopropyl alcohol. This particle-bound DNA could then be used directly as a PCR template. The significance of this is that although blood is the universal source of DNA, it is a complex mixture of cells, proteins, lipids, carbohydrates and other low molecular-weight compounds. All of these adversely affect most of the chemical analyses one would wish to perform on blood. For example, haem groups in the haemoglobin released from red blood cells bind to the DNA polymerase enzyme used in the PCR amplification of DNA. Even a 1% by volume concentration of whole blood can inhibit PCR. Only the leukocytes (white blood cells) in blood contain DNA, the erythrocytes do not. From Table 1.2 we find that the erythrocytes exceed the number density of leukocytes by a ratio of about 700 : 1. Xie *et al.* [60] were able to extract a sufficient number of leukocytes from whole blood to allow PCR to be performed, without the use of an expensive (and often unstable) coating of a leukocyte-specific antibody such as CD45 on the magnetic particles.

1.4.7 Surface Forces (Cell Patterning)

The forces considered so far can be used to sort or selectively move suspended cells and other bioparticles when they are suspended in solution. In this way it is possible to direct cells to specific locations on the surface of a substrate or within a specified location of a fluidic channel. The reliable localization of cells in a fluidic device is important in the fabrication of cell-based biosensors, such as those to be used for the detection of toxic substances or for high throughput screening of new pharmaceutical drugs. Such localization is also required for studies of cell–cell communication and signal transduction pathways in artificial neuronal networks.

Choi *et al.* [61] have demonstrated that by coating the floor of a fluidic channel with specific cell adhesion molecules (CAMs), target cells can be induced to roll over the channel floor surface. With slanted ridges in the channel floor (as used in hydrophoresis [11]) cells that roll along the surface remain attached to the surface at the ends of the ridges and so roll into the adjacent trench. Cells that do not bind to the adhesion molecules follow a different flow trajectory at the ends of the ridges and do not enter the trenches. Thus, by coating the surface of the ridges with P-selectin, HL60 and K562 leukaemia cells were separated through a combination of hydrophoresis and cell rolling [61]. HL60 cells express a high level of a receptor ligand that binds to P-selectin and so rolled into the trenches, whereas K562 cells do not express this ligand and were directed away from the trenches.

Lai *et al.* [62] have demonstrated that magnetically tagged cells can be patterned on a microstructured ferromagnetic thin film by controlling the magnetic domain walls in the film. Once the magnetic domain walls are formed by an applied magnetic field, no magnetic field is then needed to maintain the magnetization directions and positions of the domain walls. The local magnetic forces, which results from the high stray field produced by the magnetic domain walls in their remanent states, attract the cells to selected positions. Without this remanence the tagged cells would not remain attracted to the ferromagnetic film. Thus, in general, unless the cells strongly adhere to the surface they will drift back into the bulk solution once the manipulating force is removed. Protocols have been developed to pattern single cells onto defined areas by attaching (printing) peptide molecules or proteins onto the substrate. The peptide or protein is chosen so as to bind to a CAM of a target cell. CAMs are protein receptors located on a cell membrane surface and are responsible for the specific binding of a cell with other cells or with its surrounding extracellular matrix. This process is called cell adhesion. CAMs mostly fall into four main protein categories (immunoglobulins, integrins, cadherins and selectins) and typically consist of an intracellular part that interacts with the cytoskeleton, a part that spans across the cell membrane and an extracellular component that has a high specificity of binding to chemical ligands of other CAMs or of the extracellular matrix. The strategy to pattern cells of a specific type is therefore to modify the surface of a substrate with cell-adhesion ligands. Examples of this include the attachment of peptides onto supported lipid bilayers [63], of peptides onto gold patterned SiO_2 substrates [64] and the patterning of multiple antibodies along the cross-section of a single microfluidic channel [65].

1.5 Combining Dielectrophoresis with other Forces

In section 1.4 we have considered applications of the individual forces listed in Table 1.1 – but advantages can result by applying some of them together in a hybrid device. Most of the applications of dielectrophoresis described in this book are performed in fluidic devices and so we will find examples of where hydrodynamic effects such as laminar flow and hydrophoresis play an integral part. As described in section 1.2, microfluidic channels are characterized by having large ratios of their interior surface area to volume. These surfaces carry a net electric charge and attract counterions from the bulk fluid in the channel, to create what is known as a thin electrical double layer, more fully described in the next chapter and in Chapter 12 (section 12.7). This thin layer carries a net charge (ρ), which can respond to an applied DC electric field or low-frequency AC field. If the field is applied along the axis of the channel, the laminar of fluid next to the channel wall is accelerated as a result of the volume coulombic force (ρE). Adjacent fluid annuli are accelerated by the momentum transfer caused by viscous forces, until the velocity gradient approaches zero across the whole microchannel. A charged fluid layer effectively 'drags' its adjacent fluid layer along, until finally the entire fluid in the microchannel moves at a uniform velocity. This is known as electro-osmosis (see section 12.8) and can be used to advantage in what is known as insulator-based DEP (iDEP), described fully in section 10.4.2 of Chapter 10. A combination of an applied DC and AC field in iDEP can also add the extra force of electrophoresis, so that the net movement of suspended particles can depend on its electrophoretic and DEP mobility, as well as the direction and magnitude of the electroosmotic fluid flow.

Optical effects have also been incorporated in the form of so-called optoelectronic tweezers or light-induced dielectrophoresis. Optical forces are not used directly, as in an optical tweezer, instead light beams illuminate photoconductive structures to switch them on as virtual electrodes. Single cells can be manipulated in a massively parallel fashion across a large area with either laser beam generated multispot diffraction patterns, or using incoherent light sources such as LEDs or micromirror displays. Recent notable examples of this include the use of optically addressable carbon nanotube electrodes [66] and the dielectrophoretic patterning of cells using diffraction images and spin-coated organic titanium oxide photoconductors [67]. Xu *et al.* [68] have reviewed the progress in developing dielectrophoretically tunable optofluidic devices, such as adaptive lenses, optical attenuators and single pixel displays.

Examples of where dielectrophoresis has been combined in microfluidic devices with magnetic forces, acoustic forces and surface-wetting effects will now be described.

1.5.1 Hybrid Dielectrophoresis-Magnetophoresis

Issadore *et al.* [69] have described a hybrid microfluidic chip that can independently and simultaneously trap and move particles using both electric and magnetic fields. Below an array of metal pixels used for dielectrophoretic manipulation, two sets of perpendicularly oriented metal wires were electrically addressed to produce local high-gradient magnetic fields. The system was tested using lipid vesicles with internalized magnetic nanoparticles. The elastic deformation properties of such vesicles were examined by holding a vesicle in place by a dielectrophoretic force and applying a magnetic force. James *et al.* [70] used magnetophoresis to bring magnetic particles down to an array of electrodes, where dielectrophoresis could then be used for their accurate lateral placement. The objective was to develop a system similar to flow cytometry in which a high throughput of flowing antibody labelled magnetic particles can be optically interrogated for captured fluorescent target analytes. A detection level of 50 ppb of a surrogate biotoxin (ovalbumin) in a raw milk sample was demonstrated. Lyuksyutov *et al.* [71] and Kauffmann *et al.* [72] have demonstrated that the lateral movement and assembly of small particles can be achieved by diamagnetic levitation and pulsed electric field induced dielectrophoresis. By positioning two neodymium-iron-boron permanent magnets to have opposing magnetic fields, a value of 700 $N^2A^{-2}m^{-3}$ was generated for the factor $(B.\nabla)B$ (i.e., some 20-fold larger than that calculated for a single magnet in Example 1.4). This combination of magnetophoretic and dielectrophoretic manipulation can be applied to manipulate a range of diamagnetic particles, including cells, fluorescent beads and droplets of aqueous, oil and alcohol solutions, for example. Jung *et al.* [73] reported a 20 000-fold enrichment of circulating nucleated cells from peripheral blood using a hybrid device comprising a ferromagnetic wire array to create a local high-gradient magnetic field (when exposed to an external magnetic field) together with interdigitated electrodes to create a dielectrophoretic force. By applying an external magnetic flux of 0.3 T and a 2 MHz sinusoidal voltage of 4 Vp-p to the interdigitated electrodes, paramagnetic red blood cells in deoxygenated blood were driven by magnetophoresis in the opposite direction to dielectrophoretically transported white blood cells [73]. The design and simulated functioning of a hybrid magneto- and dielectrophoresis microfluidic device for the continuous collection of magnetically tagged particles has been described by Blaire *et al.* [74]. In this device magnetic beads are trapped using an interdigitated micromagnet array to create high-gradient magnetic fields and then released by dielectrophoresis using an interdigitated array of indium tin oxide (ITO) electrodes.

1.5.2 Hybrid Dielectrophoresis-Acoustophoresis

Dielectrophoresis can provide accurate particle manipulation over short distances and length scales, whereas acoustophoresis can operate over much larger distances and scales. This complementary relationship has been used by Lock *et al.* [75] to employ travelling acoustic waves to bring particles suspended in a large volume container down onto an electrode array, where travelling dielectrophoresis can then be used to concentrate the particles in a chamber of much smaller volume. Wiklund *et al.* [76] describe how the competition between long-range ultrasonic forces, short-range dielectrophoretic forces and viscous drag forces can provide flexible and gentle manipulation functions of individual cells in microfluidic devices. A system that integrates a bulk lead zirconate titanate (PZT) slab, to produce acoustic forces, with substrate patterned microelectrodes for dielectrophoresis manipulation of particles has also been described by Ravula *et al.* [77].

1.5.3 Hybrid Dielectrophoresis Electrowetting

An effect referred to as liquid-dielectrophoresis (L-DEP) can be demonstrated by placing a droplet of fluid (volume 1~2 μL) at one end of a track formed between two parallel coplanar electrodes over which a thin layer of dielectric has been deposited. On application of an AC voltage (e.g., 100 kHz, 700 V) across these electrodes a liquid 'finger' projects from the deposited droplet and moves rapidly along the electrode track [48, 78]. On removing the applied voltage, this finger of liquid breaks up into small droplets of spacing and individual volume (pL~nL) predicted by Rayleigh's instability criterion [79]. These liquid droplets can then be transported and mixed with other reagents or droplets using dielectrophoretic forces – referred to as droplet-dielectrophoresis (D-DEP) and also involving aspects of the EWOD effect as described by Jones [48]. Chugh and Kaler [79] have examined the integration L-DEP and D-DEP as a means to provide rapid and automated sample handling and bioassays, without the need for pumps or valves, on a compact chip-based platform.

These developments lead naturally to the concept of being able to transport liquid droplets that contain particles such as cells; to then separate a target cell from other cells inside the droplet by dielectrophoresis; and by EWOD to split the droplet into two smaller ones, one of which contains the isolated target cell. Valley *et al.* [80] have made significant progress along these lines by unifying optoelectrowetting and optoelectronic tweezers on the same chip. This enables the transportation of aqueous droplets by electrowetting, as well as manipulation of individual particles within those droplets by dielectrophoresis. In this device a liquid droplet is sandwiched between a top Teflon-ITO electrode and a bottom ITO electrode that is coated with a photosensitive layer of a-Si:H, an insulating layer of Al_2O_3 and a Teflon layer. When an external voltage is applied between the two ITO electrodes, in the absence of incident light, the electric field primarily exists in the highly resistive a-Si:H layer. However, upon illumination, the conductivity of the a-Si:H layer increases dramatically and causes the electric field to drop across a combination of the dielectric (oxide and Teflon) and liquid layers. If the majority of the field drops across the dielectric layer, then the droplet will experience a net optoelectrowetting force towards the illuminated region. However, if the electric field drops mainly across the liquid layer, electric field gradients will exist in the liquid (depending on the spatial localization of the light beam) and particles within the droplet will experience a dielectrophoretic force. As a demonstration, the device was used to select a single cell out of a mixture of cells and then encapsulate it in its own aqueous droplet.

1.6 Summary

The singular feature that distinguishes dielectrophoresis from the other techniques listed in Table 1.1 is that it selectively manipulates a target cell based on its intrinsic dielectric polarizability. As will be described in later chapters, two parameters contribute to this polarizability – namely the effective electrical capacitance and conductance of the cell. These properties in turn depend on such features as: the size and shape of the cell; cell surface topography associated with blebbing or microvilli; the integrity of the cytoplasmic membrane; internal features such as cytoplasm conductivity and the nucleus-cytoplasm volume ratio. All of these features contribute to the frequency spectrum of the cell's electrical impedance, which can be examined over a wide range of electrical frequencies and be uniquely characteristic of a particular cell type. The most common methods currently used to quantitatively characterize or purify cell populations are flow cytometry (FACS) or magnetic bead-coupled cell separation (MACS). These methods are dependent on the existence of specific cell surface antigens and the formulation or availability of high affinity probes to these antigens. Dielectrophoresis does not require the use of fluorescent or magnetic labels – although using antibody-coated dielectric beads could be used to increase the yield or purity of target cell separation.

Magnetophoresis compares unfavourably with dielectrophoresis on two counts as a cell separation method. The electrode geometry required to produce a high value for the parameter (E.∇)E is straightforward to fabricate for a microfluidic device for dielectrophoresis, whereas quite complicated magnetic geometries are required to give high values for (B.∇)B. Also, with the exception of the separation of deoxygenated erythrocytes from leukocytes, the separation of cells based solely on differences of their intrinsic diamagnetic properties is not a practicable proposition. The situation where target cells can be subjected to the opposite polarity of dielectrophoretic force to other cells can certainly not be achieved for leukocytes using magnetophoresis.

A significant negative aspect of dielectrophoresis is that it operates only over a short distance (< 300~500 μm) from an electrode array. Acoustophoresis can operate over significant distances into a fluidic channel, but is not good at small scale manipulations of cells. The discrimination of cell type separation by acoustophoresis is based on differences in cell size, density and compressibility and this is less sensitive than discrimination based on dielectric polarizability. Optophoresis also operates over a wide scale of distances, but selective cell separation will depend on differences in the internal refractive index of the cell and will not be as sensitive as dielectrophoresis in sensing differences in membrane properties, for example. However, because dielectrophoresis is easily incorporated into a microfluidic device, its major drawback in terms of its short range of effectiveness can be offset by combining it with either acoustophoresis or optophoresis and taking advantage of the special features and advantages that are lacking in these two other complementary methods.

1.7 References

1 Andersson, H. and van den Berg, A. (2003) Microfluidic devices for cellomics: A review. *Sensors and Actuators B-Chemical* **92**, 315–325.

2 Stone H. A., Stroock A. D. and Ajdari, A. (2004) Engineering flows in small devices: Microfluidics towards a lab-on-a-chip. *Annu. Rev. Fluid Mech.* **36**, 381–411.

3 Squires, T. M. and Quake, S. R. (2005) Microfluidics: Fluid physics at the nanoliter scale. *Rev. Mod. Phys.* **77**, 977–1026.

4 Abkarian, M., Faivre, M., Horton, R. *et al.* (2008) Cellular-scale hydrodynamics. *Biomed. Mater.* **3**, 034011.

5 Bhagat, A. A. S., Bow, H., Hou, H. W. *et al.* (2010) Microfluidics for cell separation. *Med. Biol. Eng. Comput.* **48**, 999–1014.

6 Autebert, J., Coudert, B., Bidard, F. C. *et al.* (2012) Microfluidic: An innovative tool for efficient cell sorting. *Methods* **57**, 297–307.

7 Wilding, P., Kricka, L. J., Cheng, J. *et al.* (1998) Integrated cell isolation and polymerase chain reaction analysis using silicon microfilter chambers. *Anal. Biochem.* **257**, 95–100.

8 Mohamed, H., Turner, J. N. and Caggana, M. (2007) Biochip for separating fetal cells from maternal circulation, *J. Chromatogr. A* **1162**, 187–192.

9 Mohamed, H., Murray, M., Turner, J. N. and Caggana, M. (2009) Isolation of tumor cells using size and deformation, *J. Chromatogr. A* **1216**, 8289–8295.

10 Berger, S. A., Talbot, L. and Yao, L. S. (1983) Flow in curved pipes. *Annu. Rev. Fluid Mech.* **15**, 461–512.

11 Choi, S. and Park, J.-K. (2007) Continuous hydrophoretic separation and sizing of microparticles using slanted obstacles in a microchannel. *Lab Chip* **7**, 890–897.

12 Song, S. and Choi, S. (2013) Design rules for size-based cell sorting and sheathless cell focusing by hydrophoresis. *J. Chromatography A*, **1302**, 191–196.

13 Yang, S., Undar, A. and Zahn, J. D. (2006) A microfluidic device for continuous, real time blood plasma separation. *Lab on a Chip* **6**, 871–880.

14 Giddings, J. C. (1966) A new separation concept based on a coupling of concentration and flow nonuniformities, *Separ. Sci. Technol.* **1**, 123–125.

15 Giddings, J. C. (1993) Field-flow fractionation: Analysis of macromolecular, colloidal, and particulate materials. *Science* **260**, 1456–1465.

16 Roda, B., Zattonia, A., Reschigliana, P. *et al.* (2009) Field-flow fractionation in bioanalysis: A review of recent trends. *Anal. Chim. Acta* **635**, 132–143.

17 Yamada, M. and Seki, M. (2005) Hydrodynamic filtration for on-chip particle concentration and classification utilizing microfluidics. *Lab Chip* **5**, 1233–1239.

18 Yamada, M., Kano, K., Tsuda, Y. *et al.* (2007) Microfluidic devices for size-dependent separation of liver cells. *Biomed. Microdev.* **9**, 637–645.

19 Yamada, M., Nakashima, M. and Seki, M. (2004) Pinched flow fractionation: continuous size separation of particles utilizing a laminar flow profile in a pinched microchannel. *Anal. Chem.* **76**, 5465–5471.

20 Davis, J. A., Inglis, D. W., Morton, K. J. *et al.* (2006) Deterministic hydrodynamics: Taking blood apart. *Proc. Natl. Acad. Sci. USA* **103**, 14779–14784.

21 Holm, S. H., Beech, J. P., Barrett, M. P. and Tegenfeldt, J. O. (2011) Separation of parasites from human blood using deterministic lateral displacement. *Lab Chip* **11**, 1326–1332.

22 Di Carlo, D., Irimia, D., Tompkins, R. G. and Toner, M. (2009) Continuous inertial focusing, ordering, and separation of particles in microchannels. *Proc. Natl. Acad. Sci. USA* **104**, 18892–18897.

23 Lee, W. C., Bhagat, A. A. S., Huang, S., Van Vliet, K. J. *et al.* (2011) High-throughput cell cycle synchronization using inertial forces in spiral microchannels. *Lab Chip* **11**, 1359–1367.

24 Martel, J. M. and Toner, M. (2012) Inertial focusing dynamics in spiral microchannels. *Phys. Fluids* **24**, 032001.

25 Bruus, H. (2012) Acoustofluidics 7: The acoustic radiation force on small particles. *Lab Chip* **12**, 1014–1021.

26 Hawkes, J. J. and Coakley, W. T. (2001) Force field particle filter, combining ultrasound standing waves and laminar flow. *Sensors and Actuators* **75**, 213–222.

27 Petersson, F., Aberg, L., Sward-Nilsson, A-M. and Laurell, T. (2007) Free flow acoustophoresis: Microfluidic-based mode of particle and cell separation. *Anal. Chem.* **79**, 5117–5123.

28 Laurell, T., Petersson, F. and Nilsson, A. (2007) Chip integrated strategies for acoustic separation and manipulation of cells and particles. *Chem. Soc. Rev.* **36**, 492–506.

29 Yang, A. H. and Soh, H. T. (2012) Acoustophoretic sorting of viable mammalian cells in a microfluidic device. *Anal. Chem.* **84**, 10756–10762.

30 Lenshof, A., Ahmad-Tajudin, A., Järås, K. *et al.* (2009) Acoustic whole blood plasmaphoresis chip for prostate specific antigen microarray diagnostics. *Anal. Chem.* **81**, 6030–6037.

31 Yeo, L. Y. and Friend, J. R. (2009) Ultrafast microfluidics using surface acoustic waves. *Biomicrofluidics* **3**, 012002.

32 Wang, Z. and Zhe, J. (2011) Recent advances in particle and droplet manipulation for lab-on-chip devices based on surface acoustic waves. *Lab Chip* **11**, 1280–1285.

33 Ashkin, A. (1980) Applications of laser radiation pressure. *Science* **210**, 1081–1088.

34 Ashkin, A., Dziedzic, J. M., Bjorkholm, J. E. and Chu, S. (1986) Observation of a single gradient force optical trap for dielectric particles. *Optics Lett.* **11**, 288–290.

35 Ashkin, A., Dziedzic, J. M. and Yamane, T. (1987) Optical trapping and manipulation of single cells using infrared laser beams. *Nature* **330**, 769–771.

36 MacDonald, M. P., Spalding, G. C. and Dholakia, K. (2003) Microfluidic sorting in an optical lattice. *Nature* **426**, 421–424.

37 Padgett, M. and Di Leonardo, R. (2011) Holographic optical tweezers and their relevance to lab on chip devices. *Lab Chip* **11**, 1196–1205.

38 Bernabini, C., Holmes, D. and Morgan, H. (2010) Micro-impedance cytometry for detection and analysis of micron-sized particles and bacteria. *Lab Chip* **11**, 407–412.

39 Han, X., van Berkel, C., Gwyer, J. *et al.* (2011) Microfluidic lysis of human blood for leukocyte analysis using single cell impedance cytometry. *Anal. Chem.* **84**, 1070–1075.

40 Shapiro, H. M. (2003) *Practical Flow Cytometry*, 4th edn. John Wiley & Sons, Inc., Hoboken, NJ.

41 Fu, A. Y., Spence, C., Scherer, A. *et al.* (1999) A microfabricated fluorescence-activated cell sorter. *Nature Biotechnol.* **17**, 1109–1111.

42 Fu, A. Y., Chou, H-P., Spence, C. *et al.* (2002) An integrated microfabricated cell sorter. *Anal. Chem.* **74**, 2451–2457.

43 Dittrich, P. S. and Schwille, P. (2003) An integrated microfluidic system for reaction, high-sensitivity detection, and sorting of fluorescent cells and particles. *Anal. Chem.* **75**, 5767–5774.

44 Cho, S. H., Chen, C. H., Tsai, F. S. *et al.* (2010) Human mammalian cell sorting using a highly integrated micro-fabricated fluorescence-activated cell sorter (μFACS). *Lab Chip* **10**, 1567–1573.

45 Pollack, M. G., Fair, R. B. and Shenderov, A. D. (2000) Electrowetting-based actuation of liquid droplets for microfluidic applications. *Applied Phys. Lett.* **77**, 1725–1726.

46 Davoust, L., Fouillet, Y., Malk, R. and Theisen, J. (2013) Coplanar electrowetting-induced stirring as a tool to manipulate biological samples in lubricated digital microfluidics. Impact of ambient phase on drop internal flow pattern. *Biomicrofluidics* **7**, 044104.

47 Shabani, R. and Cho, H. J. (2013) Flow rate analysis of an EWOD-based device: How important are wetting-line pinning and velocity effects? *Microfluidics and Nanofluidics* **15**, 587–597.

48 Jones, T. B. (2002) On the relationship of dielectrophoresis and electrowetting. *Langmuir* **18**, 4437–4443.

49 National Physics Laboratory (1995) *Kaye and Laby Tables of Physical and Chemical Constants*, 16th edn, www.Kayelaby.npl.co.uk (accessed 5 November 2016).

50 Jones, T. B. (1995) *Electromechanics of Particles*, Cambridge University Press, Cambridge.

51 Iwasaka, M., Miyakoshi, J. and Ueno, S. (2001) Magnetophoresis of diamagnetic cells and microorganisms in a culture medium. *IEEE Trans. Magn.* **37**, 2644–2646.

52 Han, K.-H. and Frazier, B. (2005) Diamagnetic capture mode magnetophoretic microseparator for blood cells. *J. Microelectromech. Syst.* **14**, 1422–1431.

53 Takayasu, M., Duske, N., Ash, S. R. and Friedlaender, F. J. (1982) HGMS studies of blood cell behavior in plasma. *IEEE Trans. Magn.* **MAG-18**, 1520–1522.

54 Hatch, G. P. and Stelter, R. E. (2001) Magnetic design considerations for devices and particles used for biological high-gradient magnetic separation (HGMS) systems. *J. Mag. Mag. Mat.* **225**, 262–276.

55 Goya, G. F., Berquo, T. S., Fonseca, F. C. and Morales, M. P. (2003) Static and dynamic magnetic properties of spherical magnetite nanoparticles. *J. Appl. Phys.* **94**, 3520–3528.

56 Pamme, N. and Wilhelm, C. (2006) Continuous sorting of magnetic cells via on-chip free-flow magnetophoresis. *Lab Chip* **6**, 974–980.

57 Lai, M.-F., Chen, C.-Y., Lee, C.-P. *et al.* (2010) Cell patterning using microstructured ferromagnetic thin films. *Appl. Phys. Lett.* **96**, 183701.

58 Pankhurst Q. A., Connolly J., Jones S. K. and Dobson J. (2003) Applications of magnetic nanoparticles in biomedicine. *J. Phys. D: Appl. Phys.* **36**, R167–R181.

59 Gijs, M. A. (2004) Magnetic bead handling on-chip: New opportunities for analytical applications. *Microfluid. Nanofluid.* **1**, 22–40.

60 Xie, X., Nie, X., Yu, B. and Zhang, X. (2007) Rapid enrichment of leucocytes and genomic DNA from blood based on bifunctional core-shell magnetic nanoparticles. *J. Magn. Magn. Mater.* **311**, 416–420.

61 Choi, S., Karp, J. M. and Karnik, R. (2012) Cell sorting by deterministic cell rolling. *Lab Chip* **12**, 1427–1430.

62 Lai, M.-F., Chen, C.-Y., Lee, C.-P. *et al.* (2010) Cell patterning using microstructured ferromagnetic thin films. *Appl. Phys. Lett.* **96**, 183701.

63 Svedhem, S., Dahlborg, D., Ekeroth, J. *et al.* (2003) *In situ* peptide-modified supported lipid bilayers for controlled cell attachment. *Langmuir* **19**, 6730–6736.

64 Vieseh, M., Vieseh, O., Martin, M. C. *et al.* (2007) Short peptides enhance since cell adhesion and viability on microarrays. *Langmuir* **23**, 4472–4479.

65 Didar, T. F., Foudeh, A. M. and Tabrizian, M. (2012) Patterning muliplex protein microarrays in a single microfluidic channel. *Anal. Chem.* **84**, 1012–1018.

66 Huang, K.-W., Wu, Y.-C., Lee, J.-A. and Chiou, P.-Y. (2013) Microfluidic integrated optoelectronic tweezers for single-cell preparation and analysis. *Lab Chip* **13**, 3721–3727.

67 Yang, S.-M., Tseng, S.-Y., Chen, H.-P. *et al.* (2013) Cell patterning via diffraction-induced optoelectronic dielectrophoresis force on an organic photoconductive chip. *Lab Chip* **13**, 3893–3902.

68 Xu, S., Ren, H. and Wu, S.-T. (2013) Dielectrophoretically tunable optofluidic devices. *J. Phys. D: Appl. Phys.* **46**, 483001.

69 Issadore, D., Franke, T., Brown, K. A. *et al.* (2009) High-voltage dielectrophoretic and magnetophoretic hybrid integrated circuit-microfluidic chip. *J. Microelectromech. Syst.* **18**, 1220–1225.

70 James, C. D., McClain, J., Pohl, K. R. *et al.* (2010) High-efficiency magnetic particle focusing using dielectrophoresis and magnetophoresis in a microfluidic device. *J. Micromech. Microeng.* **20**, 045015.

71 Lyuksyutov, I. F., Naugle, D. G. and Rathnayaka, K. D. D. (2004) On-chip manipulation of levitated femtodroplets. *Appl. Phys. Lett.* **85**, 1817–1819.

72 Kauffmann, P., Chetouani, H., Pham, P. *et al.* (2009) Magnetophoretic and dielectrophoretic actuations coupled with diamagnetic trapping in air and liquids. *Sensor. Lett.* **7**, 470–474.

73 Jung, J., Seo, S.-K., Joo, Y.-D. and Han, K.-H. (2011) Label-free continuous lateral magneto-dielectrophoretic microseparators for highly efficient enrichment of circulating nucleated cells from peripheral blood. *Sensors and Actuators B* **157**, 314–320.

74 Blaire, G., Masse, A., Zanini, L. F. *et al.* (2013) Hybrid Bio-Mag-MEMS combining magnetophoresis and dielectrophoresis. *Eur. Phys. J. B* **86**, Art. no. 165.

75 Lock, G. M., Markx, G. H. and Pethig, R. (2002) Manipulation of particles in liquid media, EPO Patent 1202811-A1.

76 Wiklund, M., Günther, C., Lemor, R. *et al.* (2006) Ultrasonic standing wave manipulation technology integrated into a dielectrophoretic chip. *Lab Chip* **6**, 1537–1544.

77 Ravula, S. K., Branch, D. W., James, C. D. *et al.* (2008) A microfluidic system combining acoustic and dielectrophoretic particle preconcentration and focusing. *Sensors and Actuators B* **130**, 645–652.

78 Jones T. B., Gunji M., Washizu M. and Feldman M. J. (2001) Dielectrophoretic liquid actuation and nanodroplet formation. *J. Appl. Phys.* **89**, 1441–1448.

79 Chugh, D. and Kaler, V. I. S. (2008) Leveraging liquid dielectrophoresis for microfluidic applications. *Biomed. Mater.* **3**, 034009.

80 Valley, J. K., NingPei, S., Jamshidi, A. *et al.* (2011) A unified platform for optoelectrowetting and optoelectronic tweezers. *Lab Chip* **11**, 1292–1297.

2

How does Dielectrophoresis Differ from Electrophoresis?

2.1 Introduction

In Chapter 1 the methods of electrophoresis and dielectrophoresis were placed together as electrokinetic effects, with only brief descriptions of their nature. The intended objective of this book is to create an increased awareness of how dielectrophoresis can contribute to the biomedical sciences as both a research and practical tool. For those of us already engaged in these pursuits, the guidance and involvement of those trained in the molecular and life sciences is greatly desired and in most cases considered to be essential. However, published works on dielectrophoresis are mostly written by and for researchers having backgrounds in engineering or the physical sciences. Many of these publications are thus largely unhelpful in addressing the 'so what, who cares' questions of interest and relevance to those trained in the life and medical sciences. The similarity of the terms *electrophoresis* and *dielectrophoresis* is also not helpful in discouraging the impression that the latter represents a relatively esoteric extension of the former. The purpose of this chapter is therefore to describe in broad terms not only how the two subjects do share common foundations, but also how the special features of dielectrophoresis lend to it the promise of providing important contributions to cell biology in general and, in particular, to such areas as drug discovery, medical diagnostics and regenerative medicine.

Chapter 1 began by deciphering the origin and hence probable meaning of the compound word *electrophoresis*. We concluded, based on Greek translations, it implies something to do 'with electricity' and also 'with carrying things around'. We can therefore as a working definition assume that dielectrophoresis refers to an object carried by a dielectric effect. This was certainly the intended interpretation of Herbert Pohl who coined this term, with his statement [1]: 'The motion of suspensoid particles relative to that of the solvent resulting from polarization forces produced by an inhomogeneous electric field is defined as "dielectrophoresis".'

But what do we mean by a 'dielectric' effect? The prefix 'di-' translates from the Greek to mean 'across' as in to cut across or block. Thus, a *dielectric* can be thought of as a nonmetallic conductor, to include gases, liquids and solids that exhibit characteristic interactions with electric, magnetic and electromagnetic fields. Such interactions include the storage and dissipation of electric energy, where the dynamics and polarization of electric charges are fundamental aspects of this and are characterized by macroscopic properties such as permittivity, dielectric loss and electrical breakdown.

It took some time for the definition proposed by Pohl in 1951 to gain wide recognition. Twenty-six years later the title of a paper [2] was changed (despite the vociferous dissent of the senior author) to include the words '*measurements using nonuniform electric field effects*' because an eminent referee in the field of electromagnetics insisted that there was no such word as 'dielectrophoresis'! Pohl went on to write the seminal book on *his* subject [3]. A summary provided by an unknown reviewer on the book's dust cover concludes with the prophetic statement: 'A far wider range of potential applications exists than Professor Pohl has been able to include. The book should thus provide stimulating reading for imaginative research workers in the physical, medical and biological sciences.'

The two subjects of this chapter fall under the general category of 'electrokinetics'. It is well appreciated by molecular and biomedical scientists that the kinetics involved in the various forms of electrophoresis result from the interaction of the electrical charge carried by a particle and an applied electric field. Less familiar, and the subject of this book, are the electrokinetic effects that do not depend on the nature or even presence of a charge carried by a particle, but do depend on its dielectric properties. By bringing into consideration the

Dielectrophoresis: Theory, Methodology and Biological Applications, First Edition. Ronald Pethig.

dielectric properties of a particle we introduce a new set of theories, technologies and applications. These are all introduced in this chapter by describing

- common aspects as well as important differences of the principles of electrophoresis and dielectrophoresis;
- differences between their methods of measurement;
- different information that each technique can provide of the physico-chemical properties of cells and other bioparticles.

As in Chapter 1, the material is presented at two levels of detail. The style of the main text is largely descriptive, with the more theoretical details provided separately in 'boxes'. Worked examples are also included as a means to highlight the practical significances of the theory.

A common feature of electrophoresis and dielectrophoresis is that the motion of a particle is induced by its interaction with an applied electric field. A discussion of the concept of an electric field is thus a good starting point.

2.2 Electric Field

In most applications of electrophoresis the applied electric field E does not vary significantly as a function of space (i.e., it can be considered *uniform* in magnitude) and is generated using a *direct current* (DC) voltage applied to two electrodes. A schematic of how such a field can be produced is given in Figure 2.1.

Although the practical aspect of generating an electric field appears from Figure 2.1 to be straightforward, the concept involved is more subtle and takes us to the core of our subject. The convention is to define the electric field E at a point in space in terms of the electric force F_e it will exert on a test electrical charge q at that point, divided by the charge q. The force F_e is variously known as the *electric* force, *electrostatic* force or *coulombic* force.

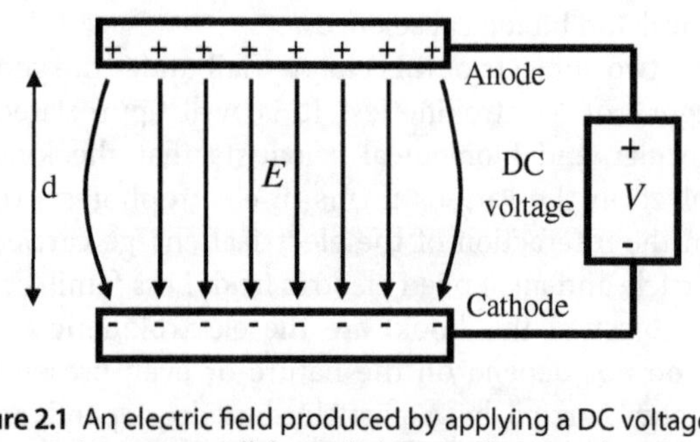

Figure 2.1 An electric field produced by applying a DC voltage difference *V* to two parallel metal plates. In the central section, away from the ends of the plates, the electric field E is uniform, has a magnitude of *V*/d volts per metre and is directed away from the positive voltage potential (anode) towards the negative potential (cathode).

We assume that the electric field already exists and that the test charge is used to measure the field. As analogies to this we can use a thermometer to test a temperature field and a mass suspended from a spring balance to test a gravitational field. The electric field is expressed mathematically as:

$$E = \frac{F_e}{q} \tag{2.1}$$

The electric field E at a point in space is thus equal to the electric force per unit charge experienced by a test charge at that point. In SI units the unit of force is 1 newton and the unit of charge is 1 coulomb (C), so that from Equation (2.1) the unit of electric field strength is 1 newton per coulomb (1 N/C). Recognizing that electric field strength or intensity is a force acting on a unit charge is important. For example, as shown in Chapter 6, it is central to understanding the concepts of electric displacement and internal field in a dielectric, as well as the derivation of an important parameter in dielectrophoresis – namely the Clausius–Mossotti factor.

Students trained in the physical and engineering sciences will often express electric field strengths in units of volts per metre rather than newtons per coulomb. This follows because the potential energy per unit charge associated with the test charge q is defined as the potential V at that point. V is a scalar quantity having magnitude but not direction. E is a vector field that possesses both magnitude and direction – it points in the direction of the greatest rate of decrease of the potential V. Mathematically the relationship between E and V is written as:

$$E = -\nabla V \tag{2.2}$$

where the *nabla* symbol ∇ denotes the gradient operator used in vector calculus (the name for this symbol is related to its shape being similar to an ancient Greek harp). ∇V is also read as *grad V* and the gradient refers to the rate of change of the potential V with distance along a particular direction. As given by Equation (2.2) the electric field vector E thus has units of volts per metre and points directly away from a positive potential towards a less positive or negative potential. Referring to Figure 2.1, the uniform field produced in the central region between the plate electrodes (away from the nonuniform fringing fields at their ends) has a magnitude given by V/d. As an example, for electrode plates spaced 1 cm apart and with a potential difference of 10 V applied to them, the field E produced has a magnitude of 1000 V/m (or 1000 N/C). Fields of between 500 and 8000 V/m are used in electrophoresis devices to separate cells in a cell mixture, according to differences in their electrophoretic mobilities.

We normally know or can calculate the magnitude of the electric field at any point of interest. By rearranging

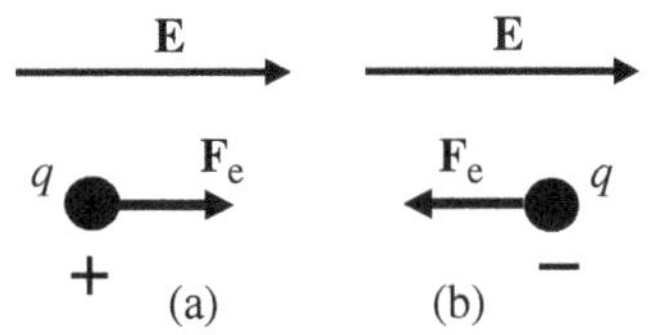

Figure 2.2 Force F_e exerted on a test charge q by an electric field E. (a) If q is a positive charge, F_e is directed along the same direction as E. (b) For the case where q is negative, F_e and E act in opposing directions. (Reproduced with permission of Wiley.)

Equation (2.1) the corresponding force exerted on a point test charge is given by:

$$F_e = qE \tag{2.3}$$

As q is a scalar quantity, having either positive or negative magnitude but no direction, then from Equation (2.3) the electric force F_e like E is also a vector. As shown in Figure 2.2, if q is a positive charge the force F_e acts in the same direction as E; if q is negative then F_e and E act in opposite directions.

2.3 Electrophoresis

Equation (2.3) was formulated for a *point* test charge because in general the electric force acting on it could vary at different points. However, we have noted that practical applications of electrophoresis mostly involve a uniform electric field, such as the one shown in Figure 2.1. In this situation, or where the dimension of the test particle is small compared to the length scale of a variation of the field strength, the electric field E can be considered to be the same in magnitude and direction at all points on the particle. Thus, we can dispense with the concept of a *point* charge and instead consider situations where an electric force acts on a real physical object, such as a cell or other bioparticle.

Measurement of the electrophoretic responses of cells freely suspended in solution represented one of the earliest of analytical methods to study the surface properties of cells and especially membrane surface charge [4,5]. Electrophoresis has also been developed as a preparative method to separate cell subpopulations from cell mixtures [6–9]. In methods known as free flow electrophoresis or microelectrophoresis, an electric field is established in a rectangular or cylindrical cuvette or chamber, so that with the aid of a microscope either the time taken for a cell to move a defined distance can be determined, or the deviation of a cell from a vertical fluid flow path (caused by applying an electric field perpendicular to the flow stream) be measured [10–12]. In the method known as laser Doppler electrophoresis, cells are induced to move through a laser beam and their velocities are determined by the extent to which light scattered from each cell is shifted in frequency [13,14]. An important aspect of such methods is that care is taken to ensure electro-osmotic flow of the liquid does not influence the measurements. Determinations of cell motion must be made in what is termed the stationary layer of fluid in the closed chamber, where the fluid flow is found to remain zero after application of the electric field.

All eukaryotic cells so far studied exhibit a *negative* electrophoretic mobility under normal physiological conditions – they move towards the anode and hence in the opposite direction to that of the applied field [15]. According to Figure 2.2 this informs us that mammalian cells carry a net *negative charge.* The number of negatively charged chemical groups per unit area of the external surface of a cell's plasma membrane exceeds the number of positively charged groups. These groups are most commonly associated with the carboxyl (COOH) and amino (NH_2) side chains of proteins and other molecules incorporated into the membrane structure. As a rough estimate there are about 10~20 acidic and basic groups per square nanometre of the membrane surface. An acidic carboxyl group can dissociate to give an ionized, negatively charged, COO^- group and a mobile proton (H^+), whereas a basic amino group can accept a proton to produce a positively charged NH_3^+ group. If the acidity of the solution around a cell is increased, for example by lowering the pH below pH 7, there will be a relatively large number of free protons in the solution. The probability that a proton will neutralize an ionized carboxyl group ($COO^- + H^+ \rightarrow COOH$) will be greater than the tendency for an ionized NH_3^+ group to give up its proton to the solution. On the other hand, raising the pH above pH 7 will have the opposite effect because there will be a relatively low concentration of free protons. There will be an increased tendency for carboxyl groups to give up a proton and become negatively charged and for NH_3^+ groups to give up their protons to the solution and become an uncharged NH_2 group. At a certain pH of the solution, depending on the relative number per unit area of carboxyl and amino side groups, the net charge on a membrane surface will be zero. The electrophoretic mobility of the cell will also be zero. This is known as the *isoelectric point*, given the symbol pI. In an early study of this effect Coulter [4] determined the isoelectric point for sheep red blood cells to occur at pH 4.6 (the cells moving towards the anode and then towards the cathode above and below this pH value, respectively). A quantitative analysis of the ionization of acidic and basic groups as a function of pH is provided by the Henderson–Hasselbalch equation derived in Box 2.1.

Although we may assume that mammalian cells carry a net, uncompensated, negative charge under normal physiological conditions at pH 7, bacteria may carry

Box 2.1 The Henderson–Hasselbalch Equation

According to the classical definition of acid and base, H_3O^+ is *acidic* (it can donate a proton) and OH^- is *basic* or alkaline (it can accept a proton). The generalized expression for the dissociation of an acid into a proton H^+ and its anion A^- is:

$$HA \leftrightarrow H^+ + A^-$$

Acids such as HCl dissociate completely and this reaction goes to completion:

$$HA \leftrightarrow H^+ + A^-$$

Other acids such as acetic acid dissociate only partially. In such cases, an acid dissociation equilibrium is established with a significant amount of undissociated acid [HA] being present. The acid dissociation constant K_a is defined as:

$$K_a = \frac{[H^+][A^-]}{[HA]}$$

Taking the log of both sides of this equation, we obtain:

$$\log K_a = \log[H^+] + \log\frac{[A^-]}{[HA]}, \text{ or}$$

$$-\log[H^+] = -\log K_a + \log\frac{[A^-]}{[HA]}$$

Substituting pH for $-\log$ [H^+] and pK_a for $^-\log K_a$ we obtain:

$$pH = pK_a + \log\frac{[A^-]}{[HA]}, \text{ or}$$

$$pH = pK_a + \log\frac{[proton\ acceptor]}{[proton\ donor]}$$

This is the Henderson–Hasselbalch equation, which permits the calculation of the degree of dissociation of an acid, given the pH of the solution and the pK_a of the acid.

either a net negative or positive charge [15, 16]. Depending on their relative contents of ionizable carboxyl side groups (aspartic acid, glutamic acid) and amino groups (arginine, histidine, lysine) protein particles may also carry a net negative or positive charge at pH 7. Examples of acidic proteins with pI values below 7.0 include pepsin, casein, insulin and albumin – and basic proteins with pI values above 7.0 include cytochrome-c, lysine and salmine. Haemoglobin and myoglobin have pI value close to 7. Nucleic acid particles (DNA, RNA) at pH 7 carry a net negative charge associated with the ionized phosphate groups in their polynucleotide chains.

Example 2.1 Net Charge on a Cell Membrane

1. Estimate, under normal physiological conditions at pH 7.2 the sign and magnitude of the net, uncompensated, charge density on a cell that contains on average 11.6 carboxyl and 1.93 amino groups per square nanometre of membrane surface. The pK_a values for the carboxyl and amino groups can be taken as 4 and 11, respectively.
2. Estimate the surface charge density at pH 3.4.

Solution 2.1

1. Using the Henderson–Hasselbalch equation derived in Box 2.1 we can write:

$$\log\frac{[COO^-]}{[COOH]} = 7.2 - 4 = 3.2$$

where COO^- and $COOH$ act as the proton acceptor and donor, respectively. The ratio of ionized (negatively charged) to nonionized (uncharged) carboxyl groups = antilog 3.2 = 1585. We can also write:

$$\log\frac{[NH_2]}{[NH_3^+]} = 7.2 - 11 = -3.8$$

The ratio of charged to uncharged amino groups = antilog 3.8 = 6310. We can therefore assume that almost 100% of the carboxyl and amino groups carry a charge. The net membrane charge is therefore (2.93 – 11.6) electronic charges per square nanometre = $-9.67 \times (1.6 \times 10^{-19})$ C/nm^2 = -1.55×10^{-6} C/cm^2.
2. At pH 3.4 we can assume that all of the amino groups (1.93 per nm^2) carry a positive charge. The ratio of charged to uncharged carboxyl groups = antilog (3.3 – 4) = 0.2. Therefore of the 11.6 carboxyl groups per nm^2, 1.93 per nm^2 carries a negative charge. Thus, the density of negative COO– charges exactly balances the density of positive $NH3^+$ charges. The cell membrane carries no net charge. This corresponds to the cell's isoelectric point.

Example 2.2 Electric Force Acting on a Cell

Estimate the magnitude of the electric force F_e that will act on a mammalian cell of diameter 10 μm in an electric field of 1000 V/m.

Solution 2.2 We will assume that mammalian cells typically carry an uncompensated negative charge density of ~1.5 μC/cm^2. A spherical cell of diameter 10 μm has a surface area ($4\pi R^2$) of ~3×10^{-6} cm^2 and so will carry a net charge of about -4.5×10^{-12} C. From Equation (2.3) the force F_e acting on this cell when subjected to a field of 1000 N/C is given by:

$$F_e = (-4.5 \times 10^{-12}\ \text{C}) \times (1000\ \text{N/C}) = -4.5 \times 10^{-9}\ \text{N}$$

The negative sign indicates that this force acts in the opposite direction to that of the electric field. To place the magnitude of this force into context, it is very small compared to the earth's gravitational force of 1.5 N that acts on an apple of mass ~150 g.

Example 2.3 Sedimentation Force Acting on a Cell
Estimate the force of sedimentation acting on a cell of diameter 10 μm when suspended in an aqueous electrolyte.

Solution 2.3 Two forces act on the cell, namely the gravitational force (weight) acting downwards and the buoyant force acting upwards. The buoyant force relates to Archimedes' principle:

> *A body that is partly or entirely submerged in a fluid is buoyed upwards by a force equal in magnitude to the weight of the displaced fluid.*

For a spherical particle of radius R of mass density γ_1 suspended in a fluid of mass density γ_2, the net force F_s acting on the particle is given by:

$$F_s = \frac{4}{3}\pi R^3 (\gamma_1 - \gamma_2)\, g$$

where g is the gravitational acceleration vector (9.815 m/s^2). The density of a weak electrolyte is ~1010 kg/m^3 and that of a typical cell ~1050 kg/m^3. For $R = 5$ μm:

$$F_s \approx (4\pi/3)(5\times10^{-6})^3(40)\times 9.815\,\text{kg m/s}^2 = 2\times10^{-13}\,\text{N}$$

The estimated electric force of 4.5×10^{-9} N obtained for Example 2.2 is thus some 20 000 times larger than the sedimentation force acting on the cell. It is also useful to compare the electric force with the randomizing Brownian force ($kT/2R$) experienced by particles when suspended in solution. For a cell of radius 5 μm the Brownian force will have a magnitude of around 4×10^{-16} N. The electric force is thus 10 million times greater than the randomizing force the cell will experience from thermally induced motions of surrounding water molecules.

The electric field E appearing in Equation (2.3) is generated by charges such as those shown induced in the parallel plate electrodes in Figure 2.1. A test charge brought into this field will generate its own electric field, but this field will not act on the test charge. This follows from the physical principle that a body cannot exert a net force on itself. If this were not so, we could for example lift ourselves off the ground by standing in a bucket and tugging with our hands on its handle! A familiar example, where a body sets up a field around itself and a second body responds to it, is the gravitational field set up by the Earth that exerts a gravitational force on other masses. Particles accelerated to Earth under the action of the gravitational field attain a terminal velocity, corresponding to the situation where there is a balance between the gravitational force and an opposing force (air friction) generated by its interaction with the surrounding atmosphere. A cell suspended in a liquid and subjected to an electric field will attain a steady-state velocity when the retarding fluid viscous force builds up to exactly counteract the electric force.

Viscous forces oppose the motion of one portion of a fluid relative to another (gases and liquids are classified as fluids). This results in the tendency for a boundary layer of fluid to remain in intimate contact with the surface of a body as it moves through a fluid. A phenomenon known as laminar flow is created around a body that moves relatively slowly through a fluid and an example of this is shown in Figure 2.3. As shown schematically in Figure 2.3(a) the flow speeds of intermediary fluid layers between the moving solid surface and the bulk fluid decreases uniformly as a function of distance away from body surface. The fluid layer immediately in contact with the body surface moves with the same velocity as the body – we say there is the condition of zero slip at the body-fluid interface. In principle, at some distance away from the body surface there will also be zero slip between the outermost fluid laminar layer carried by the moving body and its 'interface' with the bulk fluid. However, it is commonly the case that a body and its associated laminar layers of fluid 'part company' from the bulk fluid at a so-called 'slip plane'. Each elemental fluid layer carried by a moving body in a fluid is in a state of continuously increasing shear strain, caused by the fluid velocity

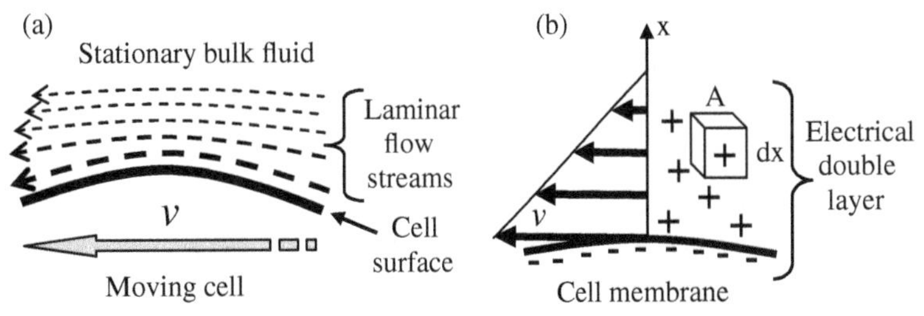

Figure 2.3 (a) A schematic of the laminar fluid layers moving with a cell relative to the stationary bulk fluid. The laminar boundary layer in contact with the cell surface moves at the same velocity v as the cell. (b) The velocity v of each fluidic laminar relative to the bulk fluid decreases linearly with distance x from the cell surface. A fluid volume Adx contains a positive ion charge density $\rho(x)$ associated with the electrical double layer.

Box 2.2 Application of Stokes' Law

The viscous fluid force $\mathbf{F}_s$ that will resist an applied force $\mathbf{F}_a$ of acceleration experienced by a spherical particle is given by Stokes' Law:

$$\mathbf{F}_S = 6\pi\eta R v \tag{2.4}$$

where R and $\boldsymbol{v}$ are the particle's radius and velocity, respectively and η is the dynamic viscosity of the fluid medium. The steady-state (terminal) velocity is attained when this force exactly balances the applied force (i.e., when $\mathbf{F}_S = \mathbf{F}_a$) and is thus given by:

$$F_S = F_a/(6\pi\eta R) \tag{2.5}$$

This is an application of Newton's First Law of Motion, which states that 'A body acted on by no net force moves with constant velocity (which may be zero) with zero acceleration.'

gradient across it. To counteract this distorting effect a viscous force (a shear stress) is generated in proportion to the rate of change of shear strain (known as the strain rate) given by the velocity gradient (dv/dx). The dynamic viscosity η of the fluid is defined as the ratio of the shear stress (force F per unit area A) to the strain rate:

$$\eta = \frac{Shear\ stress}{Strain\ rate} = \frac{F/A}{dv/dx}$$

The force F acting on each element of fluid moving with the body is thus given by:

$$F = \eta A \frac{dv}{dx}$$

The total viscous force acting on a moving body is the summation (integration) of all of the forces acting on each fluid element that moves with the body. For the case of a spherical body of radius R moving with velocity v the viscous drag force is given by Stokes' Law. This law is used in Box 2.2 to derive an expression for the steady-state velocity of a spherical particle under the influence of an accelerating force and a viscous drag force.

The concept of an electrophoretic mobility μ_e is used to quantify electrophoresis and is defined as the ratio of the steady-state velocity v and the applied field E:

$$\mu_e = \frac{v}{E} \tag{2.6}$$

with units of m^2 per volt second ($m^2\ V^{-1}s^{-1}$). In a data base comprising 288 types of eukaryotic cells, μ_e values from 0.5×10^{-8} to $3.5 \times 10^{-8}\ m^2 V^{-1}s^{-1}$ are reported [15]. The most studied cell type is the human red blood cell, having an established μ_e value of $1.1 \times 10^{-8}\ m^2 V^{-1}s^{-1}$. A simple way to calibrate electrophoresis equipment is to first make measurements using red blood cells as a standard reference. The μ_e value for a cell is commonly expressed in units of μm/s per volt/cm and in this form provides a useful insight of the magnitude of steady-state velocities and cell migration distances we are likely to observe in cell electrophoresis measurements (e.g., a red blood cell will travel 0.44 mm after 40 seconds exposure to a field of 1000 V/m in a stationary fluid). Purification of heterogeneous cell samples by electrophoresis should be possible for μ_e differences of 10–20% if the equipment used can distinguish mobility values within $\pm 0.05 \times 10^{-8}\ m^2$ per volt second.

Example 2.4 Stokes' Law has Limited Applicability to Cell Electrophoresis

A cell, of diameter 10 μm and carrying a net charge of -4.5×10^{-12} C, is suspended in an aqueous electrolyte at pH 7 and exposed to a uniform field of 1000 V/m. Estimate, taking into account the fluid viscous drag force given by Stokes' Law, the electrophoretic mobility μ_e of the cell. How does this estimate compare with typical values obtained experimentally?

Solution 2.4 Aqueous electrolytes have dynamic viscosities $\eta \sim 10^{-3}$ Pa s (1 pascal = 1 Pa = 1 N/m^2). When the accelerating force is equal to the electric force given by Equation (2.3), then from Equation (2.5) in Box 2.1 the steady-state velocity v for a cell of radius 5 μm, net charge -4.5×10^{-12} C, in a 1000 V/m field is given by:

$$v = \frac{qE}{6\pi\eta R}$$
$$= [(-4.5\times10^{-12}\ C)(10^3\ NC^{-1})]/[6\pi(10^{-3}\ Nm^{-2}\ s)(5\times10^{-6}\ m)] = -47.8\ mm/s.$$

The corresponding electrophoretic mobility value is given by Equation (2.6):

$$\mu_e = v/E = -47.8\times10^{-3}/10^3 = -4.8\times10^{-5}\ m^2\ V^{-1}\ s^{-1}$$

This value is more than 1000 times larger than that found experimentally for cells maintained at physiological pH [15]. We need to re-examine our application of Stokes' Law in calculating electrophoretic mobilities.

Stokes' Law, employed in the analysis presented in Box 2.1 to derive Equation (2.6), can be used to calculate the viscous drag force acting on an electrically *uncharged* spherical particle. However, as our analysis in Example 2.4 indicates, it is not applicable to the case of an electrically *charged* particle moving in an electrolyte under the influence of an electric field. This is because Stokes'

Box 2.3 Electrical Double-Layer Influence on Electrophoretic Mobility

As shown in Figure 2.3 the relative velocity (with respect to the bulk fluid) of the laminar fluid layers decreases linearly as a function of distance x from the membrane-fluid interface. In Chapter 12 (Equation 12.3), the shear stress τ exerted on each fluid layer is given by:

$$\tau = \eta dv/dx \tag{2.7}$$

At the steady-state condition of electrophoresis, the velocity of each fluid layer will be constant. The shear force and electric force acting on each fluid volume element Adx containing an ion charge density $\rho(x)$ must therefore be equal and opposite:

$$\eta A\frac{dv}{dx} = \mathrm{E}\rho(x)\,A\,dx \quad \text{or} \quad \mathrm{E}\rho(x) = \eta\frac{d^2v}{dx^2}$$

Replacing $\rho(x)$ using the Poisson equation (Equation (3.30) of Chapter 3):

$$-\varepsilon_o\varepsilon_m \mathrm{E}\frac{d^2\phi(x)}{dx^2} = \eta\frac{dv}{dx}$$

As boundary conditions for the integration of this equation, we can assume that as x tends to infinity the charged cell is effectively screened by the counterions (*i.e.*, the electric potential $\phi(x)$ tends to zero) and that the velocity of a fluid element Adx is zero. We also define an *electrokinetic* potential ζ, generally known as the *zeta* potential, as the potential at the boundary between the stationary bulk fluid and the surface layer of fluid moving with the cell (ζ will be less than the potential exactly at the surface of the cell membrane). On integration, we obtain the result:

$$-\varepsilon_o\varepsilon_m \mathrm{E}\varsigma = \eta v$$

This is the so-called Helmholtz–Smoluchowski equation and provides the following expression for the electrophoretic mobility μ_e:

$$\mu_e = \frac{v}{\mathrm{E}} = \frac{\varepsilon_o\varepsilon_m\varsigma}{\eta} \tag{2.8}$$

Law does not take into account the influence of the electrical double layer associated with a charged particle. As described in Chapter 12 counterions are attracted to uncompensated surface charges on a particle and are distributed into the laminar fluid layer that moves with the particle. This situation is shown schematically in Figure 2.3(a). Because they carry a charge of opposite polarity to that of the net charge on the particle, the electric forces acting on the counterions act against the electric force given by Equation (2.3). The overall scheme of forces acting on a charged particle is shown in Figure 2.4.

A derivation of the Helmholtz–Smoluchowski equation, used to relate the electrophoretic mobility with surface charge carried by a particle, is presented in Box 2.3. (An earlier theory of Helmholtz assumed that the counterions are located in a monolayer next to the

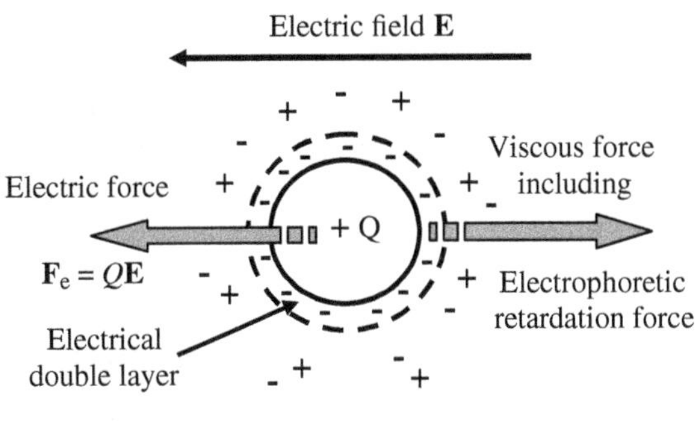

Figure 2.4 The steady-state velocity of a charged particle in an electric field is established when the electric force $\mathbf{F}_e$ is exactly balanced by the viscous drag force that includes the retardation resulting from the interaction of the field and the counterions in the electrical double layer.

uncompensated charges on the particle surface, but Smoluchowski later included the more diffuse nature of the electrical double layer.) Equation (2.8) derived in Box 2.3 works well for cases where the electrical double-layer thickness κ^{-1} (defined in Chapter 12 as the Debye screening length) is much less than the particle radius R. This applies to the case for cells suspended in aqueous electrolytes because the effective value for κ^{-1} is at most a few nanometres.

Example 2.5 Application of the Helmholtz–Smoluchowski Equation

The reported electrophoretic mobility of rat kidney cells is $-1.26 \times 10^{-8}\,\mathrm{m^2\,V^{-1}s^{-1}}$ when suspended in 150 mM NaCl at 306 K [15, 17]. Determine, using Equation (2.8), the corresponding value for the zeta potential.

Solution 2.5 Rearranging Equation (2.8) given in Box 2.3 the zeta potential ζ is:

$$\varsigma = \frac{\mu_e\eta}{\varepsilon_o\varepsilon_m}$$

in which ε_o is the permittivity of free space (8.85×10^{-12} F/m) and ε_m is the relative permittivity of the suspending medium. For an aqueous electrolyte $\varepsilon_m = 79$ and the dynamic viscosity $\eta = 10^{-3}$ Pa s (1 pascal = 1 Pa = 1 N/m^2). From the equation above:

$$\varsigma = (-1.26\times10^{-8}\,\mathrm{m^2\,V^{-1}\,s^{-1}})\times(10^{-3}\,\mathrm{N\,m^{-2}\,s})/(8.85\times10^{-12}\,\mathrm{m^{-1}}\times79)$$
$$= -18\times10^{-3}\,\mathrm{N\,m\,V^{-1}\,F^{-1}} = -18\,\mathrm{mV}.$$

(This result is obtained by noting that a farad has units of C/V and using the equivalence N/C = V/m in defining an electric field.)

The value for the zeta potential ζ obtained in Solution 2.5 corresponds to the electrostatic potential a small distance beyond the cell surface at the hydrodynamic plane of shear. No satisfactory theory appears to exist to calculate the corresponding potential at the charged membrane surface itself. However, based on a free radical quenching technique to probe the surface charge of ascites cells, a value of 5.7 mV was deduced as the difference between the cell surface potential (−20.3 mV) and the zeta potential of −14.6 mV [18]. Assuming a value close to −24 mV for the surface potential of a rat kidney cell, we can estimate the surface charge density using theories first developed by Gouy and Chapman and later extended by Grahame [19]. For cells suspended in a monovalent electrolyte such as NaCl the uncompensated charge density σ (μC/cm^2) on the cell membrane is given to a good approximation by:

$$\sigma = 12\sqrt{C}\frac{q\psi}{2kT}$$

in which q is the absolute magnitude of the charge on an electron (1.6×10^{-19} C), ψ is the membrane surface potential, C the molar concentration of ions in the bulk fluid and k Boltzmann's constant (1.602×10^{-23} J/K). Substituting $C = 150$ mM, $\psi = -24$ mV, $T = 306$ K into this expression we obtain a value for the cell surface charge density of $-1.8\ \mu$C/cm^2.

The electrophoretic mobility given by the Helmholtz–Smoluchowski equation (2.8) in Box 2.3 does not include any details regarding the size or shape of the particle. The separation by electrophoresis of particles suspended in solution therefore relies solely on differences in their surface charge densities. However, if the particles are immersed within a gel or other porous matrix, physical hindrance or entanglement effects can lead to separations that depend on particle size and shape, as well as surface charge density differences. This method is commonly used to separate bioparticles such as DNA and RNA, which have similar charge-to-mass ratios and thus exhibit similar electrophoretic mobilities.

2.4 Induced Surface Charge and Dipole Moment

Our discussion of the electrophoretic mobility of a macroscopic particle has taken no account of how an established uniform electric field might be disturbed as a result of inserting such a particle into it. This is a problem, presented in classical texts on electrostatics [e.g., 20, 21] that can be traced back to Maxwell's first formulation of the concept of an electromagnetic field [22]. The relevant theory is presented in detail in Chapter 10, but for our present purposes we can consider the imaginary exercise of first establishing a uniform field in a fluid. We then insert into this field an uncharged spherical particle whose electrical properties (conductivity and permittivity) match *exactly* those of the fluid. The insertion of the particle, requiring the displacement of fluid of equal volume, geometry and dielectric properties as the particle, will not be 'perceived' by the field. The uniform nature of the field will remain unaltered. However, consider the cases shown in Figure 2.5. If the test particle (e.g., a viable cell) is less conducting than the surrounding fluid the field and current flux lines will tend to concentrate in the fluid and avoid the particle by skirting around its surface. The counter effect occurs if the particle, such as a dying cell with an impaired membrane resistance, is more conductive than the surrounding fluid. The flux lines are now deflected into the cell interior and avoid the fluid near the particle. For these two cases and, as depicted in Figure 2.5, electric charges are induced at the interface between the fluid and the cell membrane.

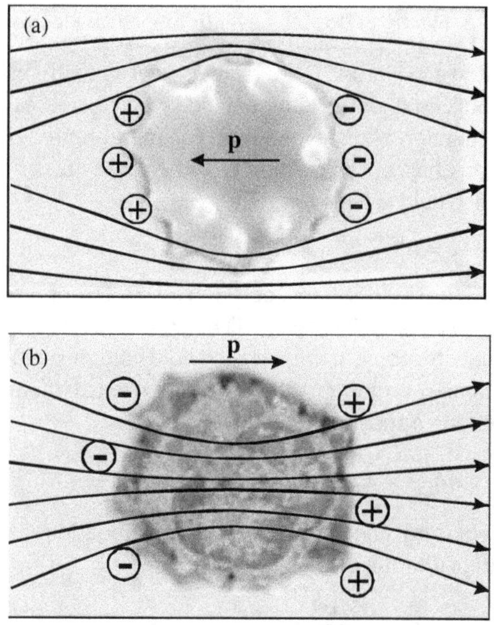

Figure 2.5 (a) A cell with an intact and viable plasma membrane will resist passive ion flow and appear to an imposed DC electric field as an electrically insulating particle. The resulting ionic currents will skirt around the cell to seek more conductive paths in the surrounding electrolyte. Induced charges appear at the membrane-electrolyte interface to produce a dipole moment **p** orientated against the applied field. (b) A cell with an impaired plasma membrane, suspended in a poorly conducting fluid, can appear as a conducting particle. The field will then penetrate into the cell interior and induced charges will lend to the cell a dipole moment aligned in the same direction as the field.

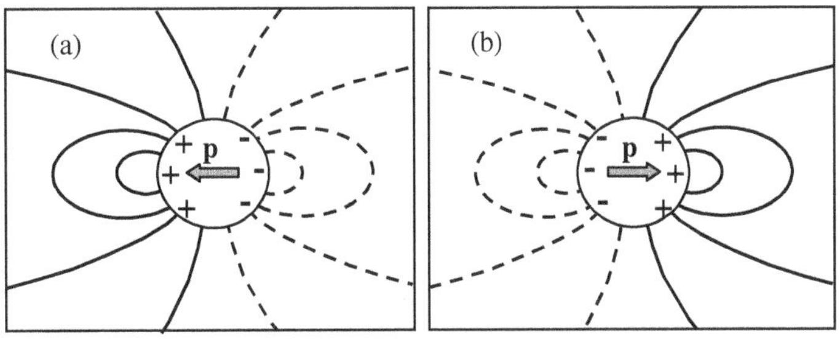

Figure 2.6 Equal and opposite surface charges are induced on a particle in an electric field, and form a dipole moment **p** that generates positive (solid lines) and negative (dotted lines) electric potential distributions. The dipole moment orientations and potential distributions shown in (a) and (b), respectively, correspond to the polarizations of the cells shown in Figures 2.5(a) and (b), respectively.

As described in greater detail in Chapters 5 and 10, the induced charges are distributed as equal numbers of positive and negative charges and polarized so as to occupy opposite sides of the particle to form a dipole moment. The net magnitude of the additional charge taken up by the particle is thus zero and will not influence the electrophoretic mobility in a *uniform* electric field. In a uniform field, the coulombic forces acting on each half of the dipole will be equal and of opposing direction. The induced dipole moments generate distributions of electric potentials as shown in Figure 2.6. Superposition of each potential distribution onto the original uniform field will produce the corresponding field distortions depicted in Figure 2.5.

The magnitude of the dipole moment p induced by the field E is determined by the effective polarizability α of the particle's material. The polarizability is defined as the *induced dipole moment per unit volume in unit field.* For cases where the particle consists of a material that is homogeneous, isotropic and whose polarization is linearly proportional to the applied field, the induced moment p is thus given by:

$$\mathrm{p} = \alpha v \mathrm{E} \tag{2.9}$$

where v is the volume of the particle. For the case of a spherical particle of radius R and, as described in detail in Chapter 6, the polarizability factor α can be presented as:

$$\alpha = 3\varepsilon_0\varepsilon_m[CM] \tag{2.10}$$

The factor [CM], generally known as the Clausius–Mossotti factor, depends on the electrical properties (conductivity σ and permittivity ε) of the particle and its surrounding fluid and is bounded by the values:

$$-0.5 < [CM] < 1.0 \tag{2.11}$$

Negative $[CM]$ values arise where the particle is less polarizable than the fluid (i.e, its σ and / or ε value is less than the fluid's) and corresponds to the situations shown in Figure 2.5 (a) and Figure 2.6 (a). The situations shown in Figure 2.5 (b) and Figure 2.6 (b) correspond to a positive $[CM]$ value where the particle is *more* polarizable than its surrounding medium. Substitution of Equation (2.10) into Equation (2.9), together with the volume for a spherical particle of radius R, gives the following expression for p:

$$\mathrm{p} = 4\pi R^3\varepsilon_o\varepsilon_m[CM]\,\mathrm{E} \tag{2.12}$$

Example 2.6 Magnitude of Dipole Moment Induced in a Cell

A cell of diameter 10 μm is suspended in an aqueous electrolyte and exposed to a uniform field of 1000 V/m. Calculate the magnitude and direction (with respect to the applied field) of the dipole moment induced in the cell for the two values of the Clausius–Mossotti factor: [CM] = 0.5 and [CM] = −0.2.

Solution 2.6 The induced dipole moment is given by Equation (2.12):

$$\mathrm{p} = 4\pi R^3\varepsilon_o\varepsilon_m[CM]\,\mathrm{E}$$

$$\mathrm{p} = 4\pi(5\times10^{-6}\,\mathrm{m})^3\times 79\,(8.854\times10^{-12}\,\mathrm{F/m})\;[\mathrm{CM}]\,10^3\,\mathrm{V/m}$$

Noting that farad volt ≡ coulomb, then

$$\mathrm{p} = 1.1\times10^{-21}[\mathrm{CM}]\,\mathrm{Cm}$$

For [CM] = 0.5, p = 5.5 × 10^{-22} C m (aligned in the *same* direction as the field). For [CM] = −0.2, p = −2.2 × 10^{-22} C m (aligned *against* the field direction).

Historically, the magnitude of a dipole moment has been measured in debye units; 1 debye (1 D) = 3.34 × 10^{-30} coulomb metre and so is of the order of magnitude of a fraction of electronic charge multiplied by the order of molecular dimensions. Water, considered to be one of the more polarizable of small molecules, has O–H bond lengths of ~0.1 nm and dipole moment of 1.86 D, giving a polarization charge +0.33e on each hydrogen and −0.66e on the oxygen atom (where e is the electronic charge of 1.6 × 10^{-19} C). A myoglobin protein molecule exhibits a dipole moment of ~150 D [23]. Thus, expressed in debye units we gain an appreciation of the magnitudes of the polarized charges and molecular sizes involved. The dipole moment of 5.5 × 10^{-22} C m, calculated in Example

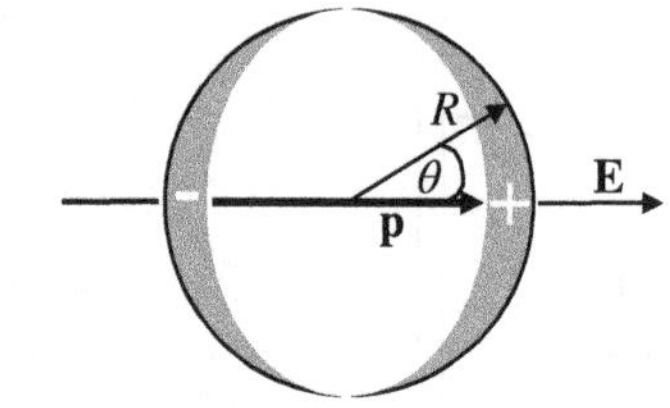

Figure 2.7 The induced surface charge density (shown as grey shading) on a spherical particle is distributed as a cosine function of the angle θ between the radius vector R and the direction of the induced dipole moment p aligned with the field E.

2.6 for a 10 μm diameter cell is equivalent to 1.6×10^8 D and thus represents a gigantic combination of charge and dimension compared to a water or protein molecule. We can estimate the magnitude of polarized charge Q for the cell using the definition of a dipole moment p:

$$p = Qd \tag{2.13}$$

and treating Q as a localized charge on opposite sides of a cell of diameter d (see Figure 2.8). From Equation (2.13), a dipole moment of 5.5×10^{-22} C m for a 10 μm diameter cell corresponds to $Q = 5.5 \times 10^{-17}$ C. This is *smaller* in magnitude, by a factor of ~10^5, than the typical net surface charge of -4.5×10^{-12} C carried by a cell at physiological pH (see Example 2.2). However, this net physiological charge is distributed uniformly around the cell surface and so produces a net dipole moment of zero, whereas the induced charges of $+5.5 \times 10^{-17}$ C and -5.5×10^{-17} C are each located on opposite sides of the cell. The induced charges are not in fact localized as point charges, but are distributed around the cell surface according to a cosine function of the angle between the field direction and the radial vector from the centre of the

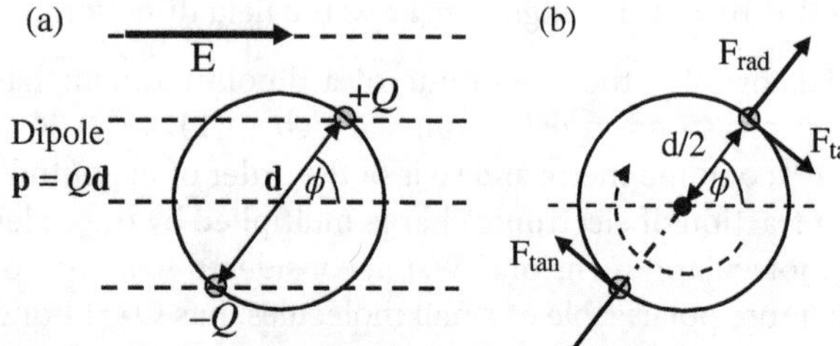

Figure 2.8 (a) A spherical particle with a dipole moment **p** is shown oriented at an angle ϕ to an imposed uniform electric field E. (b) The coulombic force experienced by each charge of the dipole is shown resolved into its radial and tangential components. The radial components ($F_{rad} = \pm QE\cos\phi$) cancel each other, but the tangential components ($F_{tan} = QE\sin\phi$) combine to produce a net rotational torque ($T = pE\sin\phi$).

spherical cell. The induced surface charge distribution is shown schematically in Figure 2.7.

2.5 Dielectrophoresis

If a polarized cell is aligned with a *uniform* electric field, as shown in Figure 2.7, the induced positive and negative charge distributions experience the same magnitude of coulombic force but in opposing directions. In this situation the net electric force acting on the polarized particle is thus zero. However, if the particle's moment p is oriented at an angle ϕ with the uniform field E, as shown in Figure 2.8, the particle will experience a net electrical force. This is not a translational force, but one that causes the particle to rotate about its axis at right angles to the plane of the field and induced moment. The radial force (F_{rad}) shown in Figure 2.8(b) that acts on the positive charge $+Q$ is given by the product of the charge and the field component $E\cos\phi$ acting along the radial direction. The net radial force $(+QE - QE)\cos\phi$ acting along the axis of the moment p cancel out each other. The particle will not be induced to move laterally. The component of force (F_{tan}) at right angles to the radial force that acts on the positive charge is given by $F_{tan} = +QE\cos(90° - \phi) = QE\sin\phi$. This force acts on what can be thought of as a lever of length d/2 connected to the centre of the particle. The torque T of a force about a point is given by the product of the force magnitude and the lever arm length and so for the positive charge $T^+ = (QE\sin\phi)d/2)$. The torque T^- acting on the negative charge of the dipole has the same magnitude and its direction is such as to add to T^+. There is thus a total torque $T = QdE\sin\phi = pE\sin\phi$ acting on the particle. This torque will act to align the moment p along the field direction, at which situation $\sin\phi = 0$ and $T = 0$. We can therefore assume that under equilibrium conditions the dipole moment induced in a particle is aligned with the local external electric field. We should note, though, that a polarized particle of molecular dimension will be 'buffeted' by thermal randomizing forces (Brownian motion) so that angle ϕ will be small but always changing. From this we can deduce that although a cell might exhibit an electrophoretic mobility in a uniform field, this will depend on the net surface charge associated with ionized chemical groups on the cell membrane, but will not be influenced by the existence of an induced dipole moment.

As shown in Figure 2.9, for a cell exposed to a *nonuniform* field the algebraic sum of the translational forces acting on the positive and negative elements of the induced dipole moment is not zero. The resultant translational force is known as the *dielectrophoretic* force and the mathematical expression for this is derived in Box 2.4. The force is found to be proportional to the product of the local field E and its gradient ∇E.

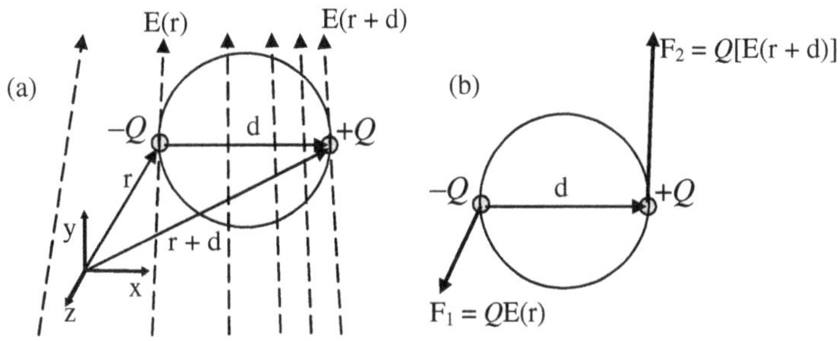

Figure 2.9 (a) The component charges of a dipole are shown located in a field gradient (illustrated by the unequal spacing of the field lines). (b) Because the field has a gradient, the component charges of the dipole experience different coulombic forces F_1 and F_2. The vector difference between F_1 and F_2 is the dielectrophoretic force acting on the dipole.

Example 2.7 Magnitude of the Field Factor (E·∇)E
A particle is located a radial distance of 40 μm from the inner electrode in the electrode arrangement shown in Figure 2.10. Derive a value for the field E and the field factor (E·∇)E at this location for the case where $r_1 = 150$ μm, $r_2 = 750$ μm and a voltage of 5 V is applied to the inner electrode.

Solution 2.7 The spherical electrode geometry shown in Figure 2.10 can be approximated as a spherical capacitor composed of an inner sphere of radius r_1 and an outer concentric shell of radius r_2. The potential $V(r)$ at a point r ($r_2 \geq r \geq r_1$) when a voltage V is applied to the inner electrode and the outer one is grounded is given by:

$$V(r) = \frac{V\, r_1\,(r_2 - r)}{r\,(r_2 - r_1)}$$

Box 2.4 Dielectrophoretic Force

Consider Figure 2.9, which shows the general case of a polarized small sphere located in a nonuniform electric field. The component charges, $+Q$ and $-Q$, of the sphere's induced dipole moment will experience different electric forces. The net difference between these forces is defined as the dielectrophoretic force F_{DEP}.

The force $\mathbf{F}_{DEP}$ is given by:

$$F_{DEP} = Q\,[E(r + r) - E\,(r)]$$

On performing a Taylor series expansion of $E(r + d)$ about r, and taking the effective length of the dipole shown in Figure 2.9 as $d = 2R$, (i.e., $\mathbf{p} = Q2R$) then:

$$F_{DEP} = Q\left[E(r) + 2R\cdot\frac{\partial E}{\partial r} + \cdot 2R^2\cdot\frac{\partial^2 E}{\partial r^2}\cdot\cdot + \frac{(2R)^n}{n!}\cdot\frac{\partial^n E}{\partial r^n} + \cdots\right] - QE(r)$$

If the sphere's diameter $2R$ is much smaller than the scale of the field nonuniformity $\delta^n E/\delta r^n$, we can ignore the terms where n > 1, to give to a good approximation:

$$F_{DEP} = Q2R\cdot\frac{\partial E}{\partial r} = (\mathrm{p}\cdot\nabla)E \tag{2.14}$$

Substituting into this equation the expression given in Equation (2.12) for p:

$$F_{DEP} = 4\pi R^3\varepsilon_o\varepsilon_m\,[CM]\,(E\cdot\nabla)E \tag{2.15}$$

As ∇ is a differential operator with respect to the space coordinates, it operates only on what follows it. Thus (E·∇)E is interpreted as the scalar (or dot) product of the vector E and the vector ∇E. As shown in Figure 2.12, the dielectrophoretic force acts along the direction of ∇E and not along the field lines.

Box 2.5 Refinement of Dielectrophoretic Force Equation

From Equation (2.15) in Box 2.4 we have:

$$F_{DEP} = 4\pi R^3\varepsilon_o\varepsilon_m\,[CM]\,(E\cdot\nabla)E$$

This can be simplified by employing the vector transformation [24]:

$$(A\cdot\nabla)B = \nabla(A\cdot B) - (B\cdot\nabla)A - A\times(\nabla\times B) - B\times(\nabla\times B)$$

In our case A = B = E. We have also, in Equation (2.2), defined E as the gradient of a scalar quantity. This is a necessary and sufficient condition to classify E as an irrotational field, so that $\nabla\times E = 0$ [24, p. 149]. The vector transformation of (E·∇)E thus leads to the identity:

$$2(E\cdot\nabla)E = \nabla E^2$$

Validation of this relationship for the field generated by coaxial cylindrical electrodes is given by Equations (3.45) and (3.46) in Chapter 3, and can also be checked for the fields produced by various forms of charge distributions given in Table 3.2.

Equation (2.15) for the dielectrophoretic force can thus be written as:

$$F_{DEP} = 2\pi R^3\varepsilon_o\varepsilon_m\,[CM]\nabla E^2 \tag{2.16}$$

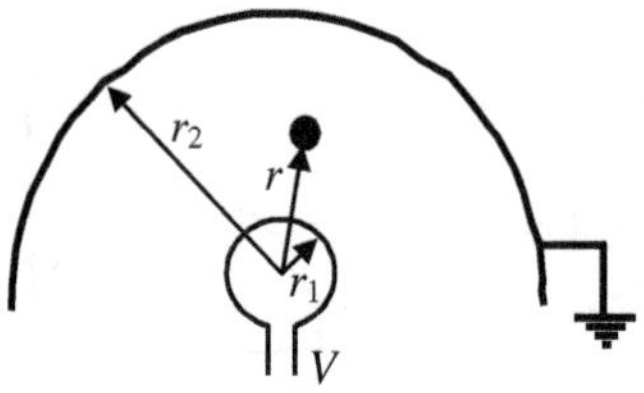

Figure 2.10 A particle located at radial distance r in a nonuniform electric field generated by electrodes of spherical geometry and radii of curvature r_1 and r_2. A voltage V is applied to the inner electrode and the outer one is grounded to earth potential. The product of the field strength and field gradient (**E.∇E**) acting on the particle is given by Equation (2.18).

The field at this point is given by:

$$\mathrm{E} = -\nabla V = \frac{r_1 r_2 V}{r^2 (r_2 - r_1)} \mathrm{r}_o$$

where $\mathrm{r_o}$ is the unit radial vector. For $r_1 = 150$ µm, $r_2 = 750$ µm, $r = 190$ µm, $V = 5$ V, from this equation:

$$\mathrm{E} = 2.6 \times 10^4 \,\mathrm{V/m}.$$

$$\nabla \mathrm{E} = \frac{\partial \mathrm{E}}{\partial r} = -\frac{2 r_1 r_2 V}{r^3 (r_2 - r_1)} \mathrm{r}_o \tag{2.17}$$

The negative sign indicates that the field gradient ∇E decreases with increasing radial distance r from the inner electrode. From these expressions for E and ∇E:

$$(\mathrm{E} \cdot \nabla)\mathrm{E} = -\frac{2 r_1^2 r_2^2 V^2}{r^5 (r_2 - r_1)^2} \mathrm{r}_o \tag{2.18}$$

For $r_1 = 150$ µm, $r_2 = 750$ µm, $r = 190$ µm, $V = 5$ V, from Equation (2.18):

$$(\mathrm{E}\nabla)\mathrm{E} = -7.1 \times 10^{12} \quad \mathrm{V^2/m^3}$$

Values for E and −(E·∇)E as a function of the radial distance r from the inner electrode are given in Figure 2.11.

Example 2.8 Magnitude of the Dielectrophoretic Force

Consider the particle in Example 2.7 to be a cell of diameter 10 µm suspended in an aqueous medium. Calculate the dielectrophoretic force F_{DEP} acting on the cell for the two values of the Clausius–Mossotti factor: $[CM] = 0.5$ and $[CM] = -0.2$.

Solution 2.8 From Equation (2.15) in Box 2.4:

$$\mathrm{F}_{DEP} = 4\pi R^3 \varepsilon_o \varepsilon_m [CM] (\mathrm{E} \cdot \nabla)\mathrm{E}$$

For $R = 5$ µm, $\varepsilon_\mathrm{m} = 79$ (noting $\varepsilon_\mathrm{o} = 8.854 \times 10^{-12}$ F/m) and using the value for (E·∇)E of $-7.1 \times 10^{12}\,\mathrm{V^2/m^3}$ obtained in Example 2.7:

$$\mathrm{F}_{DEP} = -7.8 \times 10^{-12}\,[\mathrm{CM}]\,(\mathrm{V\,F.V/m = N})$$

For $[CM] = 0.5$: $\mathrm{F}_{DEP} = -3.9 \times 10^{-12}$ N.

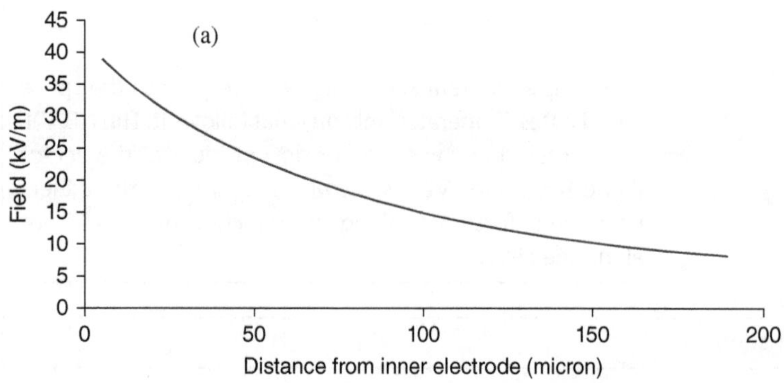

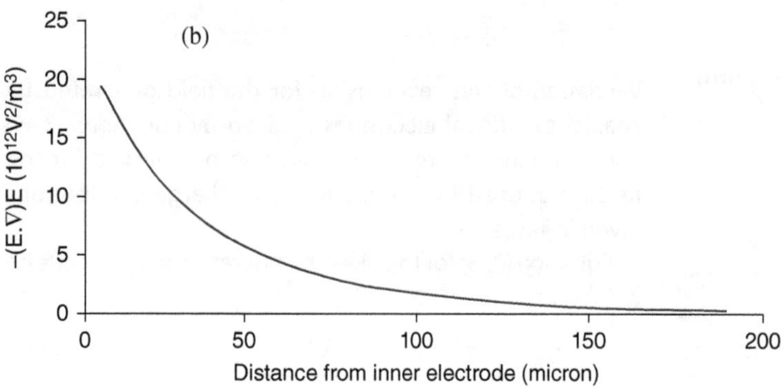

Figure 2.11 (a) Value of the field E with radial distance from the inner electrode of the geometry shown in Figure 2.9, with $r_1 = 150$ µm, $r_2 = 750$ µm and an applied voltage of 5 V. (b) The corresponding value of the field factor −(E·∇)E (note the negative sign). (Reproduced with permission of Wiley.)

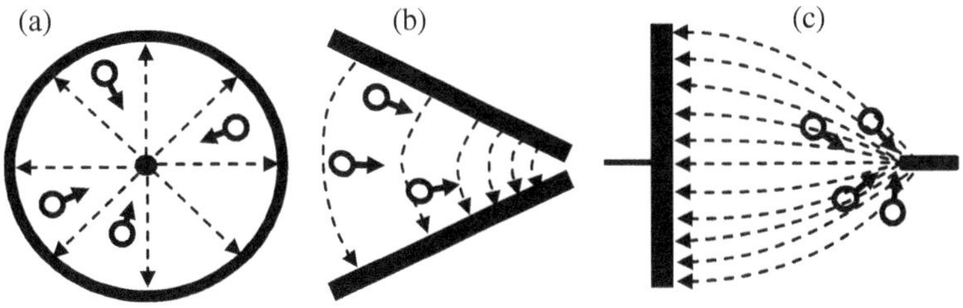

Figure 2.12 Particles moving by positive dielectrophoresis for different electrode geometries. (a) Particles move along the ∇E vector parallel to the electric field lines. (b) Particles move along the ∇E vector perpendicular to the field. (c) Particles move along the ∇E vector at a small angle to the field. For all three geometries, positively charged particles exhibiting electrophoresis would move along the field line directions.

The magnitude of this dielectrophoretic force is about 20 times larger than the sedimentation force acting on the cell (see Example 2.3) and 10 000 times larger than the randomizing Brownian force. The minus sign indicates that this force will direct the cell along a radial direction *towards* the inner electrode – i.e., *up* the electric field gradient. This is termed *positive* dielectrophoresis.

For $[CM] = -0.2$: $F_{DEP} = 1.6 \times 10^{-12}$ N. This positive value indicates that the dielectrophoretic force is such as to direct the cell along a radial direction *away from* the inner electrode – i.e., *down* the electric field gradient. This is termed *negative* dielectrophoresis.

From Box 2.4 we have the following expression for the dielectrophoretic force:

$$F_{DEP} = 4\pi R^3 \varepsilon_o \varepsilon_m [CM](E \cdot \nabla)E$$

From this equation we have confirmation of the following:

- The dielectrophoretic force is zero if the field is uniform (i.e., if $\nabla E = 0$).
- The dielectrophoretic force is ponderomotive. With all other factors remaining constant the larger the particle volume the greater will be the dielectrophoretic force acting on it.
- Electrode geometry is an important experimental design consideration in the control of the field factor (E.∇)E. From Examples 1.7 and 1.8 we find that a value for this factor of 10^{16} V^2/m^3 can result from a modest applied voltage of 1 V and produce a significant dielectrophoretic force on a cell. (E.∇)E has dimensions of V^2/m^3 and so its magnitude can be increased by a suitable scaling down of the electrode dimensions. For example, a fixed value of the dielectrophoretic force can be achieved for a hundredfold reduction of the applied voltage with a thousandfold reduction of the characteristic scale of the electrodes.
- The polarity (positive or negative) of the dielectrophoretic force is determined by the particle's polarizability relative to that of the surrounding medium, as expressed by the Clausius–Mossotti factor $[CM]$. If $[CM]$ is positive (i.e., the particle is more polarizable than the suspending medium) the force is directed up a field gradient towards an electrode. This provides the condition for *positive* dielectrophoresis. If $[CM]$ is negative (the particle is less polarizable than the medium) the force is directed down a field gradient away from an electrode. This provides the condition for *negative* dielectrophoresis.
- The direction of motion of a particle induced by a dielectrophoretic force is determined by the field gradient vector ∇E. The concept of the scalar product (E.∇)E is discussed in Chapter 10, but for our present purposes it is sufficient to state that the direction of motion is determined by the direction of ∇E, which as shown in Figure 2.12 depends on the electrode geometry. For example, for a cyclindrical or spherical geometry ∇E is parallel to the field E, but for a funnel design such as that shown in Figure 2.12(b) ∇E acts at right angles to the field lines.

These facts highlight the following differences between electrophoresis and dielectrophoresis:

- Electrophoresis can be performed using fields that are either uniform or nonuniform, but dielectrophoresis requires the application of a large field gradient.
- The electrophoretic force is zero, but the dielectrophoretic force is not, if the particle carries no net fixed charge. Dielectrophoresis depends on the existence of field-induced charges that lead to the cell having the properties of a dipole moment.
- Particle size does not influence electrophoretic mobility, but the dielectrophoretic force is directly dependent on particle size (as shown in Chapter 10).
- Electrode geometry is not of major consideration for enhancing electrophoresis, but is important in the design of dielectrophoresis devices, where high field strengths and specific forms of field nonuniformity are often important.

However, an important distinguishing feature of dielectrophoresis is revealed in Equation (2.16), where the field factor (E·∇)E is shown to relate to the *square* of the applied voltage. Thus, dielectrophoresis is independent of the polarity of the applied field direction (because $(-1)^2 = (+1)^2 = 1$). Alternating current (AC) as well as direct current (DC) voltages can be used, whereas meaningful electrophoresis experiments can only be performed using DC fields. This conclusion is drawn not

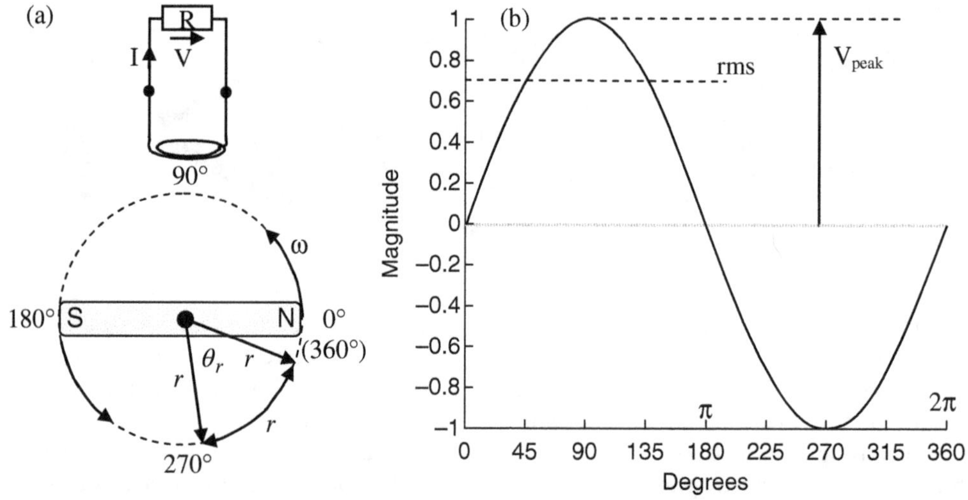

Figure 2.13 (a) A simple form of an AC generator consists of a coil of wire placed above a rotating bar magnet. A current I and voltage V is generated when the coil experiences a change in magnetic field, cycling through positive and negative peaks when the north and south pole, respectively, pass close to it. The rate of rotation of the magnet can be expressed as either cycles or radians per second, where one radian θ_r is defined as the angle subtended by an arc of length equal to the radius. The circumference of a circle is $2\pi r$, so that 2π radians is equivalent to 360°. (b) If the bar magnet is rotated at a constant rate the generated voltage (and current) takes the form of a sine wave, with a repeat wavelength of 360° (2π radians). The root mean square (rms) value of the sinusoid is given by the peak value divided by the square root of two, and is equal to $0.707\,V_{pk}$.

solely from the use of the particular electrode geometry shown in Figure 2.10, but as stated in Box 2.5 is applicable to all the fields that can be generated using electrodes. An alternating current is one in which the flow of charge periodically reverses direction, whereas in a direct current the charges flow consistently in one direction. The abbreviations AC and DC are also used to indicate whether a voltage or electric field is either alternating or direct, respectively. A simple concept, based on Faraday's discovery of magnetoelectric induction, describing the generation of an AC current and voltage is shown in Figure 2.13.

From Box 2.5 we have an alternative expression to Equation (2.15) for the dielectrophoretic force:

$$F_{DEP} = 2\pi R^3 \varepsilon_o \varepsilon_m \,[CM]\, \nabla E^2 \tag{2.19}$$

in which the magnitude of the field E is the root mean square (rms) value as derived in Box 2.6 and also defined in Figure 2.13. Using the rms field value can be useful because this equation also gives the dielectrophoretic force for the case of an applied DC field. AC voltmeters and ammeters show the rms value of the voltage and current and the rms value is also used to specify the household AC mains supply (e.g., 120 V in the United States, 230 V in Europe and many other countries). Equation (2.19) is the one most often presented in publications on dielectrophoresis and in this form emphasizes the square-law dependence of the force upon the applied field E. This means that the direction of the dielectrophoretic force is independent of the polarity of the voltage applied to an electrode, as for example the inner one shown in Figure 2.10. The voltage can be AC or DC. On the other hand, electrophoresis produces a particle motion whose direction depends on the polarity of the charge on the particle and also upon the polarity of the applied voltage. Reversal of the voltage polarity and hence also of the field, reverses the direction of electrophoretic motion. Above a certain frequency, the inertia of the particle will bring it to a halt.

But what is the advantage of using an AC voltage to achieve dielectrophoresis? The particle radius R does not depend on the field frequency and even if the suspending medium permittivity ε_m and field factor ∇E^2 did change appreciably (which they do not over the frequency range usually employed) this would not be of much use or interest. The only factor in Equation (2.19) that can change with frequency is $[CM]$. In Chapter 10 it will be shown that $[CM]$ depends on the effective conductance and electrical capacitance of the particle and that under appropriate experimental conditions a cell can be observed to change from positive to negative dielectrophoresis simply by altering the frequency of the applied field. An example of this is shown in Figure 2.14 for yeast cells, using electrodes of similar geometry to that shown in Figure 2.10.

The effect shown in Figure 2.14, where a transition from negative to positive dielectrophoresis occurs as the field frequency is increased from the kHz range of

Box 2.6 The RMS Value of a Sinusoidal AC Waveform

The usefulness of the root mean square (rms) value of a sinusoidal current or voltage waveform is that it gives the same heating effect $(I_{DC})^2R$ or $(V_{DC})^2/R)$ as the equivalent DC power dissipated in a resistor R. Consider a sinusoidal voltage of constant radian frequency $\omega = 2\pi f$, where f is the frequency (cycles / second):

$$V(t) = V_{pk}\sin(\omega t)$$

The periodic time constant T is the reciprocal of the frequency, so that $T = 2\pi/\omega$. At any instant in time the power dissipated is $V^2(t)/R$, and the time-averaged value of this over a complete cycle of the sinusoidal voltage is given by:

$$\left\langle \frac{V(t)^2}{R} \right\rangle = \frac{1}{T}\int_0^T \frac{V_{pk}^2}{R}\sin^2(\omega t)\,dt = \frac{1}{R}\frac{V_{pk}^2}{2T}$$

$$\left[t - \frac{1}{2\omega}\sin(2\omega t)\right]_0^T = \frac{V_{pk}^2}{2R}$$

This is defined as the equivalent heating effect of the rms voltage, so that:

$$\frac{V_{rms}^2}{R} = \frac{V_{pk}^2}{2R}, \quad \text{or} \quad V_{rms} = \frac{V_{pk}}{\sqrt{2}} = 0.7071 \times V_{pk}$$

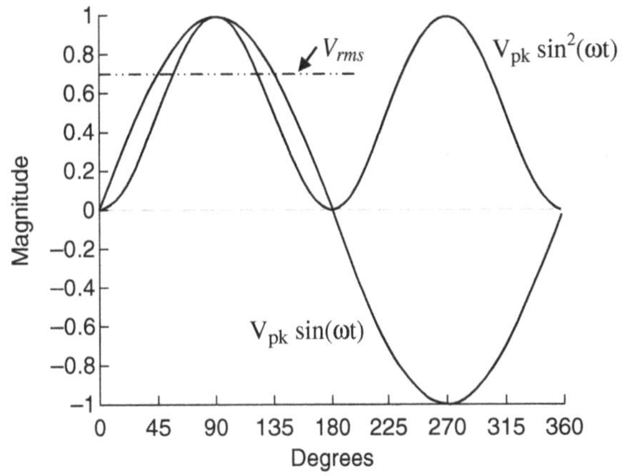

This figure shows plots of the AC voltage waveform $V(t) = V_{pk}\sin(\omega t)$ and the square of this function (for $V_{pk} = 1$ volt). The root mean square voltage (V_{rms}) is indicated.

frequencies to the MHz range, can be employed as a check on cell viability. When suspended in an appropriate medium, the high resistance to passive ion flow across the plasma membrane of a viable cell lends to it the properties of a particle of low polarizability having a negative [*CM*] value. For cells suspended in a weakly conducting medium, this can result in negative dielectrophoresis. As the field frequency is increased to the MHz region, the capacitance of the membrane can act to electrically bypass its high resistance so that the electric field penetrates into the conducting cytoplasm. In a weakly conducting medium the [*CM*] value can change from a negative to a positive value so that the cell exhibits positive dielectrophoresis. A cell with a compromised plasma membrane, resulting from apoptosis, necrosis or cell damage, will generally not exhibit a negative dielectrophoretic response at low frequencies. As discussed in Chapter 11, this can be used to separate dead cells from viable cells, or to monitor by dielectrophoresis the effect of cytotoxic agents on a cell population. Changes in the surface morphology of a cell (e.g., appearance of blebs, microvillii or extra membrane folding) either through apoptosis or cell differentiation, will influence the membrane capacitance and along with it the frequency at which the crossover from negative to positive dielectrophoresis occurs. Changing the conductivity of the cell suspending medium will also change the value of [*CM*] and an example of this is shown in Figure 2.15. With the applied field frequency fixed at 10 kHz, the same cells can either be collected at the electrode or repelled from them into a 'field cage', simply by changing the conductivity of the suspending medium. Forcing cells to gather together by negative dielectrophoresis can be used to manufacture artificial cell structures, for example.

The experimental quantification of cell electrophoresis is commonly achieved either by monitoring the cell movement through a microscope or by detecting the back scattering of laser light from the cells.

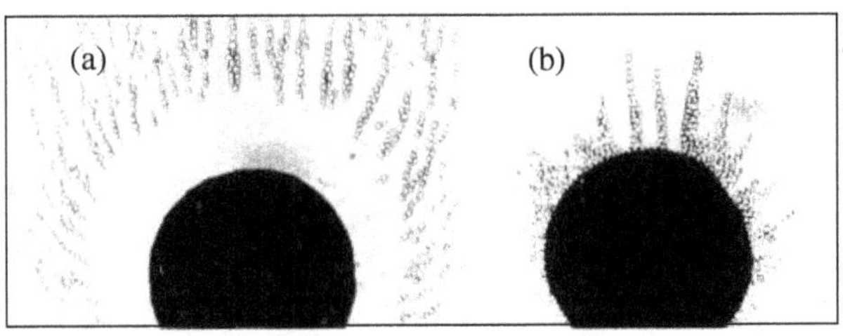

Figure 2.14 Viable yeast cells suspended in 280 mM mannitol of conductivity 40 mS/m. (a) Repelled from an electrode by negative dielectrophoresis with an applied 1 kHz field. (b) Collecting at an electrode by positive dielectrophoresis at 10 MHz [25]. (Reproduced with permission of the Institute of Physics.)

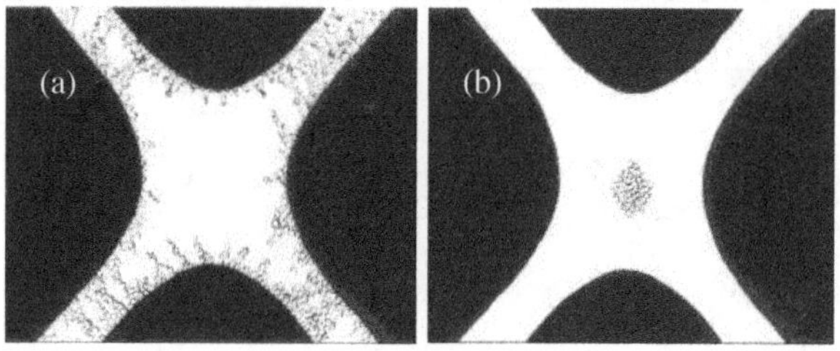

Figure 2.15 Viable yeast cells suspended in 280 mM mannitol with a 10 kHz voltage signal applied to the quadrupole electrodes. (a) Positive dielectrophoretic collection with mannitol conductivity 0.36 mS/m. (b) Negative dielectrophoretic collection into a *field cage* with mannitol conductivity 17 mS/m [26]. (Reproduced with permission of the Institute of Physics.)

Dielectrophoresis can also be monitored at the single cell level and has been used to determine the so-called DEP crossover frequency where the dielectrophoresis (DEP) motion makes the transition from negative to positive DEP. A simple way to monitor the dielectrophoretic response of many cells in suspension is shown in Figure 2.16. The absorbance of a light beam directed through a gap between a set of opposing electrodes (e.g., through the central axis of the four electrodes shown in Figure 2.15) will be influenced by the presence of cells acting as light scattering centres. The reduction of the number of cells in the bulk solution, as a result of their being attracted to the electrodes by positive dielectrophoresis, will result in an increased intensity of the light transmitted through the electrode array. An example of this is given in Figure 2.16 for measurements at 10 Hz, which is at a low enough frequency for an electrophoretic response to be detected. The cells are steadily attracted to the electrodes by positive dielectrophoresis, but they are also pushed back and forth at a regular rate of ten times a second by the electric force associated with their permanent charge. This is detected by a slow increase of optical transmittance (decrease of absorbance) with a superimposed oscillatory perturbation at 10 Hz. With increasing frequency, the inertia of the cell mass damps out the electrophoretic response – the cell cannot respond fast enough to changes of the field polarity.

2.6 Summary

The main contrasts between electrophoresis and dielectrophoresis can be summarized as follows.

2.6.1 Electrophoresis

The direction of electrophoretic motion of a particle depends on the polarity (positive or negative) of the fixed electric charge that it carries, as well as on the polarity of the applied electric field. A DC field is employed and is usually uniform, but can in principle be nonuniform. The size and shape of the particle is not a limiting or controlling factor – electrophoresis can readily be observed for macroscopic objects such as cells and charged colloidal particles, as well as much smaller particles such as proteins, nucleic acids, molecular and atomic ions. Counterions, which form an electrical double layer with the fixed charge on the particle, influence the magnitude of the steady-state electrophoretic velocity. Application of Stokes' Law to describe the viscous drag force overestimates this velocity because it does not account for the electric force that acts on the counterions in the

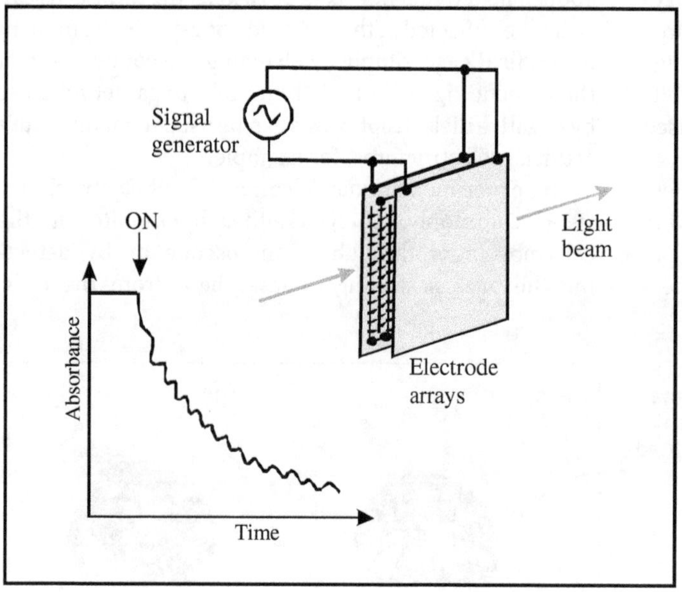

Figure 2.16 A simple method to observe the dielectrophoretic behaviour of cells suspended between two electrode arrays involves monitoring changes in the intensity of a light beam passing through the suspension. The optical absorbance, after applying a 1 Hz voltage signal (ON) to a suspension of *Micrococcus luteus* exhibits oscillations of magnitude associated with electrophoresis [27].

laminar fluid layer that moves with the particle. The effective charge on the particle, as given by the zeta potential at the hydrodynamic shear surface between the particle-associated fluid layer and the bulk solution, is employed in the Helmholtz–Smolchowski equation to give an accurate electrophoretic mobility value.

2.6.2 Dielectrophoresis

Dielectrophoresis does not require the particle to possess a fixed electric charge, but instead relies on the generation of induced surface charges related to the intrinsic dielectric properties of the particle and its surrounding medium. These induced charges lend to the particle the properties of a large dipole moment. The direction of dielectrophoretic motion does not depend on the polarity of the applied electric field – both DC and AC fields can be used – but the field employed must be nonuniform. The size and shape of the particle is a controlling factor for dielectrophoresis, because this determines the magnitude of the induced dipole moment. Significant efforts are required to observe the dielectrophoretic motion of molecular-sized objects. In AC applications of dielectrophoresis, Stokes' Law can be employed to estimate the steady-state velocity of a cell or other large particle and use of the Helmholtz–Smolchowski equation is not relevant. However, as will be described in Chapter 11, for cases where the electrical double-layer thickness is comparable to or greater than the size of submicron particles, the dielectrophoretic response can depend on the fixed particle charge.

2.7 References

1 Pohl, H. A. (1951) The motion and precipitation of suspensoids in divergent electric fields. *J. Appl. Phys.* **22**, 869–871.
2 Pohl, H. A. and Pethig, R. (1977) Dielectric measurements using non-uniform electric field (dielectrophoretic) effects. *J. Phys. E: Sci. Instrum.* **10**, 190–193.
3 Pohl, H. A. (1978) *Dielectrophoresis: The Behaviour of Neutral Matter in Nonuniform Electric Fields*, Cambridge University Press, Cambridge.
4 Coulter, C. B. (1920) The isoelectric point of red blood cells and its relation to agglutination. *J. Gen. Physiol.* **3**, 309–323.
5 Mehrishi, J. N. and Bauer, J. (2002) Electrophoresis of cells and the biological relevance of surface charge. *Electrophoresis* **23**, 1984–1994.
6 Hannig, K. (1982) New aspects in preparative and analytical continuous free-flow cell electrophoresis. *Electrophoresis* **3**, 235–243.
7 Heidrich, H. G. and Hannig, K. (1989) Separation of cell population by free-flow electrophoresis. *Methods Enzymol.* **171**, 513–531.
8 Roman, M. C. and Brown, P. R. (1994) Free-flow electrophoresis as a preparative separation technique. *Anal. Chem.* **66**, 86–94.
9 Eggleton, P. (1998) Separation of cells using free flow electrophoresis, in *Cell Separation: A Practical Approach* (eds D. Fisher, G. E. Francis and D. Rickwood). Oxford University Press, Oxford, pp. 213–252.
10 Bangham, A. D., Heard, D. H., Flemans. R. and Seamen, G. V. F. (1958) An apparatus for microelectrophoresis of small particles. *Nature* (London) **182**, 642–644.
11 Fuhrmann, G. F. and Ruhenstroth-Bauer, G. (1965) Cell electrophoresis employing a rectangular measuring cuvette, in Cell Electrophoresis (ed. E. J. Ambrose). J. & A. Churchill Ltd, London, pp. 22–25.
12 Lukiewicz, S. and Korohoda, W. (1965) Some recent advances in the techniques of cell microelectrophoresis, in Cell Electrophoresis (ed. E. J. Ambrose). J. & A. Churchill Ltd, London, pp. 26–33.
13 Uzgiris, E. E. and Kaplan, J. H. (1974) Study of lymphocyte and erythrocyte electrophoretic mobility by laser doppler spectroscopy. *Anal. Biochem.* **60**, 455–461.
14 Josefowicz, J. Y. (1979) Electrophoretic light scattering and its application to the study of cells, in *Methods of Cell Separation*, vol. **2** (ed. N. Catsimpoolas). Plenum Press, New York, NY, pp. 67–91.
15 Slivinsky, G. G., Hymel, W. C., Bauer, J. and Morrison, D. R. (1997) Cellular electrophoretic mobility data: A first approach to a database. *Electrophoresis* **18**, 1109–1119.
16 Jucker, B. A., Harms, H. and Zehnder, A. J. B. (1996) Adhesion of the positively charged bacterium Stenotrophomonas. *J. Bacteriol.* **178**, 5472-5479.
17 Price, J. A. R., Pethig, R., Lai, C.-N. *et al.* (1987) Changes in cell surface charge and transmembrane potential accompanying neoplastic transformation of rat kidney cells. *Biochim. Biophys. Acta* **898**, 129–136.
18 Pethig, R., Gascoyne, P. R. C., McLaughlin, J. A. and Szent-Györgyi, A. (1984) Interaction of the 2,6-dimethoxysemiquinone and ascorbyl free radicals with Ehrlich ascites cells: A probe of cell-surface charge. *Proc. Nat. Acad. Sci. USA* **81**, 2088–2091.

19 Grahame, D. C. (1947) The electrical double layer and the theory of electrocapillarity. *Chem. Rev.* **41**, 441–501.
20 Maxwell, J. C. (1873) *A Treatise on Electricity and Magnetism*, Clarendon Press, Oxford. (Third edition republished by Oxford University Press, New York, NY, 1998.)
21 Abraham, M. (1932) *The Classical Theory of Electricity and Magnetism*, Hafner, New York, NY. (English translation of German eighth edition.)
22 Maxwell, J. C. (1865) A dynamical theory of the electromagnetic field. *Phil. Trans. Roy. Soc. London* **155**, 459–512.
23 South, G. P. and Grant, E. H. (1972) Dielectric dispersion and dipole moment of myoglobin in water. *Proc. R. Soc. Lond.* **A328**, 371–387.
24 Hsu, H. P. (1984) *Applied Vector Analysis*, Harcourt Brace Jovanovich, San Diego, CA.
25 Wang, X.-B., Huang, Y., Burt, J. P. H. *et al.* (1993) Selective dielectrophoretic confinement of bioparticles in potential energy wells. *J. Phys. D: Appl. Phys.* **26**, 1278–1285.
26 Huang, Y and Pethig, R. (1991) Electrode design for negative dielectrophoresis. *Meas. Sci. Technol.* **2**, 1142–1146.
27 Burt, J. P. H., Al-Ameen, T. A. K. and Pethig, R. (1989) An optical dielectrophoresis spectrometer for low-frequency measurements on colloidal suspensions. *J. Phys. E: Sci. Instrum.* **22**, 952–957.

3

Electric Charges, Fields, Fluxes and Induced Polarization

3.1 Introduction

A dielectric can be defined as a material that is a poor conductor of electricity and is capable of supporting an electrostatic field. Other definitions that are given sometimes also include words to the effect that a dielectric becomes *polarized* when exposed to an electric field. Thus, introductions to the theoretical concepts of dielectric phenomena [e.g., 1] often commence with the scheme shown in Figure 2.1 of Chapter 2, where a voltage potential difference is applied across two parallel metal plates located in a vacuum. The applied potential difference generates a distribution of equal and opposite *free* electrical charges on the faces of the plates (denoted by symbols +, − in Figure 2.1). The next step is to remove the external voltage source and to replace it with a voltage measuring device, such as an electrometer. The only factor that controls the voltage across the metal plates is the density of free charge on the surfaces of the plates. To ensure that this voltage remains constant and is not affected by the measurement of it, a voltmeter is designed to have a very high resistance to electric current flow. We can therefore ignore any leakage of charge through the electrometer that might reduce the density of positive and negative free charges on the plates. The voltage potential difference across the plates with vacuum between them is measured as V_0. The exercise is then performed of inserting a solid slab of dielectric material into the space between the plates, whilst noting the reading on the electrometer. It is found, as depicted in Figure 3.1, that the voltage reading falls to a new value V_1. The distance d between the plates remains fixed, so we can assume that the electric field (V/d) between the plates has also dropped from E_0 to E_1. The ratio E_0/E_1 (or V_0/V_1) is defined as the relative permittivity ε_r (also called the dielectric constant) of the dielectric material:

$$\frac{E_0}{E_1} = \varepsilon_r \tag{3.1}$$

In Chapter 2 we noted that electrical potential V is defined in terms of potential energy per charge. So, for the potential difference across the plates to have fallen, the number of effective free charges on each plate must also have fallen. But positive and negative charges on the plates could not have neutralized each other by leaking through the electrometer or through the electrically insulating dielectric – so what has happened? To answer this we need to revisit the concepts of electric charge and field discussed in Chapter 2 and, in particular to understand the relationship between the electric field at the surface of a charged surface and the number and distribution of the charges on that surface. This is discussed in sections 3.2.2 and 3.3.2.4, where we find that the uniform field E_0 between two oppositely charged metal plates in vacuum is given by:

$$E_0 = \frac{\sigma}{\varepsilon_o} \tag{3.2}$$

where σ is the uniform free charge density per unit area on each metal plate and ε_o is the permittivity of vacuum. From Equation (3.2) the *apparent* free surface charge density (σ_1) on each plate when the dielectric is fully inserted is equal to $E_1\varepsilon_0$. Employing Equation (3.1) to substitute for E_1, the reduction $\Delta\sigma$ in apparent free surface charge density that occurs on inserting the dielectric between the plates is calculated to be:

$$\Delta\sigma = \sigma - \sigma_1 = \varepsilon_o E_0 - \frac{\varepsilon_o E_0}{\varepsilon_r} = \varepsilon_o E_0\left(1 - \frac{1}{\varepsilon_r}\right)$$
$$= \sigma\left(1 - \frac{1}{\varepsilon_r}\right)\ \mathrm{C\,m^{-2}}. \tag{3.3}$$

Clearly, even though free charges cannot leak through the dielectric, the material property we have defined as the relative permittivity ε_r is capable of neutralizing some of the free surface charge applied to the metal plates. The physical process responsible is the appearance of *polarization* charges produced by the dielectric. These polarization charges do not contribute to the measured voltage across the capacitor, but each one neutralizes a free charge on an electrode surface. We say that a *free* electronic charge has become a *bound* charge. The basic

Dielectrophoresis: Theory, Methodology and Biological Applications, First Edition. Ronald Pethig.

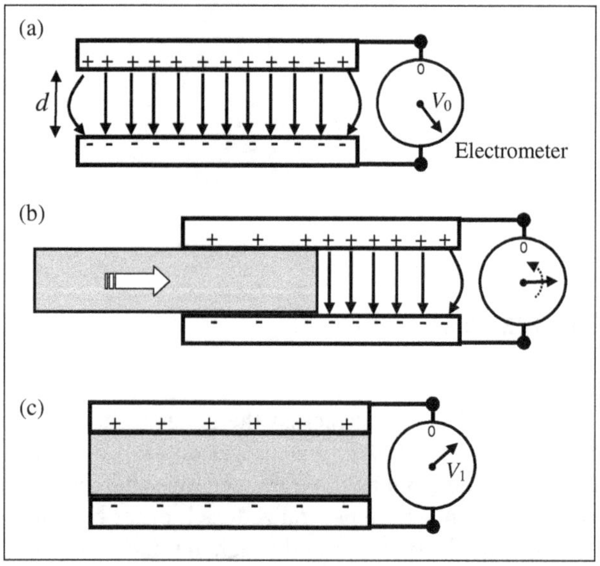

Figure 3.1 (a) An electrometer gives a voltage reading of V_0 across two parallel metal plates in vacuum, distance d apart, which have been electrically charged according to the scheme shown in Figure 2.1 of Chapter 2. (b) As a dielectric slab of relative permittivity $\varepsilon_r = 2$ is inserted between the plates the voltage reading falls. (c) When the dielectric slab is fully inserted between the plates, the voltage reading is $V_0/2 = V_1$. From Equation (3.3) this corresponds to a reduction by one-half of the effective charges on the plates – but no charge leakage has occurred through either the electrometer or the dielectric slab.

nature of this process is revealed in the derivation of Equation (3.34) and later in this chapter. It is an importance aspect of dielectrophoresis.

The practical application of the electrode system shown in Figure 3.1 is as a capacitor (also called a condenser) – a device for storing electrostatic potential energy. As described in section 3.5, this ability is quantified as its capacitance C, the amount of charge it can store at each electrode per unit of voltage applied across the electrodes. (We specify 'at each electrode' because the bound charges at opposite electrodes are equal in magnitude but opposite in polarity. The *net* charge stored in the capacitor is thus zero.) The stored charge at each electrode is the polarization charge of the dielectric, the amount of which is directly proportional to the reduction of voltage shown in Figure 3.1. The ratio C_1/C_0 of the capacitance for a dielectric inserted between the electrodes to that with no dielectric (i.e., vacuum) is thus given by:

$$\frac{C_1}{C_0} = \frac{V_0}{V_1} = \varepsilon_r$$

This provides another definition of relative permittivity ε_r.

3.2 Charges and Fields

3.2.1 Early Investigations of Electrostatic Interactions and Conduction

Electrostatic interactions involve particles that carry an electrical charge and are stationary (i.e., static). Mass and electric charges are fundamental properties of matter. Unlike mass, which is generated through the Higgs effect, we do not know what it is that endows a particle with electric charge. For some time, though, the properties and behaviour of charge have been observed and discussed. Thales of Miletus, from whose school came Socrates and who is known as one of the seven wise men of Greece, is said to have been the first (~590 BC) to describe the electrical effects created by the frictional rubbing of amber (Greek name *elektron*). Thales and other Greek and Roman writers mention that 'when a vivifying heat is applied to amber it will attract straws, dried leaves and other light bodies in the same way that a magnet attracts iron' [2]. We now understand this effect in terms of the amber acquiring a negative electric charge after being rubbed with fur (which gains a positive charge). When a piece of negatively charged amber is placed near, but not in touch with, another particle it can *induce* charges in it. No net charge is transferred to this second particle – instead two equal distributions of charge are created of opposite polarity, with the positive distribution of charges located nearest to the negatively charged amber. Being nearer to the negative charges on the amber, the induced positive charges are attracted to the amber more strongly than the more distant induced negative charges are repelled. A net electrostatic force of attraction towards the amber is therefore exerted on the second particle. From Chapter 2 we recognize this as a *dielectrophoretic* force. On taking the charged amber away, the induced charges in the second particle disappear.

In the eighteenth and nineteenth centuries, electrostatic interactions were studied using pith balls suspended by silk threads in arrangements such as those shown in Figure 3.2. (In Britain the pith was commonly obtained from the stems of elderberry bushes.) Typically, in such an experiment a polished glass rod is rubbed with a silk cloth. On bringing this rubbed rod near a pith ball, the ball moves towards the rod but after touching and thereby electrifying the ball it is repelled. The same sequence of effects is observed on bringing a piece of sealing wax rubbed with silk to touch a pith ball. However, on bringing an electrified piece of sealing wax towards a pith ball electrified by touching a rubbed glass rod, the electrified ball is attracted to the sealing wax and not repelled from it. Likewise, if two suspended pith balls are electrified by touching them with a rubbed glass rod they are repelled from the rod and from each other, but on

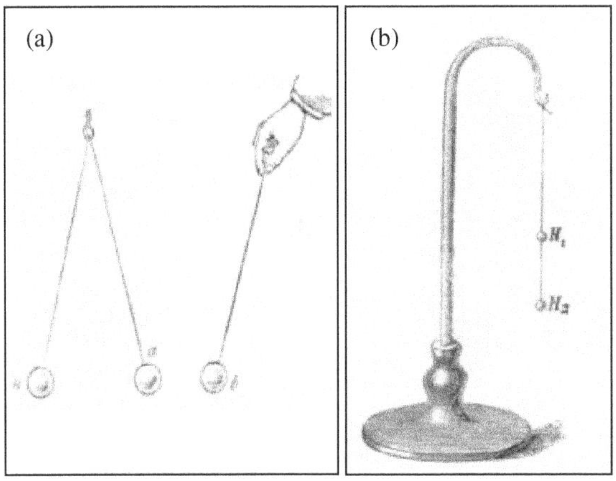

Figure 3.2 (a) Using pith balls to demonstrate electrical attraction and repulsion. (b) Using two suspended pith balls to demonstrate electrical conduction and insulation. When the balls are connected by a metal wire, electrification of the uppermost ball is carried to the lower ball (and is of the same electrical polarity). If the lower ball is suspended from the upper one by a silk thread instead of a metal wire, and the experiment is repeated, the lower ball will exhibit no sign of electrification because silk acts as an insulator. (Reproduced from R. M. Walmsley (1904) *Electricity in the Service of Man*, Cassell & Co., London, with permission of Octopus Publishing Group.)

bringing near a rubbed piece of sealing wax they are strongly attracted to it. Materials of the same kind rubbed with the same material repel each other, but between a glass rod and sealing wax rubbed with silk, for example, there is an attractive force. Likewise, a glass rod is attracted to the silk cloth or pad used to electrify it and so on.

Through such experiments it was deduced that the nature of electrification (charging with electricity) depends on the body being rubbed and on the rubbing material. The treatment of the surface of a material was also found to be important. For example, rubbing a polished glass rod with silk was found to give it a positive charge, whereas the charge produced on a ground glass rod was negative. The materials given in Table 3.1 are so arranged that any one of them becomes positively charged when rubbed by any material listed after it. Two kinds of electrification were thus identified – that of the rubbed glass, which was termed *vitreous*, and that of the rubbed sealing wax, which was termed *resinous*. Benjamin Franklin (1706–1790) recommended that these two kinds of electrification be called *positive* and *negative*, respectively. Therefore, positively charged bodies are those that exhibit the same properties as glass rubbed with silk, and negatively charged bodies are those that exhibit properties of the opposite kind.

The main evidence available before the twentieth century for a body being charged was the force it exerted on other bodies, whether that force is one of attraction or repulsion. (Modern methods include studying the tracks revealed in a liquid hydrogen bubble chamber of particles moving in a uniform magnetic field.) It was established from the early investigations that two positive or two negative charges repel each other, whilst a positive and a negative charge attract each other. In 1785 Charles Augustin de Coulomb (1736–1806) used a torsion balance (independently invented by John Michel (1724–1793) but usually known as Coulomb's torsion balance) to verify that 'the repulsive force between two small globes charged with the same kind of electricity is in the inverse ratio of the square of the distance of their centres'. He then extended this law 'to the attraction of opposite electricities' [3] and also found that the forces were proportional to the amounts of electrification of the globes. Coulomb's Law can be written mathematically as:

$$\mathrm{F} = k\frac{|q_1 q_2|}{r^2} \tag{3.4}$$

In this equation F is the *magnitude* of the force that each of two point charges q_1 and q_2, distance r apart, exert on the other. The *direction* of the force is along the line joining their centres. The parameter k is a constant whose numerical value depends on the units of measurement employed. As for all modern texts, we are using the international SI system of units (*Système international d'unités*). In this system the unit of force is one newton (N), the unit of charge is one coulomb (C) and length is given in the unit of one metre (m). The value of k in

Table 3.1 Vitreous and resinous electrification. The materials given in this table are arranged in such an order that they become positively charged when rubbed by any material listed after it. Thus, glass rubbed with cat fur becomes negatively charged, but is positively charged when rubbed with silk. Resins are negatively charged when rubbed with either cat fur or silk. (Reproduced from R.M. Walmsley (1904) *Electricity in the Service of Man*, Cassell & Co., London, p. 60.)

+ Catskin or fur	The hand	Ebonite
Wool	Wood	Resins
Glass	Sulphur	Guttaperche
Ivory	Flannel	Metals
Silk	Cotton	Guncotton
Rock crystal	Shellac	

Equation (3.4) is given as $k = 1/(4\pi\varepsilon_0)$ where ε_0 is the permittivity of free space ($8.854 \times 10^{-12}\,C^2\,N^{-1}\,m^{-2}$). To a very close approximation k therefore has a value of $9 \times 10^9\,N\,m^2\,C^{-2}$. The absolute value $|q_1q_2|$ of the product of the two charges is taken in Equation (3.4) because although they can be either positive or negative the magnitude of the force F always has a positive value. When the charges are the same in polarity, either positive or negative, each charge exerts a repulsive force on the other (see Figure 2.2 in Chapter 2). When the charges have opposite polarities the forces are ones of attraction. Coulomb's Law describes the interaction of two point charges located in vacuum and a modification of Equation (3.4) is required to take into account the effect of any molecular matter that may exist between the charges. The modification is the inclusion of what was once termed the *specific inductive capacity*, but now known as the dielectric constant or relative permittivity (ε_r) of the medium between the charges. By including the factor ε_r we take into account the kind of effect described by Equation (3.3) where polarization of the medium partially shields the charges from one another. In SI units the full mathematical form of Coulomb's Law is thus given by:

$$F = \frac{1}{4\pi\varepsilon_o\varepsilon_r}\frac{|q_1q_2|}{r^2} \qquad (3.5)$$

If the charges are located in a vacuum, then $\varepsilon_r = 1$ and Equation (3.5) is equivalent to Equation (3.4) when using SI units.

The distinction between an electrical conductor and a poorly conducting material (an insulator) were investigated using the second arrangement shown in Figure 3.2. A typical experiment to confirm the electrical conducting properties of a metal would be to suspend a pith ball by means of a silk thread and a second one below it by means of a metal wire. By touching the upper ball with a rubbed glass rod it becomes positively electrified and is consequently repelled by the glass rod. The bottom pith ball is also repelled from the glass rod, even though it has not been touched by it. Furthermore, both electrified balls are able to attract small objects and are attracted to electrified rubbed sealing wax. No electrical charge has been given to the lower ball by direct contact with the glass rod, yet it shows the same properties as the upper ball that has been electrified by direct contact. It follows that electricity from the upper ball must have passed to the lower one – the metal wire has conducted the electricity between them. If the lower ball is suspended from the upper one by a silk thread, instead of a metal wire and these experiments are repeated, the lower ball will exhibit no signs of being electrified. The silk thread behaves as an insulator – it is a poor conductor of electricity.

Experiments to demonstrate that the presence of an electrified body is sufficient to induce signs of electrification of a neighbouring conductor without making contact with it were performed using an arrangement such as that shown in Figure 3.3. An insulated sphere is charged (say positively) and an uncharged, insulated, metal rod is placed near it. Negative electrification is found on the end of the rod a nearest to the sphere, whilst an equal amount of positive electrification is found on the end b farthest from the sphere. Careful examination of the charge distributions show that it is no longer uniform around the sphere but that there is a distinctly greater charge density on the side nearest to the rod. This is shown approximately by the dotted line drawn around the sphere. The dotted lines around the rod show the approximate charge distributions at ends a and b. Lines of electric force are drawn in the lower schematic of Figure 3.3. The convention is to draw a line of electrical force leading away from a positive charge and to terminate at a negative charge. The negative electrification on the end a of the metal rod shown in Figure 3.3 indicates that a

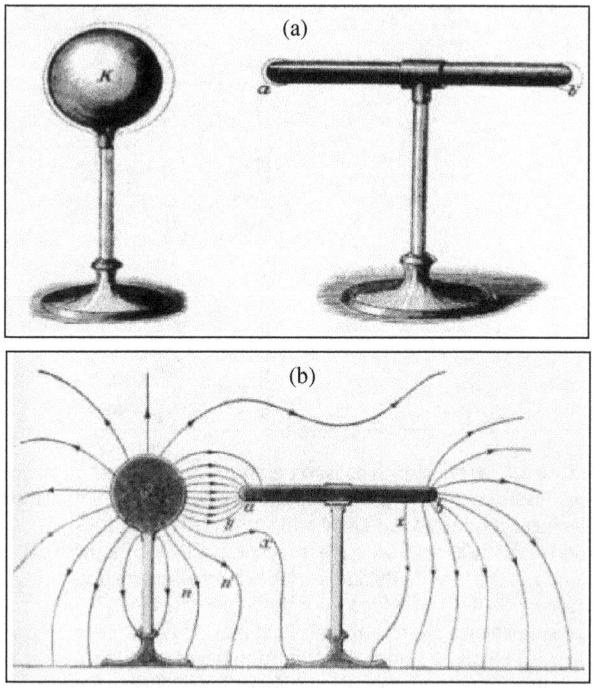

Figure 3.3 (a) The induction of electric charges has historically been demonstrated by bringing an insulated, uncharged, metal rod to within a short distance of an insulated, positively charged, metal sphere. It is found that the end *a* of the rod nearest the sphere is electrified with negative charges, whilst end *b* farthest from the sphere is electrified with positive charges. (b) The electric lines of force starting from *b* are entirely distinct from those ending at *a*. The two sets are equal in number because no net charge has been transferred to the rod, and therefore the sum of all the positive electrifications (or the lines starting from *b*) must be equal to the sum of all the negative electrifications (or the lines ending at *a*. (Reproduced from R. M. Walmsley (1904) *Electricity in the Service of Man*, Cassell & Co., London, with permission of Octopus Publishing Group.)

certain number of lines end there, whilst the positive electrification on end b similarly indicates that an *equal* number of lines set out from that end. One of the fundamental properties of an isolated electrical conductor is that no electrical force can be permanently formed within it. Otherwise the free electrical charges within it would be in motion and form a net current of electricity. This does not correspond to an electro*static* situation. Hence the force lines starting from b are entirely distinct from those ending at a. The two sets are equal in number because no charge has been given to the rod, either positive or negative and so the sum of all the positive electrifications (or the lines starting from b) must be equal to the sum of all the negative electrifications (or the lines ending at a). In Figure 3.3, nine lines have been drawn at each end of the rod, leaving thirteen lines emanating from the sphere, which do not run on to the rod. If the metal rod with its induced charges is now withdrawn some distance away from the charged sphere, the rod will show no signs of electrification, whilst the sphere will be restored to its original uniformly charged condition. Twenty-two lines of force are shown emanating from the sphere in Figure 3.4, with their negative ends terminating at 'earthing or grounding points' on the table top and other distant surfaces not shown in the figure.

Coulomb's Law, as given by Equation (3.5), describes the electrostatic interaction between two point charges. However, when two or more charges act at the same time on another one, the total electrostatic force acting on it is the vector sum of each individual force exerted by the other charges. This is known as the *principle of superposition of forces* and examples of this are given below.

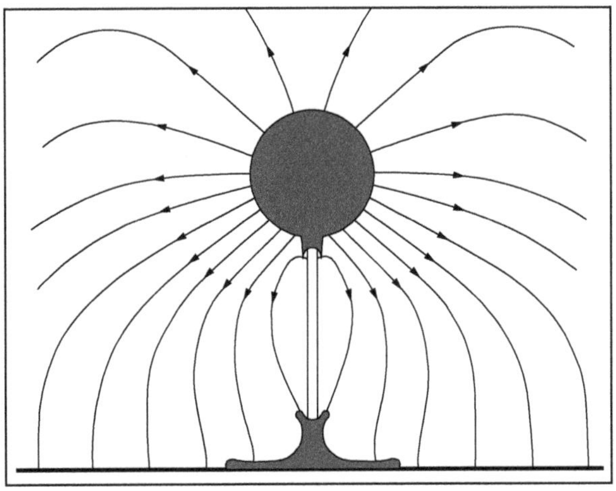

Figure 3.4 Lines of electric force are shown emanating from the positively charged sphere of Figure 3.3 after removal of the insulated metal rod. These lines terminate at negative charges distributed on the table top or other distant surfaces. (Reproduced from R. M. Walmsley (1904) *Electricity in the Service of Man*, Cassell & Co., London, with permission of Octopus Publishing Group.)

Example 3.1 Lines of Force around Three Charges
The lines of electric force associated with three point charges are shown in Figure 3.5. What can be deduced about the polarities of the charges and their relative magnitudes?

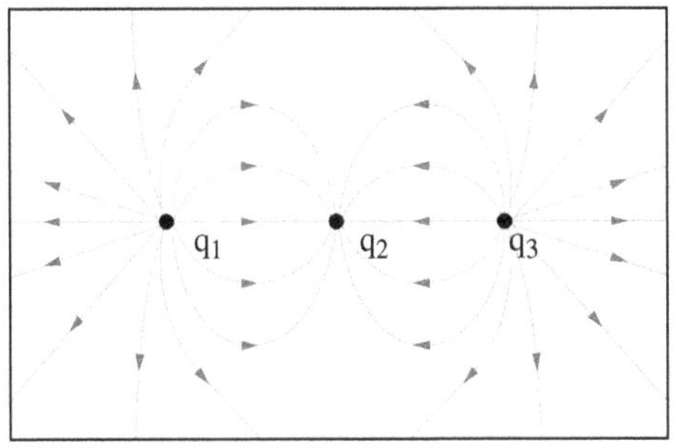

Figure 3.5 Three point charges, spaced 1 mm apart along a straight line, are shown with their associated electric force lines.

Solution 3.1 The lines of force are directed away from charge q_1 and q_3 and so they are of positive polarity. The lines of force are directed into charge q_2, which is thus of negative polarity. We could draw in many more lines of force in Figure 3.5 and so we have no *quantitative* detail to allow us to judge the relative magnitudes of the three charges. The symmetrical distribution of the force lines would suggest that charges q_1 and q_3, are of similar magnitude and that they *may* have a larger magnitude than q_2.

Example 3.2 Electrostatic Force Calculation

1. Calculate the net electrostatic force acting on charge q_3 in Figure 3.5 from q_1 and q_2. The charges are equally spaced in air along a straight line and are of sizes much smaller than their spacing apart of 1 mm. Assume that $q_1 = q_3 = 1.4\,\text{pC}$; $q_2 = -1.0\,\text{pC}$.
2. What is the force acting on q_2 from q_1 and q_3?

Solution 3.2

1. The forces F_{13} and F_{23} acting on q_3, due to q_1 and q_2, respectively, both act along the line joining the charge centres. The magnitude of the force F acting on q_3 is thus the sum of these two forces, given by $F = F_{13} + F_{23}$ (see Figure 3.6). Because they can be treated as point charges located in air, we can use Coulomb's Law in the form of Equation (3.5) and to a very good approximation take $\varepsilon_r = 1.0$:

$$F_{13} = \frac{1}{4\pi\varepsilon_o}\frac{|q_1 q_3|}{r^2} = (9\times10^9\ \text{N m}^2\ \text{C}^{-2})$$
$$(1.4\times10^{-12}\ \text{C})(1.4\times10^{-12}\ \text{C})/(2\times10^{-3}\ \text{m})^2$$
$$= 4.4\times10^{-9}\text{N} = 4.4\ \text{nN}$$

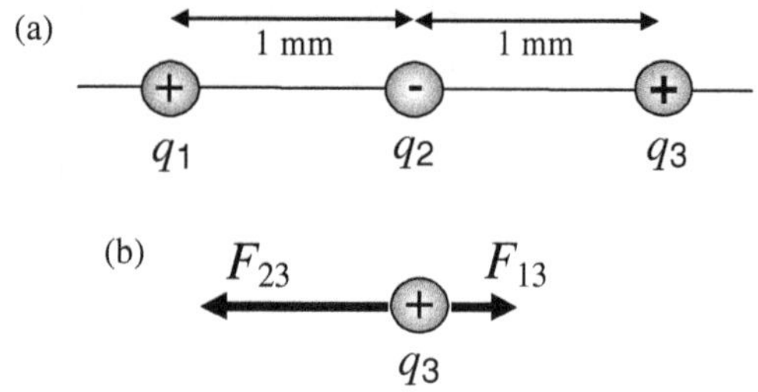

Figure 3.6 (a) Linear arrangement of the three charges shown in Figure 3.5. (b) The electrostatic force on charge q_3 is given by the vector sum of the forces F_{13} and F_{23} acting on it by q_1 and q_2, respectively.

This force repels q_3 along the line to the right.

$$F_{23} = \frac{1}{4\pi\varepsilon_o}\frac{|q_2q_3|}{r^2} = (9 \times 10^9\ \text{N m}^2\ \text{C}^{-2})$$
$$(-1 \times 10^{-12}\ \text{C})(1.4 \times 10^{-12}\ \text{C})/(10^{-3}\ \text{m})^2$$
$$= -1.26 \times 10^{-8}\ \text{N} = -12.6\ \text{nN}$$

This force attracts q_3 to the left. The force acting on q_3 is thus given by:

$$F = F_{13} + F_{23} = 4.4 + (-12.6) = -8.2\ \text{pN}$$

The negative sign indicates that this force acts to the left in Figure 3.6.

2. The net force acting on q_2 is zero, because the forces of repulsion due to q_1 and q_3 are of equal magnitude and act in opposing directions.

Examples 3.1 and 3.2 consider the superposition of electrostatic forces acting along a straight line. The vector sum of the forces was obtained by simply adding the magnitudes of the forces exerted on one charge by the others. If all the charges were not located along a straight line, we would have had to take into account the direction of each individual force. To do this we can make use of the concept of a *unit vector* $\widehat{\mathbf{r}}$, which points along the line joining two of the point charges whose electrostatic interaction is being considered. The concept of a unit vector is explained in Box 3.1 and an example of its use is given in Example 3.3. By using the unit vector $\widehat{\mathbf{r}}$ we can write Equation (3.5) in a form that expresses Coulomb's Law as a vector equation:

$$F = \frac{1}{4\pi\varepsilon_o\varepsilon_r}\frac{q_1q_2}{r^2}\hat{\mathbf{r}} \tag{3.6}$$

Example 3.3 Electrostatic Vector Force Calculation Charge q_2 in Figure 3.6 is moved, in a direction at right angles to the line joining q_1 and q_3, to the location shown in Figure 3.7. Calculate the net electrostatic force acting on charge q_3 as a result of the interactions from q_1 and q_2.

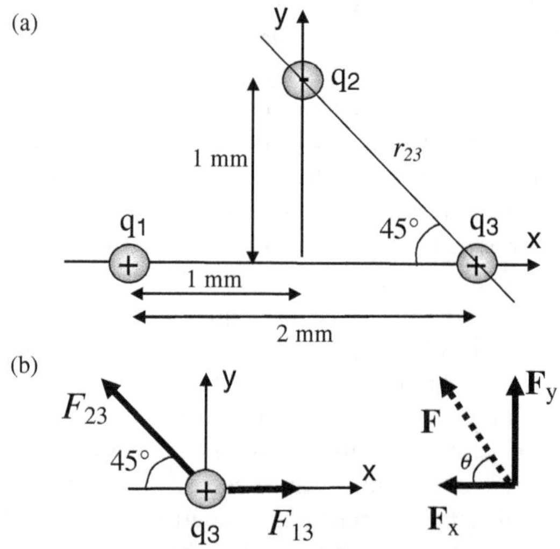

Figure 3.7 (a) The arrangement of the three charges for Example 3.3. (b) The electrostatic force on charge q_3 is the vector sum of the net forces $\mathbf{F}_x$ and $\mathbf{F}_y$ acting along the *x*-axis and *y*-axis, respectively. The angle θ is given by the magnitude of the inverse tangent function $\tan^{-1}(\mathbf{F}_y/\mathbf{F}_x)$.

Solution 3.3 The net force **F** acting on q_3 is the vector sum of the two forces due to q_1 and q_2, given by $\mathbf{F} = \mathbf{F}_{13} + \mathbf{F}_{23}$ (see Figure 3.7). We will denote the plane containing all three charges as the *xy*-plane and assign the line directed from charge q_1 to q_3 to be the positive *x*-axis.

The magnitude of F_{13} is given by:

$$F_{13} = \frac{1}{4\pi\varepsilon_o}\frac{|q_1q_3|}{(r_{13})^2} = (9 \times 10^9\ \text{N m}^2\text{C}^{-2})$$
$$(1.4 \times 10^{-12}\ \text{C})(1.4 \times 10^{-12}\ \text{C})/(4 \times 10^{-3}\ \text{m})^2$$
$$= 1.1\ \text{nN}$$

As depicted in Figure 3.7, F_{13} is a force that acts along the positive *x*-axis and has no component along the *y*-axis.

To calculate F_{23} we require the distance r_{23} between charges q_2 and q_3 in Figure 3.7. We calculate r_{23} using Pythagoras' theorem to find the hypotenuse of a right-angled triangle:

$$r_{23} = \{(1 \times 10^{-3}\ \text{m})^2 + \{(1 \times 10^{-3}\ \text{m})^2\}^{1/2} = 1.41\ \text{mm}$$
$$F_{23} = \frac{1}{4\pi\varepsilon_o}\frac{|q_2q_3|}{(r_{23})^2} = (9 \times 10^9\ \text{N m}^2\ \text{C}^{-2})$$
$$(-1 \times 10^{-12}\ \text{C})(1.4 \times 10^{-12}\ \text{C})/(1.41 \times 10^{-3}\ \text{m})^2$$
$$= -6.3\ \text{nN}$$

F_{23} is a vector force of attraction directed from q_3 towards q_2, acting at 45° to the (negative) *x*-axis as shown in Figure 3.7. F_{23} therefore has a component $F_{23}\text{Cos}45° = 6.3\text{Cos}45° = 4.45$ nN acting along the negative *x*-axis and a component $F_y = F_{23}\text{Cos}45° = 4.45$ nN acting along

Box 3.1 Unit Vectors

The concept of a *unit vector* has just one purpose, namely to act as a directional pointer. It has a magnitude of 1.0 and no units of measurement. In a two-dimensional (*xy*-plane) it is the convention to define $\boldsymbol{\hat{\imath}}$ as the unit vector that points along the positive direction of the *x*-axis, and $\boldsymbol{\hat{\jmath}}$ as the unit vector pointing along the positive direction of the *y*-axis.

Thus, for example, a force $\mathbf{F}_x$ of 0.5 N acting along the negative direction of the *x*-axis is written as $\mathbf{F}_x = -0.5\boldsymbol{\hat{\imath}}$ N, whereas $2\boldsymbol{\hat{\jmath}}$pN represents a force of 2 piconewtons directed along the positive direction of the *y*-axis. On extending to three-dimensional space, we define the unit vector that points along the positive *z*-axis (orthogonal to the *xy*-plane) as $\hat{k}$.

Consider the case of a particle subjected to a fluidic drag force F_x acting along the *x*-axis of a channel, as well as an electrophoretic force F_y acting along the *y*-axis. The total force $\mathbf{F}_T$ acting on the particle is the vector sum of these two forces, given by:

$$\mathbf{F}_T = F_x\boldsymbol{\hat{\imath}} + F_y\boldsymbol{\hat{\jmath}}$$

This emphasizes that adding two vectors requires a geometrical procedure of adding together two quantities that have both magnitude and direction. This procedure is not the same as adding together two scalar quantities, such as in the sum $7 + 2 = 9$.

The vector form of Coulomb's Law is given in Equation (3.6) as:

$$F = \frac{1}{4\pi\varepsilon_o\varepsilon_r}\frac{|q_1q_2|}{r^2}\hat{\mathbf{r}}$$

The unit vector $\hat{r}$ is equal to the displacement vector from one charge to the other, divided by the scalar distance *r* between the charges. In the *xy*-plane we can thus express $\hat{r}$ as $(x\boldsymbol{\hat{\imath}} + y\boldsymbol{\hat{\jmath}})/r$. If the line joining the charges is directed at an angle θ to the x-axis, the displacement vector is given by $(r\mathrm{Cos}\theta + r\mathrm{Sin}\theta)$. Hence, the unit vector $\hat{r}$ given in Equation (3.6) can also be written as:

$$\hat{r} = \mathrm{Cos}\theta\hat{\imath} + \mathrm{Sin}\theta\boldsymbol{\hat{\jmath}}.$$

the positive *y*-axis. The magnitude and polarity of the net vector force along the *x*-axis is given by $F_{13} - F_{23}\mathrm{Cos}\,45° = 1.1 - 4.45 = -3.35$ nN. The resultant electrostatic force **F** acting on charge q_3 can thus be written in terms of the unit vector notation described in Box 3.1 as:

$$\mathbf{F} = F_x\hat{\imath} + F_y\hat{\jmath} = -3.35\,\hat{\imath} + 4.45\hat{\jmath}\text{ nN}$$

The angle θ shown in Figure 3.7 is the angle whose tangent is given by the ratio F_y/F_x:

$$\theta = \tan^{-1}(F_y/F_x) = \tan^{-1}(4.45/3.35) = 53°$$

(The notation *arctan* is also sometimes used for the inverse tangent function $\tan^{-1}$.)

In these examples we have calculated the forces of interaction between separated charges. To do this we used Coulomb's Law, which was formulated on the basis of experimental observations. But we have not examined the physical process responsible for such interactions. What exactly is the mechanism by which charges in free space become 'aware' of each other? Michael Faraday (1791–1867) introduced the concept of lines of electric force in free space, such as those depicted in Figure 3.5. He envisaged that these lines of force represent the direction of a vector, rather like the velocity of an incompressible fluid, whose magnitude is inversely proportional to the cross-section of a narrow tube formed by such lines. But, of course, this is an imaginary concept. So, what is going on in the empty space between the charges?

3.2.2 Electric Fields

René Descartes (1596–1650), regarded as the first modern philosopher, gave particular thought to how magnetic, electric and gravitational influences can be transmitted through space. He was greatly influenced by Johannes Kepler (1571–1630) who had demonstrated the importance of mathematics in bringing clarity and certainty to the study of natural phenomena. A major problem for natural philosophy was to understand how actions are transmitted between bodies that are not in contact. Typical examples would be the observed interaction of magnets, or how the position and phases of the moon influence the extent of the fall and rise of tides. To interpret these as *occult* influences, or examples of *action at a distance*, would be contrary to Kepler's teaching. Instead, Descartes considered that one body could only act on another one through the actions of pressure and impact and for this to occur the bodies would have to be contiguous. A medium of some sort had to connect the separate bodies physically and to describe this he introduced the concept of *aether*. The space, the aether, between separated bodies was envisaged to consist of continually moving particles. Since no spaces would exist for moving particles to move into, the movement of a single particle of the aether would be incorporated into the motion of an entire closed chain of particles in the form of vortices.

Unlikely as it may seem now, this fanciful concept was taken seriously and to good advantage. For example, the Swiss mathematician Johann Bernouilli (1710–1790) considered all space to be

> permeated by a fluid aether, containing an immense number of excessively small whirlpools. The elasticity which the aether appears to possess, and in virtue of which it is able to transmit vibrations, is really due to the presence of these whirlpools; for, owing to centrifugal force, each whirlpool is continually striving to dilate, and so presses against the neighbouring whirlpools. [3, p. 95]

James Clerk Maxwell (1831–1879) employed a model of the form postulated by Descartes in his monumental achievement [4–7] of unifying the magnetic and electrical phenomena deduced experimentally by Michael Faraday (1791–1867). Maxwell makes the following summary statements in his 1865 paper [7]:

- 'The most obvious mechanical phenomenon in electrical and magnetic experiments is the mutual action by which bodies in certain states set each other in motion while still at a sensible distance from each other.'
- 'The theory I propose may therefore be called a theory of the electromagnetic field, because it has to do with the space in the neighbourhood of the electric or magnetic bodies, and it may be called a dynamical theory, because it assumes that in that space there is matter in motion, by which the observed electromagnetic phenomena are produced.'

In his model the motion of the particles constituted an electric current, the centrifugal force acting on them represented the electromotive force, and the pressure of the particles on each other corresponded to the tension or potential of the electricity. The particles themselves were capable of being magnetized. Having obtained the equation of motion of his system of particle vortices, Maxwell proceeded to determine the rate of propagation of electromagnetic disturbances through it – and thereby proved that light is an electromagnetic wave [7]. A major concept he employed was that magnetic energy represents the kinetic energy of a medium and that electric energy is the energy of strain of the same medium. 'By this conception electromagnetic theory was brought into such close parallelism with the elastic-solid theories of the aether, that it was bound to issue in an electromagnetic theory of light' [3, p. 255].

The final summary statement by Maxwell in his 1865 paper [7] provides the solution to our question as to how an isolated electric charge senses the existence of another charge: 'The electromagnetic field is that part of space which contains and surrounds bodies in electric or magnetic conditions.'

An isolated charge therefore interacts with the electric field created by other charges. Maxwell employed the vortices model of Descartes as an aid to develop equations to describe the electrical and magnetic phenomena established previously by Johann Carl Gauss (1777–1855), André-Marie Ampère (1775–1836) and Faraday. Whether or not the concept of aether was a valid one did not matter. Einstein's theory of special relativity in 1905 had its origin in efforts by Poincaré and Lorentz to determine the relative motion of the earth and the aether, but its existence was already in doubt following experiments such as those of Michelson and Morley in 1887 [8]. By 1900 Poincaré was able to ask [9] 'Our aether – does it really exist?' Although on the fringes of current thought, the concept of aether is still evoked [10], with the vortices formed by rotating electron-positron dipoles, it forms no part of modern dielectrics.

In Chapter 2, Equation (2.1), the magnitude of an electric field E at a point in space is defined as being equal to the electric force experienced by a unit test charge at that point. Based on this relationship, the electrostatic force F acting on a point test charge q_2 is:

$$\mathrm{F} = q_2\mathrm{E} \tag{3.7}$$

If the field E is generated by a single point charge q_1, then from inspection of Coulomb's Law given by Equation (3.5) the magnitude of the electric field experienced by charge q_2 at a distance r away from it is:

$$\mathrm{E} = \frac{1}{4\pi\varepsilon_o\varepsilon_r}\frac{|q_1|}{r^2} \tag{3.8}$$

The convention is that the electric field of a point charge points away from a positive charge, but points toward a negative charge. At the end of section 3.2.1 we described how Faraday introduced the concept of lines of electric force, examples of which are given in Figures 3.5 and 3.8. These lines of force were taken to represent the direction of a vector, rather like the velocity of an incompressible fluid, whose magnitude is inversely proportional to the cross-section of a narrow tube formed by such lines. This was used as a mathematical tool by Maxwell, to show that rather than representing fluid flow the lines of force show the direction of the vector field **E** at any point in space. The concept of a 'line of electric force' is thus replaced with the concept of an 'electric field line'. An indication of the field magnitude is provided by the spacing of these lines. Where the field is strong the lines are bunched together and they are spaced far apart where the field is weak. For the case of a point positive charge, its field radiates out with spherical symmetry and from Equation (3.6) the magnitude and direction of the vector field **E** are given by:

$$\mathrm{E} = \frac{1}{4\pi\varepsilon_o\varepsilon_r}\frac{q_1}{r^2}\hat{\mathrm{r}} \tag{3.9}$$

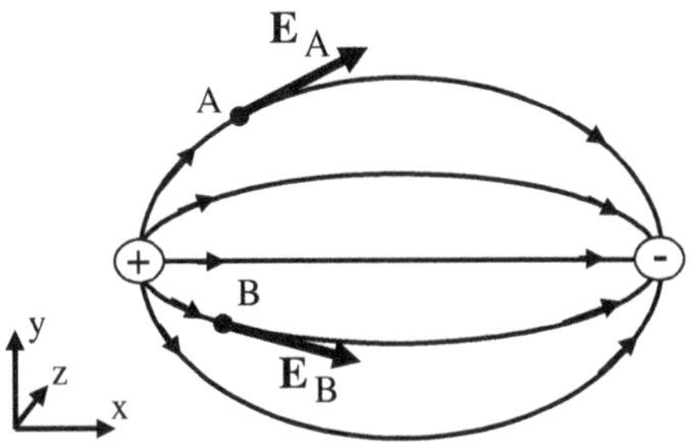

Figure 3.8 Some of the field lines are drawn between a positive and negative charge. The direction of the electric field vector **E** at point (e.g., *A* or *B*) is given by the tangent to the field line at that point. The magnitude of the field will vary along a field line, and have components E_x, E_y, E_z, along the x-, y- and z-directions.

where $\hat{r}$ is the unit radial vector. The more general case of field lines produced by two or more charges in three-dimensional space is shown in Figure 3.8. The direction of the electric field vector at any point along a field line is given by the tangent at that point. Apart from the special case of a uniform electric field, the magnitude of the field varies from point to point along a field line. In general, therefore, a vector field **E** in three-dimensional space has components E_x, E_y, E_z, along the orthogonal *x*-, *y*- and *z*-axes:

$$E = E_x\hat{i} + E_y\hat{j} + E_z\hat{k} \tag{3.10}$$

where $\hat{i}$, $\hat{j}$, and $\hat{k}$ are the unit vectors along the *x*-, *y*- and *z*-axes, respectively. The components E_x, E_y, and E_z may exhibit gradients (i.e., each varying as a function of distance along the *x*-, *y*- and *z*-axes, respectively). This is important for us, because as given by Equation (1.5) of Chapter 1, and in Box 2.4 of Chapter 2, the dielectrophoretic force depends on the gradient of the electric field. This gradient is expressed mathematically as the vector ∇E (variously known as 'grad' E or 'del' E) where the symbol ∇ functions as a differential operator, such that:

$$\nabla E = \left(\hat{i}\frac{\partial}{\partial x} + \hat{j}\frac{\partial}{\partial y} + \hat{k}\frac{\partial}{\partial z}\right) E = \hat{i}\frac{\partial E}{\partial x} + \hat{j}\frac{\partial E}{\partial y} + \hat{k}\frac{\partial E}{\partial z} \tag{3.11}$$

Although Faraday's lines of electric force cannot be interpreted as the lines of flow of a liquid, Maxwell was able to show that a hydrodynamic analogy can be made by defining an electric 'displacement' field **D** in a medium of specific inductive capacity (i.e., dielectric constant or relative permittivity) ε_r as:

$$\mathbf{D} = \varepsilon_0\varepsilon_r\mathbf{E} \text{ (units of coulomb per square metre)} \tag{3.12}$$

In equation (3.12) the relative permittivity ε_r is a number – it is dimensionless. The factor $\mathbf{D}/\varepsilon_0$ therefore has dimensions of an electric field (i.e., force per unit charge). Thus, in Maxwell's view, electric charges no longer appear as the *centres* of force envisaged by Faraday, but rather as sources of *flux* of force.

The term 'displacement' was used to describe the imagined subtle movement (shift) of *bound* charges within each molecular element of a dielectric material in response to an imposed electric field. In modern terms we know that a dielectric consists of an electrically neutral atomic structure, where the total negative charge of all the electrons exactly balances out the positive charges of the nuclei about which they orbit. Although these charges cannot migrate through the dielectric to produce a conventional electric current, they can exhibit subtle displacements in an electric field. Electrons will attempt to minimize their potential energy by distorting their orbital paths against the electric field (i.e., towards the anode) and the positive nuclei will try to move along the field (towards the cathode). We call this the polarization of the *bound* charges within the dielectric. Collectively, these microscopically small displacements create a displacement current density $d\mathbf{D}/dt$, in analogy to a conventional electric current density $j = d\sigma/dt$ caused by the long range flow of free charges. The charges shown on the metal plates in Figure 3.1 are *free* charges. If the electrometer is replaced with a metal wire, the free charges on the plates will discharge momentarily through the wire as an electric current, but will not flow through the dielectric. Instead, the electric circuit is completed by a momentary displacement current $d\mathbf{D}/dt$ caused by the depolarization of the bound charges within the dielectric. The displacement current is confined to the dielectric. From this qualitative description it becomes apparent that the steady reduction of the charges on the metal plates shown in Figure 3.1, as the dielectric is steadily inserted, is related to the displacement field **D** and the collective polarizations of all the bound charges. A more quantitative description of this effect follows in this chapter.

Example 3.4 Electric Field of a Point Charge
Find the magnitude and direction of the electric field at a distance of 1 cm away in air from a small metal sphere of diameter 1 μm that carries a negative charge of −10 nC. If the medium surrounding the sphere is changed to an aqueous solution, what effect will this have?

Solution 3.4 The charged metal sphere is very much smaller than the distance to the point of determination of the field, so we can treat it as a point charge and use Equation (3.8) to calculate the field. As shown in Figure 3.9 the field has spherical symmetry about the point charge. We do not require the *x, y, z* coordinates of the location for which the field value is wanted – only the radial distance

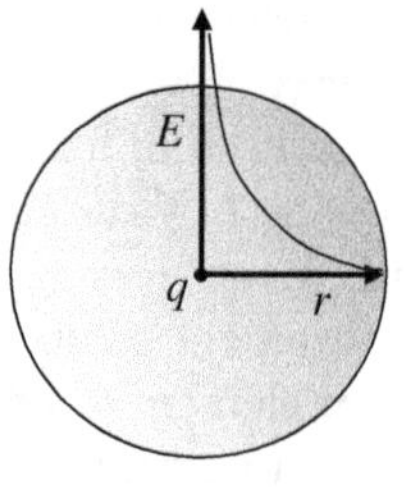

Figure 3.9 The electric field E of a point charge q falls off with radial distance r in all directions as a function of $1/r^2$.

from the point source is required.

$$E = \frac{1}{4\pi\varepsilon_o\varepsilon_r}\frac{q_1}{r^2}\hat{\mathrm{r}} = (9\times10^9\,\mathrm{N\,m^2C^{-2}})\,\frac{1}{\varepsilon_r}$$
$$(-10\times10^{-9}\,\mathrm{C})/(10^{-2}\,\mathrm{m})^2\,\hat{\mathbf{r}}$$
$$= -9\times10^5\frac{1}{\varepsilon_r}\,\hat{\mathrm{r}}\ \mathrm{N/C}$$

In air $\varepsilon_r = 1$ and $E = -9 \times 10^5$ N/C and is directed towards the charged sphere, against the direction of the unit vector $\hat{\mathrm{r}}$. If the sphere is immersed in an aqueous medium, the value for ε_r in Equation (3.8) changes from $\varepsilon_r = 1$ to $\varepsilon_r \approx 80$. The resulting electric field magnitude can be calculated as $E \approx -1.13 \times 10^4$ N/C, where the minus sign again indicates that the field points in the opposite direction to the unit vector $\hat{\mathbf{r}}$.

Example 3.5 Electric Field of a Line of Charge
A glass rod of length $2l$ is given a uniformly distributed positive charge $+Q$ by rubbing it with a silk cloth. Derive an equation to calculate the electric field of this rod in air at a radial distance r perpendicular to the centre of the rod.

Solution 3.5 We align the rod along the x-axis, as shown in Figure 3.10, and take the origins of the x- and y-axes at the centre of the rod. The electric field is calculated at a position P located distance y along the y-axis.

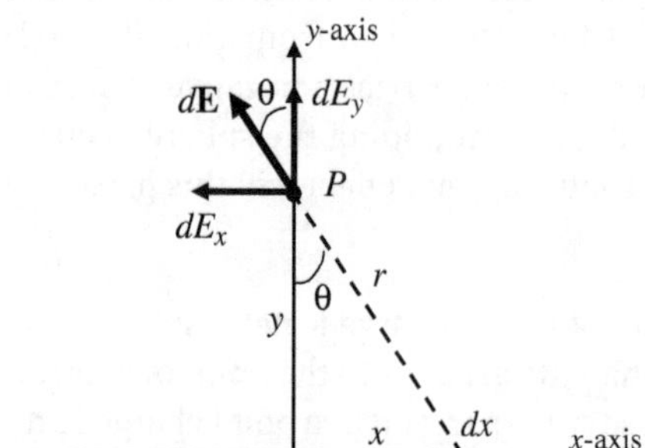

Figure 3.10 Procedure to find the field of a charged rod at a point along the perpendicular axis that bisects it.

We can define a linear charge density λ for the rod as the total charge $+Q$ divided by its length $2l$ ($\lambda = +Q/2l$). An infinitesimal segment dx at a point x along the rod will thus carry a charge $dQ = \lambda dx = +Qdx/2l$. We use Equation (3.9), with $\varepsilon_r = 1$, to calculate the vector field element $d\mathbf{E}$ produced by the elemental (point) charge dQ at point P:

$$dE = \frac{1}{4\pi\varepsilon_o}\frac{Qdx}{2l}\frac{1}{r^2}\hat{\mathrm{r}} \tag{3.13}$$

where r is the distance from point x to location P and $\hat{\mathrm{r}}$ is the radial unit vector. This vector field element has x- and y-components given by:

$$dE_y = d\mathbf{E}\mathrm{Cos}\,\theta;$$
$$dE_x = d\mathrm{ESin}\,\theta \text{ with } \mathrm{Cos}\,\theta = y/r \text{ and } \mathrm{Sin}\,\theta = x/r$$

where $r = (x^2 + y^2)^{\frac{1}{2}}$

The total field $\mathbf{E}$ at P produced by the whole rod is obtained by summing up all the $d\mathbf{E}$ elements. We do this by integrating Equation (3.13) for all x values $-l \le x \le +l$, but before ploughing through this exercise it is a good idea to check if the symmetry of the problem can lead to a shortcut. From inspection of Figure 3.10 we deduce that the summations of all the contributions to dE_x should add up to zero, because dE_x values for $0 \le x \le +l$ all point along the negative x-axis and are balanced for $-l < x \le 0$ by the dE_x values, which all point along the positive x-axis. This result applies *only* along the y-axis with its origin at the *midpoint* of the rod. Away from this axis of symmetry the total field will have both x- and y- components.

A net E_y component does exist because the dE_y values for $-l \le x \le +l$ all point along the positive y-axis. The summation of all the dE_y values to give E_y is obtained as follows:

$$E_y = \int_{-l}^{+l} dECos\theta dx = \int_{-l}^{+l} dE\frac{y}{r}dx$$
$$= \frac{1}{4\pi\varepsilon_o}\frac{Qy}{2l}\int_{-l}^{+l}\frac{dx}{r^3} = \frac{1}{4\pi\varepsilon_o}\frac{Qy}{2l}\int_{-l}^{+l}\frac{dx}{(x^2+y^2)^{3/2}}$$
$$= \frac{1}{4\pi\varepsilon_o}\frac{Qy}{2l}\left[\frac{1}{y^2}\frac{x}{(x^2+y^2)^{1/2}}\right]_{x=-l}^{x=+l}$$
$$= \frac{1}{4\pi\varepsilon_o}\frac{Q}{y(l^2+y^2)^{1/2}}$$

The total field vector $\mathbf{E}$ has no component E_x and so acts along a radial direction perpendicular to the centre of the charged rod and is expressed as:

$$E = \frac{1}{4\pi\varepsilon_o}\frac{Q}{y(l^2+y^2)^{1/2}}\hat{j} \tag{3.14}$$

If position P is moved further and further from the rod, we can expect the rod to appear more and more as a point charge. This should lead the field to have an inverse y^2 dependency as in Equation (3.9). We can check this by taking $y \gg l$ so that the square of l can be neglected in the denominator of Equation (3.14). This gives:

$$\mathrm{E} = \frac{1}{4\pi\varepsilon_o}\frac{Q}{y^2}\hat{j}$$

This indeed shows that at a distance far enough away from a charged rod its field is close to that of a point charge. Using the relationship $\lambda = Q/2l$ to express Equation (3.14) in terms of the linear charge density λ:

$$\mathrm{E} = \frac{1}{2\pi\varepsilon_o}\frac{\lambda}{y(1+y^2/l^2)^{1/2}}\hat{j} \quad (3.15)$$

We can now enquire what the field value is at a distance so close to the rod that, from the perspective of a test charge at P, the rod appears to be of near infinite length. In this case, the term y^2/l^2 is much smaller than unity and Equation (3.15) simplifies to:

$$\mathrm{E} = \frac{1}{2\pi\varepsilon_o}\frac{\lambda}{y}\hat{j} \quad (3.16)$$

The electric field close to a long line of smoothly distributed charge thus varies as the inverse of the radial distance from that line. At a distance very much greater than the length of the line of charge, the field varies with the inverse square of the distance. Between these two limiting situations, the field along a radial line bisecting the rod is given by Equation (3.15).

Example 3.6 Electric Field of a Charged Ring and Circular Disk

1. Find an expression to calculate the field on the axis of a ring of charge.
2. Use this result to find the field applied to a dielectric material, of relative permittivity ε_r, placed on top of a uniformly charged circular disk.

Solution 3.6

1. Field of a Uniformly Charged Ring in Air

The procedure for calculating the field along the axis of a charged ring of radius a is shown in Figure 3.11(a). We assume that the ring carries a positive charge Q and that it lies in the xz-plane, with the y-axis orthogonal to this. The origins of the x-, y- and z-axes are taken to be the centre of the ring. We divide the ring into infinitesimal elements ds such that the charge dQ on each element acts as a point positive charge.

The vector field $d\mathbf{E}$ produced at point P by each charge dQ of an element ds is:

$$d\mathrm{E} = \frac{1}{4\pi\varepsilon_o}\frac{dQ}{r^2}\hat{\mathrm{r}} \quad (3.17)$$

where r is the distance from an element ds to location P of value

$$r = \sqrt{y^2 + a^2}$$

The fields produced by diagonally opposite dQ charges on the ring cancel out within the xz-plane, but are additive along the y-axis. The components $d\mathrm{E}_x$ and $d\mathrm{E}_z$ of $d\mathbf{E}$ are thus zero, but the y-component $d\mathrm{E}_y$ along the y-axis component is given by:

$$d\mathrm{E}_y = d\mathbf{E}\,\mathrm{Cos}\theta = d\mathbf{E}\,y/r = \frac{1}{4\pi\varepsilon_o}\frac{dQ}{(y^2+a^2)}\frac{y}{\sqrt{y^2+a^2}}$$
$$= \frac{1}{4\pi\varepsilon_o}\frac{ydQ}{(y^2+a^2)^{3/2}}$$

The total y-component E_y of the field is found by summing up (integrating) all of the $d\mathrm{E}_y$ contributions around the ring:

$$\mathrm{E}_y = \int d\mathrm{E}_y = \int \frac{1}{4\pi\varepsilon_o}\frac{ydQ}{(y^2+a^2)^{3/2}}$$
$$= \frac{1}{4\pi\varepsilon_o}\frac{y}{(y^2+a^2)^{3/2}}\int dQ = \frac{1}{4\pi\varepsilon_o}\frac{Qy}{(y^2+a^2)^{3/2}}$$

The total vector field $\mathbf{E}$ along the y-axis of the ring of charge shown in Figure 3.11 is:

$$\mathrm{E} = \frac{1}{4\pi\varepsilon_o}\frac{Qy}{(y^2+a^2)^{3/2}}\hat{j} \quad (3.18)$$

2. Field in a Dielectric above a Uniformly Charged Circular Disk

Our approach is to recognize from Figure 3.11(b) that the free charge distribution on a circular disk can be viewed as being equivalent to a large number of concentric rings, of width dr and free charge dQ. We have the solution for the field along the axis of symmetry of a charged ring in air, so to find the field for a charged disk all we do is sum up all the contributions from the concentric system of rings that make up the total disk area. We also need to take into account that above the disk we have a medium of relative permittivity ε_r.

In Example 3.5 we denoted a linear charge density (dQ/dx C/m) using the symbol λ. For free charge distributed over a surface we will denote the charge density per unit area (dQ/dA C/m^2) by the symbol σ. (The symbol ρ will be used later in this chapter to signify a volume charge density per unit volume (dQ/dV C/m^3.)

In Figure 3.11(b) an elemental ring of charge dq has a circumference $2\pi r$, width dr and an area $dA \approx 2\pi rdr$. The surface charge density on an elemental ring is:

$$dQ = \sigma dA = \sigma 2\pi rdr$$

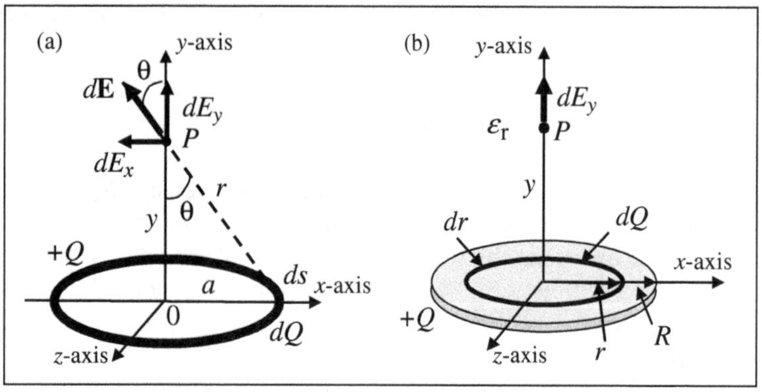

Figure 3.11 (a) Procedure to find the field at a point *P* along the axis of a charged ring. (b) Adopting the result for a charged ring to find the field in a dielectric medium above a uniformly charged disc of radius *R*.

Employing the result obtained in part 1 of this solution for the component dE_y of the field along the axis of a charged ring, then for each elemental ring in the disk:

$$dE_y = \frac{1}{4\pi\varepsilon_o\varepsilon_r}\frac{ydQ}{(y^2+r^2)^{3/2}} = \frac{1}{4\pi\varepsilon_o\varepsilon_r}\frac{y(2\pi\sigma rdr)}{(y^2+r^2)^{3/2}}$$

The total field E_y applied to the dielectric medium above the complete disk is given by integrating this expression for dE_y for all the elemental rings from the disk's centre to its outermost radius *R*:

$$E_y = \int_0^R \frac{1}{4\pi\varepsilon_o\varepsilon_r}\frac{y(2\pi\sigma rdr)}{(y^2+r^2)^{3/2}} = \frac{\sigma y}{2\varepsilon_o\varepsilon_r}\int_0^R \frac{rdr}{(y^2+r^2)^{3/2}} = \frac{\sigma y}{2\varepsilon_o\varepsilon_r}\left[-\frac{1}{(y^2+r^2)^{1/2}}\right]_{r=0}^{r=R}$$

i.e.

$$E_y = \frac{\sigma y}{2\varepsilon_o\varepsilon_r}\left[-\frac{1}{(y^2+R^2)^{1/2}}+\frac{1}{y}\right] \tag{3.19}$$

We can now enquire as to what the field will be if we increase the size of the disk, whilst maintaining the surface charge density σ constant, to the situation where the disk's radius *R* is very much larger than the distance *y* to the field measurement point *P*. In the limit, where $R \gg y$, from Equation (3.19) we have:

$$E = \frac{\sigma}{2\varepsilon_o\varepsilon_r} \tag{3.20}$$

This informs us that the field applied to the dielectric, in the central area (away from the edges) of a very large sheet of charge σ, does not vary as a function of the distance from the sheet. In other words, the field is uniform.

Equation (3.20) in Example 3.6 allows us to examine the electric field generated between two oppositely charged plates separated by a thin dielectric. This situation is shown in Figure 3.12.

The two plates carry the same magnitude of charge and so the fields E_{Ai} and E_{Bi} generated by them within the dielectric are equal in magnitude, given by Equation (3.20):

$$E_{Ai} = E_{Bi} = \frac{\sigma}{2\varepsilon_o\varepsilon_r}$$

The field produced by the upper plate *A* is directed away from the positive charges of that plate, both within the dielectric and in the surrounding air. Within the dielectric E_{Ai} is thus pointing towards the negatively charged lower plate B. The field E_{Bi} is directed towards the negative charges on plate *B*. Within the dielectric, therefore, fields E_{Ai} and E_{Bi} are additive, so that the resultant field E_i in the dielectric between the plates is given by:

$$E_i = E_{Ai} + E_{Bi} = \frac{\sigma}{\varepsilon_o\varepsilon_r} \tag{3.21}$$

This result applies only to regions within the dielectric that are not close to the edges of the plates, where the 'fringing' field is nonuniform. As shown in Figure 3.12, away from the edges, the fields E_{Ao} and E_{Bo} that extend away from the outer surfaces of the plates cancel each other, so that the external field is zero. The field generated by the oppositely charged plates is constrained mostly within the dielectric between them. Apart from

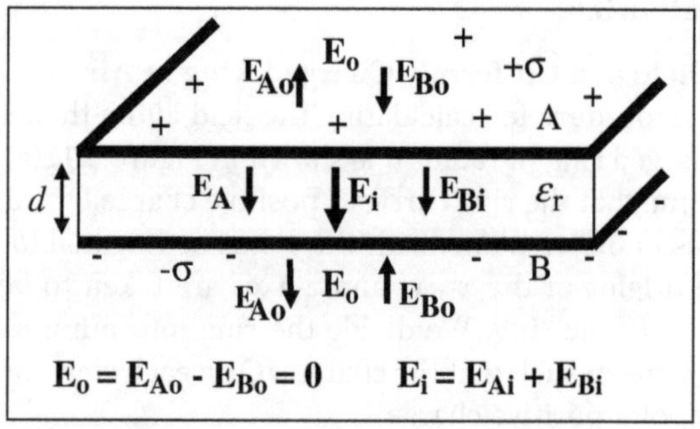

Figure 3.12 Plates (A and B) with equal and opposite charge densities ($+\sigma$ and $-\sigma$) are separated by a thin dielectric of thickness *d* and relative permittivity ε_r. The field $\mathbf{E}_i$ in the dielectric between the plates is the vector sum of the fields $\mathbf{E}_{Ai}$ and $\mathbf{E}_{Bi}$. The field $\mathbf{E}_o$ outside the plates is zero because the fields $\mathbf{E}_{Ao}$ and $\mathbf{E}_{Bo}$ are opposed.

where a 'fringing' field occurs between the edges of the two plates, the field is zero elsewhere. If we draw an imaginary surface that completely surrounds the two oppositely charged plates, the electric flux entering or leaving this closed surface area is zero. A net flux emerges from a closed surface surrounding the positively charged plate and enters a closed surface around the negatively charged plate. The relationship between the net electric flux 'flowing' across a closed surface and the net charge enclosed within this surface is given by a law formulated by Johann Carl Gauss. This law is important because it provides the means to determine the electric potentials and fields produced by electrodes in a more straightforward way than applying Coulomb's Law. It also helps us to understand what is happening in the process shown in Figure 3.1 that leads to the conclusion expressed by Equation (3.3).

3.3 Gauss's Law

Gauss's Law states that the net electric flux Φ through any closed surface that surrounds a defined volume is proportional to the net charge located within that volume. Two types of flux can be considered, namely the E-field flux Φ_E and the D-field flux Φ_D. In this book Maxwell's D-field flux, given by Equation (3.12), will mainly be employed. (The difference between the fields D/ε_0 and E will be discussed more fully in Chapter 6.) For any infinitesimal element of area dA on a closed surface the D-field flux through it is equal to the product $D_p dA$, where D_p is the component of the **D** vector normal (at right angles) to the surface of dA. The total flux Φ_D through a surface that totally surrounds a system of charges is obtained by adding up all the values of $D_p dA$ on that surface. This total D-flux is equal to the total *free* charge Q_{free} that is enclosed. In shorthand, Gauss's Law is expressed as:

$$\Phi_D = Q_{free} \tag{3.22}$$

Example 3.7 Electric Flux through a Small Surface Area

A surface element of dimensions 1 mm × 1 mm is exposed, in a vacuum, to an electric field vector **E** of magnitude 5×10^4 N/C.

1. Calculate the flux through this surface element when it is perpendicular to **E**.
2. Calculate the flux when the surface element is oriented at 40° to **E**.
3. Calculate the flux when the surface element lies parallel to **E**.

At this small scale the field can be considered to be uniform.

Solution 3.7 The three situations to be analysed are shown in Figure 3.13. The small surface element, of area $A = 10^{-6}\,\text{m}^2$, is not a closed Gaussian surface for which an inside and outside can be defined. We define a positive flux through a surface element as an outward flux, where the angle between the normal to this surface and the vector field is between 0° and 90°. In Figure 3.13 this is defined using the concept of a *normal unit vector* **n**, so that our small surface element can be defined to have a vector area **A** = A**n**. In this format the magnitude A and also the orientation of the element's surface area is provided. This information is required in order to evaluate the electric flux Φ passing through such an elemental vector area. From Figure 3.15 it is evident that the maximum E-field or D-field flux value occurs when the angle θ between the normal unit vector **n** and the vector **E** is zero. When the plane of the elemental area lies parallel to the field lines (i.e., $\theta = 90°$) no field lines pass through the area and the flux is zero. From the geometry of the situation we can see that the component of the vector area **A** along the direction of **E** varies as $\cos\theta$.

1. With the surface oriented perpendicular to **E**, $\theta = 0°$, the E-field flux Φ_E is:

$$\Phi_E = \mathbf{E}\mathbf{A}\cos 0° = (5 \times 10^4\,\text{N/C})(10^{-6}\,\text{m}^2) \times 1 = 5 \times 10^{-2}\,\text{N.m}^2/\text{C}$$

The corresponding **D**-field in vacuum is ε_o**E**, so that the D-field flux is given by:

$$\Phi_D = \varepsilon_o \mathbf{E}\mathbf{A}\cos 0° = (8.854 \times 10^{-12}\,\text{C}^2\,\text{N}^{-1}\,\text{m}^{-2})(5 \times 10^{-2}\,\text{N.m}^2/\text{C}) = 4.4 \times 10^{-13}\,\text{C}$$

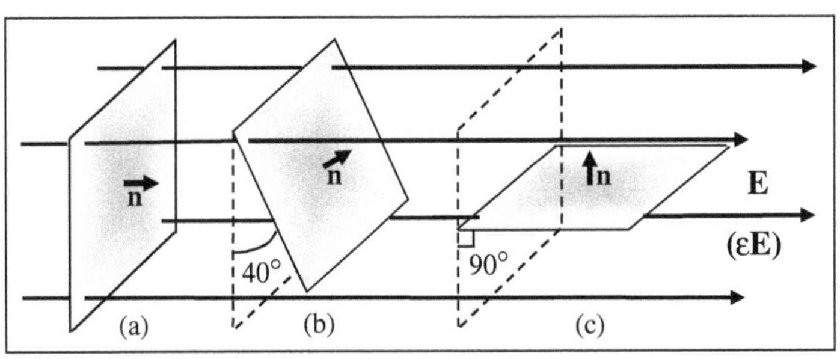

Figure 3.13 A small square surface area is shown oriented (a) perpendicular to a uniform electric field **E**, (b) at 40° to **E**, (c) parallel to **E**. Four electric field lines (**D**-lines if written as ε**E**) are drawn. The small arrow labelled **n** represents the normal unit vector directed at right angles to the plane of the square surface area.

Note how the unit (coulomb) of the D-field flux is easier to comprehend than those ($N.m^2/C$) of E-field flux. The temporal change of a D-field flux ($d\Phi_D/dt$) has units of coulomb / second, or ampere, the unit of electrical current. This is Maxwell's displacement current and from Gauss's Law we can deduce that it will relate to a change of the concentration of free charge in the system under study.

2. With the element's surface area oriented at 40° to **E**:

$$\Phi_E = \mathbf{E}A\cos 40° = (5\times10^4\,\mathrm{N/C})(10^{-6}\,\mathrm{m}^2)\times 0.766 = 3.8\times10^{-2}\,\mathrm{N.m^2/C}$$

and $\Phi_D = \varepsilon_o \mathbf{E}A\cos\theta = (8.854\times10^{-12}\,\mathrm{C^2\,N^{-1}m^{-2}})(3.8\times10^{-2}\,\mathrm{N.m^2/C}) = 1.7\times10^{-13}\,\mathrm{C}$

3. When the element's surface lies parallel to **E**, $\theta = 90°$, the E-field flux is:

$$\Phi_E = \mathbf{E}A\cos 90° = (5\times10^4\,\mathrm{N/C})(10^{-6}\,\mathrm{m}^2)\times 0 = 0$$

The D-field flux Φ_D is also zero.

The direction of the field vector **E** shown in Figure 3.13 implies that a net *positive* charge is located to the *left* of the surface element. If this situation is changed by replacing the charge on the left with a net negative charge, the direction of **E** is reversed. The angle between the normal unit vector **n** and **E** is now (180°−θ). Since cos(180°−θ) = −cosθ the magnitude of the fluxes calculated in (a) and (b) above will be the same, but 'flow' in the opposite direction. The results for (c) will still be zero.

In principle, Gauss's Law can be used in two ways. The spatial distribution of free charges can be determined by measuring the electrical field everywhere around them, or knowing the charge distribution we can determine the generated electric field. The second application is the more important for our purposes, as, for example, in determining the electric fields produced by electrified electrodes or surfaces.

A simple validation of Gauss's Law is to calculate the flux created by a positive free point charge Q. Around this charge we construct an imaginary 'Gaussian' surface in the form of a concentric sphere of radius R, as shown in Figure 3.14(a). **D**-field lines radiate from the charge and because of the symmetry of the arrangement they pass through the imaginary sphere at right angles to its surface. For every infinitesimal surface area dA on the sphere's surface the electric field magnitude E is the same and given by Equation (3.8) as:

$$\mathrm{E} = \frac{1}{4\pi\varepsilon_o\varepsilon_r}\frac{Q_{free}}{R^2}\ \text{(newton per coulomb)}$$

The magnitude of the corresponding D-field in a dielectric medium of relative permittivity ε_r is:

$$\mathrm{D} = \varepsilon_o\varepsilon_r E = \frac{1}{4\pi}\frac{Q_{free}}{R^2}\ \text{(coulomb per square metre)}$$

The total electric flux Φ_D emerging from the spherical surface is the product DA, where A is the total surface area of the sphere ($A = 4\pi R^2$):

$$\Phi_D = \mathrm{D}A = \frac{1}{4\pi}\frac{Q_{free}}{R^2}(4\pi R^2) = Q_{free}\ \text{(coulomb)} \tag{3.23}$$

This is the result stated in Equation (3.22) for the form of Gauss's Law dealing with free charges. If the positive charge is replaced with a negative one of the same magnitude, the convention is to visualize the 'flow' of electric flux as *entering* the imaginary spherical surface and terminating at the negative charge. For this case the flux Φ_D is defined to be negative. We can think of a positive charge as *source* of flux and a negative charge as a flux *sink*. The result obtained in Equation (3.23) indicates that the total flux is independent of the radius of the Gaussian sphere. We can demonstrate this by introducing another, smaller, spherical surface around the charge, as shown in Figure 3.14(a). Field lines passing through the inner sphere of radius r also pass through the larger sphere of radius R. According to Coulomb's Law, the field $\mathbf{E}_r$ at any point on the surface of the innermost sphere is *larger* than the field $\mathbf{E}_R$ on the outer sphere by an amount proportional to $(R/r)^2$. However, the elemental area dA_r on

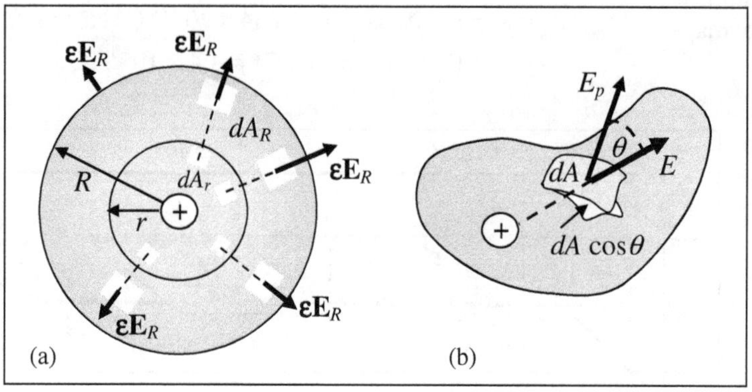

Figure 3.14 (a) Electric flux (D = εE) lines passing through two concentric spherical surfaces centred on a positive charge. The same magnitude of flux passes through the surface of both spheres. (b) On an irregular surface the projection of a surface element dA onto a spherical surface at the same point is $dA.\cos\theta$. The electrical flux ($\varepsilon E_p dA$) through an element on the irregular surface translates to a flux ($\varepsilon EdA.\cos\theta$) through a corresponding spherical surface element.

projection from the outer sphere is *smaller* than dA_R by the factor $(r/R)^2$. The two fluxes given by the products $\mathbf{D}_r.dA_r$ and $\mathbf{D}_R.dA_R$ are thus equal. The radius of a constructed imaginary Gaussian sphere is of no significance when determining the total flux. We can use this fact to accommodate situations where the field varies from point to point over a closed surface that is irregular rather than spherical.

The Gaussian surface in Figure 3.14(b) is irregular in shape. The magnitude of the normal field E_p and hence flux D_p, at an elemental surface dA varies over this surface. As shown in Figure 3.14(b) we now construct a corresponding spherical surface element at the same point as dA. The area of this spherical element will be smaller than dA and have the value $dA.\cos\theta$, where θ is the angle between E_p and the field E emerging from the spherical element. The two fluxes $D_p dA$ and $DdA.\cos\theta$ are thus equal in magnitude. This procedure can be repeated over the whole surface, where each elemental area dA is projected onto a corresponding spherical surface. If we sum up all of these fluxes (the mathematical equivalent of performing a surface integral, represented by the symbol $\int_S$) the total flux through the irregular surface will give the same result as Equation (3.23), namely:

$$\Phi_D = \int_S D\cos\theta dA = \int_S D_p dA = \int_S \varepsilon_o \varepsilon_r E_p dA = Q_{free} \tag{3.24}$$

The procedure of summing up the fluxes through a distribution of spherical surface elements of different radii may appear to be a dubious approach to verifying Equation (3.24). However, such doubt can be removed by showing that all of the elemental surfaces $EdA.\cos\theta$ can be projected onto a smaller closed spherical surface around the charge Q.

A particular reason for drawing an irregular Gaussian surface would be to surround an irregular array of single point charges. To accommodate this situation we can employ the principle of superposition, which for this purpose states that the net flux of a collection of charges is the same as the sum of the fluxes produced by each charge on its own. For a number N of free charges Q_i in a volume V surrounded by a closed Gaussian surface S, Equation (3.24) takes the form:

$$\Phi_D = \int_S D_p dA = \int_S \varepsilon_o \varepsilon_r E_p dA = \sum_{i=1}^{N} Q_i = \int_V \rho_e dV \tag{3.25}$$

where ρ_e is the density of free charge contained in a volume element dV inside the closed surface. The symbol $\int_V$ represents the mathematical process of integrating (adding up) all the charged volume elements in the total volume V.

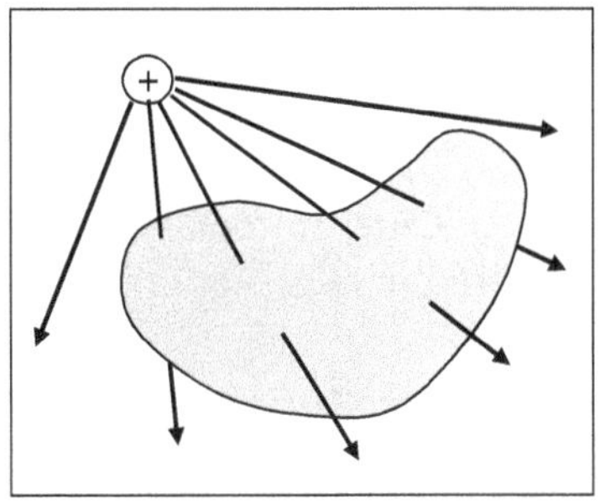

Figure 3.15 A closed Gaussian surface is shown, which surrounds a volume containing zero charge. Field lines from an external charge that enter this surface will exit the surface at another point. The *net* flux through the closed surface is zero. An electric field line can only start at a source (positive charge) and terminate at a sink (negative charge).

It follows from Equations (3.24) and (3.25) that, if the positive charge shown in Figure 3.14 is removed, or if negative charges are introduced so as to cancel the positive charges, the net flux through the Gaussian surface is zero. The presence of charges *outside* a closed surface that contains zero charge will not produce a net flux through that surface. As demonstrated in Figure 3.15, the field lines from an external point charge that enter the Gaussian surface will emerge from it at another point. Field lines can begin or end in a volume of space only when there is a net charge in that volume.

3.3.1 Alternative Forms of Gauss's Law

Equations (3.22–3.25) relate the total **D**-field flux through a closed surface to the total enclosed *free* charge. It is usually the case that we know the concentration of free charges, for example those located on a capacitor plate electrode or transferred as static charge from one surface to another. The corresponding useful form of Gauss's Law is to define the total flux as given by Equations (3.22) or (3.25). However, in some cases it is helpful to consider the total charge, comprising both the free charges and the bound charges. In this case an alternative form of Gauss's Law is expressed in terms of the E-field flux as:

$$\Phi_E = \int_S E_p dA = \frac{1}{\varepsilon_o} \sum_{i=1}^{N} Q_i = \frac{Q_T}{\varepsilon_o} \tag{3.26}$$

where Q_T is the *total* charge, free plus bound, contained within the Gaussian surface.

Equations (3.25) and (3.26) are known as the integral forms of Gauss's Law (they involve surface and volume integrations, $\int_S$ and $\int_V$). An alternative way to express Gauss's Law is the differential form, involving the gradient, or differential operator, ∇, introduced in Chapter 1

Box 3.2 Laplace's Equation

Laplace's equation has important applications in electrostatics, gravitation, and in analyses of steady-state heat flow and fluid flow. For Cartesian coordinates *x, y, z* in space it takes the general form:

$$\nabla^2 u = \partial^2 u/\partial x^2 + \partial^2 u/\partial y^2 + \partial^2 u/\partial z^2 = 0$$

$\nabla^2 u$ is called the *Laplacian* of *u*, and the theory for finding solutions for it is called *potential theory*. Practical problems involve boundary conditions for a particular volume or region *T* of space with a well defined boundary surface *S* or a region defined by some curve such as a circle. Two commonly used boundary surfaces are those having cylindrical or spherical symmetry. For a rod *T* at a fixed potential surrounded by a cylinder *S*, the coordinates *r*, θ and *z* (see figure) are related to *x, y, z* by $x = r\cos\theta$, $y = r\sin\theta$, and $z = z$. The *Laplacian* has the form:

$$\nabla^2 u = \frac{\partial^2 u}{\partial r^2} + \frac{1}{r}\frac{\partial u}{\partial r} + \frac{1}{r^2}\frac{\partial^2 u}{\partial \theta^2} + \frac{\partial^2 u}{\partial z^2}$$

For spherical symmetry (a ball *T* surrounded by a sphere *S*) the spherical coordinates *r*, θ and ϕ (see opposite) are related to *x, y, z* by $x = r\cos\theta\sin\phi$, $y = r\sin\theta\sin\phi$, and $z = r\cos\phi$, so that the Laplacian has the form:

$$\nabla^2 u = \frac{1}{r^2}\left[\frac{\partial}{\partial r}\left(r^2\frac{\partial u}{\partial r}\right) + \frac{1}{\sin\phi}\frac{\partial}{\partial \phi}\left(\sin\phi\frac{\partial u}{\partial \phi}\right) + \frac{1}{\sin^2\phi}\frac{\partial^2 u}{\partial \theta^2}\right]$$

Where the boundary condition (*e.g.*, electrostatic potential) on a sphere *S* is independent of angle θ (i.e., has cylindrical symmetry), the solution $u(r, \phi)$ will also be independent of θ. Thus, $\partial^2 u/\partial\theta^2 = 0$ and Laplace's equation takes the form:

$$\nabla^2 u = \frac{\partial}{\partial r}\left(r^2\frac{\partial u}{\partial r}\right) + \frac{1}{\sin\phi}\frac{\partial}{\partial \phi}\left(\sin\phi\frac{\partial u}{\partial \phi}\right) = 0$$

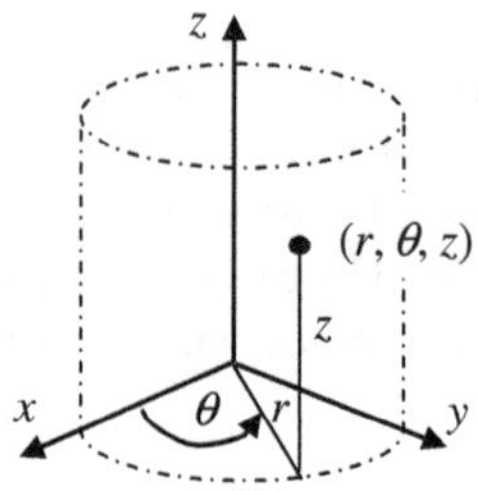

Cylindrical coordinates

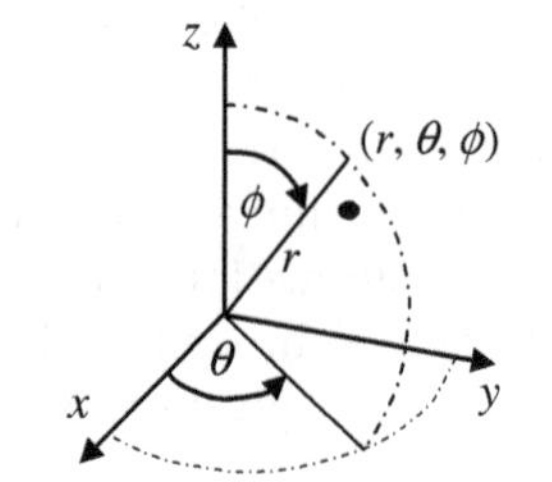

Spherical (polar) coordinates

and further described in Equation (3.11). The differential form of Equation (3.22) is:

$$\nabla \cdot \mathbf{D} = \rho_e \tag{3.27}$$

where ρ_e is the free charge density and $\nabla\cdot\mathbf{D}$ is the *divergence* of the D-field. The notation $\nabla\cdot\mathbf{D}$ is variously verbalized as div **D**, grad dot D, or del dot D. If we consider a vector D-field in three-dimensional space, having components $\mathrm{D_x}$, $\mathrm{D_y}$, $\mathrm{D_z}$, along the orthogonal *x*-, *y*- and *z*-axes, we can express vector **D** as:

$$\mathbf{D} = \mathrm{D_x}\hat{\imath} + \mathrm{D_y}\hat{\jmath} + \mathrm{D_z}\boldsymbol{k}$$

where $\hat{\imath}$, $\hat{\jmath}$ and $\hat{k}$ are the unit vectors along the *x*-, *y*- and *z*- axes, respectively. To calculate $\nabla\cdot\mathbf{D}$ we take the components of the differential operator ∇, apply them to the components of **D** and sum the results, to obtain the non-vector (scalar) version of Gauss's Law:

$$\nabla \cdot \mathrm{D} = \frac{\partial \mathrm{D}_x}{\partial x} + \frac{\partial \mathrm{D}_y}{\partial y} + \frac{\partial \mathrm{D}_z}{\partial z} = \rho_e \tag{3.28}$$

In terms of the hydrodynamic analogy we have used to describe the 'flow' of electric flux, we can envisage $\nabla\cdot\mathbf{D}$ to represent the amount of flux flowing out of, or into, a certain volume element. This is obtained by adding up all the sources of flux (positive charges) and subtracting all of the sinks of flux (negative charges) inside that volume element. The corresponding differential form of Equation (3.26) for the E-field is:

$$\nabla \cdot \mathrm{E} = \frac{\partial \mathrm{E}_x}{\partial x} + \frac{\partial \mathrm{E}_y}{\partial y} + \frac{\partial \mathrm{E}_z}{\partial z} = \frac{\rho_T}{\varepsilon_o} \tag{3.29}$$

where ρ_T is the total charge density (free plus bound). In Chapter 1, Equation (1.2), we define the electric field as the negative gradient of the voltage potential ($\mathrm{E} = -\nabla V$). Substituting this relationship into Equation (3.29), we obtain:

$$\nabla^2 V = -\frac{\rho_T}{\varepsilon_o} \tag{3.30}$$

This is known as Poisson's equation. For the case where the total charge is zero, we have Laplace's equation:

$$\nabla^2 V = 0 \tag{3.31}$$

Poisson's and Laplace's equations are valid *only* for electrostatics [11, 12]. They are modified when we have time-varying fields. In Chapters 5, 6 and 9 we find that solutions to Laplace's equation are important in the development of dielectrophoresis theory. Details given in Box 3.2 provide the first steps in this process. The E-field form of Gauss's Law, Equation (3.26), holds a particularly important position in electromagnetics – it is the first

Box 3.3 Maxwell's Equations

James Clerk Maxwell not only developed the concept of a displacement flux and showed that an electromagnetic disturbance travels through free space at the speed of light – in 1865 he also demonstrated that the basic principles of electromagnetism discovered by himself and others can be reduced to four equations. These are now known as Maxwell's equations, which in their integral and differential forms are:

1. $\int_S EdA = \frac{Q_T}{\varepsilon_o}; \nabla \cdot E = \frac{\rho}{\varepsilon_o}$ (Gauss's Law for electric fields).
2. $\int_S BdA = 0; \nabla \cdot B = 0$ (Gauss's Law for magnetic fields).

This second equation reflects the fact that single magnetic 'charges' or magnetic monopoles do not exist. The surface integral of a magnetic field over a Gaussian surface is always zero because the surface can enclose neither a net source nor sink of a magnetic field. A magnetic field line always takes the form of a closed loop. An electric field starts at a source (positive charge Q) and ends at a sink (negative charge).

3. $\int_l Bdl = \frac{J}{\varepsilon_o c^2} + \frac{1}{c^2}\frac{d\Phi_E}{dt}; \nabla \times B = \frac{J}{\varepsilon_o c^2} + \frac{1}{c^2}\frac{\partial E}{\partial t}$.

This is the generalized Ampère's Law, which states that the conduction current J and the displacement current in a closed circuit both act to produce a magnetic field. In this equation, c is the speed of light $c^2 = \frac{1}{\mu_o \varepsilon_0}$.

4. $\int_l EdA = -\frac{d\Phi_B}{dt}; \nabla \times E = -\frac{\partial B}{\partial t}$.

This is Faraday's Law, which states that a time-varying magnet flux or field creates an electric field.

In the forms written above, these equations apply to electric and magnetic fields in free space. When a material is present, the permittivity and permeability of free space (ε_o and μ_o, respectively) are replaced by the material's dielectric and magnetic properties, $\varepsilon_o\varepsilon_r$ and $\mu_o\mu_r$, respectively. Maxwell's equations inform us that a point electric charge at rest creates a static electric field but no magnetic field, whereas a point charge moving at a constant velocity produces both an electric and magnetic field. To produce an electromagnetic wave, the point charge must be accelerated.

of the four famous Maxwell's equations. These equations are described in Box 3.3.

3.3.2 Applications of Gauss's Law

Careful inspection of Figure 3.1 indicates that the excess, stationary, positive and negative charges are shown distributed along an outer edge of the two metal plates, rather than uniformly inside the metals. Was this a careless execution of the drawing, or an intentional attempt to reflect reality? Gauss's Law can provide an answer. In Figure 3.16 a Gaussian surface *S* is shown constructed so as to enclose almost the entire internal volume of an isolated solid metal rod. In our discussion of Figure 3.3, depicting electric lines of force (field lines) formed outside the ends

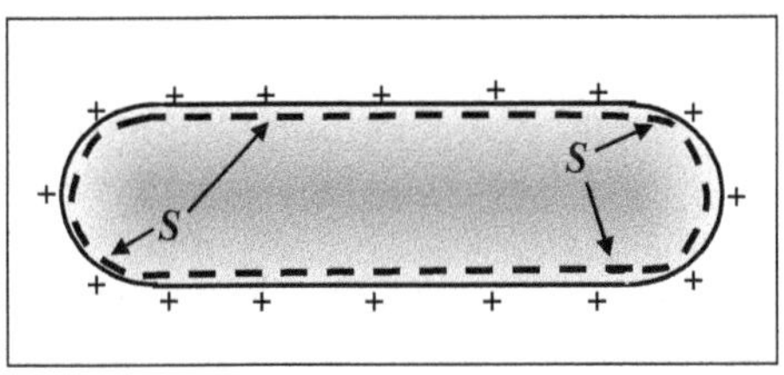

Figure 3.16 A Gaussian surface *S*, depicted as the closed dotted-line, is constructed within an isolated metal rod that has been electrified with excess positive charges. Because no electric field can permanently exist within the rod, Gauss's Law informs us that none of the charges can exist within the rod and must all be distributed at the rod's surface.

of an electrified and isolated metal rod, it was stated that no electrical force can permanently exist within the rod. This corresponds to the statement that no electric field can exist at any point within an isolated metal conductor that has been electrified or is subjected to an imposed constant electric field. The atomic structure of an electrical conductor is such that the electrons occupying the outermost atomic orbitals are delocalized – they are not bound by an electrostatic force of attraction to any one particular atomic nucleus. They can be thought of as a *gas* of free electrons. On imposing an electric field in a conductor that does not form part of a continuous electric circuit, its delocalized electrons are displaced (leaving behind an effective net positive charge) until they produce an internal field that exactly counterbalances the applied one. A continuous electric current is not produced in an electrified isolated conductor. Thus, if the electric field within the Gaussian surface shown in Figure 3.16 is zero, from Gauss's Law given by Equation (3.24) we can deduce that the net charge density *within* it is also zero. The electronic charges induced to move in an electrified metal conductor and the corresponding deficit of electronic charges that appear as positive charges, must therefore be located solely at the *surface* of the conductor. This also applies to situations such as that shown in Figure 3.1 where two metal plates have been electrified by application of a voltage potential difference to them. The surface distribution of charges shown in the drawing of Figure 3.1 was a conscious attempt to depict this fact.

It is also instructive to explore how Equations (3.25) and (3.26) can assist an understanding of the effect shown in Figure 3.1, where insertion of a dielectric slab between two charged metal plates leads to an apparent reduction of the free charge. In Figure 3.17, two cylindrical Gaussian surfaces have been constructed in a region close to the boundary between the positively charged metal plate and dielectric of Figure 3.1. Cylinders, rather than spheres and the orientation of them have been chosen to take advantage of their symmetry and the nature of the electric field produced within the dielectric. The field E given by Equation (3.21) obtained in Example 3.6(b) is uniform and directed at right angles to the interface between the metal plate and the dielectric. This field is therefore parallel to the sides of the cylinders, corresponding to the situation shown in Figure 3.13(c) and so no net flux passes through the sides of either cylinder. Cylinder 1 in Figure 3.17 has one circular end in the metal, where the field is zero. No flux passes through this end of cylinder 1. The total flux through the surface of Gaussian cylinder 1 occurs across the end located in the dielectric (total area $A = \int_S dA = \pi r^2$). For cylinder 1 the D-field form of Gauss's Law gives:

$$\int_S DdA = \int_S \varepsilon_o\varepsilon_r EdA = \varepsilon_o\varepsilon_r E\int_S dA$$
$$= \varepsilon_o\varepsilon_r E(\pi r^2) = \sigma_{free} \tag{3.32}$$

For the E-field form of Gauss's Law we will define the total charge density as:

$$\sigma_T = (\sigma_{free} + \sigma_{bound})$$

For Gaussian cylinder 1 the E-field form of Gauss's Law gives:

$$\int_S EdA = E(\pi r^2) = \frac{1}{\varepsilon_r}(\sigma_{free} + \sigma_{bound}) \tag{3.33}$$

From Equations (3.32) and (3.33) we can derive the relationship:

$$(\sigma_{free} + \sigma_{bound}) = \frac{1}{\varepsilon_r}\sigma_{free}$$

i.e.

$$\sigma_{bound} = \sigma_{free}\left(\frac{1}{\varepsilon_r} - 1\right) \tag{3.34}$$

This result, when compared with Equation (3.3), shows that $\sigma_{bound} = -\Delta\sigma$. In other words the reduction $\Delta\sigma$ of the free charge density on the metal plate that is observed on inserting a dielectric slab does not result from a leakage away of free charge, but from a fraction of the free charge being *neutralized* by the *creation* of an induced bound charge of equal magnitude and opposite polarity. But where does this bound charge reside in the dielectric? The answer is given by inspecting the Gaussian cylinder 2 shown in Figure 3.17, which is located totally within the bulk of the dielectric. The net flux through the surface of this cylinder is zero. All of the E-flux through the top of the cylinder exits through the bottom. Therefore, no charge (bound or free) can exist within the volume enclosed by the surface of cylinder 2 and by extension no bound charge can exist anywhere else within the bulk of the dielectric. The important conclusion arising from this is that: *the polarization charges produced by the dielectric can exist only as bound charges at the surfaces of the dielectric that directly face the metal electrode plates.*

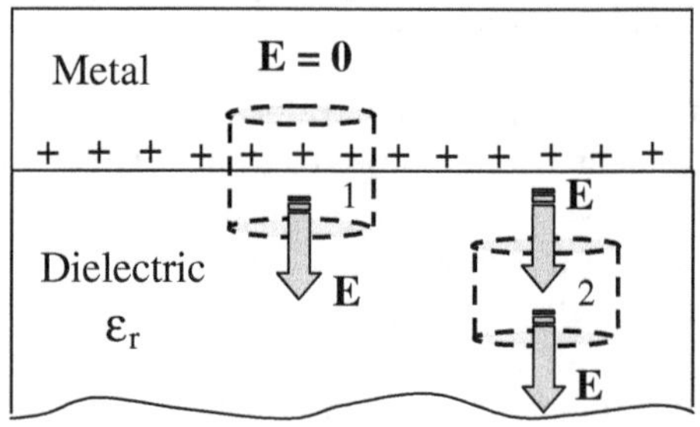

Figure 3.17 A Gaussian cylinder 1 is constructed at the interface between a dielectric and a metal plate carrying a distribution of positive *free* charges. The only flux passing through this cylinder's surface is that which emerges through the end in the dielectric. This flux is generated by the free and induced bound charges enclosed within its surface. No net flux passes through the Gaussian cylindrical surface 2, and so it contains zero charge. The induced bound charge must therefore be located at the surface of the dielectric in direct contact with the metal.

Further aspects of polarization and induced bound charge within dielectrics are considered later in this chapter, after first describing other useful applications of Gauss's Law.

3.3.2.1 Field Created by a Long Linear Charge Distribution

We will examine again, using Gauss's Law, the scheme shown in Figure 3.10 of a distribution of positive charge on a long metal rod or wire. The first step is to consider the shape of the Gaussian surface to be constructed around the distributed charge and to accomplish this we need to examine the symmetrical characteristics of the system to be analysed. From Example 3.5 we know that at locations away from the ends of the rod the field is directed radially outwards at right angles to the wire. The component of the field parallel to the central axis along the rod is zero. If we rotate the wire about this axis, the field at a fixed distance away from the wire does not change in value. The radial and rotational symmetry of the field indicates that the Gaussian surface should take the form of a coaxial cylinder, as depicted in Figure 3.18.

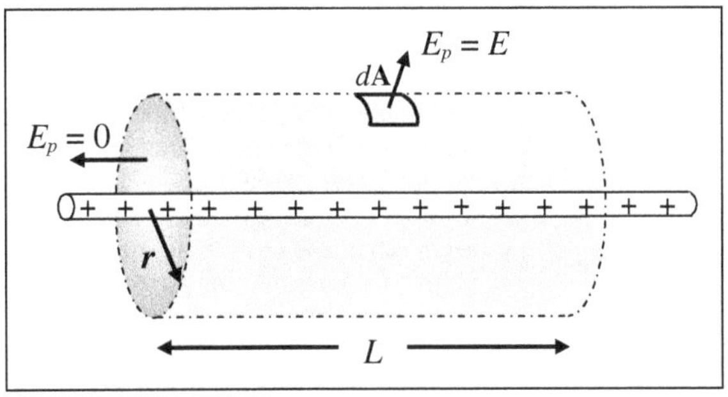

Figure 3.18 A Gaussian surface, in the form of a coaxial cylinder of radius r and length L, is drawn around a section of a long, positively, charged rod.

The electric field produced by the charged rod can be obtained from Gauss's Law, in the form of Equation (3.24), using the relationship:

$$\int_S \mathrm{E}_p dA = \frac{1}{\varepsilon_o \varepsilon_r} \sum_{i=1}^{N} Q_i \tag{3.35}$$

In this equation $\int_S \mathrm{E}_p dA$ represents the sum of the product ($\mathrm{E}_p dA$) taken over the whole surface of the Gaussian cylinder shown in Figure 3.18. We have noted that the component of the field parallel to the central axis along the rod is zero. Thus, at each point on the two ends of the cylinder the product ($\mathrm{E}_p dA$) is zero, because E_p (being directed parallel to the rod's axis) is zero. The only contribution to $\int_S \mathrm{E}_p dA$ over the Gaussian surface is thus from the cylindrical wall, of length L and radius r. The value for E_p everywhere over this surface is E and the total surface area $\int_S dA = 2\pi rL$. The left-hand side of equation (3.35) can thus be written as:

$$\int_S \mathrm{E}_p dA = \mathrm{E}(2\pi rL) \tag{3.35a}$$

For the right-hand side of Equation (3.35)

$$\frac{1}{\varepsilon_o \varepsilon_r} \sum_{i=1}^{N} Q_i = \frac{1}{\varepsilon_o \varepsilon_r} \sum_{l=0}^{l=L} \lambda dl = \frac{\lambda L}{\varepsilon_o \varepsilon_r} \tag{3.35b}$$

where λ is the linear charge density per unit length of the rod. Inserting the results given by Equations (3.35a) and (3.35b) in Equation (3.35) we have:

$$\mathrm{E}(2\pi rL) = \frac{\lambda L}{\varepsilon_o \varepsilon_r}$$

or

$$\mathrm{E} = \frac{\lambda}{2\pi \varepsilon_o \varepsilon_r r} \tag{3.36}$$

This gives, but with less effort, the same result obtained in Example 3.5, Equation (3.16), for the magnitude of the field of an infinite length of charge.

3.3.2.2 Field Created by a Charged Metal Sphere

We have demonstrated, with the use of Figure 3.16, that all of the excess charge Q given to a solid metal conductor must be located on its surface – no charge or electric field can exist inside it. This is the principle by which a 'Faraday cage' operates. Sensitive electronic systems can be protected from interference by external electric fields by enclosing them in a metal cage. At any point outside the surface of a positively charged sphere we know, from Equations (3.8) and (3.9), the field is directed radially outwards from the centre of the sphere and depends only on the distance r from the centre. Thus, the field has the same value at any point on a spherical surface that is concentric with the charged sphere. A spherical Gaussian surface of radius r, constructed concentrically around a charged metal sphere of radius R as shown in Figure 3.19(a), will take advantage of this spherical symmetry of the field.

In Equation (3.35) the expression $\int_S \mathrm{E}_p dA$ now represents the sum of the product ($\mathrm{E}_p dA$) taken over the whole surface of the Gaussian sphere shown in Figure 3.19 and has the value $4\pi r^2\mathrm{E}$. The left-hand side of equation (3.35) can thus be written as:

$$\int_S \mathrm{E}_p dA = 4\pi r^2 \mathrm{E} \tag{3.37a}$$

For the right-hand side of Equation (3.35)

$$\frac{1}{\varepsilon_o \varepsilon_r} \sum_{i=1}^{N} Q_i = \frac{Q}{\varepsilon_o \varepsilon_r} \tag{3.37b}$$

where Q is the total excess charge on the metal sphere. From Equation (3.35), using the results given by Equations (3.37a) and (3.37b), we obtain:

$$\mathrm{E} = \frac{1}{4\pi \varepsilon_o \varepsilon_r} \frac{Q}{r^2} \ (r > \mathrm{R}) \tag{3.38}$$

This is the same result as Equation (3.8), which was obtained using Coulomb's Law. Inside the charged sphere ($r < R$) the field is zero, but outside the sphere ($r > R$) the field varies as $1/r^2$ as if all of the excess charge Q distributed on the surface is concentrated as a point charge at the centre of the metal sphere. The same result is achieved for a hollow spherical shell of metal. No charge exists in the interior of a hollow shell, so no field can exist there either. The variation of the field as a function of distance from the sphere's centre is shown in Figure 3.19(a).

From this result we can also conclude that the excess charge is uniformly distributed on the surface of the metal sphere. If it were not uniformly distributed, then the field would not vary at every point outside its surface as $1/r^2$ – but in some places as some other power law of the distance r. We could anticipate this conclusion by

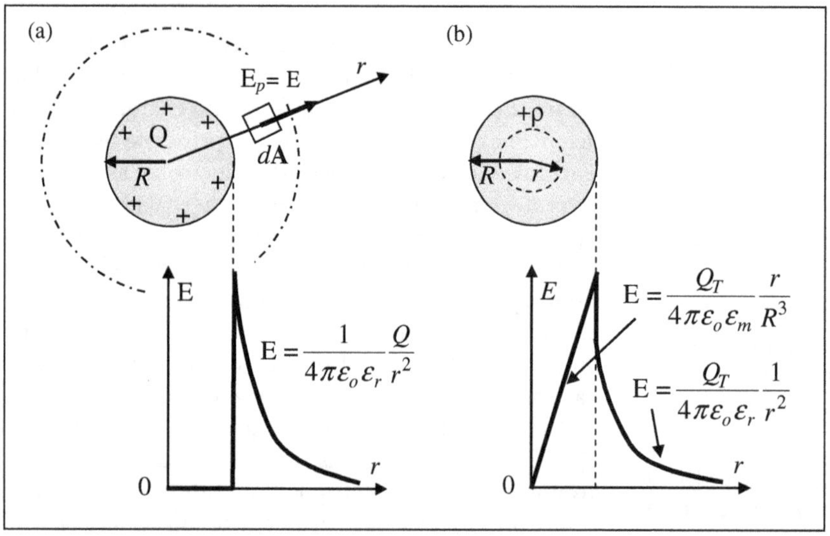

Figure 3.19 (a) A spherical Gaussian surface (dashed line) is shown constructed concentrically around a positively charged metal sphere of radius *R*. Inside the metal sphere the field *E* is zero, but outside the sphere ($r > R$) the field 'jumps' to the value it would have if the metal sphere and its excess charge *Q* were to be squashed into a point source at the centre. (b) A spherical Gaussian surface is constructed concentrically around a dielectric sphere containing a uniform distribution of trapped positive charges. Inside the dielectric sphere the field *E* is proportional to the distance from its centre. The field outside is equal to that produced if the sphere and its charge took the form of a point source.

imagining a metal sphere with excess electrons on its surface. These charges repel each other equally in all directions, until equally spaced apart and thus collectively at their minimum potential energy. If electrification of the metal sphere is accomplished by removing electrons from its atomic lattice, this removal of negative charges from otherwise electrically neutral matter will lend to the electrification the properties of an excess of positive charges. The most energetic electrons in the metal's atomic structure will rearrange themselves so that the positively charged electronic 'vacancies' are uniformly distributed apart by their mutual electrostatic repulsions. Excess negative or positive electronic charges are thus always distributed uniformly over spherical metal surfaces, or away from the edges of flat and cylindrical surfaces.

By extension, we can use this result for a metal sphere to deduce the field generated by a solid metal cylinder. For an infinite metal cylinder with charge per unit length λ, the field at any point outside the cylinder's surface is given by Equation (3.36), as if all the surface charge was located along the central axis of the cylinder. The field is zero inside the metal cylinder.

3.3.2.3 Field Created by a Charged Dielectric Sphere

For a sphere composed of a dielectric material (i.e., an electrically insulating material) we can assume that excess charges (not induced ones) are localized (trapped) throughout the material, with a corresponding uniform volume charge density ρ_e (C/m^3). Because an insulting material is unable to conduct electricity to any significant extent, a large electric field can be sustained across and within it without disturbing an electrostatic state where all excess charges are assumed to be stationary. If the charge is uniformly distributed within a spherical particle of radius *R*, we can explore the field within it by constructing an internal Gaussian sphere of radius *r* ($r < R$) as shown in Figure 3.19(*b*).

The appropriate form of Equation (3.24) with which to apply Gauss's Law is:

$$\int_S E_p dA = \frac{1}{\varepsilon_o \varepsilon_m}\int_V \rho_e dV \tag{3.39}$$

where ε_m is the relative permittivity of the dielectric material. For an internal Gaussian sphere of radius *r*, the left-hand side of this equation can be written as $\int_S E_p dA = 4\pi r^2 E$. To elucidate the right-hand side of Equation (3.39) it is convenient to define the total charge carried by the dielectric sphere as Q_T, such that:

$$Q_T = \frac{4}{3}\pi R^3 \rho_e$$

For a spherical Gaussian surface of radius *r* within the charged dielectric sphere, the enclosed internal charge q_i is given by:

$$q_i = \rho_e\left(\frac{4}{3}\pi r^3\right) = Q_T\frac{r_3}{R^3}$$

Gauss's Law in the form of Equation (3.39) is thus given as:

$$4\pi r^2 E = \frac{Q_T}{\varepsilon_o \varepsilon_m}\frac{r^3}{R^3}$$

to give

$$E = \frac{Q_T}{4\pi\varepsilon_o\varepsilon_m}\frac{r}{R^3}\ (r < R) \tag{3.40}$$

The field within a charged dielectric sphere is thus proportional to the distance *r* from the centre of the sphere. At the centre point of the sphere there is no charge, the value for *r* is zero and so as given by Equation (3.40) the field E is zero. For a Gaussian spherical surface constructed beyond the surface of the charged dielectric sphere, the enclosed charge is Q_T and we can use the result of Equation (3.38) to give the external field as:

$$\mathrm{E} = \frac{1}{4\pi\varepsilon_o\varepsilon_r}\frac{Q_T}{r^2}\ (r > \mathrm{R}) \tag{3.41}$$

where ε_r is the relative permittivity of the medium surrounding the charged dielectric sphere. The variation of the electric field from the centre of the dielectric sphere to beyond its surface is shown in Figure 3.19(*b*).

3.3.2.4 Field between Oppositely Charged Parallel Electrodes

Figure 3.20 depicts a dielectric material situated between a top metal plate that carries a positive charge density per unit area $+\sigma$ and a bottom plate with an opposite charge density $-\sigma$. The thickness of the dielectric is small in comparison to the dimensions of the plates, but this relationship is distorted in Figure 3.20 in order to show the construction of two Gaussian surfaces in the form of cylinders with their sides parallel to the direction of the uniform field E. Because opposite charges attract each other, the free charge density distributions, $+\sigma$ and $-\sigma$, are shown as close as they can get to each other, namely distributed along the interfaces between the metal plates and the dielectric.

A *positive* flux of magnitude $\varepsilon_o\varepsilon_r\mathrm{E}A$ flows *out* of the top Gaussian cylinder through the end, of area A, located in the dielectric. There is no flux through the sides of the cylinder, or in the end located in the metal. The *positive* free charge enclosed within the top Gaussian cylinder is $+\sigma A$. From Gauss's Law we have $\varepsilon_o\varepsilon_r\mathrm{E}A = \sigma A$, and so

$$\mathrm{E} = \frac{\sigma}{\varepsilon_o\varepsilon_r} \tag{3.42}$$

This is the same result given by Equation (3.21), obtained using Coulomb's Law and from superposition of the fields created by the charges on each metal plate. The charge on the bottom plate did not enter into the evaluation of Equation (3.42), but we can check this result by performing Gauss's Law for the bottom Gaussian cylinder. A *negative* flux $(-\varepsilon_o\varepsilon_r\mathrm{E}A)$ flows *out* of the bottom Gaussian cylinder, which encloses a *negative* charge $-\sigma A$. From Gauss's Law we have $-\varepsilon_o\varepsilon_r\mathrm{E}A = -\sigma A$, which gives the same result as equation (3.42).

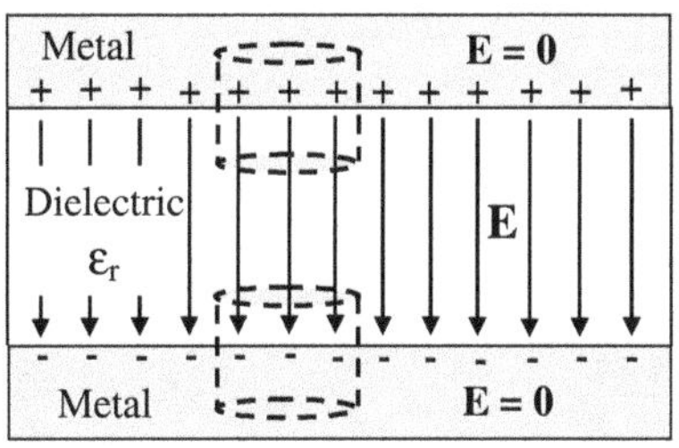

Figure 3.20 The electric field in a thin dielectric between oppositely charged ($\pm\sigma$), large area, metal plates can be derived by applying Gauss's Law to either the top or bottom Gaussian cylinder shown in this figure. In each case we obtain $\mathrm{E} = \sigma/\varepsilon_o\varepsilon_r$.

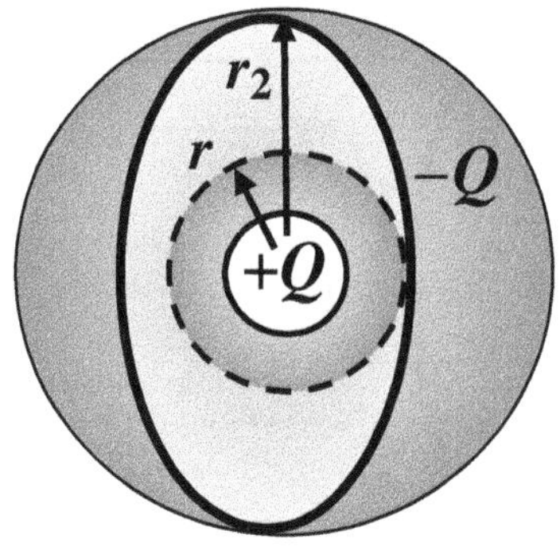

Figure 3.21 A concentric Gaussian sphere, radius r, is constructed to calculate the field produced in the dielectric occupying the space between two spherical metallic shells. The inner shell, radius r_1, has a total charge of $+Q$. The outer shell has radius r_2 and total charge $-Q$.

3.3.2.5 Field between Oppositely Charged Concentric Spherical Electrodes

We construct a spherical Gaussian surface, radius r, between the two spherical conducting shells shown in Figure 3.21. The inner shell carries a total charge $+Q$ on its surface, whilst the outer shell carries an equal but opposite charge $-Q$.

In section 3.3.2.2 it was shown that the field inside a solid or hollow metal sphere is zero (the 'Faraday cage' principle). The outer shell with charge $-Q$ does not therefore contribute to the field E generated in the space between the outer and inner shells. The field is created by the charge +Q on the inner shell and is determined by calculating the total flux through the Gaussian sphere. The surface integral to be used in Gauss's Law is $\int_S dA = 4\pi r^2$ and the field E in the dielectric between the coaxial electrodes is given by:

$$\int_S \varepsilon_o\varepsilon_r\mathrm{E}dA = \varepsilon_o\varepsilon_r\mathrm{E}(4\pi r^2) = Q\ (r_2 > r > \mathrm{r}_1)$$

to give

$$\mathrm{E} = \frac{Q}{4\pi\varepsilon_o\varepsilon_r r^2}\ (r_2 > r > \mathrm{r}_1) \tag{3.43}$$

3.3.2.6 Field between Oppositely Charged Coaxial Electrodes

We construct a cylindrical Gaussian surface, radius r and length L, in the dielectric material that separates and supports two coaxial electrodes shown in Figure 3.22. The inner cylindrical electrode has a surface charge density $+\lambda$ (C/m) whilst the outer electrodes has the form of a cylindrical metallic shell and has charge density $-\lambda$.

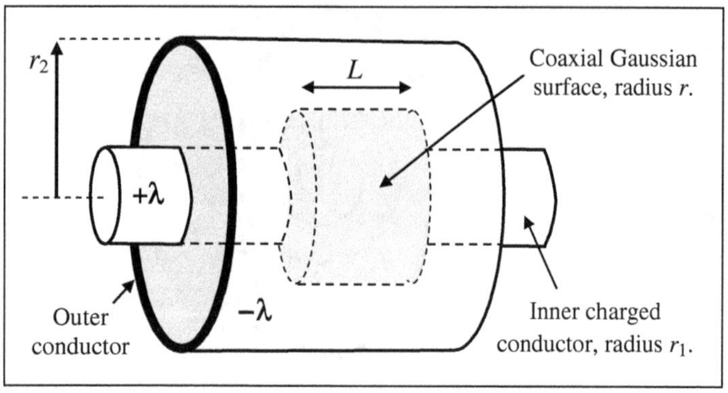

Figure 3.22 The field in the dielectric between two oppositely charged coaxial conductors is found by constructing a coaxial, cylindrical, Gaussian surface around the inner conductor.

At sufficient distances from the ends of the coaxial electrodes, the field generated by the inner one is directed radially towards the outer electrode and has no component parallel to the central axis (see section 3.3.2.1). The flux through the surface of the Gaussian cylinder is thus totally normal to its surface and no flux passes through its two ends. The surface integral to be used in Gauss's Law is $\int_S dA = 2\pi rL$ and the field E in the dielectric between the coaxial electrodes is given by:

$$\int_S \varepsilon_o\varepsilon_r \mathrm{E} dA = \varepsilon_o\varepsilon_r \mathrm{E}(2\pi rL) = \lambda L \; (r_2 > r > \mathrm{r}_1)$$

to give

$$\mathrm{E} = \frac{\lambda}{2\pi\varepsilon_o\varepsilon_r r} \; (r_2 > r > \mathrm{r}_1) \tag{3.44}$$

A summary is given in Table 3.2 of the electric fields created for the various charge distributions considered in this section. The magnitude of the dielectrophoretic force depends on the product $(\mathrm{E}\cdot\nabla)\mathrm{E}$ of the applied field and its gradient (see Equation (2.15) in Box 2.4 of Chapter 2). If the field is uniform (i.e., does not depend on position r) then $\nabla\mathrm{E}$ is zero and the dielectrophoretic force is thus also zero. From Table 3.2 we see this is the situation for the field produced by a charged metal plate, or between two oppositely charged plates. This will not be the case in regions near the edges of the plates, because the fields produced in such areas will not be uniform.

A valuable principle can be learnt from Table 3.2, namely that the electric field and its gradient vary as an inverse function of the size of an electrode. Consider the case of concentric spherical electrodes. For distances r greater than the radius r_1 of the inner conductor and less than the radius of the outer coaxial conductor, the product $(\mathrm{E}\cdot\nabla)\mathrm{E}$ is proportional to the factor $\frac{Q^2}{r^5}$. A very large

Table 3.2 The magnitudes of the electric field E and field gradient ∇E are given for a point free charge and various arrangements of charged conducting surfaces. The field is determined at distance r from the charge or charged surface (magnitude Q, λ or σ) in a medium of relative permittivity ε_r. The parameter $\frac{1}{4\pi\varepsilon_o}$ is written as k.

Distribution of charge	Field at point r	Field gradient
Point free charge Q. Eq. (3.8)	$\mathrm{E} = \frac{kQ}{\varepsilon_r}\frac{1}{r^2}$	$\nabla\mathrm{E} = -\frac{2kQ}{\varepsilon_r}\frac{1}{r^3}$
Flat metal plate (σ). Eq. (3.20)	$\mathrm{E} = \frac{\sigma}{2\varepsilon_o\varepsilon_r}$	zero
Parallel metal plates ($\pm\sigma$). Eq. (3.42)	$\mathrm{E} = \frac{\sigma}{\varepsilon_o\varepsilon_r}$	zero
Charge Q on metal sphere of radius R. Eq. (3.38)	E = zero ($r < R$)	∇E = zero ($r < R$)
	$\mathrm{E} = \frac{kQ}{\varepsilon_r}\frac{1}{r^2} \; (r > R)$	$\nabla\mathrm{E} = -\frac{2kQ}{\varepsilon_r}\frac{1}{r^3} \; (r > R)$
Charge Q in dielectric sphere, radius R, relative permittivity ε_m. Eqs (3.40) and (3.41)	$\mathrm{E} = \frac{kQ}{\varepsilon_m}\frac{r}{R^3} \; (r < R)$	$\nabla\mathrm{E} = \frac{kQ}{\varepsilon_m}\frac{1}{R^3} \; (r < R)$
	$\mathrm{E} = \frac{kQ}{\varepsilon_r}\frac{1}{r^2} \; (r > R)$	$\nabla\mathrm{E} = -\frac{2kQ}{\varepsilon_r}\frac{1}{r^3} \; (r > R)$
Charged (λ) long wire. Eq. (3.36)	$\mathrm{E} = \frac{2k\lambda}{\varepsilon_r}\frac{1}{r}$	$\nabla\mathrm{E} = -\frac{2k\lambda}{\varepsilon_r}\frac{1}{r^2}$
Concentric spherical electrodes ($\pm$Q). Eq. (3.43)	$\mathrm{E} = \frac{kQ}{\varepsilon_r}\frac{1}{r^2} \; (r_2 > r > r_1)$	$\nabla\mathrm{E} = -\frac{2kQ}{\varepsilon_r}\frac{1}{r^3}$
Coaxial cylindrical electrodes. Eq. (3.44)	$\mathrm{E} = \frac{2k\lambda}{\varepsilon_r}\frac{1}{r} \; (r_2 > r > r_1)$	$\nabla\mathrm{E} = -\frac{2k\lambda}{\varepsilon_r}\frac{1}{r^2}$

increase of the value for (E·∇)E is achieved by reducing the radius of the inner conductor, whilst maintaining the charge on the inner conductor to the same sort of level. For example, a value of $r = 2$ mm is permitted for an inner conductor radius of 1 mm, but can be reduced to $r = 0.2$ mm for an inner conductor of radius 0.1 mm. Such tenfold reduction of r corresponds to a 10^5-fold increase of (E·∇)E.

In Box 2.5 of Chapter 2 a simplification of (E·∇)E was given as:

$$2(\mathrm{E}\cdot\nabla)\mathrm{E} = \nabla \mathrm{E}^2$$

Validation of this relationship is evident in Table 3.2. For example, for the case of coaxial cylindrical electrodes, we find that

$$2(\mathrm{E}\cdot\nabla)\mathrm{E} = 2\left(\frac{2k\lambda}{\varepsilon_r}\frac{1}{r}\right)\left(-\frac{2k\lambda}{\varepsilon_r}\frac{1}{r^2}\right) = -8\left(\frac{k\lambda}{\varepsilon_r}\right)^2\frac{1}{r^3} \tag{3.45}$$

The value for $\nabla \mathrm{E}^2$ is calculated from the following procedure:

$$\nabla \mathrm{E}^2 = \frac{\partial}{\partial r}\left(\frac{2k\lambda}{\varepsilon_r}\frac{1}{r}\right)^2 = 4\left(\frac{k\lambda}{\varepsilon_r}\right)^2\frac{\partial}{\partial r}\left(\frac{1}{r^2}\right) = -8\left(\frac{k\lambda}{\varepsilon_r}\right)^2\frac{1}{r^3} \tag{3.46}$$

This gives the same result as Equation (3.45) and provides the validation we sought.

3.3.3 Summary Guidelines in Applying Gauss's Law

The examples given in this last section provide us with the following guidelines in using Gauss's Law to find the electric field at any point:

The important first step is to choose the Gaussian surface. This surface must be a *closed* surface. If the field at a particular point is to be found, this point must lie on it. The Gaussian surface may be real, such as the face of a solid body, or an imaginary one that is constructed partly or wholly in space or another medium.

The integral $\int_S \mathrm{E}_p dA$ over a closed Gaussian surface is readily evaluated analytically if the Gaussian surface and the charges within it have some form of geometric symmetry. Otherwise, some form of numerical computation is required. For example, if the charge distribution possesses a spherical or cylindrical symmetry, the Gaussian surface should take the form of sphere or cylinder, respectively.

In the integral $\int_S \mathrm{E}_p dA$, E_p is the perpendicular component of the *total* field E at each point on the Gaussian surface. The total field may result from charges that exist within and outside the Gaussian surface. If the vector E_p is directed outwards from the interior of the closed Gaussian surface and has the same magnitude at every point on the surface, then $\mathrm{E}_p = \mathrm{E}$ and $\int_S \mathrm{E}_p dA = \mathrm{E}A$. If E_p is directed inwards and is constant over the surface, then $\int_S \mathrm{E}_p dA = -\mathrm{E}A$. If no net charge at all exists within the Gaussian surface then $\int_S \mathrm{E}_p dA$ is zero.

A closed Gaussian surface can be deconstructed into separate surfaces and the integral $\int_S \mathrm{E}_p dA$ evaluated for each one of them. The integral over the whole closed surface is equal to the sum of the integrals over the separate surfaces. The following guidelines can assist in the choice of Gaussian surface: If the total field at every point is tangential to the Gaussian surface, then E_p is zero and $\int_S \mathrm{E}_p dA$ is zero. This fact can be used to choose the orientation of the sides or ends of a cylindrical Gaussian surface. If the total field is zero at every point on a surface, such as within a metal, then $\int_S \mathrm{E}_p dA$ is also zero.

3.4 Induced Dielectric Polarization

The result given by Equation (3.34), together with the conclusion drawn from applying Gauss's Law to the scheme shown in Figure 3.17, enable us to understand what is happening in Figure 3.1 in terms of the distributions of the free and induced bound charges. During the stages before and after the insertion of the dielectric slab between the charged metal plates, the total number of charges per unit area of the plates remains unchanged. This charge cannot dissipate through either the electrometer or the dielectric insulator. However, a proportion of these charges are neutralized (bound) by the appearance of polarization charges at the surfaces of the dielectric that interface with the metal plates. The amount of the free charge density so bound is directly proportional to the observed reduction of the electrometer reading shown in Figure 3.1. If A is the surface area of each plate, an induced charge $-A\sigma_{bound}$ appears on the dielectric surface next to the positively charged plate (the anode) and $+A\sigma_{bound}$ appears on its opposite surface next to the cathode. Two electric charges of opposite polarity, $\pm q$, separated by a distance d, represent a dipole of moment $\mathrm{p} = q\mathrm{d}$. As depicted in Figure 3.23 this electric dipole moment is symbolized by a vector pointing from the negative to the positive charge.

The dielectric slab of thickness d therefore possesses an induced dipole moment $p = (A\sigma_{bound} \times d)$. The macroscopic polarization P of a dielectric material is defined as the average induced dipole moment per unit volume. For the dielectric slab shown in Figure 3.1 the polarization charge is:

$$P = p/(\text{volume}) = (\mathrm{A}\sigma_{bound} \times d)/Ad = \sigma_{bound}. \tag{3.47}$$

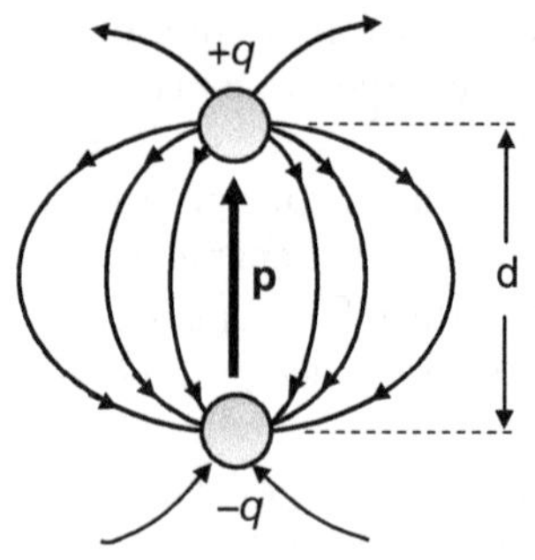

Figure 3.23 Two electric charges, of polarity $+q$ and $-q$, separated by a distance d, represent a dipole of moment $p = qd$, and is represented as a vector p directed from the negative to the positive charge.

The polarization P is therefore the induced (bound) charge densities that appear on the faces of the dielectric slab. The distribution of free and bound charges on the electrode plates, together with the polarization (bound) charges on the dielectric, are shown in Figure 3.24. The polarization vector **P**, which corresponds in magnitude to the surface charge density bound at the electrodes by the polarized dielectric, is shown directed along the direction of the applied field.

From Equation (3.3) we have the bound charge ($\Delta\sigma$) on an electrode plate as

$$\sigma_{bound} = \sigma_{free}\left(1 - \frac{1}{\varepsilon_r}\right) = \sigma_{free}\left(\frac{\varepsilon_r - 1}{\varepsilon_r}\right)$$

and from Equation (3.42) the field E imposed on the dielectric is

$$\mathrm{E} = \frac{\sigma_{free}}{\varepsilon_o \varepsilon_r}$$

Equation (3.47) can thus be written as:

$$P = \sigma_{bound} = \sigma_{free}\left(\frac{\varepsilon_r - 1}{\varepsilon_r}\right) = \mathrm{E}\varepsilon_o\varepsilon_r\left(\frac{\varepsilon_r - 1}{\varepsilon_r}\right) = (\varepsilon_r - 1)\,\varepsilon_o\mathrm{E} \tag{3.48}$$

The quantity $(\varepsilon_r - 1)$ is termed the electric susceptibility χ_e of the dielectric material, so that Equation (3.48) is often expressed as:

$$P = \chi_e \varepsilon_o E \tag{3.49}$$

The concept of electric susceptibility χ_e is illustrated by writing

$$\chi_e = \frac{P}{\varepsilon_o \mathrm{E}}$$

which corresponds to the ratio of the bound charge density to the free charge density. We have thus been able to relate the macroscopic, bulk, polarization P of the dielectric slab to a microscopic property characterized by its relative permittivity ε_r and to the average electric field E inside the dielectric. The *local* field acting on an individual molecule in the bulk of the dielectric is considered in Chapter 6, leading to the defining of a purely molecular quantity known as the *molar polarization*. The *microscopic* mechanisms responsible for the *macroscopic* polarization of a dielectric material are also described in Chapter 7. For our present discussion of induced polarization it is sufficient to consider the basic concept shown in Figure 3.25.

Viewed at the microscopic level the bulk polarization of a dielectric can be viewed as the collective interaction of the induced dipole moments of each of atom or molecule. Figure 3.25 gives an exaggerated depiction of this, where each induced microscopic dipole is shown aligned with the internal field E. In the bulk of the dielectric the positively charged end of a microscopic dipole is cancelled by a neighbouring induced negative charge. Uncompensated induced positive charges appear only at the dielectric surface facing the metal plate that carries negative free charge density (the cathode), whilst uncompensated induced negative charges appear only at the surface facing the plate that carries the positive free charge density (the anode). From the definition given by

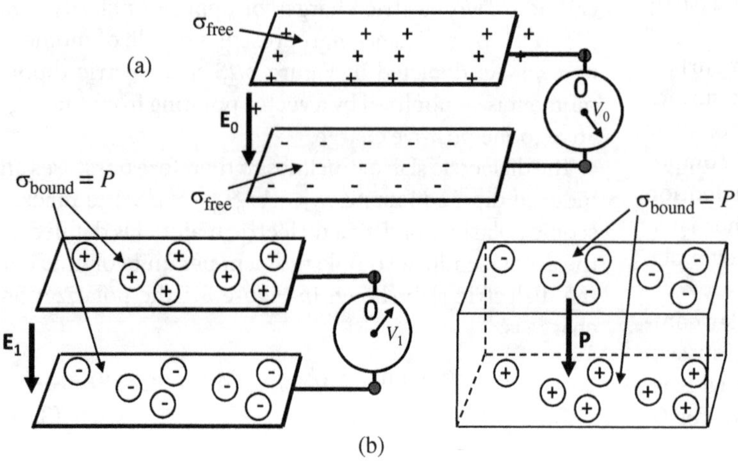

Figure 3.24 The initial and final stages of Figure 3.1 are simplified to show the distribution of the free and bound charge densities on electrodes of surface area *A*. (a) Positive and negative free charge densities are shown on the anode and cathode, respectively, before insertion of the dielectric slab. (b) On insertion of the dielectric, the voltage falls because bound charges induced at the surface of the dielectric partially neutralize the free charges. This lends to the dielectric slab the properties of a macroscopic dipole moment of magnitude $p = (A\sigma_{bound} \times d)$.

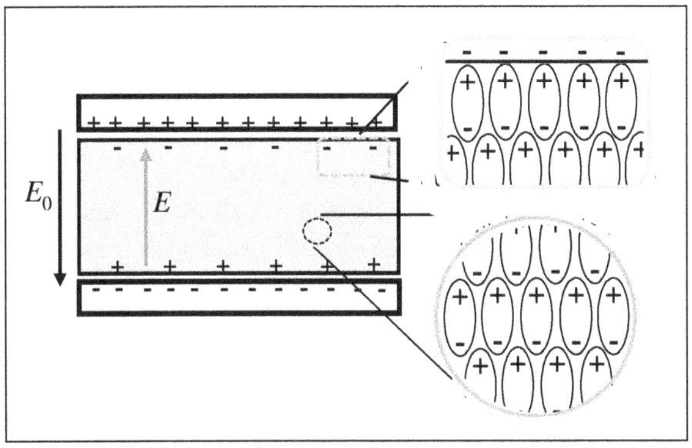

Figure 3.25 In the bulk of a polarized dielectric (enlarged circle) the induced molecular dipoles tend to align with the internal field *E*. Charges at each end of an induced dipole are neutralized by neighbouring induced dipoles. The net induced charge per unit volume *within* the dielectric bulk is thus zero. Uncompensated dipole charges at the face of the dielectric (enlarged rectangle) next to an electrode appear as bound charges that partially neutralize the free charges on the electrode.

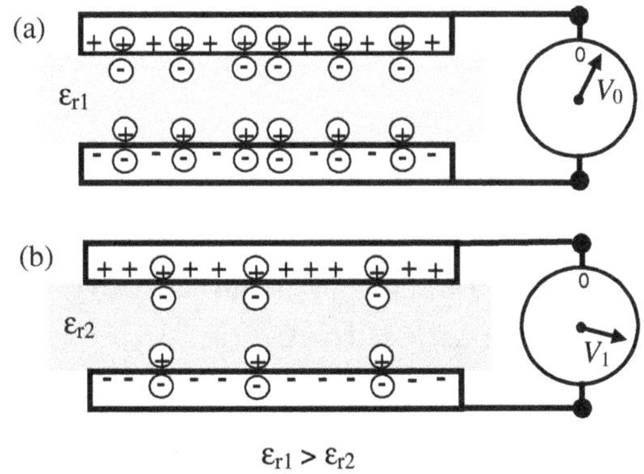

Figure 3.26 (a) An electrometer gives a voltage reading of V_0 across two charged metal plates in an aqueous solution ($\varepsilon_r \approx 80$). (b) On inserting a dielectric sheet ($\varepsilon_r = 2.5$) the voltage decreases because the induced charges at the surface of the dielectric sheet counteract some of those induced at the solution-electrode interface. The polarization (induced dipole moment per unit volume) in the dielectric sheet opposes that of the aqueous dielectric.

Equation (3.12) for the displacement vector D, Equation (3.48) can be given in the form:

$$D = P + \varepsilon_o E \tag{3.50}$$

If a dielectric medium such as a gas or liquid of relative permittivity ε_{r1}, rather than a vacuum already existed between the electrode plates shown in Figure 3.1, on inserting another dielectric between the electrodes so as to replace the first one the observed polarization is given by:

$$P = (\varepsilon_{r2} - \varepsilon_{r1})\varepsilon_o E \tag{3.51}$$

where ε_{r2}, is the relative permittivity of the replacement dielectric. For the case where $\varepsilon_{r1} < \varepsilon_{r2}$, as for example where a plastic sheet of relative permittivity $\varepsilon_{r2} \approx 2.5$ is inserted to replace an aqueous medium ($\varepsilon_{r2} \approx 80$) between electrically charged electrodes, the apparent polarization of the second dielectric will appear to be negative with respect to the one it has replaced. This is manifested as an increase of the voltage V_1 above the initial voltage V_0. This effect arises because of an increase of the free charge density on the electrode surfaces, as depicted in Figure 3.26.

3.5 Capacitance

From Table 3.2 we find that the electric field at any point in the region of a charged conductor is proportional to the density of the charge. In particular, for the case of two parallel plate electrodes, situated in a vacuum with equal and opposite *free* charge densities $\pm\sigma_o$, Equation (3.42) gives the uniform field between the plates as:

$$E_o = \frac{\sigma_o}{\varepsilon_o}$$

As depicted in Figure 3.1, if the plates are separated by a uniform distance *d*, a voltage difference V_o can be measured between the plates of magnitude given by:

$$V_o = E_o d$$

If the space between the electrodes is now filled with a dielectric of relative permittivity ε_r the voltage falls, as shown in Figure 3.1, to V_1:

$$V_1 = E_1 d = \frac{\sigma_1}{\varepsilon_o \varepsilon_r} d \tag{3.52}$$

where σ_1 is the new density of free charges on the plates, having been reduced from σ_o by an amount given by Equation (3.3) due to the appearance of induced bound charges at the dielectric's interfaces with the electrodes. To increase V_1 back to the original value V_o will require an increase of free charge on the plates above the initial value σ_o. From a practical electronics perspective the ratio of 'stored' charge *Q* in the device to the applied voltage *V* has increased. This ratio is defined as the *capacitance C* of the device (capacitor):

$$C = \frac{Q}{V} \tag{3.53}$$

The voltage difference *V* is directly proportional to the field E, which in turn is directly proportional to the charge on the electrodes. Thus, Equation (3.52) can be used to derive the capacitance of a pair of parallel plate electrodes, each of area A spaced distance d apart and containing a dielectric of relative permittivity ε_r, as:

$$C = \frac{A\sigma_1}{V_1} = \frac{A\varepsilon_o\varepsilon_r}{d} \tag{3.54}$$

3.6 Divergence Theorem and Charge Density Relaxation Time

Before giving a formal description of the divergence theorem, also known as Gauss's theorem, we can enquire as to why we should be interested in it. In Chapters 6 and 7 we learn that interfacial charges accumulate on the surface of a dielectric particle when immersed in another medium and subjected to an imposed electric field. As depicted in Figure 6.12 these charges are distributed non-symmetrically around the particle so that it assumes the properties of a macroscopic dipole moment. The magnitude and polarity of this induced dipole moment, as well as the relaxation time for the buildup and decay of the interfacial charges, dictate the dielectric and dielectrophoretic characteristics of the particle. Application of the divergence theorem can provide an insight into the mechanism controlling the relaxation time.

The divergence theorem relates the flux of a vector field through a surface to the behaviour of that vector field in the volume element enclosed by that surface. In formal language the theorem states [13]: *Let T be a closed bounded region in space whose boundary is a piecewise smooth surface S. Let* F*(x, y, z) be a vector function that is continuous and has continuous first partial derivatives in some domain containing T. Then*

$$\int_v \nabla \cdot \mathrm{F} dv = \int_s \mathrm{F} \cdot \mathrm{n} dA \tag{3.55}$$

where n *is the outer unit normal vector of S, pointing to the outside of S.*

A formal proof of this theorem is given by Kreyszig [13], which can be interpreted as stating that the outward flux of a vector field through a closed surface is equal to the volume integral of the divergence over the region enclosed by the surface. More simply put – the sum of all the sources minus the sinks within a defined region gives the net flow out of or into that region. This situation is commonly referred to as the *equation of continuity*.

Consider an electrical current density J flowing across a surface S. This surface encloses a volume V containing a charge density ρ, which we will assume can vary as a function of time. By charge density we mean the ratio $\Delta q/\Delta v$, where Δq is the charge contained within a volume element Δv. The current across the surface S is the rate at which charge crosses it. We will take this current as being positive if the current density vector J forms an angle of less 90° with the positive normal vector n (see Figure 3.13 for the case of an electric field). Also, there are no sources or sinks of charge, reflecting the fact that charge, like mass, can neither be instantaneously created nor destroyed. The integral taken over the surface S of the normal component of J therefore equals the net change of the initial charge density ρ contained within it. We can express this relationship as:

$$\int_s \mathrm{J} \cdot \mathrm{n} da = -\int_v \frac{\partial \rho}{\partial t} dv \tag{3.56}$$

This equation can be simplified by applying the divergence theorem of Equation (3.55); so that

$$\int_s \mathrm{J} \cdot \mathrm{n} da = \int_v \nabla \cdot \mathrm{J} dv$$

and Equation (3.56) becomes:

$$\int_v \left(\nabla \cdot \mathrm{J} + \frac{\partial \rho}{\partial t} \right) dv = 0$$

Both terms in the function enclosed in the brackets of this equation are spatially continuous, so the only way that the volume integral can be zero is for the function itself to be zero, i.e.,

$$\frac{\partial \rho}{\partial t} + \nabla \cdot \mathrm{J} = 0 \tag{3.57}$$

This result has the same form as the analogous case of the flow of a fluid, which is called the *condition for the conservation of mass* or the *continuity equation of a compressible fluid* [13, p. 455]. In our case, Equation (3.57) describes the *condition for the conservation of electrical charge* or the *continuity equation of electrical current flow*.

We can now proceed to investigate how the charge density ρ varies with time. We will assume that the medium containing this charge obeys Ohm's Law. This is most recognizable as the relationship $V = IR$, but in terms of the current density and electric field is given as $\mathrm{J} = \sigma_m \mathrm{E}$, where σ_m is the conductivity of the medium. Equation (3.57) can thus be written as:

$$\frac{\partial \rho}{\partial t} + \nabla \cdot \sigma_m \mathrm{E} = 0$$

From Equation (3.29) we also have for a homogeneous medium of relative permittivity ε_m

$$\nabla \cdot \mathrm{E} = \frac{\rho}{\varepsilon_o \varepsilon_m}$$

so that Equation (3.57) has the form:

$$\frac{\partial \rho}{\partial t} + \frac{\sigma_m \mathrm{E}}{\varepsilon_o \varepsilon_m} \rho = 0$$

The solution to this differential equation is given by:

$$\rho = \rho_o \exp\left(-\frac{\sigma_m}{\varepsilon_o \varepsilon_m} t\right) \quad (3.58)$$

where ρ_o is the initial charge density at time $t = 0$. Thus, the charge density decays to 1/e (i.e., to 37%) of its original magnitude ρ_o (which may be positive or negative) in a time τ given by

$$\tau = \frac{\varepsilon_o \varepsilon_m}{\sigma_m} \quad (3.59)$$

This is known as the characteristic relaxation time describing the decay of the charge density to zero and can also be expressed as a characteristic frequency $f_c = 1/(2\pi\tau)$. This result demonstrates an important theorem in electrostatics, namely that within a normal dielectric medium (i.e., not an *ideal* dielectric exhibiting zero conductivity) a permanent distribution of free charge cannot exist for an indefinite period of time. If we make the analogy that a dielectric with an initial local region of stored charge density behaves as a microscopic resistor-capacitor (*rc*) element, of effective area ΔA and thickness Δd, then $r = \Delta d/(\sigma \Delta A)$ and $c = \varepsilon_o \varepsilon_r \Delta A/\Delta d$. Inserting these identities into Equation (3.59) leads to the relationship $\tau = rc$, which students of basic electronics will recognize as the characteristic time constant for the charging and discharging of a capacitor. However, a more appropriate analogy is to consider the heterogeneous capacitor depicted in Figure 7.10, with our bounding surface S touching the interface between the two different dielectric media. It is shown in Chapter 7 that this form of heterogeneous system exhibits interfacial (Maxwell–Wagner) polarization with a relaxation time given by Equation (7.28):

$$\tau_{MW} = \varepsilon_o \frac{\varepsilon_1' d_2 + \varepsilon_2' d_1}{\sigma_1 d_2 + \sigma_2 d_1}$$

For the situation $d_1 \approx d_2$ and where region 2 is an aqueous electrolyte in contact with a cell membrane material (region 1), then $\varepsilon_2 < \varepsilon_1$ and $\sigma_2 \gg \sigma_2$ and the predicted relaxation time is close to that given by Equation (3.59). Of particular relevance to the dielectrophoretic behaviour of a biological cell is the relaxation time derived for the interfacial polarization of a spherical particle suspended in a dielectric medium, given by Equation (9.7) and of the form:

$$\tau = \varepsilon_o \frac{\varepsilon_p + 2\varepsilon_m}{\sigma_p + 2\sigma_m} \quad (3.60)$$

We will find in Chapter 9 that below ~1 MHz the dielectric properties of a viable mammalian cell are dominated by the conductivity (almost zero) and capacitance of its plasma membrane. The applied electric field hardly penetrates into the cell interior. From Table 9.3 a typical value for the relative permittivity of the membrane material is ~6, which can be taken as the value for ε_p in Equation (3.60). A cell thus appears as an insulating particle so that the value for $2\sigma_m$ is much larger than σ_p. We also have $2\varepsilon_m$ (~160) as being significantly larger than ε_p. The approximate value for the relaxation time of the interfacial polarization, known as the dielectric β-dispersion, for a mammalian cell is thus $\tau \approx \varepsilon_o \varepsilon_m/\sigma_m$. The aqueous buffers used for the suspending medium in dielectric and dielectrophoretic measurements with cells typically have an ε_m value close to 80 and conductivities of 20~50 mS/m. A value for τ of around 15~35 ns can therefore be predicted, corresponding to a characteristic frequency ($[1/(2\pi\tau]$) of 5~10 MHz. The characteristic frequency observed for the β-dispersion of blood cells is typically in this range, as shown in Figure 9.8 for the case of a suspension of red blood cells.

3.7 Summary

Two important facts have been described in this chapter:

An ideal metal conductor is unable to sustain an internal electrostatic field. The conduction electrons in the bulk of a metal respond to an applied field in a way that exactly cancels an applied electric field. If a metal is electrically charged, either positively or negatively by connecting it to the positive or negative terminal of a battery, for example, the deficiency or excess of electronic charges are located solely on the *surface* of the metal.

An ideal dielectric is unable to conduct electrical current and can support a large electrostatic field. A dielectric is electrically polarized in an electric field, manifesting itself as the appearance of bound polarization charges on the surface of the dielectric. A polarized dielectric takes on the form of an electric dipole characterized by an induced dipole moment per unit volume P, related to the bulk relative permittivity ε_r and the average field E within the dielectric by the relationship:

$$\mathrm{P} = (\varepsilon_r - 1)\varepsilon_o \mathrm{E}$$

The polarization vector **P** corresponds in magnitude to the *bound* surface charge density induced on an electrode surface by the polarized dielectric and is directed along the direction of the applied field.

A nonideal dielectric, or a region of nonvanishingly small conductivity within an ideal dielectric, cannot support a permanent distribution of free charges. An example of direct relevance to the dielectric and dielectrophoretic behaviour of biological cells in suspension is interfacial polarization arising from the field-driven accumulation of charges at the cell's surface. On removal of the field this charge decays back to zero with a characteristic relaxation time τ. Details and the consequences of this are described more fully in Chapters 6, 7 and 9.

These concepts follow on directly from the nature of electric charges and how, through the electric fields they produce, are able to exert a force on other distant charges. An isolated charge interacts with the electric field E created by other charges but is not influenced by its own field.

Two other important concepts have also been described in this chapter, namely:

The net electric flux of force ($D = \varepsilon_o \varepsilon_r E$) through a closed surface surrounding a distribution of charges in a medium of permittivity $\varepsilon_o \varepsilon_r$ is equal to the net magnitude of the enclosed charges. This result is given by Gauss's Law and is an equivalent way of expressing the empirical relationship obtained by Coulomb that describes the interaction of two point charges located in a vacuum.

Gauss's Law provides a simple way to calculate the electric field produced by various uniform distributions of charge on conducting surfaces. Examples of this include the field produced by a uniform distribution of charge along a wire or on a flat surface and between charged flat, spherical or cylindrical electrodes. The fields and field gradients produced by such charge distributions are given in Table 3.2. The charge distributions are assumed to be infinite in extent – the field relationships given in Table 3.2 do not apply in regions close to an electrode edge.

3.8 References

1 Fröhlich, H. (1958) *Theory of Dielectrics: Dielectric Constant and Dielectric Loss*, 2nd edn, Oxford University Press, Oxford.

2 Mottelay, P. F. (1922) *Bibliographical History of Electricity and Magnetism*, Charles Griffin & Co., Ltd., London, p. 7.

3 Whittaker, E. (1951) *A History of the Theories of Aether and Electricity: The Classical Theories*, Thomas Nelson & Sons Ltd, London, p. 57.

4 Maxwell, J. C. (1861) On physical lines of force, Part I: The theory of molecular vortices applied to magnetic phenomena. *Phil. Mag.* **21**, 161–175.

5 Maxwell, J. C. (1861) On physical lines of force, Part II: The theory of molecular vortices applied to electric currents. *Phil. Mag.* **21**, 281–291; 338–348.

6 Maxwell, J. C. (1861) On physical lines of force, Part III: The theory of molecular vortices applied to statical electricity. *Phil. Mag.* **21**, 17–24.

7 Maxwell, J. C. (1865) A dynamical theory of the electromagnetic field. *Phil. Trans. Roy. Soc.* **155**, 459–512.

8 Michelson, A. A. and Morley, E. W. (1887) On the relative motion of the Earth and the luminiferous aether. *Phil. Mag.* **24**, 449–463.

9 Whittaker, E. (1953) *A History of the Theories of Aether and Electricity: The Modern Theories 1900–1926*, Thomas Nelson & Sons Ltd., London, p. 30.

10 Tombe, F. D. (2010) The double helix theory of the magnetic field. *General Science Journal* 1–17.

11 Poisson, S.-D. (1813) Remarques sur une l'équation qui se présente dans la théorie des attractions des spheroides. *Nouveau bulletin des sciences, par la Société Philomatique de Paris* **3** (1812–1813), 388–392.

12 Laplace, P.-S. (1785) Théorie des attractions des spheroides et de la figure de la terre. *Mémoires de l'Académie des Sciences* pp. 113–174.

13 Kreyszig, E. (1999) Advanced Engineering Mathematics. 8 edn, John Wiley & Sons, Inc., New York, NY, pp. 505–512.

4

Electrical Potential Energy and Electric Potential

4.1 Introduction

An enduring memory for this author is of a filmed lecture given several times by the famous physicist, Richard Feynman, in which he used a 'swinging ball of death' to demonstrate the principle of conservation of energy. He took hold of a heavy brass ball suspended by a long chain and walked backwards with it until the chain was taut and the ball touched the end of his nose. He released the ball, allowing it to swing across the auditorium and over the heads of the students, who began gasping in horror when it gathered momentum on a return path back towards his face. Feynman stood still and remained so even as the gasps grew louder and more frantic. But of course, he was not at all concerned – he knew that the kinetic energy gained by the ball would not exceed the gravitational potential energy it had gained when lifted to touch his nose. On its return the ball would not smash into his face, but come to rest just in front of it before swinging back again. In doing this he illustrated the reversible conversion between potential and kinetic energy, which is a characteristic of a *conservative force*, as described more fully in Box 4.1. Examples of nonconservative forces include air resistance and friction. Their effect in reducing the kinetic energy of a body is not reversible. Feynman's 'ball of death' experienced frictional air resistance and this is why on its return path it came to rest just in front of his nose.

On bringing the ball from its hanging-down, resting, position up to his nose, Feynman's hands exerted an equal but opposite force to that acting to accelerate the ball downward in the earth's gravitational field. This force is the weight mg of the ball, where m is the ball's mass and g is the acceleration due to gravity. Close to the surface of the Earth the gravitational force can be assumed constant. Lifting the ball a height h above its equilibrium position increased the ball's gravitational potential energy by an amount $U = mgh$. At the moment of releasing the ball its velocity v was zero and hence its kinetic energy $1/2(mv^2)$ was also zero. Assuming negligible air resistance, when the ball executed its swing back down to its initial resting position it would at that moment have lost the added potential energy and converted all of it into kinetic energy. This increase in the ball's kinetic energy from its initial resting state was thus also equal to the work done on the ball by Feynman's hands. As explained in Box 4.1, this is an expression of the *work-energy theorem*, which states that *the work done by the net force on a body is equal to the change in the body's kinetic energy*. At any instant during its swinging back and forth, the algebraic sum of the ball's potential and kinetic energy remained constant. As the potential energy increased by a certain number of joules, the kinetic energy decreased by the same amount and *vice versa*. This is why Feynman knew he could stand stock still during his demonstration.

Feynman's lecture concerned the action of the earth's gravitational field on an object having mass. An electrical force shares the same property as a gravitational force in being a reversible, conservative, force. The purpose of this chapter is to describe how the concepts of a conservative force and the work-energy theorem, widely used in the subject of mechanics, are also relevant to understanding an electrokinetic effect such as dielectrophoresis.

4.2 Electrical Potential Energy

The concepts of work and potential energy described in Box 4.1 will now be applied to examples of the interaction of electric fields with charged bodies, commencing with the situations described in Chapter 2. In Figure 4.1, a positively charged particle Q is located in a uniform electric field $\mathbf{E}_x$. There are no components of the field acting along the y- and z-axes.

The field exerts a force of magnitude QE_x on the positively charged particle and this force is directed along the positive x-axis and remains constant irrespective of the particle's location in the field. From Equation (4.1) given in Box 4.1, the positive work W done by the field when the

Dielectrophoresis: Theory, Methodology and Biological Applications, First Edition. Ronald Pethig.

Box 4.1 Work, Potential Energy and the Work-Energy Theorem

In the SI system the unit of work is the *joule*, with dimensions of newton-metre (1 J = 1 N.m). The unit of work is thus the product of the unit of force and the unit of distance. If a constant force F acts on body so that it moves a distance s in a straight line along the direction as the force, the magnitude of the work *W* done on the body by the force is given by:

$$W = Fs \tag{4.1}$$

For the general case, where the force is not constant and the body moves at an angle θ to it, the work done is calculated by summing up the product F$\cos\theta.ds$ for each incremental distance *ds* taken along the total displacement between points *a* and *b*. Mathematically this is given by the equation:

$$W = \int_a^b F\cos\theta\, ds \tag{4.2}$$

In this equation, if *b* = *a*, the integral is zero, meaning that when the end and start points are the same the total work is zero, no matter how tortuous the total path has been. The work done can be positive, negative or zero. When the displacement is at 90° to the force, $\cos\theta = 0$ and so the work done is zero. For example, the work done on Feynman's brass 'ball of death' by the tension in the pendulum arm is zero, because this force has no component in the direction of the ball's motion.

The work W_{ab} done by a conservative force in displacing a body from point *a* to point *b* is equal to the negative of the change ΔU in the body's potential energy (i.e., $W_{ab} = -\Delta U = -(U_b - U_a)$. In other words, a conservative force acts to push an object towards a lower potential energy. Note how the gravitational force acts on the 'ball of death' described in the introduction to this chapter. The gravitational force does positive work when the ball is released and its potential energy decreases as it falls and then performs negative work when the ball swings back up again as the potential energy increases.

The *work-energy theorem* states that the change in kinetic energy ($\Delta KE = KE_b - KE_a$) following the displacement of a body from point *a* to point *b* is equal to the total net work W_{ab} done on that body. We thus have the relationship $W_{ab} = (KE_b - KE_a)$, which, taken together with the expression $W_{ab} = -(U_b - U_a)$, leads to the result:

$$KE_a + U_a = KE_b + U_b \tag{4.3}$$

Thus, if the only work done on a body is accomplished by a *conservative* force the sum of the kinetic energy and the potential energy remains constant.

particle moves a distance *s* in the field direction is given by:

$$W = Fs = QEs$$

This *positive* work by the field results in a *reduction* of the particle's electrical potential energy *U*. An analogy is the reduction of potential energy of a rock sinking to the bottom of a lake as a result of the work done on it by the gravitational field. To move the positively charged particle in the opposite direction, back to its original location where it had a higher potential energy, will require work to be done *against* the electrostatic force – so that the field performs *negative* work. (This is analogous to working against gravity in order to bring the rock back up to the surface of the lake.) If we replace the positive charge on the particle with negative charge, its potential energy will increase when it moves in the direction of the field along the positive *x*-axis and will decrease when it moves against the field. (This is analogous to exchanging the rock for an air bubble of the same size and shape in water.)

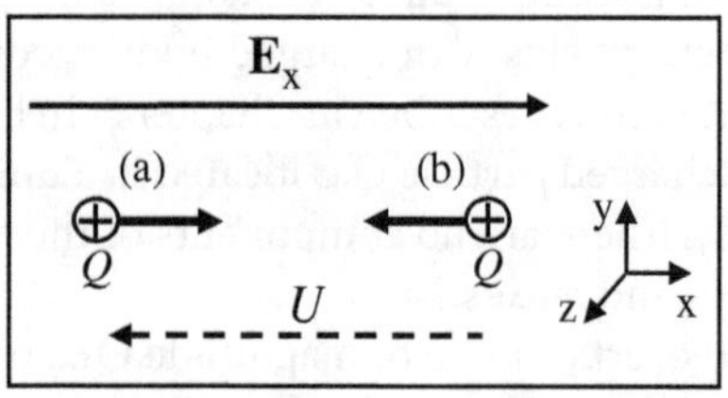

Figure 4.1 (a) When a positively charged particle *Q* moves in the direction of a uniform electric field $\mathbf{E}_x$, the field does positive work on the particle and its potential energy *U* decreases. (b) If the particle moves in the opposite direction, against the field, the field does negative work and the particle's potential energy increases.

To illustrate the conservative nature of an electrostatic force, Figure 4.1 is reconfigured to the form of Figure 4.2 to reinforce the analogy between a particle moving under the action of an electrical force and a gravitational force.

In Figure 4.2, a positively charged particle is located at position *a* in a uniform field E generated between two parallel plate electrodes. The electrostatic force QE is constant and acts on the particle so as to accelerate it towards the bottom electrode (cathode). This is analogous to the particle having mass *m* with a constant gravitational force *mg* acting to accelerate it downwards. The potential energy of a particle at a height *x* and acted on by a gravitational force is *mgx* and so by analogy the potential energy of a positively charged particle at a location *x* and acted on by an electrical force QE is *QEx*. The work

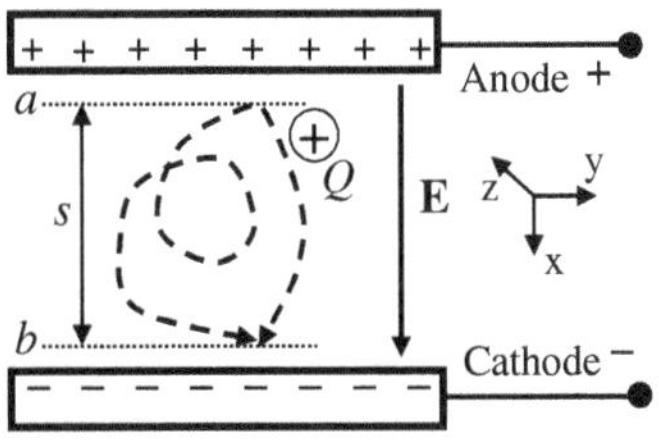

Figure 4.2 The dotted lines show two possible paths that can be taken by a positive charge Q moving from location *a* to location *b* in a uniform electric field E established between two charged parallel electrodes. Although the paths differ, the work done by the field is the same and equal to QEs. In moving from *a* to *b* the electrical potential energy of the particle *decreases* by the amount QEs.

W_{ab} done by the electric field is given by Equation (4.2) in Box 4.1 and is independent of the path taken by the charged particle in moving from location *a* to location *b* in Figure 4.2. The change in potential energy ΔU is given by the relationship:

$$W_{ab} = -\Delta U$$

where $\Delta U = U_a - U_b = QE(a - b)$. In Figure 4.2, location *b* is farther than location *a* along the *x*-axis, so that $(a - b)$ is a negative quantity (equal to $-s$). The change in electrical potential energy ΔU is thus negative and equal to $-QEs$.

Example 4.1 Work Done and Electrical Potential Energy

An extra force is applied to a particle carrying a charge of 10^{-9} C so as to move it *slowly* against the direction of a uniform field of magnitude 10^4 N/C. The particle is moved a distance of 1 cm against the field direction.

1. Has the particle lost or gained electrical potential energy after this displacement?
2. Calculate the change in electrical potential energy of the particle.
3. Why has it been specified that the particle is moved *'slowly'*?

Solution 4.1

1. The extra force slightly exceeds and is opposite to that exerted by the electric field and so does positive work on the particle – equivalent to saying that the field does negative work ($-W$). From the definition $W = -\Delta U$ we deduce that because the work done is negative the particle has *gained* potential energy.
2. The work done is given by:

$$W = \text{force} \times \text{displacement}.$$

The displacement is $-d$ (i.e., is in a direction opposite to that of the field), so that:

$$W = (QE) \times -d = -(10^{-9}\ \text{C})(10^4\ \text{NC}^{-1}) \times 10^{-2}\ \text{m} = -10^{-7}\ \text{J}$$

3. The particle is specified as moving *'slowly'* so that any kinetic energy given to it can be neglected in calculating the change in potential energy.

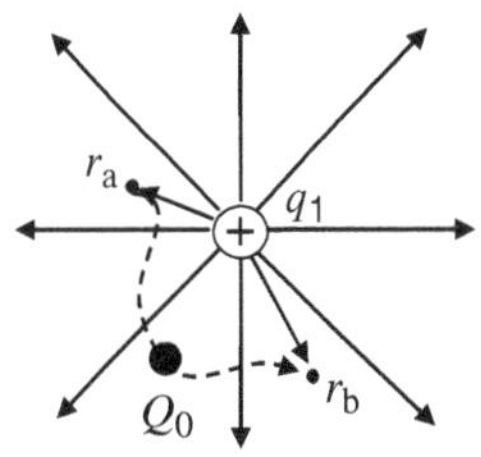

Figure 4.3 The work done on the test charge Q_0 in moving it from location *a* to location *b*, in the radial electric field produced by charge q_1, depends on the difference between the radial distances r_a and r_b and not on the path taken by Q_0, or whether the field is a uniform one.

The concept of electrical potential energy can be applied to the general case of charged particles in an arbitrary electric field. The field need not be uniform. For example, consider the case shown in Figure 4.3 of a point charge Q_0 that is moved slowly in the vicinity of a stationary point charge q_1.

An electric force is a conservative one so the change in potential energy in moving Q_0 from location *a* to location *b* will not depend on the path taken. The field around the single point charge q_1 has radial symmetry and so we can consider an equivalent path to be along a radial line from *a* to *b* as shown in Figure 4.3. This can also be deduced from Equation (4.2) in Box 4.1 because the factor $\cos\theta.ds$ is equivalent to a small displacement d*r* along a radial field line.

The magnitude of the radial field E_r at a distance from charge q_1 is given by Equation (3.8) of Chapter 3:

$$E_r = \frac{1}{4\pi\varepsilon_o\varepsilon_r}\frac{q_1}{r^2}$$

The force acting on charge Q_0 is given by:

$$F_r = Q_0E_r = \frac{1}{4\pi\varepsilon_o\varepsilon_r}\frac{Q_0q_1}{r^2} \tag{4.4}$$

This force is not constant along the displacement path and so calculation of the work done W_{ab} requires the following integration:

$$W_{ab} = \int_a^b F_r dr = \int_a^b \frac{1}{4\pi\varepsilon_o\varepsilon_r}\frac{Q_0q_1}{r^2}$$
$$= \frac{Q_0q_1}{4\pi\varepsilon_o\varepsilon_r}\left(\frac{1}{r_a} - \frac{1}{r_b}\right) \tag{4.5}$$

This work done by the field is equal to the negative of the change in the potential energy, so that

$$W_{ab} = -\Delta U = -(U_b - U_a) \tag{4.6}$$

Equations (4.5) and (4.6) are compatible if we define

$$U_a = \frac{1}{4\pi\varepsilon_o\varepsilon_r}\frac{Q_0 q_1}{r_a} \tag{4.7}$$

and

$$U_b = \frac{1}{4\pi\varepsilon_o\varepsilon_r}\frac{Q_0 q_1}{r_b} \tag{4.8}$$

Equation (4.7) describes the potential energy of the system of *both* charges when Q_0 is at a radial distance r_a from the stationary charge q_1. When r_a is an infinite distance away the potential energy U_a is zero. We can therefore interpret Equation (4.6) as the work required to bring a test charge Q_0 from rest at infinity up to a radial distance r_a from the stationary charge q_1. If Q_0 and q_1 are of the same polarity (either both positive or both of them negative) then an added force must be applied to oppose the repulsive electrostatic force between the two charges. In this case there is an *increase* of the potential energy of the charges, taken as a pair. If Q_0 and q_1 are of opposite polarity the field of q_1 does *positive* work on Q_0 and the potential energy *decreases.* Equation (4.8) can be used to describe the situation where Q_0 remains stationary and q_1 is the test charge at a radial distance r_b.

In general we are interested in the case where a test charge Q_0 is displaced in an electric field created by a distribution of point charges $q_1, q_2, q_3, \ldots q_i$ (e.g., along an electrode edge) at distances $r_1, r_2, r_3, \ldots r_i$ from the test charge. The electric field at each location along the displacement path taken by q_i will be the vector sum of the radial fields created by each of the point charges q_i. The total work done on Q_0 by this field will be the sum of these contributions. From Equation (4.7) the electrical potential energy associated with the test charge Q_0 located at a specific point is given by:

$$U = \frac{Q_0}{4\pi\varepsilon_o\varepsilon_r}\left(\frac{q_1}{r_1} + \frac{q_2}{r_2} + \frac{q_3}{r_3} + \cdots + \frac{q_i}{r_i}\right)$$

$$= \frac{Q_0}{4\pi\varepsilon_o\varepsilon_r}\sum_i \frac{q_i}{r_i} \tag{4.9}$$

where $r_1, r_2, r_3, \ldots r_i$ are the distances from the fixed point charges $q_1, q_2, q_3, \ldots q_i$ to the location of the test charge Q_0. When Q_0 is displaced the distances r_1, r_2, $r_3, \ldots$ change accordingly. The work done by the field on Q_0 to give such displacement is equal to the negative of the potential energy difference between the start and finish points of this displacement.

The total electrical potential energy for the general case of a collection of point charges is given by the following modification of Equation (4.9):

$$U = \frac{1}{4\pi\varepsilon_o\varepsilon_r}\sum_{i<j}\frac{q_i q_j}{r_{ij}} \tag{4.10}$$

We can interpret Equation (4.10) as the result of initially having a collection of charges widely separated apart from each other. This establishes our reference for zero electrical potential energy. The charges are then brought together so that the distance between q_1 and q_2 is r_{12} and so on for the general situation where q_i and q_j are separated by the distance r_{ij}. The electrostatic interactions between all possible pairing of the charges are added. The inequality expressed as $i < j$ and attached to the summation symbol Σ ensures that we do not include a charge interacting with itself and also that we include any one pair of charges only once.

Example 4.2 Work Done in Assembling a Distribution of Charges

We revisit Example 3.2 and Figure 3.6 of Chapter 3, in the form of Figure 4.4. Three charges q_1, q_2, q_3 are shown spaced in air along a straight line. Their sizes are much smaller than their spacing apart (i.e., they can be treated as point charges). We are given that: $x = y = 1$ mm; $q_1 = q_3 = 1.4$ pC; $q_2 = -1.0$ pC.

Calculate the work performed when q_3 was brought from infinity to its location alongside q_2 and q_3 as shown in Figure 4.4.

Solution 4.2 The potential energy of the assembly of three charges is given by Equation (4.10) as:

$$U = \frac{1}{4\pi\varepsilon_o\varepsilon_r}\left(\frac{q_1 q_2}{r_{12}} + \frac{q_1 q_3}{r_{13}} + \frac{q_2 q_3}{r_{23}}\right) \tag{4.11}$$

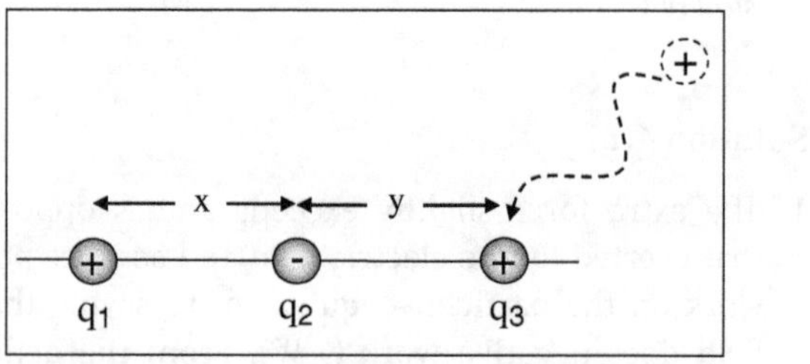

Figure 4.4 A positive charge q_3 (+1.4 pC) is brought from 'infinity' to the location shown alongside charges q_2 (−1.0 pC) and q_1 (+1.4 pC). What is the work required to achieve this action and has the net potential energy of the system of charges increased or decreased?

The potential energy associated with q_1 and q_2 when q_3 is located at infinity is calculated from Equation (4.11) using the values $r_{13} = r_{23} = \infty$:

$$U_\infty = \frac{1}{4\pi\varepsilon_o\varepsilon_r}\left(\frac{q_1q_2}{r_{12}}\right) \tag{4.12}$$

The difference ΔU between Equations (4.12) and (4.11) equates to the negative of the work W done by the field created by q_1 and q_2 on bringing q_3 to its location from infinity. We can express this relationship as:

$$\begin{aligned} W = -(U - U_\infty) &= -\frac{1}{4\pi\varepsilon_o\varepsilon_r}\left(\frac{q_1q_3}{r_{13}} + \frac{q_2q_3}{r_{23}}\right) \\ &= -\frac{q_3}{4\pi\varepsilon_o\varepsilon_r}\left(\frac{q_1}{r_{13}} + \frac{q_2}{r_{23}}\right) \end{aligned} \tag{4.13}$$

As expected, this gives the equivalent result to Equation (4.9). We have $q_1 = q_3 = 1.4\,\text{pC}$; $q_2 = -1.0\,\text{pC}$; $r_{13} = 2\,\text{mm}$; $r_{23} = 1\,\text{mm}$. Thus, from Equation (4.13):

$$\begin{aligned} W &= -\frac{q_3}{4\pi\varepsilon_o\varepsilon_r}\left(\frac{q_1}{r_{13}} + \frac{q_2}{r_{23}}\right) \\ &= -\frac{1.4\times10^{-12}\,\text{C}}{4\pi\varepsilon_o\varepsilon_r}\left(\frac{1.4\times10^{-12}\,\text{C}}{2\times10^{-3}\,\text{m}} - \frac{1.0\times10^{-12}\,\text{C}}{1\times10^{-3}\,\text{m}}\right) \\ &= -\frac{1.4\times10^{-12}\,\text{C}}{4\pi\varepsilon_o\varepsilon_r}\left(-\frac{0.6\times10^{-12}\,\text{C}}{2\times10^{-3}\text{m}}\right) \\ &= (9\times10^{9}\,\text{Nm}^2\text{C}^{-2})(4.2\times10^{-22}\,\text{C}^2\text{m}^{-1}) \\ &= 3.78\times10^{-12}\,\text{Nm (joules)} \end{aligned}$$

Positive work has been performed by the net field of q_1 and q_2 in bringing q_3 into position. The attractive force that q_2 exerts on q_3 is thus greater than the repulsive force exerted by the far more distant q_1 and the net electrical potential energy has *decreased.* (Bringing like charges together increases their potential energy; bringing unlike charges together decreases their potential energy.)

4.3 Electrical Potential

The electrical potential V at any point in an electrical field is defined as the potential energy U *per unit charge* associated with a test charge Q_0 located at that point in the field:

$$V = \frac{U}{Q_0}\ \text{N·m/C} \tag{4.14}$$

In the SI system of units the unit of electrical potential is the volt:

1 volt = 1 joule per coulomb.

In Chapter 2 the concept of an electrical field was described as the force *per unit charge* that the field exerts on a test charge Q_0. An electric field E thus has units of N/C, which is equivalent to (N.m/C)/m. From Equation (4.14), a potential V has units of N.m/C. In other words, the unit of electric field can be expressed as *one volt per metre.* This is the unit commonly used by electrical and electronic engineers and will be adopted in this book.

Expressing Equation (4.6) on a per unit charge basis we have:

$$\begin{aligned} \frac{W_{a-b}}{Q_0} &= -\frac{\Delta U}{Q_0} = -\left(\frac{U_b}{Q_0} - \frac{U_a}{Q_0}\right) \\ &= -(V_b - V_a) = V_a - V_b \end{aligned} \tag{4.15}$$

We interpret this equation as stating: 'The potential difference ($V_a - V_b$) between location a and location b is equal to the work done by an electric field in moving a body of unit charge from a to b in this field.'

A potential difference ($V_a - V_b$) becomes a useful concept if we can define V_a as a reference potential. Although the Earth's surface is on average charged negatively with respect to the atmosphere and this charge fluctuates as a result of precipitation and lightning strikes, the Earth serves as a reasonably stable reference potential for practical electrical devices. Electrical engineers commonly employ the Earth's potential (also known as ground potential) as a fixed reference and use the expression 'voltage' when referring to the potential difference *with respect to* ground at a particular part of an electrical circuit.

For the theoretical calculations in this book we will employ a physicist's usual practice of using as our reference the zero potential associated with a test charge at rest and located an infinite distance away from any other charge. Being at rest ensures that the test charge has zero kinetic energy. Being an infinite distance away from any other charges ensures that the test charge is not exposed to the influence of the electrical field from any other charge and is at a location of zero potential. Using this convention, where V_∞ is zero, we can rearrange Equation (4.15) to the form:

$$\frac{W_{\infty-b}}{Q_0} = -(V_b - V_\infty) = -V_b \tag{4.16}$$

This equation provides another way to define electrical potential: 'the potential V_b at location b is equal to the **negative** of the work done by an electric field in moving a body of unit charge from infinity to location b in this field.'

From Equation (4.13), which relates to bringing a test charge q_3 from infinity to a location b in a field created by charges q_1 and q_2, we can calculate the potential at the location of q_3 using the following equation:

$$V_b = \frac{-W}{q_3} = \frac{1}{4\pi\varepsilon_o\varepsilon_r}\left(\frac{q_1}{r_{13}} + \frac{q_2}{r_{23}}\right) \tag{4.17}$$

This equation informs us that a body having positive charge will have a positive potential when placed in an electric field created by a distribution of positive charges. A body of negative charge placed in the field created by a distribution of negative charges will also have a positive potential. This is equivalent to stating that the potential energy of a system of like charges increases as they are brought closer together. An added external force is required to bring a charged body at rest from infinity towards a field created by charges of the same polarity as itself. This is equivalent to the field of the other charges performing negative work on the body. On the other hand, a positive charge brought into the field created by a distribution of negative charges will find itself in a location of negative potential. The same result occurs for a negative charge brought into the field of positive charges. The field does positive work when a charged body moves from infinity towards a collection of charges of opposite polarity to that of the body.

The potential at any point in space due to a collection of charged spherical electrodes, each treated as a point charge, is given by the algebraic sum of the potentials of each charge. This involves evaluation of the more general form of Equation (4.17):

$$V(r) = \frac{1}{4\pi\varepsilon_o\varepsilon_r}\sum_i \frac{Q_i}{r_i} \tag{4.18}$$

In this equation the potential V is evaluated at a distance r_i from each charge Q_i and all of the values are added together algebraically. A positive charge Q_i produces a positive contribution to the overall potential and if Q_i is negative its contribution is a negative potential. An arbitrary value can be used for the positive test charge used to define the work done to bring it from infinity to the location of evaluation. A simple example is two charges Q_i of equal magnitude and opposite polarity that form an electric dipole. This is equivalent to designating $q_2 = -q_1$ in Equation (4.17). Figure 4.5 shows how the potential varies with distance from a dipole formed of two point charges. Contours that join potentials of the same value can be drawn as shown in Figure 4.5 and these are known as equipotential contours – or equipotential surfaces if drawn in three-dimensional space.

If instead of a collection of point charges we have a continuous distribution of charge along a wire (λ C/m), or over a surface (σ C/m^2), or within a volume (ρ C/m^3) the

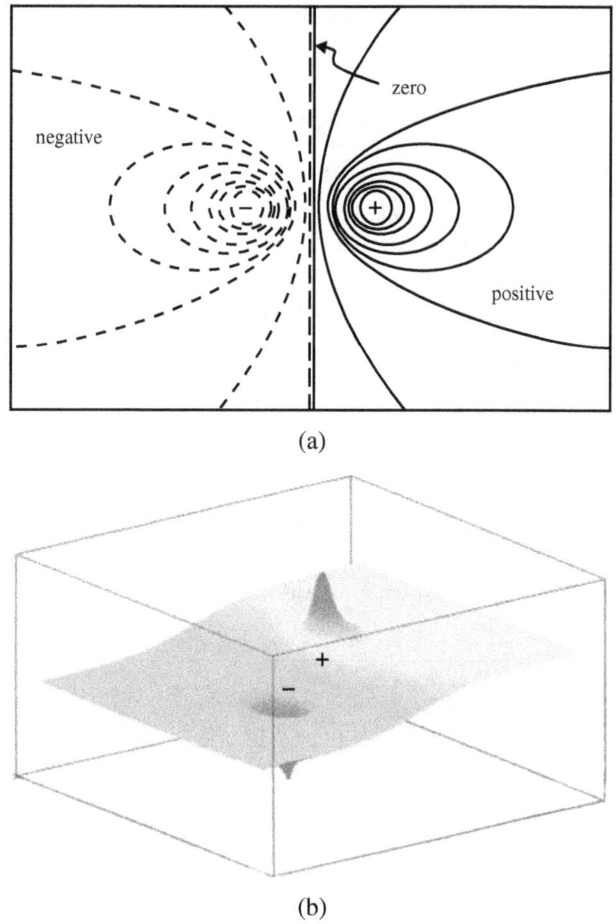

Figure 4.5 (a) A 2D plot of the equipotentials of a dipole. Along the line equidistant from both charges the potential is zero. The potential is also zero at infinite distances from both charges. (b) A 3D plot of the variation of the potential in the vicinity of a point dipole.

charge is divided up into small elements dq and the summation sign Σ in Equation (4.18) becomes an integration sign av

$$V(r) = \frac{1}{4\pi\varepsilon_o\varepsilon_r}\int \frac{dq}{r} \tag{4.19}$$

where r is now the distance from the charge element dq to the point of evaluation of the potential V.

Equations (4.18) and (4.19) can be used to find the potential due to a collection of charges or a uniform distribution of charge. However, in some cases we have the situation where these quantities are not known, but we do know or can calculate the electric field E. In this case we are able to calculate the potential *difference* between two locations. The force on a test charge Q_o used to probe the potential is given by $\mathbf{F} = Q_o\mathbf{E}$. From Equations (4.2) and

(4.15) the potential difference ($V_a - V_b$) between locations a and b is given by the relationship:

$$(V_a - V_b) = \int_a^b Eds = \int_a^b E\cos\theta ds \tag{4.20}$$

If the integral on the right-hand side of Equation (4.20) is positive, the electric field has performed positive work on a positive test charge as it moves from location a to location b. This means that the potential energy per unit charge has decreased and V_a is greater than V_b. The reverse is true if the integral is negative in value. This is consistent with the convention shown in Figure 4.2 where movement of the test charge in the *same* direction as the field vector **E** is equivalent to moving in the direction of *decreasing* potential V. Movement *against* the field vector is the same as progressing in the direction of an *increasing* potential.

Example 4.3 Potential at a Point in a Field

1. With reference to Figure 4.4, calculate the potential at the location of charge q_3, for the case $x = y = 1$ mm. How does the potential at q_3 change as distance x tends to zero but the distance $(x + y)$ between q_2 and q_2 remains fixed at 2 mm?
2. Calculate the potential at the location of charge q_3 when distance x remains fixed at 1 mm and distance y increases to 1 cm. How does the potential at q_3 change as distance x remains fixed at 1 mm, but distance y is extended beyond 10 m?

Solution 4.3

1. The potential at the location of q_3 shown in Figure 4.4, with $x = y = 1$ mm, is equal to the negative of the work W done per unit charge by the field in the movement of q_3 from infinity to that location. We have calculated this work done in Example 4.2, obtaining the result:

$$W = 3.78 \times 10^{-12}\ \text{Nm}$$

The potential at q_3 is equal to the *negative* of this work done per unit charge:

$$V = -\frac{W}{q_3} = (-3.78 \times 10^{-12}\ \text{Nm})/(1.4 \times 10^{-12}\ \text{C})$$
$$= -2.7\ \text{Nm/C} = -2.7\ \text{V}$$

The potential has decreased in magnitude on bringing it from infinity. This agrees with the conclusion expressed in Solution 4.2 that the attractive force of the negative charge q_2 dominates over the repulsive force associated with the more distant q_1.

We can determine how the potential V at q_3 changes as a function of distances x and y using Equation (4.17):

$$V = \frac{1}{4\pi\varepsilon_o\varepsilon_r}\left(\frac{q_1}{r_{13}} + \frac{q_2}{r_{23}}\right)$$
$$= \frac{1}{4\pi\varepsilon_o\varepsilon_r}\left(\frac{q_1}{(x+y)} + \frac{q_2}{y}\right) \tag{4.21}$$

As q_2 is brought closer and closer to q_1 (i.e., $x \to 0$) but the position of q_3 remains fixed with respect to q_1, Equation (4.21) tends to the form:

$$V = \frac{1}{4\pi\varepsilon_o\varepsilon_r}\left(\frac{q_1 + q_2}{y}\right)$$

With $q_1 = 1.4$ pC; $q_2 = -1.0$ pC; $y = 2$ mm:

$$V = \frac{1}{4\pi\varepsilon_o\varepsilon_r}\left(\frac{(1.4 \times 10^{-12}\ \text{C} - 1 \times 10^{-12}\ \text{C})}{2 \times 10^{-3}\ \text{m}}\right)$$
$$= (9 \times 10^9 \text{Nm}^2\text{C}^{-2})(2 \times 10^{-10}\ \text{Cm}^{-1}) = 1.8\ \text{V}$$

This result informs us that when q_3 is located at a distance that is far greater than the separation between q_2 and q_1, the combination of q_2 and q_1 acts as a net positive charge (approaching a magnitude 0.4 pC as the distance between q_2 and q_1 becomes very small). An external force is required to bring the positive charge q_3 from infinity to a location 2 mm from q_2 and q_1. The field of q_2 and q_1 performs negative work and the potential at the location of charge q_3 has a positive value.

The potential reverses polarity, from +1.8 V to −2.7 V, as the distance x increases from a very small value to 1 mm. This indicates that at some value of x the potential at q_3 is zero. From Equation (4.21) this situation occurs when:

$$\left(\frac{q_1}{(x+y)} + \frac{q_2}{y}\right) = 0 \tag{4.22}$$

For our case where $q_1 = 1.4$ pC, $q_2 = -1.0$ pC and $(x + y) = 2$ mm, the solution for Equation (4.22) occurs when $x = 0.57$ mm and $y = 1.43$ mm. When charges q_1, q_2 and q_3 are arranged as shown in Figure 4.4 with these values for x and y, the potential at the location of q_3 is zero. No work at all was required to bring q_3 from infinity to this location!

2. The potential V at q_3 when it is located 1 cm from q_2 can be calculated using Equation (4.17) with

$q_1 = 1.4\,\text{pC}, q_2 = -1.0\,\text{pC}, r_{13} = 1.1\,\text{cm}, r_{23} = 1.0\,\text{cm}$:

$$V = \frac{1}{4\pi\varepsilon_o\varepsilon_r}\left(\frac{q_1}{r_{13}} + \frac{q_2}{r_{23}}\right)$$

$$= \frac{1}{4\pi\varepsilon_o\varepsilon_r}\left(\frac{1.4\times10^{-12}\ \text{C}}{1.1\times10^{-2}\ \text{m}} - \frac{1\times10^{-12}\ \text{C}}{1\times10^{-2}\ \text{m}}\right)$$

$$= (9\times10^9\ \text{Nm}^2\ \text{C}^{-2})\left(\frac{0.3\times10^{-14}\ \text{C}}{1.1\times10^{-4}\ \text{m}}\right)$$

$$= 0.25\ \text{V}$$

The positive value of this potential indicates that the negative charge q_2 now acts to screen q_3 partially from the repulsive force of q_1. As q_3 is moved further away from q_2, but keeping the distance between q_2 and q_1 fixed at 1 mm, we approach the situation where $r_{13} \approx r_{23}$ and V is given by:

$$V \approx \frac{1}{4\pi\varepsilon_o\varepsilon_r}\left(\frac{q_1 + q_2}{r_{23}}\right)$$

When q_3 is located 10 m from q_2 we have:

$$V \approx (9\times10^9\ \text{Nm}^2\text{C}^{-2})\left(\frac{0.4\times10^{-12}\text{C}}{10\ \text{m}}\right)$$

$$= 3.6\times10^{-4}\ \text{V}$$

For practical purposes we can therefore take distances greater than 10 m as being a good approximation to an 'infinite' distance away from charges q_1 and q_2.

Example 4.4 Potential outside a Charged Metal Sphere

1. Calculate the potential at a point 100 μm in air outside the surface of a copper sphere of radius 10 μm , which carries a total charge of 10 pC.
2. Calculate the value of this potential if the charged sphere is now immersed in an aqueous electrolyte.

Solution 4.4

1. To solve this problem we make use of the result obtained in section 3.3.2.2 of Chapter 3, where Gauss's Law was applied to find the field inside and outside of a charged (solid or hollow) metal sphere of radius R. The field inside the sphere is zero, but for distances ($r > R$) outside the sphere the field is given by equation (3.38):

$$\text{E} = \frac{1}{4\pi\varepsilon_o\varepsilon_r}\frac{Q}{r^2}$$

This is the 'Inverse Square Law' given by Equation (3.8) obtained using Coulomb's Law. Outside the sphere the field varies as $1/r^2$. Thus, all of the charge Q distributed on the sphere's surface acts as a point charge Q at the sphere's centre. The potential at any point r outside the sphere is thus equal to the negative of the work done by the field in the movement of a unit test charge from 'infinity' to the location r measured from the centre of the sphere. We will use Equation (4.17) for the case where the unit test charge is brought towards a single point charge Q. The potential at point $r > R$ is thus given by:

$$V_r = \frac{1}{4\pi\varepsilon_o\varepsilon_r}\frac{Q}{r} \tag{4.23}$$

Inside the copper sphere ($r < R$) the electric field is zero (see section 3.3.2.2 and Figure 3.19). No work is therefore required to move a test charge within the copper sphere and the potential inside and on the surface is constant and fixed at the value given by Equation (4.21) with $r = R$. If we assume that the charged sphere is located in air, then $\varepsilon_r = 1.0$ and $1/(4\pi\varepsilon_o\varepsilon_r) = 9\times10^9\ \text{Nm}^2\text{C}^{-2}$. In Equation (4.23) we also have $\text{Q} = 10\,\text{pC}$ and $\text{r} = 110\,\mu\text{m}$ (we take r as the distance from the centre of the sphere, *not* from its surface). We thus calculate the potential at a point 100 μm in air (i.e., $\varepsilon_r = 1.0$) outside the surface of the copper sphere as:

$$V = (9\times10^9\ \text{Nm}^2\text{C}^{-2})\left(\frac{10\times10^{-12}\ \text{C}}{110\times10^{-6}\ \text{m}}\right) = 818\ \text{V}$$

The potential at the surface and within the copper sphere is given by setting $r = 10\,\mu\text{m}$ in Equation (4.23) and is 9000 V.
2. On immersing the charged sphere in an aqueous electrolyte the value of the relative permittivity of the surrounding medium changes from $\varepsilon_r = 1.0$ to $\varepsilon_r \approx 80$. From Equation (4.23) the potential at a location 100 μm from the surface is ~818 V/80 = 10.2 V.

Example 4.5 Equipotential Surfaces around a Spherical Electrode

A spherical metal electrode of radius $R = 20\,\mu\text{m}$, immersed in an aqueous medium, is electrically energized at a potential of −10 V with respect to a reference zero potential located an infinite distance away from it. At what distances d from the surface of the electrode has the potential increased to values of −8 V, −6 V, −4 V and −2 V?

Solution 4.5 At all points within the electrode and at its surface, the potential is fixed at −10 V. For locations outside the electrode surface, Equation (4.23) gives an

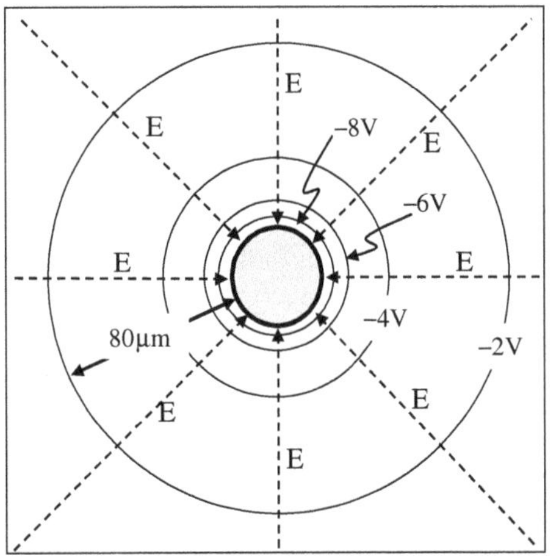

Figure 4.6 A cross-section is shown of the equipotential surfaces (−8 V, −6 V, −4 V and −2 V) around a spherical electrode of radius 20 μm, electrified to a potential of −10 V with respect to a zero reference at infinity. The radial field lines E intersect the spherical equipotential surfaces at right angles.

inverse relationship between the potential and the distance r from the electrode centre:

$$\frac{V_r}{V_R} = \frac{R}{r}$$

Setting $V_R = -10\,\text{V}$ and $R = 20\,\mu\text{m}$, electric potential values of −8 V, −6 V, −4 V and −2 V, respectively, occur at distances d ($d = r - R$) of 5 μm, 13.3 μm, 30 μm and 80 μm, respectively. Each equipotential lies on a spherical surface that is co-centric with the electrode. Cross-sections of these equipotential surfaces are shown in Figure 4.6, together with the radial field lines that point towards the centre of the electrode.

Figure 4.6 demonstrates an important relationship between equipotential surfaces and electric field lines. If a point test charge is moved over an equipotential surface, the electric field can perform no work on it. There can therefore be no component of the field tangential to an equipotential surface. Thus, a field line will always cross an equipotential contour at right angles, or stated another way: *equipotential surfaces and electric field lines are always mutually orthogonal.*

Example 4.6 Relationship between Electrode Surface Charge Density and the Potential Difference between Parallel Electrodes

Derive an expression that relates the charge density per unit area on parallel electrodes, of the form shown in Figure 4.2, to potential difference between the electrodes. Assume that the width of the electrodes is much greater than the distance d between them, so that the field generated between them can be considered uniform.

Solution 4.6 The work W_x done by the field E in moving a test charge q from the cathode to a location x above the cathode is given by:

$$W_x = -q\text{E}x$$

From Box 4.1 this work done is equal to the negative of the change in potential energy U, so that:

$$U_x - U_c = q\text{E}x$$

where U_c is the potential at the cathode. The potential V at any location is the potential energy per unit charge, so that:

$$V_x - V_c = \text{E}x$$

When the distance x is equal to the distance d between the cathode and the anode, V_x is the potential V_a at the anode, so that:

$$V_a - V_c = \text{E}d$$

The electric field between parallel electrodes is thus equal to the potential difference between the electrodes divided by the distance between them:

$$\text{E} = \frac{(V_a - V_c)}{d} \tag{4.24}$$

From Equation (3.21) of Chapter 3, the field in the dielectric between two parallel electrodes is:

$$\text{E} = \frac{\sigma}{\varepsilon_o \varepsilon_r}$$

where σ is the charge per unit area on the anode (or $-\sigma$ on the cathode). From Equation (4.24) we derive the relationship between the electrode surface charge density and the potential difference between the electrodes (anode and cathode) as:

$$\sigma = \frac{\varepsilon_o \varepsilon_r (V_a - V_c)}{d} \tag{4.25}$$

Example 4.7 Potential close to a Long Linear Distribution of Charge

Determine the potential at a location r that is close enough to a line of positive charge density λ for this charge distribution to be considered of infinite length.

Solution 4.7 In section 3.3.2 the electric field E at a distance r from an effectively infinitely long linear charge distribution was found to have only a radial component given by Equation (3.16). This equation can be simplified to the form:

$$\text{E} = \frac{\lambda}{2\pi\varepsilon_o\varepsilon_r r} \tag{4.26}$$

We can now derive the potential difference between the potential at r and a reference V_b by performing the integration given by Equation (4.20):

$$(V_a - V_b) = \int_a^b Eds$$

The field has only a radial component, so $Eds = Edr$. From Equation (4.26) we thus have:

$$(V_a - V_b) = \int_{r_a}^{r_b} \frac{\lambda dr}{2\pi\varepsilon_o\varepsilon_r r} = \frac{\lambda}{2\pi\varepsilon_o\varepsilon_r}\log_e\frac{r_b}{r_a} \quad (4.27)$$

We now have the interesting exercise of defining V_b more clearly. We have usually defined a zero potential reference to be a test charge at rest at infinity. However, inspection of Equation (4.27) reveals that if we choose V_b to be zero when r_b is set at infinity, then V_a has an infinite value for all distances r_a! This situation arises because we have assumed that our linear charge density λ extends to infinity – otherwise Equation (4.26) is not valid (see Example 3.5). However, we can avoid this problem by defining the potential V_b to be zero at an arbitrary radial distance of r_0 from the linear charge distribution. The potential at a radial distance r, with respect to this reference zero, is thus given by:

$$V(\mathrm{r}) = \frac{\lambda}{2\pi\varepsilon_o\varepsilon_r}\log_e\frac{r_0}{r} \quad (4.28)$$

The linear charge density λ is positive, so that for radial distances less than r_0 the potential is positive and for distances further away than r_0 the potential is negative. When $r = r_0$ the factor $\log_e(r_0/r)$ in Equation (4.28) is equal to $\log_e 1$ (i.e., zero) and so the potential is zero. It is important to note that Equation (4.28) gives a potential value *relative* to our artificial reference zero at r_0 – we do not obtain the *absolute* value of the potential at any location. This is clear from Equation (4.27), which shows that we are determining the *change* of potential $(V_a - V_b)$ in moving from location b to a and have chosen V_b to be zero at a location given $b = r_0$.

Example 4.8 Potential Close to a Charged Metal Rod Determine how the potential varies with radial location r outside a long charged metal rod, far away from the ends of the rod.

Solution 4.8 As shown in the derivation of Equation (3.36), Equation (4.26) also describes the potential close to a charged long metal rod or wire. The field within and at the surface of a metal rod is zero. Equation (4.28) can thus be modified by defining the radius R of the rod to be the radial distance $r = r_0$ at which $V(r)$ is designated as the reference zero. The potential at radial distances r outside a charged metal rod of radius R carrying a linear charge density λ (C/m) is thus given by the relationship:

$$V(\mathrm{r}) = \frac{\lambda}{2\pi\varepsilon_o\varepsilon_r}\log_e\frac{R}{r} \quad (r \geq R) \quad (4.29)$$

For radial distances $r \leq R$, measured from the central axis of a metal rod, the electric field is zero. The potential is constant and the same as the reference value designated as zero at the surface of the rod. Again, as for Example 4.7 it is important to appreciate that we do not obtain absolute values of the potential using Equation (4.29). The rod carries an electrical charge and so its absolute potential value cannot in reality be zero. We have simply chosen, for mathematical convenience, to designate the potential of the rod as our reference zero level.

4.3.1 Molecular Electrical Potential Surface

Molecules are formed from atomic nuclei and their electrons. The electrons create a negative electrical potential in the space surrounding the molecule, whilst the positively charged nuclei create a positive potential. The nuclei can be considered as point charges, whereas the electrons form a diffuse cloud of negative charge. The electrical potential $V(r)$ at each point r in space around a molecule can be represented as a combination of Equations (4.18) and (4.19):

$$V(r) = \frac{1}{4\pi\varepsilon_o\varepsilon_r}\left[\sum_A \frac{Z_A}{|R_A - r|} - \int \frac{\rho(r')}{|r' - r|}dr'\right] \quad (4.30)$$

In this equation Z_A is the charge on nucleus A, located at R_A and $\rho(r)$ is the electronic charge density function [1, 2]. The charge density function is a measure of the probability of finding an electron at any location and is quantified in units of electrons / bohr. The atomic radius of the hydrogen atom ($Z = 1$) is given by the most probable radius of orbit of its single electron. This is known as the Bohr radius and has a value of 5.3×10^{-11} m. A contour in three dimensions of $\rho(r) = 0.002$ electrons per cubic bohr is normally assumed to encompass at least 95% of the electronic charge of a molecule and represents a physically reasonable representation of its shape and dimensions [1].

Knowledge of the electrostatic potential surface of a molecule has become an effective tool for interpreting and predicting the reactivity of a molecule towards other chemical ligands. An approaching *electrophile* will tend to head towards those surface regions of the molecule where $V(r)$ attains its most negative values (i.e., local minima of the electrostatic surface) because this is where the molecule's electrons are most dominant and potentially available to form valence bonds with other atoms.

4.4 Electrostatic Field Energy

From Equation (4.16) the electrical potential (defined as the potential energy per unit charge) at a point in an electrical field is equal to the work done against the field E in bringing a unit charge from infinity to that point. We can thus write the energy of a point charge q_2 in the field of a single point charge q_1 as:

$$U = q_2 V_{21}$$

where V_{21} is the potential at q_2 due to q_1. We can also say that the potential is equal to the work returned by the system as one of these charges is taken back to infinity. The two charges therefore have a mutual and reciprocal energy, so that

$$U = q_1 V_{12} = \frac{1}{2}(q_1 V_{12} + q_2 V_{21})$$

Bringing another charge q_3 into the field increases the energy:

$$U = q_2 V_{21} + q_3(V_{31} + V_{32})$$

The mutual energies of charge pairs also allow us to express this as:

$$U = \frac{1}{2}(V_{12} + V_{13})q_1 + \frac{1}{2}(V_{21} + V_{23})q_2 + \frac{1}{2}(V_{31} + V_{32})q_3$$

Extending this logic to a complete and closed system of n charges:

$$U = \frac{1}{2}\sum_{i=1}^{n}\sum_{j=1}^{n} V_{ij}q_i = \frac{1}{2}\sum_{i=1}^{n} V_i q_i (i \neq j)$$

where V_i is the potential q_i due to the remaining $n - 1$ charges in the closed system. If we designate V_o as the potential arising from the charges being uniformly distributed with a volume density ρ throughout a dielectric, the energy can be expressed as a volume integral:

$$U = \frac{1}{2}\int_v V_o \rho \mathrm{d}v \tag{4.31}$$

This integral must be carried out over a volume that encloses all of the charge.

Charges are rarely distributed throughout the bulk volume of a dielectric, but appear as a surface charge density σ on conductors that form, for example, the electrodes of a capacitor. In this case the increase in the potential energy of the field results from work done by the electromotive force of a battery in building up this surface charge and can be evaluated from a surface integral of the form:

$$\delta U = \int_s V_o \delta\sigma \mathrm{d}a$$

Stratton [3, p. 110] makes the analogy of the energy stored in the field resembling the potential energy stored within an extended spring. The elastic energy of an ideal spring is equal to $(kx^2)/2$ where k is the force constant of the spring and x is the extension or compression of the spring length. We will find in section 4.4.1 that the energy stored in a capacitor is equal to $Q^2/2C$. With Stratton's analogy the charge Q is equivalent to x and $1/C$ to the force constant k. The energy of an inhomogeneously stressed elastic medium such as a spring is concentrated principally in regions of greatest strain. In this case the elastic energy per unit volume has a very definite sense. Although the analogy of the electrostatic to the elastic field is not a close one, Stratton considers it plausible to suppose that the electrostatic energy is localized in the more intense regions of the field and distributed with a *density*:

$$\delta u = \mathrm{E} \cdot \delta\mathrm{D}$$

where D is the displacement vector $\mathrm{D} = \varepsilon_o \varepsilon_r \mathrm{E}$. The total energy stored in the field can thus be obtained from the volume integral:

$$U = \int_v \int_0^{\mathrm{D}} \mathrm{E} \cdot \delta\mathrm{D}\mathrm{d}v$$

where, following the form of Equation (3.50), the increment δu is integrated from the initial state $\mathrm{D} = 0$ without an applied field to the final state $\mathrm{D} = \varepsilon_o \mathrm{E} + \mathrm{P}$. If the dielectric medium is isotropic and linear in its response to an applied field, then ε_r (and hence D) may possibly be a function of location but will be independent of E. In which case the integration of $\mathrm{E} \cdot \delta\mathrm{D}$ for increments δu is equal to $(\varepsilon_o \varepsilon_r \delta\mathrm{E}^2)/2$, so that

$$U = \frac{1}{2}\int_v \varepsilon_o \varepsilon_r \mathrm{E}^2 \mathrm{d}v \tag{4.32}$$

The integral in this case must be carried out over a volume that encompasses all regions where the field E exists. The electric field energy (joules) *per unit volume* is thus given by:

$$u = \frac{1}{2}\varepsilon_o \varepsilon_r \mathrm{E}^2 \tag{4.33}$$

4.4.1 Potential Energy of a Charged Capacitor

From the *work-energy theorem* we can derive the potential energy of a charged capacitor by evaluating the work required to give it a final charge (per electrode) Q and a potential difference V across its electrodes. If we assume that during this charging process we have attained a charge q and potential difference v, then from Equation (3.53) at this stage we have $v = q/C$, where C is the capacitance of the capacitor. The increment of work dW required to add another elemental charge dq is

$$dW = v dq = \frac{q dq}{C}$$

The total work W required to fully charge the capacitor is given by integrating these increments:

$$W = \int_0^W dW = \int_0^Q \frac{1}{C} q dq = \frac{Q^2}{2C} \tag{4.34}$$

This is the work done by an electromotive force derived from a battery. It is also the work done by the field between the capacitor electrodes on the charges during the discharging of the capacitor. If we define the totally discharged state of the capacitor to be the reference zero of its potential energy, then from the *work-energy theorem* Equation (4.34) defines the potential energy U of the charged capacitor:

$$U = \frac{Q^2}{2C} = \frac{1}{2} C V^2 \tag{4.35}$$

where V is the potential difference across the electrodes when the capacitor is at its full charge Q per electrode. A capacitor device can therefore store both charge and energy. If the task of the capacitor is to transfer charge around an electrical circuit, the work required to do this is inversely proportional to the capacitance C. This is reflected in assigning to a capacitor an equivalent electrical resistance known as its reactance X_c, where $X_c = 1/(\omega C)$. If the function is to store energy, then this task also increases with C. We can think of this energy as being stored in the field between the capacitor's electrodes. The energy stored *per unit volume* is then the energy density u. For a parallel plate capacitor with plate area A and separation d, the volume containing the field (neglecting fringing at the electrode edges) is Ad. The energy density is thus:

$$u = \frac{\frac{1}{2}(CV^2)}{Ad} = \frac{\frac{1}{2}(A\varepsilon_o\varepsilon_r/d)V^2}{Ad}$$
$$= \frac{1}{2}\varepsilon_o\varepsilon_r\frac{V^2}{d^2} = \frac{1}{2}\varepsilon_o\varepsilon_r \mathrm{E}^2$$

which gives the same result as Equation (4.33) for the energy density of any electric field.

4.4.2 Energy of a Dielectric Particle in a Field

We will now perform the imaginary procedure described by Stratton [3, p. 112] of inserting an uncharged and non-conducting dielectric particle into a dielectric medium in which an electric field E_m has already been established. The system of external charges that produce this field are kept constant during this procedure. The particle has relative permittivity ε_p, whilst the medium is assumed to be isotropic and linear so that its relative permittivity ε_m is either constant or varies as a scalar function of position. The objective is to evaluate the potential energy of the dielectric particle as a function of its location in the field.

Based on Equation (4.32) and with the displacement vector given by $\mathrm{D}_m = \varepsilon_o\varepsilon_m\mathrm{E}_m$, the initial energy U_m of the system (i.e., the total work done in establishing the field E_m) is given by the following integral, evaluated over all space:

$$U_m = \frac{1}{2}\int_v \mathrm{E_m} \cdot \mathrm{D}_m \, dv$$

As shown in a similar exercise performed in Chapter 6 (see Box 6.1), the resulting polarization of the particle produces a (dipole) field that modifies the original field. We designate this modified field at any point as E, so that the difference $\mathrm{E}_p = (\mathrm{E} - \mathrm{E}_m)$ is the field resulting from the particle polarization. The energy U_2 of the field in this new state is given by:

$$U_2 = \frac{1}{2}\int_{V_p+V_m} \mathrm{E} \cdot \mathrm{D} \, dv$$

where V_p and V_m are the volumes occupied by the particle and the medium outside the particle, respectively. The change in energy $U = (U_2 - U_m)$ is equal to the energy of the particle in the external field E_m and is given by:

$$U = \frac{1}{2}\int_{V_p+V_m} (\mathrm{E} \cdot \mathrm{D} - \mathrm{E_m} \cdot \mathrm{D_m}) \, dv \tag{4.36}$$

This also represents the work performed in the action of inserting the particle into the field E_m. Equation (4.36) can be written in the form:

$$U = \frac{1}{2}\int_{V_p+V_m} \mathrm{E} \cdot (\mathrm{D} - \mathrm{D_m}) dv + \frac{1}{2}\int_{V_p+V_m} (\mathrm{E} - \mathrm{E_m}) \cdot \mathrm{D_m} \, dv \tag{4.37}$$

The particle carries no net charge and the original system of charges producing E_m remains unchanged, so the first of the integrals in Equation (4.37) is zero, to give:

$$U = \frac{1}{2}\int_{V_p+V_m} (\mathrm{E}-\mathrm{E_m})\cdot\mathrm{D_m}\,dv$$
$$= \frac{1}{2}\int_{V_p} (\mathrm{E}-\mathrm{E_m})\cdot\mathrm{D_m}\,dv + \frac{1}{2}\int_{V_m} (\mathrm{E}-\mathrm{E_m})\cdot\mathrm{D_m}\,dv$$

Using a theorem of vector fields that states that the intergral over all space of the scalar product of an irrotational vector and a solenoid vector is zero (which applies here because $\nabla\times\mathrm{E}_m = 0$ and $\nabla\cdot(\mathrm{D}-\mathrm{D}_m) = 0$), Stratton [3, p. 113] shows that this expression can be reduced to one in terms of an integral, not over all space, but *over its own volume alone*:

$$U = \frac{1}{2}\int_{V_p} (\mathrm{E}\cdot\mathrm{D_m} - \mathrm{E_m}\cdot\mathrm{D})dv$$

Inserting the identities $\mathrm{D}_m = \varepsilon_o\varepsilon_m\mathrm{E}_m$ and $\mathrm{D} = \varepsilon_o\varepsilon_p\mathrm{E}$ into this equation we obtain the final result:

$$U = \frac{1}{2}\int_{V_p} (\varepsilon_\mathrm{m}-\varepsilon_\mathrm{p})\mathrm{E}\cdot\mathrm{E_m}\,dv \tag{4.38}$$

This result is particularly important for our subject of dielectrophoresis. It shows that, for the case where $\varepsilon_p > \varepsilon_m$, we can expect the introduction of the particle into the medium to result in a *negative* value of its electrostatic energy. Work will be required on the particle to *withdraw* it from the medium. Furthermore, this energy is further reduced if the field E_m increases. The particle will attempt to minimize its energy further by moving up a field gradient to a maximum value of this gradient. This describes the action of *positive* dielectrophoresis, where a particle is directed towards an electrode edge. On the other hand, if the particle's permittivity is less than that of the medium, its energy will be positive. It will move down a field gradient to search for a gradient minimum, well away from an electrode edge. Work will be required to *insert* the particle into the medium. This describes *negative* dielectrophoresis.

If, during the course of a dielectrophoresis experiment, a small change $\delta\varepsilon$ should occur in its permittivity, then $\delta\varepsilon\mathrm{E}\cdot\mathrm{E_m}$ will differ from $\delta\varepsilon\mathrm{E_m^2}$ by an infinitesimal amount and Equation (4.38) takes on the form:

$$\delta U = \frac{1}{2}\int \delta\varepsilon\mathrm{E_m^2}\,dv \tag{4.39}$$

This result shows that a small change of the permittivity of the particle relative to that of the suspending medium can be monitored as a change in the dielectrophoretic force.

Finally, looking ahead to concepts that are described in Chapters 6 and 10, the time-averaged dielectrophoretic force acting on a spherical particle of radius R is given from Equation (10.25) as:

$$\langle F_{DEP}\rangle = \frac{v}{2}\mathrm{Re}\left[\alpha^*\right]\nabla\mathrm{E}^2 \tag{4.40}$$

where $v = (4\pi R^3)/3$ is the particle's volume and $\mathrm{Re}[\alpha^*]$ is the real part of its polarizability, given from Box 6.1 as:

$$\alpha = 3\varepsilon_o\varepsilon_m\left(\frac{\varepsilon_p-\varepsilon_m}{\varepsilon_p+2\varepsilon_m}\right)\ \text{(per unit volume)}$$

The term in brackets is known as the Clausius–Mossotti factor, described in Chapter 6. If no, or negligible, energy losses occur, associated with charge conduction or dielectric dispersions in the particle or surrounding medium, we can consider the dielectrophoretic force to be a conservative force. In this case we can evoke the work-energy theorem described in Box 4.1 and equate the work done in moving a polarized particle from one point to another as being equal to the negative of the change of the particle's potential energy. From Equation (4.40) we can therefore give the time-averaged potential energy of the polarized particle as:

$$\langle U\rangle = -\frac{1}{2}\alpha\mathrm{E}^2\ \text{(per unit volume)} \tag{4.41}$$

This result is in agreement with the form of Equation (4.38). For the case where $\varepsilon_p > \varepsilon_m$ we can expect the introduction of an unpolarized particle into a medium (in which a field E has been established) to result in a negative value of its electrostatic energy. Work will be required on the particle to withdraw it from the medium and the field. The potential energy of the particle is reduced further if the field E is increased. If confronted with a field gradient, the particle will attempt to minimize its energy further by moving up the gradient to a larger value of E. It will undergo *positive* dielectrophoresis. The opposite effect, *negative* dielectrophoresis, will occur if the particle's permittivity is less than that of the surrounding medium. If there are conduction or dielectric losses in the particle or medium, although the change in potential energy will not be given exactly by Equation (4.41) the conditions for either positive or negative dielectrophoresis remain unchanged.

4.5 Summary

This chapter has dealt with concepts of importance to dielectrophoresis, namely the work done and the potential energy associated with interactions between charged particles and electric fields. A charged particle exposed to an electric field experiences an electrostatic force that can perform *work* on the particle, which relates directly to a *change* of the particle's *electric potential energy*.

If W_{ab} is the work done by the field in moving a charged particle from location *a* to location *b*, the change in potential energy ΔU is given by the relationship $W_{ab} = -\Delta U$. Thus, if *positive work* is done by the field, the potential energy of the particle *decreases.* A field does *positive* work on a positively charged particle when it moves in the direction of the field and so in this case the particle's potential energy *U decreases.* All systems try to attain an equilibrium state by minimizing its potential energy. Positively charged particles therefore have a natural tendency to move in the direction of an applied electric field. In order to move a positively charged particle in the direction against that of the field, work has to be done on the particle against the electrostatic force tending to force it in the opposite direction. This is equivalent to the field doing *negative* work on the particle and so the particle's potential energy increases. On the other hand, the field will perform *positive* work when a *negatively* charged particle moves *against* the field and its potential energy will *decrease.* Negatively charged particles have a natural tendency to move against an electric field's direction in order to minimize their potential energies. Moving a negatively charged particle in the field direction will increase its potential energy.

An important property of an electrostatic force is that it is a *conservative* force. Thus, the change in potential energy in moving a charged particle from location *a* to location *b* will not depend on the path taken. Moving the particle from location *a* to *b* and back again to location *a* will involve no net expenditure of work by the field. There will be no change in the particle's potential energy, no matter what paths are taken in the outward and return journeys.

Another important concept is *potential,* defined as the potential energy per unit particle charge. The potential of a charged particle depends on its position in an electric field. We are unable to measure the absolute magnitude of a potential, but instead define its value with respect to a reference potential. A conceptual reference zero corresponds to setting a test unit charge at rest (i.e., zero kinetic energy) at an infinite distance away from the influence of any field created by any other charge (i.e., zero potential energy). We can therefore define the potential *V* at any location in a field as equal to the *negative* of the work done by the field in moving a body of unit charge from infinity to that location. We can also define the potential difference ($V_a - V_b$) between location *a* and location *b* as equal to the work done by an electric field in moving a body of unit charge from *a* to *b* in this field.

Contours that join potentials of the same value are known as equipotential contours, or equipotential surfaces if drawn in three-dimensional space. Equipotential surfaces and electric field lines are always mutually orthogonal. Electrical potential, defined as the potential energy per unit charge, has units of N.m/C. In the SI system of units this unit of electrical potential is named the volt. In Chapter 2 the concept of an electrical field was described in terms of the force per unit charge that the field exerts on a test unit charge. The unit of electric field can thus also be expressed as *one volt per metre.*

These various concepts have been brought together in the form of the work-energy theorem to derive expressions for the energy stored in an electric field. The simple example of a parallel-plate capacitance *C* charged to a potential difference *V* between its plates gives this energy as:

$$U = \frac{Q^2}{2C} = \frac{1}{2}CV^2$$

This in turn leads to the following expression for the energy density stored in the field in the capacitor's dielectric:

$$u = \frac{1}{2}\varepsilon_o\varepsilon_r \mathrm{E}^2$$

This relationship also describes the energy density of any electric field generated by electrodes of any geometry.

The most important result of relevance to dielectrophoresis is the derivation of the energy of a dielectric particle as a function of its position in a uniform or nonuniform electric field established in another dielectric medium, namely:

$$U = \frac{1}{2}\int_{V_p} (\varepsilon_\mathrm{m} - \varepsilon_\mathrm{p})\mathrm{E}\cdot\mathrm{E_m}\,dv$$

For the situation where the particle's permittivity (ε_p) is larger than that (ε_m) of the suspending medium, the particle has a *negative* electrostatic energy. The particle will experience a *positive* dielectrophoretic force, tending to drive it towards an electrode where the local field will be at its maximum and where the particle will attain its lowest potential energy. Conversely, if the particle's permittivity is less than that of the medium, its energy will be positive and it will strive to find a local field minimum under the action of *negative* dielectrophoresis. This result summarizes the energetics of how a mixture of particles of different dielectric properties (for example subpopulations of cells) can be physically separated and isolated under the influence of a nonuniform electric field and that small changes in the dielectric properties of a cell (e.g., a change of cell membrane capacitance or viability) may also be monitored by its dielectrophoretic behaviour.

Based on concepts to be described in Chapters 6 and 10, the time-averaged potential energy per unit volume of a polarized particle can be given as:

$$\langle U\rangle = -\frac{1}{2}\alpha \mathrm{E}^2 \text{ (per unit volume)}$$

where α is the effective polarizability of the particle, determined by the magnitude and polarity of the Clausius–Mossotti factor to be described in Chapter 6.

Finally, it is important to note that our interpretation of the work-energy theorem applies to systems in which there are no energy losses. A dramatic example given here is the description of Richard Feynman's 'swinging ball of death'. The effect would not have been so dramatic if, instead of air, he had demonstrated his swinging ball in a highly viscous fluid. Frictional losses would have limited the ball's motion. Likewise for dielectrophoresis, the basic theory of which is often presented as if the dielectrophoretic force is a conservative force, where electrical conduction and dielectric relaxation losses can be ignored. This is discussed further in Chapter 10.

4.6 References

1. Sjoberg, P. and Politzer, P. (1990) Use of the electrostatic potential at the molecular surface to interpret and predict nucleophilic processes. *J. Phys. Chem.* **94**, 3959–3961.
2. Kumar, A., Gadre, S. R., Mohan, N. and Suresh, C. H. (2013) Lone pairs: an electrostatic viewpoint. *J. Phys. Chem A* **118**, 526–532.
3. Stratton, J. A. (1941) *Electromagnetic Theory*, McGraw-Hill, New York, NY.

5

Potential Gradient, Field and Field Gradient; Image Charges and Boundaries

5.1 Introduction

The phenomenon of dielectrophoresis can be broken down to two actions. The first one involves subjecting a particle to an electric field generated by applying potential differences (voltages) to a set of electrodes. This field induces a nonuniform distribution of charge on the particle's surface, in the form of an electric dipole moment, of magnitude proportional to the local field. The second action arises from how this induced dipole moment interacts with the field. If the field is uniform (i.e., it has zero gradient) the particle could attempt to minimize its potential energy by aligning its induced dipole with the field, rather like a compass needle aligning itself along the Earth's magnetic field. The particle might exhibit rotational motion, but will exhibit no lateral motion. Lateral motion (dielectrophoresis) of the particle can only occur where there is a spatial gradient of the field and the particle can move so as to minimize its potential energy. The dielectrophoretic force acting on the particle is thus proportional to the product of the local field (which lends to it the properties of a dipole) and the local field gradient (which causes the particle to move).

The purpose of this chapter is to provide some basic insights into the relationship between the electric field and the potential gradient at any point in space, as obtained by solving Laplace's equation, which was derived in Chapter 3. This leads to the description of distributions of charges in terms of their effective dipole moment (and higher order moments) and thus to the concept of a polarized particle exhibiting the properties of an induced dipole moment. The charges that give rise to this induced moment are formed at the boundary surface between a particle and its surrounding dielectric medium. For most biological applications of dielectrophoresis the surrounding medium is an aqueous electrolyte. The conditions that apply to the behaviour of D-field and E-field lines of force at the boundary surface between two dielectrics are also described, together with the concept of an electric field gradient.

5.2 Potential Gradient and Electrical Field

In the form of Equation (4.20) in Chapter 4 we have the following direct relationship between an electrical potential difference and the associated electrical field:

$$(V_a - V_b) = \int_a^b \mathbf{E}.d\mathbf{s} \tag{5.1}$$

As demonstrated in Examples 4.7 and 4.8 of Chapter 4 this relationship can be used to find the potential difference (V_a − V_b) between a location a and the reference level at b if we know the field **E** at these two locations. This involves the mathematical process of integrating the function **E**.d**s**. The *fundamental theorem of calculus* states that differentiation is the reverse process of integration. By performing the reverse procedure of differentiation it should therefore in principle be possible to derive **E** if we know the values of the potential V at each location. The quantity (V_a − V_b) can be found by summing the infinitesimal change of potential dV accompanying each infinitesimal change in location d**s** from the reference position b to the point of interest at a. This is expressed mathematically as:

$$(V_a - V_b) = \int_b^a dV$$

This calculation can also be performed as a reverse path taken from position a to b, denoted by reversing the limits on the integral sign:

$$(V_a - V_b) = -\int_a^b dV$$

From this relationship and Equation (5.1) we have the following equality:

$$-\int_a^b dV = \int_a^b \mathbf{E}.d\mathbf{s}$$

Dielectrophoresis: Theory, Methodology and Biological Applications, First Edition. Ronald Pethig.

Box 5.1 Differential Calculus and Gradient of a Scalar Field

Differential calculus concerns the study of the rates of change (gradients) of quantities expressed as mathematical functions. The differential $\frac{df(x)}{dx}$ of a function of x (e.g., $f(x) = x^2 + 4x + 3$) is obtained by evaluating by how much the function changes for a very small increment of x:

$$\frac{df(x)}{dx} = \lim_{\delta x \to 0} \frac{f(x + \delta x) - f(x)}{\delta x}$$

This procedure determines the slope or gradient of the function for each value of x. As an example we derive $\frac{df(x)}{dx}$ for $f(x) = x^2 + 4x + 3$) as follows:

$$\frac{df(x)}{dx} = \lim_{\delta x \to 0} \frac{[(x^2 + 2x\delta x + \delta x^2) + 4(x + \delta x) + 3] - (x^2 + 4x + 3)}{\delta x}$$
$$= \lim_{\delta x \to 0} \frac{[2x\delta x + \delta x^2 + 4\delta x]}{\delta x} = \lim_{\delta x \to 0} [2x + 4 + \delta x]$$

In the limit, as the increment δx tends to zero, $\frac{df(x)}{dx} = 2x + 4$.

The change in the x-direction of a scalar field $f(x, y, z)$ for an increment δ is given by:

$$\lim_{\delta \to 0} \frac{f(x + \delta, y, z) - f(x, y, z)}{\delta}\hat{\imath} = \frac{\partial f(x, y, z)}{\partial x}\hat{\imath}$$

where $\hat{\imath}$ is the unit vector along the x-direction. Similar expressions hold for incremental changes in the y- and z-directions, so that the overall spatial change or gradient of the scalar field at any point is given by:

$$\hat{\imath}\frac{\partial f}{\partial x} + \hat{\jmath}\frac{\partial f}{\partial y} + \hat{k}\frac{\partial f}{\partial z} = \nabla f$$

We may require the gradient along a specific direction, as for example in the direction of a unit vector $u = a\hat{\imath} + b\hat{\jmath}$ at a specific location x_0, y_0. Vector u will be a hypotenuse of unit length and so $a^2 + b^2 = 1$. The *directional* derivative $D_u f(x_0, y_0)$ is then defined as:

$$D_u f(x_0, y_0) = \lim_{\delta \to 0} \frac{f(x_0 + \delta a, y_0 + \delta b) - f(x_0, y_0)}{\delta}$$

The directional derivative along a unit vector u can also be written as $(\nabla f)\cdot u$.

For these two integrals to be equal, the integrands must also be equal, so that:

$$-dV = \mathbf{E}.d\mathbf{s} \tag{5.2}$$

In this equation the potential V is a scalar quantity. Plots such as those shown in Figure 4.5 are sometimes referred to as plots of the scalar potential *field*. An analogy is a plot of the temperature field for a room, obtained as contour lines that connect points of equal temperature. Values for potential and temperature have magnitudes, but not direction – they do not 'point' anywhere. A potential field (V) and temperature (T) field are thus examples of a scalar field (and are written as italics). On the other hand, the symbols E and ds are written in upper case fonts to indicate that they are vector quantities, possessing the qualities of magnitude as well as direction. Using the vector notation given in Box 3.1, we can write Equation (5.2) as:

$$-dV = (\hat{\imath}E_x + \hat{\jmath}E_y + \hat{k}E_z) \cdot (\hat{\imath}dx + \hat{\jmath}dy + \hat{k}dz) \tag{5.3}$$

where $\hat{\imath}$, $\hat{\jmath}$ and $\hat{k}$ are unit vectors along the x-, y- and z-axes, respectively. Because $\hat{\imath}$, $\hat{\jmath}$ and $\hat{k}$ are all perpendicular to each other, we have the following results for their dot products:

$$\hat{\imath}\cdot\hat{\imath} = \hat{\jmath}\cdot\hat{\jmath} = \hat{k}\cdot\hat{k} = 1 \times 1 \times \cos 0^\circ = 1$$

$$\hat{\imath}\cdot\hat{\jmath} = \hat{\imath}\cdot\hat{k} = \hat{\jmath}\cdot\hat{k} = 1 \times 1 \times \cos 90^\circ = 0$$

Equation (5.3) can thus be simplified to the form:

$$-dV = E_x\,dx + E_y\,dy + E_z\,dz$$

The x-, y- and z-components of the vector E can be written in terms of the partial derivatives of V as:

$$E_x = -\frac{\partial V}{\partial x}, \quad E_y = -\frac{\partial V}{\partial y}, \quad E_z = -\frac{\partial V}{\partial z} \tag{5.4}$$

where, for example, $\frac{\partial V}{\partial x}$ is termed the *partial* derivative of the potential for the variable x, obtained whilst keeping the values of variables y and z fixed as constants. In vector notation the field E can thus be given in terms of the partial derivatives in all three coordinate directions as:

$$E = -\left(\hat{\imath}\frac{\partial V}{\partial x} + \hat{\jmath}\frac{\partial V}{\partial y} + \hat{k}\frac{\partial V}{\partial z}\right) \tag{5.5}$$

This equation states that the resultant electric field vector is given as the negative vector additions of the gradients of the potentials along the x-axis, the y-axis and the z-axis. Expressed in this form highlights the fact that E is a vector. Thus, the gradient of a scalar function or field using vector notation is obtained by applying the *grad* or *del* vector operator denoted by the symbol ∇. For example, the gradient of V, verbalized as either grad(V) or del(V), is given by:

$$\nabla V = \left(\hat{\imath}\frac{\partial}{\partial x} + \hat{\jmath}\frac{\partial}{\partial y} + \hat{k}\frac{\partial}{\partial z}\right) V \tag{5.6}$$

This concept is described more fully in Box 5.1.

Equation (5.5) can thus be written more succinctly as:

$$\mathrm{E} = -\nabla V \tag{5.7}$$

The negative algebraic sign conforms to the convention that directs the vector E outward from a positive charge. For the case of a radial electric field, such as that shown in Figures 4.3 and 4.6, then:

$$\mathrm{E}_r = -\frac{\partial V_r}{\partial r} \tag{5.8}$$

An example of a radial field is one produced outside of a charged metal sphere, where from Equation (4.23) we have the potential at any radial distance r given by:

$$V_r = \frac{1}{4\pi\varepsilon_o\varepsilon_r}\frac{Q}{r}$$

From Equation (5.8), the *magnitude* of the radial electric field is derived as follows:

$$\mathrm{E}_r = -\frac{\partial V_r}{\partial r} = -\frac{\partial}{\partial r}\left(\frac{1}{4\pi\varepsilon_o\varepsilon_r}\frac{Q}{r}\right) = \frac{1}{4\pi\varepsilon_o\varepsilon_r}\frac{Q}{r^2}$$

This result agrees with that given by Equation (3.38). To express the fact that this is a *vector* field acting along a radial line directed from the centre of the metal sphere we employ the radial unit vector $\hat{\mathrm{r}}$:

$$\mathrm{E} = \hat{\mathrm{r}}\mathrm{E}_\mathrm{r}\frac{1}{4\pi\varepsilon_o\varepsilon_r}\frac{Q}{r^2}\hat{\mathrm{r}} \tag{5.9}$$

Figure 5.1 shows the electric field vectors, in a two-dimensional plane through the centre of a positively charged metal sphere, diminishing with radial distance from the sphere.

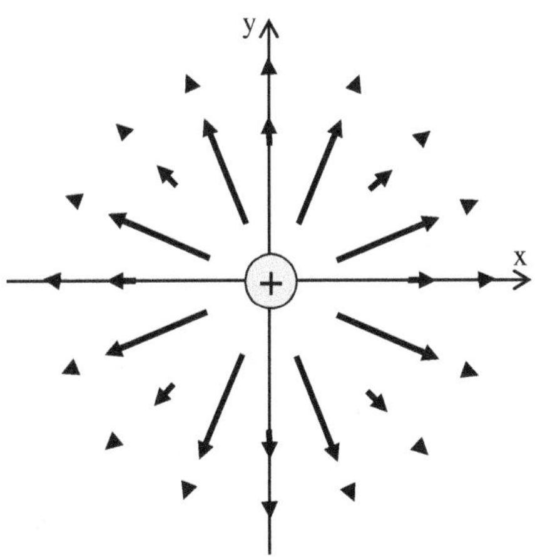

Figure 5.1 The vector field created by a positively charged metal sphere situated at the centre of an xy-plane. The base of each arrow is located at a point of interest (e.g., x_0, y_0). An arrow points in the direction of greatest decrease of the potential and its length is proportional to the magnitude of the potential gradient (i.e., the electric field) at that point. The field decreases with radial distance r as $1/r^2$.

Equations (5.6) and (5.9) enforce an important fact, namely that the gradient of V is a *vector* – it has both magnitude and direction. The minus sign in Equation (5.7) indicates that, at any point in space, the electric field vector E points towards where the potential *decreases* the most rapidly. Along an equipotential surface grad(V) is zero. A field vector is thus always orthogonal to an equipotential contour or equipotential surface.

Example 5.1 Field outside a Charged Metal Rod
Determine the components of the electrical field outside a long charged metal rod, far away from the ends of the rod.

Solution 5.1 The potential at distances r outside a charged metal rod of radius R carrying a linear charge density λ is given by Equation (4.29) derived in Solution 4.8:

$$V(\mathrm{r}) = \frac{\lambda}{2\pi\varepsilon_o\varepsilon_r}\log_e\frac{R}{r}$$

The potential and hence the field have radial symmetry about the long axis of the rod and so the field has no x-, y- or z-component and only a radial component. The magnitude of this field is given by:

$$\mathrm{E}_r = -\frac{\partial V_r}{\partial r} = -\frac{\partial}{\partial r}\left(\frac{\lambda}{2\pi\varepsilon_o\varepsilon_r}\log_e\frac{R}{r}\right)$$
$$= -\frac{\partial}{\partial r}\left(\frac{\lambda}{2\pi\varepsilon_o\varepsilon_r}[\log_e R - \log_e r\right)$$

$\frac{\partial}{\partial r}\left(\log_e R\right) = 0$ (because R is a constant) and $\frac{\partial}{\partial r}\left(\log_e r\right) = \frac{1}{r}$, so that:

$$\mathrm{E}_r = \frac{\lambda}{2\pi\varepsilon_o\varepsilon_r r}$$

Example 5.2 A Plot of a Vector Field
An electric field is given by the following function:

$$\mathrm{E}(x, y) = x\mathrm{u} - y\mathrm{v}$$

where u and v are unit vectors along the y- and x-directions, respectively. Draw a graphical representation of the vector field, to include locations such as (2, −1), for example.

Solution 5.2 The field is a function of only the x- and y-coordinates and can be plotted as a two-dimensional graph. Evaluation of the field for $(x, y) = (2, -1)$ gives:

$$\mathrm{E}(2, 1) = 2\mathrm{u} + 1\mathrm{v}$$

The magnitude of the field at this point is given by $\mathrm{E} = \sqrt{(2^2 + 1^2)} = \sqrt{5} = 2.24$ and is oriented with the x-axis at an angle given by $\theta = \tan^{-1}(2/1) = 63.4°$. This vector

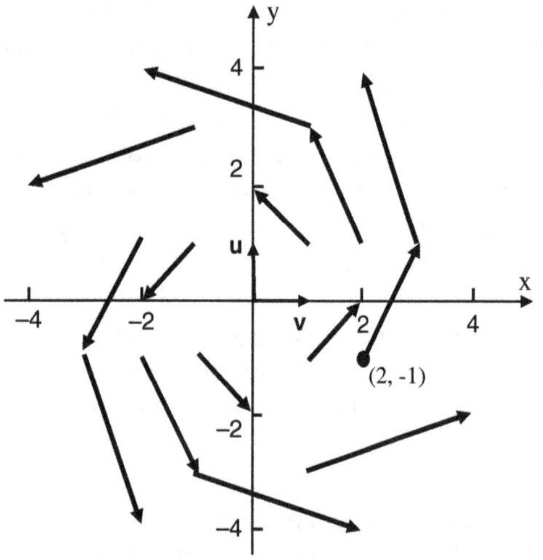

Figure 5.2 The vector field plot for the electric field E(x, y) = xu − yv, where u and v are unit vectors along the y- and x-axes, respectively.

is plotted as an arrow in Figure 5.2, together with some other vectors at other locations. We can refer to this plot as showing either a *field of vectors* or a *vector field.*

Example 5.3 A Plot of the Gradient of a Potential Field

An electrical potential is given by the following equation:

$$V(x,y) = x^2 + y^2 + 0.5xyz$$

Derive a plot showing the gradient field for this potential function, together with several equipotential contours, for the case where $z = 0$. .

Solution 5.3 In the xy-plane, for z = 0, the equipotential contours are defined by:

$$V(x,y) = x^2 + y^2 = k$$

Each contour is described by an equation for a circle, centred at the origin, of radius equal to $\sqrt{k}$. The equipotential contours for 20 V, 40 V, 60 V, 80 V and 100 V are drawn in Figure 5.3. The components of the gradient vectors at any coordinate (x, y) are given by:

$$\nabla V(x,y) = \left(\hat{\imath}\frac{\partial}{\partial x} + \hat{\jmath}\frac{\partial}{\partial y}\right)(x^2 + y^2) = 2x\hat{\imath} + 2y\hat{\jmath} \quad (5.10)$$

The gradient vectors are shown superimposed on the equipotential contours in Figure 5.3. The distance between the contours decreases as the magnitudes of consecutive equipotentials increase and this leads to an increase in the magnitude of the gradient vectors with increasing distance from the origin of the plot. The gradient vectors point in the direction of the greatest rate of decrease of the potential values and are directed orthogonal to the equipotential contours.

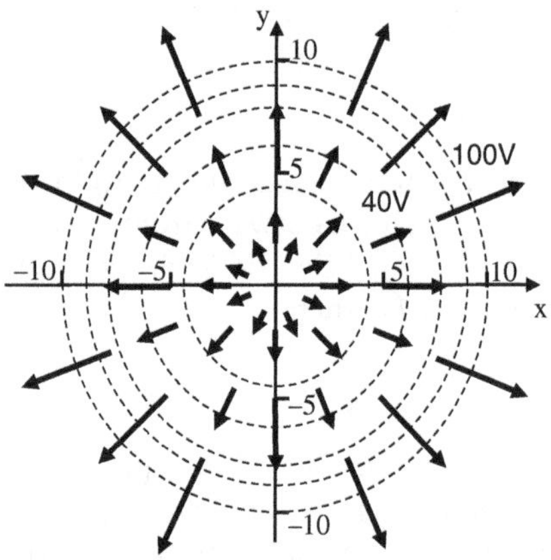

Figure 5.3 The equipotential contours (circles) and gradient vectors (arrows) are shown for the potential function $V(x, y) = x^2 + y^2$. The 40 V and 100 V equipotentials are labelled. The gradient vectors represent the magnitude of the electric field for a coordinate located at the base of an arrow. The arrow length is proportional to the magnitude of the electric field vector, the components of which are determined using Equation (5.10). The arrows point in the direction of the greatest decrease in potential at that point.

Each gradient vector shown in Figure 5.3 represents the magnitude and direction of the electric field at that location, as given by Equation (5.7) and was calculated manually using Equation (5.10). This is a tedious exercise, but a useful way to check the validity of plots obtained using mathematical software tools such as MATLAB® (MathWorks), *Maple®* (Maplesoft) or *Mathematica®* (Wolfram), for example. Relevant examples of the application of MATLAB® and *Maple®* are given in Box 5.2

5.3 Applying Laplace's Equation

The examples we have considered so far for determining electrostatic potentials and fields have involved known distributions of point charges (e.g., uniform charge distributions on spheres, large area flat plates and long wires) or electrode geometries having analytical solutions for their electric fields (e.g., parallel plate or concentric spherical electrodes, coaxial electrodes). But how do we calculate the potentials and fields produced by unknown charge distributions or arbitrary-shaped electrodes? The answer to such questions is that we are required to find a solution to the appropriate form of either Poisson's or Laplace's equation – bearing in mind that these equations are only valid for electrostatic or quasi-electrostatic situations where time-varying magnetic fields do not play a role. The field is then fully determined by the relationship $E = -\nabla V$.

Box 5.2 Gradient Field Plots using MATLAB® and Maple® Software

The following is a MATLAB program to give a contour plot of a vector field:

```
% Calculate the gradient in the xy-plane of
the potential function V(x, y) = x^2 + y^2
    v = -6:1:6;
    [x,y] = meshgrid(v);
    z = x.^2 y.^2 ;
    [px, py] = gradient(z, 2, 2,);

% Plot equipotential contours and vector
arrows (quivers) together on same plot
    contour(v,v,z)

    holdon
    quiver(v,v,px,py)
    holdoff
```

The resulting plot, with scaling, of the gradient field for the function $V(x, y) = x^2 + y^2$ is shown in Figure 5.4 (left-hand side). Although the vectors do not have the correct lengths, they do have the correct *relative* lengths compared to the other vectors in the gradient field.

The following command in *Maple*:

```
plot3d(x^2+y^2, x=-1..1, y=-1..1);
```

produces the 3D plot of the surface of the vector field shown on the right hand side of Figure 5.4. The 2D contour plot is added to show how the surface plot corresponds with the 2D gradient field plot on the left:

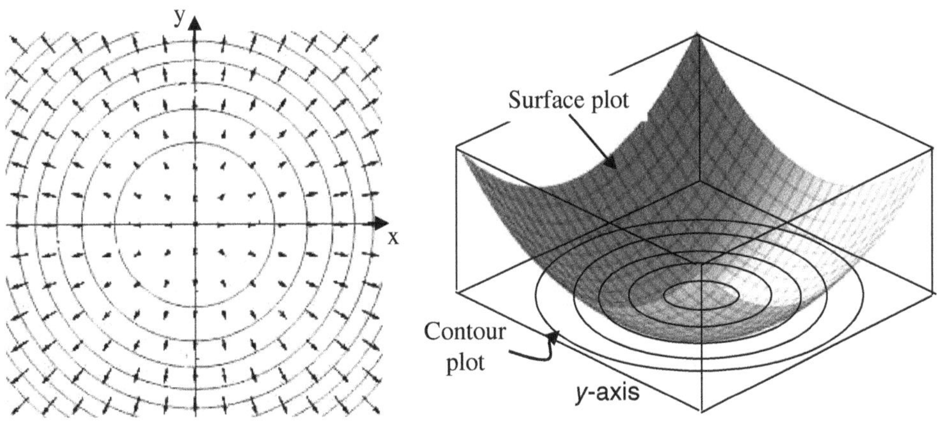

Figure 5.4

Poisson's equation was introduced in Chapter 3 and given as:

$$\nabla^2 V = \left(\frac{\partial}{\partial x^2} + \frac{\partial}{\partial y^2} + \frac{\partial}{\partial z^2}\right) V = -\frac{\rho}{\varepsilon_o}$$

For regions of space without a net charge density ρ this becomes Laplace's equation (see Box 3.2):

$$\nabla^2 V = \left(\frac{\partial}{\partial x^2} + \frac{\partial}{\partial y^2} + \frac{\partial}{\partial z^2}\right) V = 0$$

For many applications of dielectrophoresis we are concerned with solving Laplace's equation in regions of space where the potential V results from charge distributions on metal electrodes some distance away from our region of interest.

5.3.1 Laplace's Equation in One Dimension

It is instructive to consider the trivial case of where the potential V depends on only a single variable. For example, this variable could be the dimension x. In this case we write Laplace's equation as:

$$\frac{\partial^2 V}{\partial x^2} = 0$$

An analytical solution for this is readily obtained in the form

$$V(x) = ax + b$$

Values for a and b are obtained from knowledge of the imposed boundary conditions. For example, if for $x = 1$ and 10 we are given $V(x) = 3$ V and 21 V, respectively, then for $10 > x > 1$ the unique solution has $a = 2$ V,

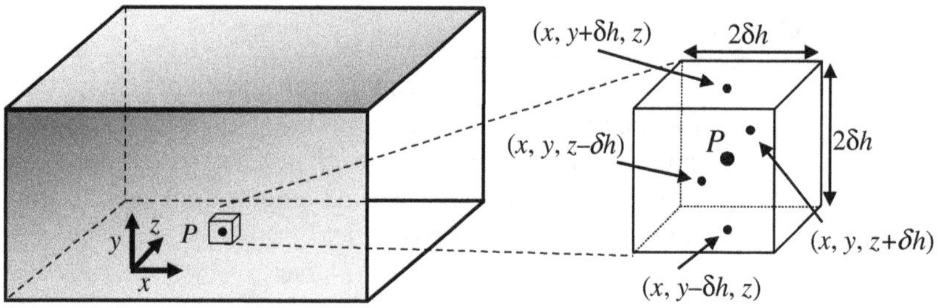

Figure 5.5 A flat metal plate is bent to form an electrode on the front face and top face of a long rectangular tube. Away from the open ends the electric potential at any point *P* inside the tube varies along the *y*- and *z*-directions, but can be considered to be independent of distance along its length (the *x*-direction). To find the potential at *P* we construct around it a Gaussian surface in the form of a small cube of side $2\delta h$.

$b = 1$ V. The general solution takes the form of a flat sloping surface. This simple example demonstrates some important properties, namely:

- Practical problems involving Laplace's equation are boundary value problems in a particular volume or region *T* having a well defined boundary surface *S* (see Box 3.2).
- At any point x the potential $V(x)$ is the average of the two potentials at equidistant points $\pm\delta$ on either side of x within the region of interest:

$$V(x) = \frac{1}{2}[V(x-\delta) + V(x+\delta)]$$

- As a consequence of this last property there are no local maxima or minima of $V(x)$ within the region of interest.

5.3.2 Laplace's Equation in Two and Three Dimensions

For the 2D situation where the potential V is independent of x, but varies with y and z, we are required to solve a partial differential equation of the form:

$$\frac{\partial^2 V}{\partial y^2} + \frac{\partial^2 V}{\partial z^2} = 0$$

For the more general 3D case:

$$\frac{\partial^2 V}{\partial x^2} + \frac{\partial^2 V}{\partial y^2} + \frac{\partial^2 V}{\partial z^2} = 0$$

These equations have far more complicated solutions than that of a sloping flat surface obtained for the one-dimensional situation. However, the three main properties listed for the one-dimensional situation also apply. In two dimensions, at any point (x, y), the potential $V(x, y)$ is the average of the potentials on equidistant points on a circle drawn around that point. Mathematically this is expressed as the summation of all the incremental potential values around the circle, divided by the circumference of that circle:

$$V(x,y) = \frac{1}{2\pi r}\oint_C V\,dl$$

In three dimensions the potential $V(r)$ is the average of the potentials on equidistant points around a sphere S of radius r. Mathematically this is expressed as the summation of all the potentials per unit elemental area around the surface of the sphere, divided by the surface area of that sphere:

$$V(r) = \frac{1}{4\pi r^2}\oint_S V\,dA$$

As an example of these properties, consider the arrangement shown in Figure 5.5, where a metal plate is bent to form the front and top faces of a long rectangular tube. The plate is charged to a known potential, as, for example, by connecting it to a terminal of a battery, with respect to a conducting surface that forms the back and bottom faces of the tube. We construct around point P a small cubic Gaussian surface, as described in section 3.3 of Chapter 3 and depicted in Figure 5.5.

A short section only of the rectangular tube is shown in Figure 5.5. By choosing locations P within the tube, far away from its ends, we simplify our analysis to that of a two-dimensional rather than a three-dimensional problem. This is sufficient to demonstrate the general principle to be outlined. The charged electrode plate generates a potential whose magnitude inside our chosen section of the tube depends only on the y- and z-coordinates in a plane that is perpendicular to the x-axis. For fixed values of coordinates y and z the potential V does not depend on location along the x-axis. In other words, the partial differential $\partial V/\partial x$ is equal to zero. From Equation (5.4) the x-component of the field is given by $E_x = -\partial V/\partial x$ and so E_x is zero. We need only consider the y- and z-components of the field.

The cubic Gaussian surface drawn in Figure 5.5, centred on location P, is assumed to contain no charge. The total electric flux through this cubic surface is therefore zero (refer to section 3.3 of Chapter 3). Furthermore, because E_x is zero there is no flux through the two faces of the cubic Gaussian surface that are parallel to the yz-plane. If we make the length $2\delta h$ of each edge of the cube small enough, we can to a good approximation assume that the fluxes through the other four faces of the cube are equal to the product of the area $(2\delta h)^2$ and normal component of the field to that surface. An analysis of this, based on application of Gauss's Law, is given in Box 5.3. The result given by Equation (5.11) can be summarized by

Box 5.3 The Potential at any Point in Space is the Average of the Potentials that Surround It

The cubic Gaussian surface drawn in Figure 5.4 contains no charge and because the x-component of the field is zero, the total flux Φ_T through the surface is given by:

$$\Phi_T = E_y(x, y + \partial h, z)(2\partial h)^2 + [-E_y(x, y - \partial h, z)(2\partial h)^2] \\ + E_z(x, y, z + \partial h)(2\partial h)^2 + [-E_z(x, y, z - \partial h)(2\partial h)^2 = 0 \quad (5.11)$$

Using the identities given in Equation (5.4) we can express the field components as:

$$E_y(x, y + \partial h, z) = -\partial V(y + \partial h, z)/\partial y = -[V(y + \partial h, z) - V(y, z)]/\partial h$$
$$E_y(x, y - \partial h, z) = -\partial V(y - \partial h, z)/\partial y = -[V(y, z) - V(y - \partial h, z)]/\partial h$$
$$E_z(x, y, z + \partial h) = -\partial V(y, z + \partial h)/\partial y = -[V(y, z + \partial h) - V(y, z)]/\partial h$$
$$E_z(x, y, z - \partial h) = -\partial V(y, z - \partial h)/\partial y = -[V(y, z) - V(y, z - \partial h)]/\partial h$$

On substituting these identities into Equation (5.11) and solving for $V(y, z)$ we obtain (after dividing throughout by $4\partial h$) the result:

$$V(y, z) = [V(y + \partial h, z) + V(y - \partial h, z) + V(y, z + \partial h) \\ + V(y, z - \partial h)]/4 \quad (5.12)$$

Therefore, the potential at a point within a small volume of space that contains no charge is equal to the average of the potentials that surround this point. In the limit as the characteristic dimension ∂h of this volume becomes infinitesimally small, this result is exact.

the statement: *in a sufficiently small volume of space that contains no charge, the potential at the centre of this volume is equal to the average of the surrounding potentials.*

A simple example of how to apply this result is to consider the case of two parallel plate electrodes and to construct in the space between them a rectangular array of square elemental cells. The location of each cell is identified using an alphanumerical code as shown in Figure 5.6. The points of intersection (nodes) of each horizontal and vertical line in the array form a rectangular grid of points. The potential value at each node in relation to the four nodes around it is given by Equation (5.12) in Box 5.3. The potentials of the nodes in the two electrodes are specified

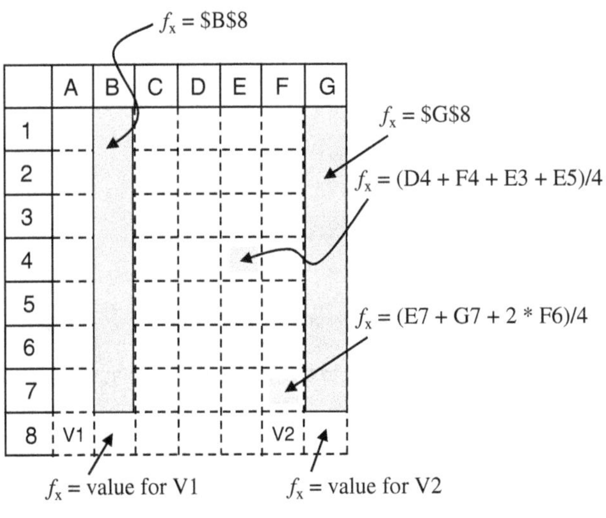

Figure 5.6 A finite-element simulation, using the Microsoft Office Excel spreadsheet, to calculate the potentials generated between two large area parallel electrodes of potentials V1 and V2. Electrode potentials of any chosen values can be inserted into the formula bar by entering them into cells B8 and G8. The function f_x is entered for each cell in the form of Equation (5.12) given in Box 5.3. The function entered for rows 1 and 7 treats the electrodes as being of large area to produce a uniform electric field.

and so these can be used as known values in a set of simultaneous equations of the form of Equation (5.12) for nodes nearest to the electrodes. An iterative method is then used by repeatedly calculating values for the potentials at all the nodes using the complete set of simultaneous equations until all the solutions converge to a single value for each node. This procedure is known as the *finite difference method* and various programming languages are available to perform it. However, most spreadsheets can automatically handle the iterative solutions of simultaneous equations for grids containing a modest number of nodes. An example of a simulation using *Microsoft Office Excel* is shown in Figure 5.7 for the simple case of two parallel plate electrodes having potentials of 7 V and 3 V. The grid of points is assumed to be located in the central section of the space between the electrodes, well away from fringing electric fields at the electrode ends. The potential down any vertical row of cells (e.g., B1–B10) is therefore assumed to be an equipotential, to give the expected uniform field between the parallel electrodes. The functions entered for the cells along rows 1 and 10 were therefore modified to be of the form shown for cell F7 in Figure 5.6. For all other cells (those in rows 2–9) functions of the form shown for cell 4E in Figure 5.6 were used.

The distribution of potentials produced in the yz-plane of the interior of the rectangular tube shown in Figure 5.5 is presented in Figure 5.8. The metal plate has been assigned a potential of +12 V and the other two sides of the tube have been grounded at zero volts. The potentials decrease steadily, from values approaching 12 V at the corner of the bent metal sheet, to zero volts at the corner formed by the other two sides of the tube. These other two sides of the tube are assumed to have conducting surfaces and to be grounded at earth potential.

Three of the square meshes (cells) close to the top left of those shown in Figure 5.8 are highlighted in Figure 5.9. By taking the differences in the potential values of adjacent

	A	B	C	D	E	F	G	H	I	J	K	L	M
1	7	6.67	6.33	6.00	5.67	5.33	5.00	4.67	4.33	4.00	3.67	3.33	3
2	7	6.67	6.33	6.00	5.67	5.33	5.00	4.67	4.33	4.00	3.67	3.33	3
3	7	6.67	6.33	6.00	5.67	5.33	5.00	4.67	4.33	4.00	3.67	3.33	3
4	7	6.67	6.33	6.00	5.67	5.33	5.00	4.67	4.33	4.00	3.67	3.33	3
5	7	6.67	6.33	6.00	5.67	5.33	5.00	4.67	4.33	4.00	3.67	3.33	3
6	7	6.67	6.33	6.00	5.67	5.33	5.00	4.67	4.33	4.00	3.67	3.33	3
7	7	6.67	6.33	6.00	5.67	5.33	5.00	4.67	4.33	4.00	3.67	3.33	3
8	7	6.67	6.33	6.00	5.67	5.33	5.00	4.67	4.33	4.00	3.67	3.33	3
9	7	6.67	6.33	6.00	5.67	5.33	5.00	4.67	4.33	4.00	3.67	3.33	3
10	7	6.67	6.33	6.00	5.67	5.33	5.00	4.67	4.33	4.00	3.67	3.33	3

Figure 5.7 An example of a finite difference simulation performed using the spreadsheet scheme shown in Figure 5.6. Potential values of 7 V and 3 V are entered into the shaded columns representing two parallel plate electrodes. The electrodes are assumed to extend way beyond the 'solution space' shown here between them.

cells, the following values for ΔV_z and ΔV_y are obtained:

$$\Delta V_z = (4.58 - 6.05) = -1.47 \text{ V}; \Delta V_y = (3.65 - 6.05) = -2.40 \text{ V}$$

The sides Δz and Δy of each square mesh depend on the size of the rectangular tube shown in Figure 5.8 and modelled by the results shown in Figure 5.8. If we take the inside dimensions (z, x, y) of the tube to be 1.1 cm × 1.0 cm, then $\Delta z = \Delta y = 0.1$ cm. We also assume that these values for Δz and Δy are small enough for the approximations to be made that

$$\frac{\Delta V_z}{\Delta z} = \frac{\partial V}{\partial z}; \quad \frac{\Delta V_y}{\Delta y} = \frac{\partial V}{\partial y}$$

	0	0	0	0	0	0	0	0	0	0	0	
12	6.01	3.65	2.54	1.92	1.53	1.24	1.01	0.80	0.60	0.40	0.20	0
12	8.40	6.05	4.58	3.62	2.95	2.43	1.99	1.59	1.20	0.81	0.41	0
12	9.54	7.57	6.11	5.04	4.21	3.53	2.93	2.37	1.80	1.22	0.62	0
12	10.18	8.58	7.27	6.20	5.32	4.56	3.85	3.15	2.42	1.65	0.84	0
12	10.60	9.30	8.17	7.19	6.32	5.53	4.75	3.95	3.08	2.13	1.09	0
12	10.91	9.87	8.92	8.05	7.25	6.47	5.68	4.81	3.83	2.70	1.40	0
12	11.16	10.35	9.58	8.85	8.15	7.44	6.68	5.80	4.73	3.42	1.82	0
12	11.38	10.78	10.20	9.63	9.07	8.47	7.79	6.96	5.89	4.43	2.46	0
12	11.59	11.19	10.80	10.41	10.01	9.57	9.05	8.38	7.42	5.95	3.60	0
12	11.80	11.60	11.40	11.20	10.99	10.76	10.47	10.08	9.46	8.35	5.99	0
	12	12	12	12	12	12	12	12	12	12	12	

Figure 5.8 The solved potential values are shown for a regular distribution of nodes across the yz-plane within the tube shown in Figure 5.5. The metal plate has a potential of 12 V and the other two conducting surfaces of the tube are grounded at zero volts. The finite difference calculations were performed using Microsoft Office Excel, using the spreadsheet format and functions shown in Figure 5.6.

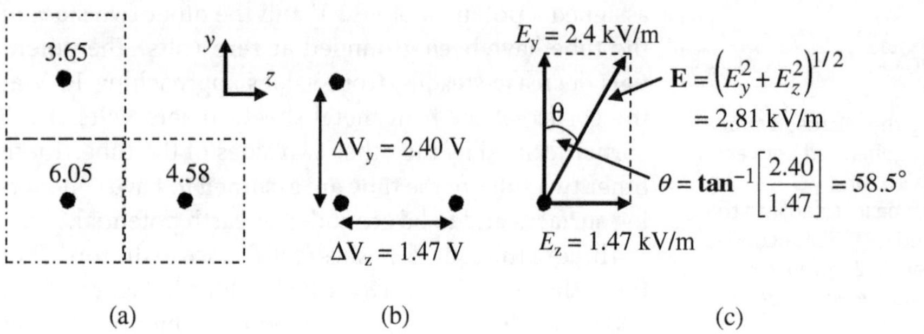

Figure 5.9 (a) Three cells are highlighted from near the top left of the spreadsheet of potential values shown in Figure 5.8. (b) Values of the corresponding potential differences ΔV_z and ΔV_y. (c) Values derived for the local field components E_y and E_z (assuming each cell is a square of side 0.1 cm) and the magnitude and orientation θ of the resultant field vector **E**.

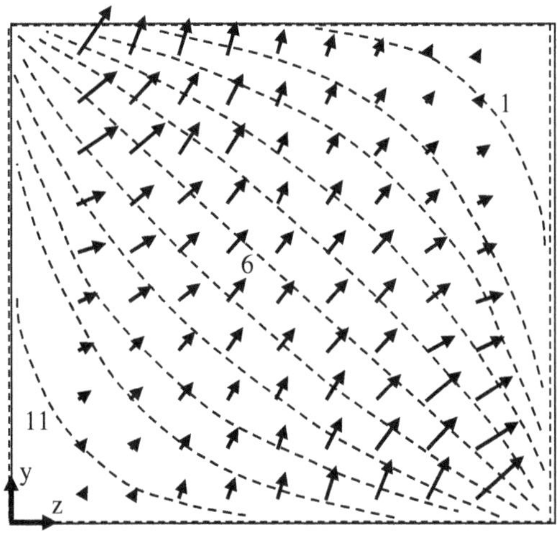

Figure 5.10 The equipotential contours (dotted lines) at 1 V intervals, corresponding to the distribution of potentials shown in the yz-plane in Figure 5.8. The resultant electric field vectors for each cell, calculated according to the procedures illustrated in Figure 5.9, are also shown. The largest field vectors occur at the top left and bottom right corners where the grounded surfaces and the 12 V biased metal plate are closest together.

From Equation (5.4) the *y*- and *z*-components of the local field vector E can then be approximated as:

$$E_y = -\frac{\Delta V_y}{\Delta y}; \quad E_z = -\frac{\Delta V_z}{\Delta z}$$

These are the relationships used to give the values for E_y and E_z shown in Figure 5.9(c). The vector addition of E_y and E_z gives the resultant field vector E:

$$E = \sqrt{E_y^2 + E_z^2}$$

This field vector is oriented at angle θ given by:

$$\theta = \tan^{-1}\frac{E_z}{E_y}$$

A map of the local field vectors, represented as arrows of lengths proportional to their magnitudes and directed at θ degrees to the *y*-axis, is shown in Figure 5.10. The bases of the arrows are located at the centre of each cell. The field vectors are superimposed onto the equipotential contours across the *yz*-plane. In close proximity to the sides of the tube, the equipotentials are nearly parallel to these surfaces. The smallest separations between the equipotential contours (and hence the largest field vectors) occur at the corners where the grounded conducting sides of the tube meet (but do not touch) the two sides formed by the charged metal plate. The field vectors cross equipotential contours at right angles.

Modelling electric potentials and fields using the finite difference method has an advantage in that the computational programs are simple to formulate [1] or are readily available as commercial software packages. A major disadvantage is that only regularly spaced nodes and well defined meshes and geometrical elements (e.g., squares, rectangles, cubes) can be used and these cannot readily be adjusted to accommodate regions that contain both fairly uniform and highly nonuniform fields. For this reason another method, known as the *finite element method*, has for some time been widely adopted. This is well described in text books [2, 3], including one that can be downloaded for free from the Internet [4]. In the finite element method the elements can be of general shapes, such as irregular triangles or tetrahedrons. Automated meshing routines give appropriately small element sizes where the field is highly nonuniform, such as at sharp electrode edges or at the ends of pin electrodes and larger elements where the electric field is relatively uniform. A widely used commercial finite element software package is COMSOL *Multiphysics*, which can be interfaced to extensive toolboxes provided by MATLAB. An example of an application of COMSOL Multiphysics is shown in Figure 5.11. An excellent summary of computer applications in electromechanics, of relevance to dielectrophoresis, is given by Hughes [5]

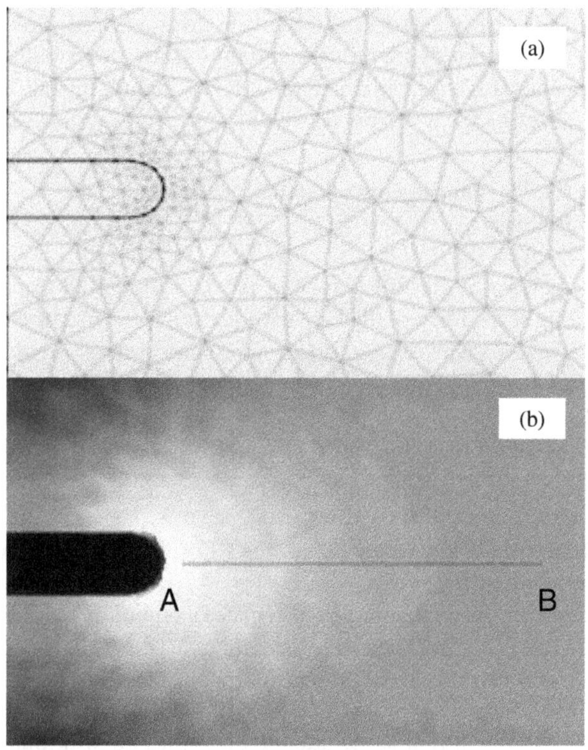

Figure 5.11 (a) An example (using COMSOL Multiphysics) where an automated meshing routine in the finite element method has reduced the sizes of the triangular elements in the region near the end of a pin electrode tip where the electric field is expected to be the largest. (b) A greyscale map of the resulting electrode field. The field is highest in the area shaded white at the electrode tip (A) and steadily decreases with distance towards point B.

Box 5.4 Solving Laplace's Equation by Separation of Variables

Consider the two dimensional form of Laplace's equation

$$\frac{\partial^2 V}{\partial x^2} + \frac{\partial^2 V}{\partial y^2} = 0$$

We *assume* that V is a product of two functions $X(x)$ and $Y(y)$ that depend only on x and y, respectively. Thus, $\partial^2 V/\partial x^2 = Yd^2V/dx^2$ and $\partial^2 V/\partial y^2 = Xd^2V/dy^2$. Substituting these equalities into the Laplace equation and dividing throughout by XY, we obtain a standard differential equation (i.e., one that does not contain *partial* differentials):

$$\frac{1}{X}\frac{d^2X}{dx^2} + \frac{1}{Y}\frac{d^2Y}{dy^2} = 0$$

For this equation to hold requires the two terms to be proportional to constants K_1 and K_2, with $K_1 + K_2 = 0$. Choosing $K_1 = -K_2 = k^2$, we have two separate linear differential equations:

$$\frac{d^2X}{dx^2} = k^2X \quad \text{and} \quad \frac{d^2Y}{dy^2} = -k^2Y$$

These have solutions of the form [7, p. 589]:

$$X(x) = Ae^{kx} + Be^{-kx} \quad \text{and} \quad Y(y) = C\sin ky + D\cos ky$$

The potential V thus takes the form:

$$V(x,y) = X(x)Y(y) = \left(Ae^{kx} + Be^{-kx}\right)\left(C\sin ky + D\cos ky\right)$$

A fit can be made with the prescribed values of the solution V or some of its derivatives on the boundary surface S or curve C by choosing any superposition of V_k with appropriate values for k. If the boundary condition involves V being prescribed on a surface the problem is called a *Dirichlet problem* and the surface a *Dirichlet surface.* If the boundary condition is given in terms of the derivative $\partial V/\partial n$ of V normal to the surface S, the problem is called a *Neumann problem*. The resulting problem is called a *mixed problem* if V is prescribed on one part of S and $\partial V/\partial n$ on the other part of S.

The method of separation of variables is not restricted to Cartesian coordinates as given here – it can also be applied for the cases of the spherical and cylindrical coordinate schemes described in Box 3.2.

and is also treated in the book by Morgan and Green [6]. The underlying numerical methods are described in various advanced mathematical texts, including that by Kreyszig [7, Ch. 19, pp. 942–988].

5.3.3 Solving Laplace's Equation by Separation of Variables

A general method for solving partial differential equations of the form of the Laplace equation is the method of *separating variables* – also known as the *product method* [7, Ch. 11, pp. 582–649]. The key step is to *assume* that the solution for the potential V is the product of functions, with each function depending on one variable only. An outline of this is described in Box 5.4.

Example 5.4 Solving Laplace's Equation in Spherical Coordinates

Derive a general solution of Laplace's equation that can be used to solve the electrostatic potential $V(r, \theta, \phi)$ in a situation where there is spherical symmetry. Restrict the analysis to cases where the prescribed boundary condition of the potential maintained on a sphere S of radius R is independent of the spherical coordinate θ.

Solution 5.4 We have two boundary conditions. The first one represents the fact that the potential on the surface of a sphere, cantered at $r = 0$ and of radius R, is maintained at a potential whose value is independent of the spherical coordinate θ. We can write this boundary condition as:

$$V(R,\theta,\phi) = f(\phi) \quad (r = R) \tag{5.13}$$

The second boundary condition recognizes the fact that at an infinite distance away from the sphere the potential will be zero. This can be written as:

$$\mathrm{Lim}_{r\to\infty} V(r,\phi) = 0 \tag{5.14}$$

From Box 3.2 the Laplace equation to be solved for $V(r, \phi)$ is:

$$\frac{\partial}{\partial r}\left(r^2\frac{\partial V}{\partial r}\right) + \frac{1}{\sin\phi}\frac{\partial}{\partial\phi}\left(\sin\phi\frac{\partial V}{\partial\phi}\right) = 0 \tag{5.15}$$

We assume a solution of the form:

$$V(r,\phi) = R(r)\Phi(\phi)$$

Employing the method of separation of variables, we substitute the equalities $\partial V/\partial r = \Phi\, dV/dr$ and $\partial V/\partial\phi = R\, dV/d\phi$ into Equation (5.15), then divide throughout by $R\Phi$ to obtain:

$$\frac{1}{R}\frac{d}{dr}\left(r^2\frac{dR}{dr}\right) + \frac{1}{\Phi\sin\phi}\frac{d}{d\phi}\left(\sin\phi\frac{d\Phi}{d\phi}\right) = 0$$

This gives the two ordinary differential equations [7, p. 638]:

$$\frac{d}{dr}\left(r^2\frac{dR}{dr}\right) = n(n+1)R \tag{5.16}$$

$$\frac{1}{\sin\phi}\frac{d}{d\phi}\left(\sin\phi\frac{d\Phi}{d\phi}\right) = -n(n+1)\Phi \tag{5.17}$$

Box 5.5 Legendre Polynomials $P_n(x)$

Named after the French mathematician Adrien-Marie Legendre (1752–1833), the Legendre polynomials are employed in the *method of power series* to solve equations (known as Legendre's equations) that cannot be solved by other methods. The Legendre polynomials are of order n, denoted by $P_n(x)$, given as [7, p 208]:

$$P_n(x) = \sum_{m=0}^{M} (-1)^m \frac{(2n-2m)!}{2^n m!\,(n-m)!\,(n-2m)!} x^{n-2m}$$
$$= \frac{(2n)!}{2^n (n!)^2} x^n - \frac{(2n-2)!}{2^n 1!(n-1)!(n-2)!} x^{n-2} + \frac{(2n-2)!}{2^n 2!(n-1)!(n-2)!} x^{n-3} \text{ etc.} \qquad [n = 0, 1, 2\ldots]$$

where $M = n/2$ or $(n-1)/2$, whichever is an integer. Simplification in evaluation of these polynomials is achieved using Rodrigues' formula [7, p 209]:

$$P_n(x) = \frac{1}{2^n n!} \frac{d^n}{dx^n}\left[\left(x^2-1\right)^n\right]$$

Thus: $P_0(x) = 1$, $P_1(x) = x$, $P_2(x) = \frac{1}{2}(3x^2-1)$, $P_3(x) = \frac{1}{2}(5x^3-3x)$

$P_4(x) = \frac{1}{8}(35x^4 - 30x^2 + 3)$, $P_5(x) = \frac{1}{8}\left(63x^5 - 70x^3 + 15x\right)$ etc.

Graphs of these polynomials are given below. They are *normalized* such that $P_n(1) = 1$ and are solutions of the Laplace equation where the imposed boundary conditions require regularity of the solutions at the boundaries. This condition is satisfied [7, p. 237] by the so-called *orthogonality* of the polynomials, such that:

$$\int_{-1}^{1} P_m(x)P_n(x)dx = 0 \ (m \neq n)$$

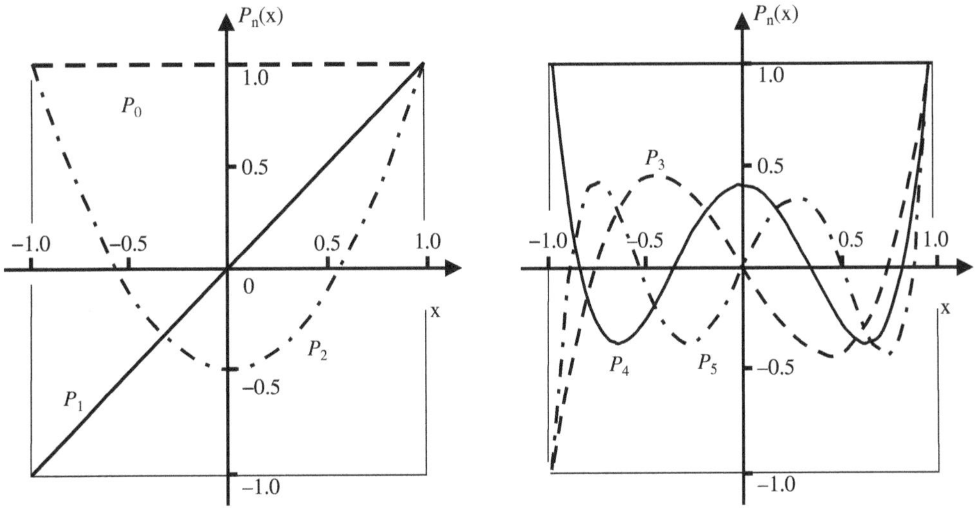

where n is so chosen [7, p. 639–640] to obtain the following two sequences of solutions $V(r, \phi) = R(r)\,\Phi(\phi)$ of the Laplace Equation (5.15):

$$V_n(r,\phi) = A_n r^n P_n(\cos\phi) \quad n = 0, 1, 2\ldots \qquad (5.18a)$$

$$V_n^*(r,\phi) = \frac{B_n}{r^{n+1}} P_n(\cos\phi) \quad n = 0, 1, 2\ldots \qquad (5.18b)$$

where $P_n(\cos\phi)$ are the Legendre polynomials in $\cos\phi$ (see Box 5.5). The absolute value of the coefficients P_n is never greater than unity and so the expanded series converges.

Example 5.5 Hemispherical Electrodes

Two hemispherical, hollow, metal electrodes of radius R are arranged as shown in Figure 5.12. The hemispheres are held in position and separated by a thin electrically insulating gasket. Find the potential inside and outside the hemispheres for the case where the upper one is maintained at a potential of 10 V and the lower one is grounded at zero volts.

Solution 5.5 We have the following boundary conditions for the potential function $f(\phi)$:

$$f(\phi) = 10 \quad (\text{top hemisphere}, r = R, 0 \le \phi \le \pi) \qquad (5.19a)$$

$$f(\phi) = 0 \quad (\text{bottom hemisphere}, r = R, -\pi \le \phi \le 0) \qquad (5.19b)$$

$$\text{Lim}_{r\to\infty} f(\phi) = 0 \quad (\text{outside of the hemispheres}) \qquad (5.19c)$$

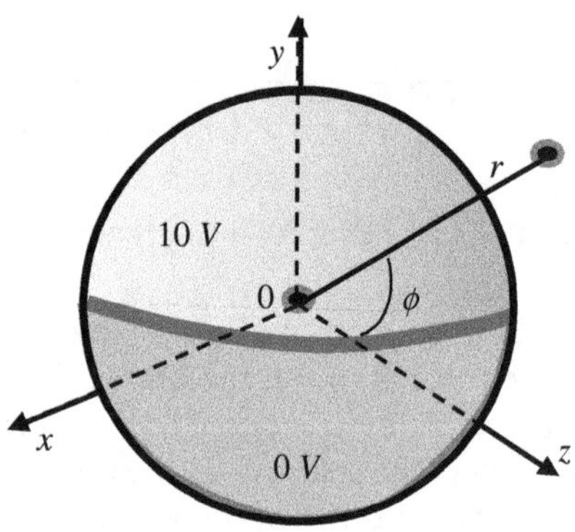

Figure 5.12 The hollow hemispherical electrode geometry analysed in Example 5.5.

Inside the Hemispheres

We know from Chapter 3 (e.g., see Table 3.3) that the electric field inside a charged hollow or solid metal sphere is zero. We can expect the same for the fields in the upper and lower internal regions of the hemispheres shown in Figure 5.12, so that the potentials throughout them will to a very good approximation be constant. It follows that the solution given by Equation (5.18a) is suitable. The coefficients B_n in Equation (5.18b) must be zero for all n – otherwise the potential would head off to infinity at $r = 0$. A solution of the Laplace Equation (5.15) can thus be obtained from the series:

$$V(r,\phi) = \sum_{n=0}^{\infty} A_n r^n P_n(\cos\phi) \tag{5.20}$$

For Equation (5.20) to satisfy the conditions given by Equations (5.19 a and b) then

$$V(R,\phi) = \sum_{n=0}^{\infty} A_n R^n P_n(\cos\phi) = f(\phi) \quad n = 0, 1, 2\ldots \tag{5.21}$$

This series corresponds to the so-called Fourier–Legendre series of $f(\phi)$ [7, p 242]. The coefficients A_nR^n are given by

$$A_n R^n = \frac{2n+1}{2}\int_{-1}^{1} f(w)P_n(w)dw. \tag{5.22}$$

in which we can set $w = \cos\phi$. The limits of integration -1 and 1 for the integral in Equation (5.22) correspond to $\phi = \pm\pi$ and $\phi = 0$, respectively. We thus have the relationships:

$$dw = -\sin\phi d\phi \tag{5.23a}$$

$$\int_{-1}^{1} f(w) = \int_{\pi}^{0} f(\phi) = -\int_{0}^{\pi} f(\phi) \tag{5.23b}$$

From Equations (5.21–5.23) and employing the expression for the Legendre polynomial P_n given in Box 5.5, we obtain:

$$A_n = \frac{2n+1}{2R^n} f(\phi)\int_{0}^{\pi} P_n(\cos\phi)\sin\phi d\phi \quad n = 0, 1, 2\ldots$$

$$= \frac{(2n+1)}{2R^n} f(\phi)\sum_{m=0}^{M}(-1)^m \frac{(2n-2m)!}{2^n m!(n-m)!(n-2m)!}\int_{-1}^{1} w^{n-2m}dw$$

where $M = n/2$ or $(n-1)/2$, whichever is an integer. In this series

$$\int_{-1}^{1} w^{n-2m}dw = \frac{2}{(n-2m+1)}$$

so that

$$A_n = \frac{(2n+1)}{R^n 2^n} f(\phi)\sum_{m=0}^{M}(-1)^m \frac{(2n-2m)!}{m!(n-m)!\,(n-2m+1)!} \tag{5.24}$$

Taking note that $(-1)^0 = 1$ and $0! = 1$, we obtain the following values for A_n:

$$n = 0:\quad A_0 = f(\phi)\frac{0!}{0!0!1!} = f(\phi)$$

$$n = 1:\quad A_1 = \frac{3f(\phi)}{2R}\frac{2!}{0!1!2!} = \frac{3}{2R}f(\phi)$$

$$n = 2:\quad A_2 = \frac{5f(\phi)}{4R^2}\left[\frac{4!}{0!2!3!} - \frac{2!}{1!1!1!}\right] = 0$$

$$n = 3:\quad A_3 = \frac{7f(\phi)}{8R^3}\left[\frac{6!}{0!3!4!} - \frac{4!}{1!2!2!}\right] = -\frac{7f(\phi)}{8R^3}$$

$$n = 4:\quad A_4 = \frac{9f(\phi)}{16R^4}\left[\frac{8!}{0!4!5!} - \frac{6!}{1!3!3!} + \frac{4!}{2!2!1!}\right] = 0$$

$$n = 5:\quad A_5 = \frac{11f(\phi)}{32R^5}\left[\frac{10!}{0!5!6!} - \frac{8!}{1!4!4!} + \frac{6!}{2!3!2!}\right] = \frac{11f(\phi)}{16R^5}$$

From Equation (5.20) the potentials within the hemispheres are thus given by:

$$V(r,\phi) = \sum_{n=0}^{\infty} A_n r^n P_n(\cos\phi)$$

$$= f(\phi)\left[P_0(\cos\phi) + \frac{3r}{2R}P_1(\cos\phi) - \frac{7r^3}{8R^3}P_3(\cos\phi) + \frac{11r^5}{16R^5}P_5(\cos\phi) - \cdots\right]$$

For the top hemisphere, ($r < R$, $0 \le \phi \le \pi$) $f(\phi) = 10$ V and, using the values given in Box 5.5 for the Legendre polynomials:

$$V(r,\phi) = 10 + 15\frac{r}{R}\cos\phi - \frac{35r^3}{8R^3}\left[5\cos^3\phi - 3\cos\varphi\right] + \frac{55r^5}{64R^5}\left[63\cos^5\phi - 70\cos^3\varphi + 15\cos\phi\right] - \cdots \tag{5.25}$$

For ϕ approaching $\pi/2$ (i.e., in the upper internal regions of the top hemisphere) the potential does not deviate much from the applied value of 10 V. Values of the potential in other regions of the top hemisphere can be obtained to the required accuracy by computing partial sums of the series given in Equation (5.25).

For the bottom hemisphere ($r < R$, $\pi/2 \leq \phi \leq 0$) $f(\phi) =$ 0 V and the potential is everywhere zero.

Outside the Hemispheres

The solution given by Equation (5.18b) for $r > R$ satisfies the boundary condition given by Equation (5.19c):

$$V_n^*(r, \phi) = \frac{B_n}{r^{n+1}} P_n(\cos\phi) \quad n = 0, 1, 2 \ldots$$

The coefficients B_n are given by:

$$B_n = \frac{2n+1}{2} R^{n+1} f(\phi) \int_0^\pi P_n(\cos\phi) \sin\phi d\phi$$

Proceeding as for Equation (5.22) we obtain the potential outside the hemispheres as

$$\begin{aligned} V(r, \phi) &= \sum_{n=0}^{\infty} B_n r^n P_n(\cos\phi) \\ &= f(\phi) \left[\frac{R}{r} P_0(\cos\phi) + \frac{3R^2}{2r^2} P_1(\cos\phi) - \frac{7R^4}{8r^4} P_3(\cos\phi) \right. \\ &\quad \left. + \frac{11R^6}{16r^6} P_5(\cos\phi) - \cdots \right] \\ &= \frac{f(\phi)R}{r} + \frac{3f(\phi)R^2}{2r^2} (\cos\phi) \\ &\quad - \frac{7f(\phi)R^4}{16r^4} \left[5\cos^3\phi - 3\cos\phi\right] + \cdots \end{aligned} \quad (5.26)$$

At large distances ($r \gg R$) from the electrodes, Equation (5.26) approximates as:

$$V(r, \phi) \approx \frac{f(\phi)R}{r}$$

which corresponds to the electrodes being treated as a point charge (see Equation (4.23) and Example 4.5)

5.3.4 Multipole Expansion of a Potential

A general scheme for deriving the potential due to a distribution of point charges is shown in Figure 5.13. A volumetric element d^3r' of charge density $\rho(r')$ is located at a distance r' along the z-axis of a polar coordinate system. The potential $V(r)$ of this charge is to be determined from an observation point P located at a distance r from the origin of the coordinate system. The length of the vector between P and $\rho(r')$ is given by $|r - r'|$.

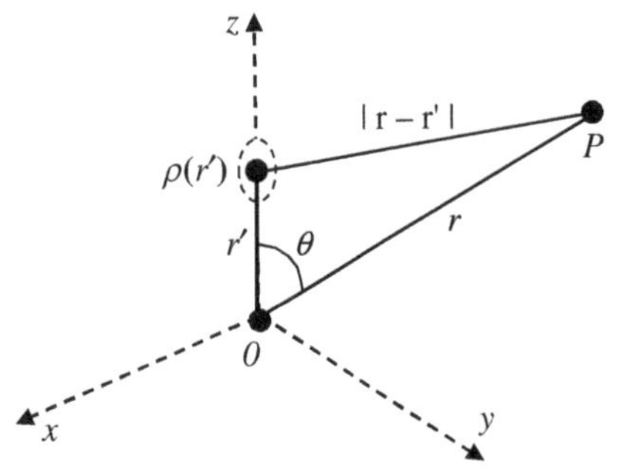

Figure 5.13 A region of charge $\rho(r')$ is located on the z-axis, distance r' from the origin 0 of a polar coordinate system. The potential $V(r)$ is to be determined from the observation point P located at distance r from the origin. The length of the vector between P and the charge is given by $|r - r'|$.

Following the form of Equation (4.18) we can write for the potential $V(r)$:

$$V(r, \theta) = \frac{1}{4\pi\varepsilon_0\varepsilon_r} \int \frac{\rho(r')}{|r - r'|} d^3r' \quad (5.27)$$

Referring to Figure 5.13, from the cosine rule for the sides of a triangle:

$$\begin{aligned} |r - r'| &= \sqrt{r^2 + r'^2 - 2rr'\cos\theta} \\ &= r\sqrt{\left(1 + \frac{r'^2}{r^2} - 2\frac{r'}{r}\cos\theta\right)} \end{aligned}$$

and

$$\frac{1}{|r - r'|} = \frac{1}{r} \frac{1}{\sqrt{\left(1 + \frac{r'^2}{r^2} - 2\frac{r'}{r}\cos\theta\right)}}$$

The function r'/r inside the square root of this equation is a *generating function* of the Legendre polynomials [7, pp. 209–210] so that:

$$\frac{1}{|r - r'|} = \frac{1}{r} \sum_{n=0}^{\infty} P_n(\cos\theta) \left(\frac{r'}{r}\right)^n \quad (r > r') \quad (5.28)$$

The coefficients of r'/r are thus the Legendre polynomials in $\cos\theta$ (see Box 5.5). Equation (5.27) becomes:

$$V(r, \theta) = \frac{1}{4\pi\varepsilon_0\varepsilon_r} \sum_{n=0}^{\infty} \frac{1}{r^{n+1}} \int \rho(r') P_n(\cos\theta) r'^n d^3r' \quad (5.29)$$

5.3.4.1 $n = 0$ Term (*Monopole Term*)

From Box 5.5 we have $P_n(\cos\theta) = 1$, so that for $n = 0$

$$V(r, \theta) = \frac{1}{4\pi\varepsilon_0\varepsilon_r} \frac{1}{r} \int \rho(r') d^3r'$$

This is equivalent to result given by Equation (4.19). The integral $\int \rho(r')d^3r'$ is the net total charge q of the volumetric charge distribution and so

$$V(r,\theta) = \frac{q}{4\pi\varepsilon_0\varepsilon_r}\frac{1}{r} \tag{5.30}$$

This describes the coulomb potential of a point charge q located at the origin, where $r = 0$. We say that a point charge at the origin represents a *singularity* because this implies an infinite charge density and potential. The $n = 0$ term is called the *monopole term.*

If instead of the charge being concentrated within a small volume element, it is distributed continuously along a length l of the z-axis with a density $\rho(z)$, the potential at a sufficiently great distance from the origin can be expressed as:

$$V(r,\theta) = \sum_{n=0}^{\infty} V_n = \frac{1}{4\pi\varepsilon_0\varepsilon_r}\sum_{n=0}^{\infty}\frac{1}{r^{n+1}}\int_0^l \rho(z)z^n dz$$
$$\int P_n(\cos\theta) \quad (r > l)$$

The first term of the expansion of this series (given by $P_0 = 1$) is:

$$V_0 = \frac{1}{4\pi\varepsilon_0\varepsilon_r}\frac{1}{r}\int_0^l \rho(z)dz = \frac{q}{4\pi\varepsilon_0\varepsilon_r}\frac{1}{r} \tag{5.31}$$

where q is now the total charge along the line.

5.3.4.2 $n = 1$ Term (*Dipole Term*)

The net charge q in Equation (5.31) can have either a positive or negative value. The net charge could also be zero, so that:

$$q = \int_0^l \rho(z)dz = 0$$

The $n = 0$ term is thus zero, so that the dominant term in Equation (5.29) for $r \gg l$ is:

$$V_1 = \frac{1}{4\pi\varepsilon_0\varepsilon_r}\frac{P_1(\cos\theta)}{r^2}\int_0^l \rho(z)zdz = \frac{p}{4\pi\varepsilon_0\varepsilon_r}\frac{\cos\theta}{r^2}$$

The quantity

$$p = \int_0^l \rho(z)zdz$$

is the *dipole moment* of the charge distribution located at points z along the z-axis. In general we can write

$$V_n = \frac{1}{4\pi\varepsilon_0\varepsilon_r}p^{(n)}\frac{P_n(\cos\theta)}{r^{n+1}} \tag{5.32}$$

and define

$$p^{(n)} = \int_0^l \rho(z)z^n dz \tag{5.33}$$

as an axial multipole of nth order. The moment p of a point charge q is thus $p = q$, whilst that for a dipole is $p = qd$, where d is the distance between the dipole's constituent charges $+q$ and $-q$.

Based on Equations (5.30–5.33) we can make the following two statements:

> At sufficiently large distances relative to the radius of a sphere that encompasses an arbitrary distribution of charge, the potential of these charges may be represented approximately by the coulomb potential of a point charge located at the origin.
>
> The potential of an arbitrary distribution of charge along a line is identical, outside a sphere whose diameter coincides with the line, with the potential of multipoles located at the origin.

For $n = 1$ the corresponding term in the potential series given by Equation (5.32) is

$$V_1 = \frac{p}{4\pi\varepsilon_0\varepsilon_r}\frac{\cos\theta}{r^2}$$

In Box 5.6 it is shown that a dipole potential V_1 is produced by locating point charges $+q$ and $-q$ along a section of the z-axis.

Example 5.6 Dipole Moment of an Assembly of Point Charges

Five point charges are located as follows: charge $3q$ at $(1.5d\hat{\imath}, 0, 0)$; $3q$ at $(0, -2d\hat{\jmath}, 0)$; q at $(0, 2d\hat{\jmath}, 0)$; $-q$ at $(d\hat{\imath}, 0, 0)$ and $-4q$ at $(0, -d\hat{\jmath}, 0)$, where $\hat{\imath}$ and $\hat{\jmath}$ are the unit vectors along the x- and y-axes, respectively. Calculate the moment of this charge distribution.

Solution 5.6 For an assembly of n point charges, the moment p is given by

$$p = \sum_{i=1}^{n} q_i r_i = qd(4.5\hat{r} - 6\hat{\jmath} + 2\hat{\jmath} - 1\hat{\imath} + 4\hat{\jmath}) = 3.5qd\,\hat{r}$$

The five point charges thus result in a dipole moment of magnitude 3.5 qd directed along the x-axis. An example of relevance to the dielectric properties of protein molecules is the surface distribution of charges for the cytochrome-c molecule shown in Figure 8.17(b). These charges contribute to the permanent dipole moment of this protein.

Box 5.6 Dipole Moment

The potential of a linear charge distribution can only be represented by a dipole if the net charge is zero and if the point of observation P is a great distance away relative to the length of the charged line. The simplest arrangement is to locate a point charge $+q$ at a position l on the z-axis and an equal but opposite charge $-q$ at the origin, as depicted in the figure below. According to Equations (4.18) and (5.32) the potential at a distant location r is

$$V = \frac{q}{4\pi\varepsilon_0\varepsilon_r}\left(\frac{1}{r_1} - \frac{1}{r}\right) = \frac{ql}{4\pi\varepsilon_0\varepsilon_r}\frac{\cos\theta}{r^2} + \textit{higher order terms}$$

The factor ql is the dipole moment $p^{(1)}$ of this charge configuration. We now perform the mental exercise of reducing the distance l to zero but at the same time increase q so that ql remains constant. In the limit as $l \to 0$ a double-point *singularity* is generated with a potential given everywhere (except at the origin) by:

$$V = \frac{p}{4\pi\varepsilon_0\varepsilon_r}\frac{\cos\theta}{r^2} = -\frac{p}{4\pi\varepsilon_0\varepsilon_r}\frac{\partial}{\partial z}\left(\frac{1}{r}\right)$$

The field of this dipole is cylindrically symmetrical about the z-axis, as depicted in Figure 5.14. In any meridian plane the radial and transverse components, E_r and E_θ, of the field strength are given by:

$$E_r = -\frac{\partial\phi}{\partial r} = \frac{1}{2\pi\varepsilon_0\varepsilon_r}\frac{p\cos\theta}{r^3}$$

$$E_\theta = -\frac{1}{r}\frac{\partial\phi}{\partial\theta} = \frac{1}{4\pi\varepsilon_0\varepsilon_r}\frac{p\sin\theta}{r^3}$$

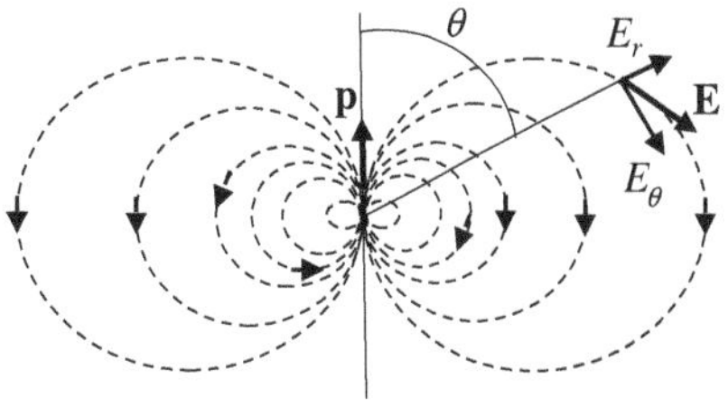

Figure 5.14 Electrical field lines in a meridian plane through the axis of a point dipole of moment **p**. The field **E** is cylindrically symmetrical about the axis of the dipole and at any point is the vector sum of the radial and transverse components, E_r and E_θ, respectively.

Although we have created a *point* dipole moment, it is in fact a *vector* p directed, for our particular case here, along the z-axis.

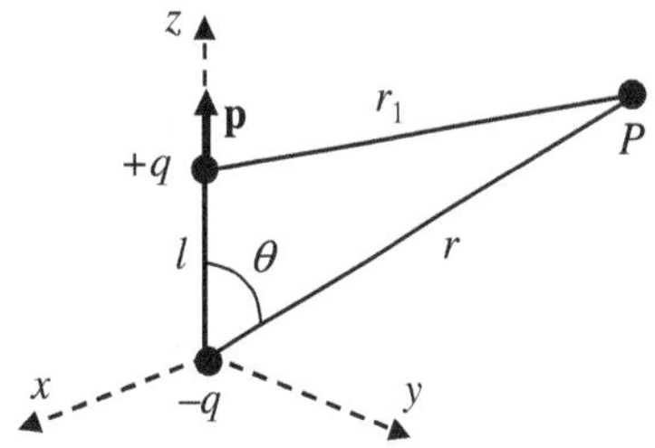

From Chapter 4 the potential energy U of a system consisting of a positive charge q at a location a in an external field of potential V(a) is given by

$$U = qV(a)$$

An easy way to determine the potential of a dipole in an external field is to sum the potential energies of its two point charges. Consider a dipole composed of a point charge $+q$ located at a, separated distance l from a charge $-q$ located at b. The potential energy of the total system is

$$U = qV(a) - qV(a)$$

To create a *point* dipole we bring the two charges closer and closer together. As l tends to zero and at the same time the net charges q tend to infinity, so that the moment ql remains fixed, then

$$U = qdV = ql.\nabla V = -\mathbf{p}\cdot\mathbf{E} = -pE\cos\theta \tag{5.34}$$

where θ is the angle made by the dipole and the external field **E**.

From the concepts described in Chapter 4 the force F exerted on the dipole by the external field is equal to the negative gradient of the potential energy U for a fixed orientation θ of the dipole, so that

$$\mathrm{F} = \nabla(\mathrm{p}\cdot\mathrm{E})_{\theta\mathrm{constant}} \tag{5.35}$$

In a uniform field the net force on the dipole is zero. However, at a fixed location in the field a change in orientation θ will lead to a change of the dipole's potential energy. A torque T will be exerted on the dipole so as to minimize this potential energy and this is given by

$$\mathrm{T} = -\frac{\partial U}{\partial\theta} = -pE\sin\theta \tag{5.36}$$

In vector notation, the vector product of two vectors A and B is called the *cross product*, denoted by A × B. This vector product produces a vector with a direction perpendicular to both **A** and **B** and of magnitude equal to $AB\sin\theta$. Equation (5.36) is therefore expressed vectorially as

$$\mathrm{T} = \mathrm{p}\times\mathrm{E} \tag{5.37}$$

5.3.4.3 The Quadrupole and Octupole

A quadrupole is constructed by locating a negative dipole moment $ql_0 = -p^{(1)}$ at the origin. A second moment

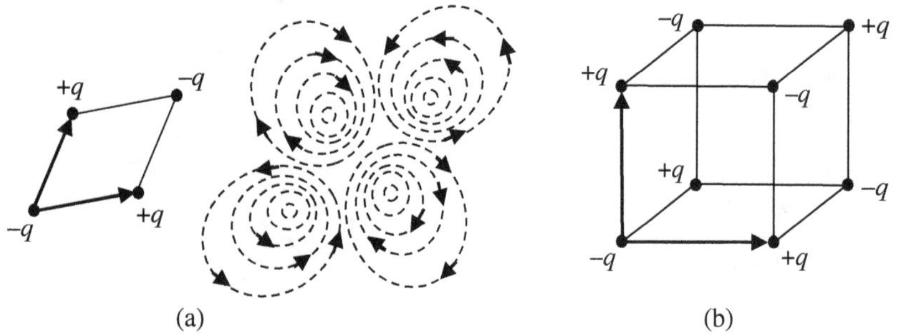

Figure 5.15 Schematic representations of (a) a quadrupole and (b) an octupole.

of equal magnitude but opposite polarity $+p^{(1)}$ is then constructed and displaced from the first moment by a small distance l_1. If the axes of both dipoles and the displacement l_1 are directed along the z-axis, this is known as an *axial quadrupole moment* and is defined as the product

$$p^{(2)} = 2\left(p^{(1)}l_1\right) = 2(ql_0l_1)$$

The *point* quadrupole is generated by letting $l_0 \to 0$, $l_1 \to 0$, $q \to \infty$ whilst maintaining (ql_0l_1) constant. The contribution V_2 to an overall potential from the quadrupole term is

$$V_2 = \frac{1}{4\pi\varepsilon_0\varepsilon_r}p^{(2)}\frac{P_2(\cos\theta)}{r^{2+1}} = \frac{1}{8\pi\varepsilon_0\varepsilon_r}\frac{1}{r^3}\left(3\cos^2\theta - 1\right)$$

A quadrupole can more generally be generated by displacing the second dipole moment from the first one in an arbitrary direction. This is depicted in Figure 5.15. Whereas the dipole moment is a vector, the moment of a quadrupole is a tensor of second rank.

Multipoles of higher order are constructed by placing a multipole of order $n-1$ and of negative moment $-p^{(n-1)}$ at the origin and then placing in an arbitrary direction near it an equal positive multipole $+p^{(n-1)}$. This is shown for an octupole ($n = 3$) in Figure 5.15. We can now add a third statement (theorem) to the two already given, based on Equations (5.30–5.33): 'The potential generated outside an uncharged sphere, by an arbitrary distribution of charges within it, is identical with the potential of a system of multipoles located at its centre.'

In Chapter 10 we will make use of this theorem to refine the equation for the dielectrophoretic force acting on a polarizable particle subjected to an inhomogeneous field. Stratton's classic text, *Electromagnetic Theory* [8, pp. 172–183] can be recommended to those readers wishing to find a more detailed description than that given here of the concept of multipoles. Jones [9] in his book *Electromechanics of Particles* also gives an excellent treatment of how the forces and torques experienced by particles subjected to electric and / or magnetic fields can be understood in terms of multipole theory.

Example 5.7 A Macroscopic (Classic) Dipole Moment

A macroscopic dipole (rather than a point dipole) such as that shown in Figure 4.5 consists of two equal and opposite charges separated a finite distance apart. Figure 5.16 shows a dipole located on the z-axis, with charge $+q$ at $z = d/2$ and $-q$ at $z = -d/2$. Derive a general expression for the potential of this dipole at a distance r from the origin.

Solution 5.7 In Figure 5.16 we define r^+ and r^- to be the vector distances, respectively, from charge q^+ and q^- to the observation point P. The potential $V(r, \theta)$ at point P due to charges q^+ and q^- is

$$V(r,\theta) = \frac{q}{4\pi\varepsilon_0\varepsilon_r}\left(\frac{1}{r^+} - \frac{1}{r^-}\right) \tag{5.38}$$

From the cosine rule for the sides of a triangle and from the procedure that gave Equation (5.28):

$$r^+ = \sqrt{r^2 + \frac{d^2}{4} - 2r\frac{d}{2}\cos\theta}$$

so that

$$\frac{1}{r^+} = \frac{1}{r}\sum_{n=0}^{\infty} P_n(\cos\theta)\left(\frac{d}{2r}\right)^n$$

We also have

$$r^- = \sqrt{r^2 + \frac{d^2}{4} - 2r\frac{d}{2}\cos(\pi - \theta)}$$

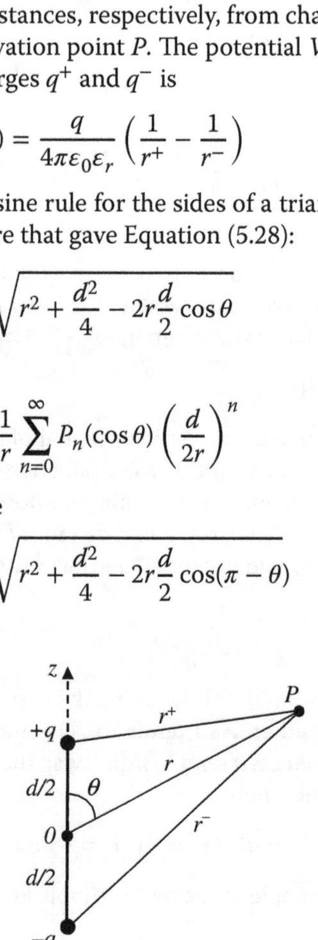

Figure 5.16 Figure for Example 5.7.

so that

$$\frac{1}{r^-} = \frac{1}{r}\sum_{n=0}^{\infty} P_n(\cos\pi - \theta)\left(\frac{d}{2r}\right)^n$$

Using the values for P_n given in Box 5.5 and noting that $\cos(\pi-\theta) = -\cos\theta$, then working through Equation (5.38) we find that the terms with even values for n cancel, leaving only terms with odd values for n in the series expansion, so that:

$$\begin{aligned} V(r,\theta) &= \frac{q}{4\pi\varepsilon_0\varepsilon_r}\left[\frac{d}{r^2}\cos\theta + \frac{d^3}{8r^4}\left(5\cos^3\theta - 3\cos\theta\right)\right] \\ &+ \frac{q}{4\pi\varepsilon_0\varepsilon_r}\left[\frac{d^5}{128r^6}\left(63\cos^5\theta - 70\cos^3\theta + 15\cos\theta\right)\right. \\ &\left. + \textit{higher terms}\right] \end{aligned} \quad (5.39)$$

Unless the distance r from the dipole is much greater than the size d of the dipole, the potential $V(r, \theta)$ of a macroscopic dipole consists of contributions from higher order multipoles. For the case where $r \gg d$, then

$$V(r,\theta) = \frac{qd}{4\pi\varepsilon_0\varepsilon_r}\frac{\cos\theta}{r^2} = \frac{p}{4\pi\varepsilon_0\varepsilon_r}\frac{\cos\theta}{r^2}$$

so that, to a good approximation, the potential and field of a macroscopic dipole are the same as those described for a 'mathematical' dipole in Figure 5.14 and Box 5.6.

Example 5.8 Dipole Moment of a Cell with Induced Surface Charge

A spherical cell of radius R, located at the origin of a rectangular system of coordinates, has an induced surface charge density σ given by $\sigma = k\cos\theta$, where k is a constant and θ is the spherical coordinate shown in Box 3.2. Find the dipole moment component of the potential at distances far away from the cell.

Solution 5.8 With reference to the spherical coordinate system shown in Box 3.2, an interesting consequence follows from the fact that the surface charge density depends only on angle θ. The components of the dipole moment along the x- and y-directions are both zero, because the same magnitude of the surface charge will exist at locations (x, y, z) and $(-x, -y, z)$. Components of the dipole lie only along the z-axis.

From section 5.3.4.2 the dipole component in the potential at a location $r \gg R$ is:

$$V_1 = \frac{p}{4\pi\varepsilon_0\varepsilon_r}\frac{\cos\theta}{r^2} \quad (5.40)$$

The component of the vector from the origin to a point on the cell's surface with charge density σ is $R\cos\theta$. The dipole moment p is thus given by the iterative integral

$$p = k\int_0^{2\pi}\int_0^{\pi}(\cos\theta)(R\cos\theta)(R^2\sin\theta)d\theta d\phi$$

Noting that

$$\int \sin^n\theta\cos^m\theta d\theta = \frac{\sin^{n+1}\theta\cos^{m-1}\theta}{n+m} + \frac{m-1}{n+m}\int \sin^n\theta\cos^{m-2}\theta d\theta$$

then

$$p = \frac{4}{3}k\pi R^3$$

From Equation (5.40) the dipole component of the potential is thus:

$$V_1 = \frac{kR^3}{3\varepsilon_0\varepsilon_r r^2}\cos\theta$$

Example 5.9 Electric Field of a 'Pure' Dipole

Derive an expression for the electric field of a *pure* dipole.

Solution 5.9 A pure dipole arises from a system of charges where only the dipole term in the multipole expansion is not zero. From section 5.3.4.2 the potential of a pure dipole of moment p, expressed in spherical coordinates, is:

$$V = \frac{p}{4\pi\varepsilon_0\varepsilon_r}\frac{\cos\theta}{r^2}$$

The electric field of this pure dipole is obtained by finding the negative gradient of its potential. Expressed in spherical coordinates (see Box 3.2) and denoting $\hat{r}, \hat{\theta}, \hat{\phi}$ as unit vectors, we obtain:

$$\begin{aligned} \mathbf{E} = -\nabla V &= -\frac{\partial V}{\partial r}\hat{r} - \frac{1}{r}\frac{\partial V}{\partial\theta}\hat{\theta} - \frac{1}{r\sin\theta}\frac{\partial V}{\partial\phi}\hat{\phi} \\ &= \frac{p}{4\pi\varepsilon_0\varepsilon_r}\frac{1}{r^3}\left[2\cos\theta\hat{r} + \sin\theta\hat{\theta}\right] \end{aligned} \quad (5.41)$$

Conversion of the unit vectors is made through the following standard formulae:

$$\hat{r} = \sin\theta\cos\phi\hat{x} + \sin\theta\sin\phi\hat{y} + \cos\theta\hat{z}$$

$$\hat{\theta} = \cos\theta\cos\phi\hat{x} + \cos\theta\sin\phi\hat{y} - \sin\theta\hat{z}$$

If we set $\phi = 0$, then

$$\begin{aligned} \mathbf{E} &= \frac{p}{4\pi\varepsilon_0\varepsilon_r}\frac{1}{r^3}\left[2\cos\theta\left(\sin\theta\hat{x} + \cos\theta\hat{z}\right) + \sin\theta\left(\cos\theta\hat{x} - \sin\theta\hat{z}\right)\right] \\ &= \frac{p}{4\pi\varepsilon_0\varepsilon_r}\frac{1}{r^3}\left[3\sin\theta\cos\theta\hat{x} + \left(3\cos^2\theta - 1\right)\hat{z}\right] \end{aligned} \quad (5.42)$$

Referring to Figure 5.16 shown in Example 5.7, the dipole moment $\mathbf{p}$ has its centre at the origin and points along the z-axis. θ is the angle between p and the unit vector $\hat{r}$ in the radial direction along r and the unit vector $\hat{\theta}$ makes an angle of $(\theta + \pi/2)$ with p. We can therefore also express Equation (5.41) in a coordinate-free form by noting that

$$\mathrm{p} \cdot \hat{r} = \mathrm{p}\cos\theta \quad \text{and} \quad \mathrm{p} \cdot \hat{\theta} = \mathrm{p}\cos(\theta + \pi/2) = -\mathrm{p}\sin\theta$$

The dipole moment **p** is given by the vector sum of its radial and angular components:

$$\mathrm{p} = (\mathrm{p} \cdot \hat{\mathrm{r}})\hat{\mathrm{r}} + (\mathrm{p} \cdot \hat{\theta})\hat{\theta} = \mathrm{p}\cos\theta\hat{\mathrm{r}} - \mathrm{p}\sin\theta\hat{\theta} \tag{5.43}$$

Inserting Equation (5.43) for p into Equation (5.41) we obtain

$$\begin{aligned} \mathrm{E} &= \frac{1}{4\pi\varepsilon_0\varepsilon_r}\frac{1}{r^3}[2 \cdot \hat{\mathrm{r}}(\mathrm{p} \cdot \hat{\mathrm{r}})\hat{\mathrm{r}} - \mathrm{p} + (\mathrm{p} \cdot \hat{\mathrm{r}})\hat{\mathrm{r}}] \\ &= \frac{1}{4\pi\varepsilon_0\varepsilon_r}\frac{1}{r^3}[3(\mathrm{p} \cdot \hat{\mathrm{r}})\hat{\mathrm{r}} - \mathrm{p}] \end{aligned} \tag{5.44}$$

5.4 Method of Image Charges

The *method of image charges*, also known as the *method of images* or the *method of mirror charges*, is a useful tool for solving problems in electrostatics. Basically, imaginary point electric charges are introduced within a conductor to replace a distribution of charges along the conductor's surface. This procedure can greatly simplify the task, for example, of calculating the force that a distribution of charges on a conductor exerts on other charges. The validity of this method depends on two so-called *uniqueness theorems*. The First Uniqueness Theorem states:

- In a specified volume T bounded by a surface S, the Laplace equation has a unique solution when the potential V is given on the boundary S.

If we do not know the potential at the boundary, but can specify the charge distributions on various conducting surfaces, we can use the Second Uniqueness Theorem:

- The electric field within a volume T is uniquely specified by the total net charge on each conductor surrounding this volume and by the charge density within T.

A simple example of the application of the First Uniqueness Theorem is given in Figure 5.17. A stationary point charge $+Q$ is shown located a short distance d above the surface of a grounded metal plate. Because the metal plate is grounded, this represents a boundary where V is zero. The potential also drops to zero as we approach an infinite distance above the metal surface. In terms of the First Uniqueness theorem we have a defined

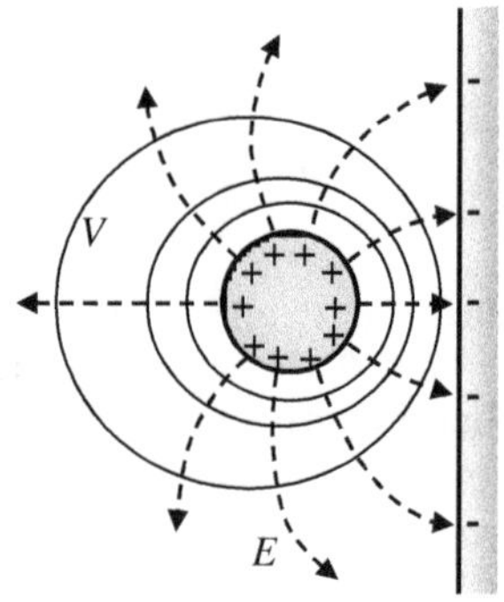

Figure 5.17 The field lines from a point charge meet an electrically conducting surface, such as a metal, at right angles. This arises because there can be no component of the electrical field parallel to the conductor's surface. The induced (negative) charges on the conductor form an equipotential surface.

volume T above the surface of the metal with specified (zero) potentials at the boundaries of this volume. We can therefore uniquely determine the potential at every point within this volume. We need find only one physical situation that replicates these boundary conditions to be sure that this represents the unique solution. We can deduce what this solution is by simple logic. We know that an electric field cannot be sustained along the surface of the metal and so the distribution of induced negative charges at the metal boundary will form an equipotential surface. As shown in Figure 5.18, this equipotential surface can be generated by replacing the metal plate with a point charge $-Q$ located at a distance d inside the space previously occupied by the metal. We are in effect treating

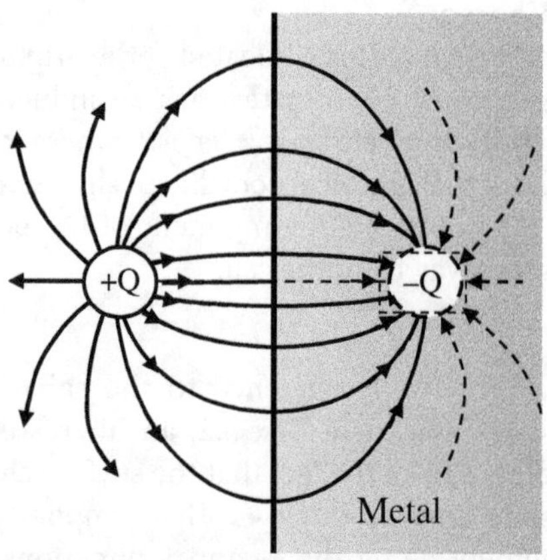

Figure 5.18 The equipotential surface of the metal surface shown in Figure 5.17 can be replaced with an imaginary point charge $-Q$ located inside it. The imaginary charge and the external charge are equidistant from the conductor's surface. The imaginary charge produces the same field outside the metal as that shown in Figure 5.17. Calculation of the force on the real charge $+Q$ from the induced distributed charge on the metal's surface is now simplified to that of using Coulomb's Law for two point charges of equal but opposite polarity.

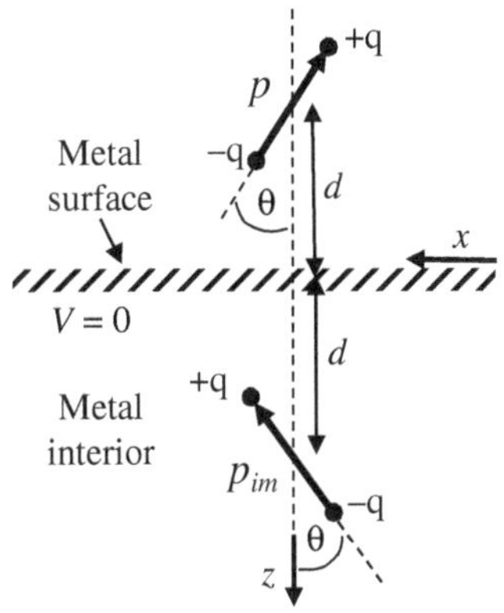

Figure 5.19 A 'real' dipole moment p is shown located distance d above the surface of a grounded metal plane and makes an angle θ with the perpendicular axis to this plane. The image dipole p_{im} makes an angle 2θ with the real dipole and its magnitude is the same as p.

the metal surface as a mirror. The image charge that is produced generates the same field outside the conductor as that shown in Figure 5.17. Calculation of the attractive force F acting on the real charge $+Q$ from the distributed negative charges it induces on the metal surface is now simplified to the application of Coulomb's Law for two point charges, of opposite polarity, separated by the distance $2d$. The result is:

$$F = -\frac{1}{4\pi\varepsilon_o}\frac{Q^2}{4d^2}$$

This attractive force is directed normal to the metal surface.

An extension of this example is to consider the case of a dipole of moment p situated a distance d above the surface plane of a metal at an angle θ with the perpendicular axis to this plane. This is illustrated in Figure 5.19. We take the coordinates of this dipole as (0, d, 0). The mirror image dipole p_{im} is located at (0, $-d$, 0) and has the same magnitude as the real dipole moment. The angle between the dipole and its image is 2θ, so that the dipole's image has the same z-component but opposite components in the x- and y-directions. A dipole moment with Cartesian components (p$\sin\theta$ $\cos\theta$, p$\sin\theta$ $\sin\theta$, p$\cos\theta$) thus has an image moment of components ($-$p$\sin\theta$ $\cos\theta$, $-$p$\sin\theta$ $\sin\theta$, p$\cos\theta$).

Equation (5.35) gives the force acting on a pure dipole in an electric field as

$$F = (p\cdot\nabla)E$$

The real dipole experiences the field E produced by its image dipole. From Equation (5.44), with a spherical coordinate system centred on the image dipole, this field is

$$E = \frac{1}{4\pi\varepsilon_0\varepsilon_r}\frac{1}{r^3}[3(p\cdot\hat{r})\hat{r} - p]$$

We set $\phi = 0$ and substitute for the unit vectors to obtain

$$\begin{aligned} E &= \frac{p_{im}}{4\pi\varepsilon_0\varepsilon_r}\frac{1}{r^3}[2\cos\theta\,(\sin\theta\hat{x} + \cos\theta\hat{z}) \\ &\quad + \sin\theta\,(\cos\theta\hat{x} - \sin\theta\hat{z}) \\ &= \frac{p_{im}}{4\pi\varepsilon_0\varepsilon_r}\frac{1}{r^3}\left[3\sin\theta\cos\theta\hat{x} + \left(3\cos^2\theta - 1\right)\hat{z}\right] \end{aligned}$$

Working through the calculation of (p$\cdot\nabla$)E we obtain the following result for the force acting on the real dipole by the field of its mirror image:

$$F = -\frac{1}{4\pi\varepsilon_o\varepsilon_r}\frac{3p^2}{16d^4}\left(1 + \cos^2\theta\right) \tag{5.45}$$

A dipole approaching the surface of a metallic plane therefore experiences an attractive force towards the metal. However, this force falls of as $(1/d)^4$ and so for modest distances from a metal surface the attractive force acting on a particle as a result of its image dipole moment is likely to be less than that arising from its surface charge, the force of which falls off as $(1/d)^2$. This is an important consideration for dielectrophoresis manipulations of biological cells, which typically possess a net negative surface charge much larger than the induced charges of its induced dipole moment.

A dipole also experiences a torque T in the plane perpendicular to the metal surface given by Equation (5.37):

$$\begin{aligned} T &= p \times E \\ &= \frac{1}{4\pi\varepsilon_0\varepsilon_r}\frac{1}{(2d)^3}[3p\cos\theta p \times \hat{r} - p \times p_{im} \\ &= \frac{p^2}{4\pi\varepsilon_0\varepsilon_r}\frac{1}{(2d)^3}(3\cos\theta\sin\theta - \sin 2\theta)\,\hat{x} \\ &= \frac{p^2}{4\pi\varepsilon_o\varepsilon_r}\frac{1}{8d^3}\left(\frac{1}{2}\sin 2\theta\right)\hat{x} \end{aligned} \tag{5.46}$$

The surface charge on a biological cell is uniformly distributed on the membrane surface and does not give rise to a dipole moment. The field of the image charge will not therefore produce a torque on the cell, but the image of its induced dipole moment will do so.

5.4.1 Polarized Particle near an Electrode or Insulator Surface

In a dielectrophoresis experiment particles are suspended in a fluid and subjected to an electric field. This field is created using electrodes that are often embedded into one or more insulating substrates that form the

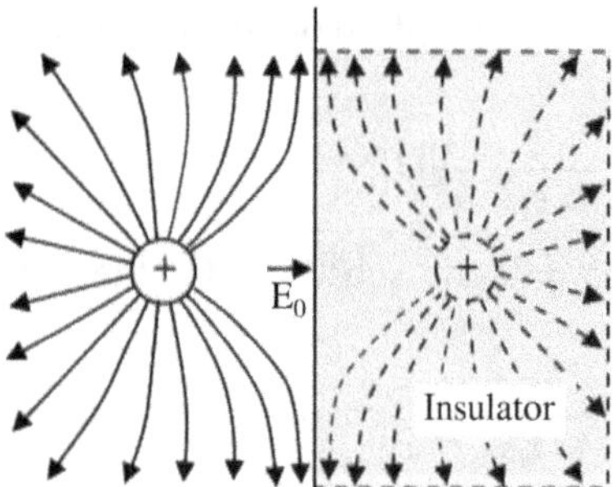

Figure 5.20 To satisfy the boundary conditions at the surface of an insulator the field of the image charge within the insulator must oppose that of the true external charge. The real charge experiences a repulsive coulombic force from the imaginary image charge.

chamber walls containing the suspended particles. The electric boundary conditions to be described in section 5.6 of this chapter must be satisfied at both the metal (or conducting) and insulator surfaces of the chamber walls. The important boundary conditions of relevance to image charges are:

- At the surface of an ideal conductor there can be no component of the field parallel to its surface. Such a wall must act as an equipotential surface. This condition is satisfied for the charge images shown in Figures 5.18 and 5.19.
- A wall composed of an insulator will not represent an equipotential surface – there will be a component of the field parallel to its surface. However, the normal component of the field at the surface of an insulator is continuous. The charge images shown in Figures 5.18 and 5.19 do not satisfy this situation.

In Chapter 6 it is shown that a particle polarized by an imposed electric field assumes the form of electric dipole moment. For a spherical particle the field produced by this dipole takes the form of Equation (5.42), which gives field components parallel (x-direction) and normal (z-direction) to the wall surface shown in Figure 5.19. As a polarized particle approaches the surface of either a conducting or insulating surface, the total field (E_T) at position r is given by:

$$E_T(r) = E_o + E_{dp} + E_{imag}$$

where E_o is the applied local field, E_{dp} is the dipole field of the polarized particle and E_{imag} is the field produced by the image charges. If the polarized particle is approaching a conducting surface, the boundary condition at the surface ($z = 0$) is given by:

$$E_T(z = 0) \cdot \hat{t} = 0$$

where $\hat{t}$ is the unit vector in a direction tangential (in this case also parallel) to the conducting surface. We have already deduced that the form of the image dipole shown in Figure 5.19 satisfies this boundary condition. If the polarized particle approaches an insulating surface, the boundary condition at the surface is given by:

$$E_T(z = 0) \cdot \hat{n} = E_{oz}$$

This condition is satisfied when the field produced by the image dipole exactly opposes the field of the 'real' dipole. A simple demonstration of this is shown in Figure 5.20 for the case of a positive point charge located near the surface of an insulator.

The corresponding situation for a dipole approaching an insulator is shown in Figure 5.21, from which it is evident that a coulombic repulsion force exists between the true dipole and its image dipole. On the other hand, an attractive force exists between a dipole and its image when approaching a conducting surface.

5.5 Electric Field Gradient

The electric field map shown in Figure 5.10 is confined to a two-dimensional plane. The map does not change along the x-axis normal to the yz-plane. In Figure 5.10 the surface of the electric field map is shown in a region where the field reaches its maximum value at the junction between the metal sheet biased at 12 V and the grounded surface of the tube. Paths can be identified in

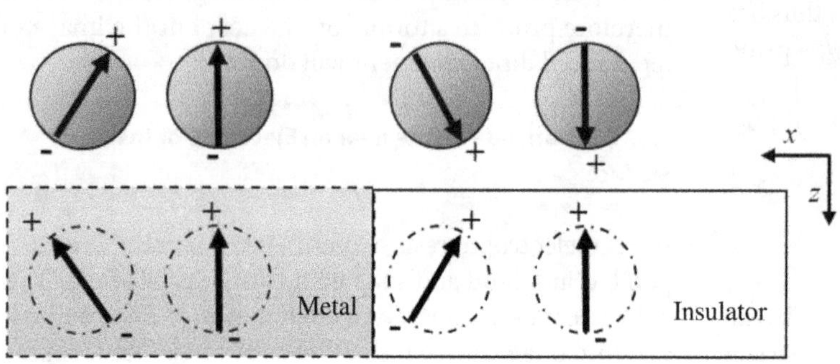

Figure 5.21 The directional sense of the image dipole for a polarized particle approaching a metal surface is opposite to that created when it approaches an insulating surface. The positive dielectrophoretic (DEP) force that causes a polarized particle to be directed towards an electrode is enhanced by the attractive coulombic force with its image. On the other hand, the effect of a negative DEP force that repels a polarized particle from an electrode towards an insulating surface is reduced by a repulsive force with its image dipole.

Box 5.7 The Gradient of an Electric Field is a 3 × 3 Tensor

An electric field vector such as $E(x, y, z) = 3\hat{\imath} + 4\hat{\jmath} + 2\hat{k}$ (where $\hat{\imath}$, $\hat{\jmath}$ and $\hat{k}$ are unit vectors along the x-, y- and z-directions, respectively) can be represented by the following 3 × 1 (column) matrix:

$$\mathbf{E} = \begin{bmatrix} 3 \\ 4 \\ 2 \end{bmatrix}$$

Multiplying a 1 × 3 (line) matrix with a 3 × 1 (column) matrix gives a result that is equivalent to a vector dot product. For example:

$$\begin{bmatrix} x_1 & x_2 & x_3 \end{bmatrix} \begin{bmatrix} y_1 \\ y_2 \\ y_3 \end{bmatrix} = x_1y_1 + x_2y_2 + x_3y_3 = x \cdot y$$

Let G = ∇E be the gradient of the field E. The components of dE are:

$$dE_x = \frac{\partial E_x}{\partial x}dx + \frac{\partial E_x}{\partial y}dy + \frac{\partial E_x}{\partial z}dz$$

$$dE_y = \frac{\partial E_y}{\partial x}dx + \frac{\partial E_y}{\partial y}dy + \frac{\partial E_y}{\partial z}dz$$

$$dE_z = \frac{\partial E_z}{\partial x}dx + \frac{\partial E_z}{\partial y}dy + \frac{\partial E_z}{\partial z}dz$$

This can be represented as the following matrix product:

$$\begin{bmatrix} dE_x \\ dE_y \\ dE_z \end{bmatrix} = \begin{bmatrix} \frac{\partial E_x}{\partial x} & \frac{\partial E_x}{\partial y} & \frac{\partial E_x}{\partial z} \\ \frac{\partial E_y}{\partial x} & \frac{\partial E_y}{\partial y} & \frac{\partial E_y}{\partial z} \\ \frac{\partial E_z}{\partial x} & \frac{\partial E_z}{\partial y} & \frac{\partial E_z}{\partial z} \end{bmatrix} \begin{bmatrix} dx \\ dy \\ dz \end{bmatrix} = G \cdot \begin{bmatrix} dx \\ dy \\ dz \end{bmatrix} \quad (5.47)$$

The gradient G of a vector electric field can thus be represented as a 3 × 3 tensor. A tensor is a matrix, but not all matrices are tensors. The three components of any vector can be multiplied by any 3 × 3 matrix. In order to classify as a 3 × 3 *tensor*, the three numbers that result from this multiplication must represent (as they do in the above equation) the components of a *vector*.

the yz-plane along which the electric field gradient is zero – the field remains constant in value. No work is required to move an electrically charged particle along such a path and its potential energy would remain constant. However, the natural tendency for a charged particle in an electric field is to move so as to reduce its potential energy. A positively charged particle will move up the E-field surface and choose the steepest field gradient possible at each point of its route. The lengths and directions of the arrows shown in Figure 5.22 denote the magnitude and direction, respectively, of the local field gradient vector. From the mathematical description given in Box 5.7, the components dE_y and dE_z of the field gradient in the yz-plane are given in terms of the partial derivatives of the field along the direction of the y-axis and the z-axis:

$$dE_y = \frac{\partial E_y}{\partial y}dy; \quad dE_z = \frac{\partial E_z}{\partial z}dz$$

As shown in the inset of Figure 5.22, the field gradient ∇E (grad E) is equal to the vector sum of dE_y and dE_z. This result gives the direction of the steepest field gradient at that particular location x, y. For example, if the field is constant in the z-direction, then $dE_z = 0$ and the steepest gradient lies normal to this direction – namely along the y-direction on the E-field surface. For the case where the incremental changes of the local field are equal in the y- and z-directions, ∇E is directed at 45° to the y- and z-axes when projected onto the E-field surface.

A positively charged particle will follow a path that at each point represents the steepest field gradient. Conversely, a negatively charged particle will seek to minimize its electrical potential energy by following a path of steepest gradients down the E-field surface. As shown in Figure 5.22 the overall path taken may involve changes in the y- and z-directions. An analogy is that of a rock rolling down a mountain slope, following a path that at any

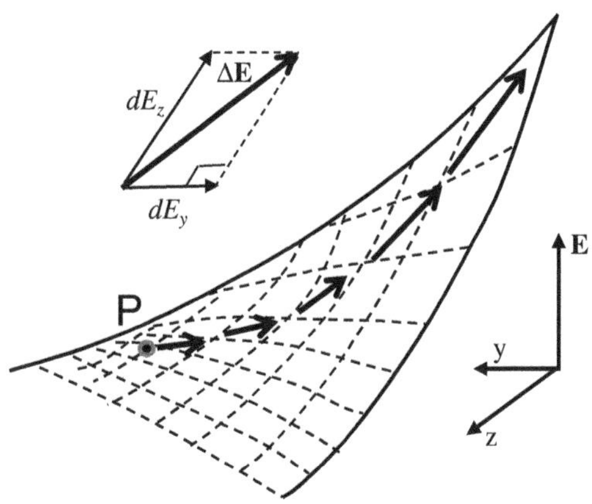

Figure 5.22 A plot of the E-field surface at one corner of the map shown in Figure 5.10 where the field reaches a maximum. The length and direction of an arrow denotes the magnitude and direction, respectively, of the local field gradient ∇E. These values are obtained by the vector addition of the field increments in the y- and z-directions. A positively charged particle at location P will move along a path that at each point follows the steepest field gradient.

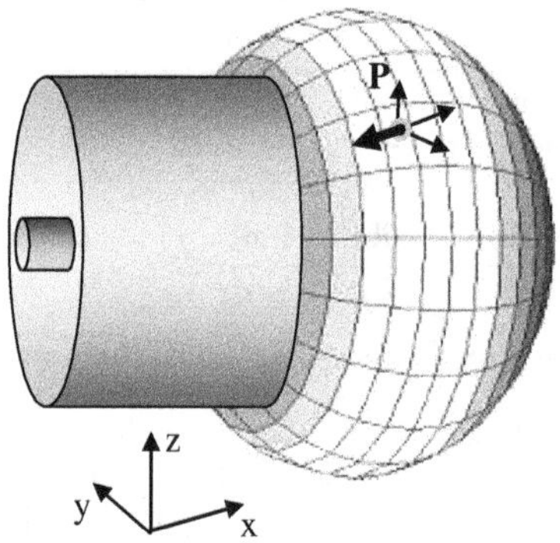

Figure 5.23 A point P is shown on a three-dimensional E-field surface created at the open end of a coaxial electrode. The direction of greatest rate of change of the electric field is given by the sum of the three partial derivatives, in the *x*, *y* and *z* directions, of the mathematical function describing the shape of the surface. As described in Box 5.7 the resultant gradient vector takes the form of a 3 × 3 tensor.

point leads to the greatest rate of decrease of its potential energy in the Earth's gravitational field. In summary, the electric field gradient has magnitude and direction – it is a vector as denoted by each arrow in Figure 5.22.

The concept of an electric field gradient, when we are confined to a two-dimensional surface, is not too difficult to visualize. However, the usual practical situation in dielectrophoresis involves the spatial manipulation of particles in electric fields that vary in three-dimensional space. An example is the semispherical field generated at the open end of a coaxial electrode system, as depicted in Figure 5.23. Visualization of a local field gradient in such a three-dimensional field surface can in fact be more difficult than the mathematics involved in deriving it! From Equations (5.5) and (5.6) the electric field is a vector quantity given by:

$$\mathbf{E}(x, y, z) = \left(\hat{\imath}\frac{\partial}{\partial x} + \hat{\jmath}\frac{\partial}{\partial y} + \hat{k}\frac{\partial}{\partial z}\right) V$$

In Box 5.7 the gradient of this field is shown to be represented by a 3 × 3 tensor. It is a vector equal to the vector sum of $d\mathrm{E}_x$, $d\mathrm{E}_y$ and $d\mathrm{E}_z$ for each coordinate direction and 'points' along the direction of greatest field gradient at that point. A charged particle on this E-field surface will 'sense' the direction of this greatest field gradient and will move along it so as to reduce its potential energy.

5.6 Electrical Conditions at Dielectric Boundaries

The boundary surface *S* between two dielectrics (1) and (2) of relative permittivity ε_{r1} and ε_{r2}, respectively, is shown in Figure 5.24. The two dielectrics are not perfect insulators and have electrical conductivities of σ_{c1} and σ_{c2}, respectively. We will imagine that a very thin transition region exists at the boundary, within which the permittivity and conductivity change rapidly but *continuously*. The potential functions ϕ_1 and ϕ_2 are thus continuous across the boundary, together with their first derivatives with respect to distance (i.e., fields E_1 and E_2 are continuous). The electric force flux ($\mathrm{D} = \varepsilon\mathrm{E}$) and current density ($\mathrm{J} = \sigma_c\mathrm{E}$) normal to *S* are thus also continuous functions across the boundary.

For the Gaussian cylinder drawn across the boundary in Figure 5.24(b) no flux passes through the sides parallel to the normal vector **n**, but only through its two circular ends of area *A*. Taking the positive direction of the normal vector to be from dielectric (1) to dielectric (2), on applying Gauss's Law in the form of equation (3.32) the change of the normal component of the flux D across the boundary is given by:

$$\begin{aligned}\int_S \mathrm{D}\cdot \mathrm{n}\,dA &= (\mathrm{D}_{n2} - \mathrm{D}_{n1})\cdot \mathrm{n} \\ &= \left(\varepsilon_o\varepsilon_2\mathrm{E}_{n2} - \varepsilon_o\varepsilon_1\mathrm{E}_{n1}\right) = \sigma_{free} \qquad (5.48)\end{aligned}$$

At the boundary surface there is thus an abrupt change in the *normal component* of D. The magnitude of this change is equal to the surface charge density at the surface boundary. This charge will have migrated to the boundary surface through one or both dielectrics, or have been introduced (e.g., injected) by some other means. If the boundary surface is uncharged, the normal component of D does not change across the boundary, so that $\mathrm{D}_{n2} = \mathrm{D}_{n1}$ to give:

$$\varepsilon_2\mathrm{E}_{n2} = \varepsilon_1 E_{n1} \qquad (5.49)$$

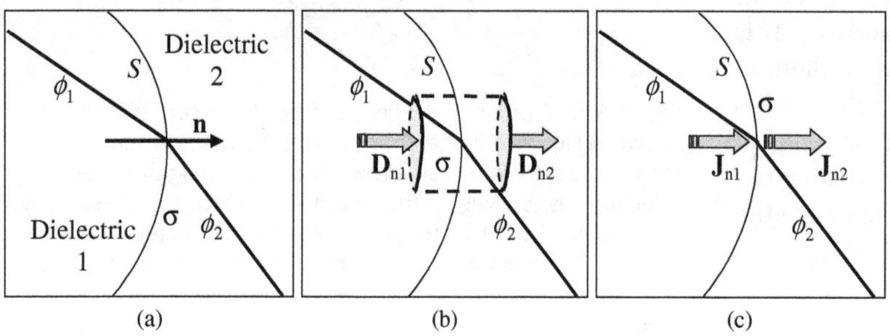

Figure 5.24 (a) A boundary surface *S* with charge density σ is shown between two dielectrics of relative permittivity ε_{r1}, ε_{r2} and conductivity σ_{c1}, σ_{c2}, respectively. The potential functions ϕ_1 and ϕ_2 are continuous across the boundary. The vector **n** is drawn normal to *S*. (b) Flux ($\mathrm{D} = \varepsilon\mathrm{E}$) normal to *S*, through the two ends of a cylindrical Gaussian surface, is continuous across the boundary. (c) The current density ($\mathrm{J} = \sigma_c\mathrm{E}$) normal to *S* is continuous across the boundary.

If one or both of the dielectrics exhibit an electrical conductivity σ_c, the current density ($J_n = \sigma_c E_n$) is equal to the flow of free charge to or across the boundary surface *S*. The change in the normal component of the net current density normal to the surface is given by:

$$(J_{n2} - J_{n1}) \cdot n = (\sigma_{c2}E_{n2} - \sigma_{c1}E_{n1}) = -\frac{\partial \sigma_{free}}{\partial t} \quad (5.50)$$

The time *t* is a common factor for D and J, related to the angular frequency ω of the voltage signal $v(t) = V_o(\sin \omega t)$ applied to the dielectrics. The sine wave is a periodic waveform derived from a radius vector **r** rotating around the origin. In the notation of complex algebra a rotating radius vector is given as

$$r(t) = r_o e^{i\omega t} = r_o(\cos \omega t + i \sin \omega t) \quad (5.51)$$

where *i* signifies $\sqrt{-1}$. The time derivative of an exponential function gives the same exponential function (i.e., $\partial(e^{i\omega t})/\partial t = i\omega e^{i\omega t}$) and so $\partial/\partial t$ can be replaced with $i\omega$. From Equations (5.48) and (5.50) we have the two equations:

$$\varepsilon_2 E_{n2} - \varepsilon_1 E_{n1} = \sigma_{free}$$

$$\sigma_{c2}E_{n2} - \sigma_{c1}E_{n1} = -i\omega\sigma_{free}$$

The requirement for zero charge buildup at the surface boundary can be written as:

$$\begin{bmatrix} \varepsilon_2 & \varepsilon_1 \\ \sigma_{c2} & \sigma_{c1} \end{bmatrix} \begin{bmatrix} E_{n2} \\ -E_{n1} \end{bmatrix} = 0$$

which is satisfied by

$$\begin{vmatrix} \varepsilon_2 & \varepsilon_1 \\ \sigma_{c2} & \sigma_{c1} \end{vmatrix} = 0,$$

or

$$(\varepsilon_2\sigma_1 - \varepsilon_1\sigma_2) = 0 \quad (5.52)$$

The conditions at *S* for the *tangential* components of the fields E_1 and E_2 are found by replacing the Gaussian cylinder with a rectangular path as shown in Figure 5.25.

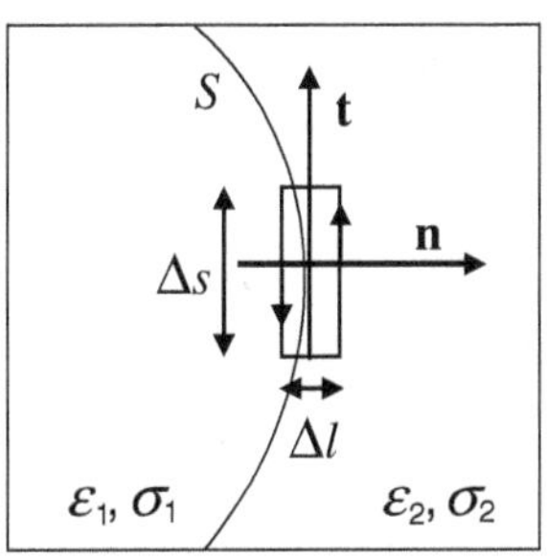

Figure 5.25 The conditions at *S* for the *tangential* components of the fields E_1 and E_2 are found by replacing the Gaussian cylinder with a rectangular path as shown in this figure.

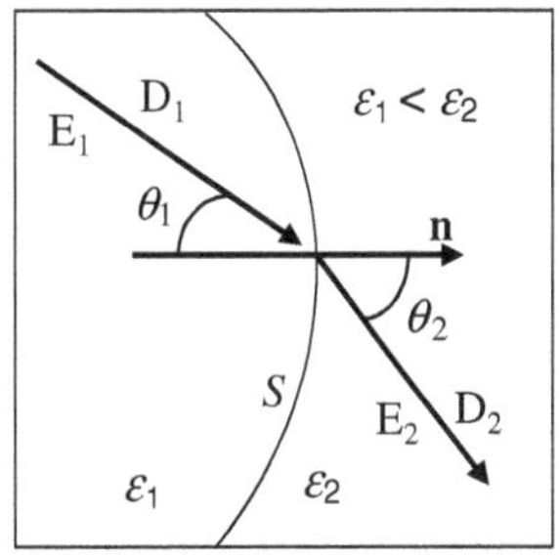

Figure 5.26 The boundary conditions (continuity of the normal component of D and of the tangential component of E) at the surface *S* between to perfect dielectrics results in the law of refraction for the D- and E-lines of force: $\tan\theta_1/\tan\theta_2 = \varepsilon_{r1}/\varepsilon_{r2}$.

The sides of this rectangle are of length Δs and are located either side of the boundary surface *S*. The ends of the rectangle that traverse *S* are each of length Δl.

For one complete circuit of the rectangular path shown in Figure 5.25 there is zero potential difference and so from Equation (5.1) we have

$$\oint E ds = 0$$

To a good approximation this integral can be approximated as

$$(E_1 \cdot t_1 + E_2 \cdot t_2)\Delta s + (E_1 \cdot n_1 + E_2 \cdot n_2)\Delta l = 0$$

where t_1,t_2 and n_1, n_2 are the unit tangential and normal vectors, respectively. In the limit as Δs and Δl tend to zero, $E_1.t_1$ and $E_2.t_2$ become tangents to *S*, so that:

$$(E_{t1} - E_{t2}) \cdot t = 0 \quad (5.53)$$

Thus, the tangential component of a field **E** at the surface boundary is continuous. It follows that the tangential component of current density J is also continuous.

The continuity of the normal component of D and the tangential component of E at a boundary surface gives rise to a particular law of refraction for ideal dielectrics. As shown in Figure 5.26, let θ_1 and θ_2 be the angles between the D-field and E-field force 'lines' and the normal vector n to the boundary surface *S*. The boundary conditions given by Equations (5.48) and (5.53) can thus be written as:

$$|D_1|\cos\theta_1 = |D_2|\cos\theta_2 \quad (5.54a)$$

and

$$|E_1|\cos\theta_1 = |E_2|\cos\theta_2 \quad (5.54b)$$

But

$$D = \varepsilon_o \varepsilon_r E$$

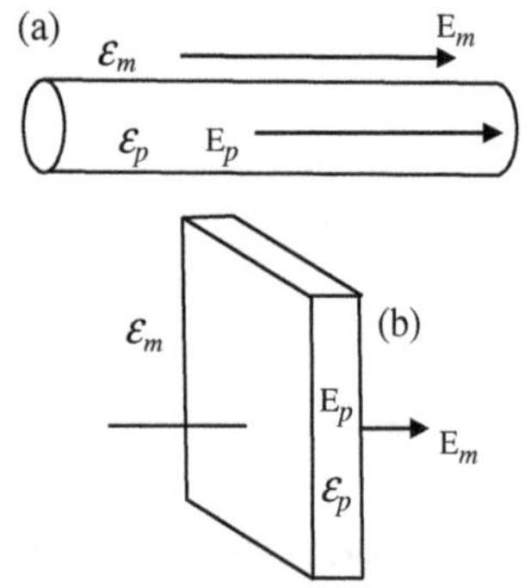

Figure 5.27 (a) A long, thin dielectric rod with its axis parallel to an external electric field. (b) A large, flat dielectric plate with its major faces perpendicular to an external electric field.

and so from Equations (5.54)

$$\frac{\tan\theta_1}{\tan\theta_2} = \frac{\varepsilon_1}{\varepsilon_2}$$

Thus, the D- and E-field lines are deflected further from the normal to the surface if they cross from one dielectric to another for which the permittivity is greater.

The boundary conditions given by Equations (5.49) and (5.53) are applied in Figure 5.27 to the cases of a long thin dielectric rod and a large flat dielectric plate. From Equation (5.53), which states that the tangential component of the externa field E_m is continuous across the interface between the rod and the external medium, then the field E_p inside the rod is given by:

$$E_p = \frac{\varepsilon_m}{\varepsilon_p} E_m \quad (5.55)$$

The field inside the rod is thus equal to the external field E_m. In Chapter 7, with Equations (7.32) and (7.33) the concept of a depolarizing factor A is described. In general, the field inside any particle is of the form:

$$E_p = E_m - E_{dp} \quad (5.56)$$

where E_{dp} is the depolarizing field arising from the polarization charges on the particle, given in terms of the induced polarization P by:

$$E_{dp} = \frac{AP}{\varepsilon_o} \quad (5.57)$$

From Equations (5.55–5.57) it is clear that for the case of a long thin rod $A = 0$. From Equation (5.49), which states that the normal component of the externa displacement D is continuous across the faces of the flat dielectric plate, the internal field is given by:

$$E_p = \frac{\varepsilon_m}{\varepsilon_p} E_m \quad (5.58)$$

The field inside the plate is thus reduced by the factor $\varepsilon_p/\varepsilon_m$ relative to the external field E_m. In this case, for the flat plate $A = 1$. As discussed further in Chapter 7, the particle geometries of a long thin rod and a large flat plate give the extreme values for A, since the values of A for all other particle shapes lie between 0 and 1. For the important case of a spherical particle, $A= 1/3$.

5.7 Summary

The electric potential at a point within a small volume of space that contains no charge is equal to the average of the potentials that surround this point. In the limit as the characteristic dimensions of this volume become infinitesimally small, this is an exact statement and forms the basis for modelling electric potentials and fields using the finite difference and finite element methods.

The gradient of a scalar function or field using vector notation is obtained by applying the *grad* or *del* vector operator denoted by the symbol ∇. Thus, the gradient of a potential V is given by:

$$\nabla V = \left(\hat{\imath}\frac{\partial}{\partial x} + \hat{\jmath}\frac{\partial}{\partial y} + \hat{k}\frac{\partial}{\partial z}\right) V$$

where $\hat{\imath}, \hat{\jmath}$ and $\hat{k}$ are unit vectors along the x-, y- and z-axes, respectively.

In vector notation an electric field E is defined as:

$$E = -\left(\hat{\imath}\frac{\partial V}{\partial x} + \hat{\jmath}\frac{\partial V}{\partial y} + \hat{k}\frac{\partial V}{\partial z}\right) = -\nabla V$$

If the field E is uniform, then everywhere $\nabla E = 0$. Because the dielectrophoretic force acting on a particle is proportional to ∇E, we have the practical requirement that the field is nonuniform. If the nonuniform field is confined to a two-dimensional plane, the field gradient at any point (x, y) is given by a 2×2 tensor of the form:

$$\nabla\mathbf{E} = \begin{bmatrix} \frac{\partial E_x}{\partial x} & \frac{\partial E_x}{\partial y} \\ \frac{\partial E_y}{\partial x} & \frac{\partial E_y}{\partial y} \end{bmatrix}$$

For the more general case of a nonuniform field in three-dimensional space, the field gradient at any point (x, y, z) is given by a 3×3 tensor:

$$\nabla E = \begin{bmatrix} \frac{\partial E_x}{\partial x} & \frac{\partial E_x}{\partial y} & \frac{\partial E_x}{\partial z} \\ \frac{\partial E_y}{\partial x} & \frac{\partial E_y}{\partial y} & \frac{\partial E_y}{\partial z} \\ \frac{\partial E_z}{\partial x} & \frac{\partial E_z}{\partial y} & \frac{\partial E_z}{\partial z} \end{bmatrix}$$

The field gradient ∇E at any point on an E-field surface has magnitude and direction, with the direction being that of the steepest field gradient at that point. A charged particle, or a particle experiencing dielectrophoresis, will minimize its electrical potential energy by moving along (or against) the direction of ∇E.

In practical dielectrophoresis experiments, the particles being manipulated can often approach the surface of a metal electrode. The particle will have assumed the properties of a dipole moment and also carry a net charge. These two attributes induce charges at the surface of the electrode, to exert a small but sometimes not negligible attractive coulombic force on the particle. A useful technique to aid calculation of this attractive force is to apply the method known variously as the *method of image charges*, or the *method of mirror charges*. The calculation is simplified by treating the metal surface as a mirror, so that imaginary charges inside the metal mirror the spatial locations of the real external charges (but reversing the electrical polarity of each real charge). The unspecified distribution of induced charges at the metal surface is replaced by a system of point charges of known polarity and location.

Finally, the following conditions apply at the boundary surface S between two dielectrics:

1. The potential functions ϕ are continuous everywhere in the dielectrics, including at the boundary between them. Thus, at the boundary surface S in Figure 5.24 we have

$$\phi_1 = \phi_2$$

2. The normal component of the displacement flux density ($D = \varepsilon E$) is continuous across the boundary surface, so that

$$\varepsilon_2 E_{n2} - \varepsilon_1 E_{n1} = \varepsilon_2 \frac{\partial \phi_2}{\partial r} - \varepsilon_1 \frac{\partial \phi_1}{\partial r} = \sigma_{free}$$

3. The normal component of the current density ($J = \sigma E$) is continuous across the boundary surface, so that

$$\sigma_{c2} \frac{\partial \phi_2}{\partial r} - \sigma_{c1} \frac{\partial \phi_1}{\partial r} = -i\omega\sigma_{free}$$

where ω is the radian frequency of the applied voltage signal.

4. The tangential component of a field E and a current density J is continuous at a surface boundary.

The application of these dielectric boundary conditions can be used to define a depolarization factor, which quantifies by how much the field inside a polarized particle is reduced relative to the external polarizing field as a result of the induced polarization charges on the particle's surface.

5.8 References

1 Binns, K. J. and Laurenson, P. J. (1963) *Analysis and Computation of Electric and Magnetic Field Problems*, Pergamon Press, Oxford.

2 Zienkiewicz, O. C. Taylor, R. L., Zhu, J. Z. (2005) *The Finite Element Method: Its Basis and Fundamentals*, 6 edn. Butterworth-Heinemann, Amsterdam.

3 Reddy, J. N. (2005) *An Introduction to the Finite Element Method*, 3 edn. McGraw-Hill, New York, NY.

4 Humphries, S. (2014) *Finite-Element Methods for Electromagnetics* (eBook), CRC Press, Boca Raton, FL.

5 Hughes, M. P. (2003) *Nanoelectromechanics in Engineering and Biology*, CRC Press, Boca Raton, pp. 239–277.

6 Morgan, H. and Green, N. G. (2003) *AC Electrokinetics: Colloids and Nanoparticles*, Research Studies Press Ltd, Baldock, pp. 191–210.

7 Kreyszig, E. (1999) *Advanced Engineering Mathematics*, 8 edn, John Wiley & Sons, Inc., New York, NY.

8 Stratton, J. D. (1941) *Electromagnetic Theory*, McGraw-Hill, New York, NY.

9 Jones, T. B. (1995) *Electromechanics of Particles*, Cambridge University Press, Cambridge.

6

The Clausius–Mossotti Factor

'Wohl dem, der seiner Väter gern gedenkt'
('Blessings on him who gladly remembers his forefathers')

J. W. V. Goethe

6.1 Introduction

Scientific papers dealing with the dielectric or dielectrophoretic properties of bioparticles often refer to the Clausius–Mossotti (*CM*) factor to describe their polarizabilities. For the case of a spherical particle of permittivity ε_p, suspended in a medium of permittivity ε_m, the *CM* factor is given as:

$$CM = \frac{\varepsilon_p - \varepsilon_m}{\varepsilon_p + 2\varepsilon_m} \quad \text{or} \quad CM = \frac{\varepsilon_p^* - \varepsilon_m^*}{\varepsilon_p^* + 2\varepsilon_m^*} \tag{6.1}$$

The version given on the left corresponds to the case of an *ideal* dielectric sphere suspended in an *ideal* dielectric medium. By *ideal* we mean that neither the particle nor medium exhibit ohmic conduction of mobile charges. Such conduction gives rise to energy loss (I^2R) in the form of heat. The other version, involving asterisk superscripts, is known as the 'complex' *CM* factor and takes into account the conduction and dielectric energy losses of the suspended particle and the surrounding medium. This particular aspect is described in section 6.4 of this chapter.

The form of what we shall term the *macroscopic CM* factor can be derived by employing Laplace's equation with the boundary conditions described in section 5.6. The method described in Box 6.1 follows the procedure described in classical textbooks of electricity and

Box 6.1 Polarization of a Dielectric Sphere Suspended in a Medium

A uniform field E_o, directed along the positive *x*-axis, is established in a homogeneous, ideal dielectric, medium of relative permittivity ε_m. An uncharged, ideal dielectric, sphere of radius *R* and relative permittivity ε_p is then inserted into this medium. The field produced by the polarized sphere will be symmetrical around the *x*-axis. Thus, using polar coordinates with the centre of the sphere at the origin and θ the angle between *r* and the *x*-axis, the Laplace equation simplifies to:

$$\nabla^2\phi = \frac{\partial^2\phi}{\partial r^2} + \frac{x}{r}\frac{\partial\phi}{\partial r} + \frac{1}{r^2}\frac{\partial^2\phi}{\partial\theta^2} + \frac{1}{r^2}\frac{\cos\theta}{\sin\theta}\frac{\partial\phi}{\partial\theta} = 0$$

For $r \gg R$ the field is not altered by polarization of the sphere, so that $\mathbf{E}_\infty = \mathbf{E}_o$ and $\phi_\infty = -E_o r\cos\theta$. The potential is finite everywhere within the sphere ($r < R$). The boundary conditions described in Chapter 5 will also apply, so that for $r = R$:

$$\phi_i = \phi_o \quad \text{and} \quad \varepsilon_p\frac{\partial\phi_i}{\partial r} = \varepsilon_m\frac{\partial\phi_o}{\partial r}.$$

Solutions of the Laplace equation are called spherical harmonics and employ Legendre functions [2, Ch. IX, pp. 194–231]. Ignoring higher order terms of $\cos\theta$ the solutions have the form:

$$\phi_o = \left(\frac{A}{r^2} - Br\right)E_o\cos\theta; \quad \phi_i = \left(\frac{C}{r^2} - Dr\right)E_o \tag{6.2}$$

where ϕ_o is the potential outside the sphere. For the internal potential ϕ_i we assume $\theta = 0°$, so that $\cos\theta = 1$. The constants have the following values:

$$A = \frac{\varepsilon_p - \varepsilon_m}{\varepsilon_p + 2\varepsilon_m}R^3; \; B = 1; \; C = 0; \; D = \frac{3\varepsilon_m}{\varepsilon_p + 2\varepsilon_m}$$

Inserting these values into Equation (6.2) the following solutions are obtained

$$\phi_o = \left(\frac{\varepsilon_p - \varepsilon_m}{\varepsilon_p + 2\varepsilon_m}\frac{R^3}{r^3} - 1\right)E_o r\cos\theta; \quad \phi_i = -\frac{3\varepsilon_m}{\varepsilon_p + 2\varepsilon_m}rE_o$$

The field inside the sphere is given by $E_i = -\nabla\phi_i = \frac{3\varepsilon_m}{\varepsilon_p+2\varepsilon_m}E_o$.

From Equation (3.51) the polarization **P** of the sphere is

$$P = (\varepsilon_p - \varepsilon_m)\varepsilon_o E_i = 3\varepsilon_0\varepsilon_m\left(\frac{\varepsilon_p - \varepsilon_m}{\varepsilon_p + 2\varepsilon_m}\right)E_o \tag{6.3}$$

Dielectrophoresis: Theory, Methodology and Biological Applications, First Edition. Ronald Pethig.

magnetism [e.g., 1]. The exercise is to evaluate the induced polarization of a spherical particle when it is inserted into a polarized dielectric medium. The polarization **P** of the sphere, which, from Chapter 3, is interpreted as an induced dipole moment per unit volume, is obtained by evaluating the field generated *within* the sphere. The result, given in Equation (6.3), is:

$$\mathrm{P} = 3\varepsilon_0\varepsilon_m\left(\frac{\varepsilon_p - \varepsilon_m}{\varepsilon_p + 2\varepsilon_m}\right)\mathrm{E}_o$$

where $\mathbf{E}_o$ is the initial uniform field within the medium of relative permittivity ε_m before the uncharged dielectric sphere of relative permittivity ε_p is embedded in it. The induced dipole moment **p** of a dielectric sphere of radius R is given by multiplying **P** by the sphere's volume:

$$\mathrm{p} = \frac{4}{3}\pi R^3\mathrm{P} = 4\pi\varepsilon_o\varepsilon_m R^3\left(\frac{\varepsilon_p - \varepsilon_m}{\varepsilon_p + 2\varepsilon_m}\right)\mathrm{E}_o \qquad (6.4)$$

Evaluation of the polarization P in Equation (6.3) of Box 6.1 was achieved using the relationship:

$$\mathrm{P} = (\varepsilon_p - \varepsilon_m)\,\varepsilon_o\mathrm{E}_i$$

which links the macroscopic phenomenon to processes at the molecular level. We can avoid making this association by noting that the derivation in Box 6.1 of the constant A in Equation (6.2) provides the following result for the component of the potential outside the sphere that is specifically associated with its polarization:

$$\phi_o = \frac{A}{r^2}\cos\theta = \left(\frac{\varepsilon_p - \varepsilon_m}{\varepsilon_p + 2\varepsilon_m}\right)\frac{R^3}{r^2}\mathrm{E}_o\cos\theta$$

This component adds to the original potential $\phi = -\mathrm{E}_o r\cos\theta$, which was initially established within the dielectric medium. Comparing ϕ_o to the potential of a macroscopic dipole, given by Equation (5.39), namely:

$$V(r,\theta) = \frac{\mathrm{p}}{4\pi\varepsilon_o\varepsilon_m}\frac{\cos\theta}{r^2}$$

we have the following equality

$$\frac{\mathrm{p}}{4\pi\varepsilon_o\varepsilon_m} = \left(\frac{\varepsilon_p - \varepsilon_m}{\varepsilon_p + 2\varepsilon_m}\right)R^3\mathrm{E}_o$$

This confirms that, from the perspective of the surrounding medium, the polarized sphere has assumed the properties of a macroscopic dipole of magnitude equal to that given by Equation (6.4). Furthermore, we can make an important observation that the constant A obtained for the solution of Laplace's equation described in Box 6.1 provides a value for the effective macroscopic moment **p** given by

$$\mathrm{p} = 4\pi\varepsilon_o\varepsilon_m\mathrm{A}$$

From this we can appreciate why the Clausius–Mossotti factor given by Equation (6.1) is so important in dielectrophoresis. It determines the magnitude and polarity of the induced dipole moment of the suspended particle and so dictates its behaviour in a nonuniform electric field. It is difficult to measure the dielectrophoretic (DEP) force acting on a particle directly, but a useful reference for determining relative DEP responses is a metallic conducting sphere suspended in air, so that $\varepsilon_p \gg \varepsilon_m$. This gives the maximum positive value of 1.0 for the *CM* factor in Equation (6.4). As another reference particle we can use an air bubble suspended in water, for which $\varepsilon_p = \sim 1.0$ and $\varepsilon_m = \sim 80$. In this case the *CM* factor has a value of −0.49, which is close to the maximum negative value it can attain, namely −0.5. For *positive* values of *CM* the induced dipole moment p is directed in the *same* direction as the field E and the particle experiences a DEP force that directs it *towards high field regions* at electrode edges. For *negative* values of *CM* the induced dipole moment p is directed in the *opposite* direction to the field and the particle experiences a DEP force that directs it *away from high field regions.* For the case where $\varepsilon_p = \varepsilon_m$ the *CM* factor in Equation (6.4) is zero. The particle is not polarized and, because the volume that it displaces has the same dielectric properties as the fluid medium, its presence does not perturb the external field E_o. The practical applications of observing this zero polarization situation include being able to determine the dielectric properties of the particle (the dielectric properties of the suspending medium are known and can be controlled); monitoring subtle changes of the dielectric properties of the particle (resulting from apoptosis or cell differentiation, for example); separating particles in a mixture for which the condition $\varepsilon_p^* = \varepsilon_m^*$ for each particle type occurs at a different frequency of the applied field.

In a paper published in 2013 in a reputable peer-reviewed journal, it is stated:

> The Clausius–Mossotti relation is named after the Italian physicist Ottaviano-Fabrizio Mossotti, whose 1850 book analyzed the relationship between the dielectric constants of two different media, and the German physicist Rudolf Clausius, who gave the formula explicitly in his 1879 book in the background not of dielectric constants but of indices of refraction.

Apart from correctly identifying the relevant scientists, this statement (which mirrors others like it in the literature) completely misrepresents their work! But does this matter? On the more important issue the authors of this paper are correct in stating that: 'In physics, this relation connects the relative permittivity of a dielectric to the polarizability of the atoms or molecules constituting the dielectric.' Thus, this refers to a *molecular* rather than *macroscopic* form of the *CM* relation, implying an understanding at the microscopic level of the relationship $\varepsilon_r = \mathbf{D}/\varepsilon_0\mathbf{E}$ given by Equation (3.12). For both the macroscopic and molecular forms of the *CM* relation, evaluations are required of the local field experienced by a test charge or polar molecule in a dielectric medium. A review of the work that has been directed to achieve this can assist in interpreting the dielectrophoretic response of a cell compared to that of a polar bioparticle, for example.

According to the advice of James Clerk Maxwell: 'It is of great advantage to the student of any subject to read the original memoirs on that subject, for science is always most completely assimilated when it is in its nascent state' [2, preface]. Because the *CM* factor is so central to the understanding and application of dielectrophoresis it is important that we understand the physical principles on which it is based. To approach such understanding we can follow Maxwell's advice and so what follows is a revue of the historic development of the theories that led to the derivation of what we will refer to as the Clausius–Mossotti–Lorentz relation. Apart from learning some important dielectric principles, readers of this chapter may also decide for themselves whether Mossotti wrote a book that 'analyzed the relationship between the dielectric constants of two different media' and that Clausius 'gave the formula explicitly in his 1879 book in the background not of dielectric constants but of indices of refraction.' :)

6.2 Development of the Clausius–Mossotti–Lorentz Relation

6.2.1 Siméon Denis Poisson and George Green

The early mathematical concepts that were applied to the theory of dielectrics largely originate from the work on magnetism by Poisson [3]. He adopted Coulomb's concept that two types of magnetic fluid, positive and negative, could arise from the decomposition of a neutral fluid. These magnetic fluids are not able to pass from one element to the next in a magnetic body, but are confined in their movements to their own individual element. If we suppose that an amount m of positive magnetic fluid is located at a point (x, y, z), then the magnetic intensity, defined as the force exerted on a unit magnetic pole, will diminish as the reciprocal distance from that point. The magnetic intensity (i.e., force) acting on a unit magnetic pole at (x', y', z') will thus have components:

$$-m\frac{\partial}{\partial x'}\left(\frac{1}{r}\right), -m\frac{\partial}{\partial y'}\left(\frac{1}{r}\right), -m\frac{\partial}{\partial z'}\left(\frac{1}{r}\right),$$
$$\text{where } r = \left[(x'-x)^2 + (y'-y)^2 + (z'-x)^2\right]^{1/2}$$

If equal quantities of the two magnetic fluids are displaced from each other within a magnetic element, the components of the magnetic intensity at (x', y', z') will be the negative derivatives with respect to x', y', z' of the function

$$\mathrm{X}\frac{\partial}{\partial x}\left(\frac{1}{r}\right) + \mathrm{Y}\frac{\partial}{\partial y}\left(\frac{1}{r}\right) + \mathrm{Z}\frac{\partial}{\partial z}\left(\frac{1}{r}\right)$$

where the vector [XYZ] is the magnetic moment of the element. If it were possible, this moment could be determined from the torque required to maintain the element at a fixed angle of orientation to the direction of the external magnetic force acting on it. To determine the magnetic intensity produced by the whole magnetic body, Poisson performed (by parts) the following triple integration:

$$V = \iiint \left(\mathrm{X}\frac{\partial}{\partial x} + \mathrm{Y}\frac{\partial}{\partial y} + \mathrm{Z}\frac{\partial}{\partial z}\right)\left(\frac{1}{r}\right) dx\, dy\, dz$$

to obtain the result

$$V = \iint \frac{1}{r}(I.dS) - \iiint \frac{1}{r}\mathrm{div}I\, dx\, dy\, dz \tag{6.5}$$

where I is the moment per unit volume of the body, more generally called the *magnetization*. This result by Poisson shows that the magnetic intensity produced in the external space of a magnetic body is the same as would be produced if the body were covered with a layer of magnetic surface-charge density $(I.dS)$ per element dS, together with a counter distribution of density div$I$ throughout its internal volume. Poisson also considered the magnetic intensity **F** at a point in a small cavity excavated within a magnetic body. For a small spherical cavity he obtained the result

$$\mathrm{F} = \mathrm{grad}V + \frac{4\pi}{3}\mathrm{I} \tag{6.6}$$

where I is the magnetization in the body at the location of the cavity. In what became known as *Poisson's Law of Induced Magnetism*, he also considered the magnetism induced in magnetizable metals such as soft iron when subjected to the magnetic field of an approaching permanent magnetic. Such metals were assumed to contain a large number of small spherical elements, with each sphere behaving as a perfect conductor of magnetic fluid so that its internal magnetic intensity is zero. In order to counteract the magnetic intensity F each sphere acquires a magnetic moment. We will see that these magnetic concepts were taken over into the theory of dielectric polarization. In particular, evaluation of the local field intensity acting on a molecule within a dielectric medium is central to the formulation of the Clausius–Mossotti factor.

Noting the arbitrary nature of the algebraic sign used in Equation (2.2) to define the relationship between the electric potential and field, the parameter V in Equations (6.5) and (6.6) takes the same form as Equation (4.19) and represents the sum of all the magnetic or electric charges in a field divided by their respective distances from a given point. This parameter was for the first time given the name *potential* by an enigmatic part-time mathematician, George Green, in his essay of 1828 [4]. Green (1793–1841), a miller's son, attended just one year of schooling in Nottingham, England, spanning the ages of 8 and 9. Yet he mastered the 'Continental' form of calculus devised by Leibnitz (rather than Newton's form) and generalized the theories of Laplace and Poisson at a time when they were not even taught at Cambridge! As posed by Grattan-Guinness [5], not least of the mysteries surrounding Green is how he knew about Poisson's paper published in Paris at the end of 1826 with sufficient time to study it before his *Essay* appeared in April 1828! The *Essay* was largely printed at his own expense, with the help of 52 supporting subscribers (most of whom would not have understood any of it) and so it *vanished from sight* [5]. His status in life as a full-time miller rather than mathematician was dramatically illustrated during the Reform Bill riots (1829), a year after his *Essay* was published. Green defended his mill with a musket whilst a daughter handed him the ammunition. On his visit to England in June 1930, Albert Einstein asked to visit the mill – which is now a museum known as Green's Windmill in the district of Sneinton, Nottingham.

On p. 16 of his *Essay,* Green states:

> Before proceeding to make known some relations which exist between the density of the electric fluid at the surfaces of bodies, and the corresponding values of the potential functions within and without those surfaces, the electric fluid being confined to them alone, we shall in the first place, lay down a general theorem which will afterwards be very useful to us. This theorem may be thus enunciated:
>
> Let U and V be two continuous functions of the rectangular co-ordinates x, y, z, whose differential coefficients do not become infinite at any point within a solid body of any form whatever; then will
>
> $$\int dx\,dy\,dz\,U\delta V + \int d\sigma U\left(\frac{dV}{dw}\right) = \int dx\,dy\,dz\,V\delta U + \int d\sigma\,V\left(\frac{dU}{dw}\right)$$
>
> the triple integrals extending over the whole interior of the body, and those relative to $d\sigma$, over its surface, of which $d\sigma$ represents an element: dw being an infinitely small line perpendicular to the surface, and measured from this surface towards the interior of the body.

This *general theorem* in fact is his now celebrated formula, called *Green's Theorem,* making the connection between surface and volume integrals. Whereas Poisson considered Equation (6.5) to be mainly a simplifying exercise, Green realized that his own theorem was of importance in relating properties inside bodies to properties on their surfaces and *vice versa*. Green's Theorem has found applications well beyond its initial intended importance in magnetism and electricity. For example, in quantum field theory Feynman's diagrams are essentially Green's mathematics in graphic form.

Being mindful of Maxwell's advice that: 'It is of great advantage to the student of any subject to read the original memoirs on that subject' [5, preface], rather than paraphrasing the relevant content of Green's *Essay* [4], which influenced the work of Clausius some 50 years later, it would therefore seem appropriate to reproduce it *verbatim*. The following are articles 4 and 5 (pp. 19–22) of the *Essay*:

> **4.** We will now proceed to determine some relations existing between the density of the electric fluid at the surface of a body, and the potential functions thence arising, within and without this surface. For this, let $\rho\, d\sigma$ be the quantity of electricity on an element $d\sigma$ of the surface, and V, the value of the potential function for any point p within it, of which the coordinates are x, y, z. Then, if V' be the value of this function for any other point p' exterior to this surface, we shall have
>
> $$V = \int \frac{\rho d\sigma}{\sqrt{[(\xi - x)^2 + (\eta - y)^2 + (\varsigma - z)^2]}}$$
>
> ξ, η, ζ being the coordinates of $d\sigma$, and
>
> $$V' = \int \frac{\rho d\sigma}{\sqrt{[(\xi - x')^2 + (\eta - y')^2 + (\varsigma - z')^2]}}$$

the integrals relative to $d\sigma$ extending over the whole surface of the body.

It might appear at first view, that to obtain the value of $V'V$ from that of V, we should merely have to change x, y, z, into x', y', z': but, this is by no means the case; for, the form of the potential function changes suddenly, in passing from the space within to that without the surface. Of this, we may give a very simple example, by supposing the surface to be a sphere whose radius is a and centre at the origin of the coordinates; then, if the density ρ be constant, we shall have

$$v = 4\pi\rho a \quad \text{and} \quad V' = \int \frac{4\pi a^2 \rho}{\sqrt{x'^2 + y'^2 + z'^2}}$$

which are essentially distinct functions.

With respect to the functions V and V′ the general case, it is clear that each of them will satisfy Laplace's equation, and consequently

$$0 = \delta V \quad \text{and} \quad 0 = \delta' V'$$

moreover, neither of them will have singular values; for any point of the spaces to which they respectively belong, and at the surface itself, we shall have

$$\bar{V} = \bar{V}'$$

the horizontal lines over the quantities indicating that they belong to the surface. At an infinite distance from this surface, we shall likewise have

$$V' = 0$$

5. To convince ourselves, that there does exist such a function as we have supposed U to be; conceive the surface to be a perfect conductor put in communication with the earth, and a unit of positive electricity to be concentrated in the point p'; then the total potential function arising from p′ and from the electricity it will induce upon the surface, will be the required value of U. For, in consequence of the communication established between the conducting surface and the earth, the total potential function at this surface must be constant, and equal to that of the earth itself, i.e. to **zero** (seeing that in this state they form but one conducting body). Taking, therefore, this total potential function for U, we have evidently $0 = \bar{U}, 0 = \delta U$, and $U = {}^1/_r$ for those parts infinitely near to p'. As moreover, this function has no other singular points within the surface, it evidently possesses all the properties assigned to U in the preceding proof.

Again, since we have evidently $U'g= 0$, for all the space exterior to the surface, the equation (4) art. 4 gives

$$0 = 4\pi(\rho) + \frac{d\bar{U}}{dw'}$$

where (ρ) is the density of the electricity induced on the surface, by the action of a unit of electricity concentrated in the point p'. Thus, the equation (5) of this article becomes

$$\bar{V} = -\int d\sigma(\rho)V$$

This equation is remarkable on account of its simplicity and singularity, seeing that it gives the value of the potential for any point p', within the surface, when V, its value at the surface itself is known, together with (ρ) the density that a unit of electricity concentrated in p' would induce on this surface, if it conducted electricity perfectly, and were put in communication with the earth.

Amongst these important concepts described in these extracts from his Essay, Green has described the important boundary conditions used in Box 6.1, namely that the potential varies continuously across a boundary surface and that at a great distance from a charged surface its potential is zero. These concepts were exploited by Clausius in his considerations of how the local forces acting on a polarizable particle in a dielectric translate to the macroscopic polarization of a whole dielectric body. Of particular relevance to this in Green's *Essay* is the following part of article 9 [4, pp. 34, 35]:

> Suppose we have a hollow, and perfectly conducting shell, bounded by any two closed surfaces, and a number of electrical bodies are placed, some within and some without it, at will; then, if the inner surface and interior bodies be called the interior system; also, the outer surface and exterior bodies the exterior system; all the electrical phenomena of the interior system, relative to attractions, repulsions, and densities, will be the same as would take place if there were no exterior system, and the inner surface were a perfect conductor, put in communication with the earth; and all those of the exterior system will be the same, as if the interior one did not exist, and the outer surface were a perfect conductor, containing a quantity of electricity, equal to the whole of that originally contained in the shell itself, and in all the interior bodies.
>
> This is so direct a consequence of what has been shown in articles 4 and 5, that a formal demonstration would be quite superfluous, as it is

> easy to see, the only difference which could exist, relative to the interior system, between the case where there is an exterior system, and where there is not one, would be in the addition of a constant quantity, to the total potential function within the exterior surface, which constant quantity must necessarily disappear in the differentials of this function, and consequently, in the values of the attractions, repulsions, and densities, which all depend on these differentials alone. In the exterior system there is not even this difference, but the total potential function exterior to the inner surface is precisely the same, whether we suppose the interior system to exist or not.

Thus, the work of Green in generalizing the concepts developed by Poisson and giving to them a more solid theoretical basis, had by 1828 endowed to the subject of electrostatics a highly sophisticated status.

6.2.2 Faraday, Mossotti, Clausius, Maxwell, Lorenz and Lorentz

Amongst Michael Faraday's many important contributions to magnetism and electricity were his demonstrations that the effect of a magnetic force on magnetic materials was analogous to that of an electrostatic force on a dielectric [6]. In particular, he recognized that dielectric bodies receive a molecular induction under the influence of an external charged body, or in other words that their molecules acquire an electrical polarity. Physicists had previously considered such molecules to be indifferent to electric induction, their intervention consisting only in the imposing of a passive resistance to the dispersal of electricity.

The analogies between the effects of magnetic and electrostatic forces were taken forward by Ottaviano-Fabrizio Mossotti. Born in 1791 in Novara, Italy, Mossotti fled to London in 1823 fearing arrest because of his liberal attitudes and membership of a secret society. He then taught astronomy at the University of Buenos Aires from 1827 to 1835 before returning to Italy in 1840 with an academic position at Pisa University. In 1848, he fought as captain of the troops of the two Tuscan universities in the war for independence from Austria and shortly before his death in 1863 he was elected senator of the Kingdom of Italy [7]. Despite these activities he set himself the goal to translate into an analytical expression the electrical induction of molecules under the influence of the local 'internal field' acting on them and to then ascertain how this influenced the distribution of the electricity in the dielectric [8]. He acknowledged that the 'subtle and abstruse analysis' ('una analisi sottile ed astrusa') he required for this had already been formulated by Poisson [3] in his theory of magnetic induction.

From the insights of Faraday and Poisson, Mossotti was able to suppose that each dielectric molecule contained corpuscles charged vitreously (i.e., positive) and also corpuscles charged resinously (i.e., negative) [8]. In the absence of an electric field the corpuscles are arranged so as to neutralize each other, but under the action of a local 'internal' field they are separated so that the entire molecule takes the form of an 'electric doublet' (i.e., dipole). This charge separation is achieved without their escaping the space of each molecule. As depicted in Figure 3.25, each molecule is polarized as a whole, with one side acquiring a positive charge and the opposite side an equal but negative charge. Mossotti may also have been the first to consider the phenomenon of electrical breakdown at the molecular level. He states that 'the charge separation is all the greater the greater the external action and in the dielectric corpuscles it grows to the point where the electricity, having over-accumulated in the extremes, overflows in leaps and sparks from one molecule to the others' ('irrompe per salto e con scintilla dalle une molecole alle altre') [8, p. 51].

It is of interest (and somewhat amusing) to read Maxwell's account of Mossotti's work:

> Thus, when Mossotti observed that certain quantities relating to electrostatic induction in dielectrics had been shewn by Faraday to be analogous to certain quantities relating to magnetic induction in iron and other bodies, he was enabled to make use of the mathematical investigation by Poisson relative to magnetic induction, merely translating it from magnetic language into the electric, and from French into Italian [9].

Does this appear to be less than fulsome praise of Mossotti's achievement? Elsewhere [1, Art. 62, p. 70] Maxwell writes:

> He assumes the existence within the dielectric of small conducting elements, capable of having their opposite surfaces oppositely electrified by induction, but not capable of losing or gaining electricity on the whole, owing to their being insulated from each other by a non-conducting medium. This theory of dielectrics is consistent with the laws of electricity, and may be actually true. If it is true, the specific inductive capacity of a dielectric may be greater, but cannot be less, than that of a vacuum. No instance has yet been found of a dielectric having an inductive capacity less than that of a vacuum, but if such should be discovered, Mossotti's physical theory must be abandoned, although his

formulae would all remain exact, and would only require us to alter the sign of a coefficient.

The most important accomplishment of James Clerk Maxwell was to demonstrate that Faraday's theory of lines of force, as well as alternative theories that invoked action at a distance, could be replaced by the concept of an electromagnetic field. In his famous paper of 1865 [10] describing the propagation of light as an electromagnetic phenomenon and the theory to calculate its speed ($c^2 = 1/(\varepsilon_o \mu_o)$), Maxwell introduces the concept of displacement current as follows:

> In a dielectric under the action of electromotive force, we may conceive that the electricity in each molecule is so displaced that one side is rendered positively and the other negatively electrical, but that the electricity remains entirely connected with the molecule. The effect of this action on the whole dielectric mass is to produce a general displacement of electricity in a certain direction. This displacement does not amount to a current, because when it is has attained to a certain value it remains constant, but it is the commencement of a current, and its variations constitute currents in the positive or the negative direction according as the displacement is increasing or decreasing. In the interior of the dielectric there is no indication of electrification, because the electrification of the surface of any molecule is neutralized by the opposite electrification of the surface of the molecule in contact with it; but the bounding surface of the dielectric, where the electrification is not neutralized, we find the phenomena which indicate positive or negative electrification.

As outlined in Box 6.2 the displacement current conceived by Maxwell has units of current density, but takes the form of a time-varying electric field rather than a 'conventional' current of electric charge. However, as with 'conventional' conduction current, a displacement current generates an associated magnetic field as given by Ampère's Circuital Law. Maxwell added this displacement current to Ampère's Law in his famous equations (see Boxes 3.3 and 6.2).

Maxwell defined the displacement flux by the relationship given in Equation (3.12):

$$D = \varepsilon_o \varepsilon_r E$$

where E is the field within the dielectric. As noted in Chapter 3, because the relative permittivity ε_r is a

Box 6.2 Displacement Current

The diagram in this box depicts the charging of a parallel-plate capacitor with current i_c.

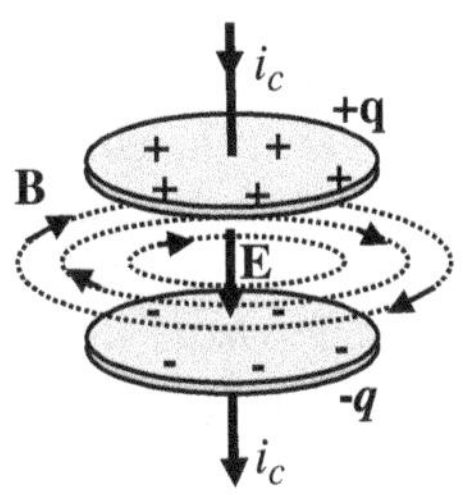

The capacitance C of the capacitor is given by $C = A\varepsilon/d$, where A is the area of the circular plates, d is their distance apart and ε ($\varepsilon = \varepsilon_o \varepsilon_r$) is the permittivity of the dielectric medium between the plates. The charge q at any instant is given by $q = Cv$, where v is the instantaneous potential difference between the plates. Neglecting field fringing, the uniform field between the plates is given by $E = v/d$. We can derive the following expression for the instantaneous charge:

$$q = Cv = \frac{A\varepsilon}{d} Ed = \varepsilon EA = \varepsilon \Phi_E$$

where Φ_E is the E-field flux through the electrode surface (see Example 3.7). The charging current at any instant is equal to dq/dt, so that

$$i_c = \frac{dq}{dt} = \varepsilon \frac{d\Phi_E}{dt}$$

To provide continuity of current between the current leads and through the dielectric in the capacitor, Maxwell conceived of the concept of a displacement current i_D:

$$i_D = \varepsilon \frac{d\Phi_E}{dt},$$

with a corresponding displacement current density j_D

$$j_D = \frac{i_D}{A} = \frac{\varepsilon}{A} \frac{d\Phi_E}{dt} = \frac{\varepsilon}{A} \frac{AdE}{dt} = \varepsilon \frac{dE}{dt}$$

Maxwell introduced the displacement current into Ampère's Circuital Law to give the third of what are universally referred to as *Maxwell's equations*:

$$\oint B \cdot dl = \mu \left(i_C + \varepsilon \frac{d\Phi_E}{dt} \right)$$

where μ is the magnetic permeability of the medium. The displacement current can thus be considered as the source of the magnetic field B created between the capacitor plates shown in the above diagram.

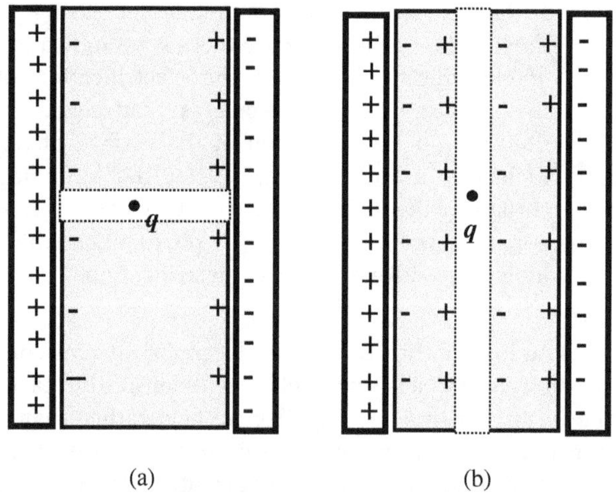

Figure 6.1 Imaginary exercises to understand the difference between the electric displacement field D/ε_o and the (internal) field E within the dielectric. (a) A unit charge placed inside a channel drilled through the dielectric at right angles to the electrodes experiences a force due to the *net free* charge on the electrodes, namely the internal field E. (b) A unit charge placed in a narrow channel parallel to the electrodes experiences a force (D/ε_o) due to the combination of the *net free* charge on the electrodes (E) *and* the induced charge (P/ε_0) on the surface of the dielectric, namely P/ε_0. D/ε_o is thus the field produced by the free charge density on the electrodes in the *absence* of the dielectric.

number, D/ε_o has dimensions of an electric field. It is important to understand the difference between D/ε_o and E and this is achieved by performing the two imaginary exercises described in Figure 6.1. With a drill of infinitely small diameter we make a channel of infinitely small cross-section through the *polarized* dielectric at right angles to the electrode plates. We now place a *unit* positive charge into this channel as shown in Figure 6.1 (a). What will be the force exerted on this test charge? According to the definition of an electric field given by Equation (2.1) the force acting on this unit charge is the *internal* field E, since it experiences only the field of the *net free* charge on the plates. For the second imaginary exercise, shown in Figure 6.1(b), we bore out an infinitely narrow channel through the dielectric that runs parallel to the plates. What will be the force exerted on a unit test charge placed in this channel? Close inspection of Figure 6.1(b) shows that this force, which we call the electric displacement D/ε_o is equal to the field produced by the sum of the *net free* charge on the electrodes and the *induced* charge on the surface of the dielectric. The charge induced on the surface of the dielectric is the polarization P defined in Chapter 3 and from Equation (3.42) the field produced inside our imaginary empty channel by this surface charge is P/ε_0. The general relationship between D and E, which assumes *nothing* about the dependency of the polarization on the field, is thus:

$$D/\varepsilon_o = E + P/\varepsilon_0 \tag{6.7}$$

We thus identify the displacement force D/ε_o as equal to the field produced by the charge density on the plates in the *absence* of the dielectric. With reference to Figure 3.1 and Equation (3.2) we can make the identity $D/\varepsilon_o = (V_0/d)\cdot n = E_0$ (where n is the unit vector normal to the electrode surface). We have also deduced that the internal field $E = (V_1/d)\cdot n = E_1$. In Equation (3.1) we have defined $\varepsilon_r = E_0/E_1$, so that on inserting our identified expressions for E_0 and E_1 we find ourselves in agreement with the relationship proposed by Maxwell, namely: $D = \varepsilon_o\varepsilon_r E$.

Maxwell died in 1879, the same year that Clausius published volume two of his book *The Mechanical Theory of Heat*, in which he gives a mathematical treatment of dielectric media [11]. Rudolf Julius Emanuel Clausius, born in 1822 in Koszalin, Poland, is better known for his derivation of the Second Law of Thermodynamics and for introducing the concept of entropy. In 1865 he concluded that the energy of the universe is constant and that the entropy of the universe tends to a maximum. He shares with Mossotti the distinction of being a war hero! In 1870, during the Franco-Prussian War, he was wounded in battle (having organized an ambulance corps) and was awarded the Iron Cross.

Like Mossotti, Clausius assumed that within a dielectric there are slightly conducting molecular corpuscles, which are separated from the others by nonconductive spaces, so that under the action of a local 'internal field' the electricity moves only within the individual corpuscles but cannot pass from one to the other. He also introduced the hypothesis that the corpuscles are electrically polar in advance ('schon im Voraus electrischpolar sind') [11, p. 66]. Clausius poses the question as to whether the deflection of the polar particles by an external electric force is limited by an elastic restoring force proportional to their deflection, or in contrast whether the mutual cohesive interactions of the polar particles result in a sufficiently large frictional force as to limit their deflection and then their return to random orientations on removal of the external force. In his mathematical treatment, Clausius adopts the fundamental equations derived by Poisson [3] and their later generalization by Green [4] and so everywhere in these previous works he translates 'north and south magnetization fluid' to be 'positive and negative electricity'. For spherically conducting corpuscles Clausius derives the following interesting relationship [11, p. 94]:

$$K = \frac{1+2g}{1-g} \tag{6.8}$$

where K is the specific inductive capacity (i.e., $K = \varepsilon_r$) of the dielectric medium and g is a simplifying factor relating the induced electric moment of a corpuscle and the electric intensity produced by the whole dielectric body. It is implicit that g is a function of the mass density ρ of the dielectric. From simple manipulation of Equation (6.8) we can thus derive the relationship:

$$\left(\frac{\varepsilon_r - 1}{\varepsilon_r + 2}\right)\frac{1}{\rho} = const \tag{6.9}$$

This is the key result that leads to what became known as the Clausius–Mossotti relation. Clausius was apparently unaware of the mathematical relationship between the refractive index and the density of a medium formulated by the Danish mathematician Ludwig Lorenz and presented by him at meetings of the Royal Danish Academy of Sciences in Copenhagen in 1869 and 1875 [12]. Also, independently of this the Dutch physicist Hendrik A. Lorentz presented the same result at a conference in Amsterdam in 1878 [13]. Of the similarity of their names and conclusions he later made the comment; 'which is certainly a curious case of coincidence' [14, p. 145]. Lorentz later shared the 1902 Nobel Prize in Physics with Pieter Zeeman for the discovery and theoretical explanation of the Zeeman effect. The relationship found by Lorenz and Lorentz (expressed in SI units) is:

$$\frac{n^2 - 1}{n^2 + 2} = \frac{N\alpha}{3\varepsilon_o} \tag{6.10}$$

where n is the refractive index and N the number of molecules per unit volume of a medium, with α the mean polarizability of the molecules. Lorenz and Lorentz published their works again in 1880 [15, 16] and so Equation (6.10) is generally known as the Lorentz–Lorenz formula. We can equate N to $N_A\rho/M$, where ρ is the density of the medium, M is the molecular weight of the atoms or molecules and N_A is the number of atoms or molecules in a mole of the medium substance – namely Avogadro's constant (6.02 × 10^{23}). Substituting for N in Equation (6.10) we obtain for the *molar refraction* P_M:

$$P_M = \frac{n^2 - 1}{n^2 + 2}\frac{M}{\rho} = \frac{N_A\alpha}{3\varepsilon_o} \tag{6.11}$$

The molar refraction is found, within limits, to be additive. We can understand this by noting that the ratio $M:\rho$ in Equation (6.11) has units of volume, so that P_M and α also have units of volume. We can thus treat P_M as a molar volume to provide an approximate measure of the actual volume (without free space) of the polarizable electronic clouds of the atoms or molecules in one gram mole, as distinct from the apparent volume given by M/ρ. As such P_M should be independent of temperature. Thus, known values of the molar refraction of various atoms and chemical bonds can be assigned to other molecular structures and their molar refractions predicted to reasonable accuracy. Examples of the additivity of polarizability volumes can be found in Table 6.1 and are based on extensive measurements by Vogel [17] of the refractive dispersion of the sodium D-line emission. The sodium D-line occurs at a frequency of 5.1 × 10^{14} Hz, which is low enough to ensure that the polarizability values include both the atomic and electronic polarizations described in Chapter 7.

Table 6.1 Values of the polarizability volume α (cm^{-3}) for some atoms and chemical groups [17].

Atomic			
H (in CH_2)	1.028	N (tertiary aliphatic amines)	2.744
C (in CH_2)	2.591	N (tertiary aromatic amines)	4.243
O (ethers)	1.764	S (sulphides)	7.921
O (acetals)	1.607	S_2 (disulphides)	16.054
O (carbonyl)	2.122	Cl	5.844
Group			
CH_2	4.647	OH (alcohols)	2.546
CH_3	5.653	SH (thiols)	8.757
CO (ketones)	4.601	NH_2 (primary aliphatic amines)	4.438
COO (esters)	6.200	NH (secondary aliphatic amines)	3.610
COOH	7.226	NH (secondary aromatic amines)	4.678

Note: The values given in this table are in cgs units, and are sometimes expressed in ångström units ($Å^3 = 10^{-24}$ cm^3). In SI units α ($F.m^2$) = $\alpha_{cgs}/(4\pi\varepsilon_o)$, so that α (cm^3) ≈ 9 × 10^{15} × α ($F.m^2$).

In 1865 Maxwell had deduced that the specific inductive capacity (i.e., relative permittivity) is 'equal to the square of the index of refraction divided by the coefficient of magnetic induction' [10, p. 501]. Therefore, unless a transparent material possesses magnetic properties, such as thin films developed using nanotechnology [18, 19], we have $\varepsilon_r = n^2$. Inserting this relationship into Equation (6.10) we obtain the result:

$$\frac{\varepsilon_r - 1}{\varepsilon_r + 2} = \frac{N\alpha}{3\varepsilon_0} = \rho\frac{N_A\alpha}{3M\varepsilon_0} \tag{6.12}$$

For any particular material the factor $(N_A\alpha/M\varepsilon_o)$ is a constant and so Equation (6.12) gives the result predicted by Clausius in the form of Equation (6.9). It also follows from Equation (6.11) that we can define the *molar polarization* P_M as:

$$P_M = \frac{\varepsilon_r - 1}{\varepsilon_r + 2}\frac{M}{\rho} = \frac{N_A\alpha}{3\varepsilon_o}$$

The molar polarization is a purely microscopic quantity related directly to the electrical properties of the molecules that form the dielectric material. We can also define the *volume polarization* P_v as

$$P_v = \frac{P_M\rho}{M} = \frac{\varepsilon_r - 1}{\varepsilon_r + 2}$$

Lorentz appears to be the first to draw attention to the fact that the relation between relative permittivity and the density ρ of a dielectric, expressed as Equation (6.9), is 'a formula corresponding to one that was given long ago by Clausius and Mossotti' [14, p. 145]. Lorentz was certainly the first to attempt an evaluation of the local electric force acting on a polarizable particle in the bulk of a polarized dielectric. He performed this exercise as part of his theory of the *propagation of light in a system of molecules* [14, Ch. IV, pp. 133–167]. He adopted the reasoning by which Kelvin came to distinguish between the magnetic force and the magnetic induction – namely by defining these as forces exerted on a pole of unit strength, placed inside a small cavity within a magnetized body. The magnetically polarized parts of the body outside the cavity turn their poles more or less towards it and thus produce on its walls a certain distribution of magnetism.

'Because the formulae for the field produced by a variable electric moment are less simple than those which determine the action of a constant molecular magnet' Lorentz recognized that 'The formulae however much resemble each other if the point for which the field of a particle is to be determined, lies at a distance from it that is small compared with the wave-length. In this case the field can be approximately considered as an electrostatic one, such as would exist if the electric moment did not change in the course of time' [14, pp. 137–138].

He therefore anticipated that the local electrical force would be similar to the magnetic case given by Equation (6.8) and first derived by Poisson [3]. Because he had introduced electrostatic units 'of such a kind that we get rid of the larger part of such factors as 4π and $\sqrt{4\pi}$, by which the formulae were originally encumbered' [14, p. 2], Lorentz anticipated and in fact found that the local field would be of the form

$$E_l = E + \left(\frac{1}{3} + s\right) P \tag{6.13}$$

Prophetically, Lorentz states that the parameter 's is a constant which it will be difficult exactly to determine' [14, p. 138].

To arrive at this result Lorentz envisaged that the polarizable particle (e.g., an atom or molecule) is enclosed within a spherical cavity 'whose dimensions are infinitely small in a physical sense, and we conceive for a moment, all other particles lying within this surface to be removed.' As indicated in Figure 6.2, this volume should thus be microscopically small compared to the distance between the electrodes, but large enough to contain a sufficient number of particles so that the dielectric properties within the cavity are uniform and the same as the bulk dielectric. Extending the concept developed by Mossotti and Clausius, Lorentz considered that each particle contains a single electron, which is displaced from its position of equilibrium by the local electric force $\mathbf{E}_l$ acting on it.

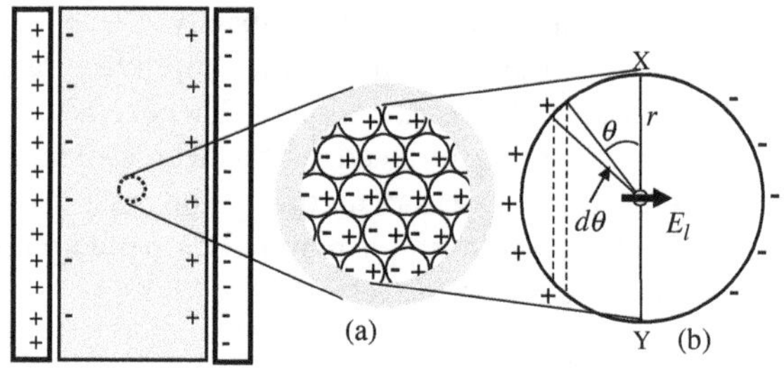

Figure 6.2 To determine the local field E_l acting on a polarizable particle in a dielectric we consider the forces acting on a unit charge at the centre of an imaginary spherical region of radius *r*. The volume of this sphere is small compared to the distance between the electrodes, but large enough to contain a sufficient number of particles so as to exhibit the bulk permittivity. The presence and polarization of this imaginary cavity does *not* affect the field in the dielectric outside it.

The distribution of induced charges on the surface of the cavity, due to the polarization of the outside portion of the dielectric body, will exert a force $\mathbf{E}_1$ that must be added to the electric force **E** generated by the electrodes. For a spherical cavity of radius a and θ the angle between the radius drawn towards an element of charge on its surface and the polarization **P**, Lorentz gives the force $\mathbf{E}_1$ as

$$E_1 = \frac{1}{4\pi a^2} \int |P| \cos^2 \theta d\theta = \frac{1}{3} P$$

The derivation of this result is given more fully in Box 6.3 and employs SI units rather than the *ad hoc* system devised by Lorentz. If the particles that have been removed from the cavity are now restored to their original places, their induced electric moments will produce a third force $\mathbf{E}_2$. Lorentz shows, for a system of particles having a regular cubical arrangement, that $\mathbf{E}_2 = 0$. This proof is reproduced in Box 6.4 and can also be applied with a certain degree of approximation to isotropic bodies in general, such as glass, fluids and gases (Box 6.5). However, in general we should write $\mathbf{E}_2 = s\mathbf{P}$. This term is included in Equation (6.13).

The induced moment p is directly proportional to the local force, so that

$$p = \alpha E_l \tag{6.14}$$

The constant of proportionality α is thus the moment induced by a field of unit intensity and represents the *polarizability* per polarizable particle of the dielectric. For gases, liquids and some solid dielectrics to good approximation $s = 0$ in Equation (6.13), so that (in SI units)

$$E_l = E + \frac{P}{3\varepsilon_0} \tag{6.15}$$

Box 6.3 Derivation of the Local (Lorentz) Field in a Polarized Dielectric

Figure 6.2 shows an imaginary spherical cavity within a polarized dielectric medium. The *local* force $\mathbf{E}_L$ on a unit charge at the sphere's centre has three components:

$$\mathbf{E}_L = \mathbf{E} + \mathbf{E}_1 + \mathbf{E}_2$$

$\mathbf{E}$ is the force due to the free charge density on the electrodes (identified in the main text as equal to $\mathbf{D}/\varepsilon_0$). $\mathbf{E}_1$ is the force arising from the polarization of the dielectric and for ease of calculation can be divided into two parts. $\mathbf{E}_{1a}$ due to the induced charge density $\mathbf{P}$ on the external surface of the dielectric (given by $\mathbf{E}_{1a} = -\mathbf{P}/\varepsilon_0$) and $\mathbf{E}_{1b}$ the Coulombic force due to charges induced on the surface of the small spherical cavity. As indicated in Figure 6.2(b) this force is determined by dividing the surface of the cavity into rings that are parallel with the electrodes. For a ring bounded by angles θ and $\theta + d\theta$ to the equatorial plane XY, the polarization perpendicular to the left-hand hemispherical surface is $\mathbf{P}\sin\theta$. For a sphere of radius r the total charge Q on the ring is thus given by:

$$Q = 2\pi r\cos\theta \times rd\theta \times \mathrm{P}\sin\theta = 2\pi r^2\mathrm{P}\sin\theta\cos\theta.d\theta$$

The Coulombic force exerted by this charged elemental ring on a *unit* charge at the centre of the sphere is equal to:

$$\frac{2\pi r^2\mathrm{P}\sin\theta\cos\theta.d\theta.\sin\theta}{4\pi\varepsilon_0 r^2} = \frac{\mathrm{P}\sin\theta\cos\theta.d\theta.\sin\theta}{2\varepsilon_0}$$

Thus, the total force ($\mathbf{E}_{1b}/2$) from the induced charges on the left-hand hemisphere is:

$$\frac{\mathrm{E}_{1b}}{2} = \frac{\mathrm{P}}{2\varepsilon_0}\int_0^{\pi/2}\sin^2\theta\cos\theta.d\theta.$$

$$= \frac{\mathrm{P}}{2\varepsilon_0}\int_0^{\pi/2}\sin^2\theta.d\sin\theta = \frac{\mathrm{P}}{6\varepsilon_0}$$

The induced (negative) charges on the right-hand hemispherical surface produce an added force $\mathbf{E}_{1b}/2 = \mathbf{P}/(6\varepsilon_0)$, which is directed from left to right, so that $\mathbf{E}_{1b} = \mathbf{P}/(3\varepsilon_0)$.

For gases and liquids, where the molecules all move independently of each other, $\mathbf{E}_2$ is close to zero. This should also be the case for solid amorphous dielectrics. Lorentz also showed that $\mathbf{E}_2 = 0$ for crystals of cubic lattice structure (see Boxes 6.4 and 6.5).

For gases, liquids and some solid dielectrics, the local field acting on a polarizable particle in a polarized dielectric is thus to good approximation given by:

$$\mathrm{E}_L = \mathrm{E} + \mathrm{E}_{1a} + \mathrm{E}_{1b} = \frac{\mathrm{D}}{\varepsilon_o} - \frac{\mathrm{P}}{\varepsilon_o} + \frac{\mathrm{P}}{3\varepsilon_0}$$

$$= \left(\mathrm{E} + \frac{\mathrm{P}}{\varepsilon_o}\right) - \frac{\mathrm{P}}{\varepsilon_o} + \frac{\mathrm{P}}{3\varepsilon_0} = \mathrm{E} + \frac{\mathrm{P}}{3\varepsilon_0}$$

From Equation (3.48) we have $\mathrm{P} = (\varepsilon_r - 1)\varepsilon_o\mathrm{E}$. On substituting for P, we obtain the following expression for the internal field – known as the Lorentz field:

$$\mathrm{E}_l = \left(\frac{\varepsilon_r + 2}{3}\right)\mathrm{E} \tag{6.16}$$

For all dielectric media $\varepsilon_r > 1$. We therefore conclude that the local, microscopic, field is always *larger* than the macroscopic field applied to the dielectric. This result, which at first sight appears nonintuitive, arises because inside our envisaged cavity within the bulk of a dielectric, the polarizable charge within an atom or molecule

Box 6.4 Lorentz's Proof that E_2, as Cited in Box 6.3, is Zero for a Cubic Lattice

The following is the proof given by Lorentz [14, Note 55, p. 308] that the force $\mathbf{E}_2$ (cited in Box 6.3 and the main text) is zero for the case of the medium within the spherical cavity of Figure 6.2 having a cubic lattice structure.

In the case of a cubical arrangement all the particles within the sphere may be said to have equal electric moments $\mathbf{p}$. Taking the centre of the spherical cavity as origin of coordinates, we have for the force exerted in the direction of x by a particle situated at the point (x, y, z), at a distance r from the centre,

$$\frac{\mathrm{p}_x}{4\pi}\cdot\frac{3x^2 - r^2}{r^5},\ \frac{\mathrm{p}_y}{4\pi}\cdot\frac{3xy}{r^5},\ \frac{\mathrm{p}_z}{4\pi}\cdot\frac{3xz}{r^5}$$

But the sums

$$\sum\frac{3x^2 - r^2}{r^5},\ \sum\frac{3xy}{r^5},\ \sum\frac{3xz}{r^5}$$

are zero when extended to all the particles within the sphere. For the second and third sum this is immediately clear if we take the axes of coordinates parallel to the principal direction of the cubical arrangement. Further, for axes of this direction,

$$\sum\frac{3x^2 - r^2}{r^5} = \sum\frac{3y^2 - r^2}{r^5} = \sum\frac{3z^2 - r^2}{r^5}$$

Showing that each of these expressions must be zero, because their sum is so.

Box 6.5 The Field due to Dipoles within an Isotropic Dielectric Sphere is Zero

The field due to individual polar molecules within the spherical region shown in Figure 6.2 can be obtained by taking the average of the fields due to all the dipoles inside the sphere. From Equation (5.44) the field at a distance r from a dipole is:

$$\mathrm{E} = \frac{1}{4\pi\varepsilon_0\varepsilon_r}\frac{1}{r^3}[3(\mathrm{p}\cdot\hat{\mathrm{r}})\hat{\mathrm{r}} - \mathrm{p}]$$

The spatial average of the x-component of the field is given by

$$\langle E_x\rangle = \frac{1}{4\pi\varepsilon_o\varepsilon_r}\sum\left[\frac{3\left(p_x x^2 + p_y xy + p_z xz\right)}{r^5} - \frac{p_x}{r^3}\right]$$

In an isotropic dielectric the spatial averages of the fields in the x-, y- and z-components are equal and so

$$\langle x^2\rangle = \langle y^2\rangle = \langle z^2\rangle = \frac{\langle r^2\rangle}{3};\ \langle xy\rangle = \langle yz\rangle = \langle zx\rangle = 0$$

The average of the fields due to individual dipoles inside the sphere is therefore zero. This result will not hold if there are molecular interactions between the dipoles, such as results from hydrogen-bonding in water, or for polymers such as proteins with oriented molecular side groups.

experiences the applied macroscopic field *plus* the field due to the polarization of the dielectric medium around the cavity. It exactly mirrors the result deduced for magnetization by Poisson in the form of Equation (6.6) and taken forward by Mossotti, Clausius and Lorentz for the case of dielectric polarization.

If N is the number of polarizable particles per unit volume, then from equations (6.14) and (6.16), the induced moment per unit volume (i.e., the polarizability P) is given by:

$$\mathrm{P} = N\mathrm{p} = N\alpha\mathrm{E}_l = N\alpha\left(\frac{\varepsilon_r + 2}{3}\right)\mathrm{E} \tag{6.17}$$

From Equation (3.48)

$$\mathrm{P} = (\varepsilon_r - 1)\,\varepsilon_o\mathrm{E}$$

Using this expression to eliminate P/E from Equation (6.17) we obtain the result given by Equation (6.12), which we will now refer to as the Clausius–Mossotti–Lorentz (CML) relation:

$$\frac{\varepsilon_r - 1}{\varepsilon_r + 2} = \frac{N\alpha}{3\varepsilon_0} \tag{6.18}$$

If we cannot assume $s = 0$ in Equation (6.13) the CML relation takes the form

$$\frac{\varepsilon_r - 1}{\left[1 + a\left(\varepsilon_r - 1\right)\right]} = \frac{N\alpha}{\varepsilon_0} \tag{6.19}$$

where

$$a = \frac{1}{3} + s \tag{6.20}$$

with s being a numerical parameter that is a measure of interactions with neighbouring polarized particles that might increase the local field in the dielectric.

6.2.3 Peter Debye

By 1910 it was found that gases and liquids could be assigned to one of two broad classes of dielectric. The first class, mainly substances having relatively low values for ε_r, were found to have relatively constant values for their polarizability α as a function of frequency and temperature. The second class possessed higher ε_r values than expected from their refractive index and α was found to decrease with increasing temperature. From our discussion of Equation (6.11) this temperature effect was not expected. An explanation was proposed by the Dutch-American physical chemist Peter J. W. Debye, who received the Nobel Prize in Chemistry in 1936 partly for this work. Debye developed further the hypothesis of Clausius that some molecules possess a permanent dipole moment. The total polarizability α_T of a molecule may thus comprise contributions from an induced polarization α_i, resulting from distortion of their electronic charge distribution, as well as from reorientation of their permanent dipole moment **p** under the influence of an external field [20, 21]. As shown in Chapter 7, where orientation polarization is discussed, the total polarizability α_T of a polar molecule takes the form:

$$\alpha_T = \alpha_i + \frac{\mathrm{p}^2}{3kT} \tag{6.21}$$

For molecules that possess a permanent dipole moment and thus a total polarizability given by Equation (6.21), we can write

$$\frac{\varepsilon_r - 1}{\varepsilon_r + 2} = \frac{1}{3\varepsilon_o}\sum N\left(\alpha_i + \frac{\mathrm{p}^2}{3kT}\right) \tag{6.22}$$

Debye [21, pp. 9–11] derived this equation using the same derivation of the cavity field described by Lorentz [14, pp. 138–139]. The factor α_i denotes the electronic polarizability of the molecule, **p** its permanent dipole moment and N its number density (molecules/ml). The summation Σ includes all types of molecule existing in the dielectric sample. The Boltzmann constant k and absolute temperature T arises, as described in Chapter 7, from the application of Boltmann statistics to determine the probability of finding a dipole aligned along the direction of an applied electric field. When a field is applied to a medium containing polar molecules, their dipole moments experience a torque tending to align them with the field. This orienting tendency is opposed by thermal agitation. If θ is the angle between a dipole moment $\mathbf{p}$ and the field direction, the component of the moment in the field direction is $\mathbf{p}\cos\theta$. In Chapter 7 (see Box 7.3) it is shown that under normal conditions the thermal average of $\cos\theta$ is given by

$$\langle\cos\theta\rangle = \frac{\mathrm{pE}}{3kT} \qquad (6.23)$$

If the local electric field E_L acts on a single polar molecule in a polar liquid, then over a period of time it will assume an average moment m in the direction of the field given by contributions from an induced moment and an orientation polarization:

$$\mathrm{m} = \alpha_i \mathrm{E}_L + \mathrm{p}\langle\cos\theta\rangle \qquad (6.24)$$

where $\mathbf{p}$ is the 'real' intrinsic dipole moment value for the molecule.

6.3 Refinements of the Clausius–Mossotti–Lorentz Relation

Equations (6.18) and (6.22) are the simplest forms of the Clausius–Mossotti–Lorentz relation but they apply to two different situations. Equation (6.18) works very well for solids and nonpolar liquids and solids, where relative permittivity values are low and the right-hand side of the equation does not approach a value of 1.0. However, for polar liquids (e.g., water with $\varepsilon_r \approx 80$ at room temperature) Equation (6.22) offers the possibility that the number density and polarizability of the molecules can be such that the right-hand side of the equation can reach a value of unity. In this case ε_r attains a value of infinity, so that even a small applied field would result in an infinite polarization. Saturation of this polarization would in fact maintain it at a finite value, but instead the dielectric would become spontaneously polarized to the ferroelectric state. Viewed another way, according to Equation (6.22) polar liquids with high ε_r values should be close to a Curie point – a temperature below which the dielectric exhibits ferroelectric behaviour – and so be very sensitive to a change of pressure, temperature or electric field strength. Instabilities of this form have not been observed for polar liquids.

This problem was addressed by Lars Onsager, a Norwegian-American physical chemist, winner of the 1968 Nobel Prize in Chemistry and a doctoral student of Peter Debye. He gives the following amusing account [22]: 'how about the dielectrics that Debye had done? I was too lazy to go to the libraries and sat down and worked it out and lo and behold it came out quite different!' . . . 'And some years later in 1935 it was plainly high time to publish it.'

In his model, Onsager replaced the cavity used by Lorentz and Debye with a much smaller one [23]. The cavity has the same radius as the polar molecule of interest, which has a permanent intrinsic moment (when isolated as in the gas phase) of value $\mathbf{p}_o$. When located in an electric field the component $\mathbf{m}$ of its induced and orientation moment in the field direction is given by Equation (6.24), i.e.

$$\mathrm{m} = \alpha_i \mathrm{E}_L + \mathrm{p}_o\langle\cos\theta\rangle$$

Onsager now considers the effect of introducing a rigid dipole of moment $\mathbf{m}$ and radius R into a cavity of radius R, located within a dielectric medium of relative permittivity ε_m. To simplify the calculations, the moment $\mathbf{m}$ is treated as a point dipole located at the centre of the molecule. No external field is applied and so the medium surrounding the cavity is not polarized.

As shown in Box 6.6, the field of this dipole polarizes the surrounding dielectric and gives rise to a *reaction field* $\mathbf{E}_r$ within the cavity. This in turn enhances both the permanent and the induced dipole moment components of $\mathbf{m}$. Thus, although a molecular dipole cannot directly contribute to its own local field, it can do so indirectly by inducing polarizations in neighbouring molecules. However, according to Onsager [23] this reaction field is always parallel to the moment $\mathbf{m}$ and so will exert no torque and contribute nothing to the dipole orientation polarization of a polar liquid. Thus, only a part of the local Lorentz field employed in the derivation of Equation (6.22) contributes to the orienting torque of the permanent dipoles. As a consequence of including the reaction field it is implicit in the result obtained by Lorentz [14, pp. 138–139] and repeated by Debye [21, pp. 9–11], that the dipole moment can partly orient itself. This is equivalent to a person placing each foot into a bucket and lifting themselves off the floor and is why Equation (6.22) predicts the possibility of a dielectric 'catastrophe' in the form of a ferroelectric transition.

As outlined in Box 6.6, the local field $\mathbf{E}_L$ consists of the *cavity* field and the *reaction* field. The cavity field $\mathbf{E}_c$ is that part of the local field that remains unaltered if the molecule has its dipole moment removed and only this contribution should be included in the procedure to derive Equation (6.22). The reaction field $\mathbf{E}_r$ is the component of the local field that appears when the permanent

Box 6.6 Onsager's Modification of the Lorentz Field

Consider a molecular dipole of intrinsic moment **p** and radius *R*. In an electric field the component **m** of its induced and orientation moment in the field direction is given by Equation (6.24). A dipole of moment **m** is placed in an empty spherical cavity of radius *R*, located in an *unpolarized* medium of relative permittivity ε_m. This dipole is considered to be a point dipole located at the centre of the cavity. The potential ϕ must everywhere satisfy Laplace's equation and the following boundary conditions:

i) $\phi(r,\theta) - \dfrac{\text{m}\cos\theta}{r^2}$
is a continuous function for $0 \le r \le \infty$.

ii) $\left(\dfrac{\partial \phi_i}{\partial r}\right) = \varepsilon_m \left(\dfrac{\partial \phi_o}{\partial r}\right)$ at $r = R$

ϕ_i and ϕ_o are the potentials inside and outside the cavity, respectively.

The following solutions of the form of Equation (6.4) are obtained:

$$\phi_o = \left(\frac{\text{m}^*}{\varepsilon_m r^2}\right)\cos\theta;\ \phi_i = \frac{\text{m}\cos\theta}{r^2} - \text{E}_r r\cos\theta$$

with

$$\text{m}^* = \frac{3\varepsilon_m}{2\varepsilon_m + 1}\text{m, and } \text{E}_r = \frac{2(\varepsilon_m - 1)}{2\varepsilon_m + 1}\frac{\text{m}}{R^3}$$

m* is the effective moment of the material inside the cavity as viewed from *outside* it and E_r is the *reaction field* within the cavity that arises from the polarization induced in the medium *outside* the cavity by this moment's field.

The moment m is now removed from the cavity and a uniform field E_o applied. The field E_c inside this empty cavity is given by E_i derived in Box 6.1, with $\varepsilon_p = 1$:

$$\text{E}_c = \frac{3\,\varepsilon_m}{1 + 2\varepsilon_m}\text{E}_o$$

The local (i.e., Lorentz) field E_L acting on the dipole inside the cavity is given by

$$\text{E}_L = \text{E}_r + \text{E}_c = \frac{2(\varepsilon_m - 1)}{2\varepsilon_m + 1}\frac{\text{m}}{R^3} + \frac{3\varepsilon_m}{2\varepsilon_m + 1}\text{E}_o \tag{6.25}$$

dipole moment is restored and because its direction is parallel to the dipole it exerts no torque on **m** and should not be included in the determination of the orientation polarization. The local field acting on the dipole in its cavity is given by Equation (6.25) in Box 6.6 as

$$\text{E}_L = \frac{2(\varepsilon_m - 1)}{2\varepsilon_m + 1}\frac{\text{m}}{R^3} + \frac{3\varepsilon_m}{2\varepsilon_m + 1}\text{E}_o$$

Substituting into Equation (6.24) the value of $\mathbf{E}_L$ given by Equation (6.25) we obtain the relationship

$$\text{m}\left[1 - \frac{2(\varepsilon_m - 1)}{(2\varepsilon_m + 1)}\frac{\alpha_i}{R^3}\right] = \text{p}_o\langle\cos\theta\rangle + \frac{3\varepsilon_m}{(2\varepsilon_m + 1)}\alpha_i\text{E}_o \tag{6.26}$$

The polar molecule is embedded in a spherical cavity of radius equal to its own radius. From the concept of molar refraction discussed in association with Equations (6.10) and (6.11) and the fact that α_i has units of volume, we can make the following substitution:

$$\frac{\alpha_i}{R^3} = \frac{n^2 - 1}{n^2 + 2} \tag{6.27}$$

Employing Maxwell's result that the relative permittivity is equal to the square of the index of refraction, we can also identify $\varepsilon_{m\infty} = n^2$ as the value of the dielectric's relative permittivity at frequencies in the optical range (~5×10^{14} Hz). Equation (6.27) then becomes

$$\text{m}\left[\frac{3\left(2\varepsilon_m + \varepsilon_{m\infty}\right)}{\left(2\varepsilon_m + 1\right)\left(\varepsilon_{m\infty} + 2\right)}\right] = \text{p}_o\langle\cos\theta\rangle + \frac{3\varepsilon_m}{\left(2\varepsilon_m + 1\right)}\alpha_i\text{E}_o \tag{6.28}$$

which can be written as

$$\text{m} = \text{p}\langle\cos\theta\rangle + \frac{\varepsilon_m\left(\varepsilon_{m\infty} + 2\right)}{\left(2\varepsilon_m + \varepsilon_{m\infty}\right)}\alpha_i\text{E}_o \tag{6.29}$$

As shown in Box 6.7 this relationship was used by Onsager [23] to derive Equation (6.34) that relates the product of the number density and intrinsic dipole moment of polar molecules in a polar liquid to its static (or low frequency) relative permittivity value ε_m and the value $\varepsilon_{m\infty}$ determined by measurement at optical frequencies of the refractive index *n*, where $\varepsilon_{m\infty} = n^2$.

Replacing the number density *N* with $N_A\rho/M$, where N_A is Avogadro's constant, ρ the density of the pure polar liquid and *M* the molecular weight of the polar molecules, then from Equation (6.34) Onsager's theory

Box 6.7 Onsager's Equation for the Orientation Polarization of a Polar Liquid

The parameter **p** in Equation (6.29) represents the effective dipole moment (orientational and induced) *inside* the cavity and from Equation 6.28) is given by:

$$\mathrm{p} = \frac{(2\varepsilon_m + 1)(\varepsilon_{m\infty} + 2)}{3(2\varepsilon_m + \varepsilon_{m\infty})}\mathrm{p}_o \tag{6.30}$$

where $\mathbf{p}_o$ is the intrinsic, 'real', value of the moment. However, to evaluate the polarization of the polar liquid we require the dipole moment $\mathbf{p}^*$ as viewed from *outside* the cavity. From Equation (6.23) the mean orientation of $\mathbf{p}^*$ is given by

$$\langle \cos\theta \rangle = \frac{\mathrm{p}^*\mathrm{E}_o}{3kT} \tag{6.31}$$

Let N be the number of dipoles per unit volume in the polar liquid. The polarization P per unit volume is thus $\mathrm{P} = N\mathrm{m}$, so that from Equations (6.29) and (6.31) we have

$$\mathrm{P} = N\mathrm{m} = N\left[\frac{\mathrm{pp}^*}{3kT} + \frac{\varepsilon_m(\varepsilon_{m\infty} + 2)}{(2\varepsilon_m + \varepsilon_{m\infty})}\alpha_i\right]\mathrm{E}_o \tag{6.32}$$

If we assume that the volume of the liquid equates to the sum of the volumes of the polar molecules, then $(4\pi R^3 N)/3 = 1$. From Equation (6.27) we thus have

$$\alpha_i = R^3\left(\frac{n^2 - 1}{n^2 + 2}\right) = \frac{3}{4\pi N}\left(\frac{\varepsilon_{m\infty} - 1}{\varepsilon_{m\infty} + 2}\right)$$

Substituting this value for α_i into Equation (6.32) and employing the relationship given by Equation (3.48) that $\mathrm{P} = (\varepsilon_m - 1)\varepsilon_o\mathrm{E}_o$, we obtain:

$$(\varepsilon_m - 1) = N\frac{\mathrm{pp}^*}{3\varepsilon_o kT} + \frac{3\varepsilon_m(\varepsilon_{m\infty} - 1)}{(2\varepsilon_m + \varepsilon_{m\infty})} \tag{6.33}$$

Onsager [23] shows that $\mathrm{p}^* = \frac{3\varepsilon_m}{2\varepsilon_m + 1}\mathrm{p}$

so that from Equation (6.30) $\mathrm{p}^* = \frac{\varepsilon_m(\varepsilon_{m\infty}+2)}{(2\varepsilon_m + \varepsilon_{m\infty})}\mathrm{p}_o$

to give from Equation (6.33) the following relationship:

$$\frac{N\mathrm{p}_o^2}{9\varepsilon_o kT} = \frac{(\varepsilon_m - \varepsilon_{m\infty})(2\varepsilon_m + \varepsilon_{m\infty})}{\varepsilon_m(\varepsilon_{m\infty} + 2)^2} \tag{6.34}$$

results in the following relationship for the orientation polarization of a pure polar liquid:

$$\frac{N_A\mathrm{p}_o^2}{9\varepsilon_o kT} = \frac{(\varepsilon_m - \varepsilon_{m\infty})(2\varepsilon_m + \varepsilon_{m\infty})}{\varepsilon_m(\varepsilon_{m\infty} + 2)^2}\frac{M}{\rho} \tag{6.35}$$

Onsager's theory certainly overcomes the problem of dielectric instability inherent with Debye's Equation (6.22), but it not without possible sources of error. For example, because the cavity is assumed to contain only one molecule, short range interactions with other molecules are neglected. Long range interactions, however, are included within the evaluation of the reaction field. Wilson [24] also highlighted the neglect of molecular anisotropy in Onsager's theory and to accommodate this he proposed that Equation (6.35) should be modified to the form where $\varepsilon_{m\infty} = n'^2$, with n' being the refractive index corresponding to the polarizability along the axis of the dipole. Kirkwood [25] suggested a modification to take account of orientation hindrance arising from electrostatic interactions and short-range intermolecular forces. Fröhlich and Sack [26] infer that Onsager's model should hold only if the relative permittivity is less than 9. These authors also fault Debye's application of the Lorenz field in that it implies the polarization cannot follow the rotation of the dipole, whereas Onsager's theory is faulted because it errs too much the other way – the reaction field is assumed to follow the dipole orientation and so does not exert any force on the dipole [26]. Following Onsager's method to determine the local field, Fröhlich and Sack surround the selected dipolar molecule by a cavity of volume $1/N$ (where N is the number of molecules per unit volume) and treat the outside as a continuous medium whose dielectric properties are the same as the macroscopic dielectric properties to be calculated. As with Onsager's method, they also split the local field into two parts: (i) the cavity field, which is obtained on the assumption that the dipole has been removed from the cavity; (ii) the reaction field, which is the change of the field at the position of the dipole through its action upon the surrounding dipoles. However, Fröhlich and Sack calculate the cavity field slightly differently from Onsager. Onsager removes not only the dipole from the cavity in order to determine the cavity field, but also the whole molecule. The relative permittivity of the cavity is thus unity. Fröhlich and Sack remove the point dipole quality of the molecule but leave the rest of the molecule behind, so that the relative permittivity of the cavity assumes a value given by the refractive index – namely $\varepsilon_{m\infty}$. The cavity field given in Box 6.6 thus becomes

$$\mathrm{E}_c = \frac{3\varepsilon_m}{2\varepsilon_m + \varepsilon_{m\infty}}\mathrm{E}_o$$

The treatment of the reaction field $\mathbf{E}_r$ by Fröhlich and Sack [26] also differs from that of Onsager, who assumes that the reaction field attains the value corresponding to the situation where the dipole remains in a given direction for a time that is long compared to its orientation relaxation time τ. Instead, Fröhlich and Sack describe their cavity field $\mathbf{E}^{FS}{}_r$ by its dynamic properties (i.e., by its time-dependent differential equation) as

$$\tau \frac{dE_r}{dt} = -E_r + E_r^{FS} \cdot u$$

where **u** is the unit vector in the direction of the dipole and the field reaches its equilibrium value as an exponential function of the form $\sim e^{-t/\tau}$. Unless the dipole remains infinitely long in the same direction, then $E_r^{FS} > |E_r|$. For a spherical cavity, Fröhlich and Sack obtain the following expression for the reaction field:

$$E_r^{FS} = \frac{2(\varepsilon_m - \varepsilon_{m\infty})}{\varepsilon_{m\infty}(2\varepsilon_m + \varepsilon_{m\infty})} \frac{m}{R^3}$$

which can be compared to that obtained (see Box 6.6) by Onsager:

$$E_r = \frac{2(\varepsilon_m - 1)}{2\varepsilon_m + 1} \frac{m}{R^3}$$

The final expression obtained by Fröhlich and Sack [26] for the static relative permittivity is

$$\varepsilon_m = \varepsilon_{m\infty} + \frac{3\varepsilon_m}{(2\varepsilon_m + \varepsilon_{m\infty})} \frac{p^2}{R^3 kT} \times \left[1 + \frac{1}{3} \frac{(\varepsilon_m - \varepsilon_{m\infty})^2}{\varepsilon_m \varepsilon_{m\infty}}\right]$$

This gives values for ε_m roughly half-way between those obtained using Debye's and Onsager's models. Because only one molecule occupies the cavity, short range interactions have been neglected.

Other proposed modifications to Onsager's model include that by Abbott and Bolton [27] who assumed that the polar molecule can be represented as a prolate ellipsoid and evaluated the reaction field of a point dipole lying on the axis of a cavity of this form. Buckingham [28], however, found this approach to be inconsistent for three reasons: (i) the total moment of a particular molecule was assumed to be parallel to the permanent moment. This is not so because the induced moment contributes to the mean moment in the direction of the field by a similar order of magnitude to that from the permanent moment; (ii) the polarizability of the ellipsoid was assumed to be uniform, whereas its actual value in the direction of the dipole will not in general be equal to the mean; (iii) the following relationship

$$\varepsilon_m - 1 = \frac{4\pi \langle m \rangle}{VE}$$

where $\langle m \rangle$ is the mean value of the component of the moment **m** of a molecule in the direction of the field E and V is the molecular volume, is not valid for nonspherical, ellipsoidal, molecules. It is only correct if the moment corresponds to that of a point dipole. While the field of a uniformly polarized sphere satisfies this requirement, that of a ellipsoid does not. This error becomes apparent when it is realized that ε_m is a macroscopic property of a substance and is determined by observing the polarization of a large specimen and not that of a single molecule. Buckingham devises a theory to apply Onsager's theory to a molecular model, consisting of an optically isotropic ellipsoid, to derive the orientation polarization, which does not contain these three deficiencies [28]. He derives the following relationship:

$$\frac{\varepsilon_m}{\varepsilon_m(2\varepsilon_m + 1)} - \frac{n^2 - 1}{9} \times \sum \left[\frac{1}{[\varepsilon_m - (\varepsilon_m - 1)A_i]\,[\varepsilon_m + (n^2 - 1)A_i]}\right] = \left[\frac{1 + (n^2 - 1)A_a}{\varepsilon_m + (n^2 - 1)A_a}\right]^2 \frac{4\pi p^2}{9kTV}$$

where $\frac{4\pi ab^2}{3V} = 1$; a, b and c are the semiaxes of an ellipsoid and parameter A is the internal field function (also known as the depolarization factor) described in the discussion of interfacial polarization in section 7.5 of Chapter 7. Buckingham finds that his result yields dipole moment values, which are in better agreement with experiment than those calculated by the theory of Abbott and Bolton. He concludes: 'Since these authors claim that their equation is superior to that of Onsager when applied to non-associating liquids, the theory presented in this paper should also be an improvement on Onsager's' [28].

An obvious application of Buckingham's equation would be in the evaluation of the dipole moments of proteins. As discussed in Chapter 8, this approach does not appear to have been adopted.

6.4 The Complex Clausius–Mossotti Factor

The complex form of the CM factor, as given in Equation (6.1), is required when time dependent electric fields are used. If an alternating voltage source, instead of a constant voltage, is applied to the plate electrodes of Figure 3.1, the dielectric will still be polarized. However, the relationships

$$P(t) = (\varepsilon_r - 1)\varepsilon_o E(t) \tag{6.36}$$

as the time dependent form of Equation (3.48) and

$$D(t) = \varepsilon_o \varepsilon_r E(t) \tag{6.37}$$

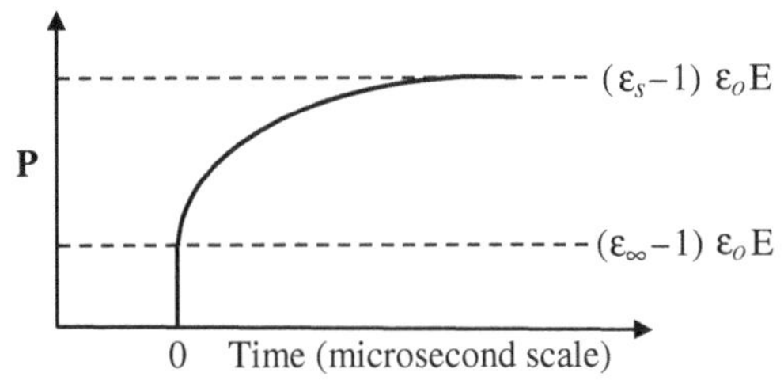

Figure 6.3 Time dependence of polarization P of a dielectric on application of a constant electric field E. Electronic polarization of magnitude $(\varepsilon_\infty - 1)\varepsilon_o$E occurs almost instantaneously, but inertia of molecular dipoles slows down the polarization as it approaches the steady-state level of $(\varepsilon_s - 1)\varepsilon_o$E.

given by Equation (3.12), will need modification to a more general form. The reason for this is that the displacement charges associated with the polarization exhibit inertia. This inertia becomes apparent when an electric field is suddenly applied to a dielectric. As illustrated in Figure 6.3 the polarization P takes time to reach its final value. Under normal conditions the time-dependent polarization P(t) will be proportional to the time-dependent field E(t), but the ratio P(t)/E(t) will vary as a function of the alternating field frequency ω. If the applied voltage frequency is low, say 1 Hz or lower, the polarization and displacement given by Equations (6.36) and (6.37) will faithfully follow the time variation of the field. If the dielectric comprises permanent molecular dipoles, this corresponds to their moments reorienting exactly in phase with the alternating field. The parameter ε_r will then accurately equate to the relative permittivity used so far in this book – referred to as the static relative permittivity. As the frequency is increased to (say) 1 MHz, the situation may be such that that there is a significant lag between the response of the dipoles and the change of the field magnitude and polarity. Finally, as the frequency is increased to (say) 100 MHz the dipoles may not be able to respond to the applied field at all. The effective relative permittivity is then equivalent to the value $\varepsilon_{m\infty}$ in Equation (6.28), which remains constant in value as the frequency is increased up to the optical range and before resonance effects occur (see Chapter 7). In Figure 6.3 the timescale involved with electronic polarization appears as an instantaneous response.

Equations (6.36) and (6.37) may still be used and the polarization curve shown in Figure 6.3 taken into account, if the relative permittivity ε_r is represented as a *complex* permittivity (ε_r^*), written as

$$\varepsilon_r^* = \varepsilon' - i\varepsilon'' \tag{6.38}$$

where $i = \sqrt{-1}$, ε' is the real component (Re) and ε'' is the imaginary component (Im). We can understand the reason for this by considering a sinusoidal voltage waveform, which is the most common one used to produce a time-varying field in a dielectric. It is shown in Chapter 10, Box 10.3 that a sinusoidal field can be mathematically expressed as

$$\begin{aligned} \mathrm{E}(t) &= \mathrm{Im}\left[\mathrm{E}_o e^{i\omega t}\right] = \mathrm{Im}\left[\mathrm{E}_o\left(\cos\omega t + i\sin\omega t\right)\right] \\ &= \mathrm{E}_o \sin\omega t. \end{aligned} \tag{6.39}$$

The factor E_o represents the peak magnitude of the field and remains constant with time t, whilst ω is the radian frequency – also expressed as $\omega/2\pi$ cycles per second and given in SI derived units of hertz (Hz). If this time-varying field is applied for a sufficient length of time the polarization will vary with the same periodicity, but at the onset of molecular inertia effects it will lag behind the field. We can represent this as a phase lag φ, such that

$$\begin{aligned} \mathrm{D}(t) &= \mathrm{D}_o \sin\left(\omega t - \varphi\right) \\ &= \mathrm{D}_o\left(\sin\omega t \sin\varphi - \cos\omega t\cos\varphi\right) \\ &= \mathrm{D}'\sin\omega t - \mathrm{D}''\cos\omega t \end{aligned}$$

with $\mathrm{D}' = \mathrm{D}_o\sin\varphi$, $\mathrm{D}'' = \mathrm{D}_o\cos\varphi$. The general forms of Equations (6.36) and (6.37) become

$$\mathrm{P}(t) = \left(\varepsilon_r^* - 1\right)\varepsilon_o \mathrm{E}(t) \tag{6.40}$$

$$\mathrm{D}(t) = \varepsilon_o\varepsilon_r^*\mathrm{E}(t) \tag{6.41}$$

For example, we can write Equation (6.41) as

$$\begin{aligned} \mathrm{D}(t) &= \varepsilon_o\varepsilon_r^*\mathrm{Im}\left(\mathrm{E}_o e^{i\omega t}\right) = \varepsilon_o\left(\varepsilon' - i\varepsilon''\right)E_o\sin\omega t \\ &= \varepsilon_o\left(\varepsilon'^2 + \varepsilon''^2\right)^{1/2}E_o\sin(\omega t - \varphi) \end{aligned}$$

We can define a loss tangent as

$$\tan\varphi = \frac{\mathrm{D}_o\sin\varphi}{\mathrm{D}_o\cos\varphi} = \frac{\mathrm{D}''}{\mathrm{D}'} = \frac{\varepsilon''}{\varepsilon'}$$

to give the ratio of the out-of-phase to the in-phase components of D and this in turn relates to the ratio of the energy loss per cycle and the energy stored per cycle. At a low frequency where there is no appreciable phase lag, the value of ε'' is zero and ε' equates to ε_r. The energy required to establish the electric field during the first half-cycle of the sine wave is completely recovered during the next half-cycle. However, if D and E exhibit a phase difference ε'' will have a finite value and there is a net loss of energy related to the heat produced by the displacement current density $\partial\mathrm{D}/\partial t$. At very high frequencies, where the molecular dipoles have no time at all to respond to changes of the field, the parameter ε'' again has a zero value, there is no energy loss and ε' equates to $\varepsilon_{r\infty}$. Typical variations of ε' and ε'' as a function of frequency are shown in Figure 6.4.

Energy loss in a nonideal dielectric occurs from both a 'conventional' conduction current j_c and a displacement

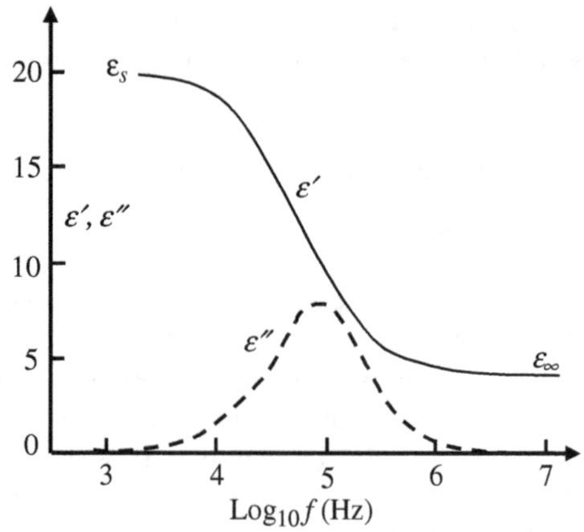

Figure 6.4 Typical variations of the dielectric parameters ε' and ε'' in Equation (6.38) as a function of frequency. The value of ε' at low frequencies corresponds to the static relative permittivity ε_s and at high frequencies approaches the value ε_∞ derived from optical refraction measurements.

current $\partial D/\partial t$, so that from Ohm's Law ($j = \sigma E$) we have the relationship

$$j_c + \frac{\partial D}{\partial t} = \sigma_c E + \varepsilon_o \varepsilon_r \frac{\partial E_o e^{i\omega t}}{\partial t} = \sigma_c E + i\omega\varepsilon_o\varepsilon_r E_o e^{i\omega t}$$
$$= (\sigma_c + i\omega\varepsilon_o\varepsilon_r)E$$

The two forms of energy loss, conduction current and displacement current, are given by the following relationships, respectively:

$$W_\sigma = \frac{1}{2}\sigma_c E^2 \quad \text{and} \quad W_\varepsilon = \frac{1}{2}\omega\varepsilon_o\varepsilon'' E^2 \tag{6.42}$$

As described by 'Willy' Wagner [29], a complex conductivity σ^* can thus be defined as

$$\sigma^* = \sigma' + i\sigma'' = \sigma_c + i\omega\varepsilon_o\varepsilon_r \tag{6.43}$$

A finite value for ε'' can be seen to have the same effect as a conductivity of $\omega\varepsilon_o\varepsilon''$, so that ε'' can be equated to $\sigma'/\omega\varepsilon_o$. From Equation (6.38) the absolute complex permittivity (units of F/m) can thus be expressed as

$$\varepsilon^* = \varepsilon_o\varepsilon_r^* = \varepsilon_o(\varepsilon' - i\varepsilon'') = \varepsilon_o\varepsilon' - i\frac{\sigma'}{\omega} \tag{6.44}$$

From Equations (6.43) and (6.44) it is apparent that at low frequencies as $\omega \to 0$ then $\sigma^* \to \sigma_c$, whilst at high frequencies as $\omega \to \infty$ then $\varepsilon^* \to \varepsilon_o\varepsilon'$. At low frequencies the dielectric (and hence dielectrophoretic) properties of a particle and medium are thus dominated by conduction processes, whilst at high frequencies dielectric polarization processes are more important. Defining both a complex permittivity and conductivity is thus useful and they can be expressed in terms of each other through the relationship

$$\sigma^* = i\omega\varepsilon_o\varepsilon^* \tag{6.45}$$

The complex form of the Clausius–Mossotti (*CM*) factor, given in Equation (6.1) to describe the polarization of a spherical particle of relative permittivity ε_p and conductivity σ_p suspended in a medium of relative permittivity ε_m and conductivity σ_m, can thus be given as either:

$$CM = \frac{\varepsilon_p^* - \varepsilon_m^*}{\varepsilon_p^* + 2\varepsilon_m^*} \quad \text{or} \quad CM = \frac{\sigma_p^* - \sigma_m^*}{\sigma_p^* + 2\sigma_m^*} \tag{6.46}$$

The dipole moment value given by Equation (6.4) for the electrostatic case can now be given for the case of sinusoidal fields of the form $E = \text{Im}\left[E_o e^{i\omega t}\right]$ as either

$$p = 4\pi\varepsilon_o\varepsilon_m R^3 \left(\frac{\varepsilon_p^* - \varepsilon_m^*}{\varepsilon_p^* + 2\varepsilon_m^*}\right) E \tag{6.47a}$$

or

$$p = 4\pi\varepsilon_o\varepsilon_m R^3 \left(\frac{\sigma_p^* - \sigma_m^*}{\sigma_p^* + 2\sigma_m^*}\right) E \tag{6.47b}$$

The MATLAB program in Box 6.9 models the complex *CM* factor for a spherical particle suspended in an aqueous electrolyte. The results produced for the frequency variations of the real and imaginary components of the *CM* factor are given in Figure 6.5.

The MATLAB program presented in Box 6.9 specifies the particle conductivity as 10 mS/m, whilst the conductivity of the electrolyte is much lower at 0.1 mS/m. At low frequencies, as $\omega \to 0$ and $\sigma^* \to \sigma_c$, we thus expect from Equation (6.46) that the *real* value of the *CM* factor will have a positive value, in agreement with the result shown in Figure 6.5. The particle's permittivity is specified to be eight times less than that of the electrolyte and so based on Equations (6.44) and (6.46) we can predict, as shown in Figure 6.5, that at high frequencies the *CM* factor will have a negative *real* value. This means that the effective dipole moment of the spherical particle will reverse polarity as the field frequency is increased from 1 kHz to 1 MHz. We interpret the dipole moment as the distribution of the algebraic sum of the bound and free charges at the interface between the particle and the surrounding fluid medium. This *interfacial* polarization process is described more fully in Chapter 7. Changes of this algebraic sum of the charges occurs at a characteristic response (or relaxation) time τ, which for the general multipole case is given as [30]

$$\tau = \frac{n\varepsilon_p + (n+1)\varepsilon_m}{n\sigma_p + (n+1)\sigma_m} \tag{6.48}$$

Box 6.8 Derivation of the Real and Imaginary Components of *CM*

The complex Clausius–Mossotti (*CM*) function is obtained by substituting into Equation (6.46) the complex permittivity ε^* given by Equation (6.44):

$$CM = \frac{(\varepsilon_p - i\sigma_p/\omega) - (\varepsilon_m - i\sigma_m/\omega)}{(\varepsilon_p - i\sigma_p/\omega) + 2(\varepsilon_m - i\sigma_m/\omega)} = \frac{\omega(\varepsilon_p - \varepsilon_m) - i(\sigma_p - \sigma_m)}{\omega(\varepsilon_p + 2\varepsilon_m) - i(\sigma_p + 2\sigma_m)}$$

Multiplying the numerator and denominator by the complex conjugate of the denominator we obtain:

$$CM = \frac{\omega^2(\varepsilon_p - \varepsilon_m)(\varepsilon_p + 2\varepsilon_m) + (\sigma_p + 2\sigma_m)(\sigma_p - \sigma_m)}{\omega^2(\varepsilon_p + 2\varepsilon_m)^2 + 2(\sigma_p + 2\sigma_m)^2} + i\frac{\omega(\varepsilon_p - \varepsilon_m)(\sigma_p + 2\sigma_m) - (\varepsilon_p + 2\varepsilon_m)(\sigma_p - \sigma_m)}{\omega^2(\varepsilon_p + 2\varepsilon_m)^2 + 2(\sigma_p + 2\sigma_m)^2}$$

On substituting into this expression the characteristic relaxation (response) time τ given by Equation (6.49), we obtain for the real component [30]:

$$\text{Re}[CM] = \left[\left(\frac{\omega^2\tau^2}{1+\omega^2\tau^2}\right)\left(\frac{\varepsilon_p - \varepsilon_m}{\varepsilon_p + 2\varepsilon_m}\right) + \left(\frac{1}{1+\omega^2\tau^2}\right)\left(\frac{\sigma_p - \sigma_m}{\sigma_p + 2\sigma_m}\right)\right]$$

whilst the imaginary component simplifies to the form:

$$\text{Im}[CM] = \frac{3\omega(\varepsilon_p\sigma_m - \varepsilon_m\sigma_p)}{\omega^2(\varepsilon_p + 2\varepsilon_m)^2 + (\sigma_p + 2\sigma_m)^2}$$

---

where ε_p and ε_m are the absolute (rather than relative) permittivity values. For the case of the dipole approximation ($n = 1$) the response time is thus

$$\tau = \frac{\varepsilon_p + 2\varepsilon_m}{\sigma_p + 2\sigma_m} \tag{6.49}$$

As shown in Box 6.8 the interfacial charging process can be taken into account in the derivation of the real component of the *CM* factor. The method is simply to substitute either of the expressions for σ^* and ε^* given by Equations (6.43) and (6.44) into Equation (6.46) and incorporate the expression for the relaxation time given by Equation (6.49). The expression obtained for the real component is:

$$\text{Re}[CM] = \left[\left(\frac{\omega^2\tau^2}{1+\omega^2\tau^2}\right)\left(\frac{\varepsilon_p - \varepsilon_m}{\varepsilon_p + 2\varepsilon_m}\right) + \left(\frac{1}{1+\omega^2\tau^2}\right)\left(\frac{\sigma_p - \sigma_m}{\sigma_p + 2\sigma_m}\right)\right] \tag{6.50}$$

This is the same result, but in a different form, to the more general one obtained by Molinari and Viviani [31] and Benguigui and Lin [32] that includes the transient DEP response. For our purposes the form of

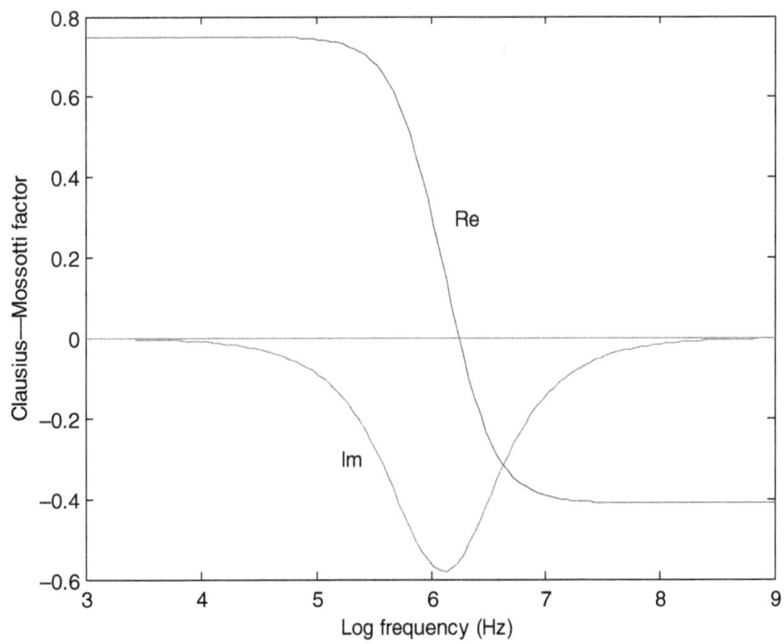

Figure 6.5 Plots of the real (Re) and imaginary (Im) components of the Clausius–Mossotti factor as a function of frequency, created using the MATLAB program in Box 6.9. At low and high frequencies the imaginary component approaches a value of zero.

Equation (6.50) is the more helpful because we can readily deduce from it that as $\omega \to 0$ (i.e., $\omega\tau \ll 1$) we have for the electrostatic case and at low frequencies:

$$\mathrm{Re}[CM] \approx \frac{\sigma_p - \sigma_m}{\sigma_p + 2\sigma_m} \tag{6.51}$$

For the situation as $\omega \to \infty$ ($\omega\tau \gg 1$) we then have at high frequencies:

$$\mathrm{Re}[CM] \approx \frac{\varepsilon_p - \varepsilon_m}{\varepsilon_p + 2\varepsilon_m} \tag{6.52}$$

Figure 6.5 was generated for a spherical particle suspended in an aqueous electrolyte, with the following conductivity and permittivity parameters: $\sigma_p = 10$ mS/m; $\sigma_m = 1$ mS/m; $\varepsilon_p = 10\varepsilon_o$ and $\varepsilon_m = 80\varepsilon_o$. Equations (6.51) and (6.52) predict that at low and high frequencies the real values for the *CM* factor should be 0.75 and −0.41, respectively. The plot shown in Figure 6.5 reproduces these predictions. An example is given in Figure 6.6 of where the particle's conductivity is less than that of the surrounding medium, but with a permittivity larger than the medium. From the specified dielectric parameters ($\sigma_p = 0.1$ mS/m; $\sigma_m = 1$ mS/m; $\varepsilon_p = 60\varepsilon_o$ and $\varepsilon_m = 20\varepsilon_o$) the predicted low and high frequency Re[*CM*] values should be −0.43 and 0.4, respectively. These predictions are mirrored in the curve shown for the real part of the *CM* factor in Figure 6.6.

Equations (6.47a and b) inform us that the induced dipole moment **p** is a complex quantity having real and imaginary components. If the reference axis for the real component Re(**p**) is, as shown in Figure 6.7, aligned with the direction of the applied field $\mathbf{E}_o$ we can define a phase angle φ as $\tan\varphi = \mathrm{Im(p)/Re(p)}$.

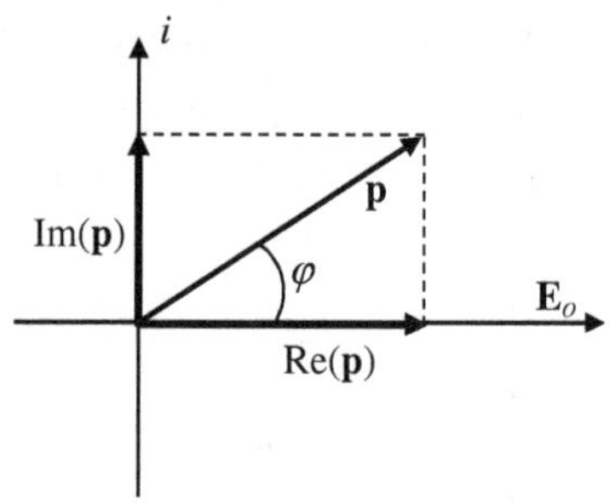

Figure 6.7 The induced dipole moment p consists of a real component Re(p) along the direction of the applied field E_o, and an imaginary component Im(p) directed along the imaginary axis *i*. The magnitude of the moment **p** is given by the vector sum of Re(p) and Im(p). In this diagram p *leads* the field by the phase angle φ

For the case where $\mathrm{p}^* = \mathrm{Re(p)} + i\mathrm{Im(p)}$, the moment p leads the applied field $\mathbf{E}_o$ by θ degrees. The magnitude of the moment is given by

$$\mathrm{p} = \sqrt{(\mathrm{Re(p)})^2 + \mathrm{Im(p)}^2)}$$

The MATLAB program given in Box 2.9 was modified to calculate and plot the magnitude of the moment and its phase angle as a function of frequency. Results obtained using it are shown in Figures 6.8 to 6.11.

Finally, from the analysis of the electrical polarization of a dielectric ellipsoid, the expression given by Equation (6.4) for the induced dipole moment p of a

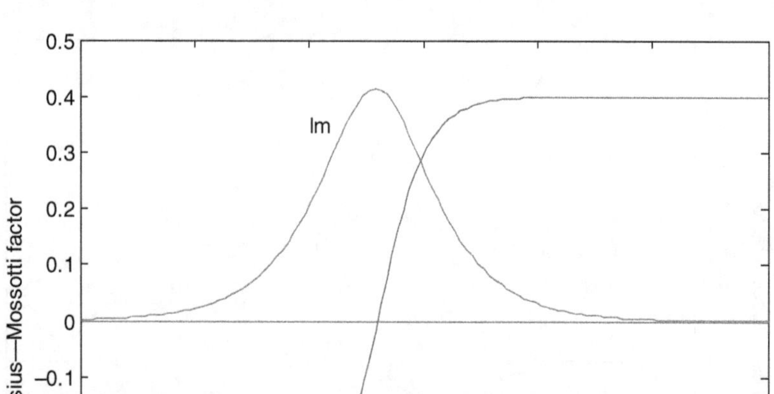

Figure 6.6 Plots of the real and imaginary components of the Clausius–Mossotti factor as a function of frequency. The MATLAB program in Box 6.9 was modified to give values for σ_p and σ_m of 0.1 mS/m and 1 mS/m, respectively, together with values for ε_p and ε_m of $60\varepsilon_o$ and $20\varepsilon_o$, respectively. At the low and high frequencies the real part of the *CM* factor has values predicted by Equations (6.50) and (6.51) where the imaginary component approaches zero.

Box 6.9 MATLAB Program to Plot the Complex Clausius–Mossotti Factor

Lines 10–14 of the following MATLAB program assign permittivity and conductivity values (in SI units) to a spherical particle and its fluid suspending medium. Lines 16–17 specify that we wish to generate 120 angular frequency values (w) between 1 kHz and 1 GHz and to calculate the real and imaginary values of the complex Clausius–Mosotti factor at each of these frequency values. These calculations are performed in lines 20–24. The format of the plots and labelling of the axes and graph is specified in lines 26–29.

```
1 %  CMfactor.m
2
3 % The particle is assigned a conductivity kc1 and permittivity kp1.
4 % The suspending medium has conductivity kc2 and permittivity kp2.
5 % ki (i=1,2) complex permittivity of particle, medium.
6 % pO: permittivity of free space.
7 % All parameters in SI units (metres, S/m, F/m).
8 % -----------------------------
9
10 pO=8.854e-12;
11 kc1=10e-3;
12 kp1=10*pO;
13 kc2=10e-4;
14 kp2=80*pO;
15
16 f=logspace(3,9,120);
17 zeroline=f-f;
18 w=2*pi*f;
19
20 k1=kp1-i*kc1 ./w;
21 k2=kp2-i*kc2 ./w;
22 cmf=(k1-k2) ./(k1+2*k2);
23 rm=real(cmf);
24 im=imag(cmf);
25
26 plot(log10(f),rm,'-', log10(f), im,'-', log10(f),zeroline,'-');
27 text(6.2,0.3,'Re'), text(5.2,-0.3,'Im')
28 xlabel('Log Frequency (Hz)')
29 ylabel('Clausius-Mossotti Factor')
```

spherical particle can be generalized to the general case of an ellipsoidal particle having semiaxes a, b and c:

$$\mathrm{p_x} = \frac{4\pi abc}{3}\varepsilon_o\varepsilon_m\left(\frac{\varepsilon_p - \varepsilon_m}{\varepsilon_m + A_x(\varepsilon_p - \varepsilon_m)}\right)\mathrm{E}_x \quad (6.53)$$

where p_x is the induced polarization along one of the axes with the applied field parallel to this axis. The parameter A_x is the depolarization factor described in Chapter 5, Equation (5.57). This factor is described more fully in Chapter 7 and values for A_x for various sizes of oblate and prolate spheroids are shown in Figure 7.11. For a

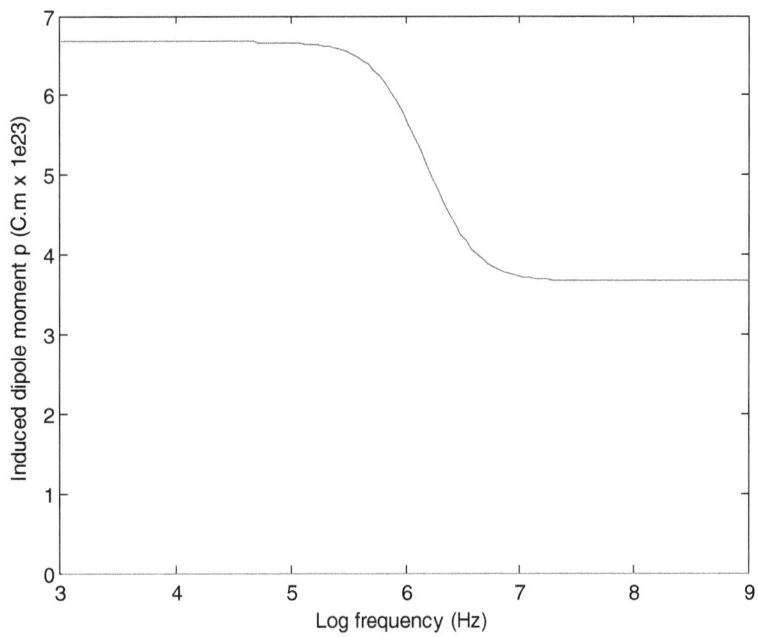

Figure 6.8 The magnitude of the induced dipole moment p as a function of frequency, obtained using a modified version of the MATLAB program in Box 6.9, for the same particle and medium conductivities that produced the *CM* factor values given in Figure 6.5.

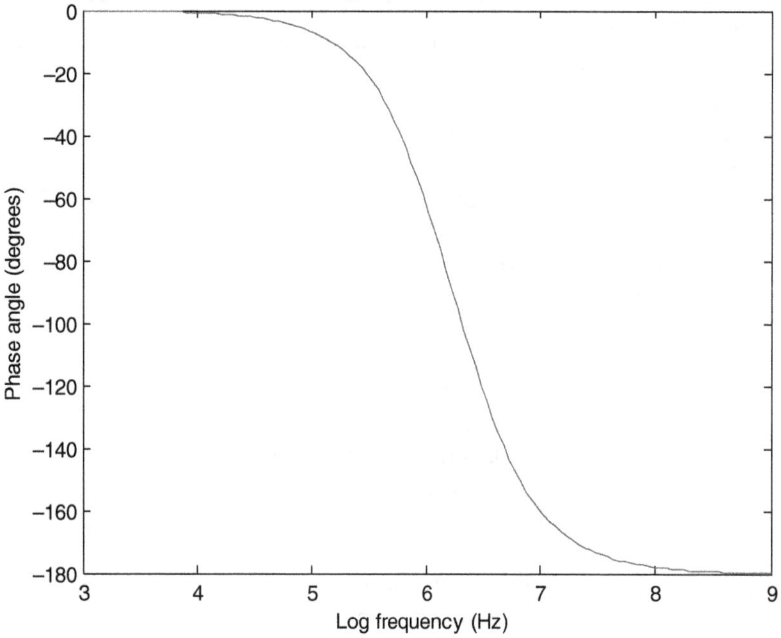

Figure 6.9 At low frequencies the moment **p** of Figure 6.8 is in phase with the applied electric field. As the frequency is increased the moment increasingly *lags* the field and at high frequencies attains a maximum phase difference of −180°.

spherical particle ($a = b = c = R$), the depolarization parameter has the same value for all directions of the field and orientation of the particle, so that $A_x = A_y = A_z = 1/3$. Substituting this value into Equation (6.53) leads to the result:

$$p = 4\pi\varepsilon_o\varepsilon_m R^3 \left(\frac{\varepsilon_p - \varepsilon_m}{\varepsilon_p + 2\varepsilon_m}\right) E_o$$

in agreement with Equation (6.4). The rod-shaped particle shown in Figure 5.25 is directed with its long (major) axis along the field direction and in this case we find that $A_x = 0$. A straight DNA molecule, or a virion such as the Tobacco Mosaic Virus, can be approximated to be a prolate spheroid with its major axis much larger than its two other axes (e.g, $a \gg b = c$). If the molecule or virion is aligned with the field, then the depolarizing factor along its major axis is close to zero, so that Equation (6.53) simplifies to the form:

$$p_x \approx \frac{4\pi a^2 b}{3}\varepsilon_o\varepsilon_m \left(\frac{\varepsilon_p - \varepsilon_m}{\varepsilon_m}\right) E_x \quad (6.54)$$

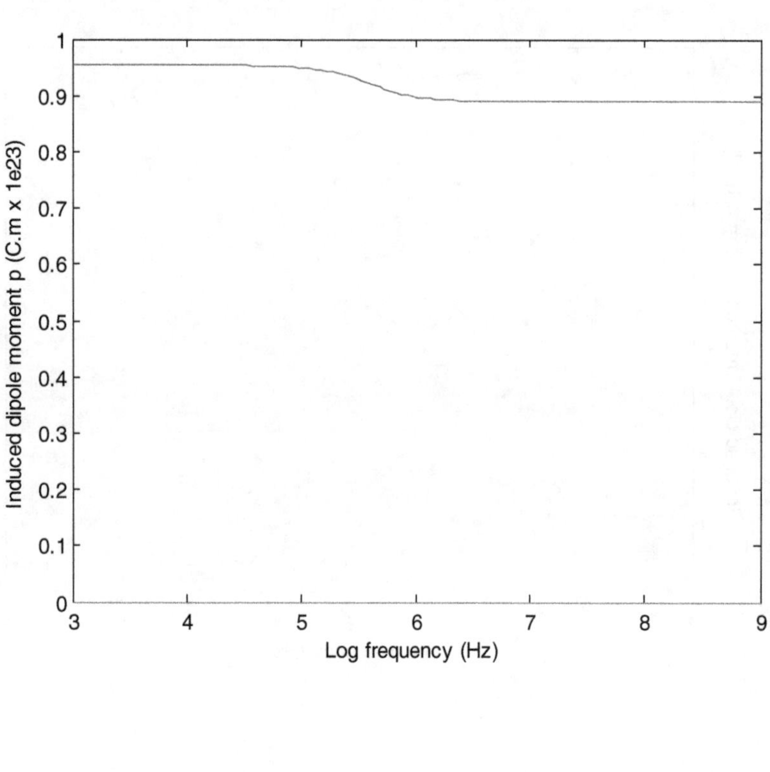

Figure 6.10 Frequency variation of the magnitude of the induced moment corresponding to the real and imaginary components of the Clausius–Mossotti factor shown in Figure 6.6 ($\sigma_p = 0.1$ mS/m; $\sigma_m = 1$ mS/m; $\varepsilon_p = 60\varepsilon_o$ and $\varepsilon_m = 20\varepsilon_o$).

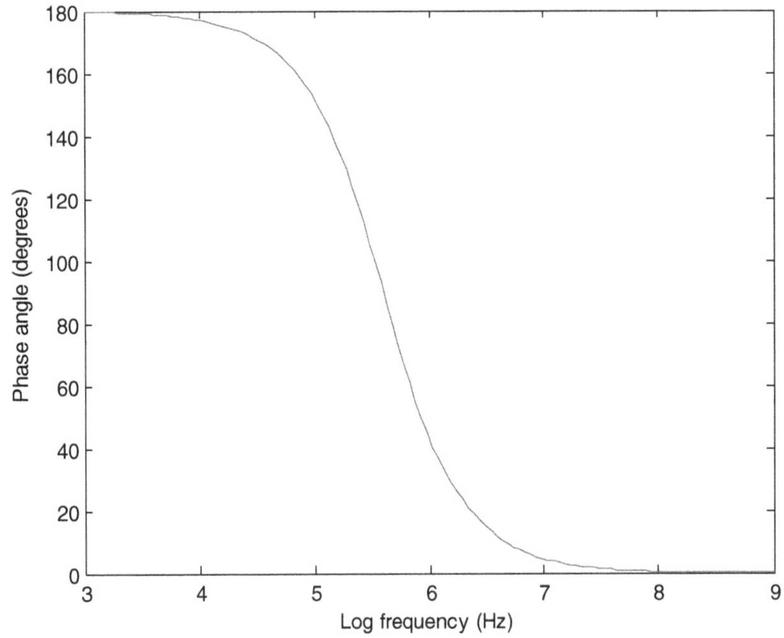

Figure 6.11 At low frequencies the moment **p** of Figure 6.10 *leads* the applied electric field by 180°. As the frequency is increased this leading phase angle decreases until at high frequencies the induced moment is exactly in phase with the applied field.

However, some care should be taken in adopting Equation (6.54) for prolate spheroidal particles, because it is not always the case that it will have its major axis directed along the field direction. The orientational torque acting on nonspherical particles will in general vary as a function of frequency. The orientation that leads to a stable (i.e., minimum potential energy) situation may not be with the major axis in alignment with the field direction, but instead where a minor axis assumes this position. In this case the relevant depolarization factor tends to a value of 1 rather than zero. This is discussed fully by Stratton [33] and Jones [34].

6.5 Summary

There are two functions in the dielectrics literature known as the Clausius–Mossotti factor. They take on the same mathematical form and use the same symbols, but in fact apply to two quite distinct problems.

The version of principal relevance to dielectrophoresis is referred to in this chapter as the *macroscopic* Clausius–Mosssotti factor. It arises when dealing with the polarization of a particle embedded in a medium whose dielectric properties differ from that of the particle. A uniform electric field $\mathbf{E}_0$ has already been established in the medium and it is assumed that for distances r far away from the particle's centre its polarization field does not alter the imposed uniform field. This is a standard problem, described with solutions in the classical electrostatics literature for more than 100 years, which involves solving Laplace's equation for specified boundary conditions. An example is given in Box 6.1 for the case of a spherical particle having a *macroscopic* static relative permittivity ε_p, suspended in a medium of relative permittivity ε_m. The first order solution of Laplace's equation (i.e., ignoring terms containing $\cos^2\theta$ and higher, where θ is the angle between the vector distance r and the field direction) gives the potential as:

$$\phi = \left(\frac{K_1}{r^2} - K_2 r\right) \mathrm{E}_o \cos\theta;$$

where K_1 and K_2 are constants. The factor $\phi = \frac{K_1}{r^2}\mathrm{E}_o \cos\theta$ describes the field of a dipole and this term appears only outside the spherical particle, with

$$K_1 = \frac{\varepsilon_p - \varepsilon_m}{\varepsilon_p + 2\varepsilon_m} R^3$$

where R is the radius of the particle. The solution *outside* the particle gives $K_2 = 1$ for $r \gg R$. *Inside* the particle there is no dipole field (i.e., K_1 is zero) and

$$K_2 = \frac{3\varepsilon_m}{\varepsilon_p + 2\varepsilon_m}$$

The lines of constant electric force and the resultant fields created inside and outside the particle are plotted out in Figure 6.12. For the case where the effective polarizability of the particle exceeds that of the surrounding medium, the induced charges result in a dipole moment collinear with the external field and the external field appears to be drawn into the particle. However, the induced surface charges act in a way to decrease the

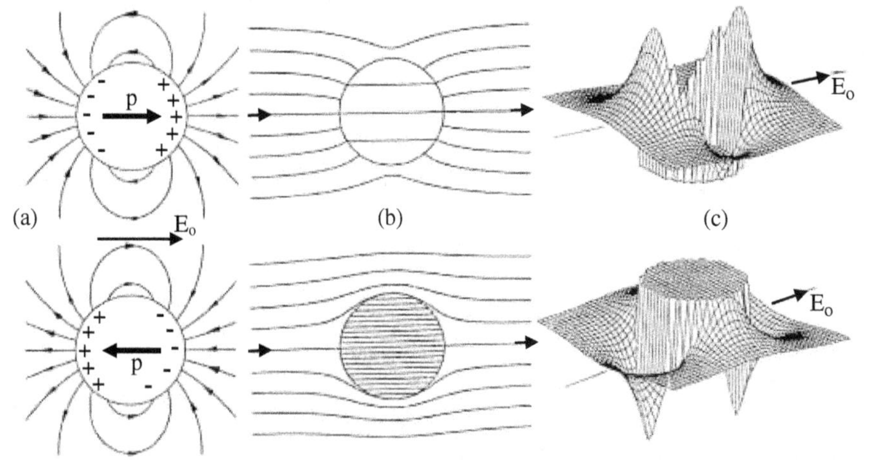

Figure 6.12 (a) The dipole moment p induced in a particle and the lines of constant electric force produced by surfaces charges induced on a spherical particle in a uniform field E_o, for the case (*top*) where the particle's polarizability is greater or (*bottom*) less than the surrounding medium. (b and c) The resultant field patterns to show how the external field is distorted into the surface of a particle that is more polarizable than the medium, to give an internal field less than the external field. For a particle less polarizable than the medium, the external field is distorted around the particle and the internal field is greater than the external field.

internal field of the particle. When the particle polarizability is less than the surrounding medium, the external field skirts around the particle and the internal field is increased.

On the reversal of an electric field to the system of particle plus external medium, the induced charges do not instantaneously rearrange themselves to produce a reversed dipole moment, but respond with a characteristic relaxation time. This needs to be taken into account when using alternating electric fields and is accomplished by defining a complex *CM* factor that leads to the generalized expression given by Equations (6.47).

The other version of the Clausius–Mossotti factor that appears in the dielectrics literature is the *molecular* CM factor and to distinguish this from the *macroscopic CM* factor we have referred to it in this chapter as the Clausius–Mossotti–Lorentz relation. This has its origins in efforts to translate the theories of Poisson and Green so as to relate the macroscopic polarizability and relative permittivity of a dielectric body to the local field that acts directly on the polarizable elements within it. The approach to this problem is to enclose a single polarizable element or a collection of such elements within an *imaginary* cavity and to calculate the internal cavity field.

Lorentz found that the local field $\mathbf{E}_l$ is related to the externally applied field **E** by the relationship (see Equation (6.13)):

$$\mathrm{E}_l = \mathrm{E} + \left(\frac{1}{3} + s\right) \mathrm{P}/\varepsilon_o$$

in which s is a constant 'which it will be difficult exactly to determine' [14, p. 138]. The form of the *CM* factor given by Equation (6.1) is only obtained by equating s to zero. Formulation of the *molecular* CM factor is thus not an exact procedure. We have discussed it in some detail in this chapter because it may provide insights into how to formulate a theory correctly to describe the dielectrophoretic behaviour of molecular-sized particles that possess a permanent dipole moment.

In this chapter we have also introduced the concepts of complex permittivity (ε^*) and complex conductivity (σ^*). The complex permittivity description of a dielectric material is employed to take into account the fact that it is not a *perfect* insulator, but exhibits energy losses associated with an electric field-induced movement of free charges (e.g., a leakage current) or the relaxation of permanent dipoles. The term *complex* is a mathematical term to indicate that $\varepsilon_r{}^*$ is a quantity having real and imaginary components, expressed as

$$\varepsilon_r^* = \varepsilon' - i\varepsilon''$$

where $i = \sqrt{-1}$ and ε' is the real component (Re) in phase with the applied sinusoidal voltage signal. The factor $-i\varepsilon''$ with its minus sign indicates that the imaginary (Im) component ε'' *lags* the voltage signal by a phase angle of 90°. $\varepsilon_r{}^*$ is thus a *vector* quantity, whereas ε' and ε'' are *scalar* quantities. The absolute values of ε' and ε'' are equal to $\varepsilon_o\varepsilon'$ and $\varepsilon_o\varepsilon''$(Farad m^{-1}), respectively. As given by Equation (6.42) the parameter ε'' quantifies the energy loss *per cycle* of the AC voltage applied to the dielectric. If the material is held between two parallel electrodes of area A and separation d, its capacitance C is given by

$$C = \frac{A\varepsilon_o\varepsilon'}{d} \text{ (Farad)}$$

If we now wish to introduce a leakage current effect or some other type of conductivity, we require the admittance Y or complex capacitance C^* of the material:

$$\mathrm{Y} = C^* = i\omega C = \frac{i\omega A}{d}\varepsilon_o\varepsilon_r^* = i\omega\frac{A}{d}\varepsilon_o(\varepsilon' - i\varepsilon'')$$
$$= \frac{A}{d}\varepsilon_o(\omega\varepsilon'' + i\omega\varepsilon') \text{ (Siemens)} \qquad (6.55)$$

Complex conductivity is employed to describe a conductive material that also exhibits the properties of a capacitor, possibly as a result of the buildup of charge (interfacial polarization) at sites of heterogeneity in the material. Complex conductivity is expressed as

$$\sigma^* = \sigma' + i\sigma''$$

where the real component σ' is proportional to the energy loss per second (power loss). The corresponding admittance **Y** or complex conductance G^* of the material is:

$$Y = G^* = \frac{A}{d}\sigma^* = \frac{A}{d}(\sigma' + i\sigma'') \tag{6.56}$$

Comparing the real and imaginary components of equations (6.55) and (6.56) we find the following relationships

$$\sigma' = \omega\varepsilon_o\varepsilon'' \text{ and } \sigma'' = \omega\varepsilon_o\varepsilon'$$

The complex conductivity and complex permittivity can thus be expressed in terms of each other through the relationship

$$\sigma^* = i\omega\varepsilon_o\varepsilon_r^*$$

The complex form of the Clausius–Mossotti (*CM*) factor, given in Equation (6.1) to describe the polarization of a spherical particle of relative permittivity ε_p and conductivity σ_p suspended in a medium of relative permittivity ε_m and conductivity σ_m, can thus be given as either:

$$CM = \frac{\varepsilon_p^* - \varepsilon_m^*}{\varepsilon_p^* + 2\varepsilon_m^*} \quad \text{or} \quad CM = \frac{\sigma_p^* - \sigma_m^*}{\sigma_p^* + 2\sigma_m^*}$$

The real and imaginary components of the *CM* factor are derived in Box 6.8. The real component $\text{Re}[CM^*]$ is the one of direct relevance to dielectrophoresis. As shown in Chapter 10, electrorotation and travelling wave dielectrophoresis depend on the imaginary component $\text{Im}[CM^*]$. In our treatment of complex conductivity for the analysis of Debye-type relaxations in a material, the DC conductivity is considered to be negligible. If this is not the case, Grant [35] has described a method for including its effect.

Finally, from the expression given by Equation (6.53) for the polarization of an ellipsoid, the complex *CM* factor for an ellipsoidal particle is given by:

$$CM = \left(\frac{\varepsilon_p^* - \varepsilon_m^*}{\varepsilon_m^* + A(\varepsilon_p^* - \varepsilon_m^*)}\right)\text{E}_x \tag{6.57}$$

where the parameter A is the depolarization factor described in Chapter 5, Equation (5.57).

6.6 References

1 Abraham, M. (1907) *Theorie der Elecktrizität*, B. G. Teubner, Leipzig, 3rd ed., pp. 157–166. English translation of Abraham M. (1932) *The Classical Theory of Electricity and Magnetism*, 8th edn. Hafner, New York, NY, pp. 79–80.

2 Maxwell, J. C. (1873) *A Treatise on Electricity and Magnetism*, Vol. 1. Clarendon Press, Oxford.

3 Poisson, S.-D. (1826) Mémoire sur la théorie du magnétisme. *Mémoires de l'Académie des Sciences de l'Institut de France (1824)* **5**, 247–338.

4 Green, G. (1828) *An Essay on the Application of Mathematical Analysis to the Theories of Electricity and Magnetism*, T. Wheelhouse, Nottingham.

5 Grattan-Guinness, I. (1995) Why did George Green write his essay of 1828 on Electricity and Magnetism? *American Mathematical Monthly* **102**(5), 387–396.

6 Faraday, M. (1839–1855) *Experimental Researches in Electricity*, 3 volumes, Taylor & Francis, London (reprinted in 2 volumes, 1965, Dover Publications, Inc., New York, NY).

7 Liberti, L. (1995) Ottaviano Fabrizio Mossotti: The Youth Years (1791–1823). B.Sc. thesis. Imperial College, http://www.lix.polytechnique.fr/~liberti/maths-history/mossotti/mossotti.html (accessed 20 November 2016).

8 Mossotti, O. F. (1850) Discussione analitica sull'influenza che l'azione di un mezzo dielettrico ha sulla distribuzione dell'electricitá alla superficie di più corpi elettrici disseminati in esso. *Memorie di Matematica e Fisica della Societá Italiana delle Scienze (Modena)* **24**(II), 49–74.

9 Maxwell, J. C. (1871) Remarks on the mathematical classification of physical quantities. *Proc. London Mathematical Soc.* **3**(34), 224–233.

10 Maxwell, J. C. (1865) A dynamical theory of the electromagnetic field. *Phil. Trans. Roy. Soc. (London)* **155**, 459–512.

11 Clausius, R. (1879) *Die mechanische Wärmetheorie*, Zweiter Band, Abschnitt III: Behandlung dielectrische Medien, Friedrich Vieweg und Sohn, Braunschweig, pp. 62–97.

12 Lorenz, L. (1870) Experimentale og theoretiske Undersøgelser over Legermenes Brydningsforhold, I *Kongelige Danske Videnskabernes Selskab Skrifter* **5**,

203; Lorenz, L. (1875) Experimentale og theoretiske Undersøgelser over Legermenes Brydningsforhold, II *Kongelige Danske Videnskabernes Selskab Skrifter* **10**, 485.

13 Lorentz, H. A. (1878) Verhandelingen der Akad. v. Wet. te Amsterdam, Deel 18. [Memoirs of the Academy of Science and Law, Amsterdam, Part 18.]

14 Lorentz, H. A. (1916) *The Theory of Electrons*, 2nd edn, B. G. Teubner, Leipzig.

15 Lorentz, H. A. (1880) Über die Beziehung zwischen der Fortphlanzungsgeschwindigkeit des Lichtes und der Körperdichte. *Ann. Physik Chemie* **245**(4), 641–665.

16 Lorenz, L. (1880) Über die Refractionsconstante. *Ann. Physik Chemie* **247**(9), 70–103.

17 Vogel, A. I. (1948) Physical properties and chemical constitution: Parts X–XXIII. *J. Chem. Soc.* 607–683; 1814–1855.

18 Li, Y., Zhu, H., Gu, H. *et al.* (2013) Strong transparent magnetic nanopaper prepared by immobilization of Fe_3O_4 nanoparticles in a nanofibrillated cellulose network. *J. Mater. Chem. A.* **1**, 15278–15283.

19 Louzguine-Luzgin, D. V., Ketov, S. V., Orava, J. and Mizukami, S. (2014) Optically transparent magnetic and electrically conductive Fe–Cr–Zr ultra-thin films. *Physica Status Solidi (a)* **211**(5), 999–1004.

20 Debye, P. J. W. (1912) Einige Resultate einer kinetischen Theorie der Isolatoren, *Physikalische Zeitshrift* **13**(3), 97–100.

21 Debye, P. (1929) *Polar Molecules*, The Chemical Catalog Co., New York, NY.

22 Onsager, L. (1971) *Autobiographical commentary, in Critical Phenomena in Alloys, Magnets and Superconductors* (eds. R. E. Mills, E. Ascher and R. I. Jaffee). McGraw-Hill, New York, NY, p. xx.

23 Onsager, L. (1936) Electric moments of molecules in liquids. *J. Am. Chem. Soc.* **58**, 1486–1493.

24 Wilson, J. N. (1939) The dielectric constants of polar liquids. *Chem. Rev.* **25**, 377–406.

25 Kirkwood, J. G. (1939) The dielectric polarization of polar liquids. *J. Chem. Phys.* **7**, 911–919.

26 Fröhlich, H. and Sack, R. (1944) Dielectric properties of dipolar substances. *Proc. Roy. Soc. Lond. A* **182**(991), 388–403

27 Abbott, J. A. and Bolton, H. C. (1952) The influence of molecular shape in the electrostatic theory of polar liquids. *Trans. Faraday Soc.* **48**, 422–430.

28 Buckingham, A. D. (1953) The application of Onsager theory to spheroidal molecules. *Trans. Faraday Soc.* **49**, 881–886.

29 Wagner, K. W. (1914) Erklärung der dielektrischen Nachwirkungsvorgänge auf Grund Maxwellscher Vorstellungen. *Archiv f. Electrotechnik* **2**(9), 371–387.

30 Inoue, T., Pethig, R., Al-Ameen, T. A. K. *et al.* (1988) Dielectrophoretic behaviour of *Micrococcus lysodeikticus* and its protoplast. *J. Electrostatics* **21**, 215–223.

31 Molinari, G. and Viviani, A. (1978) Analytical evaluation of the electro-dielectrophoretic forces acting on spherical impurity particles in dielectric fluids. *J. Electrostatics* **5**, 343–354.

32 Benguigui, L. and Lin, I. J. (1982) More about the dielectrophoretic force. *J. Appl. Phys.* **53**(2), 1141–1143.

33 Stratton, J. A. (1941) *Electromagnetic Theory*, McGraw Hill, New York, pp. 211–213.

34 Jones, T. B. (1995) *Electromechanics of Particles*, Cambridge University Press, Ch. 5.

35 Grant, F. A. (1958) Use of complex conductivity in the representation of dielectric phenomena. *J. Appl. Phys.* **29**(1), 76–80.

7

Dielectric Polarization

7.1 Introduction

Figure 3.1 gives a simple way to determine the relative permittivity of a slab of dielectric, by observing the change in the reading of an electrometer as it is inserted in the gap between a pair of charged parallel electrodes. From these external observations we can deduce the permittivity of the dielectric material, but without more information we have no understanding of how this relates to polarization processes occurring at the molecular or submolecular level. Similarly, in a dielectrophoresis experiment we typically observe the induced translational motion of a particle (e.g., cell, bacteria, fluorescently labelled virus or RNA molecule) suspended between electrodes. We assume that the particle has been polarized by the electric field, but we have no direct way to relate this to the molecular or structural properties of the particle – or more importantly to biological parameters such as a cell's viability or state of differentiation. We have to advance to such levels of understanding before dielectrophoresis can reach its full potential as a tool for the biomedical sciences.

To proceed with this objective it is useful to review the basic concepts of electrostatics and dielectrics described in previous chapters. Important facts are highlighted in Chapter 3, namely that an ideal electrical conductor and an ideal dielectric behave quite differently when subjected to an electric field. In brief:

- an ideal conductor cannot sustain an internal electrostatic field;
- an ideal dielectric does not conduct electric current and can support a large internal electrostatic field.

In Figure 3.1 the ratio E_0/E_1 of the initial field E_0 between the plates to the field E_1 when the slab is fully inserted is defined as the relative permittivity ε_r of the material. If the material is an ideal metal, which exhibits no resistance to current flow, then $E_1 = 0$ and the ratio E_0/E_1 has an infinite value. In terms of Equation (3.3) there is a complete eradication of the original free charge applied to the electrode plates (i.e., $\Delta\sigma = \sigma$). From this we deduce that ideal conductors possess an infinite relative permittivity, which is equivalent to stating that ideal metals are infinitely polarizable.

Dielectric materials possess finite values of permittivity. The ratio E_0/E_1 is thus finite and so a dielectric only *partially* counteracts an applied external field. Electrons and positively charged nuclei in the molecular structure of a dielectric material are therefore exposed to a field within the dielectric medium. In Chapter 6, we found that, to a first approximation, the *local* field acting on a polarizable element in the dielectric, known as the Lorentz field, is equal to $[(\varepsilon_r+2)/3]E_1$ and so is larger than the macroscopic field E_1 set up in the dielectric. The value of ε_r is a measure of the extent to which the electric charge distributions in polarizable elements of the dielectric are displaced or 'polarized' by the *local* field. In this chapter, we will seek some quantitative understanding of such polarization at the atomic and molecular level.

7.2 Electrical Polarization at the Atomic and Molecular Levels

A basic concept of the electronic structure of atoms is that their electrons occupy what are known as electron shells. Each shell is composed of one or more subshells, which represent the different orbital path shapes that electrons have around the nucleus of their parent atom. The first, innermost shell, has one subshell called the 1s shell and can accommodate a maximum of two electrons. Hydrogen and helium atoms have this electronic structure, with hydrogen possessing one orbiting electron and helium two. The second shell has two subshells (2s, 2p), which can accommodate a maximum of eight electrons, with two in the 2s shell and six in the 2p shell. The third shell has three subshells (3s, 3p, 3d) and can accommodate a maximum of 18 electrons, with ten in the 3d shell – and so on through the periodic tables of elements. A covalent chemical bond between two atoms is formed by the sharing of unpaired electrons, one from the outer electron shell of each atom. These shared electrons enter

Dielectrophoresis: Theory, Methodology and Biological Applications, First Edition. Ronald Pethig.

an electronic orbital, which is common to both atoms, acting to reduce the repulsive force between the two positively charged nuclei. According to this scheme the hydrogen atom, with one unpaired electron, can form only one covalent bond, whilst carbon with four electrons in its outermost 2s and 2p subshells can form four bonds. A simple example of this is the methane molecule (CH_4), where one carbon atom forms covalent bonds to four hydrogen atoms.

Electrons in the outer shells have higher average energies than those in the inner shells and their electron orbitals extend farther from the nucleus. This contributes to how chemically reactive a particular atom may be in its interaction with other atoms and it determines the electrical polarizability of a molecular structure of which it forms a part.

7.2.1 Nonpolar, Polar and Ionic Bonds

In a covalent bond formed between two identical atoms, such as the C-C bond between two carbon atoms, the bonding electrons are equally shared between the atoms. Such a bond is termed *nonpolar*. For example, molecules such as Cl_2, H_2 and F_2 are nonpolar. Different atoms exhibit different tendencies for the sharing of their electrons. This tendency can be quantified in terms of their *electronegativity*, using a scale measured from a hypothetical zero to a maximum value of 4.0. The electronegativity values of some atoms of biological importance are listed in Table 7.1 and are based on the scale devised by Linus Pauling [1]. Fluorine, the most electronegative atom, is assigned an electronegativity value of 3.98.

We note from Table 7.1 that atoms located at the upper right of the Periodic Table of Elements are more electronegative and those to the lower left are least electronegative. From this we can judge that carbon disulphide (CS_2) has an almost equal sharing of its electrons when forming its C-S covalent bonds. We call such a bond a *nonpolar* bond. As a guideline, a maximum difference of $0.4 \sim 0.5$ in electronegativity values can be used to define

Table 7.1 The electronegativity values for some atoms in the Periodic Table of elements, based on the Pauling electronegativity scale [1].

I	II	III	IV	V	VI	VII	VIII
H 2.10							
			C 2.55	N 3.04	O 3.44	F 3.98	
Na 0.93	Mg 1.31			P 2.19	S 2.58	Cl 3.16	
K 0.82	Ca 1.00						

the limit for the formation of a nonpolar bond. For a C-Cl bond there is an electronegativity difference of 0.61 and thus a significant unequal sharing of electrons between the C and Cl atoms. Electronic charge on average spends more time closer to the chlorine atom (giving it a slightly negative charge $\delta-$) and less time near the carbon atom (which thus acquires a slightly positive charge $\delta+$). The C-Cl bond is termed a *polar* bond. This quality is even more pronounced in an H-F bond, where the electronegativity difference is 1.88. The polar bonds in molecules such as NH_3 and H_2O result in their possessing a permanent electric dipole moment. Such molecules will tend to align themselves with an externally applied electric field. However, where the electronegativity difference between atoms in a bond is greater than around 2.0, we approach the situation where there is complete transfer of an electron from the least to the most electronegative atom. This type of bond is termed *ionic*. Common salt (NaCl) is a good example, forming ionic crystals held together by the coulombic forces between the positively charged Na^+ and negatively charged Cl^- atoms. Other examples of an ionic solid include salts such as KCl, LiF and $MgCl_2$.

When two highly electronegative atoms form a chemical bond, this bond is usually quite unstable. This occurs in hydrogen peroxide (H-O-O-H), where the strong attractions of bonding electrons towards the two strongly electronegative oxygen atoms make it a highly reactive molecule.

7.2.2 Polarization by Electronic and Atomic Distortion

When molecules in a gas, liquid or solid are exposed to an electric field they experience an electrostatic force, causing their atomic nuclei and electronic orbitals to be slightly displaced from their normal equilibrium configurations. This effect is known as *distortion* polarization and involves two distinct contributions, known as *electronic* and *atomic* polarization.

7.2.2.1 Electronic Polarization

The basic process can be described using the idealized atomic model shown in Figure 7.1. An atom of atomic number Z is represented as a spherical cloud of electrons of radius R and total charge $-Ze$ surrounding a nucleus of charge $+Ze$. The atomic radius of the hydrogen atom ($Z = 1$) is given by the most probable radius of orbit of its single electron. This is known as the Bohr radius and has a value of 5.3×10^{-11} m. The negative electronic charge is confined to a volume $4/3(\pi R^3)$ and has a volume charge density ρ_e given by:

$$\rho_e = \frac{3Ze}{4\pi R^3} \tag{7.1}$$

An electric field E acting on an atom will tend to pull the nucleus and the electron cloud apart. This is opposed

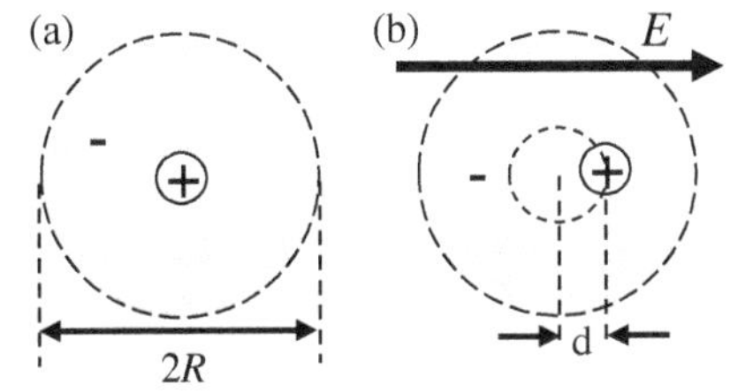

Figure 7.1 (a): An idealized atomic structure is shown of a nucleus of point charge +*Ze* surrounded by a spherical 'cloud' of orbiting electrons of total electronic charge −*Ze* and radius *R*. (b): In an electrical field *E* the opposing electrostatic forces acting on the nucleus and electron 'cloud' is balanced by their force of attraction. This results in a stable separation *d* between the nucleus and the centre of the spherical 'cloud' of electronic charge. The inner circle of radius *d* represents a Gaussian surface enclosing that fraction of negative charge involved in the attractive force with the positive nucleus.

by the electrostatic force of attraction between the point positive charge of the nucleus and the diffuse negative charge in the spherical electron cloud. The equilibrium separation distance d shown in Figure 7.1(a) corresponds to the situation where the force pulling the nucleus and electron cloud apart balances the force of their electrical attraction. However, only a fraction of the negative charge density is involved in establishing this equilibrium. We can understand this by drawing a spherical Gaussian surface of radius d around the nucleus, as shown in Figure 7.1. Outside this Gaussian surface the diffuse electron cloud can be considered as a series of concentric charged spheres, each one acting as a Faraday cage with zero internal field (see sections 3.3.2.2 and 3.3.2.5 of Chapter 3). The force of attraction with the positive nucleus, which lies just within the Gaussian surface, thus involves only the negative charge density ρ_{ed} within a sphere of radius d. This fraction of the charge is given by:

$$\rho_{ed} = Ze\left[\frac{3}{4\pi d^3}\right]\Big/\left[\frac{3}{4\pi R^3}\right] = \frac{Zed^3}{R_3} \tag{7.2}$$

The coulombic force of attraction, given by Equation (3.5), between the nucleus (positive charge Ze) and the charge given by Equation (7.2) is:

$$F_{attraction} = \frac{1}{4\pi\varepsilon_o}\frac{Ze(Zed^3/R^3)}{d^2} = \frac{(Ze)^2}{4\pi\varepsilon_o R^3}d = kd \tag{7.3}$$

The factor k on the right-hand side of this equation gives the proportionality between the restoring force and the displacement from equilibrium and acts rather like the force constant of an ideal mechanical spring that obeys Hooke's Law. When the restoring force is directly proportional to the displacement from equilibrium the system exhibits a resonant oscillation called *simple harmonic motion*. A body or system that undergoes simple harmonic motion is called a *harmonic oscillator*. The characteristic frequency of resonance of the electronic polarization effect we have described here is above 10^{14} Hz and lies within the ultraviolet (UV) frequency band.

The force acting to stretch the distance between the nucleus and the negative electronic cloud of charge is:

$$\mathrm{F}_{stretch} = Ze\mathrm{E} \tag{7.4}$$

The equilibrium separation distance d is obtained by equating the restoration force of attraction given by Equation (7.3) and the stretching force given by Equation (7.4):

$$d = \frac{4\pi\varepsilon_o R^3}{Ze}\mathrm{E} \tag{7.5}$$

The electric moment p induced by this displacement of the nucleus from the centre of the spherical electronic cloud is given by:

$$\mathrm{p} = Zed = 4\pi\varepsilon_o R^3\mathrm{E} \tag{7.6}$$

In Chapters 3 and 6, the macroscopic polarization P of a dielectric material is defined as the average induced dipole moment per unit volume. If the number density of atoms of the form shown in Figure 7.1 is N, then P is given by:

$$\mathrm{P} = N\mathrm{p} = 4\pi\varepsilon_o R^3 N\mathrm{E} \tag{7.7}$$

The corresponding value of the relative permittivity ε_r is obtained from Equation (3.48):

$$\mathrm{P} = (\varepsilon_\mathrm{r} - 1)\varepsilon_0\mathrm{E}$$

to give

$$\varepsilon_r = 1 + \frac{\mathrm{P}}{\varepsilon_0\mathrm{E}} = 1 + 4\pi R^3 N \tag{7.8}$$

Although this treatment is based on a simplistic model of hydrogen-like atoms that possess spherical electron 'shells', several important facts emerge:

- A polarized atom acts like a vibrating spring and exhibits a characteristic resonant frequency, typically ~2×10^{14} Hz in the ultraviolet region of the optical spectrum.
- At field frequencies below this resonance the magnitude of electronic polarization α_e is directly proportional to the local field E. We have the relationship: $\mathrm{P} = \alpha_e\mathrm{E}$.
- The magnitude of the electronic polarizability of an atom increases as the volume occupied by the electrons increases.

7.2.2.2 Atomic Polarization

If the molecule contains polar bonds, so that the atoms involved carry different effective charges, the nuclei are displaced with respect to one another in an electric field. This charge displacement produces an induced electric moment superimposed upon that arising from electronic polarization. This additional contribution to the overall molecular polarization is called *atomic* polarization α_a. The molecule itself need not have a permanent dipole moment – only polar chemical bonds. These polarized bonds also behave like vibrating springs and exhibit a characteristic resonant frequency in the infrared region of the optical spectrum at $\sim 2 \times 10^{13}$ Hz.

The total polarization arising from electronic and atomic distortions is given by

$$P = (\alpha_e + \alpha_a)E \tag{7.9}$$

Example 7.1 Polarizability of Solid Hydrogen
At temperatures below 14 K hydrogen forms a solid of density 0.086 g/cm^3. Estimate the value of the relative permittivity of solid hydrogen.

Solution 7.1 For simplicity we will assume that solid hydrogen takes the form of condensed hydrogen atoms (rather than condensed H_2 molecules). The atomic mass of a hydrogen atom is 1 amu, so 1 g of solid hydrogen contains Avogadro's number of atoms (i.e., 6×10^{23}). The value for N in Equation (7.6) is thus calculated to be:

$$N = (6 \times 10^{23})(8.6 \times 10^{-2})$$
$$= 5.16 \times 10^{21\ \mathrm{cm}^{-3}} = 5.16 \times 10^{27}\mathrm{m}^{-3}$$

Using this value for N, together with $R = 0.53$ nm, from Equation (7.8) we have:

$$\varepsilon_r = 1 + \frac{P}{\varepsilon_0 E} = 1 + 4\pi R^3 N$$
$$= 1 + 4\pi(5.3 \times 10^{-11})^3(5.16 \times 10^{27}) = 1.0097$$

This result implies that the electronic polarizability of our model hydrogen atom is very small. A more realistic calculation [2] of the electrical polarizability of a classical hydrogen atom in its quantum-mechanical ground state energy leads to a value that is larger by a factor of 21/4 than the result obtained using Equation (7.6). This gives a relative permittivity value $\varepsilon_r = 1.05$, compared to a value of 1.00 for vacuum. This still represents a very low polarizability.

As we advance from element to element through the periodic table, with each increase by 1 of the atomic number an electron is added to the electronic shell to neutralize the charge of an added proton. This extra electron occupies the same volume in the outermost orbital shell, but because the nuclear charge increases the electron is held more tightly by electrostatic attraction into a smaller effective atomic radius. We therefore expect the electronic polarizability to decrease from *left to right* along a row of the periodic table of elements. A noble gas occupies the end of each row (VIII) with its outermost electron shell completely occupied. The addition of another proton and electron (increasing the atomic number by 1) gives the first member (an alkali metal such as sodium or potassium) of the next row of elements. This added electron goes into an unoccupied outer shell, which represents a stepwise increase in atomic radius. In the periodic table of elements we can therefore expect the polarizability to increase down a column of elements. An example of this occurs in column IV of the table, where crystals of carbon (diamond) with Z = 6, silicon (Z = 14) and germanium (Z = 32) have relative permittivity values at room temperature of 5.5, 12 and 16, respectively. These permittivity values arise solely from electronic polarization. The much larger polarizability values of this set of atoms, compared to hydrogen, arise primarily from the fact that their electron shells consist of a combination of spherical '*s*' and dumb-bell shaped '*p*' orbitals (sp-hybrids) and not a simple spherical orbital as depicted in Figure 7.1.

7.2.3 Ionic Polarization

The lattice structure of a common ionic crystal, namely sodium chloride (NaCl), is shown in Figure 7.2. Sodium is situated at the left-hand side of the periodic table of elements and chlorine on the right. From Table 7.1 we see that sodium and chlorine atoms exhibit an electronegativity difference of 2.23, with chlorine as the more electronegative. Sodium chloride thus forms what is termed an ionic crystal. Each sodium atom carries a net positive charge and every chlorine atom a net negative charge and the electrostatic interactions between them are strong and contribute significantly to its crystal lattice structure. We can use the concept of electrical potential energy described in Chapter 4 and Equation (4.10) to derive the electrostatic energy of an ionic crystal and this is outlined in Box 7.1.

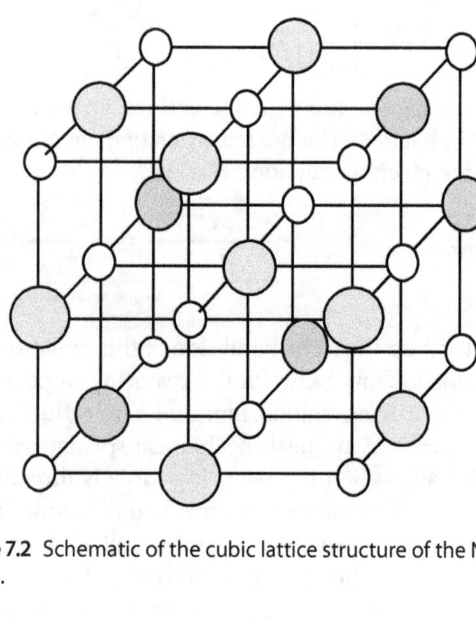

Figure 7.2 Schematic of the cubic lattice structure of the NaCl crystal.

Box 7.1 Electrostatic Potential Energy of an Ionic Crystal

The potential energy of a single Na^+ or Cl^- ion in the lattice structure shown in Figures 7.2 and 7.3 can be obtained using Equation (4.10). Each ion experiences an attractive force to ions of opposite charge and repulsion from ions of the same charge. We will focus attention on one of the Na^+ ions (we could equally well choose a Cl^- ion) and let r be the shortest distance between a Na^+ and Cl^- ion. The nearest neighbours of a Na^+ ion in the lattice are 6 Cl^- ions at a distance of $\sqrt{r}$, 12 Na^+ ions at a distance of $\sqrt{(2r)}$, 8 Cl^- ions at $\sqrt{(3r)}$, 6 Na^+ ions at $\sqrt{(4r)}$, 24 Cl^- ions at $\sqrt{(5r)}$ and so forth. From Equation (4.10) the electrostatic energy due to one ion is:

$$U = \frac{1}{4\pi\varepsilon_0\varepsilon_r}\sum_{i<j}\frac{q_iq_j}{r_{ij}} = \frac{q^2}{4\pi\varepsilon_0\varepsilon_r}\frac{1}{r} \times \left[\frac{6}{\sqrt{1}} - \frac{12}{\sqrt{2}} + \frac{8}{\sqrt{3}} - \frac{6}{\sqrt{4}} + \frac{24}{\sqrt{5}}\cdots\right]$$

The series within the square brackets converges to the value 1.7476 and is the *Madelung Constant M* for the rock salt type of crystal lattice, of which the sodium chloride crystal is a member. The value of M depends on the crystal structure. For example, for the fluorite (e.g., CaF_2 structure) $M = 2.5194$ and for the corundum (e.g., Al_2O_3) structure $M = 4.1719$.

The number of ions in 1 mole of an ionic crystal is m_wN_A, where m_w is its molecular weight and NA is the Avogadro constant. The general formula for the total molar electrostatic energy of an ionic crystal is:

$$U = m_wN_A\frac{z^-z^+q^2}{4\pi\varepsilon_0\varepsilon_r}\frac{M}{\eta}$$

Where z^- and z^+ are the number of charges on the anion and cation, respectively ($z^- = z^+ = 1$ for NaCl; $z^- = 3$, $z^+ = 2$ for Al_2O_3) and r_l is the shortest cation-anion distance for the lattice type.

The contributions to the electronic polarization of an ionic crystal are not shared equally between the cations and anions. For example, in a NaCl crystal the chlorine atoms, having gained an increased electronic charge density, contribute more to the crystal's electronic polarizability than the sodium atoms that have lost a fraction of their overall electronic charge. However, an ionic crystal exposed to a static electric field can exhibit an additional polarization effect of magnitude that significantly exceeds its intrinsic electronic polarizability.

In the lattice structure of crystalline NaCl the dipole moment (depicted by arrows in Figure 7.3) of any pair of Na^+ and Cl^- ions is completely cancelled by a neighbouring dipole. The polarization per unit volume due to the ionic nature of the crystal is zero. However, as shown in an exaggerated manner in Figure 7.3, an applied electric field exerts an electrostatic force ($F = q_iE$) on each ion, pulling the Na^+ and Cl^- ions in opposite directions. This distorts the crystal lattice and results in noncancellation of adjacent dipole moments. The net dipole moment is directed along the field direction and enhances the polarizability per unit volume of the crystal above that due to its electronic polarizability. The force acting to restore the lattice structure takes a similar form as Equation (7.3), with the spring constant k related to the compressibility of the lattice and the electrostatic forces between the Na^+ and Cl^- ions.

Yamashita [3] employed a quantum mechanical method to derive theoretical estimates of the polarizability of the LiF ionic crystal. He examined the cases for a frequency of an applied electric field that is low enough for an electrostatically induced lattice distortion to keep pace with the changing field and also for a frequency so high that no significant lattice distortion is possible. He obtained a high frequency relative permittivity value of 2.33, which can be considered to result from polarization of the electronic charge distributions around the ions. This value is not too far from the experimental one of 1.92. Based on estimated values for the Madelung constant described in Box 7.1, together with

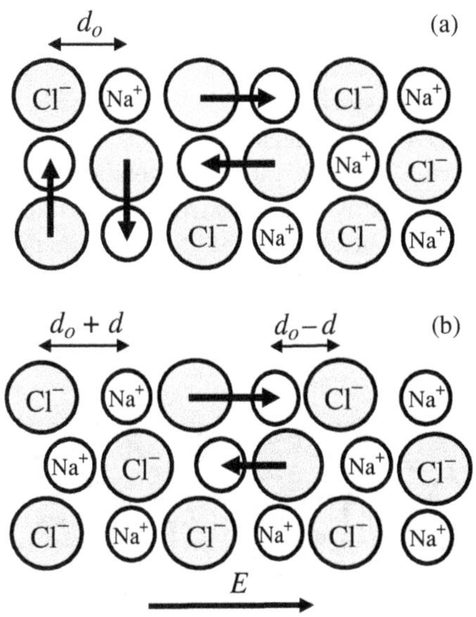

Figure 7.3 (a) In the lattice structure of crystalline NaCl the dipole moment (arrow) of any pair of Na^+ and Cl^- ions is exactly counterbalanced by a neighbouring dipole moment. The dipole moment and hence polarization per unit volume of the crystal is zero. (b) On application of an electric field E the induced displacements of the ions leads to noncancellation of adjacent dipole moments and produces a net dipole moment per unit volume.

the compressibility of the LiF lattice structure, a relative permittivity value of 9.3 was obtained for the static and low frequency situation. The fourfold increase of permittivity results from an electrostatic-induced distortion of the crystal lattice of the form depicted in Figure 7.3.

7.2.4 Polarization arising from Dipole Moment Orientation

The field-induced dipole moments shown in Figure 7.3 for an ionic solid do not respond independently to an applied field. A net dipole moment per unit volume of the ionic crystal is induced as a result of a small electrostatic-induced distortion of the lattice structure. The induced dipoles disappear when the field is removed – they are not a permanent intrinsic feature of the crystal's structure and cannot be randomly distributed or oriented.

However, some molecules possess an atomic structure that results in their having a permanent, dipole moment that can respond independently to an applied electric field. An example is fluorobenzene, shown in Figure 7.4, whose structure has the basic form of a benzene ring with a fluorine atom replacing one of the six hydrogen atoms. A benzene molecule consists of six polar C-H bonds, with each carbon atom on average carrying a net negative charge donated by its bonded hydrogen atom. The symmetrical arrangement of the six C-H bonds results in the benzene ring having on average no net dipole moment. The centre of the positive charges is at the same point as the centre of the negative charges and the polarizability of the molecule arises solely from field-induced distortions of its electron shells. As shown in Figure 7.4(b) the replacement of a hydrogen atom with fluorine gives to the fluorobenzene molecule an asymmetrical molecular electrostatic surface [4] and thus properties of a permanent dipole moment. The centres of the positive charges and negative charges are not coincident. Benzene is a *nonpolar* molecule whereas fluorobenzene is a *polar* molecule. On applying an electric field each fluorobenzene molecule experiences a torque, given by Equation (5.37) that tends to align the electric moment along the field direction. This effect is depicted in Figure 7.3(c). Above 229 K fluorobenzene is in its liquid state and each dipole moment is relatively free to rotate, leading to a relative permittivity value of around 5.5 at room temperature. In its solid state form rotational motions of the dipoles are hindered and the relative permittivity is reduced to a value near 2.5, similar to that for liquid benzene, which has no dipole moment. A universally important molecule with a permanent dipole moment is water. From Table 7.1 we see that there is an electronegativity difference of 1.34 between a hydrogen and oxygen atom, with oxygen being the more electronegative. The polarity of the two O-H bonds and the influence of the lone-pair electrons on the oxygen atom [7] give the water molecule a permanent dipole moment $\mathbf{p}_{HOH}$ (see Figure 7.5). The value of this dipole moment is evaluated by adding vectorially the two OH-bond moments, so that

$$p_{HOH}^2 = p_{OH}^2 + p_{OH}^2 + 2p_{OH}p_{OH}\cos\theta$$

in which $\theta = 104.5°$. From the charge distributions given in Figure 7.5 the bond moment $\mathbf{p}_{OH} = 5.05 \times 10^{-30}$ C m (1.515 D), to give $p_{HOH} = 6.19 \times 10^{-30}$ C m (1.855 D).

The dipole moment value given in Figure 7.5 for water is cited in both SI (C m) and debye (D) units. The

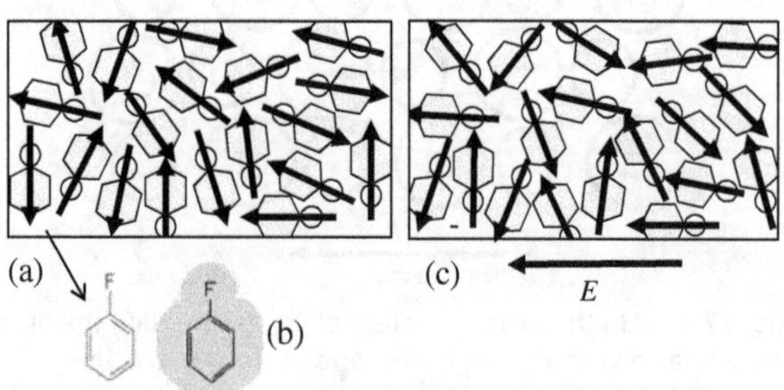

Figure 7.4 (a) Fluorobenzene molecules form a liquid consisting of independent permanent dipoles of overall random orientation. (b) The dipole moment results from the asymmetric electrostatic profile of the molecule [4]. (c) The dipoles tend to align with an applied electric field *E* and relax back to random orientations when the field is removed.

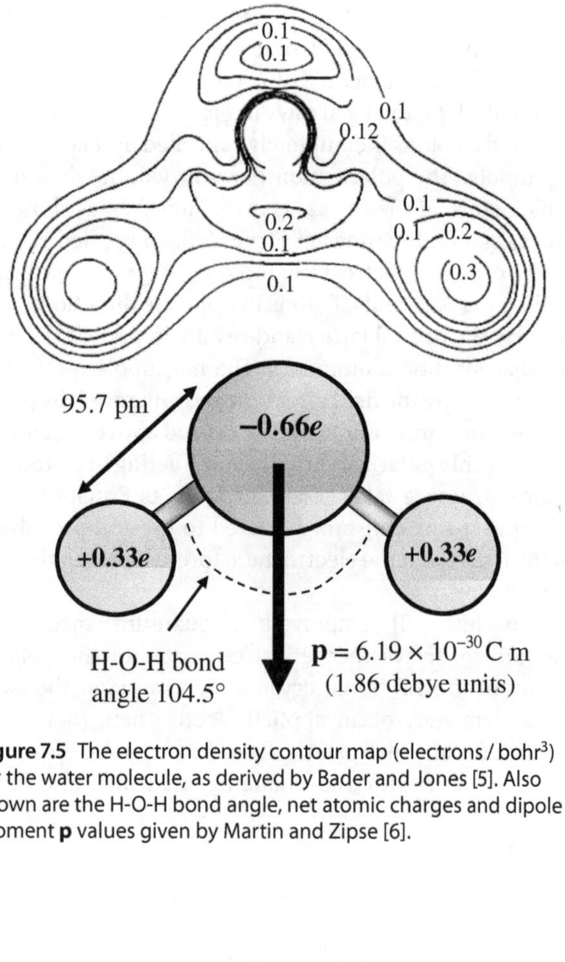

Figure 7.5 The electron density contour map (electrons / bohr³) for the water molecule, as derived by Bader and Jones [5]. Also shown are the H-O-H bond angle, net atomic charges and dipole moment **p** values given by Martin and Zipse [6].

Box 7.2 The SI and Debye Units of Electric Dipole Moment

The literature on the electrical and dielectric properties of materials contains a bewildering range of units. For example, Maxwell [22, p. 499] cites the resistance of various metals in B.A. units, which are assumed equal to 0.98677 Earth quadrants! This refers to the length of mercury column as the resistance standard established in 1862 by the British Association for the Advancement of Science, with 1 m being one ten-millionth of the distance along the meridian through Paris from the North Pole to the Equator. The system of units based on length (millimetre), mass (milligram) and time (second) proposed by Gauss in 1832, was modified to the cgs (centimetre, gram, second) system in 1874 by a B.A. committee that included Maxwell, William Thomson and a Mr Stoney (who placed on record 'that the centimetre was recommended as the unit of length against my earnest remonstrance')*. Mr Stoney's preference for the metre came to pass in the 1940s with the mks system (metre, kilogram, second). In 1960, this was replaced by the Système International d'Unités (SI), founded on seven mutually independent base units and quantities. The recommended SI measure of a dipole moment is the coulomb metre.

Coulomb's Law (Equation 3.4) differs in its cgs and SI form:

$$F = \frac{q_1 q_2}{r^2} \text{ dyne (cgs)}; F = \frac{1}{4\pi\varepsilon_o}\frac{q_1 q_2}{r^2} \text{ newton (SI)}$$

The cgs-esu unit is based on the statcoulomb (statC) and franklin (Fr) as the unit for electrical charge. By definition, two charges of 1 statC placed 1 cm apart in vacuum repel each other with a force of 1 dyne (10^{-5} N). In the SI system the coulomb is the quantity of electricity carried in 1 s by a current of 1 A. The ampère is defined in terms of the force developed between two parallel conductors placed 1 m apart in vacuum. A current of 1 A in each conductor creates a force of 2×10^{-7} newton per metre length. The following, dimensionally correct, equivalence between the statcoulomb and the coulomb is evaluated using Maxwell's equation $c^2 = 1/(\mu_o\varepsilon_o)$, where c is the speed of light in vacuo ($\sim 2.998 \times 10^8$ m s^{-1}) and μ_o is the magnetic permeability of free space defined to be $4\pi \times 10^{-7}$ H m^{-1}:

$$(1/\sqrt{4\pi\varepsilon_o}) \text{ C } \textit{is equivalent to } (10^{-1}c) \text{ statC}$$

A charge of 1 C is thus equivalent to $\sim 2.998 \times 10^9$ statC, or conversely, 1 statC is equivalent to $\sim 3.336 \times 10^{-10}$ C. This enables the following conversion to be made:

$$1 \text{ debye unit} = 10^{-18} \text{ statC cm} \approx 3.336 \times 10^{-30} \text{Cm}$$

*Report of the 43rd Meeting of the British Association for the Advancement of Science, 1873, p. 222. (Contributed by the Natural History Museum Library, London, http://biodiversitylibrary.org/page/29853091, accessed 9 December 2016.)

procedure to convert between them is outlined in Box 7.2. The debye unit was usefully defined to be of the same order as the product of an electronic charge (4.8×10^{-10} esu) and internuclear distances ($\sim 10^{-8}$ cm) within a molecule. Quoting the permanent dipole moment of haemoglobin as 480 D gives a good and immediate sense of how much larger the product of size and charge distribution for this protein molecule is compared to that of a water dipole (1.86 D). Citing these two values in SI units (1.6×10^{-27} C m and 6.2×10^{-30} C m, respectively) is somewhat tedious to write down and decipher. A yocto, with a value of 10^{-24}, is the *smallest* SI prefix that can be used in the SI system and a term such as microyoctocoulomb-metre to denote 10^{-30} C m is not permitted. Classical texts on dielectrics, such as those by Debye (1929) and Fröhlich (1958) cite dipole moment values in esu units (readily converted into debye units by removing the 10^{-18} multiplier) and many papers are still published where the debye unit is preferred over the SI unit. The debye unit will be adopted in this chapter, with conversions to SI units where considered useful.

The debye unit was introduced to honour the Dutch-American physical chemist Peter J. W. Debye and this was most likely inspired by his attendance at the 60th General Discussion of the Faraday Society, held at Oxford University in April 1934, on the subject of *The Determination and Interpretation of Dipole Moments*. Samuel Sugden (1892–1950), Professor of Physical Chemistry at Birkbeck College, London, attended this symposium, along with a strong international representation of the key figures in the subject for discussion. Sugden's paper [8], received on 28 February 1934, for printing and distribution to the registered attendees, was the only one that referred to the *debye unit* as a measure of dipole moment. Fred Fairbrother cites dipole moment values in units of 10^{-18} esu in his presented paper (received 20 March 1934) [9], but on 6 August 1934, he submitted a paper to *Nature* in which he gives dipole moments in units of Debye [10]. A report of the symposium, with the author's signature given as S. S. [11], converts all of the presented dipole moment values from esu into debye units and a footnote indicates that 1 D = 10^{-18} esu cgs. There is no doubt that Sugden authored this report and later that year he published a paper on the dipole moments of vapours using debye units [12]. The use of the debye unit became widely adopted from that time onwards.

Debye received the Nobel Prize in Chemistry in 1936, partly for his pioneering work in introducing the concept of molecular dipole moments for the elucidation of the dielectric properties of some vapours and liquids [13, 14]. As described in Chapter 6, evidence had been gathering that for a first class of gases the molar polarization remained constant with changes in temperature, but that for a second class the polarization decreased with increasing temperature. This appeared to be particularly evident for gases and liquids exhibiting large polarizabilities. In the theory of paramagnetism it was customary to assume the paramagnetic effect was due to the pre-existence of molecules that carry a permanent magnetic moment and can be oriented by a magnetic field in the same way as a small magnet. Although a molecule may not carry a net charge, Debye saw no reason why the centre of its positive charges should not be coincident with the centre of its negative charges. If this occurs, the molecule has a permanent electric moment and can exhibit polarization not only by electronic distortion, but also by field-induced orientation. Temperature-induced motions of the molecules will disturb the extent of orientation created by the field and this would account for the second class of gases and liquids where the polarization decreased with increasing temperature.

Debye developed an understanding of the effect of temperature on polarization by orientation similar to that described in Box 7.3, where the *thermal average* of the

Box 7.3 Orientational Polarization and the Langevin Function

In the absence of an electric field, the moments of a collection of dipoles will on average be distributed with the same probability over all directions in space. With an applied field, the potential energy of a dipole moment is given by Equation (5.34) as:

$$U = -\mathbf{p}\cdot\mathbf{E} = -\mathbf{pE}\cos\theta$$

where θ is the angle between the moment **p** and the field vector **E**. From Boltzmann–Maxwell statistics the probability of finding a dipole oriented in an element of solid angle $d\Omega$ (steradians) is proportional to $\exp(-U/kT)$, where k is the Boltzmann constant (1.3805×10^{-23} J/K) and T is the absolute temperature. As shown in Figure 7.6, a moment pointing in the direction of $d\Omega$ has a component $\mathbf{p}\cos\theta$ in the field direction. The thermal *average* of $\cos\theta$ is thus given by:

$$\langle\cos\theta\rangle = \frac{\int \exp(-U/kT)\cos\theta\, d\Omega}{\int \exp(-U/kT)\, d\Omega}$$

where the integrations are to be taken over all possible directions. The solid angle $d\Omega$ encompasses all moments lying between θ and $d\theta$, so that $d\Omega = 2\pi\sin\theta.\, d\theta$, to give

$$\langle\cos\theta\rangle = \frac{\int_0^{\pi} 2\pi\sin\theta\cos\theta\exp(\mathrm{pE}\cos\theta/kT)d\theta}{\int_0^{\pi} 2\pi\sin\theta\exp(\mathrm{pE}\cos\theta/kT)d\theta}$$

Let $x = \mathbf{pE}/kT$ and $y = \cos\theta$, then:

$$\langle\cos\theta\rangle = \frac{\int_{-1}^{1}\exp(xy)y\,dy}{\int_{-1}^{1}\exp(xy)dy}$$

$$= \frac{d}{dx}\log\int_{-1}^{1}\exp(xy)dy = \coth x - \frac{1}{x} = L(x)$$

The function $L(x)$ is known as the Langevin function because it resembles the result obtained by Paul Langevin in 1905 to find the mean magnetic moment of gas molecules possessing permanent magnetic moments. A plot of the Langevin function is shown in Figure 6.6. The function $\coth x$ can be expanded as a series

$$\coth x = \frac{1}{x} + \frac{x}{3} - \frac{x^3}{45} + \frac{2x^5}{945} - \cdots$$

For most practical situations the factor x ($\mathbf{pE}/kT$) $\ll 1$, so to a good approximation

$$\langle\cos\theta\rangle = \frac{1}{x} + \frac{x}{3} - \frac{1}{x} = \frac{x}{3} = \frac{\mathrm{pE}}{3kT}$$

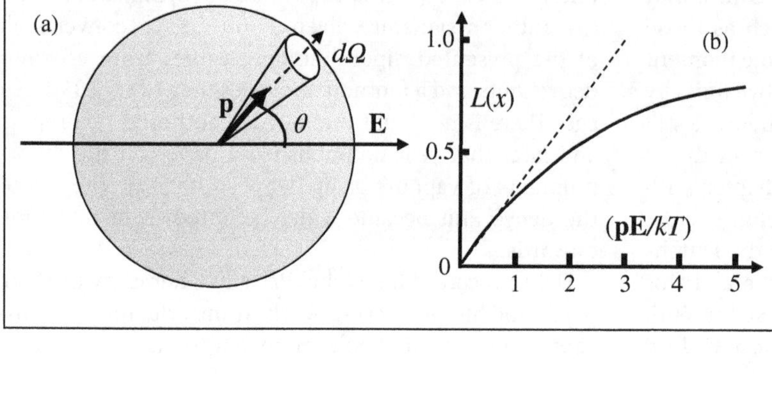

Figure 7.6 (a): A moment **p** pointing in the direction of the solid angle $d\Omega$ has a component $\mathbf{p}\cos\theta$ along the field vector **E**. (b): A plot of the Langevin function.

angle θ between an electric moment and an imposed electric field is given by the relationship:

$$\langle \cos\theta \rangle = \frac{pE}{3kT}$$

The average moment per molecule in the direction of the applied field is thus:

$$\langle p \rangle = p \langle \cos\theta \rangle = \frac{p^2 E}{3kT} \tag{7.10}$$

The magnitude of the mean orientational polarization α_o is

$$\alpha_o = \frac{p^2}{3kT}$$

The total polarizability of a gas or liquid whose molecules possesses a permanent electric moment is given by the sum of the polarizations due to electronic, atomic and orientational polarization. The total mean electric moment per unit volume is thus:

$$\langle p \rangle = \left(\alpha_e + \alpha_a + \frac{p^2}{3kT} \right) E \tag{7.11}$$

Equation (7.11) is referred to as the Debye equation. Its formulation is based on classical statistics and so assumes that the molecular system reaches an equilibrium state. The frequency of the applied field must therefore be much smaller than the rotational frequency of the molecule, so that its dipole moment can follow the field direction without appreciable lag and hence irreversible energy loss. We must also assume that any interactions between the molecules can be neglected and this restricts the application of Debye's equation to gases and vapours at low pressures. The development of dielectric theory to take into account quantum theory and in particular to include quantization of the rotational energy in accordance with the requirements of wave mechanics, was initiated in 1926 by Kronig, Manneback, Mensing, Pauli and van Vleck. The last four chapters of Debye's book describe these developments [14, pp. 125–167].

7.3 Dipole Relaxation and Energy Loss

Figure 6.3 shows how the polarization of an unpolarized dielectric composed of freely dipoles increases with time after an electric field is applied as a step function at time $t = 0$. In our timeframe (and hence frequency range) of interest we can to good approximation assume that the polarization **P** is composed of two parts, $\mathbf{P}_1$ and $\mathbf{P}_2$. The component $\mathbf{P}_1$ arises from the electronic and atomic charge displacements described in sections 7.2.2.1 and 7.2.2.2 and attains its final constant value $\mathbf{P}_1 = (\varepsilon_\infty - 1)\varepsilon_o\mathbf{E}$ almost instantly on application of the applied field **E**. $\mathbf{P}_2$ builds up at a much slower rate associated with the field-induced readjustment of the orientations of the molecular dipoles. Debye assumed [14, pp. 87–89] that this relaxation process approaches a new steady-state distribution of orientations at the exponential rate $(1 - e^{-t/\tau})$ where τ is a characteristic response time that does not vary with time t (but may depend on temperature, for example). This is equivalent to the polarization lagging behind **E** in such a way that it approaches a final value of $(\varepsilon_s - 1)\varepsilon_o\mathbf{E}$ at a rate proportional to $[(\varepsilon_s - 1)\varepsilon_o\mathbf{E} - \mathbf{P}_2]$, so that

$$\frac{dP_2(t)}{dt} = \frac{1}{\tau}[(\varepsilon_s - 1) - P_2(t)]\varepsilon_o E$$

If the field is applied as a step function at $t = 0$, when $\mathbf{P}_2 = 0$ and $\mathbf{P}_1$ instantaneously attains its value of $(\varepsilon_\infty - 1)\varepsilon_o\mathbf{E}$, then on integration of this equation

$$\mathbf{P} = \mathbf{P}_1 + \mathbf{P}_2 = [(\varepsilon_\infty - 1) + (\varepsilon_s - 1)(1 - e^{-t/\tau})]\varepsilon_o E$$

The polarization thus approaches its final steady state value asymptotically at a rate characterized by a relaxation time constant τ. If the field is suddenly removed, thermal vibrations cause the dielectric to relax asymptotically back to its nonpolarized state ($\mathbf{P} = 0$) with time constant τ.

With an applied periodic field $E = E_o e^{-i\omega t}$, then $dE/dt = -i\omega E$ and $d\mathbf{P}_2/dt = -i\omega(\varepsilon(\omega) - 1)\varepsilon_o\mathbf{E}$. We obtain the following result

$$\mathbf{P} = \mathbf{P}_1 + \mathbf{P}_2 = \left[(\varepsilon_\infty - 1) + \frac{(\varepsilon_s - 1)}{1 + i\omega\tau} \right] \varepsilon_o E$$

which corresponds to a frequency-dependent permittivity

$$\varepsilon(\omega) = \varepsilon_\infty + \frac{(\varepsilon_s - \varepsilon_\infty)}{1 + i\omega\tau} \tag{7.12}$$

in which ε_∞ is to be interpreted as the relative permittivity value obtained at high frequencies where the dipoles are unable to respond to the alternating field. (Electronic engineers may recognize this transformation from the time domain to the response in the frequency domain as the Laplace transformation of the function $f(t) = (1 - e^{-\alpha t})\cdot u(t)$ into $L\{f(t)\} = \alpha/[s(s + \alpha)]$). In the time domain, for our frequency range of interest, ε_∞ is the relative permittivity at time $t = 0$. The change in the value of $\varepsilon(\omega)$ from a constant value of ε_s at low frequencies to a constant value of ε_∞ at high frequencies is known as a dielectric dispersion, whose magnitude is defined as $\Delta\varepsilon = (\varepsilon_s - \varepsilon_\infty)$. The frequency at which ε_∞ attains its constant value is well below the frequencies where the electronic and atomic distortion resonances occur.

With reference to Equation (6.38), Equation (7.12) has real and imaginary components given by

$$\varepsilon'(\omega) = \varepsilon_\infty + \frac{(\varepsilon_s - \varepsilon_\infty)}{1 + \omega^2\tau^2} \tag{7.13}$$

$$\varepsilon''(\omega) = \frac{(\varepsilon_s - \varepsilon_\infty)\omega\tau}{1 + \omega^2\tau^2} \tag{7.14}$$

Figure 6.4 gives plots of these two parameters for the case where $\tau = (2\pi \times 10^5\,\text{Hz})^{-1} = 1.6 \times 10^{-6}$ s. Equations (7.12–14) refer to Debye-type dielectric relaxations involving the exponential function $e^{-t/\tau}$, with a single characteristic relaxation time τ. The phase difference between the polarization and the applied field leads to an absorption of energy and joule heating. The rate of conversion of electrical energy to heat in the dielectric is represented by the imaginary component ε''. The frequency ω_{pk} corresponding to maximum loss is found by differentiating Equation (7.14) with respect to ω and equating the result to zero:

$$\frac{d\varepsilon''(\omega)}{d\omega} = \frac{(1 - \omega^2\tau^2)(\varepsilon_s - \varepsilon_\infty)\tau}{(1 + \omega^2\tau^2)^2} = 0 \text{ when } \omega\tau = 1$$

The frequency of peak energy loss is thus given by $\omega_{pk} = 1/\tau$ radians per second (or $1/(2\pi\tau)$ Hz). From Equation (7.14) the corresponding peak value of ε'' is given by

$$\varepsilon''_{pk} = \frac{(\varepsilon_s - \varepsilon_\infty)}{2}$$

The width of the ε'' peak across its half-height values is 1.14 decades of frequency. The value of ε' where ε'' is at its peak value is found by inserting $\omega\tau = 1$ into Equation (7.14):

$$\varepsilon'(\omega\tau = 1) = \frac{(\varepsilon_s + \varepsilon_\infty)}{2}$$

The real component ε' represents the relative permittivity parameter ε_r given in Equation (3.3). With a constant (DC) voltage applied to its two parallel plate electrodes, a capacitor containing a dielectric slab of relative permittivity ε_r will have a capacitance given by the formula

$$C = \frac{A\varepsilon_o\varepsilon_r}{d} \text{ Farad}$$

where A is the surface area of each plate separated a distance d apart. With a sinusoidal voltage applied to the plates, the capacitance will exhibit a frequency dependence given by

$$C(\omega) = \left[\varepsilon_\infty + \frac{(\varepsilon_s - \varepsilon_\infty)}{1 + \omega^2\tau^2}\right]\frac{A\varepsilon_o}{d} = C_\infty + \frac{(C_s - C_\infty)}{1 + \omega^2\tau^2} \tag{7.15}$$

This frequency-dependent capacitance can be modelled as an equivalent electrical circuit, as shown in Box 7.4.

Another important feature of Debye-type dielectric polarizations can be found by eliminating the factor $\omega\tau$ from Equations (7.13) to obtain the relationship

$$\left[\frac{(\varepsilon' - \varepsilon_\infty) - (\varepsilon_s - \varepsilon_\infty)}{2}\right]^2 + (\varepsilon'')^2 = \left[\frac{(\varepsilon_s - \varepsilon_\infty)}{2}\right]^2$$

This equation is of the form $x^2 + y^2 = r^2$, namely, the equation of a circle of radius r. However, because ε'' can only have positive values, a plot of ε' against ε'' produces a semicircle of radius $(\varepsilon_s - \varepsilon_\infty)/2$ with the centre at $[(\varepsilon_\infty + (\varepsilon_s - \varepsilon_\infty)/2,0)$. Such a plot is known as a Cole–Cole plot [15] and an example is given in Figure 7.7.

Deviations from an ideal Debye-type single relaxation dispersion are likely to occur for macromolecules such as proteins. Factors contributing to this can arise from combinations of cooperative and isolated motions of polar side groups, for example. This gives rise to a spread of relaxation times, each one contributing to a Debye-type dispersion. For a set of closely spaced relaxation times, the resultant ε'' loss curve as a function of frequency will be much broader than the one shown in Figure 7.7. The width of such ε'' curves across their half-height values can exceed two or even three decades of frequency. If a symmetrical distribution of relaxation times occurs about a mean relaxation time, then a depressed semicircle is found in the Cole–Cole plot and to describe this Equation (7.12) can be modified to the form:

$$\varepsilon(\omega) = \varepsilon_\infty + \frac{(\varepsilon_s - \varepsilon_\infty)}{1 + (i\omega\tau)^{1-\alpha}}$$

where α has values in the range $0 \le \alpha \le 1$ [15]. For a single relaxation time $\alpha = 0$. As the situations tends to an infinite number of relaxation times, the value of α

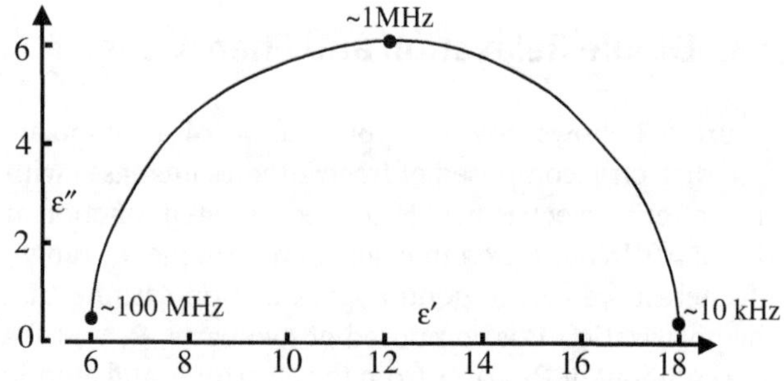

Figure 7.7 The plot of ε' versus ε'' for a Debye-type dipole relaxation process takes the form of a semicircle. This is known as a Cole–Cole plot [10].

Box 7.4 Electrical Circuit for Modelling a Debye-type Dielectric Dispersion

The following electrical circuit, composed of two capacitors and one resistor, can be used to model the Debye-type dielectric dispersion given by Equations (7.13) and (7.15). The values of the capacitors and resistor do not change as a function of frequency.

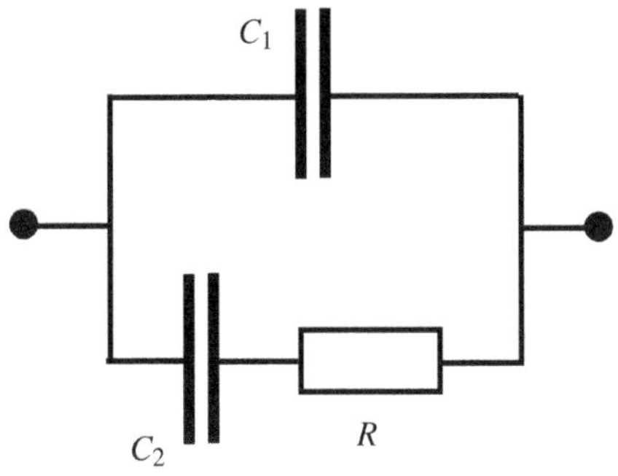

As a parallel network it is more convenient to analyse it in terms of admittance rather than impedance. From standard circuit theory its effective admittance **Y** is given by:

$$Y(\omega) = i\omega C_1 + \frac{i\omega C_2}{(1 + i\omega\tau)} = i\omega C(\omega) \text{ with } \tau = RC_2$$

$C(\omega)$ has a real (in-phase) component given by

$$C'(\omega) = C_1 + \frac{C_2}{(1 + \omega^2\tau^2)}$$

For this to be equivalent to the expression given by Equation (7.16) requires that we specify $C_1 = C_\infty$ and $C_2 = C_s - C_\infty$. Capacitance C_1 therefore represents the instantaneous polarization associated with field-induced electronic distortions and atomic displacements, with C_2 determining the magnitude of the dielectric dispersion given by $\Delta\varepsilon = (\varepsilon_s - \varepsilon_\infty)$. The imaginary, quadrature phase, component is given by:

$$C''(\omega) = \frac{\omega\tau C_2}{(1 + \omega^2\tau^2)}$$

At low frequencies ($\omega\tau \ll 1$) and high frequencies ($\omega\tau \gg 1$) the value of $C''(\omega)$ is zero. The model circuit is thus purely capacitive so that, after a sufficient length of time for the steady-state condition to be attained, the (displacement) current in the dielectric leads the applied voltage by 90°. $C(\omega)$ has a peak value at $\omega\tau = 1$. The frequency where this peak value occurs is determined by both the resistor R and capacitor C_2 ($\tau = RC_2$) but, as shown by Equation (7.14), the peak value itself is solely determined by the magnitude of C_2.

---

tends to unity. The value of α for water is typically 0.013, indicating that the water molecule relaxes with almost a single relaxation time. To describe a nonsymmetrical distribution of relaxation times, in which the distributions on the high frequency side of the principal relaxation time decreases more rapidly than those on the low frequency side, another modification can be made [16] of the form:

$$\varepsilon(\omega) = \varepsilon_\infty + \frac{(\varepsilon_s - \varepsilon_\infty)}{(1 + i\omega\tau)^\gamma}$$

where γ has values in the range $0 < \gamma \leq 1$ and results in a skewed arc-shaped plot of ε' against ε''. More generally, the $\varepsilon(\omega)$ function takes the form

$$\varepsilon(\omega) = \varepsilon_\infty + (\varepsilon_s - \varepsilon_\infty)\int_0^\infty \frac{G(\tau)}{1 + i\omega\tau} d\tau$$

where $G(\tau)$ is the fraction of molecules or other polarizable elements at a particular instant associated with a relaxation time between τ and $(\tau + d\tau)$.

The mathematical and empirical factors β and γ given above to describe the spread of relaxation times can be replaced by one having a more readily understandable physical meaning by employing mutual transformation relations known as the Kramers–Krönig relations:

$$\varepsilon'(f) - \varepsilon_\infty = \frac{2}{\pi}\int_0^\infty \frac{\varepsilon''(f)F}{F^2 - f^2} dF$$

$$\varepsilon''(f) = -\frac{2}{\pi}\int_0^\infty \frac{\varepsilon'(f) - \varepsilon_\infty}{F^2 - f^2} dF$$

In these equations, which give the relationship between a dielectric dispersion and the energy loss, F is a dummy variable supplementing the frequency f and over which either ε' or ε'' is integrated to find the other. Apart from being able to deduce the real component of a complex permittivity from the imaginary component (or the reverse of this) these relationships can be used to relate dielectrophoresis and electrorotation spectra [17, 18]. If the difference in the permittivities either side of a dielectric dispersion is $\Delta\varepsilon$, then from the Kramers–Krönig relations, we deduce that

$$\Delta\varepsilon = \frac{2}{\pi}\int_{-\infty}^\infty \varepsilon'' d(\ln f)$$

This is interpreted to mean that the total area under the dielectric loss curve of ε'' against $\log_{10} f$ should equal $(\pi\Delta\varepsilon)/(2 \times 2.3)$. Thus, the total area under the ε'' loss curve is proportional to the total concentration

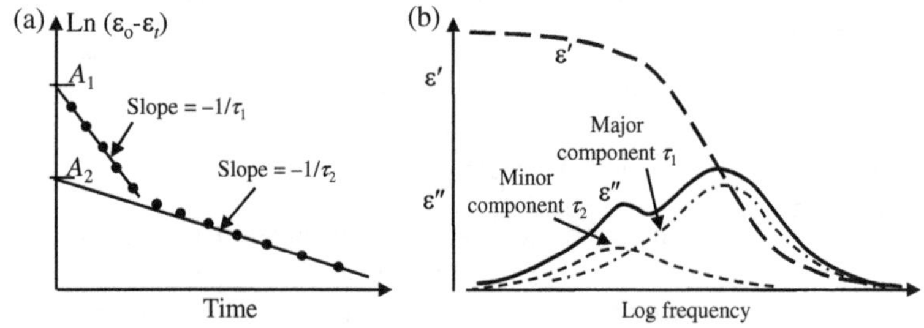

Figure 7.8 (a): Based on time domain data, a natural logarithmic plot such as this will reveal the components of a dispersion arising from two separate relaxation processes. (b) Frequency domain data analysed in the forms of Equations (7.16a) and (7.16b).

N of dipoles in the dielectric material and their dipole moment **p**, irrespective of their distribution of relaxation times (the factor 2.3 arises when the experimental data is more conveniently plotted against $\log_{10} f$ rather than the natural logarithm). Making use of Onsager's Equation (6.35) it can be shown [14] that

$$\int_{-\infty}^{\infty} \varepsilon'' d(\ln f) = \frac{\pi}{2}(\varepsilon_s - \varepsilon_\infty) = \frac{\pi}{54\varepsilon_o kT}(\varepsilon_s + 2)^2 p^2 N$$

Thus, as the distribution of relaxation times increases from that of a single relaxation time, a plot of ε'' versus $\log_{10} f$ becomes broader in extent, but the peak height (ε''_{pk}) reduces so as to maintain the total area under the curve at an unchanging value [19]. We can therefore define a parameter α' that represents the spread of relaxation times by the expression

$$\alpha' = \frac{\pi \varepsilon''_{pk}}{Area\ of\ \varepsilon''\ curve}$$

where $\pi\varepsilon''_{pk}$ would equal the area under a single-relaxation Debye-type loss curve. The value of α' falls from unity and tends to zero as the distribution of relaxation times tends to an infinite one.

A deviation from an ideal Debye-type single relaxation will appear if there are two or more dipole species with relaxation times that are close enough together for there to be an overlap of their permittivity dispersion curves and dielectric loss peaks. An example of this is described in Chapter 8 for small sugar molecules dissolved in water. The dominant contribution is the relaxation of the bulk water molecules ($\tau_1 = 1.85 \times 10^{-11}$ s), with a smaller contribution consisting of relaxation of the sugar dipoles, with a relaxation time $\tau_2 = 6.9 \times 10^{-11}$ s. In the time domain the apparent relative permittivity $\varepsilon(t)$, as a function of time after application of the voltage signal at time $t = 0$, can be represented as a combination of overlapping exponentials:

$$\varepsilon(t) = \varepsilon(o) + A_1 e^{-t/\tau_1} + A_2 e^{-t/\tau_2}$$

Taking natural logarithms of this expression we obtain the relationship

$$\ln[\varepsilon(t) - \varepsilon(o)] = \ln(A_1) - \frac{t}{\tau_1} + \ln(A_2) - \frac{t}{\tau_2}$$

The values of the coefficients A and relaxation times τ can thus be evaluated using a $\log_e$ plot such as that shown in Figure 7.8(a). In the frequency domain, the real and imaginary components of the permittivity given by Equations (7.13) and (7.14) can be given in the form:

$$\varepsilon'(\omega) = \varepsilon_\infty + \frac{\gamma(\varepsilon_s - \varepsilon_\infty)}{1 + \omega^2\tau_1^2} + \frac{(1-\gamma)(\varepsilon_s - \varepsilon_\infty)}{1 + \omega^2\tau_2^2} \quad (7.16a)$$

$$\varepsilon''(\omega) = \frac{\gamma(\varepsilon_s - \varepsilon_\infty)\omega\tau_1}{1 + \omega^2\tau_1^2} + \frac{(1-\gamma)(\varepsilon_s - \varepsilon_\infty)\omega\tau_2}{1 + \omega^2\tau_2^2} \quad (7.16b)$$

where γ is the fractional contribution of the major relaxation process having a relaxation time τ_1. The minor relaxation process has a relaxation time τ_2. Analysis of the $\varepsilon'(\omega)$ and $\varepsilon''(\omega)$ data can then proceed by finding the best fit of these expressions to that data, as shown in Figure 7.8(b).

7.3.1 Complex Conductivity

The dielectric loss factor ε'' is a measure of the energy dissipated in a dielectric as a result of the field-induced relaxation of dipole moment orientations. The energy loss is associated with the in-phase (lossy) component of the *complex conductivity* $\sigma*$ of the dielectric and not its out-of-phase (energy-storing) capacitive component. For a dielectric exhibiting a Debye-type dispersion, electrically modelled using the circuit shown in Box 7.4, the complex conductivity is derived from the magnitude of the admittance by

$$Y = \frac{A}{d}(\sigma' + i\sigma'')$$

Noting that $Y(\omega) = i\omega C(\omega)$ and $\tau = RC_2$, the following expressions can be derived from Box 7.4 for the real and imaginary components of the complex conductivity:

$$\sigma^* = \sigma' + i\sigma''$$

with

$$\sigma' = \frac{(\sigma'_\infty - \sigma'_s)\omega^2\tau^2}{(1 + \omega^2\tau^2)} \quad (7.17a)$$

$$\sigma'' = \omega C_\infty + \frac{(\sigma'_\infty - \sigma'_s)\omega\tau}{(1 + \omega^2\tau^2)} \quad (7.17b)$$

According to the equivalent circuit shown in Box 7.4, at low frequencies where $\omega\tau \ll 1$, the value of $\sigma'_s = 0$. At high frequencies σ' attains a value $\sigma'_\infty = dR/A$, where d and A are the thickness and cross-sectional area of the dielectric slab, respectively. This increase of σ' with increasing frequency mirrors the fall of ε' shown in Figure 6.4. On application of a constant amplitude sinusoidal voltage $V_0(\sin \omega t)$ to the dielectric, the electrical power loss increases from zero at very low frequencies to a final level of $V_0{}^2/R$ at frequencies beyond the dielectric dispersion. At very high frequencies for the circuit shown in Box 7.4, the reactance values of the two capacitors are negligible compared to the resistance R.

From Equations (6.44) and (7.14) the conductivity $\sigma(\omega)$ associated with dipole relaxation can be written as

$$\sigma(\omega) = \frac{(\varepsilon_s - \varepsilon_\infty)\omega^2\tau}{1+\omega^2\tau^2} = \frac{\Delta\varepsilon'\omega^2\tau}{1+\omega^2\tau^2}$$

From Equation 7.17a, we can therefore deduce that

$$\Delta\varepsilon' = \tau\Delta\sigma' \tag{7.18}$$

This relationship is a manifestation of the Kramers–Krönig relations discussed earlier in this chapter.

In measurements on biological samples, it is invariably the case that, apart from dielectric displacement currents, long-range ionic conduction also contributes to electrical energy loss. Over the frequency range of relevance to our present discussion this conductivity does not depend on frequency and can be treated as a DC conductivity by adding a parallel conductor to the circuit shown in Box 7.4. This term can be added to the expression for $\sigma(\omega)$ to give

$$\sigma(\omega) = \sigma_{dc} + \sigma(\omega) = \sigma_{dc} + \frac{\sigma_\infty\omega^2\tau^2}{1+\omega^2\tau^2} \tag{7.19}$$

According to Equation (6.44) a DC conductivity contributes to the dielectric loss factor ε'' according to the relationship $\varepsilon''(\omega) = \sigma_{dc}/\omega$. This contribution thus becomes significant as the frequency is reduced (see, for example, Figure 7.13).

7.3.2 Physical Models for Dipole Relaxation

Polar molecules in solution may be assumed to be randomly oriented and to change direction continually as a result of thermal (Brownian) motion. On application of an electric field there will be a small shift in the directional distribution of the dipole orientations. Debye has provided a very simple model to derive an estimate of the orientational relaxation time of polar molecules in solution when subjected to a time-varying electric field [14, pp. 83–85]. In this model the polar molecules are assumed to be spheres whose rotation is opposed by a frictional torque ξ. The appropriate characteristic relaxation time τ is given by

$$\tau = \frac{\xi}{2kT}$$

According to the relationship derived by Stokes, namely

$$\xi = -8\pi R^3\eta\Omega$$

where R is the sphere radius, η is the dynamic viscosity of the surrounding fluid and Ω is the angular velocity of the sphere. The relaxation time is thus given by

$$\tau = \frac{4\pi\eta R^3}{kT} \tag{7.20}$$

In the next chapter we will find that this simple model is useful for interpreting the dielectric properties of aqueous solutions of amino acids, peptides and proteins. Considering the fact that the structure of water involves hydrogen-bond associations, Debye's model predicts a surprisingly accurate value for the relaxation time of water. Using the molecular radius of the water molecule to be half the interoxygen distance of 1.4×10^{-10} m and the 293 K viscosity value as 10^{-3} kg/m s, the 293 K experimental value for τ is calculated as 8.5×10^{-12} s, in good agreement with the experimental value of 9.3×10^{-12} s [20, p. 47]. Furthermore, the ratio of the τ values for D_2O and H_2O is 1.27, which is very close to the ratio of their viscosities at 1.25. Debye states [14, p. 85] that we can expect the calculated value of τ to 'be approximately correct even for the rotation of as small a particle as a molecule' because the corresponding calculation for the mobility and diffusion of ions and molecules dissolved in liquids is accurately predicted by another equation devised by Stokes, namely Equation (12.20) and which, as shown in Chapter 12, leads to the Stokes–Einstein equation for the diffusion coefficient $D = kT/(6\pi\eta R)$.

The relaxation time τ represents the reciprocal of the mean rate coefficient of the dielectric relaxation process and as can be expected to obey an Arrhenius temperature law of the form

$$\tau = Ae^{(\Delta H/RT)} \tag{7.21}$$

where ΔH is the Arrhenius activation enthalpy per mole and A is a constant. On differentiation of this relationship we obtain

$$\frac{d(\ln\tau)}{d(1/T)} = \frac{\Delta H}{R} \tag{7.22}$$

A plot of $\ln \tau$ against $1/T$ gives a straight line of slope $\Delta H/R$. The relaxation process can be further developed

Box 7.5 Free Energy Change ΔG associated with Molecular Relaxations

A dielectric relaxation involves the reorientation of a molecule and interactions with neighbouring molecules. In Chapter 8 the dielectric relaxation of water is described in terms of the breaking, remaking and reorganization of hydrogen bonds. Assuming that these processes occur at constant temperature and pressure, we can use as a measure of the potential energy released or stored the concept of Gibbs free energy (named after Josiah Willard Gibbs, an early founder of the science of thermodynamics). Gibbs demonstrated that free-energy G is given by the relationship:

$$G = H - TS$$

where H is the heat energy (also termed enthalpy) of the molecular system, T is the absolute temperature and S is termed the entropy and provides a measure of the degree of disorder of the system. We are interested in the change of free-energy ΔG that results from a molecular relaxation and contributions to this (at constant temperature and pressure) come from the changes in heat content and entropy:

$$\Delta G = \Delta H - T\Delta S$$

For the case of water, for example, enthalpy H is released when hydrogen bonds are formed, or absorbed when they are broken. ΔH is thus equal to the overall change in bond energies. We can distinguish between a reaction in which heat is given off (an exothermic reaction) and one in which heat is absorbed (an endothermic reaction). In an exothermic reaction ΔH is negative and the system contain less energy than for its original state. In an endothermic reaction (heat absorbed) ΔH is positive and the energy of the system increases.

By convention, ΔS is positive when entropy and thus disorder, increases. The Second Law of Thermodynamics (devised by Clausius) states that the entropy of an isolated system, which is not in equilibrium, will tend to increase over time and to approach a maximum value at equilibrium. Thus, an overall positive change in entropy of a system is indicative of a spontaneous reaction. The evaporation at 100 °C of water into a gas (steam) is an example of this, because the arrangement of the water molecules becomes more disordered. A negative value for ΔH (heat given off) and a positive ΔS (increased order) tend to lead to a spontaneous reaction. An example of this is the formation of solid water (ice) from liquid water at 0 °C. A quantitative example is given in Chapter 8, where it is shown that the attraction of sodium and chloride ions in water is mainly driven by entropy and not enthalpy.

---

in terms of a chemical rate process [21, 22] and Equation (7.21) becomes

$$\tau = \frac{h}{kT} e^{-(\Delta S/R)} e^{(\Delta H/RT)} \tag{7.23}$$

where h is Planck's constant and ΔS is the molar entropy of activation for the relaxation process. The basic thermodynamic concepts involved in this approach to describing dielectric relaxation are summarized in Box 7.5.

In solids, the molecular dipoles are not free to rotate as they would be in a liquid. Fröhlich [23, 24] introduced a useful physical model of a single type of dipole for which two orientations are possible, these being alignments parallel and antiparallel to the external electric field. As a result of the crystalline structure of the dielectric material, these two orientations are separated by a potential energy barrier ΔE_b as shown in Figure 7.9. On application of a constant field **E** the energy of the site corresponding to an alignment of the dipole along the field is lowered relative to the site where the dipole is aligned against the field. The difference in potential heights of the two minima at sites 1 and 2 is $qd\mathbf{E}$, where d is the distance between the sites.

With no applied electric field the probability P_{12} of a dipole making the transition from having its positive charge located in site 2 is equal to the probability P_{21} for it having the reverse orientation. According to statistical mechanics, we can also write

$$P_{12} = \frac{e^{-(V_2/kT)}}{e^{-(V_2/kT)} + e^{-(V_1/kT)}};$$

$$P_{21} = \frac{e^{-(V1/kT)}}{e^{-(V_1/kT)} + e^{-(V2/kT)}}$$

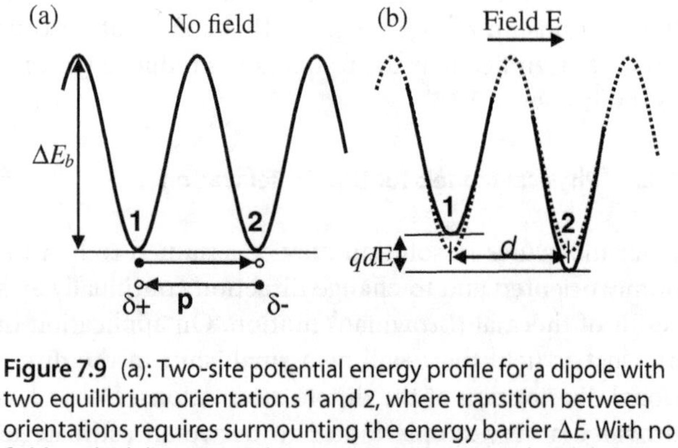

Figure 7.9 (a): Two-site potential energy profile for a dipole with two equilibrium orientations 1 and 2, where transition between orientations requires surmounting the energy barrier ΔE. With no field applied the potential energy minima are equal and there is an equal probability for both orientations of the dipole **p**. (b): With an external field applied in the direction shown, there is an increased probability of finding the dipole oriented along the direction of the field than against it (i.e., with its positive charge $\delta+$ located in site 2).

where these probabilities are normalized by making $P_{12} + P_{21} = 1$ because the dipole must be oriented in one of the two situations. On application of the field $V_1 - V_2 = qd\mathrm{E}$ and so

$$P_{12} - P_{21} = \frac{e^{(qdE/kT)} + 1}{e^{(qdE/kT)} - 1}$$

The relaxation time is given by $1/\tau = 2P_{12}$ and is therefore determined by an Arrhenius type function of the form of Equations (7.20) and (7.21). In many solids we might expect that in the absence of an external field the equilibrium positions 1 and 2 of the dipole orientations are not equal. In this case, applying absolute reaction rate theory, the molecular relaxation time is given approximately by

$$\tau = \frac{h}{kT} e^{(\delta E_b/kT)}$$

where δE is equal to the smallest energy barrier height. If there is a distribution of small barrier heights allowing for continuous fluctuations in molecular arrangements, for example, the result will be a broader distribution of relaxation times [22]. Extensions of the two-site barrier model have been developed to include both two- and three-dimensional multisite models giving more than just one single relaxation time [25,26].

In Chapter 8 it is shown that a protein molecule can be considered as a spheroid whose surface is covered with charged groups and with an interior consisting of a folded-up string of peptide dipoles and polar side groups (Figures 8.15 and 8.17). We can imagine that these various charges and dipole elements populate a large array of micro potential energy wells, giving rise to a permanent dipole moment **M** that fluctuates with time. The value of M, when the protein is suspended in water, will depend on the reaction field arising from polarization of the water molecules. From a consideration of the internal and external free energies of a system consisting of a large dielectric sphere of volume V embedded in a continuous medium, Fröhlich [24, p. 177] derived the following equation for the mean square fluctuation $\langle \mathrm{M}^2 \rangle$ of the moment M:

$$\langle \mathrm{M}^2 \rangle = 3kTV\varepsilon_o(\varepsilon_p - 1)\frac{2\varepsilon_w + 1}{2\varepsilon_w + \varepsilon_p} \tag{7.24}$$

in which ε_p is the relative permittivity of the spherical particle and ε_w is the relative permittivity of the surrounding medium (which we will assume is water). It is now possible to compute values for $\langle \mathrm{M}^2 \rangle$ based on X-ray diffraction data that provides details of the three-dimensional structure of a protein molecule. Equation (7.24) can therefore be used to calculate the permittivity of a protein molecule. This is described further in Chapter 8.

7.4 Interfacial Polarization

The electrical conditions to be satisfied at dielectric boundaries are described in Chapter 5. At a boundary surface between two dielectrics, with permittivity and conductivity values of ε_1, σ_1 and ε_2, σ_2, respectively, no charge is created at their interface if the condition given by Equation (5.52) is satisfied

$$(\varepsilon_2\sigma_1 - \varepsilon_1\sigma_2) = 0$$

It follows that a heterogeneous dielectric containing different constituents for which the condition $\varepsilon_2\sigma_1 = \varepsilon_1\sigma_2$ is not met, a 'residual' free charge will appear within the dielectric. A quantitative analysis of this effect requires application of the boundary conditions given in Chapter 5, namely that the normal components of the displacement flux density ($D = \varepsilon E$) and the current density ($J = \sigma E$) are continuous:

$$\varepsilon_2 \mathrm{E}_{n2} - \varepsilon_1 \mathrm{E}_{n1} = \varepsilon_2 \frac{\partial \phi_2}{\partial \mathrm{r}} - \varepsilon_1 \frac{\partial \phi_1}{\partial \mathrm{r}} = \sigma_{free}$$

and

$$\sigma_{c2} \frac{\partial \phi_2}{\partial \mathrm{r}} - \sigma_{c1} \frac{\partial \phi_1}{\partial \mathrm{r}} = -i\omega\sigma_{free}$$

where ω is the radian frequency of an applied alternating field. The parameter σ_{free} denotes the concentration of 'true' charges that can only appear at, or be removed from, the interface by long-range conduction. The term '$-i$' indicates that this conduction current leads the potential ϕ by a phase angle of 90°.

Maxwell [27, Part II, pp. 435–458] was the first to provide a quantitative analysis of the buildup of 'residual' free charges in heterogeneous media for the electrostatic case and Wagner [28] extended this for alternating (AC) fields. The generation of charges at the interfaces between dissimilar components of a dielectric is known as *Maxwell–Wagner interfacial polarization.* Maxwell chose for his model the case of a composite dielectric consisting of a number of plane strata, where the electric field acts in a direction normal to the strata. A simple example of this is the two-layer capacitor shown in Figure 7.10.

The heterogeneous capacitor shown in Figure 7.10 consists of two parallel layers of dielectric material of surface area A and thicknesses d_1 and d_2 located between two electrodes. The two-layer system effectively represents two capacitors connected in series, with their complex capacitance values given by

$$C_1 = \frac{A\varepsilon_o\varepsilon_1^*}{d_1} = \frac{A\varepsilon_o(\varepsilon_1' - i\sigma_1/\omega)}{d_1}; \; C_2 = \frac{A\varepsilon_o(\varepsilon_2' - i\sigma_2/\omega)}{d_2}$$

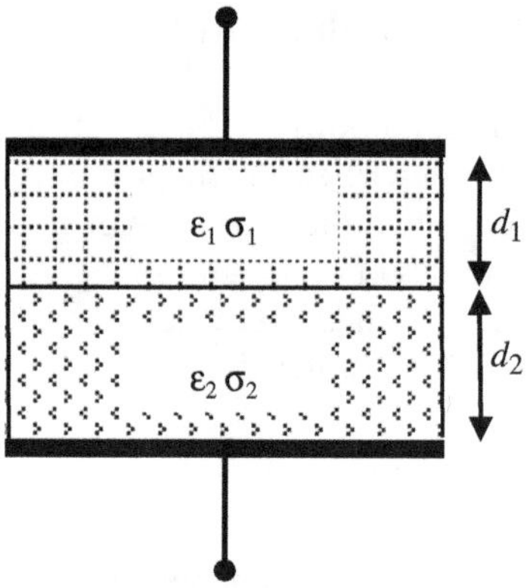

Figure 7.10 A two-layer heterogeneous dielectric between two parallel electrodes.

The total capacitance C_T is given by $\frac{1}{C_T} = \frac{1}{C_1} + \frac{1}{C_2}$, to give

$$C_T = \frac{A\varepsilon_o(\varepsilon_1'\omega - i\sigma_1)(\varepsilon_2'\omega - i\sigma_2)}{d_2(\varepsilon_1'\omega - i\sigma_1) + d_1(\varepsilon_2'\omega - i\sigma_2)}$$

On the assumption that there is an exponential buildup and decay with time of the interfacial free charge, characterized by a single relaxation time, it is sensible to expect Maxwell–Wagner interfacial polarization to behave in a similar manner to a Debye-type dielectric dispersion. This will be the case even though both dielectric materials in the composite capacitor may not themselves exhibit dipole moment relaxations. Based on Equation (7.15) we can thus expect the capacitance to vary with frequency according to a relationship of the form

$$C_T(\omega) = C_\infty + \frac{\Delta C}{1 + \omega^2\tau_{MW}^2} \tag{7.25}$$

where τ_{MW} is the characteristic relaxation time and ΔC is the difference between the capacitance C_s measured at a low frequency (the electrostatic case) and the constant value C_∞ obtained at high frequencies. The total effective capacitance can also be written in terms of an effective permittivity as

$$C_T^* = \frac{A\varepsilon_o\varepsilon_{eff}^*}{(d_1 + d_2)} = \frac{A\varepsilon_o(\varepsilon' - i\varepsilon''')}{(d_1 + d_2)}$$

Through tedious (but not complicated) algebra, the following parameters are derived:

$$\varepsilon'(\omega) = \varepsilon_\infty + \frac{(\varepsilon_s - \varepsilon_\infty)}{1 + \omega^2\tau_{MW}^2} \tag{7.26a}$$

$$\varepsilon''(\omega) = \frac{(\varepsilon_s - \varepsilon_\infty)\omega\tau_{MW}}{1 + \omega^2\tau_{MW}^2} + \frac{\sigma_{eff}}{\omega} \tag{7.26b}$$

in which σ_{eff} is the overall effective conductivity of the two-layer system and

$$\begin{aligned}
\varepsilon_s &= \frac{(d_1 + d_2)(\varepsilon_1' d_1\sigma_2^2 + \varepsilon_2' d_2\sigma_1^2)}{(\sigma_1 d_2 + \sigma_2 d_1)^2} \\
\varepsilon_\infty &= \frac{(d_1 + d_2)\varepsilon_1'\varepsilon_2'}{\sigma_1 d_2 + \sigma_2 d_1} \\
\tau_{MW} &= \frac{\varepsilon_1' d_2 + \varepsilon_2' d_1}{\sigma_1 d_2 + \sigma_2 d_1} \\
\sigma_{eff} &= \frac{(d_1 + d_2)\sigma_1\sigma_2}{\sigma_1 d_2 + \sigma_2 d_1}
\end{aligned} \tag{7.27}$$

The relaxation time constant given by Equation (7.27) equates to the time required for the interfacial free charge on the spherical particle to buildup to 1/e of its final value (i.e., the time to acquire 63.2% of total polarization).

An alternative approach is through an electric circuit model as described in Box 7.6.

These results are in agreement with those derived by Van Beek [29–31] and supported by the theoretical and experimental work of Sillars [32]. From Equations (7.26) and (7.27) it is clear that if the dielectric materials making up the composite capacitor do not exhibit dielectric dispersions (i.e., the conductivity and permittivity parameters do not vary as a function of frequency) the Maxwell–Wagner relaxation time τ_{MW} will also be independent of frequency. However, if the materials exhibit dipole relaxations in the frequency range of the interfacial polarization, Equation (7.28) assumes the more complicated frequency-dependent form

$$\tau_{MW} = \frac{\varepsilon_1'(\omega)d_2 + \varepsilon_2'(\omega)d_1}{\sigma_1(\omega)d_2 + \sigma_2(\omega)d_1} \tag{7.29}$$

which for a particular composite component i:

$$\varepsilon_i'(\omega) = \varepsilon_{i\infty} + \frac{(\varepsilon_{is} - \varepsilon_{i\infty})}{1 + \omega^2\tau_i^2}$$

$$\sigma_i(\omega) = \frac{(\varepsilon_{is} - \varepsilon_{i\infty})\omega^2\tau_i}{1 + \omega^2\tau_i^2}$$

After deriving the electrostatic case, Maxwell states [27, p. 457]:

> This investigation shews that a dielectric composed of strata of different kinds may exhibit the phenomena known as electric absorption and residual discharge, although none of the substances of which it is made exhibit these phenomena when alone. An investigation of the cases in which the materials are arranged otherwise than in strata would lead to similar results, though the calculations would be more complicated, so that we may conclude

Box 7.6 An Electric Circuit Model of Maxwell–Wagner Polarization

The following is a simple circuit to model the two-layer heterogeneous capacitor shown in Figure 7.10. As for the circuit in Box 7.4, the values of the electrical components do not change as a function of frequency.

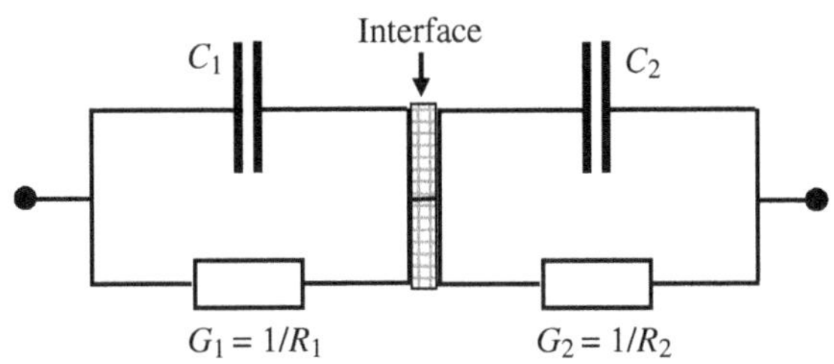

This pair of parallel networks connected in series can be analysed in terms of an overall effective admittance or impedance. From the example of Box 7.4 we know that an effective admittance **Y** of the form:

$$Y(\omega) = Y_s + \frac{(Y_s - Y_\infty)}{(1+\omega^2\tau_{MW}^2)} = Y' + iY''$$

describes a Debye-type relaxation process. The following expressions for the real and imaginary admittance components are obtained:

$$Y' = \frac{(G_1G_2 - \omega^2C_1C_2)/(G_1+G_2) - \omega^2(C_1+C_2)(C_1G_2+C_2G_1)/(G_1+G_2)^2}{(1+\omega^2\tau_{MW}^2)}$$

$$Y'' = \omega\frac{(C_1G_2+C_2G_1)/(G_1+G_2) - (C_1+C_2)(G_1G_2-\omega^2C_1C_2)/(G_1+G_2)^2}{(1+\omega^2\tau_{MW}^2)}$$

where

$$\tau_{MW} = \frac{(C_1+C_2)}{(G_1+G_2)} \tag{7.28}$$

At low frequencies ($\omega\tau_{MW} \ll 1$):

$$Y' = \frac{G_1G_2}{(G_1+G_2)},\ Y'' = 0.$$

The current is in phase with the voltage. At high frequencies ($\omega\tau_{MW} \gg 1$):

$$Y' = \frac{(C_1C_2)(G_1+G_2) - (C_1+C_2)(C_1G_2+C_2G_1)}{(C_1+C_2)^2},$$

$$Y'' = \omega\frac{(C_1C_2)}{(C_1+C_2)}$$

The current leads the voltage by 90°.

---

that the phenomena of electric absorption may be expected in the case of substances composed of parts of different kinds, even though those individual parts should be microscopically small.

Wagner [28] repeated Maxwell's analysis of a composite capacitor for the case of an applied alternating field, but also extended his model to that of small conducting spheres suspended in a dielectric medium. Each sphere was assumed to have conductivity σ_p and relative permittivity ε_p, with the surrounding dielectric medium having dielectric properties σ_m and ε_m, with $\sigma_m = 0$. Wagner derived expressions for ε' and ε'' of the same basic forms as Equations (7.26) and (7.27), with

$$\tau_{MW} = \frac{2\varepsilon'_m + \varepsilon'_p}{\sigma_p}\varepsilon_o \tag{7.30}$$

Analyses have also been obtained for spherical, ellipsoidal and cylindrical particles (σ_p, ε_p) suspended in a homogeneous medium (σ_m, ε_m). The limiting low-frequency and high-frequency permittivity and conductivity values, together with the Maxwell–Wagner relaxation time τ_{MW}, are found to depend on a parameter A_i, known variously as the *internal field function* or the *depolarization factor*, along the axis of the ellipsoid. For example, the general expression for the interfacial charging relaxation time of spheres and ellipsoids suspended in a fluid, with the particles occupying a volume fraction v, is given by

$$\tau_{MW} = \frac{\varepsilon'_m + A_i(1-v)(\varepsilon'_p - \varepsilon'_m)}{\sigma_m + A_i(1-v)(\sigma'_p - \sigma'_m)} \tag{7.31}$$

As shown by Maxwell [33, Art. 437, p. 67] for the case of the magnetization of an ellipsoid and applying this to electrical polarization, the distortion of the potential in a uniform field within a dielectric medium produced by an ellipsoidal particle with semiaxes a, b, c, depends on the parameters

$$A_a = \frac{abc}{2}\int_0^\infty \frac{ds}{\sqrt{(s+a^2)^3(s+b^2)(s+c^2)}}$$

$$A_b = \frac{abc}{2}\int_0^\infty \frac{ds}{\sqrt{(s+a^2)(s+b^2)^3(s+c^2)}}$$

$$A_c = \frac{abc}{2}\int_0^\infty \frac{ds}{\sqrt{(s+a^2)(s+b^2)(s+c^2)^3}}$$

For arbitrary values of the axes $A_a + A_b + A_c = 1$, so that for spherical particles ($a = b = c$) we have $A_a = A_b = A_c = 1/3$.

We can understand the physical significance of the parameter A_i by noting that the field inside a particle will be of the form

$$\mathbf{E}_i = \mathbf{E} - \mathbf{E}_p \tag{7.32}$$

where $\mathbf{E}$ is the field in the external medium and $\mathbf{E}_p$ is the field arising from the polarization $\mathbf{P}$ due to buildup of residual charge on the particle's surface. This polarization field is given by

$$\mathbf{E}_p = A\mathbf{P}/\varepsilon o \tag{7.33}$$

For the special case of a sphere, $a = b$ and $A_a = 1/3$. Inserting this value into Equation (7.31) for the case of a dilute suspension of noninteracting spheres (i.e., $\nu \approx 0$) and for $\sigma_2 = 0$, we obtain the result derived by Wagner for τ_{MW} as given in Equation (7.30). Within the dielectrics community the concept of Maxwell–Wagner interfacial polarization is often employed to understand the dielectric properties of materials, such as polymeric insulators, in which lossy particulate materials are assumed to be present as unwanted impurities. This corresponds, with reference Figure 6.12, to the external field being distorted into the particle surface such that the internal field is less than the external field. In this case the polarization P is positive and the parameter A_i acts as a *depolarizing* factor. For the case of the imagined spherical *cavity* shown in Figure 6.2, we have the situation where the surrounding medium is more polarizable than the particle. The charge distributions induced on the cavity surface are such that P is a *negative* quantity. From Equations (7.32) and (7.33) the field within an empty spherical cavity is

$$\mathrm{E}_i = \mathrm{E} - \mathrm{E}_{dp} = \mathrm{E} + A\mathrm{P}/\varepsilon_o = \mathrm{E} + (\mathrm{P}/\varepsilon_o)/3$$

This is the same result given by Equation (6.15) for the local field in a cavity.

The depolarizing factor along the *a*-axis of a *prolate* ellipsoid (a > b, where *a* and *b* are the principal axes) is given by

$$A_a = \frac{-1}{(a/b)^2 - 1} + \frac{a/b}{[(a/b)^2 - 1]^{3/2}} \log_e(a/b) + [(a/b)^2 - 1]^{1/2} \tag{7.34}$$

For *oblate* ellipsoids ($a < b$)

$$A_a = \frac{-1}{(a/b)^2 - 1} + \frac{a/b}{[(a/b)^2 - 1]^{3/2}} arc\cos(a/b) \tag{7.35}$$

Values for the depolarizing factor, as a function of the ratio *a*/*b*, are plotted in Figure 7.11. These equations for the depolarizing factor are used to find the field parameter f_1 in the formulation of mixture theories outlined in Box 8.2 and used to describe the dielectric properties of protein solutions and cell suspensions, for example.

Clearcut demonstrations of Maxwell–Wagner polarization in polymeric samples are difficult to find in the literature. For example, the appearance of a large dielectric dispersion, on forming a nonpolar surface layer composed of silica microspheres onto a copoloymer, is shown in Figure 7.12. This was considered [36] to be an example of Maxwell–Wagner polarization, but such interpretation is in part complicated by the Cole–Cole plots shown in Figure 7.12 being obtained at different temperatures. Large dielectric dispersions observed for suspensions of spherical colloidal particles were concluded to arise from surface ion conduction effects, rather than as a result of a Maxwell–Wagner polarization [37]. This is discussed in more detail in the next chapter in connection with interpretations of the dielectric properties of biological particles and materials. A direct way of demonstrating the buildup of residual free charge in polymers containing nonpolar inclusions, fillers, air-inclusions and micro-cracks is by their thermally stimulated discharge. An example of this is given by van Turnhout in a technique where depolarization currents are measured in an arrangement with or without an air gap between an electrode and a metallized polymer film [38]. This allowed a distinction to be made of dipolar depolarization and Maxwell–Wagner polarization in a wide range of polymers. Investigations of Maxwell–Wagner polarization remains an important aspect of the development of thin film capacitors [39].

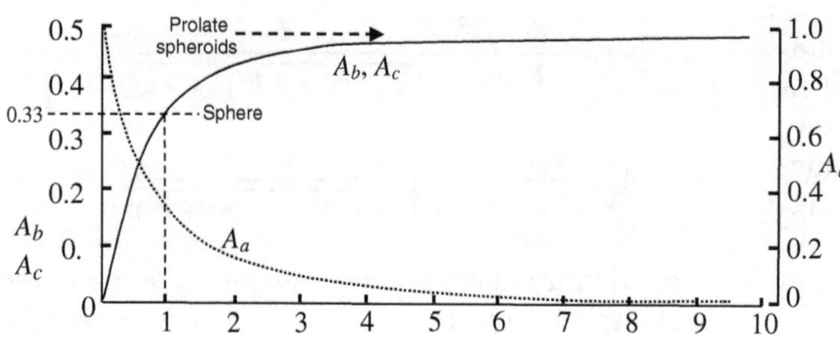

Figure 7.11 The depolarization factors A_a, A_b and A_c, for ellipsoids of revolution about the *a*-axis. These plots are derived from the tables given by Ross and Sack [34] and O'Konski [35].

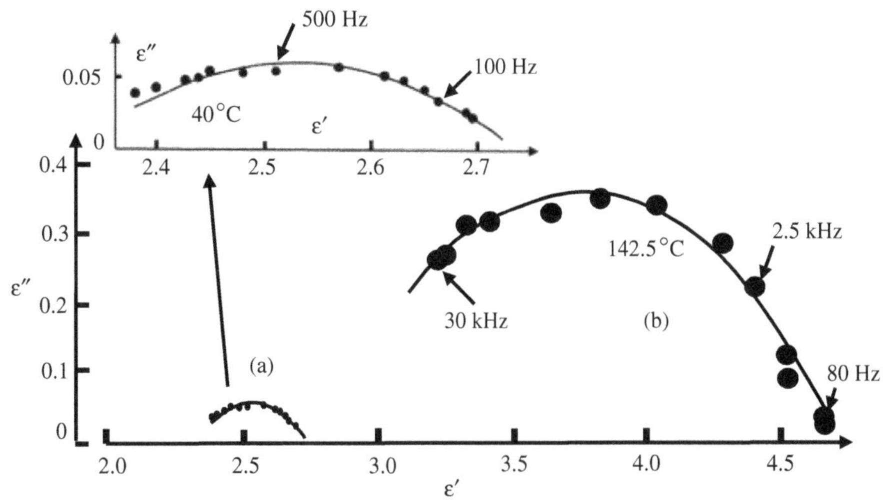

Figure 7.12 Cole–Cole plots of ε' versus ε'' for (a) polystyrene-poly(methyl) methacrylate copolymer and (b) the copolymer with a surface layer composed of micron-sized silica particles (Aerosil®). The pure copolymer exhibited a small ($\Delta\varepsilon' \approx 0.4$) dispersion, compared with that after surface treatment ($\Delta\varepsilon' \approx 1.8$). Based on Lipatov and Fabulyak [36].

7.4.1 Electrode Polarization

Interfacial polarization described by the Maxwell–Wagner effect results from free charges accumulating at the boundaries of different materials in a heterogeneous dielectric. For all dielectrics in general, charges can also accumulate at the electrodes. This modifies the field contours in the dielectric near the electrodes and results in a measured increase of the dielectric's effective capacitance. This effect is known as electrode polarization. To a first approximation electrode polarization can be represented as an impedance Z_e in series with the admittance $\mathbf{Y}(\omega)$ of the circuit shown in Box 7.6. Z_e is usually dominated by a capacitive reactance $1/i\omega C_e$, so that at very low frequencies an applied voltage can appear mainly across the electrode-sample interface and not across the dielectric sample itself. In general Z_e will have both a resistive and capacitive character and will exhibit a frequency-dependence of the form:

$$Z_e(\omega) = \frac{A}{(i\omega)^m} = \frac{A}{\omega^m}\left[\cos\left(\frac{\pi}{2}m\right) - i\sin\left(\frac{\pi}{2}m\right)\right]$$

where A and m are constants. This relationship is another manifestation of the Kramers–Krönig relations. In fact, from dielectric measurements on biological samples, the value for m typically varies from ~0.3 to 0.5 as the frequency increases from 20 Hz to 200 kHz [40–42]. A value for $m = 0.5$ (i.e., a phase angle of 45°) was derived by Warburg in his treatment of electrode polarization effects observed in electrochemical reactions at electrodes [43, 44]. In this case the diffusion of ions to and away from an electrode is a major controlling factor – hence the assumed value for m of 0.5. In dielectric measurements of the kinetics of albumin-ligand binding reactions induced by pulsed electric fields, Scheider found that lateral charge transfer along the irregular surface of the electrode was responsible for the fractional power frequency dependence of electrode polarization [45]. This emphasizes the significance of electrode surface roughness. This is evident from the fact that the capacitive reactance, $1/i\omega C_e$, dominates the electrode impedance. Roughening the surface of an electrode increases the surface area of the electrode and thus also the effective capacitance capacitive reactance C_e. This reduces the electrode's reactive impedance (Xc = $1/i\omega C_e$). A common practical method to reduce electrode impedance effects is to increase the surface area of a platinum electrode by depositing a layer of platinum black on it. This method was first adopted for electrochemical studies by Warburg [43] and validated for dielectric studies on biological materials by Schwan [46]. Another practical method is to attempt to evaluate the magnitude and frequency-dependence of the electrode polarization for a particular electrode arrangement in a measurement chamber used, for example, to contain cell suspensions. This is accomplished by filling the chamber with a known concentration (and hence conductivity) of the electrolyte or buffer and measuring its impedance (real and imaginary components) as a function of frequency.

Ishai *et al.* [47] have reviewed the fundamental properties of electrode polarization with relevance to high conductivity samples and its implications for both dielectric and impedance spectroscopy. The use of fractal electrodes to bypass the effects of electrode polarization is also described, together with a physical explanation as to the limitations of such an approach. An overview of the electrochemical and physical phenomena that influence the magnitude of electrode polarization, its implications to dielectric spectroscopy and the various approaches of either *apparatus-based compensation* or *analytical post-processing* to correct for its effects, has recently been presented by Feldman *et al.* [48].

7.5 Summary

In this chapter the basic processes involved in the polarization of an ideal dielectric have been described. By *ideal* we mean that the dielectric behaves as an electrical insulator and it is assumed that the material is homogeneous in structure and has isotropic physical properties. The dielectric may or may not contain polar molecules that possess permanent electric dipole moments capable of rotating freely and independently of one another. If the dielectric is nonpolar its electrical polarization results only from field-induced distortions of electron 'clouds' around atoms, together with displacements of atoms linked by polar chemical bonds. The time taken to reach the steady-state value of polarization is typically less than 10^{-12} s. The dielectric will exhibit a relatively low value of relative permittivity, which remains constant at its electrostatic value up to electrical frequencies of $\sim 10^{12}$ Hz.

For a dielectric whose structure consists of polar molecules, the polarization process in response to the sudden application of an electric field follows a time course shown schematically in Figure 6.3 and reproduced in Figure 7.13(a). It is assumed that the dielectric was devoid of any polarization at the time of the field's application. Following the almost instantaneous electronic and atomic polarization events, the dielectric's polarization asymptotically approaches a final steady-state value. If the dielectric's structure consists of a single type of polar molecule having a single rotational time constant, the time course of the polarization is characterized by an exponential function with a constant relaxation time τ. At this time the polarization **P** has attained $(1 - 1/e) = 63.2\%$ of its final value $(\varepsilon_s - 1)\varepsilon_o \mathrm{E}$. Strictly speaking, the polarization never reaches this final value, but after a time equal to 10τ is within 0.005% of it. The polarization process is reversible. On removal of the field, thermal agitations cause the polarization to relax asymptotically back to zero at a rate characterized by the same time constant τ. For applied fields of low alternating frequency the dipoles are able to keep in step with the changes in magnitude and direction of the field. The dielectric exhibits a relative permittivity equal or close to its electrostatic value ε'_s. The phase difference between the alternating field and the displacement current arising from the dipole reorientations is zero or very small. The energy loss, characterized by its dielectric loss parameter ε'', is zero or very small. The maximum dielectric loss occurs at a frequency ω where $\omega\tau = 1$. With increasing frequency the dipoles are less able to respond to the field and the dielectric's relative permittivity approaches its high-frequency limit ε'_∞. This change of permittivity with frequency is known as a dielectric dispersion, with magnitude defined as $\Delta\varepsilon = (\varepsilon_s - \varepsilon_\infty)$. Because at the very high frequencies the dipoles are unable to respond to the applied field's periodicity, there is no displacement current and the loss parameter ε'' falls back to zero. A typical frequency plot of $\varepsilon'(\omega)$ for a polar dielectric is shown in Figure 7.13(b). According to Equation (7.13) this is described mathematically as:

$$\varepsilon'(\omega) = \varepsilon_\infty + \frac{(\varepsilon_s - \varepsilon_\infty)}{1 + \omega^2\tau^2}$$

This relationship can also be written in terms of the dispersion's characteristic frequency, corresponding to where the loss parameter ε'' attains its peak value at $f_c = 1/(2\pi\tau)$:

$$\varepsilon'(\omega) = \varepsilon_\infty + \frac{(\varepsilon_s - \varepsilon_\infty)}{1 + f^2/f_c^2} \tag{7.36}$$

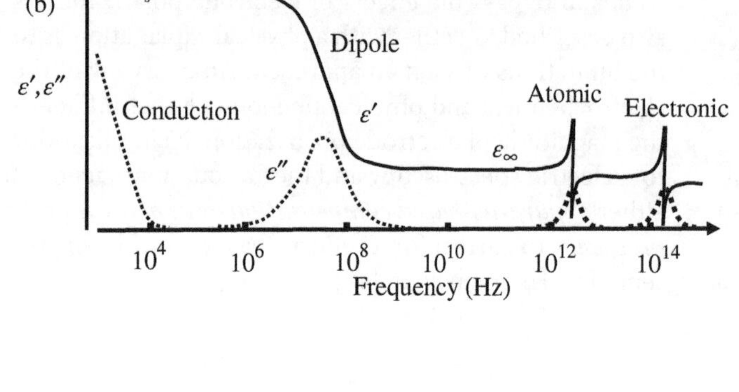

Figure 7.13 (a) The polarization of a polar dielectric as a function of time. (b) The corresponding variations of ε' and ε'' as a function of frequency, resulting (in time order) from electronic distortion, atomic displacement and relaxation of dipole orientation. There is an additional contribution to ε'' at low frequencies if the dielectric exhibits electrical conductivity.

As described in section 7.2.2, the electronic and atomic polarizations occur as resonant phenomena and exhibit distinctly different $\varepsilon'(\omega)$ profiles to that of a relaxing dipole. The energy losses are shown as $\varepsilon''(\omega)$ peaks. Also shown in Figure 7.13(b) is the energy loss associated with a dielectric exhibiting electrical conductivity.

Two other forms of polarization have also been described – namely Maxwell–Wagner interfacial polarization and electrode polarization. Interfacial polarization is the buildup of free charge at the interface between different materials in a heterogeneous dielectric, or at the surfaces of dispersed impurity particles. The main reason for including this phenomenon in this chapter is that it is an important factor to consider when analysing measurements performed on suspensions of cells or other bioparticles. This is described in detail in the next chapter. Electrode polarization (also referred to as electrode impedance) results from the buildup of free charge at the interface between a conducting dielectric and an electrode. This takes the form of an electrical double layer similar to that developed in a capacitor and has an associated capacitive reactance. It can be a dominant effect at frequencies below ~10 kHz when performing dielectric or dielectrophoretic measurements on samples that are dissolved or suspended in aqueous electrolytes of high conductivity. If the electrode impedance is large, a significant proportion of the applied field will appear across the electrode-sample interface and not across the sample itself. This can result, for example, in a dielectrophoretic response being weaker than expected.

7.6 References

1 Pauling, L. (1932) The Nature of the chemical bond, IV: The energy of single bonds and the relative electronegativity of atoms. *J. Am. Chem. Soc.* **54**, 3570–3582.

2 Bowers, W. A. (1985) The classical polarizability of the hydrogen atom. *Am. J. Phys.* **54**, 347–350.

3 Yamashita, J. (1952) The theory of the dielectric constant of ionic crystals. *Progr. Theoretical Physics* **8**, 280–290.

4 Sjoberg, P. and Polizer, P. (1990) Use of the electrostatic potential at the molecular surface to interpret and predict nucleophilic processes. *J. Phys. Chem.* **94**, 3959–3961.

5 Bader, R. F. W. and Jones, G. A. (1963) The electron density distributions in hydride molecules. I. The water molecule. *Canadian J. Chem.* **41**, 586–606.

6 Martin, F. and Zipse, H. (2005) Charge distribution in the water molecule – A comparison of methods. *J. Computational Chem.* **26**, 97–105.

7 Kumar, A., Gadre, S. R., Mohan, N., Suresh, C. H. (2013) Lone pairs: An electrostatic viewpoint. *J. Phys. Chem. A.* **118**, 526–532.

8 Sugden, S. (1934) Atomic polarization. *Trans. Faraday Soc.* **30**, 734–739.

9 Fairbrother, F. (1934) The dipole moments of the halogen hydrides in solution. *Trans. Faraday Soc.* **30**, 862–870.

10 Fairbrother, F. (1934) Determination of dipole moments in solution. *Nature* **134**(3386), 458–459.

11 S. S. (1934) Dipole moments and their interpretation. *Nature* **133**(3369), 802–803.

12 Groves, L. G. and Sugden, S. (1934) The dipole moments of vapours. Part I. *J. Chem. Soc.* **232**, 1094–1098.

13 Debye, P. J. W. (1912) Einige Resultate einer kinetischen Theorie der Isolatoren. *Physikalische Zeitshrift* **13**(3), 97–100.

14 Debye, P. (1929) *Polar Molecules*, Chemical Catalog Co., New York, NY.

15 Cole, R. H. and Cole, K. S. (1941) Dispersion and absorption in dielectrics. II. Alternating current characteristics. *J. Chem. Phys.* **9**, 341–351.

16 Davidson, D. W. and Cole, R. H. (1950) Dielectric relaxation in glycerine. *J. Chem. Phys.* **18**, 1417.

17 Pastushenko, V. Ph., Kuzmin, P. I. and Chizmadzhev, Yu. A. (1985) Dielectrophoresis and electrorotation of cells: Unified theory of spherically symmetric cells with arbitrary structure of membrane. *Biological Membranes* **5**, 65–77.

18 Wang, X-B., Pethig, R. and Jones, T. B. (1992) Relationship of dielectrophoresis and electrorotational behaviour exhibited by polarised particles. *J. Phys. D: Appl. Phys.* **25**, 905–912.

19 Sillars, R. W. (1938) The behaviour of polar molecules in solid paraffin wax. *Proc. Roy. Soc. London* **169A**, 66–83.

20 Hasted, J. B. (1973) *Aqueous Dielectrics*, Chapman & Hall, London.

21 Glasstone, S., Laidler, K. J. and Eyring, H. (1941) *The Theory of Rate Processes*, McGraw-Hill, New York.

22 Kauzmann, W. (1942) Dielectric relaxation as a chemical rate process. *Rev. Mod. Phys.* **14**, 12–44.

23 Fröhlich, H. and Sack, R. (1944) Properties of dipolar substances. *Proc. Roy. Soc. London* **182A**, 388–403.

24 Fröhlich, H. (1958) *Theory of Dielectrics*, 2nd edn, Clarendon Press, Oxford, pp. 17–21; 62–70.

25 Hoffman, J. D. (1955) Theory of dielectric relaxation for a single-axis rotator in a crystalline field. *J. Chem. Phys.* **23**, 1331–1339.

26 Meakins, R. J. (1955) The dielectric properties of urea occlusion compounds. *Trans. Faraday Soc.* **51**, 953–961.
27 Maxwell, J. C. (1891) *A Treatise on Electricity and Magnetism*, Vol. **1**, 3rd edn. Clarendon Press, Oxford.
28 Wagner, K. W. (1914) Erklärung der dielektrischen Nachwirkungsvorgänge auf Grund Maxwellscher Vorstellungen. *Archiv f. Electrotechnik* **2**(9), 371–387.
29 Van Beek, L. K. H. (1960) The Maxwell–Wagner–Sillars effect. Describing apparent dielectric loss in inhomogeneous media. *Physica* **26**, 66–68.
30 Van Beek, L. K. H., Booy, J. and Looyenga, H. (1965) Dielectric loss in heterogeneous model systems. *Appl. Sci. Res.* B **12**, 57–64.
31 Van Beek, L. K. H. (1967) Dielectric behaviour of heterogeneous systems, in *Progress in Dielectrics* (eds J. B. Birks and J. Hart), Vol. **7**, Heywood, London, pp. 71–114.
32 Sillars, R. W. (1937) The properties of a dielectric containing semiconducting particles of various shapes. *J. Inst. Electr. Eng.* **80**(484), 378–394.
33 Maxwell, J. C. (1891) *A Treatise on Electricity and Magnetism*, 3rd ed. Vol. **2**, Clarendon Press, Oxford.
34 Ross, I. G. and Sack, R. A. (1950) Solvent effects in dipole moment measurements. *Proc. Phys. Soc.* **63**, 893–906.
35 O'Konski, C. T. (1960) Electric properties of macromolecules. V. Theory of ionic polarization in electrolytes. *J. Phys. Chem.* **64**, 605–619.
36 Lipatov, Yu. S. and Fabulyak, F. Y. (1972) Relaxation processes in the surface layers of polymers at the interface. *J. Appl. Polymer Sci.* **16**, 2131–2139.
37 Schwan, H. P., Schwarz, G., Maczuk, J. and Pauly, H. (1962) On the low frequency dielectric dispersion of colloidal particles in electrolyte solution. *J. Phys. Chem.* **66**, 2626–2635.
38 van Turnhout, J. (1971) Thermally stimulated discharge of polymer electrets. *Polymer J.* **2**, 173–191.
39 Liu, G.-Z., Wang, C., Wang, C.-C. *et al.* (2008) Effects of interfacial polarization on the dielectric properties of BiFeO3 thin film capacitors. *Appl. Phys. Lett.* **92**, 122903.
40 Onaral, B. and Schwan, H. P. (1982) Linear and non linear properties of platinum electrode polarisation. Part 1: frequency dependence at very low frequencies. *Med. and Biol. Eng. and Comput.* **20**, 299–306.
41 Schwan, H. P. (1992) Linear and nonlinear electrode polarization and biological materials. *Annala Biomed. Eng.* **20**, 269–288.
42 Moussavi, M., Schwan, H. P. and Sun, H. H. (1994) Harmonic distortion caused by electrode polarisation. *Med. and Biol. Eng. and Comput.* **32**, 121–125.
43 Warburg, E. (1889) Über das Verhalten sogenannter unpolarisirbarer Elecktroden gegen Wechselstrom. *Ann. Physik und Chemie* **67**, 493–499.
44 Warburg, E. (1901) Über die Polarisationscapacität des Platins. *Ann. Physik* **6**, 125–135.
45 Scheider, W. (1977) Real-time measurements of dielectric relaxation of biomolecules kinetics of a protein-ligand binding reaction. *Ann. N. Y. Acad. Sci.* **303**, 47–56.
46 Schwan, H. P. (1963) Determination of biological impedances, in *Physical Techniques in Biological Research* (ed. W. L. Nastuk), Vol. 6. Academic Press, New York, NY, pp. 323–407.
47 Ishai, P. B., Talary, M. S., Caduff, A. *et al.* (2013) Electrode polarization in dielectric measurements: a review. *Meas. Sci. Technol.* **24**, 102001 (21 pp.).
48 Feldman, Y., Ishai, P. B. and Raicu, V. (2015) Electrode polarization, in *Dielectric Relaxation in Biological Systems* (eds. V. Raicu and Y. Feldman), Oxford University Press, Oxford, Ch. 2.2.

8

Dielectric Properties of Water, Electrolytes, Sugars, Amino Acids, Proteins and Nucleic Acids

8.1 Introduction

The scientific literature describing investigations and biomedical applications related to the electrical and dielectric properties of biological materials is vast (e.g. [1–9]). In this chapter, focus is directed on aspects of the subject of direct relevance to dielectrophoresis experiments. Such experiments are most often performed using aqueous suspensions of the particles under investigation. The dielectrophoretic force acting on the particle depends on the Clausius–Mossotti factor, which, from Chapter 6, is defined in terms of the dielectric properties of the particle as well as the suspending medium. After first reviewing the dielectric properties of pure water, the effects of adding salts, sugars, pH buffers, amino acids and proteins to water are described in this chapter. Combinations of such chemicals are added to the suspending medium to maintain the cell in a physiologically viable state. Knowledge of the extent to which these chemicals alter the dielectric properties of pure water provides the means to monitor or control the magnitude (positive or negative) of the Clausius–Mossotti factor. The experiment could also be directed towards manipulating proteins or DNA particles, for example, so the frequency-dependent polarizations of these particles are also described. The dielectric properties of cells are described in Chapter 9.

8.2 Water

The dielectric properties of pure water at 298 K (25 °C) are summarized in Figure 8.1 as the variations of ε' and ε'' with frequency and the corresponding Cole–Cole plot (see Figure 7.7). This data is based on a comprehensive and critical analysis of the published experimental data available in 1995, for the frequency range 0–1000 GHz and temperature range 0–100 °C [10]. The principal dielectric dispersion, characterized at 25 °C by a ε'' loss peak at 18 GHz, is interpreted as a Debye-type molecular dipole relaxation. The Cole–Cole plot shown in Figure 8.1 is a slightly depressed semicircle described by the modified Cole–Cole equation (7.12)

$$\varepsilon(\omega) = \varepsilon_\infty + \frac{(\varepsilon_s - \varepsilon_\infty)}{1 + (i\omega\tau)^{1-\alpha}} \tag{8.1}$$

The parameter α has a value of 0.013, corresponding to a near ideal Debye-type dielectric dispersion characterized by a single relaxation time. (Recall from section 7.3 that $\alpha = 0$ corresponds to a single relaxation time, with α tending to unity for an infinite number of relaxation times.) At 25 °C, from the Cole–Cole plot shown in Figure 8.1, $\tau = 1/(2\pi \times 18.56\,\text{GHz}) = 8.58 \times 10^{-12}$ s for pure water. At high frequencies, the main dielectric dispersion for water merges into two more dispersions and two atomic resonances. For example, at 25 °C relaxations occur at 167.8 GHz and 1.94 THz, with resonances at 4.03 THz and 14.48 THz [11]. This makes it difficult to determine an accurate value for the permittivity parameter ε_∞, which is approached asymptotically at high frequencies. A value that has been commonly adopted in the literature is $\varepsilon_\infty = 1.77$, based on the refractive index for water of $n_\infty = 1.33$ and the relationship $\varepsilon_\infty = {n_\infty}^2$ [12]. From infrared spectra obtained for water and D_2O, Zelsmann [13] concluded that the most consistent interpretations of the data were obtained by assigning the refractive index $n_\infty = 1.45$, which gives $\varepsilon_\infty = 2.1$. In any case, ε_∞ should be considered as no more than a 'fudge factor' used to obtain the best fit of the low-frequency dielectric data to a Cole–Cole semicircle plot [14]. A good example of this is the data fitting performed by Hasted [3], who derived $\varepsilon_\infty = 4.23 \pm 0.16$ and $\tau = 9.3 \times 10^{-12}$ s at 20 °C (with corresponding values of 4.20 ± 0.16 and 7.2×10^{-12} s, respectively, at 30 °C). These value for ε_∞ can be considered to be too high as a parameter to define the high-frequency 'end point' of the dielectric dispersion for water at 25 °C.

Dielectrophoresis: Theory, Methodology and Biological Applications, First Edition. Ronald Pethig.

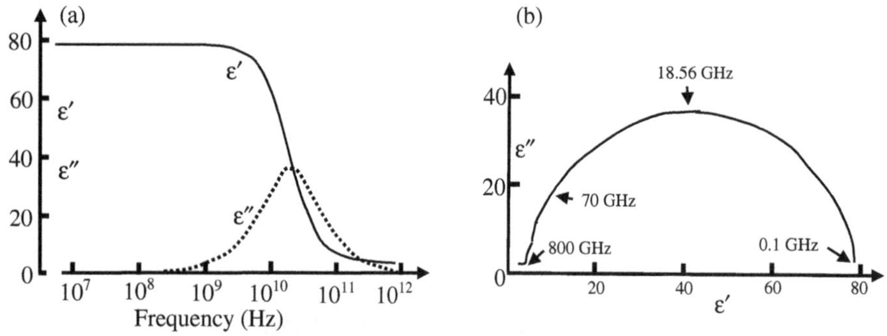

Figure 8.1 Permittivity data for pure water at 25 °C: (a) Plots of ε' and ε''; (b) The corresponding Cole–Cole plot. (Derived from Ellison *et al.* [10,11].)

The static relative permittivity value of pure water can be calculated using the function

$$\varepsilon_s(T) = A.e^{-bT} \tag{8.2}$$

with $A = 87.85306$, $b = 0.00456992$ [10]. The static permittivity values calculated using this formula for the temperature range 0–100 °C are presented in Table 8.1, together with the values and error estimates published in 1981 by the Commission on Physico-Chemical Measurements and Standards of the International Union of Pure and Applied Chemistry (IUPAC). The significant differences in these two sets of values between 80 °C and 100 °C reflect the experimental difficulties involved in measuring the permittivity at these temperatures (principally due to dissolved gases forming air bubbles) and the paucity of experimental data. The interpolations of Ellison *et al.* to derive Equation (8.3) take into account the totality of experimental data reported in the scientific literature [10]. The magnitude of the dielectric dispersion for water at 25 °C can be determined as $\Delta\varepsilon' = \varepsilon_s'-\varepsilon_\infty' = (78.37 - 2.1) = 76.17$ relative permittivity units.

Table 8.1 Values of the static permittivity of pure water and absolute error estimates between 0 °C and 100 °C, as published by the International Union of Pure and Applied Chemistry (IUPAC) in 1981, together with values calculated using Equation (8.3).

°C	ε_s IUPAC	ε_s Equation (8.3) [10]
0	87.87 ± 0.07	87.85 ± 0.03
10	83.91 ± 0.07	83.93 ± 0.03
20	80.16 ± 0.05	80.18 ± 0.03
25	78.36 ± 0.05	78.37 ± 0.03
30	76.57 ± 0.05	76.60 ± 0.03
40	73.16 ± 0.04	73.18 ± 0.03
50	69.90 ± 0.04	69.91 ± 0.03
60	66.79 ± 0.04	66.78 ± 0.03
70	63.82 ± 0.05	63.80 ± 0.03
80	61.03 ± 0.05	60.95 ± 0.03
90	58.32 ± 0.05	58.23 ± 0.03
100	55.72 ± 0.06	55.63 ± 0.03

From Equations (7.13) and (7.14) we can determine the following dielectric parameters at 25 °C:

$$\varepsilon''_{pk} = \frac{(\varepsilon_s - \varepsilon_\infty)}{2} = 38.09$$

$$\varepsilon'(\omega\tau = 1) = \frac{(\varepsilon_s + \varepsilon_\infty)}{2} = 40.27$$

The dielectric properties of water, along with its unique physico-chemical properties, are directly related to its molecular electron density distribution shown in Figure 7.5. According to the charge distribution given in this figure, each water molecule is endowed with a permanent dipole moment of 1.855 debye units (6.2×10^{-30} C m). The effective positive charge of the hydrogen atoms also allows them to interact with the lone pair electrons of the oxygen atom of a neighbouring water molecule to form a hydrogen bond. An example where a water dimer is formed in this way is shown in Figure 8.2. The linear H-O····H formation gives the most stable form of H-bond (the dotted line), with a bond energy of 5.58 kcal/mol (23.3 kJ/mol) [15,16] compared to the H-O covalent bond energy of 492 kJ/mol [17]. H-bonds are thus relatively weak and can be perturbed by thermal fluctuations of the water structure. They are made and broken very rapidly, with the lifetime of each H-bond being about 10^{-12} s. With the lone-pair electron orbitals of the oxygen each able to form a H-bond with a neighbouring water molecule and the two hydrogen atoms able to act as proton donors, a single water molecule has the potential to form four H-bonds and thus an extended 3-D network with other water molecules. According to its molecular size, liquid water should have its melting and boiling points 100 K lower than they are. Its heat of vaporization, heat of fusion and surface tension is higher than that of the comparable hydrides H_2S and NH_3, or even than those of most other common liquids. These physical properties arise directly from the strong forces of attraction created by the H-bonds between the molecules in water. Without them, water on earth would exist as a gas rather than as a liquid. The formation of the H-bond network is facilitated by cooperativity, through which the

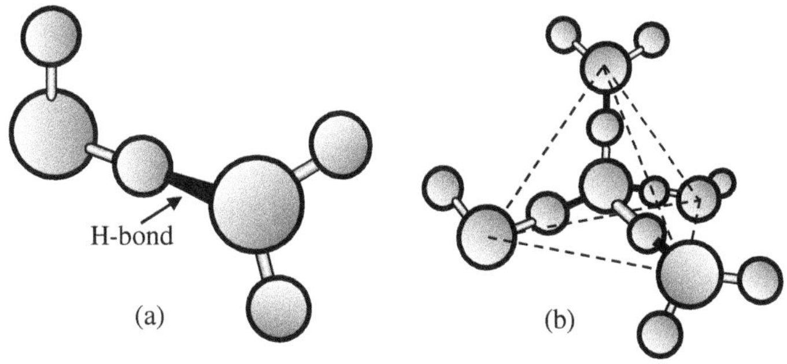

Figure 8.2 (a) The hydrogen bonding of two water molecules. (b) The tetrahedral hydrogen-bonding structure of the most common crystalline form of ice [20].

strength of a H-bond between two water molecules is enhanced by the formation of a second H-bond between one of them and a third water molecule [18, 19]. In liquid water there is a near 60% probability for a fourfold H-bonding, with a considerable content of threefold and fivefold H-bonded water molecules, to give a somewhat random network [19]. In its most common solid-state form, namely the ice structure Ih, with a density (0.9167 g cm^{-3}) at 0 °C significantly lower than that (0.9998 g cm^{-3}) of liquid water, there is almost a perfect tetrahedral structure with 80% of the water molecules having fourfold H-bonding [20]. This tetrahedral structure is shown in Figure 8.2.

The H-bonds in water are made and broken about 10^{12} times a second. The transient nature of how the covalent bonds and H-bonds may exchange places from one instant to the next can be depicted as shown below:

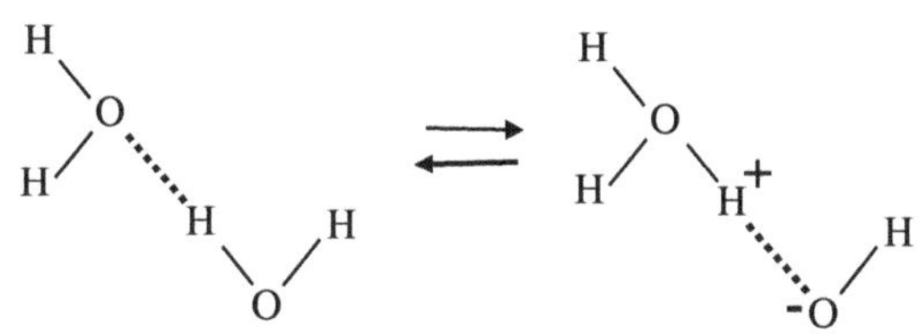

There is thus a finite (but small) probability of finding three hydrogen atoms associated with one oxygen atom to form a hydronium ion (H_3O^+), leaving another oxygen atom with only one hydrogen to form a hydroxyl ion (OH^-). The positively charged hydrogen atoms of the hydronium ion attract the electronegative, oxygen, ends of the surrounding water molecules to form a stable hydrated hydronium ion, as shown in Figure 8.3.

The dissociation (ionization) of water can therefore be written as:

$$2H_2O \leftrightarrow H_3O^+ + OH^-$$

for which the equilibrium constant is given by:

$$K_{eq} = \frac{[H_3O^+]\,[OH^-]}{[2H_2O]} = \frac{[H^+]\,[OH^-]}{[H_2O]} \tag{8.3}$$

where the brackets denote concentrations in moles per litre. To derive the final right-hand expression of this equation, we have divided the numerator and denominator by $[H_2O]$. The concentration of water remains virtually unaltered by its partial dissociation, since (at 25 °C) a litre of pure water contains only 1.0×10^{-7} M of H_3O^+ and an equal number of OH^- ions, whereas the concentration of water in a litre (1000 g) of pure water is 1000 g/L divided by the gram molecular weight (18 g/mol) – namely 55.5 M. Thus, the concentration of water is virtually a constant and it makes no real sense to include it in Equation (8.3) as if it were a variable. Equation (8.3) can thus be simplified by multiplying through by $[H_2O]$:

$$55.5\ K_{eq} = [H^+][OH^-] \tag{8.4}$$

The constant K_{eq} can be combined with the concentration of water (55.5 M) to give a constant K_w termed the ion product of water. From Equation (8.4) at 25 °C, this is given by:

$$K_w = [H^+][OH^-] = 10^{-14}$$

If $[H^+]$ for some reason increases, as when an acid substance is dissolved in water, $[OH^-]$ will decrease so as to keep the product $[H^+]\ [OH^-] = 10^{-14}$. This reaction is the basis for the pH scale, measured as a concentration of H^+ (actually H_3O^+). As described in Box 2.1, the term pH

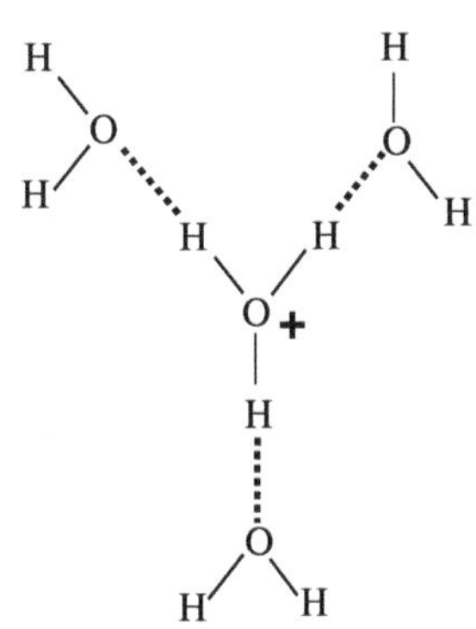

Figure 8.3 A hydronium ion in aqueous solution, surrounded by three H-bonded water molecules.

can be thought of as a shorthand term for the *negative log of hydrogen ion concentration.* At 25 °C for pure water the concentration of hydrogen ions is 10^{-7} mol, which leads us to a value of 7.0 for the pH.

Example 8.1 Onsager's Equation and the Dielectric Dispersion for Water

The following relationship, given in Box 6.7, was derived by Onsager to describe the dielectric dispersion exhibited by polar liquids:

$$\frac{N\mathbf{p}_o^2}{9\varepsilon_o kT} = \frac{(\varepsilon_m - \varepsilon_{m\infty})(2\varepsilon_m + \varepsilon_{m\infty})}{\varepsilon_m(\varepsilon_{m\infty} + 2)^2} \quad \text{(Equation 6.34)}$$

The magnitude of the dispersion is given by $\Delta\varepsilon = (\varepsilon_m - \varepsilon_{m\infty}) \equiv (\varepsilon_s - \varepsilon_\infty)$. How accurate is this formula in determining the value of $\Delta\varepsilon$ for liquid water at 25 °C?

Solution 8.1 From Equation (8.3) and Table 8.1, for pure water at 25 °C, $\varepsilon_s = 78.37$. For the commonly adopted value $\varepsilon_\infty = 1.77$ [2], the right-hand side (*RHS*) of Equation (6.34) is:

$$\frac{(\varepsilon_s - \varepsilon_\infty)(2\varepsilon_s + \varepsilon_\infty)}{\varepsilon_s(\varepsilon_\infty + 2)^2} = 10.9\,(RHS : 6.34)$$

The number density N of water molecules is given by:

$$N = \frac{N_A\,\rho}{M} = \frac{(6.02 \times 10^{23})\,(1\,gm.cm^{-3})}{18\,gm}$$
$$= 3.34 \times 10^{22}\ \text{cm}^{-3} = 3.34 \times 10^{28}\ \text{m}^{-3}$$

According to the charge distribution given in Figure 7.5, each water molecule has a dipole moment of 6.2×10^{-30} C m (1.855 debye units). The left-hand side (*LHS*) of Equation (6.34) thus equates to:

$$\frac{N\,\mathbf{p}_o^2}{9\varepsilon_o kT}$$
$$= \frac{(3.34 \times 10^{28}\,\text{m}^{-3})\,(6.2 \times 10^{-30}\,\text{Cm})^2}{9\,(8.854 \times 10^{-12}\,\text{Fm}^{-1})\,(1.38 \times 10^{-23}\,\text{JK}^{-1})\,(298\,\text{K})}$$
$$= 3.92\,(LHS : 6.34)$$

We thus find a mismatch between the theoretical prediction given by the left-hand side of Onsager's equation (6.34) and the numerical value of the right-hand side obtained using experimentally determined permittivity values. In fact, using the commonly accepted value $\varepsilon_\infty = 1.77$, the solution of Equation (6.34) leads to the value $\varepsilon_s = 31$ at 25 °C (instead of $\varepsilon_s \approx 78$).

The inadequacy found in Solution 8.1 of Onsager's equation in predicting the static permittivity of liquid water was recognized by Kirkwood [21, 22]. In Solution 8.1 no account is given of the effect of the H-bonds in liquid water. The value 1.855 debye units assigned to the dipole moment $\mathbf{p}_o$ in Equation (6.34) corresponds to an isolated water molecule in water's vapour phase. For a water molecule in liquid water, the value of $\mathbf{p}_o$ is enhanced due to induction by the dipole fields of neighbouring water molecules. This effect is taken into account in Onsager's equation (6.34). However, the formation of a tetrahedral coordination of H-bonds to its neighbours means that a field-induced reorientation of a water dipole has to be coordinated with rearrangement of its neighbours. To consider this, Kirkwood introduced an orientation correlation factor g, so that Onsager's equation (6.34) is modified to the form [21, 22]:

$$\frac{N_A\, g\, \mathbf{p}_o^2}{9\varepsilon_o kT} = \frac{(\varepsilon_m - \varepsilon_{m\infty})(2\varepsilon_m + \varepsilon_{m\infty})}{\varepsilon_m(\varepsilon_{m\infty} + 2)^2}\frac{M}{\rho} \quad (8.5)$$

The value $g = 1$ corresponds to no correlation in orientation of a dipole with its neighbours. A situation where there are cooperative orientations of the moments corresponds to g values larger than unity. For a perfect and rigid tetrahedral coordination of the H-bonds in water, about which free rotations are permitted but no bond bending, the orientation correlation parameter is given as

$$g = 1 + z\cos^2\frac{\theta}{2}$$

where θ is the H-O-H bond angle in the water molecule and z is the coordination number ($z = 4$ for a perfect tetrahedron) [20, 21]. Based on coordination numbers and intermolecular distances available for water from published x-ray scattering data, Oster and Kirkwood [22] calculated values of g for the temperature range 0 to 83 °C. They also used Equation (8.5) to calculate the static relative permittivity values, assuming an optical refractive index $n = 1.33$ (i.e., $\varepsilon_\infty = 1.77$). At 25 °C there is a very close agreement between experiment and theory, with experimental and calculated values for g of 2.68 and 2.67, respectively (corresponding to experimental and calculated valued for ε_s of 78.5 and 78.2, respectively). The difference between experiment and theory widens with increasing temperature. At 83 °C the experimental and calculated values for g are 2.49 and 2.82, respectively (corresponding to experimental and calculated valued for ε_s of 59.9 and 67.5, respectively). The experimental value of g thus decrease with increasing temperature, whereas the theoretical value increases. This is considered to reflect a partial destruction of the tetrahedral water structure with increasing temperature, along with a more pronounced bending of the H-bonds [22]. It should however be noted that the value deduced for g, based on Equation (8.5) and using experimentally determined values of ε_s, are quite sensitive to the value assumed for ε_∞. For example, adopting $\varepsilon_\infty = 2.1$ [12] gives $g = 2.35$, whereas with $\varepsilon_\infty = 4.23$ [3] gives $g = 1.0$.

As briefly discussed in Chapter 7, in the form of Equation (7.23), it is possible to describe a dielectric relaxation process as being analogous to a chemical rate process. The temperature variation of the relaxation time τ is then approximately exponential, according to the equation:

$$\tau = \frac{h}{kT} e^{-(\Delta S/R)} e^{(\Delta H/RT)}$$

where ΔS and ΔH are the molar entropy and enthalpy, respectively, of activation for the relaxation process and h is Planck's constant. If ΔS and ΔH are not dependent on temperature a graph of $\log_e(kT\tau/h)$ against $1/T$ should be a straight line of positive slope, from which ΔH can be calculated. A straight line does result between 0 and 50 °C, yielding a value for the activation enthalpy $\Delta H = 17.5$ kJ/mol [11]. This is 25% less than the H-bond energy of 23.3 kJ/mol [15, 16]. Based on computer simulations, this could be associated with the formation of a 'defect' in the water structure due to the formation of a bifurcated H-bond that gives a molecule five coordinated nearest neighbours, equivalent to it possessing five H-bonds [23]. The bifurcated H-bond lowers the potential energy barrier for reorientation of the five-fold H-bonded water and weakens the previously existing H-bonds [24]. Thus, although it is experimentally well established that the dielectric relaxation of liquid water is approximately described by a single relaxation time of ~9 ps at room temperature, corresponding to a near single exponential polarization decay of the Debye type described in Chapter 7, we should not consider this relaxation as a simple field-induced reorientation of a single water dipole moment. Instead we should consider the relaxation as resulting from the dynamics of the H-bond network and structure of water. An understanding of this remains elusive, as is amusingly (and aptly) summarized by the following statement [25]: *the literature on water structure and dynamics is vast as an ocean, with frequent whirlpools of contradictions and disputes.*

Good starting points for interested readers to 'dip their toes' into this important subject are the excellent reviews by Ohmine and Tanaka [26] and later by Agmon [25].

8.2.1 Electrical Mobility of Protons

To complete the description of the dielectric properties of pure water, we should consider its electrical conductivity. Although the dissociation of water is sometimes written as $H_2O \leftrightarrow H^+ + OH^-$ to emphasize the production of protons, we have seen that the electrostatic binding energy of the proton is so large that it has no independent existence in condensed phases such as water. A proton is generally considered to be present as hydronium, H_3O^+, which gives it an equivalent size between that of an hydrated sodium and a hydrated potassium ion. From Equation (2.6) the electrical mobility of an ion is defined as $\mu_e = v/E$, where v is the terminal speed acquired under the influence of an electric field E. To a good approximation we can assume that the terminal speed is reached when the accelerating force ($F_a = qE$) is balanced by the Stokes viscous drag force. This viscous force is directly proportional to the size of the ion and so the electrical mobility of a proton should be of the same magnitude as that of a Na^+ or K^+ ion.

Table 8.2 The electrical mobility of ions at 25 °C in dilute aqueous solution [27].

Cation	Mobility (10^{-8} m²/V.s)	Anion	Mobility (10^{-8} m²/V.s)
H^+, H_3O^+	36.2	OH^-	20.6
K^+	7.6	Cl^-	7.9
Na^+	5.2	F^-	5.7

Table 8.2 gives the electrical mobility values for various ions in water, from which it is evident that the apparent rate of migration of the H_3O^+ ion in an electrical field is significantly greater than that exhibited by Na^+ and K^+ ions.

How can we account for the anomalously high proton mobility? The accepted viewpoint is that a transport process, known as the Grotthuss mechanism, is responsible. This mechanism is named after Theodor Grotthuss (1806), who suggested that electrical conduction through water resulted from the oxygen atoms simultaneously receiving and transferring a single hydrogen atom [28]. This proposal was consistent with the concept at that time of a water molecule having the chemical formula OH instead of H_2O and that an understanding of ions in solution (let alone H-bonds) was at a very primitive level. Nevertheless, his description that throughout the conduction process 'only the water molecules located at the tip of the conducting wires will be decomposed, whereas all those located at intermediate positions will exchange their composing principles reciprocally and alternatively, without changing their nature' proved to be remarkably insightful. The modern version of the Grotthuss mechanism, is depicted in the sequence of events below. The first step involves the injection and binding of a proton into an H-bond network:

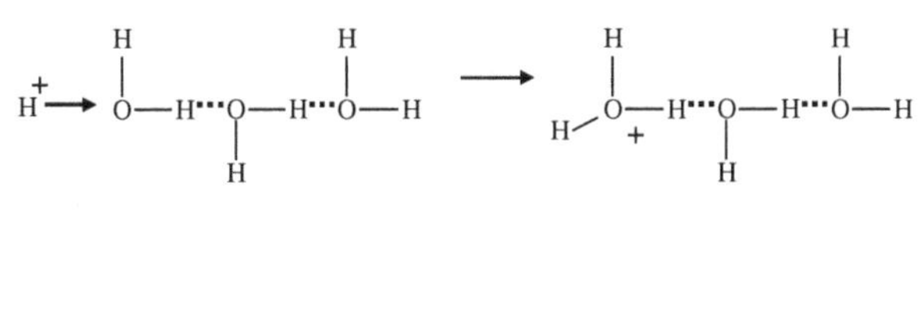

Subsequent steps involve the localized rearrangement of protons and H-bonds, followed by the release of a proton from the H-bond network:

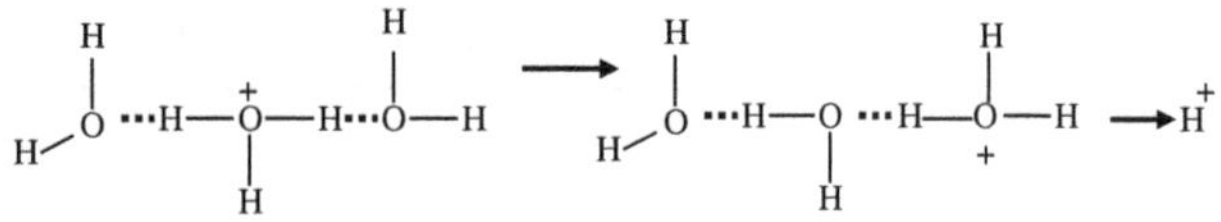

The final step is reorganization of the protons to re-establish the H-bonded water structure that existed before the injection of a proton.

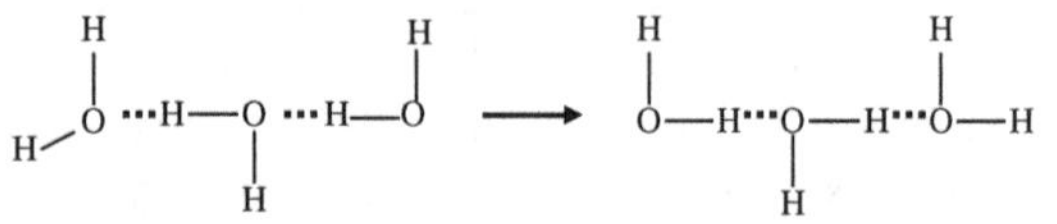

It is with this process in mind that one of the author's mentors (Albert Szent-Gyorgyi) once remarked that 'water was the only molecule he knew that could turn around with turning around!' A partial analogy for the Grotthuss mechanism is the operation of a bucket brigade in damping down a fire, where the buckets move but the people do not. Proton conduction does not involve the diffusion of either the hydronium ions or the protons themselves! A better analogy is electronic conduction along a copper wire. As an electron is injected into the cathodic end of a wire, another electron is simultaneously ejected from the end at the anode. This mode of proton transport is of relevance to bioenergetic processes that involve proton diffusion in protein complexes and the pumping of protons across cell membranes [29].

Example 8.2 Electrical Conductivity of Pure Water Derive a value for the conductivity of pure water at 25 °C.

Solution 8.2 The conductivity of pure water results solely from the electrical mobility of the H_3O^+ and OH^- ions produced in the dissociation of water and can be calculated from the formula

$$\sigma_e = q\left([H_3O^+]\mu^+ + [OH^-]\mu^-\right) \text{S/m}$$

where $[H_3O^+]$, $[OH^-]$ and $\mu+$, μ^- are the concentrations and mobility of the hydronium and OH^- ions, respectively. At 25 °C, $[H_3O^+] = [OH^-] = 10^{-7}$ M, which is equivalent to a number concentration of $(6.03 \times 10^{23} \times 10^{-7})$ ions per litre, or $6.03 \times 10^{19}\,\text{m}^{-3}$. Adopting the mobility values given in Table 8.2:

$$\sigma_\varepsilon = 1.6 \times 10^{-19}(6.03 \times 10^{19}[36.2 \times 10^{-8}] + 6.03 \times 10^{19}[20.6 \times 10^{-8}]) = 5.48\ \mu\text{S/m}$$

This result is more commonly expressed in the reciprocal terms of electrical resistivity as 18 MΩ cm. Deionized (DI) water has a conductivity of ~10 μS/m (~10 MΩ cm). Completely degassed pure water, which is then brought into equilibrium with the atmosphere, contains dissolved carbon dioxide and exhibits a conductivity of ~75 μS/m. Normal drinking water 'from the tap' has a conductivity in the range 50~80 mS/m due mainly to the presence of sodium, calcium, chlorine and phosphate ions. The conductivity of the aqueous component of a cell's cytoplasm is ~1.4 S/m [30].

8.3 Electrolyte Solutions

An electrolyte is defined as a substance that contains atoms or molecules that acquire a negative or positive charge (by gaining or losing electrons) to form ions when dissolved in a suitable solvent. We say that the substance *ionizes*. The most common solvent is water; ionizable substances include soluble salts, acids and bases. Calcium, chloride, phosphate, potassium and sodium are examples of physiologically relevant electrolytes. Polyelectrolytes are formed by the dissolution in water of macromolecules, such as polynucleic acids (e.g., DNA) and polypeptides (e.g., proteins), that contain charged functional chemical groups.

Typical steady-state external and internal concentrations of ions for a mammalian cell are given in Table 8.3. This information is helpful when performing dielectrophoresis experiments on cells. It is important to maintain the cells in a 'comfortable' physiological state and this means suspending them in an appropriate electrolyte solution. Of the ions listed in Table 8.3, potassium and chloride ions are *actively* conducted the most readily across a cell membrane and they distribute themselves not too far from what is termed a Donnan equilibrium. As described in Figure 8.4, this corresponds to the concentration product $[K^+]_{in}\cdot[Cl^-]_{in}$ inside the cell being approximately equal to the product $[K^+]_{out}\cdot[Cl^-]_{out}$ in the extracellular solution. This distribution of ions across the membrane is shown in Figure 8.4. The most concentrated free inorganic ion in the cytoplasm is K^+, which is typically 10~30 times as concentrated in the cytosol as in the extracellular fluid. The internal concentrations of free Na^+ and Cl^- are typically ten times or so less than their external concentrations. These

Table 8.3 Mammalian muscle cells typically have the following extracellular and internal (cytosol) concentrations of elemental ions and macromolecular anions (A^-).

Extracellular ion concentration (mM)	Cytosol ion concentration (mM)
Na^+ : 120; K^+: 2.5; Ca^{2+}: 2.0; Cl^-: 120	Na^+: 10; K^+: 140; Ca^{2+}: $<10^{-6}$; Cl^-: 3~4; A^-: 140

Internal	External
$[K^+][Cl^-]$: ~490 $[Na^+][Cl^-]$: ~35 $[Ca^{2+}][Cl^-]$: ~10^{-6}	$[K^+][Cl^-]$: ~300 $[Na^+][Cl^-]$: ~10^4 $[Ca^{2+}][Cl^-]$: ~240

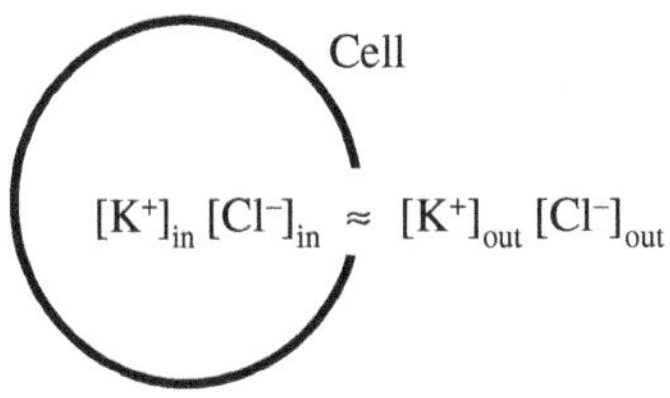

Figure 8.4 The ionic content (mM) of biological tissues is such that the potassium and chloride ions are the most concentrated and most permeable through the cytoplasm membrane. They distribute themselves close to that for an ideal Donnan equilibrium, which corresponds to the concentration product $[K^+]_{in}\cdot[Cl^-]_{in}$ inside the cell being close to the extracellular product $[K^+]_{out}\cdot[Cl^-]_{out}$.

asymmetric distributions of ions gives rise to a potential difference across the membrane, ranging from 50~90 mV, with the extracellular medium being taken as the zero reference and the cytoplasm having the negative potential. Maintenance of this potential and the ion concentration gradients requires *active* transport of ions across the membrane. An important factor in this process is the Na^+-K^+ pump, which transports Na^+ ions out of a cell and K^+ into a cell, against their electrochemical gradients. These pumps use up free energy, derived from glucose. (The chemical bond energy released by the oxidation of glucose in the cytoplasm of cells is transformed to a high energy phosphate bond in ATP. This stored energy is then used by the cell to perform various metabolic functions.) In order to maintain a cell in a 'comfortable' physiological state during an experiment it is therefore important to supply it with glucose as a chemical source of free energy.

Another important way to keep a cell 'comfortable' is to suspend it in a solution, which helps to minimize osmotic stress. Osmosis is the term given for the diffusion of water down its concentration (activity) gradient through a porous material such as a membrane. The cytoplasm of a typical cell is roughly equivalent to a 145 mM KCl solution. If we suspend a cell in a 100 mM KCl solution, water will flow down its activity gradient and into the cytoplasm, causing the cell to swell and possibly also to burst. We say that the cell is suspended in a *hypotonic* solution. Conversely, suspending the cell in a *hypertonic* solution of 200 mM KCl will cause it to shrivel, because water will flow from the cytoplasm and across the membrane to the outside. If a cell is suspended in a 145 mM KCl solution, it should neither shrink nor swell. We say that the solution is an *isotonic* one. These concepts only work if the KCl does not flow down its own concentration gradient so as to bring into equilibrium the internal and external KCL concentrations. What prevents this happening? Small molecules, such as oxygen, carbon dioxide and water are able to pass freely across the membrane, but the passage of larger molecules such as amino acids and sugars is carefully regulated. KCl is a small molecule too, but, importantly, in aqueous solution it dissociates into K^+ and Cl^- ions. As explained in Box 8.1, ions do not *passively* diffuse through a cell membrane to any significant extent. They are confronted by a huge potential energy barrier to do so and can only cross the membrane by active transport through an ion channel or pump, such as the Na^+-K^+ pump.

To understand the concept of an isotonic solution, we should think in terms of the osmolarity – also known as the osmotic concentration – of a solution. The osmolarity of a solution is a count of the number of particles dissolved in it, expressed as osmoles per litre. A 145 mM KCl molecule dissociates into two equal numbers of K^+ and Cl^- ions, so that the cytosol of a cell typically has an osmolarity of 290 mOsm/L. An osmolarity difference across the membrane equates to an activity gradient of water across it, assuming that all the osmotic solute is completely *impermeant*. Urea (CH_4N_2O) is a small molecule (molecular mass 60 g/mol) that does not dissociate in water – when dissolved it remains as a single molecule. Thus, a 290 mM solution of urea has the same osmolarity as an ideal 145 mM solution of KCl. However, suspending a cell in a 290 mM solution of urea would destroy the cell. It would rapidly swell and burst, because the small urea molecules are freely permeable through the cell membrane. With the passage of urea down its concentration gradient into the cytoplasm, the activity of the extracellular water increases. Water will now flow down its own activity gradient into the cytoplasm. If they can freely diffuse through the cytoplasm membrane, adding small, uncharged, molecules to a cell suspending medium does nothing to control the tonicity of the solution. What happens if we add glucose (molecular mass 180 g/mol) to the solution? Glucose dissolves readily in water, without dissociating or ionizing and, most importantly does not diffuse readily through a cell membrane. A solution composed of 10 mM glucose and 140 mM KCl thus has an osmotic concentration equivalent to 145 mM KCl. A cell suspended in this solution would not only have a chemical energy source, it would also neither swell nor shrink. In practice, we may wish to lower the electrical conductivity and alter the pH of the suspending medium, by adding HCl and changing the relative concentrations of KCl and glucose, for example. It is therefore of value to have an appreciation of how dissolved salts and sugars influence the dielectric properties of water and, in particular, how this may alter the value of the parameter ε_s^* in the Clausius–Mossotti factor.

Box 8.1 Cell Membranes Present an Impermeable Barrier to Passive Ion Flow

The figure in this box depicts the energy barrier to be overcome in moving a charged particle from water into a lipid membrane.

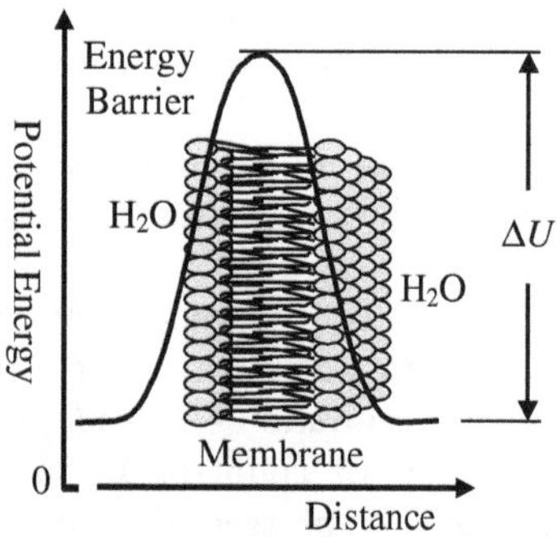

Based on the *Work-Energy Theorem* (see Box 4.1) the height ΔU of this barrier is the difference in the work required to assemble the charge in these two environments. From Equation (4.10), the potential energy change as an increment of charge δq is brought to a sphere of radius r and of charge q is given by:

$$\partial U = \frac{q\partial q}{4\pi\varepsilon_o\varepsilon_r}\frac{1}{r}$$

Thus, the total work required to proceed from zero charge to a final charge Q is:

$$U = \frac{1}{4\pi\varepsilon_o\varepsilon_r r}\int_0^Q q\partial q = \frac{Q^2}{8\pi\varepsilon_o\varepsilon_r r}$$

The value to be used for r is the ionic radius, which for Na^+ is 0.095 nm. Let $\varepsilon_r = 2.35$ for the membrane interior, an estimate based on values of 2.31 and 2.42 for stearic acid and palmitic acid, respectively (*CRC Handbook of Chemistry and Physics*, 2001). For water, assume $\varepsilon_r = 80$, so that:

$$\Delta U = U_{(\varepsilon r=2.35)} - U_{(\varepsilon r=80)} = (5.2\times10^{-19}\ \text{J}) - (1.5\times10^{-20}\ \text{J}) \approx 5\times10^{-19}\ \text{J}$$

This represents an energy barrier equivalent to 122 kT at 298 K (25 °C), nearly 100 times larger than the thermodynamic energy (3 kT/2) available to an ion in equilibrium with its environment at temperature T. We have ignored entropy changes associated with the ordering of water dipoles around a solvated ion. Accounting for this, by taking $\varepsilon_r = 35$, only slightly reduces the value of ΔU to 4.9×10^{-19} J (118 kT at 298 K). For our purposes, therefore, we can assume that a cell membrane (represented as a lipid bilayer in the figure above) presents an impermeable barrier to *passive* ion flow.

Example 8.3 Composition of an Electrolyte for a Dielectrophoresis Experiment

An investigation of the dielectrophoretic characteristics of a mammalian cell culture requires that the cells are suspended at 20 °C in an isotonic KCl solution (equivalent to 290 mOsm/L) that contains 2 mM glucose and 0.5 mM $CaCl_2$. An aqueous 0.5 mM $CaCl_2$ solution has a conductivity of 9 mS/m. The electrical conductivity of the complete aqueous cell suspending medium should be close to 60 mS/m.

1. What concentration of mannitol should be added to an initial solution, composed of 2 mM glucose, 0.5 mM $CaCl_2$ plus KCl, to give an isotonic solution of conductivity close to 60 mS/m.
2. Would it be a good idea to use urea instead of mannitol?

Solution 8.3

1. Mannitol and glucose are not electrically charged in solution and so will not contribute to the total conductivity. We therefore require enough KCl to raise the conductivity from 9 mS/m (provided by 0.5 mM $CaCl_2$) to 60 mS/m. An aqueous 4.2 mM KCl solution has a conductivity of ~51 mS/m (see Figure 8.5).

   The total osmotic concentration of the initial sugar plus salt solution is the sum of:
   - 4.2 mM KCl = 8.4 mOsm/L (dissociates into K^+ and Cl^-)
   - 0.5 mM $CaCl_2$ = 1.5 mM Osm/L (dissociates into Ca^+ and two Cl^- ions)
   - 2 mM glucose = 2 mOsm/L (glucose does not dissociate)

   This gives a total of 11.9 mOsm/L.

   We require a total osmotic concentration of 290 mOsm to be isotonic – and so need to add 290 – 11.9 = 278.1 mM of mannitol (like glucose, mannitol does not dissociate).
2. It would *not* be a good idea to use urea instead of mannitol. Urea is freely permeable through cell membranes and so cannot be used to control the tonicity of a cell suspension fluid. For example, a cell placed in 280 mM urea would rapidly swell as urea, followed by water, enter the cell down their activity gradients. Mannitol is not freely permeable through a cell membrane and can be used to control tonicity.

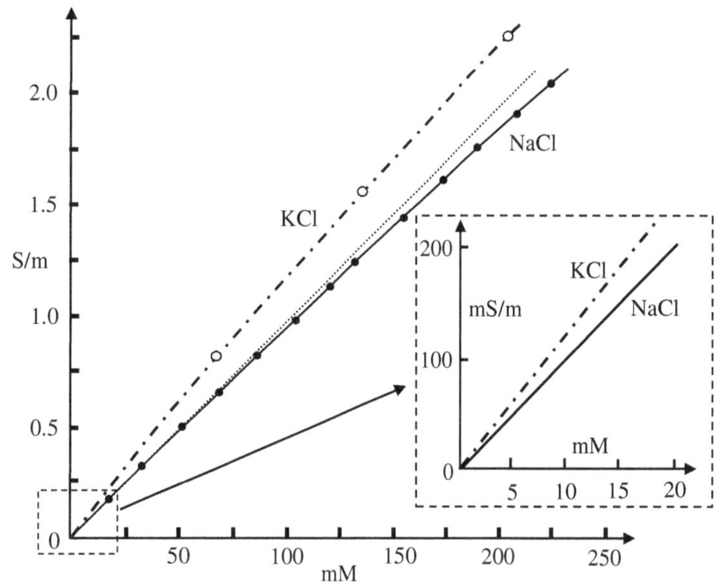

Figure 8.5 The conductivity at 20 °C of aqueous NaCl and KCl solutions is shown as a function of the dissolved salt concentration. For concentrations less than ~70 mM there is a linear relationship between conductivity and concentration. (Derived from the *CRC Handbook of Chemistry and Physics,* 87 edn, CRC Press, Boca Raton, FL, 2006–2007.)

In Example 8.3 mannitol is used to create osmotic equilibrium for a suspension of cells. Too much mannitol results in hypertonicity and the cells shrink as water flows from out of them down the water activity gradient. This effect is used in *osmotherapy* to reduce cerebral oedema by removing excess water from the intercellular and extracellular spaces in the brain. The normal osmotic concentration of blood plasma is around 280–290 mOsm/L. A hypertonic solution of mannitol (or a saline solution) can be injected into the carotid artery so as to increase the plasma osmotic concentration to around 300–320 mOsm/L. Water flows out of the brain into the blood stream in an attempt to maintain an osmotic equilibrium, decreasing the intracranial pressure. This same procedure can be used to weaken the blood-brain barrier formed by so-called tight junctions between the endothelial cells on the arterial wall. Diffusion of molecules in the blood cannot diffuse down their concentration gradient through this barrier – this can only be achieved by active transport of a selective group of molecules. However, by increasing the osmotic concentration of the blood the endothelial cells shrink and the tight junctions are stretched open. Drugs administered to the blood stream can then diffuse freely into the brain. In veterinary medicine a hypertonic solution is used to treat acute glaucoma. Serving the same function as the blood-brain barrier, there is a blood-ocular barrier. By lowering the water content of the vitreous humour, the intraocular pressure can be lowered.

8.3.1 Ions in Water

In an ionic crystal such as NaCl we can estimate, from the Van der Waals radii values given in Table 8.4, that the closest separation distance between a Na^+ and Cl^- ion is 0.4 nm. The potential energy U of attraction of this ion pair can be calculated using Equation (4.10):

$$U = \frac{1}{4\pi\varepsilon_o\varepsilon_r}\frac{Q_{Na^+}Q_{Cl^-}}{r_{(Na-Cl)}}$$

The relative permittivity of a sodium chloride crystal has a value $\varepsilon_r = \sim 6$. With this value inserted into the above equation, the potential energy of attraction of the Na^+-Cl^- ion pair is calculated to be $\sim 9.6 \times 10^{-20}$ J (i.e., 23 kT at room temperature). When a NaCl crystal is placed in water and water molecules diffuse into its bulk structure, the value of ε_r will increase and the energy of attraction will decrease. In the limit, as the value of ε_r tends to ~80, the energy of ion attraction approaches a value of 4×10^{-21} J (1.8 kT), which is only a little more than the energy of $3kT/2$ associated with thermal fluctuations. This energy of attraction is insufficient to result in stabilization of the ion pair and is the physical reason why salt crystals dissolve and dissociate in water.

Table 8.4 Atoms have a characteristic size known as their Van der Waals radius. The values given below were determined from the mechanical properties of gases and X-ray determination of atomic spacing between unbonded atoms in crystals (derived from [31]).

Atom	Radius (nm)	Atom	Radius (nm)
Hydrogen	0.12	Oxygen	0.15
Carbon	0.17	Phosphorus	0.18
Chlorine	0.17	Potassium	0.28
Nitrogen	0.16	Sodium	0.23
Magnesium	0.17	Sulfur	0.18

To understand this dissociation process more clearly we can find the relative contributions of the enthalpy and entropy changes by recognizing that the energy of attraction U is equivalent to the Gibbs free-energy of the system of the ions and their surrounding medium. From the Second Law of Thermodynamics, contributions to the free energy of the system (including a system of charges) in thermal equilibrium with its surroundings, at constant temperature and pressure, will come from its enthalpy H and entropy S according to the equation given in Box 7.4:

$$U = H - TS \tag{8.6}$$

If the enthalpy is relatively insensitive to changes of temperature, the entropy term can be approximated as $S = -\partial U/\partial T$, which from equation (4.10) gives:

$$\begin{aligned} S &= -\frac{\partial}{\partial T}\left(\frac{Q_{Na^+}\,Q_{Cl^-}}{4\pi\varepsilon_o\varepsilon_r}\frac{1}{r_{(Na-Cl)}}\right) \\ &= \frac{Q_{Na^+}\,Q_{Cl^-}}{4\pi\varepsilon_o\varepsilon_s^2\,r_{(Na-Cl)}}\frac{\partial\varepsilon_r}{\partial T} = U\frac{1}{\varepsilon_s}\frac{\partial\varepsilon_s}{\partial T} \end{aligned} \tag{8.7}$$

From Table 8.1 it is evident that Equation (8.2) provides an accurate value for ε_s over a wide temperature range and so, in the above equation, we can replace the differential with $\partial\varepsilon_s/\partial T = -b.\varepsilon_s(T)$. From Equation (8.7), with $b = 0.0046$, this gives $S = -4.6 \times 10^{-3}U$. The value of TS in Equation (8.6) at 298 K is thus $-1.37U$, which is larger than the total free energy U. This implies that the attraction of sodium and chloride ions is mainly driven by entropy and not enthalpy. Work has been expanded on the system of ions and water so as to create a more ordered system. The ions themselves are only weakly associated and so this increase in order must be associated with how the water molecules interact with the ions. This interaction involves the torques induced on the water molecules by the interactions of their dipole moments with the electric fields around the charged ions. This restricts the rotational mobility of the water molecules and creates a hydration 'shell' of oriented water molecules around an ion, as schematically depicted in Figure 8.6. The reduction of orientation mobility and dipole alignments creates the increase of order responsible for the negative value we have deduced for ST in Equation (8.7). It also results in a reduction of the local value of ε_s.

A positively charged ion orientates a neighbouring water molecule such that the negative component of its dipole is directed towards the ion. This will serve to screen the ion's positive charge and to reduce its coulomb potential. Water molecules will orientate in the opposite sense around a negatively charged ion and so also act to screen this charge. This screening is also enhanced by the tendency on average of neighbouring ions to overcome thermal vibrations and to be attracted to a countercharged ion. So, although the interaction between counter ions is sufficiently weak for their salt to dissociate and dissolve in water, the balance between electrostatic forces and thermal agitation is such that on average ions with the same charge will tend to avoid each other and those of opposite charge to spend more time near each other. This weak association of counter ions will increase as their average separation distance decreases (i.e., as the salt concentration increases). The electrolyte therefore behaves as if it were not 100% dissociated and this is reflected in the value of its activity coefficient. For example, from Table 8.5 we can deduce that a 100 mM solution of KCl will exhibit an osmolarity of 180, rather than the value of 200 mOsm if it were to act as an *ideal* ionic solution. This effect is responsible for the deviation from a linear relationship of the conductivity and salt concentration shown in Figure 8.5.

The influence of dissolved ions on the dielectric properties of water can thus be considered to result from the combination of a volumetric effect in replacing polar water molecules by nonpolar ionic particles, together with the reduced orientation mobility of water molecules in the hydration shells around the ions. The relative permittivity of an ionic aqueous solution is thus expected to fall with increasing concentration of a dissolved salt.

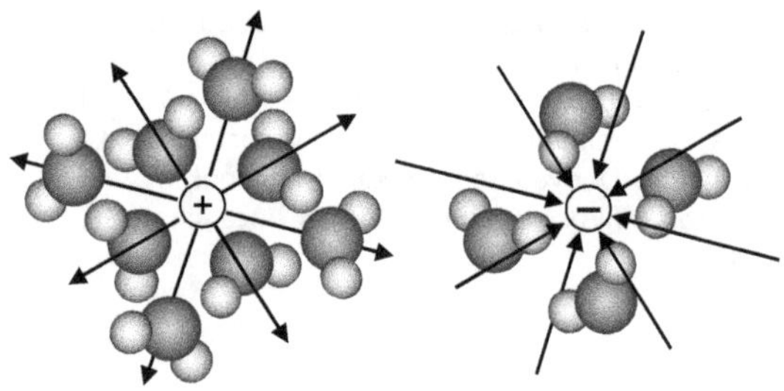

Figure 8.6 The electrostatic interactions between (left) a cation and (right) an anion with surrounding water molecule dipoles results in a 'structured hydration shell' and reduced rotational freedom of neighbouring water molecules. This represents a significant reduction of entropy of each ion-water system.

Table 8.5 Activity coefficient (defined as the ratio of the activity divided by the molal concentration) values as a function of concentration at 25 °C for some common compounds that dissociate into ions in aqueous solution. (Derived from the *CRC Handbook of Chemistry and Physics*, 87 edn, 2006–2007.)

Substance	0.01 M	0.05 M	0.1 M	0.5 M	1 M
KCl	0.901	0.816	0.768	0.649	0.604
NaCl	0.903	0.822	0.779	0.681	0.657
$MgCl_2$	0.734	0.590	0.535	0.485	0.577
$CaCl_2$	0.727	0.577	0.528	0.444	0.495
HCl	0.905	0.832	0.797	0.759	0.811

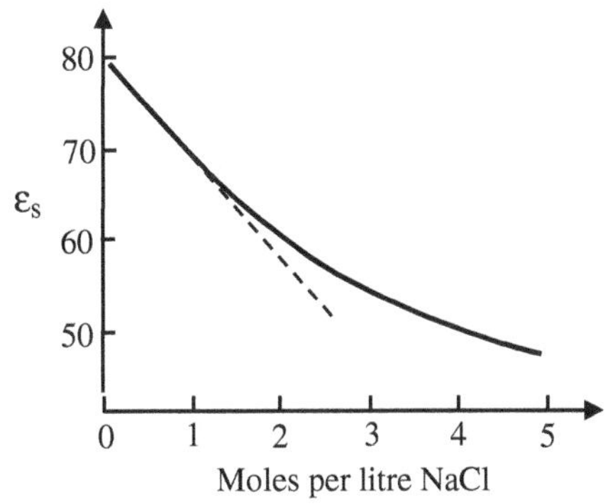

Figure 8.7 The value of the static relative permittivity ε_s of water falls with increasing concentration of dissolved NaCl (based on Hasted *et al.* [32]).

This is shown schematically in Figure 8.7 for the case of an aqueous solution of sodium chloride. It can be seen from this figure that the static relative permittivity of a NaCl solution falls from a value ~79 to less than 50 as the concentration of dissolved NaCl is increased from zero to 5 M. However, for most dielectrophoretic measurements on suspended cells we are only interested in aqueous salt solutions of concentrations less than ~150 mM. As indicated in Figure 8.7 and found experimentally by Hasted *et al.* [32], for salt concentrations less than 0.5 M a linear relationship exists between the static relative permittivity ε_s and concentration of the form:

$$\varepsilon_s = \varepsilon_w + 2\bar{\delta}c \tag{8.8}$$

where ε_w is the static relative permittivity value for pure water, c is the salt concentration in moles per litre. The factor $\bar{\delta}$ quantifies the extent to which the salt modifies the permittivity value and is given by:

$$\bar{\delta} = \frac{\delta^+ + \delta^-}{2} \tag{8.9}$$

with δ^+ and δ^- being the contributions arising from the cation and anion, respectively. Values of $\bar{\delta}$ for some salts in aqueous solution are given in Table 8.6. We can see that $\bar{\delta}$ has *negative* values, indicating that the addition of a salt to water *lowers* its permittivity. This means

Table 8.6 Values of the dielectric augmentation factor $\bar{\delta}$ in Equation (8.7) for some salts in water at 22 °C [32]. The negative values indicate that $\bar{\delta}$ acts as a *decrement*.

Salt	$\bar{\delta}$ (±1)
KCl	−5
NaCl	−5.5
LiCl	−7
HCl	−10
NaOH	−10.5
$MgCl_2$	−15

Table 8.7 Values of δ^+ and δ^- in Equation (8.8) for ions in water at 22 °C [32].

Cation	δ^+ (±1)	Anion	δ^- (±1)
Na^+	−8	Cl^-	−3
K^+	−8	F^-	−5
Li^+	−11	I^-	−7
H^+, H_3O^+	−17	SO_4^{2-}	−7
Mg^{2+}	−24	OH^-	−13

that the volume occupied by the ion and its surrounding hydration has a lower polarizability than the volume of bulk water it has displaced. In this situation the factor $\bar{\delta}$ is termed as the dielectric *decrement.* Water molecules attracted to a negative ion are less rotationally hindered than those attracted to a positive ion and so the reduction of the permittivity is much larger for cations than anions. The decrement values given in Table 8.7 are based on the assumption that the value for $\bar{\delta}$ for NaCl can be apportioned as $\delta^+ = -8$ and $\delta^- = -3$ [32]. A review of the status of our understanding up to 2001 of the physico-chemical factors that influence the dielectric decrements exhibited by aqueous electrolyte solutions has been given by Buchner and Barthel [33]. This is an ongoing research activity, with unsettled problems remaining to be solved.

Example 8.4 Relative Permittivity of KCl Solution
Derive an estimate of the room temperature relative permittivity of an aqueous 150 mM KCl solution.

Solution 8.4 From Tables 8.1 and 8.6 we obtain the values $\bar{\delta} = -5$ and $\varepsilon_w = 80.2$, respectively. The value of ε_s is calculated from Equation (8.8):

$$\varepsilon_s = \varepsilon_w + 2\bar{\delta}c = 80.2 + 2(-5 \times 0.15) = 80.2 - 1.5 = 78.7$$

As well as the effect of lowering the relative permittivity of the aqueous solvent, dissolved ions generally *decrease* the orientation relaxation time. The exception to this rule is the hydronium ion or proton, where the relaxation time *increases.* To a first approximation, this may be considered to result from the disruption by the solvated ions of the normal hydrogen-bond structure of pure water. For concentrations of dissolved ions less than 1 molar, this effect can be expressed in terms of a relaxation frequency increment δf:

$$f = f_w + c\delta f$$

where δf is the sum of the increments per mole resulting from the cation and anion, respectively. For Na^+, K^+ and Mg^{2+} the cationic increment is 0.44 (± 0.2) GHz; for Cl^- the anionic increment is 0.44 (± 0.2) GHz [34]. For H_3O^+

(H^+) instead of an increment of the relaxation frequency there is a decrement of −0.34 (± 0.2) GHz. It is doubtful that these small differences from the relaxation frequency of ~18 GHz exhibited by pure water will ever be of any practical relevance in dielectrophoresis experiments.

8.3.2 Aqueous Sugar Solutions

Sugars are carbohydrates, formed only of carbon, hydrogen and oxygen, with many of the simple kinds having the chemical formula $C_nH_{2n}O_n$ where n has values between 3 and 7. Exceptions to this formula include sucrose ($C_6H_{12}O_{11}$), mannitol ($C_6H_{14}O_6$) and deoxyribose ($C_5H_{10}O_4$). Sugar molecules are polar; they possess permanent dipole moments but do not ionize and so have neutral charge. They dissolve readily in water, not only because they are polar, but also because they can form H-bonds of the type OH····H with surrounding water molecules. The 'simple' sugars are monosaccharides, examples being glucose ($C_6H_{12}O_6$) – also known as dextrose – and fructose ($C_6H_{12}O_6$). Glucose and fructose have the same chemical formula, but glucose takes the form of a six-membered ring and fructose a five-membered ring. These molecules remain intact when dissolved in water and for 99.9% of the time also remain in their ring form. Mannitol has a linear structure. In their ring form monosaccharides can form glycoside bonds with other monosaccharides to create a disaccharide (such as sucrose, maltose and lactose). Raffinose is a trisaccharide formed by the glycoside bonding of galactose, glucose and fructose. Multiple glycoside bonding produces polysaccharides, such as starch and cellulose. When digested, saccharides are acted on by sucrase enzymes and broken down to their components. For example, sucrose breaks down to glucose and fructose.

The dielectric properties of aqueous solutions of galactose, glucose, mannose and ribose were studied over the frequency range 100 kHz to 10 GHz by a research group at the British-Dutch company, Unilever [34]. Dielectric dispersions were observed above 100 MHz, which when analysed (using the methods described in Figure 7.8) revealed three overlapping relaxation processes. For example, a 2.8 M glucose solution exhibited a broad dispersion, with the dielectric loss peaking at ~6 GHz. The dominant component of this dispersion exhibited a relaxation time of 1.85×10^{-11} s (corresponding to a loss peak at 8.6 GHz) and was considered to be the relaxation of the bulk water dipoles. A smaller contribution to the overall dispersion exhibited a relaxation time of 6.9×10^{-11} s (a loss peak at 2.3 GHz) and this was assigned to rotational relaxation of the glucose molecules. A much weak dispersion, with a relaxation time of $27 \pm 7 \times 10^{-11}$ s (a loss peak at ~0.59 GHz), was assigned to relaxations of the water of hydration around the glucose molecules. Theoretical modelling of the dielectric properties of dilute aqueous solutions of glucose and maltose reveal essentially the same phenomena, with the interesting finding that the dispersion associated with the hydration increases with the size of the sugar molecule and hence number of hydroxyl groups available to hydrogen bond to bulk water molecules [35]. From this work we can conclude that the dielectric properties of sugar solutions should be considered in terms of a three-component system, namely, bulk water molecules exhibiting their normal relaxation process, the bound hydration around each sugar molecule exhibiting a much slower relaxation process, plus the relaxing polar sugar molecules.

Dipole moment values of various sugar molecules and their influence on the dielectric properties of water were also evaluated by the research group at Unilever. This data was detailed in a company report and made available to Hasted for his book [3]. These results are presented in Table 8.8, together with those obtained by Saito *et al.* [36] and Arnold *et al.* [37]. The decrements were evaluated using a modified form of Equation (8.8):

$$\varepsilon_s = \varepsilon_w + \delta c \tag{8.10}$$

Defining the change in permittivity as $\Delta\varepsilon = \varepsilon_w - \varepsilon_s$, then $\delta = \Delta\varepsilon / c$. Hasted [3] and Arnold *et al.* [37] measured the sugar concentrations in moles per litre, whereas Saito

Table 8.8 Values for δ in Equation (8.9), based on molar concentrations of the sugar and dipole moments for aqueous sugar solutions at 25 °C.

Molecule	δ	Moment Debye units	Reference
Arabinose	–	4.3	[3, p. 197]
Fructose	−1.2[a]		[36]
Galactose	−3.28	5.3	[3, pp. 196–197]
Glucose	−4.27 −4.1[a]	4.7	[3, pp. 196–197] [36]
Mannose	−4.25	4.8	[3, pp. 196–197]
Mannitol	−2.48 −2.63	–	[3, p. 196] [37]
Myoinositol	–	5.0	[3, p. 197]
Raffinose	−8.62	–	[37]
Ribose	−2.72 −3.1[a]	5.1	[3, pp. 196–197] [36]
Sorbitol	−2.75 −2.51	–	[3, p. 196] [37]
Sucrose	−7.69 −7.5[a]	–	[37] [36]

Note: [a]Converted from data based on concentrations cited as molar fractions. The conversion procedure is described in Example 8.6.

et al. [36] employed *molar fractions*. The molar fraction (χ_i) is defined as the amount (n_i) of a constituent, given in moles, divided by the total amount (n_{tot}) of all the constituents of a solution. This is expressed by the relationship:

$$\chi_i = \frac{n_i}{n_{tot}} \tag{8.11}$$

Example 8.5 Preparing a 0.02 Molar Fraction of Glucose in Water

10 g of glucose (mol. wt. = 180 g/mol) is to be dissolved in water to obtain a 0.02 molar fraction of glucose in solution. What volume of pure water (mol. wt. = 18 g/mol) should be used to dissolve the glucose?

Solution 8.5 50 g of anhydrous glucose is equivalent to an amount of 50/180 = 0.278 mol. Equation (8.11) can be rearranged to give the required amount of water (n_w) in terms of the given quantities χ_i (0.02) and n_i (0.278 mol):

$$\chi_i = \frac{n_i}{n_{tot}} = \frac{n_i}{n_i + n_w}, \text{ so that } n_w = \frac{n_i(1-\chi_i)}{\chi_i} \tag{8.12}$$

The required amount of water is given as:

$$n_w = \frac{0.278\,(1 - 0.02)}{0.02} = 13.62 \text{ mol}$$

Now, 13.62 mol of water weighs 13.62 × 18 = 245.2 g. On the assumption of a density value of 1 g/mL for pure water, we require that the 50 g of anhydrous glucose be dissolved in 245.2 mL of pure water to obtain a 0.02 molar fraction of glucose in solution.

Example 8.6 Converting Dielectric Data from Molar Fractions to Molar Concentrations

The following data was obtained by Saito *et al.* [36] for the low frequency relative permittivity of pure water and various sugar solutions. The corresponding dielectric decrements given in Table 8.8 have been converted to be consistent with the molar concentration basis used by Hasted [3] and Arnold *et al.* [37]. Check the accuracy of these conversions.

Substance	Molar fraction	ε_r	$\Delta\varepsilon$
Pure water	–	78.4	–
Fructose	0.05	74.7	−3.7
Glucose	0.01	76.1	−2.3
Ribose	0.015	75.8	−2.6
Sucrose	0.01	74.2	−4.2

Solution 8.6 The values of the dielectric decrements given in Table 8.8 are given by

$$\delta = \frac{\Delta\varepsilon}{c}$$

where c is the sugar concentration in moles per litre. To convert the data of Saito *et al.*, to be consistent with this definition of the dielectric decrement, the amount n_w in Equation (8.11) should be equivalent to 1 litre of pure water, namely 1000 gs of water, corresponding to 1000/18 = 55.6 mol. On this basis, we can obtain the value for δ as $\Delta\varepsilon / n_i$. To obtain n_i we rearrange Equation (8.12):

$$n_i = \frac{n_w \chi_i}{(1-\chi_i)} = \frac{55.6\chi_i}{(1-\chi_i)} \tag{8.13}$$

For the case of fructose, we obtain $n_i = 2.93$ and the value for δ as −3.7/2.93 = −1.2. This result is entered into the table below, together with those obtained for glucose, ribose and sucrose. There is close agreement between the converted decrement values obtained for each of these three sugars and the other results given in Table 8.8

Sugar	χ_i (mol frac.)	n_i (Eqn 8.12)	$\Delta\varepsilon$	δ ($\Delta\varepsilon/n_i$)
Fructose	0.05	2.93	−3.7	−1.2
Glucose	0.01	0.56	−2.3	−4.1
Ribose	0.015	0.85	−2.6	−3.1
Sucrose	0.01	0.56	−4.2	−7.5

The linear relationship given by Equation (8.10) holds up to solute concentrations of no more than around 0.5 moles per litre. Arnold *et al.* [37] have explored the nonlinearity at higher sugar concentrations and analysed their results in terms of the following relationship:

$$\varepsilon_s = \varepsilon_w + \delta_1 c + \delta_2 c^2 \tag{8.14}$$

The values for δ_1 obtained for mannitol, sorbitol and sucrose are given in Table 8.8. The values for δ_2 for these three sugars are +0.13, −0.14 and −0.19, respectively. Apart from their relevance to the dielectric properties of cell suspending media, practical applications of such data include the development of dielectric sensors for determining the sugar content of foodstuffs and to detect water adulteration of honey, for example [38].

8.4 Amino Acids and Proteins in Solution

8.4.1 Amino Acids and Polypeptides

The term *amino acid* could in principle be used to refer to any compound that contains amino (–NH_2) and acidic groups. The term is however normally restricted to α-amino acids that contain an ionizable carboxyl group

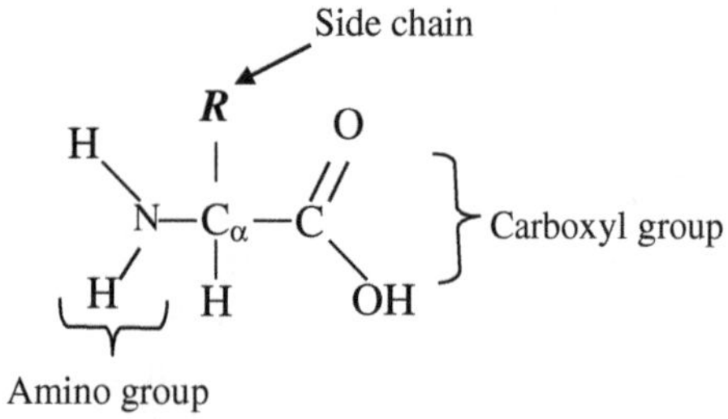

Figure 8.8 The chemical structure of an α-amino acid.

(–COOH) and can be isolated from natural sources. More than 100 have been isolated, but only 20 are commonly obtained when proteins are subjected to the hydrolytic action of boiling acid. The basic chemical structure of an α-amino acid (apart from proline) is shown in Figure 8.8. Proline possesses an *imino* group (-NH-) instead of an amino group.

The central carbon atom, called the alpha-carbon (C_α), is bonded to an amino (or for the case of proline an imino) group, a carboxyl group and a hydrogen atom. A variable chemical group *R* termed the *side chain* is also bonded to the C_α carbon. This side chain gives an amino acid its special characteristic. Glycine has the simplest side chain, namely a single hydrogen atom. This lends to glycine, with its two hydrogen atoms about the C_α carbon atom, the property of symmetry. The remaining amino acids do not possess such symmetry. They are chiral(from the Greek $\chi\varepsilon\iota\rho$ meaning hand) and so have two mirror-image (sterioisomeric) structures, designated as the right- (dextro) and left-handed (levo) form – the D and L forms, for short. Only the L forms of amino acids are found in protein molecules, but D-amino acids form part of bacterial cell walls and occur in some antibiotics. In accordance with the description of acids and bases given in Chapter 2 (Box 2.1) the acidic carboxyl group is ionized as $-COO^-$ and the basic groups are ionized as $-NH_3{}^+$ (or $=NH_2{}^+$ for the case of proline). The predominant forms of an amino acid across the pH range are shown in Figure 8.9. The doubly charged form, termed a dipolar ion or zwitterion, is predominant at neutral pH.

The chemical structures of the 20 common amino acids are given in Tables 8.9–8.11 and are classified according to whether their side chain is hydrophilic or hydrophobic.

Five of the side chains in the 20 common amino acids are ionizable and their p*K* values are given in Table 8.11. The main factors to consider when determining whether a side chain is hydrophobic or hydrophilic are:

- carbon and nonpolar groups do not readily hydrogen bond to water and are thus hydrophobic;
- oxygen and nitrogen can hydrogen bond to water and are thus hydrophilic;
- ionizable groups (e.g., $-COO^-$, $-NH_3{}^+$ or $=NH_2{}^+$) are hydrophilic;
- polar groups are hydrophilic.

The zwitterionic nature of amino acids has the consequence that their solvation by water is accompanied by a large negative change in volume, resulting from the strong electrostatic interaction between the polar water molecules and the two charged groups. Similarly, since the zwitterion represents a large dipole, neutral solutions of amino acids (which may have a negligible dc and low frequency conductivity) exhibit a high relative permittivity and absorb infrared radiation at 1580 cm^{-1}, an absorption band (in wavenumbers) characteristic of the carboxylate ion and not at 1720 cm^{-1} as would be the case for an uncharged –COOH group. The simplest α-amino acid is glycine, in which the side group R is hydrogen. The distance *d* between the centres of the positive ammonium group and the negative carboxyl group should be about 0.32 nm, so that the effective dipole moment should have a value given by

$$p = qd = (1.6 \times 10^{-19})\,(3.2 \times 10^{-10})$$
$$= 5.1 \times 10^{-29}\ \text{C m} = 15.3\ \text{debye units}$$

This value of the dipole moment compares reasonably with that of 20 debye units obtained by Wyman [39] from dielectric measurements on glycine solutions. Dunning and Shutt [40] showed further that the permittivity of glycine solutions is constant from pH 4.5 to pH 7.5 but falls sharply on either side of these pH values. The interpretation of this is that at the extremes of pH glycine possesses a single net charge only, so that the dipolar, zwitterionic, form disappears in strongly acidic or alkaline solutions.

The dipole moment per unit volume of a zwitterionic α-amino acid is larger than that of water, so that we can expect an amino acid solution to exhibit a greater static

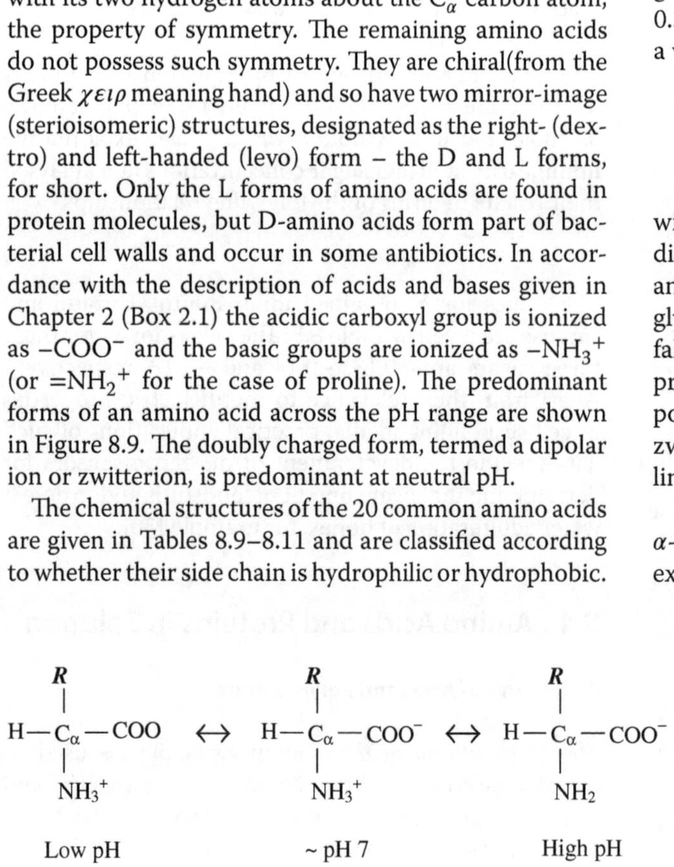

Figure 8.9 The ionic forms of an α-amino acid at various pH values. At neutral pH, amino acids exist predominately in the zwitterion (doubly ionized) form and exhibit a large dipole moment.

Table 8.9 Amino acids with hydrophobic (nonpolar) side chains *R*.

Amino acid	Side-chain structure *R*	Amino acid	Side-chain structure *R*
Alanine (Ala)	$-CH_3$	Isoleucine (Ile)	$-C(CH_3)(H)-CH_2-CH_3$
Leucine (Leu)	$-CH_2-CH(CH_3)(CH_3)$	Methionine (Met)	$-CH_2-CH_2-S-CH_3$
Phenylalanine (Phe)	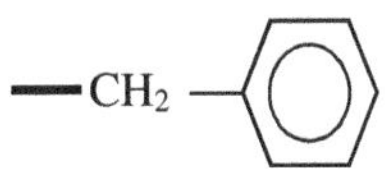	Proline (Pro)	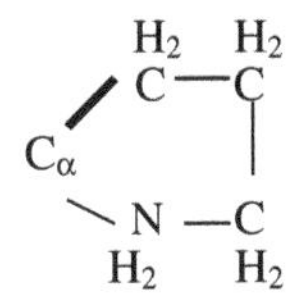
Tryptophan (Trp)	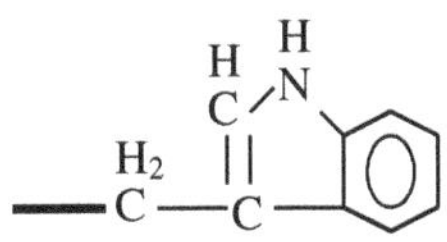	Valine (Val)	$-CH(CH_3)(CH_3)$

and low-frequency permittivity than water. This is illustrated for glycine in Figure 8.10. At room temperature, the characteristic frequency of the dielectric dispersion due to the rotation of glycine is 3.3 GHz [4], so that the dispersion overlaps with that for water shown in Figure 8.1, with a relaxation frequency ~18 GHz. The simple relaxation model given by Equation (7.20) predicts a relaxation frequency of ~12.6 GHz for glycine in water. The difference between the predicted and actual value indicates that there are significant electrostatic interactions between the amino acid and the water molecules. Qualitatively, it is not surprising the experimental finding gives a slower rotation of glycine than that calculated on the basis that only frictional forces, expected for a solid sphere rotating in a fluid and no molecular interactions occur between the solute and the solvent. Quantitatively, no straightforward theoretical treatment of the dielectric behaviour of an amino acid solution has been formulated. For practical purposes, however, it is possible to express the dielectric behaviour of amino acids in aqueous solution in the form

$$\varepsilon_s = \varepsilon_w + \delta c \tag{8.15}$$

Table 8.10 Amino acids with hydrophilic (uncharged, polar) side chains *R*.

Amino acid	Side-chain structure *R*	Amino acid	Side-chain structure *R*
Asparagine (Asn)	$-CH_2-C(=O)NH_2$	Cysteine (Cys) Glycine (Gly)	$-CH_2-SH$ $-H$
Glutamine (Gln)	$-CH_2-CH_2-C(=O)NH_2$	Serine (Ser)	$-CH_2OH$
Threonine (Thr)	$-C(OH)(H)-CH_3$	Tyrosine (Tyr)	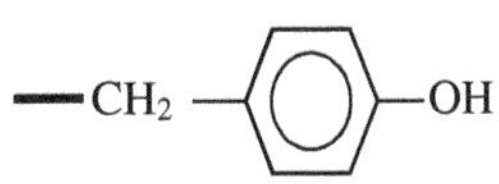

Table 8.11 Amino acids with hydrophilic (charged) side chains *R*.

Amino acid	Side-chain structure *R*	Amino acid	Side-chain structure *R*
	Positively charged (pH > pK)		Negatively charged (pH < pK)
Arginine (Arg) pK~12	— $(CH_2)_3$-NH-C(//NH_2^+)(\\NH_2)	Arginine (Arg) pK~7	— CH_2-COO^-
Histidine (His) pK~6.5	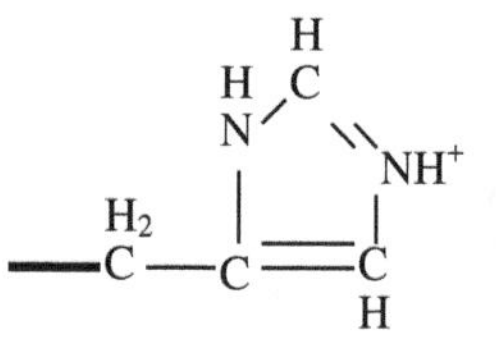	Glutamic acid (Glu) pK~4.7	— $CH_2-CH_2-COO^-$
Lysine (Lys) pK~10.5	— $(CH_2)_4-NH_3^+$		

where ε_s and ε_w are the permittivities of the solution and pure water, respectively and c is the concentration (mol/L) of the amino acid. For aqueous solutions of α-amino acids at 25 °C, δ has a value of some 26–28 permittivity units per mole for frequencies approaching 1 GHz and concentrations up to 2.5 M. Recalling from Example 8.6 that pure water has a molarity of 55.6, it is perhaps not surprising that the measured permittivity of amino acid solutions varies more or less linearly with the concentration of solute (i.e., that δ is constant), even for what are, from a biological standpoint, rather high concentrations of amino acid. These values of δ give dipole moment values for amino acid molecules that are some ten times that of an ordinary polar molecule [40,41]. This property led Kirkwood to remark that a dipolar ion is in a sense a superpolar molecule, surrounded by an intense electrostatic field [42].

The fact that amino acid solutions exhibit a dielectric *increment* at low frequencies indicates that the volume occupied by the relaxing molecule and its bound hydration exhibits a larger polarizability than the volume of bulk water it has displaced. For electrical frequencies higher than those at which the polar amino acid can respond by changing its orientation, we should expect to find a dielectric *decrement.* The volume occupied by the amino acid and its hydration has a smaller polarizability than an equivalent volume of normal bulk water. This effect is demonstrated by the dielectric properties of ε-aminocaproic acid (also known as 6-aminohexanoic acid) observed by Shepherd and Grant [43] and summarized in Figure 8.11.

From Figure 8.11 it is clear that the dielectric increment exhibited by ε-aminocaproic acid significantly exceeds the δ values of 26–28 permittivity units per mole found

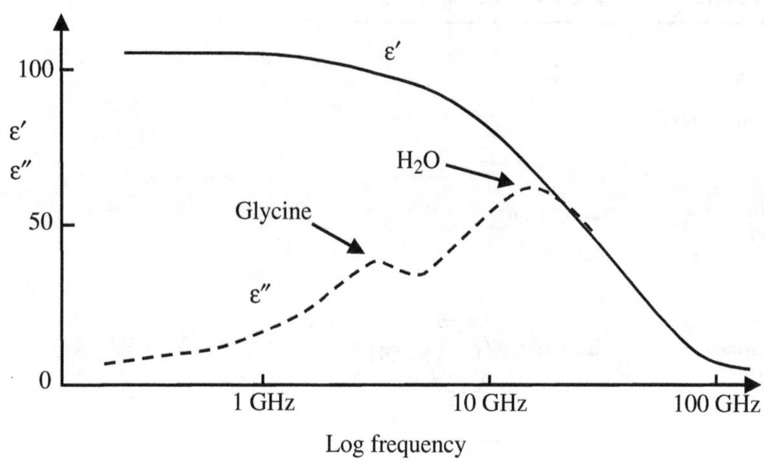

Figure 8.10 The dielectric dispersion exhibited by a 1 M aqueous solution of glycine at 20 °C. The low-frequency permittivity exceeds that of pure water (see Figure 8.1) and the ε'' loss peak at 3.3 GHz for glycine overlaps that at 18 GHz for the bulk water. (Based on [4], Figure 5.7.)

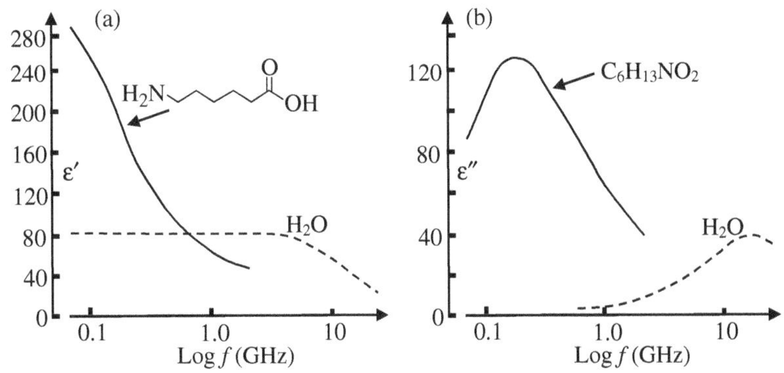

Figure 8.11 (a) A 3 M aqueous solution of ε-aminocaproic acid (chemical structure shown) exhibits a dielectric *increment* at low frequencies, but a dielectric *decrement* above 1 GHz. (b) The corresponding loss peaks for ε-aminocaproic acid (chemical formula shown) and the bulk water. (Based on measurements by Shepherd and Grant [43] at pH 7.6 and 20 °C).

for α-amino acids. This fact was first described by Wyman [44] in his dielectric experiments on various classes of amino acids. He demonstrated that not only the α-amino acids form a class having nearly similar dielectric increment values, but so also do the β- and γ-amino acids, for example. There is a regular increase of the increment δ with the number n of chemical bonds between the amino and carboxyl groups ($n = 2$ for α-amino acids, 3 for β-amino acids and so on). This is demonstrated in Figure 8.12(a), where the average values of δ for the classes of amino acids, as obtained by Wyman [44], are plotted against n, the number of chemical bonds between the amino and carboxyl groups. This demonstrates that δ increases linearly with n by a factor of about 14 for each additional chemical bond length. This behaviour is understandable in terms of these amino acids existing in aqueous solution predominantly in the form of dipolar ions. With increasing number of bonds (i.e., distance) between the charged amino and carboxyl end groups, the molecular dipole moment and hence dielectric increment δ increases in magnitude. The amino acid polypeptides can also be expected to form a homologous series and this is demonstrated in Figure 8.12(b) for the first seven peptides of glycine, where the values for δ increases linearly with the number of constituent chemical bonds n between the terminal NH_3^+ and COO^- groups.

A pH buffer commonly used to maintain physiological pH in cell culture is known as HEPES and can be added to cell suspending mediums for dielectrophoresis experiments. It is a sulfonic acid ($C_8H_{18}N_2O_4S$), which when dissolved in water becomes a zwitterion, with a measured dielectric increment value of 90 [45]. Another commonly used sulfonic acid buffer is MOPS ($C_7H_{15}NO_4S$), with a dielectric increment of 40 [45]. Concentrations of these buffers above 20 mM are not usually recommended for mammalian cell suspensions. According to Equation (8.14) we can therefore expect an increase of the permittivity to be no more than 1.8 and 0.8 permittivity units, respectively, when using HEPES and MOPS, respectively.

8.4.2 Proteins

A covalent chemical bond, known as a peptide bond, can be formed between two amino acids to form a dipeptide, involving the amino group of one amino acid and the carboxyl group of the other. As shown in Figure 8.13(a) a peptide bond leads to the release of a molecule of water in a so-called condensation reaction. Each amino acid is reduced in chemical structure to an amino acid *residue*. The carboxyl (C=O) and nitrogen atom forming the peptide bond between the two amino acid residues exhibit a resonating, partial double-bond character, as depicted in Figure 8.13(b).

Because of this resonance bonding, the six atoms of the peptide group (the two alpha carbons, the carbon, oxygen and nitrogen plus hydrogen) all lie in the same plane, as

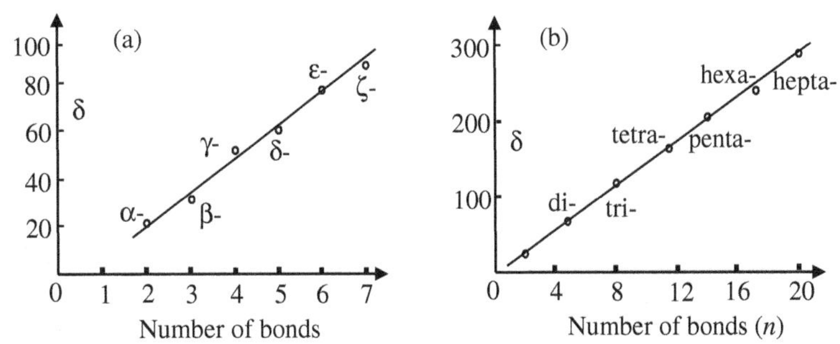

Figure 8.12 The variation of the dielectric increment δ in Equation (8.14) with the number of bonds n between the terminal charged groups of: (a) the various classes of amino acids in aqueous solution; (b) the glycine polypeptides (reproduced from Figures 3.4 and 3.5 in [5]).

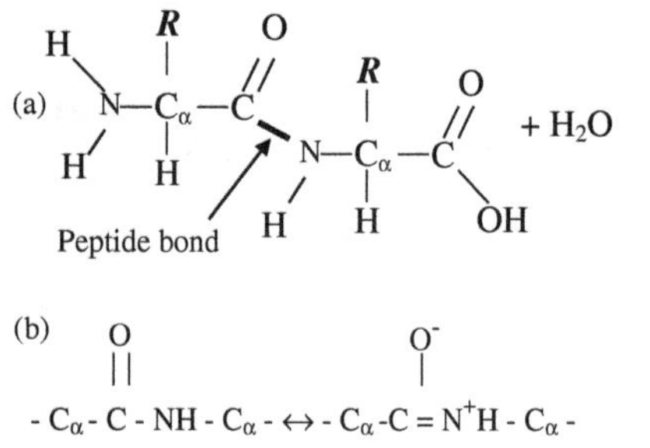

Figure 8.13 (a) The formation of a peptide bond between two amino acids, accompanied by the release of a water molecule. (b) The peptide bond has a partial, resonant, double bonding of the carbon and nitrogen atoms.

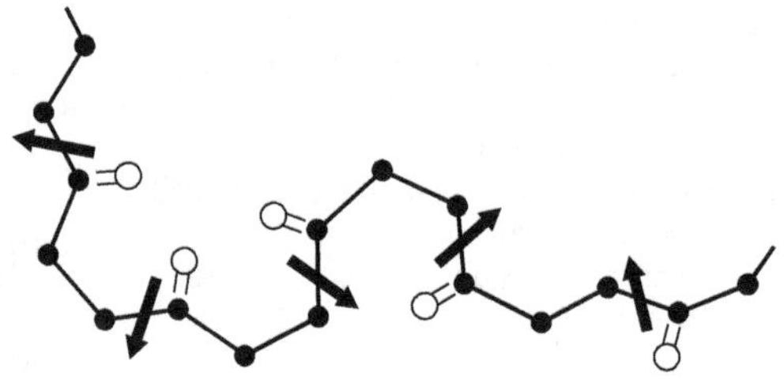

Figure 8.15 A schematic representation of a polypeptide chain as a string of connected dipole moments (arrows). Each peptide moment makes an angle of 46.7° to each C-N bond along the chain.

shown in Figure 8.14(a). Based on a simple application of molecular orbital theory, the peptide unit can be calculated to have a permanent dipole moment of 3.63 debye units directed from the oxygen to the carbon atom, at an angle of 46.7° to the C-N bond [5, Chapter 2]. This value was later revised down slightly to 3.5 debye units [46]. Independent rotation of two planar peptide groups about their connecting alpha carbon is possible. As shown in Figure 8.14(b) the relative conformation of a pair of planar peptide groups can be defined by two dihedral angles φ and ϕ, where φ is the angle of rotation about the C_α-C bond and ϕ is the angle of rotation about the N-C_α bond. Although angles φ and ϕ can theoretically both assume all the values from 0 to 360°, physically realizable conformations are in fact limited by restrictions on

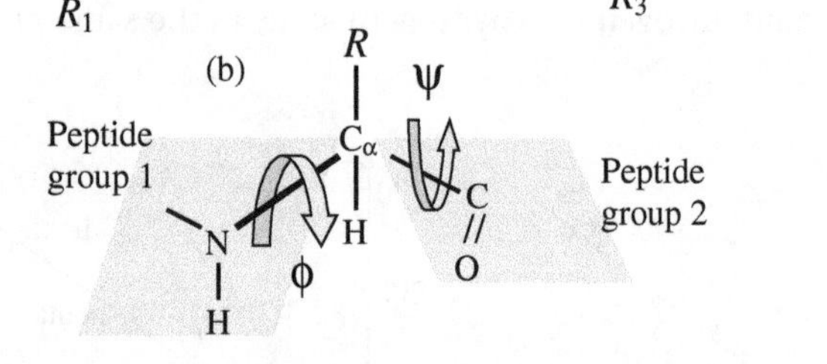

Figure 8.14 (a) The double-bond nature of the peptide bond results in the peptide group of atoms to lie in the same plane. (b) The relative orientation of two adjacent planar peptide groups is determined by the angles of rotation ϕ and φ about the connecting C_α atom.

the allowed van der Waals contact distances between different atoms in the dipeptide, especially for those atoms that are located in the side chain *R*.

Three amino acids can covalently bond together to form a tripeptide and when four do so a tetrapeptide is produced and so on to form resulting structures known as oligopeptides. Many amino acid residues bonded together form a long structure known as a polypeptide chain. With each peptide unit possessing a permanent dipole moment, polypeptide chains can be considered as strings of connected dipole moments, as shown in Figure 8.15. Proteins consist of one or more polypeptide chains. To perform their biological function (e.g., as an enzyme or a structural element such as a microfilament) proteins fold into one or more specific spatial conformations dictated by the sequence of residues in their polypeptide chains and the corresponding permitted values for the rotational angles ϕ and φ. Protein sizes range from a lower limit of around 50 to several thousand amino acid residues. An average protein contains around 300 residues. Very large aggregates can be formed from protein subunits, for example many thousand actin molecules assemble into a microfilament.

A protein's function is determined by its three-dimensional structure, which in turn is determined by the linear sequences of the amino acids in the one or more polypeptide chains from which it is composed. There are four distinct categories of a protein's structure:

- *Primary structure*: This is defined by the amino acid sequence of the polypeptide chains. A specific gene in a cell determines the primary structure of a protein. Basically, a specific sequence of nucleotides in DNA is transcribed into mRNA, which is then read by structures called ribosomes in a process that translates the mRNA code into a polypeptide chain. The sequence of a protein is unique to that protein and defines its structure and function. The primary structure is held together by the covalent peptide bonds made during the process of protein biosynthesis or translation by ribosomes. These peptide bonds provide rigidity to the

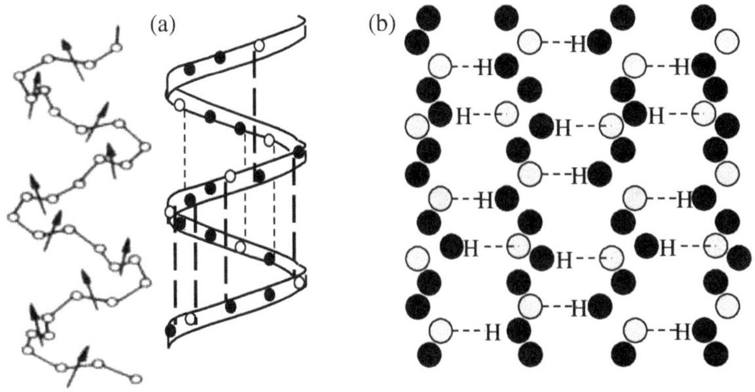

Figure 8.16 (a) The α-helix structure is stabilized by H-bonds (NH⋯O) parallel to the long axis of the helix. The axis of the helix is directed at an angle of about 56° to the C-N bond of each peptide residue, with each turn of the helix having a dipole moment of ~13 D directed parallel to this axis. (b) The β-sheet structure is stabilized by H-bonds between adjacent sections of a polypeptide chain.

protein. The primary structure can also be defined by the covalent bonding of sulphur atoms between two cysteine residues in the same or different polypeptide chains. These bonds are termed disulphide bridges.

- *Secondary structure*: This refers to the arrangement of parts of a polypeptide chain into highly regular sub-structures, the most prominent of which are the alpha helix and the beta-pleated sheet structures shown in Figure 8.16. Hydrogen bonds are responsible for stabilizing these two structures. The conformations of the amino acid residues in the alpha helix correspond to values for ϕ of −45° −50° and $\varphi = -60°$ and every turn of the helix includes 3.6 residues. Each residue participates in a hydrogen bond, so that every successive helix turn is held in place to an adjacent helix turn by three to four hydrogen bonds. With a dipole moment of ~3.5 debye units per residue, each turn of an α-helix has a dipole moment of around 13 debye units [5, Chapter 2]. The residues in a beta-pleated sheet structure have conformations with $\phi = -135°$ and $\varphi = +135°$ and is held together by hydrogen bonds. However, because water-amide hydrogen bonds are generally stronger than amide-amide hydrogen bonds, these secondary structures are stable only when the local concentration of water is sufficiently low, as for example in the fully folded protein state.
- *Tertiary structure*: This is the 3D structure of a single protein molecule, involving the spatial arrangement of the secondary structures, including the folding of parts of the polypeptide chain between α-helices and β-sheets. As depicted in Figure 8.17(a), it describes the completely folded and compacted polypeptide chain. Several polypeptide chains can be combined into a single protein molecule through ionic interactions (salt bridges) between oppositely charged ionized side chains, hydrogen bonds, hydrophobic 'bonding' interactions, disulphide bridges and intermolecular van der Waals forces between nonpolar groups. As a general rule, the hydrophilic (charged and polar) amino acid residues are located on the outside of a folded protein, with the hydrophobic residues buried inside the polypeptide structure. This is known as the 'oil drop' model for proteins and is driven by entropy (water molecules around a nonpolar surface group would be forced to form a cagelike structure having lower entropy than normal bulk water). A schematic of the cytochrome-c protein molecule is shown in Figure 8.17(b) to show the ionized chemical groups on the molecule's surface and the dipole moments associated with α-helices. This schematic is based on the X-ray crystallographic structure determined by Dickerson *et al.* [47] and shows what the authors term to be the back view of the molecule showing the two crevices that lead up to the haem group. From the locations of the charged acidic and basic groups on a protein's surface and using the method described in Example 5.6, it is possible to determine the permanent dipole moment associated with these charges [48–51]. In some proteins, the dipole moments of buried α-helices can facilitate the long-range transfer of electronic charges [52, 53].

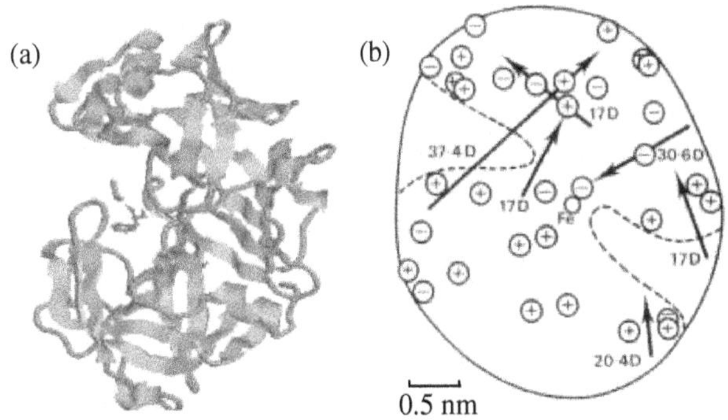

Figure 8.17 (a) The folding of a single polypeptide chain, to form the tertiary structure of a protein molecule, is stabilized by links between α-helices and β-sheets plus noncovalent interactions. (b) A schematic of the cytochrome-c molecule to show the locations of the charged side groups at pH7 and the permanent dipole moments of α-helices (reproduced from Figure 3.3 in [5]).

Table 8.12 The composition, molecular weight and isoelectric point (pI) of some proteins.

Protein	Number of Residues	Chains	Molecular weight	pI
Insulin	51	1	5808	5.4
Cytochrome-c	105	1	12 330	~10.0
Lysozyme	129	1	13 930	11.2
Myoglobin	153	1	16 890	~7.9
Pepsin	337	2	34 620	~1.0
Haemoglobin	574	4	64 500	7.1
Serum albumin	609	1	68 500	4.7
RNA polymerase II	4158	12	550 000	~7.0

- *Quaternary structure*: The forming of a complex of several protein molecules, or protein subunits, that function as part of a larger assembly or protein complex is referred to as a quaternary structure. A protein may shift between several, reversible, similar structures in performing its biological function, either as an enzyme controlling chemical reactions or as a structural element.

The numbers of constituent amino acid residues and polypeptide chains for several proteins are presented in Table 8.12. The isoelectric points of the proteins are also included in Table 8.12. Depending on their amino acid composition, different proteins have differing numbers of charged polar groups. For example (bovine) serum albumin has a net excess of acidic groups and so carries a net negative charge (equivalent to 19 electrons) at neutral pH 7. The isoelectric point (see Example 2.1) of this protein occurs at pH 4.7. Horse cytochrome-c, on the other hand, has an excess of basic groups and an isoelectric point close to pH 10. At neutral pH cytochrome-c carries a net positive charge and this is compatible with the distribution of charges shown on its surface in Figure 8.17(b). There must, however be a high degree of electrical symmetry of these surface charges, because as shown in Table 8.13 the dipole moments of small globular proteins are typically several hundred debye units. This corresponds to one and no more than two net electron charges displaced by a protein molecular diameter.

8.4.2.1 Analysis of Dielectric Dispersion

The dielectric properties of aqueous protein solutions have been studied since the 1930s and this work, up to 1942, was reviewed by Oncley [54]. As shown in Figure 8.18, a typical dispersion is observed in the MHz region (i.e., with a relaxation time of the order microseconds). The observed dielectric increment and dipole moment values were much larger than those found for amino acids

Table 8.13 The dipole moments (± ~5%) for various proteins ranked in order of increasing molecular weight.

Protein	Molecular weight	Dipole moment (D)	Reference
RNAse SA	10 500	440	[59]
Phospholipase	13 000	141	[50]
Cytochrome-c	13 000	235	[50]
Ribonuclease	13 700	350	[60][a]
		280	[50]
Lysozyme	14 300	122	[50]
Myoglobin	17 000	167	[61]
		150	[48][a]
Trypsin	23 000	271	[50]
Carboxypeptidase	34 000	637	[50]
Haemoglobin	64 000	495	[49]
		523	[61]
Serum albumin	66 000	710	[62]
Concanavaline	102 000	411	[50]

Note: [a] Dipole moment value calculated using Equation (8.24). All of the other dipole moments were calculated using Oncley's empirical equation, Equation (8.23).

and peptides and this was understood to reflect the proteins having larger molecular weights. Oncley assumed that the dispersions resulted from the Debye rotation model, where there is a competition between the electrical torque and viscous frictional force acting on the polar protein molecules. At low electrical frequencies the orienting torque acting on both the protein and water molecules is sufficient to overcome Brownian motions and frictional forces, so that the protein solution exhibits a permittivity larger than that for pure water. As the frequency is increased, to where the rate of reversal of the

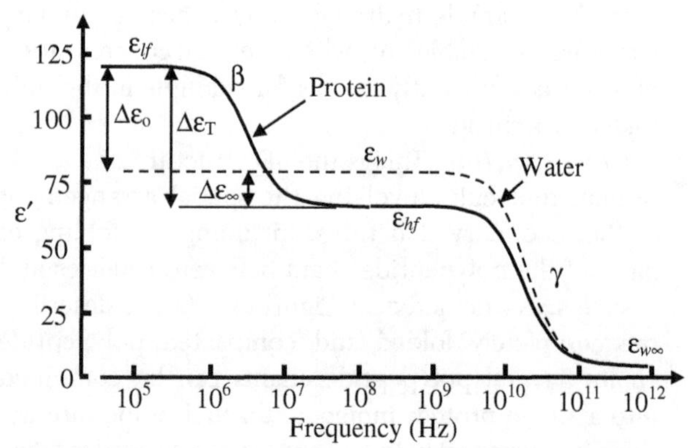

Figure 8.18 Typical dielectric dispersion exhibited by an aqueous protein solution. The dipole moment is calculated from the value of the total increment $\Delta\varepsilon_T$, equal to the sum of the increment $\Delta\varepsilon_o$ and decrement $\Delta\varepsilon_\infty$. In the literature the dispersions exhibited by the protein and water are often termed as the β-dispersion and γ-dispersion, respectively.

field approaches and then exceeds the fluctuation rate of the Brownian motions, the permittivity decreases. With a further increase of the frequency the dispersion of the solvent water molecules is observed. In the literature the dispersion exhibited by the polar protein molecule is often called the β-dispersion and the term γ-dispersion is used for the bulk water dispersion. Using the constants indicated in Figure 8.18, the β-dispersion obeys a Debye-type relaxation given by:

$$\varepsilon' = \varepsilon_{hf} + \frac{\Delta\varepsilon_T}{1 + \omega^2\tau_p^2} \tag{8.16}$$

where τ_p is the characteristic relaxation time of the protein molecule.

Analysis of the relaxation times can provide an indication of the size and shape of a protein molecule. By considering proteins to be ellipsoids of revolution, having semiaxes a, b, b, Oncley [54] assumed that there would be two characteristic relaxation times: τ_a for orientation involving rotation of the a-axis around the b-axis and τ_b for rotation of the b-axis about the a-axis. Using the results of Perrin's analysis [55] of the diffusion constants of ellipsoids in viscous media, Oncley derived these two relaxation times in terms of the ratio a/b of the semiaxes of the ellipsoid and the relaxation time τ_o of a sphere of the same volume, with τ_a given by:

$$\tau_o = \frac{4\pi\eta\, a\, b^2}{k\, T} \tag{8.17}$$

In his summary of the relaxation times of 19 amino acids and peptides, Oncley [54] found that the dielectric relaxation times for glycine and α-alanine were about 10% smaller than that predicted by Equation (8.17), but with increasing molecular size the relaxation times became increasingly larger than expected, but by never more than 50%. Whereas it might be expected that the amino acids and peptides molecules, being not much larger than the solvent water molecules, will not conform very closely to the Debye rotational model given by Equation (7.20) this should not be the case for the much larger protein molecules. Assuming that protein molecules in solution behave hydrodynamically as rigid ellipsoids of revolution generated by rotating an ellipse with semiaxes of length a and b about the a-axis and that the dipole moment is directed in a fixed direction with respect to these axes, the characteristic relaxation times are given from Perrin's work [55] as:

$$\tau_i = \frac{4\pi\eta ab^2}{kT} P_i(a/b) = \frac{3V\eta}{kT} P_i(a/b) \tag{8.18}$$

In this equation, the subscript i refers to the relaxation of either the a- or b-axis about the other, with V being the effective volume of the protein molecule. The two functions $P_i(a/b)$ depend only on the axial ratio a/b. Depending on whether ellipses are rotated about their minor or major axis, the resulting surfaces are either prolate or oblate. The situation $a > b$ corresponds to a prolate spheroid (similar to the shape of a rugby ball or American football), whereas $a < b$ corresponds to an oblate (squashed soccer ball shape) spheroid. As shown in Figure 8.19, two curves can be drawn of the possible combinations of axial ratio and the effective molecular volume V. These two curves correspond to the assumption that the measured relaxation time is associated with the rotation of one of the principal axes, with the molecular dipole moment pointing along the rotating axis. In general the dipole moment orientation does not coincide with the axis direction and a curve lying between the two shown in Figure 8.19 is the more appropriate, since the dipole moment will have a component along each axis.

An example of the application of Perrin's functions is the analysis of the dielectric data obtained for sperm whale myoglobin by South and Grant [48]. From measurements of the β-dispersion for myoglobin solutions as a function of protein concentration, fluid viscosity and temperature, a value of 30 ± 3 ns was obtained for the relaxation time τ_p in Equation (8.16) at 20 °C. According to Equation (8.17) this gives an effective molecular volume of 41 nm^3. However, from the X-ray diffraction analysis of myoglobin by Kendrew *et al.* [56] it is known that this protein is an oblate spheroid of molecular volume around 15 nm^3. From Figure 8.19(b) this implies an axial rati b/a of about 5. However, the X-ray data indicated an axial ratio of 2, which from Figure 8.19 gives the effective molecular volume to be around 35 nm^3. An increase in effective volume of ~20 nm^3 is compatible

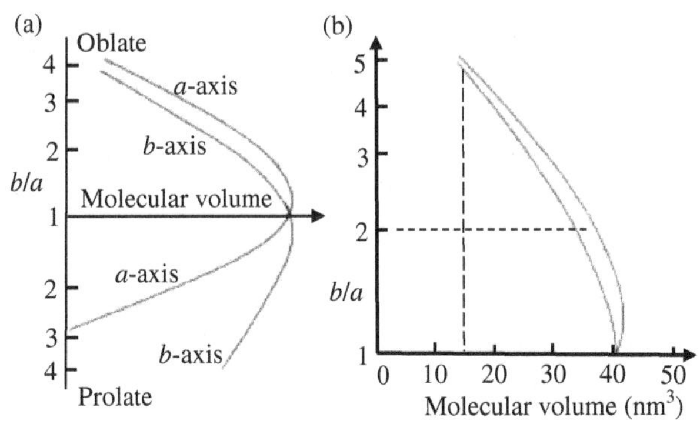

Figure 8.19 (a) Plots of the Perrin functions of axial ratio (b/a) against effective molecular volume for ellipsoids with rotations about either the a- or b-axis [55]. (b) The Perrin plot for whale myoglobin, which is an oblate spheroidal protein. The dashed lines indicate the molecular volume and axial ratio determined by X-ray diffraction [56]. An analysis of the β-dispersion gives a volume of 41 nm^3 [48].

with a hydration 'layer' of about two water molecules thickness, which rotates with the myoglobin molecule.

To derive the protein dipole moment from the observed dielectric increment $\Delta\varepsilon_T$ (see Figure 8.18) Oncley [54, 57] employed the following relationship (after conversion from esu to SI units):

$$\frac{bN_A p^2}{9000\varepsilon_o kT} = \Delta\varepsilon_T \frac{M}{c_g} \tag{8.19}$$

where c_g is the protein concentration in grams per litre and b is an empirical parameter taken to be equal to 5.8. This requires some explanation. We first note that the parameter N in Equation (6.34) in Box 6.7 is the number density per unit volume (cm^3) of polar molecules in a pure polar liquid and this is given by $N = N_A\rho/M$, where N_A is Avogadro's constant, with ρ the density (g/mL) of the pure polar liquid and M the molecular weight of the polar molecules. For Equation (8.19) we require the equivalent number density of protein molecules per cm^3 ($N_A c_g/1000M$), with M the molecular weight of the protein molecule and c_g the protein concentration in grams per litre. The introduction of the empirical parameter b is based on the examination by Wyman of dielectric measurements for 141 polar liquids, for which dipole moments had already been evaluated in their vapour state or as dilute solutions in nonpolar solvents [58]. Wyman [58] concluded that for these liquids an empirical relationship existed between their values of relative permittivity and *volume* polarization P_v of the form:

$$P_v = \frac{(\varepsilon_m + 1)}{A}$$

Expressed as a *molar* polarization, this relationship takes the form:

$$P_M = \frac{(\varepsilon_m + 1)}{A}\frac{M}{\rho} \tag{8.20}$$

For a large number of the liquids $A = \sim 8.5$, with upper and lower bounds of 11 and 6.2, respectively. This finding demonstrated conclusively the inadequacy of the Clausius–Mossotti factor to describe the molar polarization, as given by Equation (6.12). The empirical relationship of Equation (8.20) can be compared with the molar polarization derived theoretically by Onsager to take account of the reaction field created by induced polarization of the surrounding solvent dipoles:

$$P_M = \frac{(\varepsilon_m - \varepsilon_{m\infty})(2\varepsilon_m + \varepsilon_{m\infty})}{\varepsilon_m(\varepsilon_{m\infty} + 2)^2}\frac{M}{\rho} \tag{8.21}$$

For aqueous protein solutions, $\varepsilon_m \gg \varepsilon_{m\infty}$ and so from Equations (8.20) and (8.21) the following approximate relationship can be made:

$$\frac{\varepsilon_m}{A} = \frac{2\varepsilon_m}{(\varepsilon_{m\infty} + 2)^2} \text{ i.e., } A = \frac{(\varepsilon_{m\infty} + 2)^2}{2} \tag{8.22}$$

The typical value of $A = 8.5$ found by Wyman [58] corresponds to $\varepsilon_{m\infty} = 2.13$ (or a refractive index of 1.46). The upper and lower bounds for A of 11 and 6.2, corresponds to $\varepsilon_{m\infty}$ ranging between 1.63 and 2.69 (i.e., refractive indices ranging from 1.28 to 1.64). These results compare favourably with the value $\varepsilon_{m\infty} = 1.77$ (refractive index of 1.33) employed by Oster and Kirkwood [22] in evaluating the value for the correlation factor g. The factor b in Oncley's equation (8.19) thus represents an empirical way to correct for internal field effects.

As we noted in the derivation of Onsager's equation (6.35), the local field is a sum of the Lorentz cavity field and Onsager's reaction field. The reaction field arises from polarization of surrounding solvent molecules, which in turn polarizes the protein molecule and increases its dipole moment. Oncley found that the dielectric behaviour of carboxyhaemoglobin at the lower frequencies was similar to that of amino acids and peptides. For example, the low frequency dielectric increment (on a gram per mL basis) for the protein was found to be 0.33, close to that of 0.30 obtained for glycine. Oncley chose $b = 5.8$, as he states 'until more adequate theoretical grounds are available', because this value in Equation (8.19) resulted in a dipole moment value for glycine of 15 debye units. This is a value in close agreement with those obtained by methods other than dielectric spectroscopy (e.g., solvent action of neutral salts on glycine; freezing point measurements; calculation based on distance between the negative and positive charges). Oncley's measurements upon a series of solutions of crystallized horse carboxyhaemoglobin yielded consistent values of $\Delta\varepsilon_o = 0.33$, $\Delta\varepsilon_\infty = 0.11$ and a critical frequency of 1.9 MHz (i.e., $\tau = 8.4 \times 10^{-8}$ s). Based on a molecular weight value of 66 700 and application of the empirical Equation (8.19), this gave a dipole moment value of 480 debye units for that protein molecule. Oncley went on to establish the dipole moments for various other proteins: egg albumin (250 D); horse serum albumin (380 D); horse serum pseudoglobulin-γ (1100 D); edestin (1400 D) [54].

Dipole moment values for other proteins in aqueous solution are presented in Table 8.13. In this Table the order of the proteins is given according to molecular weight. There is no obvious correlation between the dipole moment value and molecular weight. This is in line with the finding of Barlow and Thornton who studied the distributions of charged groups in 32 proteins of known three-dimensional structure [63]. It was found that these proteins exhibited a wide variety of charge distributions, ranging from the highly symmetric to the highly asymmetric. In the majority of proteins that interact with a highly charged ligand, the charge distributions were relatively asymmetric and the dipoles acted in the right sense to aid ligand binding. The major contribution to protein dipoles was concluded to be from formally charged groups. The moments along the polypeptide backbone are only significant for the alpha- and beta-structures (see Figure 8.16), where the helix dipoles are aligned and tend to reduce the overall molecular dipole moment.

From Equation (8.19) Oncley's empirical equation gives the dipole moment p as:

$$p = \left(\frac{9000\varepsilon_o MkT}{bN_A} \frac{\Delta\varepsilon_T}{c_g} \right)^{1/2} \tag{8.23}$$

The calculations of the dipole moment values given in Table 8.13 were based on this relationship, with the empirical approximation that $b = 4.5$ (rather than 5.8). In some cases measurements of $\Delta\varepsilon_T$ are taken as a function of concentration c_g. If this does not produce a linear relationship, the value of the so-called *intrinsic* dielectric increment, $\Delta\varepsilon_T/c_g$, extrapolated to zero c_g is used in Equation (8.23) [e.g., 49, 50]. In another form of this relationship (taking $b = 4.5$) the dipole moment is calculated using the formula

$$p = \left(\frac{2\varepsilon_o MkT\delta}{N_A} \right)^{1/2} \tag{8.24}$$

where δ is the *specific* dielectric decrement, calculated from the slope of a linear plot of $\Delta\varepsilon_T$ versus protein concentration c measured in kg m^{-3} (mg/mL). Examples in Table 8.13 of this approach are the moment values calculated for myoglobin [48] and ribonuclease [60]. Apart from the empirical approximation that $b = 4.5$, Kirkwood's correlation factor g given in Equation (8.5) is assigned a value of unity. This is probably a good assumption, because local polarizations of the solvent water molecules are likely to cancel each other when averaged over the protein's relatively large surface area.

It should also be remarked that the theory devised by Buckingham [64] and described in Chapter 6, does not appear to have been applied to the calculation of the dipole moments of proteins. The Onsager–Kirkwood theory places the polar molecule, in the form of a point dipole rather than an assembly of charges, into a spherical cavity. A spherical cavity could be quite unsatisfactory for some of the oblate (or prolate) spheroidal proteins and the shape factors included in Buckingham's theory could provide a useful improvement on what has been applied so far to the calculation of protein dipole moments. Despite doubts that can be expressed regarding the accuracy of the moment values given in the literature, it is encouraging and also quite remarkable to find that good agreement exists between the experimental values and those computed from the known distributions of charges and bond moments in the protein structures. This fact is demonstrated in Table 8.14.

Table 8.14 Comparisons of the calculated with measured dipole moments for various proteins. The calculated value is the vectorial sum of the moment due to fixed charges at the isoelectric point and the core moment arising from the main polypeptide chain and side chains [50].

	Calculated moment (D) [50]			
Protein	**Charges**	**Core (θ)**	**Sum**	**Measured (D)**
Phospholipase	130	34 (104°)	125	141 [50]
Cytochrome-c	267	49 (138°)	233	235 [50]
Ribonuclease	295	19 (85°)	331	350 [59]
Lysozyme	127	61 (92°)	138	122 [50]
Myoglobin	247	57 (148°)	199	195 [60]
Trypsin	295	62 (95°)	296	271 [50]
Carboxypeptidase	473	152 (2°)	624	637 [50]
Concanavaline	432	22 (79°)	436	411 [50]

The calculations by Takashima and Asami [50] presented in Table 8.14 were performed at the isoelectric pH of each protein, where the effective positive and negative charges are equal. In addition to the dipole moment due to these fixed surface charges, the core protein moments were also calculated by vectorially summing main-chain and side-chain carbonyl moments. The peptide residue moments could not be used because X-ray analyses do not provide the coordinates of H atoms, making it impossible to find the direction of the NH bonds. The total computed moment values given in Table 8.14 were obtained by the vectorial sum of the fixed charge and core moments, using the calculated angle (θ) between the fixed charge and core dipole moments. The distance between the negative and positive charge centres were also computed and these were found to be very small compared to the diameter of the protein molecule. This demonstrates that the surface positive and negative charges are randomly distributed. For lysozyme, for example, the charge centre separations were found to be almost zero. It is also interesting to note from Table 8.14 that none of the core moments were found to be zero. This means, despite the presence of large segments of random coil configurations in these globular proteins, that the spatial orientations of the carbonyl bonds in the main chain and side chains maintain a certain level of regularity.

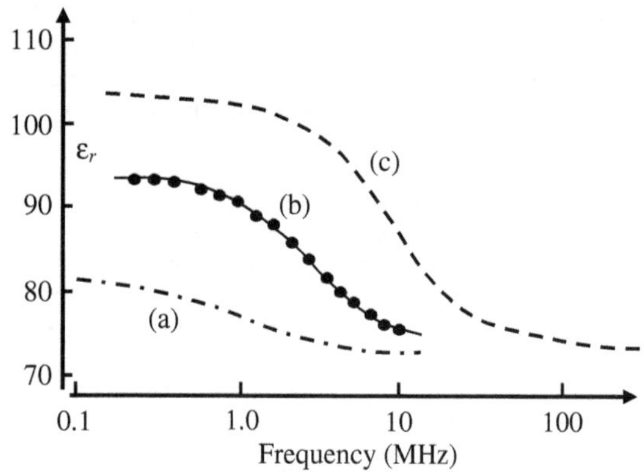

Figure 8.20 Dielectric dispersions exhibited by: (a) 20 mg/ml haemoglobin, $\Delta\varepsilon_T = 7.2$ [49], (b) 150 mg/ml myoglobin, $\Delta\varepsilon_T = 18$ [48], (c) 9.9 g/L cytochrome-c, $\Delta\varepsilon_T = 29.5$ [50].

Example 8.7 Calculating the Dipole Moment of a Protein from its β-Dispersion
The β-dispersions for three protein solutions are shown in Figure 8.20. Calculate the dipole moments of these proteins using the information provided.

Solution 8.7 Equation (8.24) can be given in the form:

$$p = \left(\frac{2\varepsilon_o kT}{N_A}\right)^{1/2} (M\delta)^{1/2} = A\,(M\delta)^{1/2}\ \text{C.m}$$

Using the following values: $\varepsilon_o = 8.854 \times 10^{-23}$ F m^{-1}, $k = 8.854 \times 10^{-23}$ J K^{-1}, $T = 298$ K, $N_A = 6.022 \times 10^{23}$, the parameter $A = 3.48 \times 10^{-28}$ C.m$^{-1/2}$. (The Tables in Appendices A and D may be used to check the veracity of these stated dimensions.) The information given in Figure 8.20, together with the gram molecular weights cited in Table 8.13, are tabulated below. To give the correct dimensions of dipole moment, the value for the increment δ are calculated in units of m^{3}g^{-1}.

Protein	M	$\Delta\varepsilon_T$	δ (m^{3}g^{-1})	$(M\delta)^{1/2}$ (m$^{3/2}$)	$A(M\delta)^{1/2}$ (C.m)	Moment (debye)
Haemoglobin	64 000	7.2	3.59×10^{-4}	1.43	1.67×10^{-27}	501
Myoglobin	17 000	18.0	1.20×10^{-4}	5.03	4.97×10^{-28}	149
Cytochrome-c	13 000	29.5	3.02×10^{-4}	1.98	6.90×10^{-28}	207

There is close agreement obtained above for the dipole moments of haemoglobin and myoglobin with the values cited in Table 8.13. However, there is a significant discrepancy between the value of 207 D given above for cytochrome-c and that (235 D) cited by Takashima and Asami [50]. The reason for this is that the value of the specific increment depended markedly on the protein concentration. When the measured increments were extrapolated to zero concentration, a value of 0.515 L g^{-1} (rather than 0.302 used above) was obtained for the intrinsic decrement [50].

Example 8.8 The Conductivity Increment accompanying the β-Dispersion for Cytochrome-c
The critical frequency (i.e., relaxation frequency) of the cytochrome-c dispersion shown in Figure 8.20 is 8 MHz. Calculate the magnitude of the conductivity change $\Delta\sigma$ over the frequency range that defines the β-dispersion for this protein.

Solution 8.8 The relationship between the permittivity increment and conductivity change is given by Equation (7.18):

$$\Delta\varepsilon' = \tau\Delta\sigma'$$

To match permittivity units (F m^{-1}) with conductivity units (S m^{-1}) this relationship takes the form:

$$\Delta\sigma' = (\varepsilon_o\Delta\varepsilon')/\tau = (8.854 \times 10^{-12}\Delta\varepsilon')2\pi f_{cr}$$

where f_{cr} is the point of inflection of the dispersion curve (i.e., the critical frequency at 8 MHz). Inserting the value $\Delta\varepsilon' = 29.5$ and $f_{cr} = 8$ MHz into this equation gives $\Delta\sigma' = 13.1$ mS m^{-1}. The reported conductivities of the cytochrome-c solutions were ~100 mS m^{-1} [50]. The conductivity increment related to the β-dispersion thus represents a 13% change of the protein solution conductivity, which is quite significant.

8.4.2.2 Protein Hydration

In one of the earliest reports of the physicochemical properties of a protein, Sorensen [65] asked 'Does crystallized egg-albumin contain water?' Following a detailed compositional analysis, he found that such samples contain about 0.22 g water per g water-free egg albumin. In later work, he gives the water content of crystallized haemoglobin as 0.35 g/g [66]. It is now known that protein molecules can have from 0.20 to 0.70 g strongly associated (bound) water per gram of protein. Through modern x-ray and neutron diffraction studies we have for some proteins a detailed knowledge of the location and bonding of much of this water. By using the latest high-resolution NMR techniques, we can even identify individual molecules of hydration water and characterize their binding sites on the protein molecule. Thus, in general and as a convenient method of classification, two kinds of water molecules associated with proteins have been identified: internal water and peripheral water. The internal water molecules, which form an integral part of a protein structure, diminish local charge-charge interactions and reduce destabilizing effects that arise from otherwise unbonded proton donors and acceptors [67]. As a rough gauge of the upper limit of internal water, we can consider the small enzyme pancreatic trypsin inhibitor (mol wt 6700), which contains four internal waters [67]. As this represents a hydration content of 1.07 wt%, we can conclude that most of the water 'bound' to proteins is associated with the protein surface. Parentheses have been placed around the word *bound* because this hydration content should not be thought of as a solid shell of water. These waters of hydration have residence times and exchange with bulk water in time scales typically ranging from tens of nanoseconds to microseconds, as well as being highly mobile in terms of reorientation and translational mobility. In what follows the terms 'bound water' and 'hydration shell' will sometimes be used, merely as a way to distinguish this component of water from the bulk water in aqueous protein solutions.

In a footnote to his paper, Oncley [57] suggests that the value of $\Delta\varepsilon_\infty$ might be used to calculate the amount of water associated with the protein molecule. This protein hydration or 'bound' water is commonly defined as the weight of water carried through the solution by unit weight of protein in sedimentation, diffusion and electrophoresis experiments. In our case we broaden this definition to include the amount of water that remains associated with a protein molecule during electric field induced rotations. Not all of this hydration should be considered to be *irrotationally* bound water – a proportion of it may be capable of hindered reorientations that contribute to the overall polarizability of a protein solution.

Oncley's logic was that the decrement $\Delta\varepsilon_\infty$ gives a measure of the extent to which at high frequencies the protein molecule and its associated water behaves as a particle of low permittivity. The volume occupied by the protein and its associated hydration has a lower polarizability than the bulk water it has displaced and so the permittivity $\varepsilon_{\rm hf}$ at high frequencies, leading up to the γ-dispersion, is therefore proportional to the remaining water. Defining v as the volume of water displaced by one gram of anhydrous protein, with $\varepsilon_{\rm w}$ and $\varepsilon_{w\infty}$ defining the dispersion of normal bulk water as shown in Figure 8.18, we have the following relationships

$$(\varepsilon_{\rm hf} - \varepsilon_{w\infty}) = (\varepsilon_{\rm w} - \varepsilon_{w\infty})(1 - vc)$$
$$\Delta\varepsilon_\infty / c = (\varepsilon_{\rm w} - \varepsilon_{w\infty})v$$

where c is the concentration of the protein in grams per litre. The partial specific volume v_{sp} of the anhydrous protein can also be defined as:

$$v_{sp} = v - w/\rho_w$$

where w is the gram mass of water bound to the protein and ρ_w is the density of the water. Using his measured value of $\Delta\varepsilon_\infty = 0.11$ per gram of protein, with $v_p = 0.75$, Oncley obtained the value for w as about 0.6 g of water per gram of carboxyhaemoglobin [57]. This approach neglects possible interactions between protein dipoles, as well as the facts that the protein molecules are not dispersed in a homogeneous dielectric and have internal fields that depend on their shapes. These aspects are treated in various so-called *mixture theories*, an excellent review of which is given by Reynolds and Hough [68]. As correctly conceived by Oncley, at frequencies above the β-dispersion the protein with its bound water can be considered as a spheroidal cavity of low permittivity suspended in a fluid of high permittivity. Two general formulae to describe the dielectric properties of mixtures of this form are formulated in Box 8.2. Equation (8.25a) is considered [68] appropriate for the case of small particles dispersed in a continuous medium and this is relevant to protein solutions. Equation (8.25b) is more appropriate for the situation where the two component volume fractions v_1 and v_2 are nearly the same [68]. Equation (8.25a) is simplified when dealing with dilute protein solutions (volume fraction v_1) where interactions between the dipole fields of the protein molecules can be neglected and the permittivity ε_2 taken to be ε_w, equivalent to that of bulk water. The internal (depolarizing) field factor f_1 is obtained using Equations (7.33) and (7.34) for the case of prolate and oblate spheroids, respectively.

A version of the mixture theory used by Buchanan *et al.* [69] for low protein concentrations takes the form

$$\varepsilon_{mx} = \varepsilon_w - \beta\, v_p(\varepsilon_w - \varepsilon_p)$$

where v_p and ε_p represent the volume fraction and permittivity of the protein, respectively and β is a parameter that depends on the axial ratio of the protein spheroid. This equation was derived directly from one developed by Fricke [70] to describe the conductivity of mixtures and incorporates the approximation that $\varepsilon_w \gg \varepsilon_p$. For an axial ratio of 1 (a sphere) $\beta = 1.5$. The following relationship was employed to find the weight of water w per gram of protein:

$$\frac{\Delta\varepsilon_\infty}{c} = \beta\left[\frac{(\varepsilon_w - \varepsilon_{p\infty})\,v_{sp} + (\varepsilon_w - \varepsilon_{w\infty})\,w}{100}\right]$$

where v_{sp} is the partial specific volume of the anhydrous protein, c the concentration in gram protein per 100 mL of solution and $\varepsilon_{p\infty}$ is the high-frequency permittivity of the protein (taken as ~2). The following total hydration values (g per g protein) were obtained: lysozyme (0.46); egg albumin (0.32); β-lactoglobulin (0.32); methaemoglobin (0.32); bovine serum albumin (0.41). Of this total hydration, amounts ranging from a third to half were estimated to be water that is irrotationally bound to the protein. A drawback of this approach is that

Box 8.2 The Permittivity of Mixtures

A general formula for the permittivity of a molecular mixture can readily be formulated, but as in most things the difficulties lie in the details. If the mixture consist of two components of permittivities ε_1 and ε_2 and occupy volume fractions v_1 and v_2 ($v_1 + v_2 = 1$), the average electric displacement $\langle D\rangle$ and average electric field $\langle E\rangle$ are given by:

$$\langle D\rangle = v_1\langle D_1\rangle + v_2\langle D_2\rangle ; \langle E\rangle = v_1\langle E_1\rangle + v_2\langle E_2\rangle$$

Assume that the permittivity of the mixture is given by

$$\varepsilon_{mx} = \frac{\langle D\rangle}{\langle E\rangle}$$ and that for each component

$$\langle D_1\rangle = \varepsilon_1\langle E_1\rangle ; \langle D_2\rangle = \varepsilon_2\langle E_2\rangle$$

From the above relationships for the average displacement and field we then have

$$\varepsilon_{mx} = \varepsilon_1 v_1 f_1 + \varepsilon_2 v_2 f_2$$

where $v_1 f_1 + v_2 f_2 = 1; f_1 = \frac{\langle E_1\rangle}{\langle E\rangle}$ and $f_2 = \frac{\langle E_2\rangle}{\langle E\rangle}$

From these relationships, we can derive two equivalent and general formulae:

$$\varepsilon_{mx} = \varepsilon_2 + (\varepsilon_1 - \varepsilon_2)\, v_1 f_1 \tag{8.25a}$$

and

$$(\varepsilon_{mx} - \varepsilon_1)\, v_1 f_1 + (\varepsilon_{mx} - \varepsilon_2)\, v_2 f_2 = 0 \tag{8.25b}$$

The difficulty in the details is finding appropriate values for f_1 and f_2, as well accounting for possible interactions between the components. Equations (8.26a and b) should theoretically give the same result, but in practice this is not so because of the different approximations often required for factors f_1 and f_2. Unfortunately, it is only possible to obtain exact values for these factors for the case of parallel slabs, or for very dilute suspensions of particles of ellipsoidal shape. Otherwise, only approximate values can be assigned to f_1 and f_2.

no knowledge is gained of the separate polarizabilities (i.e., effective permittivities) of the protein molecule itself and of its bound hydration. The derived hydration values are critically dependent on the assumed values for these two parameters [1, pp. 195–197].

The following variation of Fricke's formula

$$\frac{\Delta_\infty}{3\varepsilon_w} = \frac{\varepsilon_w - \varepsilon_{eff}}{\varepsilon_{eff} + 2\varepsilon_w} v_p \tag{8.26}$$

was employed by South and Grant [48] to derive the effective permittivity value ε_{eff} of myoglobin, using the measured value for $\Delta\varepsilon_\infty/v_p$ of 0.040 m^3/kg. Based on measurements of the relaxation time and Equation (8.18) the total volume of the protein molecule and its hydration coverage was calculated to be 40 nm^3, which from Equation (8.25) results in a value $\varepsilon_{eff} = 54$ for the effective permittivity of the myoglobin molecule. It may now be assumed that the protein molecule is made up of a sphere of volume v_p and static permittivity ε_p surrounded by a region of hydration of permittivity ε_{hs} to give a total volume V. The following equation, derived by Schwan [1, p. 197], relates these various parameters:

$$\frac{\varepsilon_{eff} - \varepsilon_{sh}}{\varepsilon_{eff} + 2\varepsilon_{sh}} = \frac{v_p}{V}\frac{\varepsilon_p - \varepsilon_{sh}}{\varepsilon_p + 2\varepsilon_{sh}}$$

South and Grant assumed that ε_p is governed by atomic and electronic polarization only, so that its value would be much less than that, ε_{hs}, of the hydration 'shell'. On this basis the value $\varepsilon_{hs} = 103$ was predicted, in close agreement to that found previously for haemoglobin [71].

However, Pitera *et al.* [72] have computed the static permittivity ε_p for four different proteins, simulated under at least two different conditions of pH, temperature, solvation, or ligand binding. These computations made use of Equation (7.24) devised by Fröhlich and involved calculations of dipole moment fluctuations per unit volume as a function of time up to 5 ns. In agreement with previous computations [73–76] it was found that the behaviour of the charged residues is the primary determinant of the effective permittivity. For example, the relative permittivity for the whole lysozyme molecule was found to be 25.7, in contrast to values of roughly 2.6 and 1.9 for the protein without charged groups or just the polypeptide backbone on its own, respectively [72]. Furthermore, only environmental changes that alter the properties of charged residues exert a significant effect on the permittivity value. In contrast, buried water molecules or ligands have little or no effect on protein dielectric properties. The permittivity values obtained for the other whole proteins were: α-lactalbumin (12.6 at pH 2.0, ~16.2 at pH 8.0; rat fatty acid binding protein (~40.7); the llama antibody heavy-chain variable domain (17.2). Simonson [77] performed molecular dynamic simulations for the two small proteins, the SH3 domain in the viral adaptor protein v-Crk (57 amino acids) and Staphylococcal nuclease (136 amino acids), obtaining relative permittivity values of 21.6 and 16.0, respectively. These relative permittivity values for the whole protein molecule, ranging from 12.6 to 40.7, brings into question the very low value assumed for anhydrous myoglobin by South and Grant and the effective value of 54 found for this molecule with its accompanying hydration [48]. Relative permittivity values of ~100 obtained for the hydration 'shells' of myoglobin and haemoglobin [48, 71] can be taken as an overestimate.

8.4.2.3 The δ-Dispersion and Protein Hydration

Small dispersions, collectively known as the δ-dispersion, have been identified in the frequency range from around 100 MHz to 1 GHz (i.e., between the β-dispersion and the γ-dispersion shown in Figure 8.18). The literature on this is extensive (e.g., 78–94) and only a broad outline of it will be presented here, concentrating on the more recent works and conclusions. We find that this remains an active research area with unsettled questions.

The first indications of an additional dispersion around 100 MHz for protein solutions were observed in Hasted's laboratory [69, 78] and later confirmed by Grant and his colleagues [79, 81, 83–85]. The general consensus is that the δ-dispersion results from the relaxations of water molecules in the hydration coverage of a protein, but the possibility of contributions from other effects should also be considered. For example, the δ-dispersion appears to encompass at least two relaxation processes [84, 85, 93] with the possibility that intraprotein motions are involved [82, 84, 91] as well as a glass transition of the protein structure [92]. It should also be taken into account that even in simple solutions, those far less complicated than protein solutions, a great variety of dielectric effects occur [86]. For protein solutions, where different polarization mechanisms overlap, it is particularly difficult to unambiguously identify the origins of specific relaxations.

Muira *et al.* [87] conclude that the δ-dispersion around 100 MHz is caused by orientation of bound water molecules on the protein surface, supplemented by fluctuations of surface polar side groups. They also suggest that although the β-dispersion results from the rotation of the protein molecule, for proteins of relatively high molecular weight there is also a contribution from the migration of counter ions on the protein's surface. In later work [88] this group also studied the freezing of globular protein solutions by microwave dielectric measurements. They identified three classes of water, namely: bulk water, which freezes at −5 °C; unfreezable water, which forms a hydration component of 0.36 g water per gram protein around the protein molecule; water firmly attached to the protein surface. The amount of the attached water remains constant with changes in temperature but its relaxation ceases below −60 °C.

Oleinikova *et al.* [91], in a paper with the interesting part title 'What can really be learned from dielectric spectroscopy of protein solutions?' found that the dielectric spectrum exhibited by ribonuclease solutions in the MHz–GHz range could be decomposed into five modes of Debye type diffusive behaviour. Whereas they give the standard interpretation of the dominant β-relaxation and γ-relaxation (protein 'tumbling' and bulk water relaxation, respectively) they make a significant departure in terms of the δ-dispersion. Instead of a single or possibly double mode of relaxation, Oleinikova *et al.* find three modes. They attribute the high frequency component (δ_3), having a relaxation time near 40 ps, to hydration water reorientation. This corresponds to what can be described as *loosely bound water*. They also declare that the existence of *tightly bound* water, often deduced from the low frequency part in the nanosecond regime (δ_1), is inconsistent with a highly mobile hydration layer observed by NMR techniques and molecular dynamics simulations. The notion that the δ-relaxation can be attributed to the exchange of hydration water molecules with bulk water molecules is also rejected, in favour of a process involving protein-water crosscorrelations. A contribution from the displacement or diffusion of counterions at the charged protein surface was also discounted. Component δ_2 near 500 MHz is discussed in terms of intraprotein motions superimposed on protein tumbling. Finally, Oleinikova *et al.* [91] found that as the protein concentration of their solutions increased from 0.5 to 6 wt%, the effective dipole moment of the protein decreased. This effect was attributed to protein-protein interactions.

Wolf *et al.* [93] report a detailed study of two concentrations of aqueous lysozyme solutions in the frequency range from 1 MHz to 40 GHz and for temperatures from 275 to 330 K. A well pronounced δ-dispersion was measured, together with detailed information on its temperature dependence. The complete broadband spectrum could be fitted using a Debye function for the β-dispersion around 10 MHz, a Cole–Cole function for the γ-relaxation of the bulk water around 20 GHz and the δ-dispersion around 100 MHz was well described by a single Cole–Cole function. The temperature dependences of the β- and γ-relaxations were found to be closely correlated. A significant temperature dependence of the dipole moment of the protein was attributed to conformational changes. Very interesting results and conclusions were obtained from measurements of the temperature dependence of the dc conductivity arising from ionic charge transport. A breakdown of the Debye–Stokes–Einstein relation was found, indicating that the dc conductivity was not completely governed by the mobility of the solvent molecules, but instead is closely connected to the dynamics of the hydration coverage around the protein. An unexpected and unexplained finding was that the ionic charge transport and the δ-relaxation were determined by identical energy barriers.

8.4.2.4 Other Interpretations of the β-Dispersion

At the time when Oncley and others were employing Debye's concept of a permanent dipole moment that relaxes according to his rotation model, a strange scenario developed. For Debye himself had appeared to disclaim such an interpretation of the β-dispersion exhibited by protein solutions! In 1928, with Falkenhagen [95], he proposed an alternative *ion atmosphere* model. In this model the protein molecule exists within a cloud of countercharge ions, so that the dispersion arises from field-induced displacement of these ions from the centre of charge of the protein. This model predicted that the dielectric relaxation time τ should vary inversely with the ionic strength of the protein solution. However, in 1972, South and Grant [48] demonstrated that the dielectric relaxation time for myoglobin was only weakly dependent on the solution conductivity (i.e., ionic strength). It was found after electrodialysis of the solution to remove small ions, leaving only two or three ions per protein molecule (hardly an 'ionic atmosphere'), that the dielectric properties of the myoglobin molecule did not change to any significant extent [48]. After Oncley's work in 1942 other models were also proposed to account for the dielectric dispersions exhibited by proteins. These included the *proton fluctuation* model of Kirkwood and Shumaker [96] where the dispersion arises from the field induced redistribution of protons between the neutral and charged basic groups on the protein surface to which protons can bind; the *structured water* model proposal by Jacobson [97] where the hydration around the protein is responsible for the dispersion and not the protein molecule itself; the *interfacial surface conductivity* model proposed by O'Konski [98]; the *Maxwell–Wagner interfacial polarization* model applied to protein solutions proposed by Schwan [1, pp. 188–189]; the *ion mobility model* proposed by Schwarz [99].

Although these various models are not without interest and are of broad relevance to biodielectrics (but seemingly not to the interpretation of the dielectric properties of proteins), it serves no purpose to describe them any further here. For those interested in more details, the paper by South and Grant [48] is highly recommended. They described each of the models in detail, a summary of which is given in Table 8.15. One by one through careful logic and experimental evidence they convincingly discounted each model and concluded that Debye's dipole moment theory and rotation model provides the best way to interpret the β-relaxation exhibited by protein solutions.

8.5 Nucleic Acids

Nucleic acids are responsible for storing the information and instructing the cell on the proteins it should synthesize. The two information-storing molecules in cells are deoxyribonucleic acid (DNA) and ribonucleic acid (RNA). Proteins are polymers constructed from 20 different monomers (the amino acids) but DNA and RNA

Table 8.15 Summary given by South and Grant [48] of the Debye rotating dipole model and alternative mechanisms proposed to describe the β-dispersion exhibited by protein solutions.

Mechanism	Quantitative theoretical treatments	
	Increment $\Delta\varepsilon_T$	Relaxation time τ
Debye rotation	$\text{const. x } \dfrac{\text{p}^2}{MT}$	$\dfrac{4\pi R^3\eta}{kT}$
Ion atmosphere [95]	$\dfrac{\text{const.}}{c^{1/2}} - \text{const.}$	$\dfrac{\text{const.}}{c_{\text{ions}} \times s_w}$
Proton fluctuation [96]	$\text{const.} \sum_i \dfrac{r_i^2}{2 + \dfrac{[\text{H}^+]}{K_i} + \dfrac{K_i}{[\text{H}^+]}}$	None available
Structured water [97]	None available	None available
Maxwell–Wagner [1]	$9v_p \dfrac{(\varepsilon_w s_p - \varepsilon_p s_w)^2}{(\varepsilon_p + 2\varepsilon_w)(s_p + 2s_w)^2}$	$\dfrac{\varepsilon_p + 2\varepsilon_w}{s_p + 2s_w}$
Surface conductivity [98]	as for Maxwell–Wagner, with s_p replaced by $s_p + \dfrac{2\lambda}{R}$	
Ion mobility [99]	$\dfrac{9}{4} 9v_p \dfrac{v_p}{\left(1 + \dfrac{1}{2}v_p\right)^2}$	$\dfrac{R^2}{2ukT}$

Note: Key to notation not already used in this chapter: $s = \sigma\varepsilon_o^{-1}$ (units of reciprocal second) where σ is the conductivity. Suffices w and p refer to the solvent (water) and protein, respectively. Characters u and λ refer to surface ion mobility and surface conductivity, respectively.

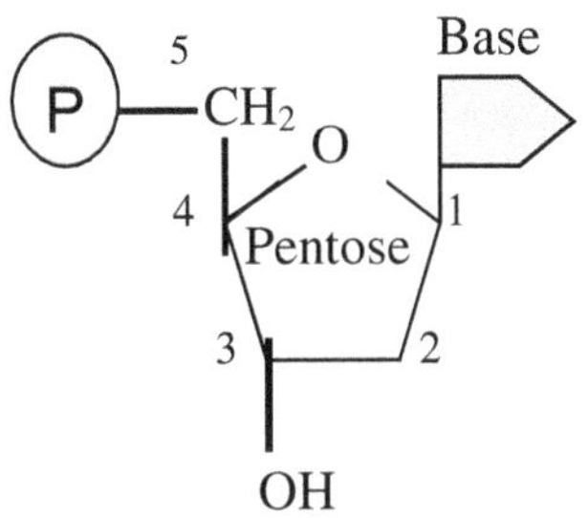

Figure 8.21 A nucleotide consists of a 'base' linked to a phosphate group by a five-carbon sugar (pentose) molecule.

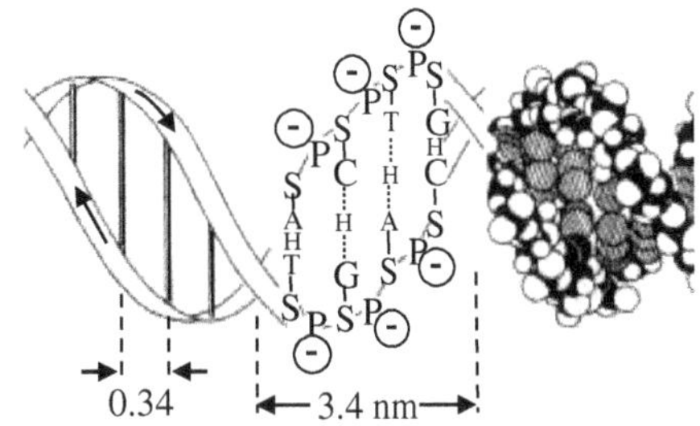

Figure 8.22 A schematic of the B-DNA, the most common form of the DNA double helix, based on Watson and Crick [100]. The two backbone strands are directed in opposite directions, with the base pairs stacked with their molecular plane perpendicular to the helical axis. The B-DNA molecule should not therefore possess a permanent dipole moment.

consist of just four monomers – called nucleotides. As shown in Figure 8.21, a nucleotide is composed of a phosphate group (P) and a 'base' linked together by a five-carbon sugar (pentose) molecule. The bases found in DNA are adenine, guanine, cytosine and thymine, conventionally abbreviated as A, G, C and T. In RNA the thymine base (T) is replaced by uracil (U). In DNA the pentose sugar molecule is *deoxyribose*, whereas in RNA it is *ribose*.

Nucleic acids consist of chains of nucleotides formed in a condensation reaction to create a *phosphodiester* bond, in which a water molecule is released. This is equivalent to the creation of a glycosidic bond between sugars or of a peptide bond between amino acids. Two nucleotides joined by such a bond forms a dinucleotide. A trinucleotide represents a single strand of DNA containing three nucleotides and so on, as more phosphodiester bonds are created. As more nucleotides are added, a long DNA single strand is produced having a defined chemical orientation. One end (the so-called $3'$ end) of a DNA strand has a free hydroxyl group (attached to carbon 3 of the sugar), whilst the other end (the $5'$ end) has a phosphate group. This orientation has important implications regarding the properties of DNA. The biologically native state of DNA is a double helix composed of two intertwined single strands of DNA. This is depicted in Figure 8.22.

The two single strands of DNA in the double helix structure proceed from carbon $5'$ to $3'$, but are directed in opposing directions, as shown in Figures 8.22 and 8.23. The two DNA strands are held together by hydrogen bonds linking their bases and only if the two strands of the helix are antiparallel can the members of each base pair fit together within the double helix. The two DNA strands can, in theory, form either a right-handed or left-handed helix, but the structure of the sugar-phosphate backbone is such that the right-handed helix is the more favourable geometry. As first proposed by Watson and Crick [100] (with their famous understatement: 'It has not escaped our notice that the specific pairing we have postulated immediately suggests a possible copying mechanism for the genetic material.') the size, shape and chemical composition of the bases dictates that base A is always paired with T, and G is paired with C. As shown in Figure 8.23 the A-T pair is held together by two hydrogen bonds and the G-C pair by three. To break the G-C pair thus requires more energy (87.9 kJ/mol) than that required to break the A-T pair (50.6 kJ/mol) [101]. This difference is reflected in the finer details of how the DNA polymer is copied. X-ray studies have determined that the stacked bases are regularly spaced 0.34 nm apart along the helix and that the length of one complete helix turn is 3.4 nm (to give ~10 pairs of bases per turn). The hydrogen bonds between the bases gives the double helix considerable stability and rigidity, but also allows the double helix a good degree of flexibility, enabling long DNA chains to coil up to form supercoils or condensed structures of very large molecular weight. The polypeptide alpha-helix shown in Figure 8.16 is far less flexible than a DNA chain because the hydrogen bonds hold together adjacent parts of the helix. An important feature to note in Figures 8.22 and 8.23 is the negative charge carried by the phosphate groups when in aqueous solution at neutral pH. These charges can be mapped on individual molecules of DNA using Kelvin probe force microscopy [102]. Unlike protein molecules, which possess either a net positive or negative charge at neutral pH depending on their amino acid content, DNA molecules have a net negative charge irrespective of their nucleotide content. A DNA molecule in aqueous solution is thus a *polyelectrolyte*, carrying a high negative charge density surrounded by an equally large cloud of positive counterions. When crystallized out of solution DNA carries with it these counterions to form a salt, such as the sodium or potassium salt of DNA. As the pH is lowered below pH 5 the DNA double helix structure begins to denature as cytosine bases, with a pK_a ~4.6, become protonated [103, 104].

Whereas the number of amino acid residues in proteins ranges from below one hundred to several

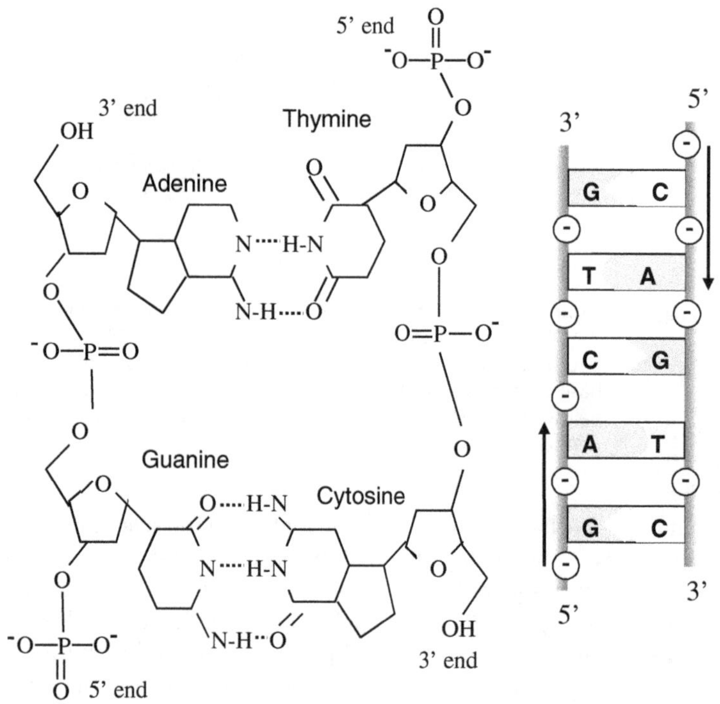

Figure 8.23 A schematic of a segment of double-stranded DNA to show the base-pair complementarity of the adenine (A) -thymine (T) and guanine (G) –cytosine (C) pairings. An important electrical feature is the negative charge carried by each phosphate group.

thousand, DNA molecules are typically very much larger. For example, the DNA molecule in the single chromosome of an *E. coli* bacterium comprises around five million base pairs – a number defining the genome size of *E. coli*. If it were to be fully stretched out this DNA would have a length of ~1.5 mm – some three orders of magnitude longer than the *E. coli* bacterium itself! How does the chromosome package itself inside this bacterium? The secret lies in the flexibility of a DNA double helix that allows it to coil and fold into a superhelix. This can be simulated by continuously twisting an elastic band and slowly bringing the ends together, so that the twisted band first forms small coils that then proceed to curl into a tight knot. Human cells (apart from red blood cells, which do not have a nucleus) contain 46 chromosomes, containing a total of 3.2×10^9 base pairs. If the DNA from all 46 chromosomes of a single human cell were to be connected and straightened out, its total length would be ~2 m! Apart from the germ cells (eggs and sperm) a typical human cell contains two copies of 22 of these chromosomes, numbered from 1 to 22 in order of diminishing physical size. Females have two X chromosomes and males one X and one Y chromosome to give a total of 46 chromosomes. The X chromosome is inherited from the mother and the Y chromosome from the father. The 22 chromosomes, plus the X and Y chromosome, can be distinguished from one another by staining with dyes that distinguish between DNA that is rich in either A-T or G-C nucleotide base pairs. Each chromosome type can be identified by the distinctive patterns of coloured bands along them and chromosomal abnormalities can be detected.

To assist in the packaging of this DNA into the nucleus of a human cell it is wrapped around protein molecules, called histones, to form structural units called nucleosomes that are spaced at regular intervals along the main DNA chain, rather like beads on a string. Arrays of nucleosomes form chromatin fibres that are then further packaged into chromosomes. This form of DNA packaging occurs in cells with a nucleus (i.e., eukaryotic cells) but not in those lacking a nucleus (prokaryotic cells) such as bacteria, where typically the total DNA forms a large circular molecule. A gene corresponds to a stretch of DNA that contains the sequential information for the production of proteins or RNA chains that have functional roles in the cell. Some stretches of DNA do not encode for proteins or RNA and, at present, a quite large percentage of this so-called 'junk' DNA has no known biological function. The entirety of the genes and noncoding sequences of DNA in a cell is called its genome. Many types of virus do not possess DNA and instead their genome consists of the coding information contained in another polynucleic acid called RNA.

RNA is very similar to DNA but differs in a few important structural details. RNA nucleotides contain ribose (DNA contains deoxyribose – a type of ribose that lacks one oxygen atom) and has the base uracil rather than thymine present in DNA. Thus, complimentary A-U pairing occurs in a RNA molecule, rather than the A-T pairing found in DNA. Also, whereas DNA takes the form in a cell of a double-stranded helix, RNA is single stranded. A RNA chain is thus much more flexible than DNA, able to fold up into a variety of three-dimensional shapes containing sections of single strand loops and double helices wherever parallel strands are able to form complimentary nucleotide base pairs. Some of the shapes that RNA molecules can adopt enable them to perform catalytic functions. Different types of RNA are central to the synthesis of proteins and are transcribed from DNA by enzymes called RNA polymerases. These enzymes bind to the DNA in the nucleus of eukaryotic cells, separate the two strands of the nuclear DNA and pair ribonucleotide bases to the template DNA strand according to the Watson–Crick base pairing shown in Figure 8.22 (with uracil replacing thymine). Thus, for example, the action of RNA polymerase is to produce a strand of RNA with a nucleotide sequence CUGA (rather than the sequence CTGA of a DNA strand if DNA polymerase had been in action). Many RNA polymerases can act on a single strand of DNA at the same time to speed up this transcription process. A type of RNA called messenger RNA (mRNA) carries coding information, obtained from the DNA template, in the form of tri-nucleotide units called codons that each code for a single amino acid. There are $4^3 = 64$ different codon combinations possible with a triplet codon of three nucleotides and all 64 codons are assigned for either amino acids or start and stop signals during translation of the mRNA code into a polypeptide sequence. Because there are only 20 common amino acids, there is some redundancy in the assignment of the mRNA triplet codons. For example, the triplet codons GGU, GGC, CGA, GGG are all used for the synthesis of glycine, with codons UAA, UGA and UAG used as instructions to stop the translation (synthesis) process. In performing this process, strands of mRNA interact in the cytoplasm with protein structures called ribosomes (in recent years ribosomes have become important targets in the search for new antibiotics to fight the emergence of drug resistant bacteria). In eukaryotic cells the mRNA is formed inside the nucleus and has to pass through pores in the nuclear membrane to locate organelles known as ribosomes in the cytoplasm. Ribosomes consist of proteins and ribosomal RNA polymers, which together act as the molecular 'machine' to read mRNA and to translate the information it carries into the production of amino acid chains that form proteins. Different types of transfer RNA (tRNA) molecules mediate this process by transferring a specific amino acid to the growing peptide chain. The different tRNA molecule can be attached to only one type of amino acid and each one contains a three base anticodon that can base pair to the corresponding codon on the mRNA chain.

The statement 'DNA makes RNA makes Protein' summarizes the so-called *Central Dogma of Molecular Biology*, first enunciated by Francis Crick [105] and which states that the sequential structural information stored in a protein cannot be transferred to another protein or to a nucleic acid. Crick used the word 'dogma' by way of a catch phrase without realizing its implied interpretation – in fact he wished his concept to be considered as an hypothesis [106]. In living systems there are three major classes of linear biopolymer, namely DNA, RNA and proteins, whose monomer sequences encode information. There are 9 conceivable direct transfers of information possible between these three classes. The transfer of information is assumed to be an error-free transfer where the molecular sequence of one biopolymer is used as a template to construct another biopolymer with a molecular sequence that depends entirely on that template. Transfers that can occur in all cells, known as general transfers, are the three cases of DNA→DNA (DNA replication), DNA→RNA (transcription) and RNA→protein (translation). Special transfers are ones that do not occur in most cells but may occur in special circumstances, such as in virus-infected cells and are the three cases of RNA→RNA, RNA→DNA and DNA→protein. A known example of the RNA-DNA transfer takes place in retroviruses, where DNA is synthesized using RNA as a template. An enzyme known as reverse transcriptase carries out this process. The human immunodeficiency virus (HIV) is a retrovirus and is the cause of AIDS. After HIV has bound to a target cell, normally one of the vital blood cells of the immune system, the RNA content of the virus and various enzymes including reverse transcriptase and ribonuclease and protease, are injected into the cell. The single strand of viral RNA genome is then transcribed into double strand DNA and integrated into a chromosome of the host cell, which can lead to possible reproduction of the virus [107].

8.5.1 Dielectric Properties of Nucleic Acids

8.5.1.1 DNA

The earliest reports of the dielectric properties of DNA are those of Jungner *et al.* [108–111] performed before 1952 and thus in ignorance of its double helix structure announced in 1953 [100]. Well defined dispersions centred at around 2 MHz were observed for relatively dilute solutions (0.01 to 0.1 wt %) of calf thymus and salmon DNA having relatively low molecular weights of $2\sim6 \times 10^5$. The magnitudes of these dispersions translated to very large molar dielectric increments, ranging from 2.9×10^5 to 2.9×10^8 [110]. This was interpreted as a Debye-type relaxation of a large dipole moment (1750 D to 56 000 D!) directed at right angles to the axis of the DNA fibre. Each constituent nucleotide was calculated to have a moment of 0.9 D, stacked parallel with each other and all pointing in the same direction [110]. Orienting effects of surrounding polar molecules, to form a *quasicrystalline* structure of the fibre molecules, were considered capable of overcoming the large repulsive Coulombic forces that would accompany such a structure [111]. Had Gunnar Jungner and his colleagues been aware of the double helix structure of DNA, they would have realized that in its native form DNA should not have a permanent dipole moment. With its two intertwined single strands of DNA running in opposing directions and the base pairs stacked with their molecular plane perpendicular to the helical axis, the dipole moments of paired nucleotides and dipoles in the main strands all cancel out. In its double helical form the DNA molecule has a highly regular and ordered internal structure – in other words it is in a crystalline state. The quasicrystalline structure postulated by Jungner *et al.* can therefore be considered as visionary. A significant departure from the conventional interpretation of the Debye relaxation model was also found by Jungner *et al.*, namely that the derived dipole moment values were *directly proportional* to the molecular weight of the DNA molecule. From Equation (8.24), according to the Debye model, the moment should vary as the square root of the molecular weight. This anomaly was recognized and discussed, but with no clear conclusions [110]. The direct relationship found between the dielectric increment and molecular weight led to a proposed procedure for monitoring the enzymatic degradation of DNA [109]. More importantly it actually provides a clue as to the mechanism of polarization. These various results and conclusions by Jungner *et al.* are remarkable considering the fact that they neither had the benefit of knowing about the helical structure of DNA nor the techniques available for its purification and preservation. Finally, it should be noted that (as far as this author can find) Jungner and his colleagues have to date reported the only dielectric measurement on the complex (nucleohistone) formed by DNA wrapping itself around a protein to assist its packaging into the cell nucleus. The magnitude of the dielectric molar increment exhibited by nucleohistone prepared from calf thymus was found to be less than the DNA on its own [108]. This effect can be interpreted in terms of the protein shielding some of the negative charge of the DNA and is a topic worthy of further study.

The existence of a permanent dipole moment directed at right angles to the long axis of the DNA molecule was disproved using an ingenious method to measure the relative permittivity of DNA strands subjected to the shearing stress of laminar flow [97, 112–114]. With this method an electric field could be applied either perpendicular or parallel to the long helical axis. According to the dipole orientation envisaged by Jungner *et al.* [110], when the DNA molecules are oriented by laminar fluid flow their dipoles should be perpendicular to the flow direction. Applying an electric field perpendicular to the flow should therefore result in the measured relative permittivity being larger than that obtained in a stationary fluid where there is random orientation of the DNA molecules. In fact, it was found that the permittivity value decreased [112, 113]. With the applied field and fluid flow parallel, the permittivity should decrease as the fluid velocity is increased. Instead, the permittivity was found to increase (which implies a moment directed *along* the DNA axis). Furthermore, the maximum decrease or increase in the dielectric constant amounted only to about 10–15% of the total dielectric increment. Jacobson [97, 112] interpreted these finding to indicate that the dipole moments producing the dielectric increment are not strongly fixed to the DNA molecule. He considered it impossible to interpret the dielectric results as the result of mechanical orientations of the complete molecule in accordance with the Debye rotation model. Instead, he thought it more likely that the dielectric properties of DNA (as well as protein) solutions are mainly due to an ordering effect on surrounding water molecules [97]. We have seen that this interpretation for protein solutions was shown by South and Grant [48] to be invalid and no evidence has been obtained to confirm it for DNA solutions.

Takashima [115–118] extended the dielectric investigations of DNA using high molecular weight ($2{\sim}3 \times 10^6$) samples and performing measurements in the frequency range 50 Hz–200 kHz. He found a dielectric dispersion centred at around 100 Hz, corresponding to DNA having dipole moment values as large as 100 000 debye. This dispersion thus occurs at a much lower frequency than that found by Jungner *et al.* at around 2 MHz. Following the convention employed to label dielectric dispersions according to the order in which they appear on an increasing frequency scale, the first (low-frequency) dispersion is termed the α-dispersion and appears in the frequency range 10 Hz to 10 kHz. The second dispersion that occurs from around 10 kHz to a few MHz, the one discovered by Jungner *et al.*, we can term as the β-dispersion. Takashima [115] found that the dipole moment and dielectric relaxation time associated with the α-dispersion were both proportional to the molecular weight of the DNA molecule (agreeing with the finding of Jungner *et al.* for the β-dispersion). He also concluded that the dipole moment is directed along the major axis, instead of perpendicular to it. A typical α-dispersion, analysed by Takashima [116] for high molecular weight DNA is shown in Figure 8.24. On heat

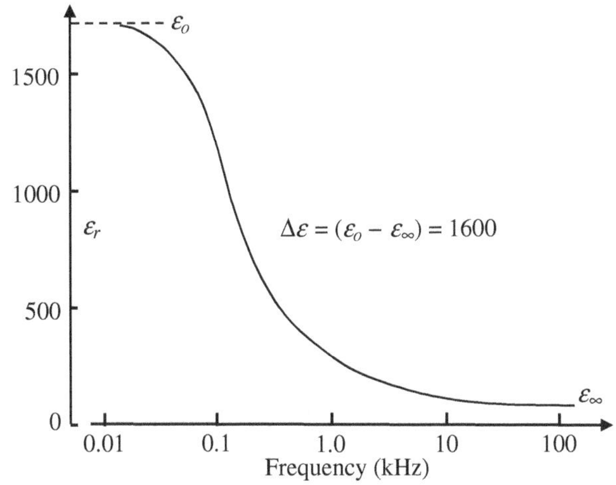

Figure 8.24 Dielectric dispersion exhibited by 0.01 wt.% concentration of salmon sperm DNA in water at neutral pH. The length of the DNA was determined to be 740 nm by birefringence measurement. The value for the low-frequency permittivity ε_o was determined from an extrapolation of a Cole–Cole plot (based on Takashima [116]).

denaturation to single strand, random coil, DNA the large dielectric increment of ~1600 permittivity units reduced to a value of about 20 units. Sonication was applied to the samples to produce DNA with smaller molecular weights and lengths, a procedure that led to the finding that the dielectric increment was proportional to the square of the average length. Unexpectedly, the relaxation time was also found to obey the same relationship. These result cast doubt on an interpretation of the DNA relaxation as being a field orienting effect as given by Debye's rotating dipole model. Takashima [116] cites unpublished work to account for these findings, which applied the counterion polarization theory proposed by Schwarz [99] to the case of an electric field directed along the major axes of cylindrical molecules. The details of this are given 13 years later in Takashima's book [7, pp. 204–209]. Basically, the counterion polarization theory predicts that the counterions that act to partially screen the negatively charged phosphate groups will be displaced along the surface of the DNA molecule under the influence of an applied electric field. The relaxation time of the resulting dielectric dispersion will depend upon the effective mobility of the ions along the macromolecule's 'surface' and for rod-shaped macromolecules will be given by

$$\tau = \frac{\pi\varepsilon_z L^2}{2uzq^2}$$

where ε_z is the effective permittivity of the surrounding ionic atmosphere of z ions per unit length, u is the (two-dimensional surface) counterion mobility, q is the charge on the ions and L is the length of the DNA molecule. The expectation that τ is proportional to L^2 was confirmed by Takashima [117].

Example 8.9 Magnitude of the Conductivity Increment for DNA Solutions

A distinctive feature of the dispersion shown in Figure 8.24 is the large value of ~1600 permittivity units derived for the dielectric increment $\Delta\varepsilon$. How does this translate to the magnitude of the conductivity change $\Delta\sigma$ that takes place over the frequency range of the dispersion?

Solution 8.9 Following the procedure in Example 8.8, we can use the following relationship to determine the magnitude $\Delta\sigma'$ of the conductivity change:

$$\Delta\sigma' = (\varepsilon_o\Delta\varepsilon')/\tau = (8.854 \times 10^{-12}\Delta\varepsilon')2\pi f_{cr}$$

The relaxation frequency f_{cr} is approximated equal to 100 Hz, so that with $\Delta\varepsilon = 1600$ this gives $\Delta\sigma' = 0.56$ μS m^{-1}. This is a very small conductivity change compared to that of the β-dispersion for cytochrome-c, analysed in Example 8.8.

To estimate the value of ε_s that defines the low-frequency boundary of the α-dispersion, Takashima [115–118] often relied on extrapolation of partial semi-circular Cole–Cole plots to frequencies below 50 Hz. This extrapolation was no longer required after the development of techniques to perform measurements down to 5 Hz [119] and 0.3 Hz [119–124]. These techniques were developed to enable dielectric measurements to be performed on conducting solutions, without errors arising from electrode polarization effects. An example of this low-frequency data is shown in Figure 8.25, where the dramatic dependence on the magnitude and relaxation time of the α-dispersion with salt concentration of the solution is clearly demonstrated.

At a meeting of the Biophysical Society, Stellwagen *et al.* [125] presented evidence for the existence of two dielectric dispersions for DNA: a large one at low frequencies and a smaller one in the MHz region. This work does not appear to have been subsequently published, but the existence of dispersions appearing at higher frequencies than the α-dispersion has been confirmed by others. The first confirmation came from the work of van der Touw and Mandel, who found dispersions for low molecular weight DNA at around 30 kHz and 160 kHz [126, 127]. The low-frequency dispersion (α-dispersion) exhibited by native DNA was found to be very sensitive to both the DNA molecular weight and ionic concentration of the solution, in accordance with Takashima's earlier findings. On sonication of the native sample to lower molecular weight fractions, the α-dispersion drastically reduced

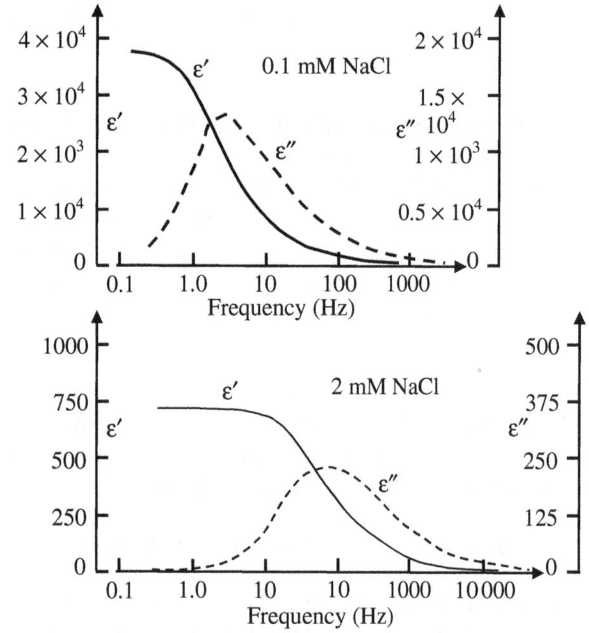

Figure 8.25 Values of the real and imaginary permittivity components, in *excess* of those values for pure water, at 25 °C for 0.05 wt.% calf thymus DNA of molecular weight 4×10^6. Large reductions of the dielectric increments and relaxation time occur with increasing salt concentration (based on Hayakawa *et al.* [121]).

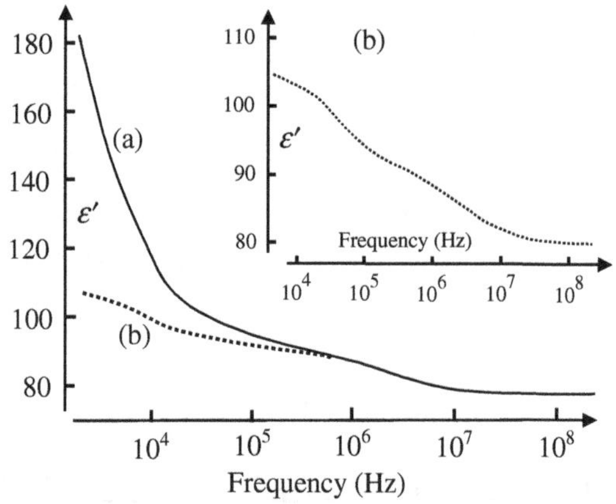

Figure 8.26 Relative permittivity of the sodium salt of calf thymus DNA in aqueous 3×10^{-4} M NaCl solution. (a) Native sample of molecular weight $\sim 4 \times 10^6$ before sonication. (b) Sonicated sample of molecular weight 3×10^5 (based on van der Touw and Mandel [126] and Mandel [127]).

in magnitude and as depicted in Figure (8.26) exhibited a high-frequency 'tail' with evidence of two small dispersions at relaxation frequencies around 30 kHz and 160 kHz. These dispersions were relatively insensitive to changes the salt concentration of the solution [127]. With the passing of time, studies that extended up to frequencies as high as 1 GHz revealed further details of dispersions hidden in the 'tail' of the α-dispersion. This work is summarized in Table 8.16.

Mandel considered the dielectric behaviour of DNA to be qualitatively similar to that exhibited in the same frequency range by synthetic charged polyelectrolytes such as polyacrylic acid [127]. The thermodynamic properties of polyelectrolyte solutions can be understood in terms of the phenomenon of *association,* where a certain fraction of the counterions stays in close neighbourhood of the macromolecular chain. Being unable to move independently through the solution, these strongly associated counterions reduce the effective charge on the polyion. Mandel described how, for synthetic linear charged macromolecules, it had become increasingly obvious that the dielectric increment was not primarily determined by the average permanent dipole moment of the macromolecular chain. The magnitudes of the observed increments would lead to extremely high values of the dipole moment, which cannot be accounted for by their molecular structure. Furthermore, the dielectric increments become vanishingly small if the charge on the macromolecule tends to zero, as can easily be checked with weak polyelectrolytes for which the charge can be controlled by addition of either strong low-molecular base or acid. Therefore, Mandel proposed that the observed dielectric effects can be attributed to large, *induced,* dipole moments originating in the distribution of strongly associated counterions around the polyions. The equilibrium distribution of these counterions along the chain is slightly disturbed by the application of an external electric field, giving rise to an induced dipole moment roughly proportional to the square of the

Table 8.16 A summary of the dielectric dispersions observed for DNA solutions that can be considered as contiguous with the β-dispersion first reported by Jungner *et al.* [108–111] and confirmed later by van der Touw and Mandel [126,127]. The dielectric increment ($\Delta\varepsilon$) values have been normalized to 1 g/L.

Author reference	$\Delta\varepsilon_1$	f_1	$\Delta\varepsilon_2$	f_2	$\Delta\varepsilon_3$	f_3
[128]	6.0	20 kHz	2.3	549 kHz	–	–
[129]	0.88	7 MHz	0.64	33 MHz	0.19	200 MHz
[130]	20.3	137 kHz	4.82	2.01 MHz	0.507	11.6 MHz

average extension of the macromolecule [127]. This conclusion mirrors Takashima's experimental finding that the dielectric increment was proportional to the square of the average length of sonicated DNA samples [116]. The field-induced polarization of ionic 'atmospheres' can also occur in normal electrolyte solutions [131]. However, owing to the small dimensions of simple ions and the spherical symmetry of the ionic atmosphere, the resulting dielectric increments are very small [132].

There is broad agreement amongst those working in the field of biodielectrics that Mandel's counterion fluctuation model explains the polarization mechanism responsible for the α- dispersion exhibited by DNA solutions. The model may also apply to the β-dispersion (referred to as the δ relaxation in some of the literature, e.g. [129]). Mandel argued that because the α-relaxation is dependent on molecular weight (i.e., proportional to molecular length) it corresponds to the migration of the more diffusely bound counterions over distances of the same order as the entire dimension of the DNA molecule. The counterions are envisaged to move freely under the influence of an external electric field along a subsection of the DNA structure until they meet a potential energy barrier. These barriers result from perturbations in the equipotentials arising from 'kinks' in the average wormlike conformational shape of the DNA molecule [126,127]. Related experimental evidence to support this scenario came from the electrophoresis studies of Ross and Scruggs [133], who established the binding order for alkali metal ions as $Li^+ > Na^+ > K^+ >$ with respective relative strengths of 1.5 : 1 : 0.8. The binding order for the divalent ions studied was found to be $Mn^{++} > Mg^{++} > Ca^{++}$. NMR studies [134, 135] had indicated that a good proportion of such ions are bound loosely to the DNA molecule as a whole, rather than to discrete sites such as the phosphate groups and are in a state of essentially complete hydration and free translational and rotational mobility. Harrington [136] had also provided evidence that short persistence lengths are maintained in high-molecular-weight DNA in aqueous NaCl solutions over a large range of ionic concentration. Because the β-relaxation is relatively independent of the molecular weight, it is reasonable to suppose that it arises from the migration of ions of the more strongly associated counterions (condensed ions [137–139]) over shorter distances bounded by smaller potential energy barriers. Apart from Mandel's own experimental and theoretical contributions to the counterion polarization model [126,127, 140, 141] other major contributors to the general theory are Manning [137–139], Oosawa [142, 143], Imai and Onishi [144], Minikata [145], Ito *et al.* [146] and Bordi *et al.* [147]. Bordi and Sarti provide a comprehensive review of the theories developed to describe the dielectric properties of polyelectrolytes [148]. Significant dielectric experiments since 1980 that confirm the broad applicability of Mandel's counterion fluctuation model include those of Charney and Lee [149], Bonicontro *et al.* [128], Saif *et al.* [129], Bakewell *et al.* [130], Bone *et al.* [150, 151], Katsumoto *et al.* [152] and Tomic *et al.* [153].

The following is a general overview of the counterion fluctuation model to describe the dielectric properties of DNA solutions. In dilute salt solution at low concentration, the DNA molecule has a rodlike conformation due to the coulombic repulsion between the regularly spaced and negatively charged phosphate groups. As the DNA concentration increases there is a transition from what is termed the *dilute* region to the *semidilute* one. The DNA molecules become entangled, with a population of counterions binding to it so as to partially neutralize the overall charge of the phosphate groups. A subpopulation of counterions is confined in an electrostatic potential well, so close to the DNA surface that the Coulombic repulsion energy between adjacent phosphate groups becomes less than the thermal energy kT. This is the counterion *condensation* process. Manning [138] introduced a dimensionless structural parameter ξ given by:

$$\xi = \frac{q^2}{4\pi\varepsilon_o\varepsilon_r kTb} \tag{8.27}$$

where q is the charge on each phosphate group and b is the average axial charge spacing along the DNA helical axis. Although not stated as such, the factor ξ is simply the ratio of the Bjerrum length ($\lambda_{B'}$) to the parameter b. The Bjerrum length is defined as the separation at which the electrostatic interaction between two elementary charges is comparable in magnitude to the thermal energy kT. The condition for condensation thus corresponds to the situation where $\xi = \lambda_B/b \geq 1$. From Figure 8.22, showing the B-form of DNA, there are two phosphate charges spaced 0.34 nm apart along the axis, so $b =$ 0.17. For an aqueous solution ($\varepsilon_r = 79$) at 25 °C (298 K), so that from Equation (8.27) $\xi = 4.16$. The condition for counterion condensation is satisfied. The fraction f of the counterions that condense is:

$$f = \frac{1 - Z^{-1}\xi^{-1}}{Z} \tag{8.28}$$

where Z is the counterion valency. The fraction of condensed counterions in an aqueous Na^+ or Mg^{2+} environment are thus 0.76 or 0.44, respectively. Counterion *condensation* is defined to be that mode of binding of counterions where the neutralized charge fraction of the DNA equals that given by Equation (8.28) over a broad concentration range [138]. An experimentally verifiable effect that confirms the existence of condensed ions is described by Saif *et al.* [129]. They reported a set of experiments on 1 g/L concentrations of DNA dialysed against deionized water and a series of saline solutions.

After seven days of dialysis against deionized water, during which the dialysis solution was changed frequently, the concentration of free bulk ions in the solution was found to have a conductivity of 5 mS/m (equivalent to a 0.5 mN NaCl concentration). Likewise, after dialysis against saline solutions of 1.5 mN, 5 mN and 15 mN, it was found in each case that the conductivity of the DNA solution increased to a value equal to that of the dialysis solution plus 5 ± 1 mS/m. This excess conductivity is a consequence of the counterion condensation.

In a DNA solution in which all counterions have the same valency, we can therefore consider three classes of ions: (i) *condensed* counterions that neutralize a fraction of the DNA phosphate charge and are spatially delocalized along the DNA molecule (this fraction is 0.76 for monovalent counterions); (ii) *diffuse* counterions that neutralize the remaining fraction (0.24) of the phosphate group charge and form the outer part of the electrical double layer shown in Figure 2.4, with a concentration that decays exponentially with distance from the DNA surface; (iii) *bulk* ions in equilibrium in the bulk solution and not influenced by the charged DNA molecules.

On the time scale of the α-dispersion (milliseconds to seconds) the DNA molecule as a whole is expected to be stationary. The relaxation mechanism most likely arises from fluctuations of the condensed counterions moving over a distance comparable with the end-to-end length of the DNA molecule, or a length L_s (parallel to the external field) between potential energy barriers caused by sudden kinks or bends in the wormlike DNA backbone [126]. For calf thymus DNA the subunit length L_s is calculated to be 62 nm at 25 °C [126, 127]. The polarizability α of the condensed counterion layer along a direction parallel to the DNA helical axis is

$$\alpha = \frac{AZ^2q^2nL_s^2}{12kT} \tag{8.29}$$

where n is the number of counterions condensed around the DNA, given by:

$$n = (1 - Z^{-1}\xi^{-1})\frac{L_s}{Zb} \tag{8.30}$$

The factor A in Equation (8.29) describes the electrostatic stability of the ionic double layer around the DNA, taking into account repulsion between the charged phosphate groups and Debye screening of these charges by the counterions. This factor is given by [126, 127]:

$$A = [1 - 2\,(Z\xi - 1)\ln(\kappa b)]^{-1} \tag{8.31}$$

where κ is the inverse of the Debye screening length described in Chapter 2. From Equation (8.29) the dielectric increment is given by the following expression:

$$\Delta\varepsilon = \frac{N\alpha}{3\varepsilon_o} \tag{8.32}$$

where the factor 1/3 accounts for the random orientation of these subunits and N is the number of subunits of length L_s per m^3, calculated to be 3.8×10^{22} g^{-3} m^{-3} [129]. Based on measured values for $\Delta\varepsilon$ there was good agreement between experimental values and those calculated using Equation (8.32). For example, for a 3.9 g/L solution of calf thymus DNA the experimental and calculated values for $\Delta\varepsilon$ were 2.23 ± 0.22 and 1.9 ± 0.6, respectively, corresponding to polarizability values of α_{exp} and α_{calc} of 6.1×10^{-33} and 5.1×10^{-33} C^2 m N^{-1}, respectively [129]. This is close to the value of 3×10^{-33} C^2 m N^{-1} obtained by Suzuki *et al.* [154] from studies of field induced orientation of λ-DNA and plasmid DNA. For the largest the three dielectric increments observed for plasmid DNA by Bakewell *et al.* [130] the polarizability was determined to be 8×10^{-30} C^2 m N^{-1}.

The origin of the α-dispersion is now considered to result from polarization *along* the DNA helix, but the case for the β-dispersion is not so clear. Most of the literature on the dielectric properties of DNA solutions has assumed that the β-dispersion involves polarization of bound counterions along shorter subunit lengths. However, there is also good evidence to suggest that for polyelectrolyte chains in general the polarization is *perpendicular* to the chains. The free counterions are considered to move within an electrostatic potential that is modulated by the presence of potential wells associated with neighbouring polyelectrolyte chains. Under the influence of an external electric field these counterions can polarize by free diffusion in three-dimensional space to a scale of the order of the distance between the chains. This concept has been reviewed by Bordi and Sarti [148]. The implication that this has for a re-evaluation of the dielectric properties of DNA solutions in the kHz to MHz frequency range has yet to evolve.

Mention should be made of an unfortunate period in the history of biodielectrics. During the 1980s, much attention was paid to the perceived biological hazards of electromagnetic radiation. For example, a sizeable literature reports nonthermal physiological effects induced by exposure to various (sometimes quite specific) wavelengths of weak electromagnetic radiation, or statistical evidence for increased suicide rates, cancer and infertility risks for those living near radio masts or overhead power lines. Most of these studies were later found to be either erroneous or irreproducible (in fact mostly both). Studies of the dielectric properties of DNA solutions did not escape this fiasco. Resonance absorptions at specific microwave frequencies were reported for DNA solutions [155–159] for which 'striking' theoretical support was obtained [160]. All of this experimental work was found to be irreproducible by the careful measurements of others [161–164].

8.5.1.2 RNA

The number of reported studies of the dielectric properties of RNA is low. Bonincontro *et al.* [165] measured the dielectric properties of isolated ribosomal RNA (rRNA) to verify the extent to which it contributes to the dispersion observed at MHz frequencies for entire *E. coli* 70S ribosomes. The bacterial ribosome, a nucleoprotein particle responsible for the decoding of messenger RNA (mRNA) into protein, sediments as 70S particles (of size 21 nm) consisting of roughly two-thirds RNA and one-third protein [166]. The 70S particle is formed of two subunits, 30S and 50S; 30S is composed of one RNA molecule and 21 ribosomal proteins, while 70S comprises two RNA and 33 proteins. Magnesium ions play an important part in the maintenance of the ribosomal activity, both *in vivo* and *in vitro*. Two distinct dielectric relaxations were observed for the 70S particle and its two subunits. The α-dispersion at around 100 KHz exhibited a strong dependence on Mg^{2+} ions, shifting to lower frequencies in the absence of the ion, consistent with a larger exposure of the RNA moity to the solvent [167]. The α-dispersion is thus considered to be associated with fluctuating counterions along the rRNA chains. The β-dispersion appears in the MHz range and is not affected by magnesium ions. At 25 °C the isolated ribosomal proteins exhibit a dispersion at ~5 MHz, with a dielectric increment $\Delta\varepsilon$ of 8.1 permittivity units, compatible with a Debye-type rotation of an average dipole moment of 730 ± 30 debye units. The activation enthalpy of 4.1 kcal mol^{-1} for the isolated proteins was less than half of that (8.7 kcal mol^{-1}) observed for the whole 70S particle. The isolated ribosomal RNA exhibits a dispersion at ~9 MHz at 25 °C with a dielectric increment of ~30 permittivity units. This dispersion was analysed according to Mandel's counterion fluctuation model [126, 127] where counterions can freely move along a subunit of the RNA, but cannot cross from one subunit to another unless it surmounts a potential energy barrier between them. The length of this subunit is closely related to the frequency of relaxation and the mobility of the counterions. From an analysis using Mandel's model, a subunit length of ~15 nm was determined [165]. The fact that this length is considerably smaller than the subunit length of 62 nm at 25 °C for calf thymus DNA [127] was taken to indicate that rRNA possesses a large flexibility. Most significantly, the relaxation time for the RNA dispersion *increased* slightly with temperature and so contrary to the case for the proteins did not obey the Arrhenius Law where the relaxation time *increases* with temperature. Mainly on the basis of this finding, Bonincontro *et al.* [165] concluded that the β-dispersion present in the MHz region in ribosome suspensions can be attributed to the proteins and *not* rRNA. However, in later studies this conclusion was retracted. On removal of some of the proteins from the 70S particle by treatment with LiCl (known to inhibit ribosome function) only the α-dispersion was modified [168]. It was thus concluded that rRNA and proteins remaining in the core of the 70S particle are mainly responsible for the β-dispersion present in the MHz region. An interesting added finding for the α-dispersion was the calculation of 86 nm for the average subunit length for polarization of the condensed counterions. This is surprisingly similar to the maximum circumference of the ribosome if it is assumed to be approximately spherical.

The dielectric properties of ribosomal RNA extracted from the 16S and 23S ribosomal subunits of *E. coli* have also been deduced from dielectrophoresis studies over the frequency range 3 kHz to 50 MHz by Giraud *et al.* [169]. This is described further in Chapter 11, but in brief a positive value for the Clausius–Mossotti factor was found between 3 kHz and 1 MHz, followed by reduction to a *negative* value as the frequency was increased above 9 MHz. This behaviour has not been observed for DNA. The low frequency behaviour corresponds to an induced dipole moment of 3300 D and a polarizability of 7.8×10^{-32} F.m^2. The behaviour above 9 MHz is consistent with the rRNA molecules exhibiting a negative polarizability with respect to the surrounding aqueous medium, equivalent to a net *negative* dipole moment of 250 D, an effective permittivity value of $78.5\varepsilon_0$ and a relatively small surface conductance of ~0.1 nS. This suggests that the rRNA samples studied had a fairly open structure accessible to the surrounding water molecules, with counterions strongly bound to the charged phosphate groups in the rRNA backbone. The finding of a net *negative* dipole moment for the hydrated rRNA could indicate that the induced dipole at the charged rRNA-water interface is antiparallel to the rRNA dipole. The polarization of the hydration layer effectively screens the rRNA molecule from the external field instead of amplifying it as predicted by Onsager's model (Equation 6.35). This effect, which can result in negative dielectrophoresis, has been reported for the case of hydrated proteins by numerical simulations [170].

Finally, protein-RNA interactions play important roles in biological processes such as gene regulation and protein synthesis, for example. Although many structures of various types of protein-RNA complexes have been determined, the mechanism of protein-RNA recognition is not clear. Ahmad and Sarai [171] performed a systematic analysis of three bulk electrostatic properties of RNA-binding proteins, namely net charge, dipole moment and quadrupole moment. These were calculated from low-resolution protein structures with only main-chain coordinates, in order to estimate how far these simple properties are able to identify RNA-binding proteins from control proteins. Their results indicate that there exists a pattern of electric moments in RNA-binding proteins, which is different from the control data as well as within the proteins binding to various types of RNAs. One type of RNA-binding proteins can be distinguished from the other on the basis of these properties with various degrees of accuracy. The dipole and quadrupole moments for proteins that bind to ribosomal RNA stand out by being larger in comparison with all other classes of protein, suggesting that the main driving force for the formation and functioning of ribosomal assembly has strong electrostatic character revealed not only by their overall charge, but also by orientations and spherical asymmetry contained in higher values of the dipole moments.

8.6 Summary

This chapter has described the dielectric properties of aqueous solutions of particular relevance to dielectrophoresis experiments on biological materials. The dielectric properties of pure water can therefore serve as our baseline. At 25 °C pure water exhibits a static relative permittivity value close to $\varepsilon_w = 78.4$ and remains at this value as the frequency of an applied electric field is increased up to 1 GHz. The static (DC) and low-frequency permittivity is temperature sensitive, for example changing to values of 80.2 and 76.6 at 25 °C and 25 °C, respectively. Beyond 1 GHz pure water exhibits a dielectric dispersion, with the associated dielectric loss parameter (ε'') having a peak value at ~18 GHz. This dispersion (known as the γ-dispersion) closely matches that expected of a Debye-type orientational relaxation of the water dipoles, characterized by a single relaxation time $\tau = 8.58 \times 10^{-12}$ s. The high-frequency tail of the γ-dispersion merges into the first of two small dispersions. At 25 °C these occur at 167.8 GHz and 1.94 THz. Atomic resonances then follow at 4.03 THz and 14.48 THz. This makes it difficult to determine an accurate value for the permittivity parameter ε_∞ that bounds the high-frequency end of the γ-dispersion. The infrared refractive index (n) for water is 1.33, which based on the relationship $\varepsilon_\infty = n^2$ gives $\varepsilon_\infty = 2.1$. Water in its vapour phase has an individual molecular dipole moment of 1.855 debye units (6.2×10^{-30} C m). When the vapour phase condenses into the polar liquid, the effective dipole moment of each water molecule is increased by the induction effect of its neighbours. This can be taken into account by employing Onsager's improvement of the Debye model, but even so the predicted value for ε_s is 31.0 at 25 °C instead of $\varepsilon_s = 78.4$. This disparity is understood in terms of tetrahedral bonding by hydrogen bonds of a water molecule to its neighbours. A field-induced reorientation of a water dipole has to be coordinated with rearrangements of its neighbours. Kirkwood introduced an orientation correlation factor g to take this effect into account.

The highest frequency of relevance to most dielectrophoresis measurements is no more than around 50 MHz, but some experiments have been reported to operate up to ~0.5 GHz. This is still below where the γ-dispersion commences. The absolute value of the permittivity for pure water at 25 °C can therefore be assumed to equal $78.4\varepsilon_o$, where ε_o is the permittivity of free space (8.854×10^{-12} farads per metre). However, this value is decreased if a salt is added to the water. This response to the presence of solvated ions arises from more than just the volume effect of replacing polar water molecules with nonpolar ionic particles. In particular, the strong electric field generated by the point charge of each ion has the effect of orienting the water molecules, hindering their rotations in response to an applied electrical field. For salt concentrations less than 0.5 M a linear relationship exists between the static relative permittivity ε_s and salt concentration of the form:

$$\varepsilon_s = \varepsilon_w + 2\bar{\delta}c, \text{ with } \bar{\delta} = (\delta^+ + \delta^-)/2$$

where c is the salt concentration in moles per litre. The factor $\bar{\delta}$ quantifies the extent to which the salt modifies the permittivity value, with δ^+ and δ^- being the contributions arising from the cation and anion, respectively. The values for $\bar{\delta}$ is *negative* and so represents a dielectric *decrement*. The addition of a salt to water *lowers* its permittivity because the volume occupied by the ion and its surrounding hydration has a lower polarizability than the volume of bulk water it has displaced. The mean value $\bar{\delta}$ is used because water molecules attracted to a negative ion are less rotationally hindered than those attracted to a positive ion.

Sugar molecules dissolve readily in water, not only because they are polar but also because they can form hydrogen bonds with surrounding water molecules. Although they are polar and uncharged, the volume occupied by a sugar molecule and its surrounding hydration has a lower polarizability than the volume of bulk water it has displaced. Aqueous sugar solutions thus also exhibit a dielectric *decrement*. For example, a 1 M glucose solution exhibits a static relative permittivity value of ~74.2 at 25 °C. This value remains constant up to a frequency of ~100 MHz where the commencement of two weak dispersions occurs before the γ-dispersion of the bulk water. The dielectric properties of sugar solutions can be understood in terms of a three-component system: (i) the rotationally hindered bound hydration around each sugar molecule; (ii) the relaxing polar sugar molecules; (iii) normal bulk water. For example, glucose solutions are found to have a small dielectric loss peak at 2.3 GHz (relaxation time $\tau \approx 0.07$ ns) assigned to rotational relaxation of the glucose dipoles. The bound water of hydration around the glucose molecules is considered responsible for a much weaker dispersion at ~0.59 GHz ($\tau \approx 0.3$ ns) with the bulk water exhibiting a slightly modified γ-dispersion loss peak at 8.6 GHz ($\tau \approx 0.02$ ns).

The zwitterionic nature of amino acids lends to them a dipole moment per unit volume exceeding that of water. Amino acid solutions thus exhibit a larger static permittivity value than water. For aqueous solutions of α-amino acids at 25 °C the dielectric *increment* per mole has a value of ~27 relative permittivity units for frequencies approaching 1 GHz and concentrations up to 2.5 M. The Debye model for dipole relaxation predicts a relaxation frequency of ~12.6 GHz for glycine in water. However, the observed characteristic frequency is 3.3 GHz. The difference between the predicted and experimental value indicates that there are significant electrostatic interactions between the amino acid and the water molecules. The molar dielectric increments of zwitterion buffers such as HEPES and MOPS are larger than that of α-amino acids, with values of 90 and 40 permittivity units, respectively. The maximum recommended concentration of these pH buffers is 20 mM for mammalian cell suspensions, so that their use should increase the permittivity above that of pure water by no more than 1~2 permittivity units (i.e., ε_s increases to ~80 at 25 °C. The net effect of adding salts and sugars (with a resulting dielectric *decrement*) and a pH buffer (with a resulting dielectric *increment*) to an aqueous cell suspending medium, for example, can in principle be calculated for theoretical modelling purposes. However, the correction required is likely to be very small and negligible in comparison with other approximations used in the modelling.

Protein solutions typically exhibit a dielectric dispersion in the MHz region (i.e., relaxation times τ_p of the order microseconds). The consensus is that this dispersion results from electric field-induced orientation relaxations of the protein's permanent dipole moment. The observed total dielectric increment $\Delta\varepsilon_T$ and corresponding dipole moment values are much larger than those found for amino acid and peptide solutions, reflecting the fact that proteins have much larger molecular weights. The total dielectric increment value $\Delta\varepsilon_T$ takes into account the fact that, at the high-frequency end of the dispersion, the permittivity value (ε_{hf}) is less than that for water because the volume occupied by the nonrelaxing protein molecule and its water of hydration has a lower polarizability than the bulk water it displaces. Following the convention for naming dispersions according to their order in the frequency scale, the dispersion exhibited by a protein solution in the 100 kHz–10 MHz range is called its β-dispersion. It obeys a frequency (ω) dependent Debye-type relaxation of the form:

$$\varepsilon' = \varepsilon_{hf} + \frac{\Delta\varepsilon_T}{1 + \omega^2\tau_p^2}$$

The protein dipole moment is calculated using the formula

$$\mathrm{p} = \left(\frac{2\varepsilon_o MkT\delta}{N_A}\right)^{1/2}$$

where M is the protein molecular weight and δ is the *specific* dielectric decrement, calculated from the slope of a linear plot of $\Delta\varepsilon_T$ versus protein concentration (mg/mL). Dipole moment values for proteins typically lie in the range 150~700 debye units, with no obvious relationship between the dielectric increment and molecular weight. The equation used to calculate the dipole moment can be modified to take into account nonspherical shapes, such as oblate or prolate spheroids.

Apart from the γ-dispersion due to the bulk water at GHz frequencies, aqueous solutions of DNA in its native (double-helix) form exhibit two other dispersions: the α-dispersion that extends from sub-hertz frequencies up to ~1 kHz; the β-dispersion lying between ~10 kHz and ~50 MHz. The α-dispersion exhibits a very large dielectric increment, which when inserted into the equations above produce dipole moment values as large as 100 000 debye units! Other characteristic of the α-dispersion for DNA are that it is very sensitive to an increase of the concentration of salts in the solution (the dielectric increment decreases and the dispersion shifts to a higher frequency) and that the dielectric increment and relaxation time is proportional to the square of the effective length of the DNA. This last result suggests that helical DNA has a large longitudinal dipole moment. However, the Debye dipole rotation model through Equation (7.20) predicts that the relaxation time should be proportional to the cube of the effective length of the molecule. An even more conspicuous observation is that the dipole moment decreases drastically on heat or acid denaturation of its double stranded structure. This result stands out because the double helix structure of native DNA should not possess a permanent dipole moment, whereas in its single strand form it might do so. In its native form DNA consists of two intertwined helical strands of DNA running in opposing directions, with the base pairs stacked in such a way that their molecular planes are perpendicular to the helical axis. Native DNA does not possess a permanent dipole moment because the moments of paired nucleotides and dipoles in the main strands should all cancel out. There is broad consensus that the α-dispersion exhibited by aqueous solutions of native DNA can be understood in terms of field-induced fluctuating migration of the more diffusely bound counterions over distances of the same order as the entire dimension of the DNA molecule. The counterions are attracted to the negatively charged phosphate groups along the DNA chains and are envisaged to move freely under the influence of an external electric field along the DNA structure until they meet a potential energy barrier. These barriers result from perturbations in the equipotentials due to 'kinks' in the average 'wormlike' conformational shape of the DNA molecule. The DNA molecule itself is *not* considered to possess a permanent dipole moment. This fluctuating ion model may also apply to the β-dispersion, where more strongly bound and localized counterions are restricted to shorter sublengths of the DNA. However, there is also good evidence to suggest that for polyelectrolyte chains other than DNA the induced polarization is *perpendicular* to the chains. Mobile counterions are considered to move within an electrostatic potential that is modulated by the presence of potential wells associated with neighbouring polyelectrolyte chains. Under the influence of an external electric field these counterions can polarize by free diffusion in three-dimensional space to a scale of the order of the distance between the chains. Clarification of the origins of the DNA β-dispersion requires further experimental exploration and analysis.

Unlike the situation for DNA, few studies of the dielectric properties of RNA have been reported. Complete assemblies of the *E. coli* 70S ribosome and its two subunits 30S and 50S, exhibit an α-dispersion at around 100 kHz and a β-dispersion at MHz frequencies. The α-dispersion is considered to be mainly associated with fluctuating counterions along the rRNA chains, whilst contributions to the β-dispersion involve both counterion fluctuations along the RNA and relaxations of the permanent dipole moments of the core proteins. There is an indication, from dielectrophoresis experiments, that the rRNA molecule exhibits a net negative dipole moment above 9 MHz. This may result from the polarization of the hydrated water molecules around the highly charged rRNA acting to screen, rather than amplify, the local electric field experienced by the rRNA.

8.7 References

1 Schwan, H. P. (1957) Electrical properties of tissue and cell suspensions, in *Advances in Biological and Medical Physics*, vol. **V** (eds J. H. Lawrence and C. A. Tobias) Academic Press, New York, NY, p 147–209.

2 Cole, K. S. (1968) *Membranes, Ions and Impulses*, University of California Press, Berkeley, CA.

3 Hasted, J. B. (1973) *Aqueous Dielectrics*, Chapman & Hall, London.

4 Grant, E. H., Sheppard, R. J. and South, G. P. (1978) *Dielectric Behavior of Biological Molecules in Solution*, Oxford University Press, Oxford.

5 Pethig, R. (1979) *Dielectric and Electronic Properties of Biological Materials*, John Wiley & Sons, Ltd, Chichester.

6 Pethig, R. and Kell, D. B. (1987) The passive electrical properties of biological systems: their significance in physiology, biophysics and biotechnology. *Phys. Med. Biol.* **32**(8), 933–970.

7 Takashima, S. (1989) *Electrical Properties of Biopolymers and Membranes*, Adam Hilger, Bristol.

8 Foster, K. R. and Schwan, H. P. (1995) Dielectric properties of tissues, in *Handbook of Biological Effects of Electro-magnetic Radiation*, 2nd edn (eds C. Polk and E, Postow). CRC Press, Boca Raton, FL, pp. 25–102.

9 Grimnes, S. and Martinsen, O. G. (2000) *Bioimpedance and Bioelectricity Basics*, 3rd edn. Academic Press, San Diego.

10 Ellison, W. J., Lamkaouchi, K. and Moreau, J.-M. (1996) Water: A dielectric reference. *J. Mol. Liq.* **68**, 171–279.

11 Ellison, W. J. (2007) Permittivity of pure water, at standard atmospheric pressure, over the frequency range 0–25 THz and the temperature range 0–100 degrees C. *J. Phys. Chem. Reference Data* **36**(1), 1–18.

12 Fröhlich, H. (1958) *Theory of Dielectrics*, 2nd edn. Oxford University Press, Oxford, p. 140.

13 Zelsmann, H. R. (1995) Temperature dependence of the optical constants for liquid H_2O and D_2O in the far IR region. *J. Mol. Structure* **350**, 95–114.

14 Powles, J. G. (1993) Cole–Cole plots as they should be. *J. Mol. Liq.* **56**, 35–47.

15 Suresh, S. J. and Naik, V. M. (2000) Hydrogen bond thermodynamic properties of water from dielectric constant data. *J. Chem. Phys.* **113**(21), 9727.

16 Rastogi, A., Yadav, S. and Suresh, S. J. (2011) Note: extraction of hydrogen bond thermodynamic properties of water from dielectric constant and relaxation time data. *J. Chem. Phys.* **135**, 086101.

17 Maksyutenko, P., Rizzo, T. R. and Boyarkin, O. V. (2006) A direct measurement of the dissociation of water. *J. Chem. Phys.* **125**, 181101.

18 Stokely, K., Mazza, M. G., Stanley, H. E. and Franzese, G. (2010) Effect of hydrogen bond cooperativity on the behaviour of water. *Proc. Natl. Acad. Sci. USA* **107**(4), 1301.

19 Henchman, R. H. and Irudayam, S. J. (2010) Topological hydrogen-bond definition to characterize the structure and dynamics of liquid water. *J. Phys. Chem. B* **114**(50), 16792–16810.

20 Eisenberg, D. and W. Kauzmann, W. (1969) *The Structure and Properties of Water*, Oxford University Press, New York, NY.

21 Kirkwood, J. G. (1939) The dielectric polarization of polar liquids. *J. Chem. Phys.* 7(10), 911–919.

22 Oster, G. and Kirkwood, J. G. (1943) The influence of hindered molecular rotation on the dielectric constants of water, alcohols and other polar liquids. *J. Chem. Phys.* **11**(4), 175–178.

23 Sciortino, F., Geiger, A. and Stanley, H. E. (1992) Network defects and molecular mobility in liquid water. *J. Chem. Phys.* **96**(5), 3857–3865.

24 Kaatze, U., Behrends, R. and Potel, R. (2002) Hydrogen network fluctuations and dielectric spectroscopy of liquids. *J. Non-Cryst. Sol.* **305**, 19–28.

25 Agmon, N. (2012) Liquid water: From symmetry distortions to diffusive motion. *Accounts Chem. Res.* **45**(1), 63–73.

26 Ohmine, I. and Tanaka, H. (1993) Fluctuation, relaxations, and hydration in liquid water. Hydrogen-bond rearrangement dynamics. *Chem. Rev.* **93**, 2545–2566.

27 Atkins, P. W. and De Paula, J. (2002) *Physical Chemistry*, 7th edn. W. H. Freeman, New York, NY, Ch. 27.

28 de Grotthuss, C. J. T. (1806) Mémoire sur la décomposition de l'eau et des corps qu'elle tient en dissolution à l'aide de l'électricité galvanique. *Annales de Chimie (Paris)* **58**, 54–74. An English translation of this paper is given by Pomès, R. (2006) *Biochim. Biophys. Acta* **1757**, 871–875.

29 Wraight, C. A. (2006) Chance and design: Proton transfer in water, channels and bioenergetic proteins. *Biochim. Biophys. Acta* **1757**, 886–912.

30 Foster, K. R., Bidinger. J. M. and Carpenter, D. O. (1976) The electrical resistivity of cytoplasm. *Biophys. J.* **16**(9), 991–1001.

31 Bondi, A. (1964) Van der Waals volumes and radii. *J. Phys. Chem.* **68**, 441–451.

32 Hasted, J. B., Ritson, D. M. and Collie, C. H. (1948) Dielectric properties of aqueous ionic solutions. Parts I and II. *J. Chem. Phys.* **16**(1), 1–21.

33 Buchner, R. and Barthel, J. (2001) Dielectric relaxation in solutions. *Annu. Rep. Prog. Chem., Sect. C* **97**, 349–382.

34 Tait, M. J., Suggett, A., Franks, F. *et al.* (1972) Hydration of monosaccharides: A study by dielectric and nuclear magnetic relaxation. *J. Solution Chem.* **1**(2), 131–151.

35 Höchtl, P., Boresch, S. and Steinhauser, O. (2000) Dielectric properties of glucose and maltose solutions. *J. Chem. Phys.* **112**(22), 9810–9821.

36 Saito, A., Miyawaki, O. and Nakamura, K. (1997) Dielectric relaxation of aqueous solution with low-molecular-weight nonelectrolytes and its relationship with solution structure. *Biosci. Biotech. Biochem.* **61**(11), 1831–1835.

37 Arnold, W. M., Gessner, A. G. and Zimmermann, U. (1993) Dielectric measurements on electro-manipulation media, Biochim. *Biophys. Acta* **1157**, 32–44.

38 Guo, W. C., Zhu, X. H., Liu, Y. and Zhuang, H. (2010) Sugar and water contents of honey with dielectric property sensing. *J. Food. Eng.* **97**(2), 275–281.

39 Wyman, J. (1934) Dielectric constants of polar solutions. *J. Am. Chem. Soc.* **56**, 536–544.

40 Dunning, W. J. and Shutt, W. J. (1938) The dielectric constants of zwitterions and polar molecules as related to p_{H}. *Trans. Faraday Soc.* **34**, 479–785.

41 Kirkwood, J. G. (1934) Theory of solutions of molecules containing weakly separated charges with special application to zwitterions. *J. Chem. Phys.* **2**(7), 351–361.

42 Kirkwood, J. G. (1939) Theoretical studies upon dipolar ions. *Chem. Rev.* **24**, 233–251.

43 Shepherd, J. C. W and Grant, E. H. (1968) Dielectric properties of amino acid solutions. I. Dispersion in aqueous ε-aminocaproic acid solutions. *Proc. Roy. Soc. Lond. A*, **307**(1490), 335–344.

44 Wyman, J. (1936) The dielectric constants of solutions of dipolar ions. *Chem. Rev.* **19**(3), 213–239.

45 Arnold, W. M. (2001) Positioning and levitation media for the separation of biological cells. *IEEE Trans. Ind. Appl.* **37**(5), 1468–1475.

46 Hol, W. G. J. (1985) The role of the a-helix dipole in protein function and structure. *Prog. Biophys. Mol. Biol.* **45**, 149–195.

47 Dickerson, R. E., Takano, T., Eisenberg, D. *et al.* (1971) Ferricytochrome *c*: I. General features of the horse and bonito proteins at 2 Å resolution. *J. Biol. Chem.* **246**, 1511–1535.

48 South, G. P. and Grant, E. H. (1972) Dielectric dispersion and dipole moment of myoglobin in water. *Proc. Roy. Soc. Lond. A* **328**, 371–387.

49 Takashima, S. (1993) Use of protein database for the computation of the dipole moments of normal and abnormal hemoglobins. *Biophys. J.* **64**(5), 1550–1558.

50 Takashima, S. and Asami, K. (1993) Calculation and measurement of the dipole moment of small proteins: use of protein data base. *Biopolymers* **33**(1) 59–68.

51 Antosiewicz, J. (1995) Computation of the Dipole Moments of Proteins. *Biophys. J.* **69**, 1344–1354.

52 Herz, T., Otto, P. and Clark, T. (2000) On the band gap in peptide α-helices. *Int. J. Quantum Chem.* **79**(2), 120–124.

53 Chen, X., Zhang, L., Sun, W. *et al.* (2010) α-helix c-terminus acting as a relay to mediate long-range hole migration in proteins. *J. Phys. Chem. Lett.* **1**(10), 1637–1641.

54 Oncley, J. L. (1942) The investigation of proteins by dielectric measurements. *Chem. Rev.* **30**, 433–450.

55 Perrin, F. (1934) The Brownian movement of an ellipsoid – the dielectric dispersion of ellipsoidal molecules. *J. Physique Radium* **5**, 497–511.

56 Kendrew, J. C., Watson, H. C., Standberg, B. E. *et al.* (1961) A partial determination by X-ray methods and its correlation with chemical data. *Nature* **190**, 666–670.

57 Oncley, J. L. (1938) Studies of the dielectric properties of protein solutions. I. Carboxyhemoglobin. *Am. J. Chem. Soc.* **60**, 1115–1123.

58 Wyman, J. (1936) Polarization and dielectric constant of liquids. *J. Am. Chem. Soc.* **58**, 1482–1486.

59 Chari, R., Singh, S. N., Yadav, S. *et al.* (2012) Determination of the dipole moments of RNAse SA wild type and a basic mutant. *Proteins* **80**, 1041–1052.

60 Keefe, S. E. and Grant, E. H. (1974) Dipole moment and relaxation time of ribonuclease. *Phys. Med. Biol.* **19**, 701–707.

61 Schlecht, P. (1969) Dielectric properties of hemoglobin and myoglobin. II. Dipole moment of sperm whale myoglobin. *Biopolymers* **8**(6), 757–765.

62 Scheider, W., Dintzis, H. W. and Oncley, J. L. (1976) Changes in the electric dipole vector of human serum albumin due to complexing with fatty acids. *Biophys. J.* **16**, 417–431.

63 Barlow, D. J. and Thornton. J. M. (1986) The distribution of charged groups in proteins. *Biopolymers* **25**(9) 1717–1733.

64 Buckingham, A. D. (1953) The application of Onsager theory to spheroidal molecules. *Trans. Faraday Soc.* **49**, 881–886.

65 Sorensen, S. P. L. (1918) Protein studies. *Hoppe-Seyler's Zeit. Physiol. Chem.* **103**(1), 1–14.

66 Sorensen, S. P. L. and Sorensen, M. (1933) The solubility and dissociation tendency of carbonoxide-haemoglobins in ammoniumsulphate solution. *Biochem. Zeit.* **258**, 16–47.

67 Edsall, J. T. and McKenzie, H. A. (1983) Water and Proteins. 2. The location and dynamics of water in protein systems and its relation to their stability and properties. *Advan. Biophys.* **16**, 15–183.

68 Reynolds J. A. and Hough, J. M. (1957) Formulae for dielectric constant of mixtures. *Proc. Phys. Soc. B* **70**, 769–775.

69 Buchanan, T. J., Haggis, G. H., Hasted. J. B. and Robinson, B. G. (1952) The dielectric estimation of protein hydration. *Proc. Roy. Soc. Lond. A* **213**, 379–391.

70 Fricke, H. (1924) A mathematical treatment of the electrical conductivity and capacity of disperse systems I. The electric conductivity of a suspension of homogeneous spheroids. *Phys. Rev.* **24**, 575–587.

71 Grant, E. H., South, G. P., Takashima, S. and Ichimura, H. (1971) Dielectric dispersion in aqueous solutions of oxyhaemoglobin and carboxyhaemoglobin. *Biochem. J.* **122**, 691–699.

72 Pitera, J. W., Falta, M. and van Gunsteren, W. F. (2001) Dielectric properties of proteins from simulation: The effects of solvent, ligands, pH, and temperature. *Biophys. J.* **80**, 2546–2555.

73 Gilson, M. K., and Honig, B. H. (1986) The dielectric constant of a folded protein. *Biopolymers* **25**, 2097–2119.

74 Nakamura, H., Sakamoto, T. and Wada, A. (1988) A theoretical study of the dielectric constant of protein. *Protein Eng.* **2**, 177–183.

75 King, G., Lee, F. S. and Warshel, A. (1991) Microscopic simulations of macroscopic dielectric constants of solvated proteins. *J. Chem. Phys.* **95**(6), 4366–4377.

76 Simonson, T. (1998) The dielectric constant of cytochrome-c from simulations in a water droplet including all electrostatic interactions. *J. Am. Chem. Soc.* **20**, 4875– 4876.

77 Simonson, T. (2013) What is the dielectric constant of a protein when its backbone is fixed. *J. Chem. Theory Comp.* **9**, 4603–4608.

78 Haggis, G. H., Buchanan, T. J. and Hasted. J. B. (1951) Estimation of protein hydration by dielectric measurements at microwave frequencies. *Nature* **167**, 607–608.

79 Grant, E. H. (1962) Electrical behaviour of egg albumen solutions at ultra-high frequencies. *Nature* **196**, 1194–1195.

80 Schwan, H. P. (1965) Electrical properties of bound water. *Ann. N.Y. Acad. Sci.* **125**, 344–354.

81 Grant, E. H. (1966) Dielectric dispersion in bovine serum albumen. *J. Mol. Biol.* **19**, 133–139.

82 Pennock, B. E. and Schwan, H. P. (1969) Further observations on electrical properties of hemoglobin-bound water. *J. Phys. Chem.* **73**, 2600–2610.

83 Grant, E. H., Mitton, B. G. R., South, G. P. and Sheppard, R. J. (1974) An investigation by dielectric methods of hydration in myoglobin solutions. *Biochem. J.* **139**(2), 375–380.

84 Essex, C. G., Symonds, M. S., Sheppard, R. J. *et al.* (1977) Five-component dielectric dispersion in bovine serum albumin solution. *Phys. Med. Biol.* **22**, 1160–1167.

85 Grant, E. H., McClean, V. E. R., Nightingale, N. R. V. *et al.* (1986) Dielectric behavior of water in biological solutions – studies on myoglobin, human low-density lipoprotein, and polyvinylpyrrolidone. *Bioelectromagnetics* **7**, 151–162.

86 Kaatze, U. (1990) On the existence of bound water in biological systems as probed by dielectric spectroscopy. *Phys. Med. Biol.* **35**, 1663–1681.

87 Pethig, R. (1992) Protein-water interactions determined by dielectric methods. *Annu. Rev. Phys. Chem.* **43**, 177–205.

88 Miura, N., Asaka, N., Shinyashiki, N. and Mashimo, S. (1994) Microwave dielectric study on bound water of globule proteins in aqueous solution. *Biopolymers* **34**, 357–364.

89 Miura, N., Hayashi, Y., Shinyashiki, N. and Mashimo, S. (1995) Observation of unfreezable water in aqueous-solution of globule protein by microwave dielectric measurement. *Biopolymers* **36**, 9–16.

90 Sun, Y. X., Ishida, T. and Hayakawa, S. (2004) Dielectric study of heat-denatured ovalbumin in aqueous solution by time domain reflectometry method. *J. Agric. Food Chem.* **52**, 2351–2357.

91 Oleinikova, A., Sasisanker, P. and Weingärtner, H. (2004) What can really be learned from dielectric spectroscopy of protein solutions? A case study of ribonuclease A. *J. Phys. Chem. B* **108**, 8467–8474.

92 Shinyashiki, N., Yamamoto, W., Yokoyama, A. *et al.* (2009) Glass transitions in aqueous solutions of protein (bovine serum albumin). *J. Phys. Chem. B* **113**, 14448–14456.

93 Cametti, C., Marchetti, S., Gambi, C. M. C. and Onori, G. (2011) Dielectric relaxation spectroscopy of lysozyme aqueous solutions: analysis of the delta-dispersion and the contribution of the hydration water. *J. Phys. Chem. B* **115**, 7144–7153.

94 Wolf, M., Gulich, R., Lunkenheimer, P. and Loidl, A. (2012) Relaxation dynamics of a protein solution investigated by dielectric spectroscopy. *Biochim. Biophys. Acta* **1824**, 723–730.

95 Debye, P. and Falkenhagen, H. (1928) Dispersion of conductivity and dielectric constants of strong electrolytes. *Physik. Zeit.* **29**, 121–132; 401–426.

96 Kirkwood, J. G. and Shumaker, J. B. (1952) The influence of dipole moment fluctuations on the dielectric increment of proteins in solutions. *Proc. Natl. Acad. Sci. USA* **38**(10), 855–862.

97 Jacobson, B. (1955) On the interpretation of dielectric constants of aqueous macromolecular solutions. Hydration of macromolecules. *J. Am. Chem. Soc.* **77**(11), 2919–2926.

98 O'Konski, C. T. (1955) Effect of interfacial conductivity on dielectric properties. *J. Chem. Phys.* **23**(8), 1559.

99 Schwarz, G. (1962) A theory of the low-frequency dielectric dispersion of colloidal particles in electrolyte solution. *J. Phys. Chem.* **66**(12), 2636–2642.

100 Watson, J. D. and Crick, F. H. C. (1953) Molecular structure of nucleic acids: A structure for deoxyribose nucleic acid. *Nature* **171**, 737–738.

101 Guerra, C. F., Bickelhaupt, F. M., Snijders, J. G. and Baerends, E. J. (2000) Hydrogen bonding in DNA base pairs: Reconciliation of theory and experiment. *J. Am. Chem. Soc.* **122**, 4117–4128.

102 Leung, C., Maradan, D., Kramer, A. *et al.* (2010) Improved Kelvin probe force microscopy for imaging individual DNA molecules on insulating surfaces. *Appl. Phys. Lett.* **97**, 203703.

103 Zimmer, Ch., Luck, G. and Venner, H. (1968) Studies on the conformation of protonated DNA. *Biopolymers* **6**, 563–574.

104 Russo, N., Toscano, M., Grand, A. and Jolibois, F. (1998) Protonation of thymine, cytosine, adenine, and guanine DNA nucleic acid bases: Theoretical investigation into the framework of density functional theory. *J. Comp. Chem.* **19**(9), 989–1000.

105 Crick, F. (1970) Central dogma of molecular biology. *Nature* **227**, 561–563.

106 Judson, H. F. (1979) *The Eighth Day of Creation: Makers of the Revolution in Biology*, Simon & Schuster, New York, pp. 335–339.

107 Greene, W. C. (1993) AIDS and the immune system. *Scientific American* **269**(3), 99–105.

108 Jungner, G., Jungner, I. and Allgén, L. G. (1949) Molecular weight determination on thymonucleic acid components by dielectric measurements. *Nature* **163**(4152), 849–850.

109 Jungner, G., Jungner, I. and Allgén, L. G. (1949) Enzymatic degradation of desoxyribose nucleic acid by crystalline desoxyribonuclease. *Nature* **164**(4180), 1009.

110 Jungner, G. and Jungner, I. (1952) On the interpretation of dielectric results on sodium thymonucleate. I. On the relation between increment, dipole moment and molecular weight. *Acta. Chem. Scandinavica* **6**, 1391–1404.

111 Jungner, G. (1952) On the interpretation of dielectric results on sodium thymonucleate. II. The dipole association of highly polymerized molecules. *Acta. Chem. Scandinavica* **6**, 1405–1416.

112 Jacobson, B. (1953) Method for obtaining streaming orientation and simultaneous determination of

dielectric properties in macromolecular solutions. *Rev. Sci. Instr.* **24**(10), 949– 954.

113 Jerrard, H. G. and Simmons, B. A. W. (1959) Dielectric studies on deoxyribonucleic acid. *Nature* **184**(4700), 1715–1716.

114 Peterlin, A. and Reinhold, Chr. (1965) The influence of laminar flow on the dielectric constant of dilute polymer solutions. *Kolloid-Zeit. u. Zeit Polymere* **204**, 23–28.

115 Takashima, S. (1963) Dielectric dispersion of DNA. *J. Mol. Biol.* **7**, 455–467.

116 Takashima, S. (1966) Dielectric dispersion of deoxyribonucleic acid. II. *J. Phys. Chem.* **70**, 1372–1380.

117 Takashima, S. (1967) Mechanism of dielectric relaxation of deoxyribonucleic acid. *Advances in Chemistry Series* **63**, 232–252.

118 Takashima, S. (1967) Effect of ions on the dielectric relaxation of DNA. *Biopolymers* **5**, 899–913.

119 Hanss, M. (1966) Mesure de la dispersion dielectrique en tres basse frequence de solutions de DNA. *Biopolymers* **4**, 1035–1041.

120 Tung, M. S., Molinari, R. J., Cole, R. H. and Gibbs. J. H. (1977) Influence of temperature and ionic strength on the low frequency dielectric dispersion of DNA solutions. *Biopolymers* **16**, 2653–2669.

121 Hayakawa, R., Kanda, H., Sakamoto, M. and Wada, Y. (1975) New apparatus for measuring the complex dielectric constant of a highly conductive material. *Jap. J. Appl. Phys.* **14**(12), 2039–2052.

122 Sakamoto, M., Kanda, H., Hayakawa, R. and Wada, Y. (1976) Dielectric relaxation of DNA solution. *Biopolymers* **15**, 879–892.

123 Sakamoto, M., Hayakawa, R. and Wada, Y. (1978) Dielectric relaxation of DNA solution. II. *Biopolymers* **17**, 1507–1512.

124 Sakamoto, M., Hayakawa, R. and Wada, Y. (1979) Dielectric relaxation of DNA solution. III. Effects of DNA concentration, protein contamination and mixed solvents. *Biopolymers* **18**, 2769–2782.

125 Stellwagen, N. C., Shirai, M. and O'Konski, C. T. (1965) Dielectric investigations of DNA solutions. Ninth Annual Biophysical Society Meeting, San Francisco, California, 24 February. Abstract.

126 van der Touw, F. and Mandel M. (1974) Dielectric increment and dielectric dispersion of solutions containing simple charged linear macromolecules: I. *Theory, Biophys. Chem.* **2**, 218–241.

127 Mandel, M. (1977) Dielectric properties of charged linear macromolecules with particular reference to DNA. *Annal. New York Acad. Sci.* **303**, 74–87.

128 Bonicontro, A., Cametti, C., Di Basio, A. and Pedone, F. (1984) Effects of ions on counterion fluctuation in low-molecular weight DNA dielectric dispersions. *Biophys. J.* **45**, 495–501.

129 Saif, B., Mohr, R. K., Montrose, C. J. and Litovitz, T. A. (1991) On the mechanism of dielectric relaxation in aqueous DNA solutions. *Biopoloymers* **31**, 1171–1180.

130 Bakewell, D. J., Ermolina, I., Morgan, H. *et al.* (2000) Dielectric relaxation measurements on 12 Kbp plasmid DNA. *Biochim. Biophys. Acta* **1493**, 151–158.

131 Debye, P. and Falkenhagen, H. (1928) Dispersion of conductability and dielectricity constants of strong electrolytes. *Phys. Zeit.* **29**, 401–426.

132 van Beek, W. M. and Mandel, M. (1978) Static relative permittivity of some electrolyte solutions in water and methanol. *J. Chem. Soc., Faraday Trans. 1* **74**, 2339–2351.

133 Ross, L. D. and Scruggs, R. L. (1964) Electrophoresis of DNA. II. Specific interactions of univalent and divalent cations with DNA. *Biopolymers* **2**, 79–89.

134 Spegt, P. and Weill, G. (1976) Magnetic resonance distinction between site bound and atmospherically bound paramagnetic counter-ions in polyelectrolyte solutions. *Biophys. Chem.* **4**, 143–149.

135 Reuben, J., Shporer, M. and Gabbay, E. J. (1975) The alkali-ion-DNA interaction as reflected in the nuclear relaxation rates of ^{23}Na and ^{87}Rb. *Proc. Nat. Acad. Sci. USA* **72**, 245–247.

136 Harrington, R. E. (1978) The optico-hydrodynamic properties of high molecular weight DNA. III. The effects of NaCl concentration. *Biopolymers* **17**, 919–936.

137 Manning, G. S. (1969) Limiting laws and counterion condensation in polyelectrolyte solutions: I. Colligative properties. *J. Chem. Phys.* **51**, 924–933.

138 Manning, G. S. (1978) The molecular theory of polyelectrolyte solutions with applications to the electrostatic properties of polynucleotides. *Quart. Rev. Biophys.* **11**, 179–246.

139 Manning, G. S. (1978) Limiting laws and counterion condensation in polyelectrolyte solutions: V. Further development of the chemical model. *Biophys. Chem.* **9**(1), 65–70.

140 Mandel, M. (1961) The electric polarisation of rod-like, charged macromolecules. *Mol. Phys.* **4**, 489–496.

141 Vreugdenhil, T., Van der Touw, F. and Mandel, M. (1979) Electric permittivity and dielectric dispersion of low molecular weight DNA at low ionic strength. *Biophys. Chem.* **10**, 67–80.

142 Oosawa, F. (1970) Counterion fluctuation and dielectric dispersion in linear polyelectrolytes. *Biopolymers* **9**, 677–688.

143 Oosawa, F. (1957) A simple theory of thermodynamic properties of polyelectrolyte solutions. *J. Polymer Sci.* **23**, 421–430.

144 Imai, N. and Onishi, T. (1959) Analytical solution of Poisson–Boltzmann equation for the two-dimensional many-center problem. *J. Chem. Phys.* **30**, 1115–1116.

145 Minikata, A. (1977) Dielectric dispersion of polyelectrolytes due to ion fluctuation. *Annal. New York Acad. Sci.* **303**, 107–120.

146 Ito, K., Yagi, A., Ookubo, N., Hayakawa, R. (1990) Crossover behavior in high-frequency dielectric relaxation of linear polyions in dilute and semidilute solutions. *Macromolecules* **23**, 857–862.

147 Bordi, F., Cametti, C., Gili, T. *et al.* (2005) Solvent quality influence on the dielectric properties of polyelectrolyte solutions: A scaling approach. *Phys. Rev. E.* **72**, 031806.

148 Bordi, F. and Sarti, S. (2015) Dielectric properties of polyelectrolytes and lipid vesicles, in *Dielectric Relaxation in Biological Systems* (eds V. Raicu and Y. Feldman), Oxford University Press, Oxford, Part 3.5.

149 Charney, E. and Lee, C.-H. (1980) Polyelectrolyte orientation in electric fields. The effect of the dielectric constant. *Macromolecules* **13**, 66–68.

150 Bone, S. and Small, C. A. (1995) Dielectric studies of ion fluctuation and chain bending in native DNA. *Biochim. Biophys. Acta* **1260**, 85–93.

151 Bone, S., Lee, R. S. and Hodgson, C. E. (1996) Dielectric studies of intermolecular interactions in native DNA. *Biochim. Biophys. Acta* **1306**, 93–97.

152 Katsumoto, Y., Omori, S., Yamamoto, D. *et al.* (2007) Dielectric dispersion of short single-stranded DNA in aqueous solutions with and without added salt. *Phys. Rev. E.* **75**, 011911.

153 Tomić, S., Babić, S. D., Vuletić, T. *et al.* (2007) Dielectric relaxation of DNA aqueous solutions. *Phys. Rev. E.* **75**, 021905.

154 Suzuki, S., Yamanashi, T., Tazawa, S. *et al.* (1998) Quantitative analysis of DNA orientation in stationary AC electric fields using fluorescence anisotropy. *IEEE Trans. Ind. Appl.* **34**(1), 75–83.

155 Swicord, M. L. and Davis, C. C. (1982) Microwave absorption of DNA between 8 and 12 GHz. *Biopolymers* **21**, 2453–2460.

156 Swicord, M. L., Edwards, G. S., Sagripanti. J. L. and Davis, C. C. (1983) Chain-length dependent microwave absorption of DNA. *Biopolymers* **22**, 2513–2516.

157 Swicord, M. L. and Davis, C. C. (1983) An optical method for investigating the microwave absorption characteristics of DNA and other biomolecules in solution. *Bioelectromagnetics* **4**, 21–42.

158 Edwards, G. S., Davis, C. C., Saffer. J. D. and Swicord, M. L. (1984) Resonant microwave absorption of selected DNA molecules. *Phys. Rev. Lett.* **53**, 1284–1287.

159 Edwards, G. S., Davis, C. C., Swicord, M. L. and Saffer. J. D. (1985) Microwave-field driven acoustic modes in DNA. *Biophys. J.* **47**, 799–807.

160 van Zandt L. L., Kohli, M. and Prohofsky, E. W. (1982) Absorption of microwave radiation by DNA double helix in aquo. *Biopolymers* **21**, 1465–1468.

161 Foster, K. R., Stuchly, M. A., Kraszewski, A. and Stuchly, S. S. (1984) Microwave dielectric absorption of DNA in aqueous solution. *Biopolymers* **23**, 593–599.

162 Foster, K. R., Epstein, B. R. and Gealt, M. A. (1987) Resonances in the dielectric absorption of DNA? *Biophys. J.* **52**, 421–425.

163 Maleev, V. Y., Kashpur, V. A., Glibitsky, G. M. *et al.* (1987) Does DNA absorb microwave energy? *Biopolymers* **26**, 1965–1970.

164 Gabriel, C., Grant, E. H., Tata, R. *et al.* (1987) Microwave absorption in aqueous solution of DNA. *Nature* **328**, 145–146.

165 Bonincontro, A., Mari, C., Mengoni, M. and Risuleo, G. (1997) A study of the dielectric properties of *E-coli* ribosomal RNA and proteins in solution. *Biophys. Chem.* **67**, 43–50.

166 Kaczanowska, M. and Ryden-Aulin, M. (2007) Ribosome biogenesis and the translation process in *Escherichia coli*. *Microbiol. Mol. Biol. Rev.* **71**(3), 477–494.

167 Bonincontro, A., Briganti, G., Giansanti, G. *et al.* (1996) Electrical conductivity and dielectric dispersion of *E. coli* 70S ribosomes and of 30S and 50S subunits: effects of magnesium. *Colloids Surfaces* **6**, 219–226.

168 Bonincontro, A., De Francesco, A. and Risuleo, G. (1999) Dielectric properties of ribosomal core particles lacking a select population of proteins. *Zeit. Naturforschung C* **54**, 569–572.

169 Giraud, G., Pethig, R., Schulze, H. *et al.* (2011) Dielectrophoretic manipulation of ribosomal RNA. *Biomicrofluidics* **5**, 024116.

170 Matyushov, D. V. (2012) Dipolar response of hydrated proteins. *J. Chem. Phys.* **136**, 085102.

171 Ahmad, S. and Sarai, A. (2011) Analysis of electric moments of RNA-binding proteins: implications for mechanism and prediction. *BMC Structural Biology* **11**, 8.

9

Dielectric Properties of Cells

9.1 Introduction

This chapter focuses on aspects of the dielectric properties of mammalian and bacteria cells of direct relevance to the prediction and interpretation of their dielectrophoretic behaviour. For a large number of cases the frequency range of relevance extends from around 50 kHz to 50 MHz, within which range the magnitude and polarity (positive or negative) of the Clausius–Mossotti factor is controlled by the Maxwell–Wagner interfacial polarization effect described in Chapter 7. For those who specialize in the dielectric properties of cell suspensions, this polarization is identified as the β-dispersion. This dielectric dispersion is controlled by the passive electrical properties of the plasma membrane and this is fully described here. Dielectrophoresis experiments that extend down to frequencies below 1 kHz begin to probe polarization effects related to the electrokinetic behaviour of counter-ions that are attracted to the net negative charge bound to the plasma membrane of mammalian cells and to the net negative charge at the surface and within the cell wall of bacteria. This gives rise at low frequencies to the appearance of the α-dispersion. The characteristic frequency of this dispersion is inversely proportional to the square of the cell radius and so is likely to have a greater influence on experiments involving suspensions of bacteria cells than for mammalian cells. There is an increasing interest in extending dielectrophoresis studies to higher frequencies. This will also be discussed here. However, it is unlikely that frequencies much above 500 MHz will be explored and so the γ-dispersion due to the relaxation of water molecules in the aqueous suspending medium will not be considered, because it will have no influence on the Clausius–Mossotti factor.

9.2 Cells: A Basic Description

Cells are the structural and functional units of all known living organisms. Some organisms, such as amoebae, algae and simple bacteria consist of a single cell but other organisms such as animals are composed of many cells – they are multicellular. Humans, for example, comprise around 10^{14} cells of typical diameter 10 μm and each of mass around 1 ng. A controlling factor limiting the size of a cell is the ratio between its outer surface area and its volume. For a given volume of nucleus, a small cell has more surface area through which to exchange nutrients, gases and other chemicals between the external and internal cell media than a large cell. There is also a limit to the biochemical processes that a nucleus can control in a cytoplasm. Current biomedical applications of dielectrophoresis are directed towards characterizing and selectively capturing blood cells or pathogenic bioparticles such as bacteria, viruses and prions. Brief descriptions of blood cells and some pathogenic bioparticles are given in Appendix L.

All living cells have a plasma membrane that encloses their contents and serves as a semiporous barrier to the outside environment. The membrane acts as a boundary, holding the cell constituents together and keeping other substances from entering. However, the plasma membrane is permeable to specific molecules, allowing nutrients and other essential elements to enter the cell and waste materials to leave it. Small molecules, such as oxygen, carbon dioxide and water, are able to pass freely across the membrane, but the passage of larger molecules (e.g., amino acids and sugars) is carefully regulated. As described in Box 8.1, the basic lipid structure of an intact cell membrane acts as an impermeable barrier to charged species such as ions. According to the accepted model, known as the *fluid mosaic model*, the plasma membrane is composed of a phospholipid bilayer. Individual lipids and proteins can move freely within the bilayer as if it was a fluid, with the membrane-bound proteins forming mosaic patterns on the membrane's outer and inner surfaces. Many diverse proteins are embedded within the phospholipid bilayer of the plasma membrane, while other proteins simply adhere to its two surfaces. Some have carbohydrates attached to their external chemical groups and are

Dielectrophoresis: Theory, Methodology and Biological Applications, First Edition. Ronald Pethig.

referred to as *glycoproteins*. The positioning of proteins on the plasma membrane is related in part to the organization of the filaments that comprise the cytoskeleton, which help anchor them in place. The cytoskeleton forms the framework of a cell. It consists of protein microfilaments and larger microtubules that support the cell, to give it its shape and help with the movement of its internal organelles. The arrangement of proteins also involves the hydrophobic and hydrophilic regions found on the surfaces of the proteins. The hydrophobic regions of the protein associate with the hydrophobic interior of the plasma membrane, whereas hydrophilic regions extend past the surface of the membrane into either the cytosol of the cell or the outer environment. Many of the transmembrane protein structures form channels and pumps through which ions can be transported.

We can distinguish between the *passive* and *active* electrical properties of cells. The passive properties are associated with the linear conductance and capacitance of cells (including excitable cells with resting membranes), whilst the active properties relate to impulse excitation and propagation of ions in the membranes of excitable cells (e.g., nerve cells). The main focus of this chapter will reflect the fact that the literature on the dielectrophoretic properties of cells has to date mainly been related with their passive electric properties. The texts by Cole [1], Takashima [2] and Hille [3] together provide a comprehensive coverage of the active electrical properties of cell membranes.

9.3 Electrical Properties of Cells

The passive electrical properties of biological tissue were scientifically explored in the 1770s by Henry Cavendish, who used his own body to complete the circuit formed by a Leyden jar connected to various lengths of glass tubes filled with salt solution! Cavendish found the electric 'velocity' (i.e., current) he experienced was directly proportional to the degree of electrification (i.e., voltage) of the Leydon jar and that salt solutions exhibit greater conducting powers than pure water. Cavendish did not publish his findings at the time, but relevant extracts from his notebooks were presented 100 years later in 1879 in a volume edited by James Clerk Maxwell titled *The Electrical Researches of Hon. Henry Cavendish*. The relationship between voltage, current and resistance had been formalized in 1825 when Georg Ohm published the results of his experiments to measure the relationship between applied voltage and current in various lengths of wire [4]. The list of early contributors to electrophysiology includes the Nobel Prize winners Max Planck (the founder of quantum mechanics) and Walther Nernst (formulator of the Third Law of Thermodynamics). Encompassed in what is still known as the Nernst–Planck equation, cells were envisaged to be surrounded by a thin layer of nonaqueous electrolyte across which ions move into or out of a cell down concentration gradients and with an electric field [5,6]. In 1902, in the absence of any knowledge of the structure of a cell and with no physicochemical evidence for the existence of an enveloping membrane, Bernstein [7] combined all of the relevant information available to him in the form of his influential membrane hypothesis:

- Cells consist of a conducting electrolyte surrounded by a thin membrane that is largely impermeable to ions.
- In its resting state, a difference in electrical potential exists across the membrane.
- On electrical stimulation of an excitable cell, there is a transient increase of the membrane permeability towards potassium ions. This results in a significant reduction in the transmembrane potential difference, assumed to be the Nernst diffusion potential for potassium ions.

It was commonly known that living tissue exhibited a relatively high DC resistivity, which decreased significantly in death and was accompanied by the release of electrolytes. A key unanswered question was whether this resulted from the release of electrolyte ions that are organically bound in live cells, or because viable cells are surrounded by an electrically insulating barrier that decomposes on cell death. The answer came from the first AC electrical impedance measurements on cells, performed by Rudolf Höber [8–10]. His experiments have been described in detail elsewhere [11] but in summary Höber provided the first experimental evidence that cells possess a resistive dielectric membrane that surrounds a conducting electrolytic interior. He determined that the conductivities of compacted red blood cells and frog muscle tissue measured at MHz frequencies were significantly higher than that measured at ~150 Hz. He deduced that, at low frequencies, the current path was around the cells, but at high frequencies, the current was able to penetrate into the conducting cell interior. The internal conductivity of a red blood cell was estimated to be close to that of a 0.2% NaCl solution. Confirmation that the MHz value represented the internal conductivity was obtained by using saponin to lyse the envelope surrounding the interior, finding that the MHz conduction current remained almost unchanged whereas the low-frequency conduction increased significantly. This was taken as evidence that the conducting interior of a red blood cell is contained within a resistive dielectric envelope ('eine dielektrische Hülle' [8, p. 237]. These discoveries were made using primitive dielectric measurement techniques. Low-frequency voltages (100~200 Hz) were generated using a hand-cranked rotating magnetic disk, whilst MHz voltages were obtained using an ignition coil and spark gap coupled via an inductor primary coil wrapped around a gas lamp tube to a bridge circuit. This circuit consisted of a glass plate capacitor, tubes containing reference salt solutions and the blood sample, a crystal detector and headphones [11].

Apart from representing a major milestone in the development of cell biology and electrophysiology (i.e., first physical evidence for the existence of a resistive cell membrane) Höber gave the first description of what we now name as the dielectric β-dispersion exhibited by cell suspensions and fresh tissue. The β-dispersion is basically a manifestation of a cell's outer membrane acting as a high resistance to direct current and low frequency alternating current, as well as possessing a large capacity for accumulating ions at the membrane surface (i.e., exhibiting Maxwell–Wagner interfacial polarization). The first studies of this are commonly attributed [e.g., 12] to much later works, such as those of Fricke and co-workers [13–15].

After Höber, the next demonstration of the existence of the β-dispersion exhibited by cells was that of a banker and pioneer of radio broadcasting (from air balloons) in Belgium, namely Maurice Philippson [16, 17]. He described how the specific impedance of the compacted red blood cells fell from a high value (3890 $\Omega\,cm^{-3}$) at 1 kHz to a low one (200 $\Omega\,cm^{-3}$) when extrapolated beyond 3.5 MHz to higher frequencies. He achieved this by measuring the potential across the sample for a range of frequencies, using a tube voltmeter, at the same time as the current. Philippson does not appear to have been aware of Höber's earlier work. Philippson devised the equivalent circuit shown in Figure 9.1(a) to describe the electrical properties of compacted cells or tissue. It consists of the protoplasm resistance R in series with a parallel combination of the membrane resistance r and capacitance C. According to this circuit the DC resistance and impedance at low frequencies is $R + r$ and tends to the value R at very high frequencies. From Philippson's results for compacted red blood cells, components R and r have estimated values of 200 Ωcm^{-3} and 3690 Ωcm^{-3}, respectively. Because Philippson could measure only the magnitude of the impedance without any phase angle determination, it was not possible for him to evaluate the membrane capacitance value C.

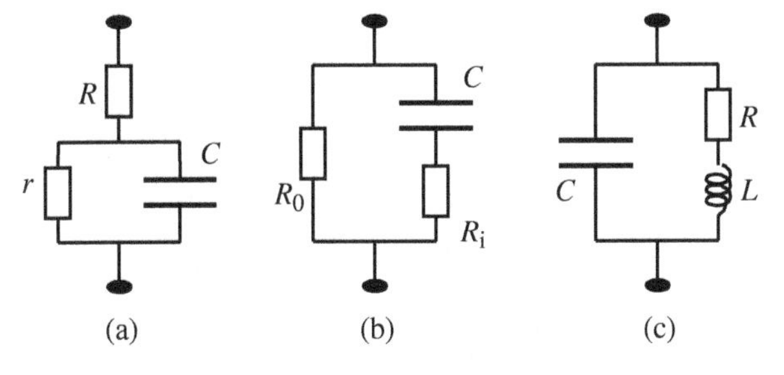

Figure 9.1 Equivalent circuits proposed for: (a) tissues and compacted red blood cells by Philippson [17]. R and r are attributed to the cytoplasm and membrane resistance, respectively and C to the membrane capacitance; (b) red blood cell suspensions proposed by Fricke and Morse [14], in which R_0 is the resistance to current flow around the cell and R_i is the cytoplasm resistance; (c) the squid giant axon by Cole and Baker to account for their discovery of an anomalous (inductive) reactance residing in the membrane structure [41]. The components R, C and L were determined to have values of 1 kΩ cm^2, 1 μF/cm^2 and 0.2 H cm^2, respectively.

The effects that Höber found were repeated by Fricke and Morse [13, 14], who obtained a cell internal resistivity of 310 Ω cm and also determined that the erythrocyte membrane possessed a capacitance of 8.1 mF/m^2. Assuming a value of 3 for the relative permittivity of the membrane material, the thickness of the erythrocyte membrane was estimated to be 3.3 nm. Although this value turned out to be an underestimate, it was the very first indication of the exquisite thinness of the cell membrane. The proposed equivalent circuit to represent an erythrocyte in suspension is shown in Figure 9.1(b). Membrane capacitance values close to 8 mF/m^2 have also been obtained for neutrophils [18], eosinophils [19], mast cells [20] and chromaffin cells [21] using the patch-clamp technique [22]. As indicated in Table 9.1 the resistance of the cytoplasm membrane is too large to be measured by conventional dielectric measurements on cell suspensions. The patch-clamp technique does not suffer such a limitation, so that Takashima *et al.* [23] were able to confirm that the membrane of a human erythrocyte behaves as a good insulator, with a specific resistance of ~2 × 10^5 Ω cm^2. Model cancer cells, such as HeLa and mouse myeloma, were found to have membrane capacitances of 19 mF/m^2 and 10 mF/m^2, respectively, with membrane specific resistances of around 10^4 Ω cm^2 [24].

Cole was the first to demonstrate that dielectric measurements can detect physical changes associated with change of cell state, by observing an increase in the total capacitance of sea urchin eggs on their fertilization [25]. He also used schematics such as Figure 2.5 to illustrate how high-frequency electric flux lines penetrate into the cell interior through the cell membrane. Average cell membrane capacitances of 8.6 mF/m^2 before fertilization and 33 mF/m^2 after fertilization were determined [26] to be associated with changes in the plasma membrane and not with the appearance of a fertilization membrane as first presumed [25]. These early dielectric measurements on cells were performed using cell suspensions or centrifuged cell compactions. Cole was the first to report measurements on single cells (sea urchin eggs) [27] and by 1941 he was able to tabulate dielectric data for a range of large single cells (plant, protozoa, marine egg, muscle and nerve) [28]. Some of this data is reproduced in Table 9.1 and remains of value today. From Table 9.1 it is

Table 9.1 Dielectric properties of cells (derived from Cole [28]) given in units commonly found in the literature, for temperatures ~20 °C. The symbol 'S' indicates single cell measurements, otherwise all results refer to those obtained for cell suspensions. The symbol ∞ indicates a resistance too high to measure accurately.

Cell type	Membrane capacitance (μF/cm²)	Membrane resistance (Ω cm²)	Internal resistivity (Ω cm)
Blood			
Erythrocyte			
chicken	0.8	∞	140
dog	0.81	∞	
human	0.8	∞	140
rabbit	0.8	∞	
Leucocyte			140
rabbit	1.0		
Nerve			
Squid			
resting	1.1 (S)	10^3 (S)	30 (S)
excited	1.1 (S)	25 (S)	90 (S)
Marine egg			
Arbacia			
unfertilized	1.1 (S)	> 100 (S)	180 (S)
fertilized	2.8 (S)	> 100 (S)	210 (S)
Cumingia			
unfertilized	2.7 (S)	∞ (S)	180 (S)
fertilized	2.3 (S)	∞ (S)	165 (S)
Plant and Protozoa			
Amoeba			
proteus			815 (S)
Bacteria			
acidophilis	0.95	∞	
Chlorella	0.33	∞	460
Nitella			
resting	0.94 (S)	2.5×10^5 (S)	87 (S)
excited	0.80 (S)	500 (S)	87 (S)
protoplasm			212 (S)
Paramecium			145 (S}
Valonia			
ventricosa	1.0 (S)	> 10^4 (S)	
Yeast	0.6	∞	

evident that resting membranes present a large resistance to passive ion flow. If we assume a membrane thickness of ~7 nm the membrane resistivities range from 10^6 Ω m and greater to immeasurably large values. Blinks [29] had also provided early confirmation of Höber's conclusion [8] that this membrane resistance value falls dramatically on death of the cell.

Measurements of the properties of nuclear membranes had to await the development of microelectrode impalement techniques (see section 9.3.1). They were limited to measurements on systems such as marine eggs or fruit fly salivary gland cells, whose nuclei measure up to 40 μm in diameter. The nuclear envelope of salivary gland cells (*Drosophila, Chironomids* and *Sciarids*) were found to have relatively low specific resistances in the range of 1 to 10 Ω cm^2, with very large capacitances ranging from 1000 to 5000 mF/m^2 [30,31]. The plasma membrane resistance and capacitance were determined to fall in the ranges 220 to 1200 Ω cm^2 and 17 to 100 mF/m^2, respectively, whilst the resistivities of the cytoplasm and nucleoplasm were measured to be about 100 Ω cm. The interesting proposal was made that the anomalous resistance and capacitance values obtained for the nucleus resulted from connections between the nuclear envelope and the endoplasmic reticulum [31]. In other words, the surface over which the resistance and capacitance were distributed was larger than that of the outer surface of the nucleus. This can be quantified by introducing a membrane topography parameter ϕ that represents the ratio of the actual membrane area of the cell or nucleus, to the membrane area ($4\pi r^2$) that would form a perfectly smooth and spherical covering of the cytoplasm or nucleoplasm. The measured membrane capacitance C_m and resistance R_m thus take the form:

$$C_m = \phi C_o; \quad R_m = (1/\phi)R_o \tag{9.1}$$

where C_o and R_o represent the values for a perfectly smooth membrane, corresponding to ϕ having a value of 1.0. In Chapter 11, when detailing the dielectrophoretic properties of cells, the concept of a membrane topography parameter ϕ is found to be useful. Based on the calculation [32] that a smooth membrane has a C_o value of 6 mF/m^2, the values of 1000 to 5000 mF/m^2 obtained [30] for nuclei correspond to a range of values for ϕ of 160 to 830. It is now known that, in all eukaryotic cells, the nuclear envelope is formed by two concentric, porated, lipid bilayer membranes, the outer one of which is contiguous with the endoplasmic reticulum. It is reasonable to consider that such a structure could lead to a large capacitance value. However, an estimate of 6.2 mF/m^2 was obtained, from dielectric measurements over the range 0.1 to 250 MHz, for the combined capacitance of the two membranes that make up the nuclear envelope of mouse lymphocytes [33]. This suggests that the outer and inner nuclear membranes are electrically connected in series, in which case the membrane exhibiting the lowest specific capacitance (the inner one) will dominate their combined, series, capacitance. The conductance of the nuclear envelope was also determined to be 15 S/cm^2, representing a relatively low specific resistance of $6.7 \times 10^{-2}\ \Omega\ \text{cm}^2$ [33]. A low resistance of $6 \times 10^{-2}\ \Omega\ \text{cm}^2$ has also been reported for the nuclear envelope of mouse liver cells, taken to indicate that the nuclear envelope is penetrated by aqueous channels [34]. The electrophysiology of the cell nucleus is an active field of research [35] and should assist a more thorough understanding of the dielectric properties of cell nuclei.

Resistivity values ranging from 310 Ω cm for the cytoplasm of erythrocytes [14] and 100 Ω cm for the cytoplasm of salivary gland cells [31] can be taken as representative of a wide range of cells. Measurements performed at frequencies up to 250 MHz on a variety of erythrocyte types give cytoplasm resistivity values of 170 to 230 Ω cm and internal relative permittivity values ranging from 50.1 to 55.1 [36]. Measurements at frequencies around 400 MHz (which should probe the cell interior) give relative permittivity values of 35 to 60 for a variety of tissue [32, 37] and resistivity values between 80 and 160 Ω cm [32]. A 'pure' physiological solution (e.g., ~150 mM NaCl) has a nominal resistivity of 70 Ω cm and relative permittivity of ~79. The increased internal resistivity and reduced permittivity values obtained for erythrocytes can be attributed in particular to dissolved haemoglobin molecules. The presence of organelles and structures such as mitochondria, the endoplasmic reticulum and the nuclear envelope will contribute to the internal dielectric properties of other cell types. Summaries of the dielectric properties of the various parts of a cell, as determined by the experimental contributions outlined above, are given in Tables 9.2 and 9.3.

In their application of cable theory to understanding signal transmission along a nerve fibre, Hodgkin and Rushton [38] assumed that the structure of a nonmedullated nerve fibre consisted of a conducting protoplasm and a thin surface membrane of high leakage resistance and large capacitance per unit area. The papers of Höber [8] and Fricke and Morse [14] were cited as the scientific sources for this assumption, together with that of Curtis and Cole who had shown that the squid giant axon possessed similar electrical properties to that of the red blood cell [39]. Hodgkin and Rushton [38] also referred to the surprising discovery [40, 41] that the squid axon exhibited an inductive reactance below 150 Hz (see Figure 9.1(c) for the proposed equivalent circuit). Following the development of the patch-clamp technique by Cole [42], an understanding of the implications of this inductive element in the membranes of nerve cells could begin, leading to the elucidation of nerve pulses in terms of

Table 9.2 The dielectric properties of the components of various types of cell as determined by various experiments (references given in square brackets).

Cell part	Sp. resistance Ω m^2	Conductivity S/m	Capacitance mF/m^2	Permittivity $\varepsilon'/\varepsilon_o$
Plasma membrane	> 1 [23,24]	< 10^{-8} [a]	8–100 [14, 23, 28, 30]	~5 [b]
Cytoplasm		0.44–1.25 [13,14, 28, 32]		35–60 [32, 36, 37]
Nuclear envelope	10^{-4}–10^{-3} [31]	< 10^{-4} [a]	10^3–5 × 10^3 [31]	
Nucleoplasm		~1.0 [30]		

Notes: [a]Calculated assuming a nominal membrane thickness of 7 nm. [b]Based on capacitance of ~6 mF/m^2 estimated for smooth plasma membrane [32].

Table 9.3 The dielectric properties at 24 °C of the components of mouse erythrocytes and lymphocytes (Asami *et al.* [34]).

Cell type and component	Conductivity S/m	Capacitance mF/m^2	Permittivity $\varepsilon'/\varepsilon_o$
Erythrocyte			
Plasma membrane	$< 10^{-5}\ \sigma_{cyt}$	7.2	5.7
Cytoplasm	0.62		59
Lymphocyte			
Plasma Membrane	$< 10^{-5}\ \sigma_{cyt}$		6.8
Cytoplasm	0.32		60
Nuclear envelope[a]	6×10^{-3} S/m (1.5×10^5 S/m^2)	6.2	28
Nucleoplasm	1.35		52

Note: [a]The double membrane structure of the nuclear envelope is modelled as a single homogeneous shell of thickness 40 nm.

'reversible alterations in sodium and potassium permeability arising from changes in membrane potential' [43] and the award in 1963 of the Nobel Prize to Hodgkin and Huxley. There were not a few who thought it unfortunate that Kenneth Cole's work had been overlooked in giving this award. In 1967 he received the US National Medal of Science for his work that led to the *sodium theory of nerve transmission.*

The membranes of nonexcitable cells (e.g., blood cells, cancer cells) do not exhibit a frequency-dependent membrane capacitance, even below 200 Hz [e.g., 23]. Nerve membranes do show such behaviour, associated with the action of membrane ion channels [e.g., 44]. The inductive effect described by Cole [40] can be eliminated by blocking the potassium channels with nerve toxins. This reveals that an increase in membrane capacitance, from a value of 1.0 to 1.23 μF/cm^2, occurs in squid axons during the so-called long action potential [44]. These active dielectric properties (i.e., inductive and capacitance changes associated with action potentials) should in principle be amenable to investigation by dielectrophoresis, but to date no such observations appear to have been reported.

9.3.1 Single Cell Measurements using Microcapillary Electrodes

Electrical activity (ECG, EMG, EEG, etc.) in living systems can be detected by placing electrodes in galvanic contact with the body or inside its tissue to measure the fields set up by electric currents flowing through the extracellular fluids [45]. Since these currents originate across cell membranes, a more direct and quantitative approach is to measure electrical events across the membrane of a single cell. This measurement is done by comparing the electric potential of one side of the membrane with that of the other side. An external sensing electrode is placed in electrical continuity with the outside of the cell and another is inserted through the membrane and into the cell's cytoplasm. The difference between these two potentials is the membrane potential V_m, given as the intracellular potential *relative* to the extracellular potential. The external potential in the suspending buffer solution is arbitrarily defined as the reference zero. A simple electrical stimulating and recording arrangement is shown in Figure 9.2.

As shown in Figure 9.2 the cell is immersed in a physiological buffer solution containing a reference electrode. Glass capillary microelectrodes, with tip diameters less than 0.1 micron and filled with an electrolyte such as 3 M KCl, can be inserted into cells with negligible damage to their membranes. The tip resistance of such microcapillary electrodes can approach values of 20~50 MΩ and so will not act as an electrical short-circuit across the membrane but can serve as a voltage probe. The first step is to insert the tip of such a recording electrode through the membrane of the cell. Before the tip of this microelectrode enters the cell, it and the reference electrode are at the same potential (taken to be reference zero). When the fine capillary tip penetrates the membrane, the cytoplasm is in continuity with the electrical connection to a voltage amplifier via a fine column of electrolyte that fills the inside of the capillary electrode (e.g.,

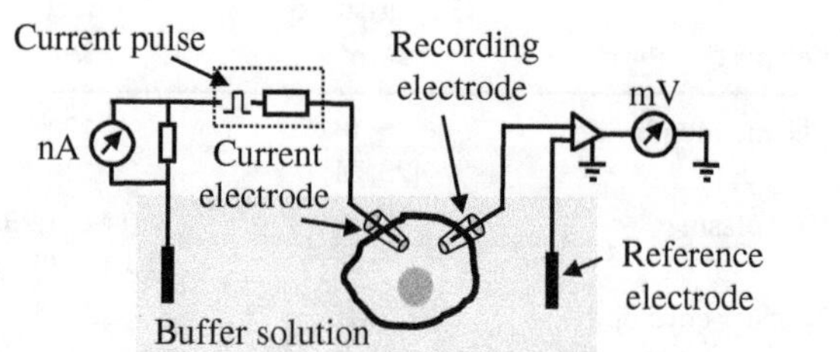

Figure 9.2 A basic system is shown for stimulating and recording the electrical properties of a cell membrane. A signal generator creates a current pulse into or out of the cell. The difference between the potential of an external electrode and one located in the cytosol gives the membrane potential V_m.

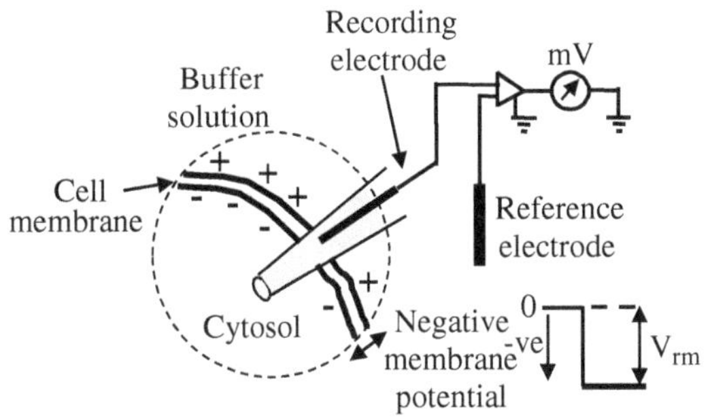

Figure 9.3 When a high-resistance glass capillary voltage probe is inserted into a cell it records a negative potential with respect to the outside of the cell. This is the resting membrane potential (V_{rest}).

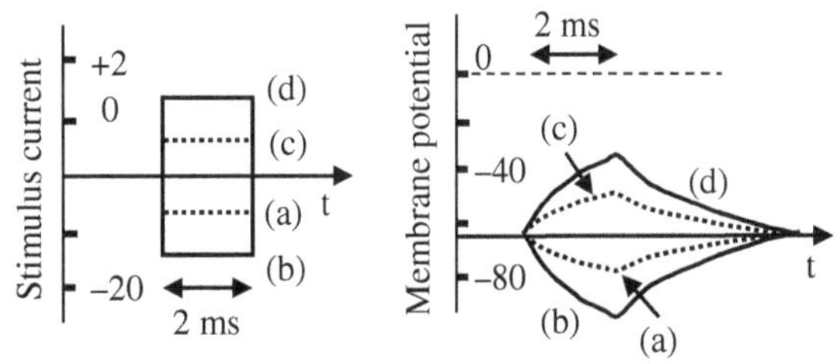

Figure 9.4 When a current pulse is applied that removes positive charge from inside a cell, hyperpolarization of the membrane occurs. The intracellular negative potential is increased (e.g., from −60 to −70 mV). A current pulse of opposite polarity will add positive charge to the inner surface of the cell membrane, causing depolarization of the membrane (e.g, from −60 to −50 mV).

a 3M solution of KCl). As the tip of the recording microelectrode is advanced, penetration of the plasma membrane is indicated by the sudden appearance of a negative potential shift of the voltage trace (see Figure 9.3). The steady negative potential recorded by the electrode tip in the cytoplasm is the resting membrane potential V_{rm}. All cells that have been investigated have a *negative* resting potential, which can be as high as −100 mV. The potential sensed by the intracellular electrode does not change as the tip is advanced further into the cell. Thus, the entire potential difference between the cell interior and cell exterior exists across the surface membrane and in the regions immediately adjacent to the inner and outer membrane surfaces.

The electrical properties of the cell membrane can be examined by causing a pulse of current to pass through the membrane so as to produce a perturbation in the membrane potential. A second microelectrode, the current electrode shown in Figure 9.2 can deliver such a current. The current from this electrode, in the form of a current pulse generated by applying a step voltage in series with a high value resistance (>1 GΩ), flows across the membrane in either the inward (buffer to cytoplasm) or the outward direction depending on the polarity of the step voltage. If the current pulse draws current to the outside, so that positive charge is removed from inside the cell via the current electrode, the potential difference across the membrane increases (hyperpolarizes). The intracellular negative potential increases in magnitude (e.g., from −60 to −70 mV). With hyperpolarization the membrane usually produces no other response than this increase of negative potential. If a current pulse is passed from the electrode into the cell, positive charge will be added to the inner surface of the cell membrane. This charge causes the potential difference across the membrane to decrease and the cell is then said to become *depolarized* (e.g., from −60 to −50 mV). These two types of response are shown in Figure 9.4.

As the strength of the outward pulse is intensified, depolarization will increase, as shown in Figure 9.4. *Excitable* cells, such as nerve, muscle and many receptor cells, exhibit a *threshold potential* at which the membrane will produce a strong active response. This is known as the *action potential* shown in Figure 9.5. The action potential is caused by the activation of membrane channels permeable to sodium, which themselves are activated by the reduction in voltage difference between the two sides of the cell membrane. The opening of the sodium channels in response to depolarization and the resulting flow of sodium ions into the cell provide an example of *membrane excitation.*

We can now appreciate that cell membranes respond to stimuli with two quite different classes of electrical behaviour – namely, *passive* and *active* behaviour:

9.3.1.1 Passive Electrical Response

This is always produced when an electric current is forced across a biological membrane, because of the electrical capacitance and conductance properties of the membrane. Passive responses occur independently of any molecular changes that open or close gated ion channels in the membrane. The resistance (reciprocal of conductance) of a cell membrane is associated with leakage pathways that allow inorganic ions to cross the membrane.

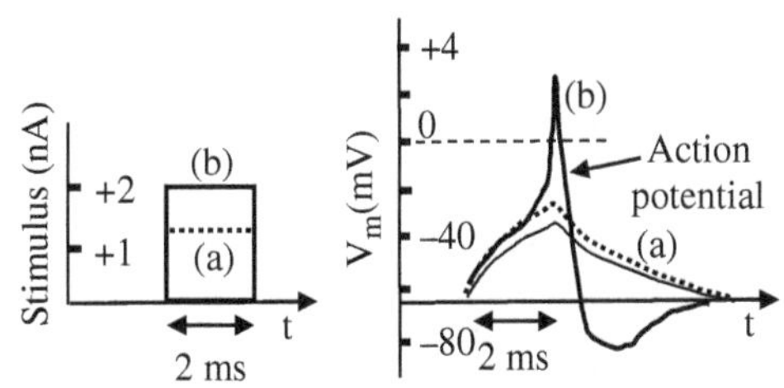

Figure 9.5 Excitable cells, such as nerve, muscle and many receptor cells, exhibit a threshold depolarization potential at which the membrane will produce a strong active response. This is known as the action potential.

The capacitance of a membrane is a measure of the extent to which the ion impermeability of the membrane leads to separation of electrical charges across the membrane.

9.3.1.2 Active Electrical Response

Such responses, known as membrane excitations, are found in excitable tissue such as nerve, muscle and sensory receptors. They depend on the opening and / or closing of numerous *ion channels* (also called *membrane channels*) in response to a stimulus. Some ion channels are *gated* (i.e., opened and shut) by changes in voltage across the membrane, while others are opened by the binding of transmitter or messenger molecules. Other channels, primarily in sensory receptor cells, are activated by specific stimulus energies such as light (photoreceptors) or mechanical strain (mechanoreceptors). When a certain group of channels selectively permeable to a certain species of ion is opened, a current may be carried across the membrane. As in the case of sodium channels, such a current normally produces a voltage signal across the membrane. The gating of ion channels is the immediate cause for nearly all electrical activity in living tissue.

9.3.1.3 Membrane Resistance

The passive resistance of a membrane is a measure of its permeability to ions. In saline solutions the resistivity of pure phospholipids is as high as $10^{13}\,\Omega$ m. This can be compared to 298 K values that range from 0.6 to 0.8 Ω m for prepared physiological solutions (buffers). The value for sea water is ~0.2 Ω m. A 4 nm thick lipid bilayer can be estimated to have a specific resistance of 40 kΩ m^2. The significantly lower resistivities of biological membranes (typically 0.01~1 Ω m^2) therefore can be assumed to arise from structures other than the lipid bilayer itself. These structures are protein-bounded aqueous pores (aquaporins) and various ion channels embedded in the lipid. The density of different channels typically range from 50 to 500 per μm^2, with conductances of 1~100 pS. However, many of these channels may not be 'open' at any given time.

If a step pulse of steady current is applied across the membrane, the membrane potential shifts by ΔV_m from the resting value. ΔV_m depends on the magnitude of the applied current ΔI and the membrane resistance R, which can be determined from Ohm's Law:

$$\Delta V_m = R\Delta I$$

Consider the two spherical cells shown in Figure 9.6 – one small and the other large. The two cells have membranes of the same *specific resistance* R_m to electric current (i.e., the same resistance per unit square area of membrane). As schematically demonstrated in Figure 9.6, for a given increment of current ΔI inserted into both cells the large cell will show a smaller increment of voltage ΔV_m. This arises because the same current will flow through a larger area of membrane. Because the input resistance of a cell (i.e., the total resistance encountered by current flowing into or out of a cell) is a function of both membrane area A and specific resistance R_m of a cell, it is useful when comparing membranes of different cells to correct for the effect of membrane area on the current density. Thus, the specific membrane resistance is calculated as:

$$R_m = RA = \frac{\Delta V_m}{\Delta I} A \ (\Omega \mathrm{m}^2)$$

9.3.1.4 Membrane Capacitance

Because they are very thin (~7 nm) and virtually impermeable to ions over most of their surface area, cell membranes can violate the principle of electroneutrality at the *microscopic* scale. Negative charges accumulated at or near one surface of a membrane will interact electrostatically over the short distance of the membrane thickness, with positive charges on the other side of the membrane. The ability of the cell membrane to accumulate and separate electric charge is called its membrane capacitance. Electronic engineers can view this situation as a very thin dielectric (the lipid bilayer) sandwiched between two conductors (electrolytes) representing the basic form of a capacitor. Cell membranes contain a lipid bilayer of about 3 nm in thickness (verified by electron microscopy) with

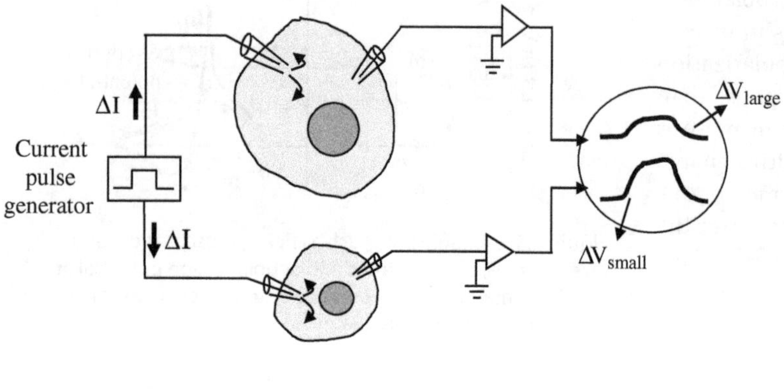

Figure 9.6 Although the membranes of the two cells shown in this diagram have the same dielectric properties, the measured input resistance ($\Delta V_m/\Delta I$) of the larger cell is lower than that of the smaller cell. To correct for the difference in current density the input resistance of a cell is often quoted as a *specific* resistance with units of Ω m^2.

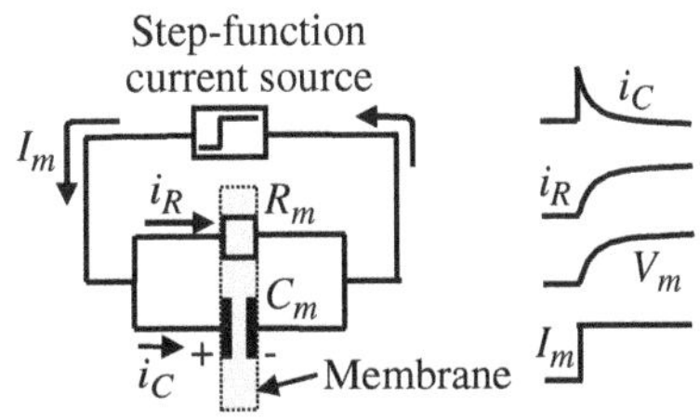

Figure 9.7 The equivalent circuit for a cell membrane can be represented as a parallel combination of the membrane resistance R_m and capacitance C_m. Time courses are shown for the resistive current i_R, capacitive current i_C, membrane potential V_m (across the membrane resistance and capacitance) on injection of a pulse of current (I_m) through the membrane.

proteins protruding on each side. As discussed in Box 8.1, a relative permittivity of $\varepsilon_r \approx 2.4$ can be assigned to this lipid phase of the membrane. The polar side chains of the proteins are predominantly exposed to the aqueous medium each side of the membrane and should contribute to an overall relative permittivity for the membrane that is larger than that of the lipids. Assuming a total thickness of ~7 nm for the combined protein-lipid structure of the membrane, together with an estimated permittivity ~$5\varepsilon_o$, the membrane capacitance for a *smooth* cell surface can be estimated as $C_m \approx 6.3\,\text{mF/m}^2$.

The equivalent circuit for a cell membrane to describe the charging and discharging of the membrane on application and then removal of a current pulse is shown in Figure 9.7. The relationship between potential V and time during the charging of the membrane capacitance is given by:

$$V(t) = V_o e^{-\tau/R_m C_m}$$

where the time to fall to $1/e$ of its initial value is the time constant given by $\tau = R_m C_m$. Having determined the membrane specific resistance R_m, the membrane capacitance can be determined from measurement of the membrane time constant (typically 0.1~10 ms). Experimental values obtained for C_m are normally larger than the theoretical value of ~6.3 mF/m^2 determined above for a smooth membrane structure. The experimental value (typically 10~30 mF/m^2) obtained correlates closely with the extent to which the area of an otherwise smooth membrane surface is increased as a result of the presence of membrane folds and proturbances, such as blebs and microvilli, for example.

Example 9.1 Trans-Membrane Ion Transfer and the Membrane Potential

Consider a mammalian cell of radius 10 µm with a membrane capacitance of 10 mF/m^2. How many ions have to be transferred across the membrane to create a resting membrane potential of minus 70 mV? Express this number of ions as a percentage of the average potassium ion concentration (150 mM) in the cytoplasm of a mammalian cell.

Solution 9.1 The total surface area of the cell is $4\pi R^2 = 4\pi 10^{-10}\,\text{m}^2$, to give the total membrane capacitance C for the cell as:

$$C = (4\pi 10^{-10}\,\text{m}^2) \times (10 \times 10^{-3}\,\text{F m}^{-2}) = 4\pi 10^{-12}\,\text{F}$$

In Chapter 2 the capacitance C of a 'device' (in this case a cell membrane) is defined as the amount of charge Q it can store at each electrode per unit of voltage V applied across the electrodes (i.e., $C = Q/V$). In Figure 3.1, the electrodes are metal plates and the charges are electrons. In Figure 9.7, the ionic solutions either side of the membrane serve as the electrodes and the charges are ions. The membrane acts as the dielectric of the capacitor. To establish a negative membrane potential (V_{mr}) requires either a flow of negative ions across the membrane into the cell, or positive (potassium) ions in the cytoplasm to flow out of the cell. To set up a potential of 70 mV will require a displacement of charge Q given by:

$$Q = V_{mr}C = (70 \times 10^{-3}\,\text{V}) \times (4\pi \times 10^{-12}\,\text{F}) = 8.8 \times 10^{-13}\,\text{C}$$

To obtain the amount of monovalent ions in moles to be transferred across the membrane we divide this result by the Faraday constant (the magnitude of charge per mole of electrons: $q/N_A = 9.65 \times 10^4$ C/mol):

$$\text{Q(moles)} = \text{Q(C)}/(9.65 \times 10^4) = 9.1 \times 10^{-18}\,\text{mol}$$

We are asked to compare this amount with the content of potassium ions in the cytoplasm. The concentration (moles per litre) of potassium ions is 150 mM. The cell volume is $(4\pi R^3)/3\,\text{m}^3 = (4\pi \times 10^{-15})/3\,\text{m}^3 = (4\pi \times 10^{-12})/3$ L. The potassium concentration $[\text{K}^+]$ in this cell volume is thus:

$$[\text{K}^+] = (4\pi \times 10^{-12})/3 \times 0.15 = 6.3 \times 10^{-13}\,\text{mol}$$

The quantity of charge to be displaced across the membrane compared with $[\text{K}^+]$ is:

$$(9.1 \times 10^{-18}\,\text{mol})/(6.3 \times 10^{-13}\,\text{mol}) = 1.44 \times 10^{-5}$$

Thus, to charge the membrane to −70 mV requires as little as 1.4×10^{-3}% of the cell's total potassium to be transferred across the membrane. The rule of electroneutrality, that the number of positive charges must equal the negative charges, remains essentially unviolated at the *macroscopic* scale within and outside the cell. An imbalance of charges exists only at the *microscopic* scale across the membrane.

9.4 Modelling the Dielectric Properties of Cells

Typical results obtained for the dielectric properties (conductivity and permittivity as a function of frequency) of an aqueous suspension of cells are shown in Figure 9.8 for the case of human erythrocytes [23]. The large increase of conductivity, first observed by Höber [8], is clearly evident and is denoted as the β-dispersion [12]. This dispersion represents the frequency dependence of the interfacial polarization at the boundary between the insulator-like plasma cell membrane and the outside conducting electrolyte. On application of an external field to a cell suspension, a transient accumulation of ions forms at the membrane surfaces. That this occurs can be verified, as Höber [8] did, by porating the membrane structure with a detergent. Also evident in Figure 9.8 is the 'tail' of a low-frequency dispersion (labelled as the α-dispersion) that merges with the β-dispersion. Studies of the α-dispersion are made difficult by electrode impedance effects but the consensus is that this dispersion arises from counter-ion conduction and fluctuations near the charged surface groups on the cell membrane. For frequencies below the β-dispersion the electric field does not penetrate the plasma membrane. Interpreting the β-dispersion using the concept of fluctuating counter-ions described in Chapter 8 for DNA and polyelectrolyte solutions can therefore be considered of relevance also for the case of suspended cells. This is discussed in more detail in section 9.5.

We now address the question as to how information regarding the dielectric properties of the various components of a cell can be extracted from characteristics such as those presented in Figure 9.8. A first approach to this is shown in Box 9.1, based on a procedure described by Maxwell for calculating the resistivity of a compound material consisting of small resistive spheres dispersed in another type of resistive medium [46]. This was extended

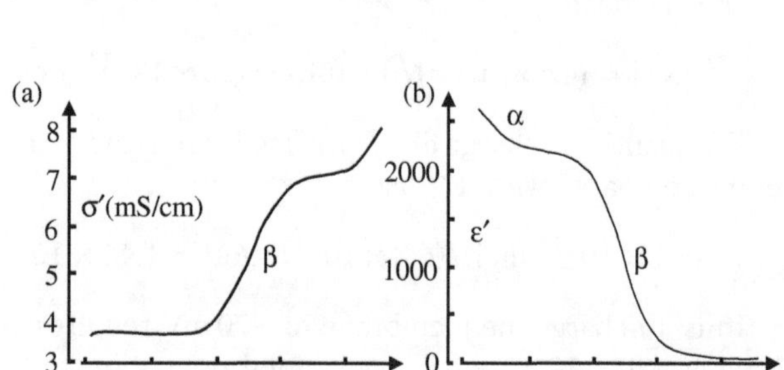

Figure 9.8 (a) Conductivity ($\sigma' = \omega\varepsilon''$) of an aqueous solution containing a 43.8% concentration by volume of human erythrocytes. (b) Relative permittivity (ε') of the same cell suspension. The 'tail' of the α-dispersion and the β-dispersion are shown (based on Takashima *et al.* [23]).

later by Wagner [47] when he defined the constituents of such a compound material in terms of the *complex* conductivities of its components. The term *effective* permittivity ε_{eff} is used in Box 9.1 to signify that a defined spherical volume of a cell suspension may be replaced conceptually with an equal volume of homogeneous *smeared-out* bulk properties, such that substitution of one volume with the other would not alter the electric field in the surrounding medium. In a dielectric measurement of a cell mixture the measured permittivity value is ε_{eff}. Equations that have been formulated to describe the permittivity or conductivity of cells and other particles dispersed in a medium are generally known as *mixture* equations.

To take into account conduction effects, Equation (9.2), derived in Box 9.1, can also be given in terms of complex permittivity values:

$$\frac{\varepsilon^*_{eff} - \varepsilon^*_m}{\varepsilon^*_{eff} + 2\varepsilon^*_m} = \nu_c \frac{\varepsilon^*_c - \varepsilon^*_m}{\varepsilon^*_c + 2\varepsilon^*_m} \tag{9.3}$$

On rearranging Equation (9.3) the effective permittivity of a spherical volume of fluid medium of complex permittivity ε^*_m, in which cells of complex permittivity ε^*_c are suspended, is given by:

$$\varepsilon^*_{eff} = \frac{2\varepsilon^*_m + \varepsilon^*_c + 2\nu_c(\varepsilon^*_c - \varepsilon^*_m)}{2\varepsilon^*_m + \varepsilon^*_c - \nu_c(\varepsilon^*_c - \varepsilon^*_m)}\varepsilon^*_m \tag{9.4}$$

Removing a spherical volume of the cell suspension and replacing it with a sphere of the same volume and with a homogeneous permittivity ε^*_{eff} will in no way alter the original field distribution in the surrounding medium. Converting the complex permittivity terms into complex conductivities using the identity $\sigma^* = i\omega\varepsilon_o\varepsilon^*$ given by Equation (6.45) leads to the same expression for ε^*_{eff} as that derived by Wagner [47]. In an impedance measurement of a cell suspension the value of ε^*_{eff} is measured directly, whilst the volume fraction ν_c and complex permittivity ε^*_m of the fluid medium can be determined by centrifuging a known volume of the cell suspension and measuring the volume and impedance of the supernatant. Equation (9.4) can in principle give the average permittivity value of the cells in the suspension. However, this equation is derived on the assumption that the number of cells per unit volume is so small that the value of ε^*_m obtained for the pure suspending fluid (without any cells) is the same as that assumed in Box 9.1 (with cells present). We are in fact assuming the relationship $\varepsilon^*_m \approx \varepsilon^*_{eff}$, implying that for a sufficiently large observation scale a heterogeneous compound material can be considered as a homogeneous one. Inserting this

Box 9.1 Permittivity of a Dilute Suspension of Cells

An external field E is applied to a fluid medium of permittivity ε_m containing a suspension of cells, each of radius R_c and permittivity ε_c. The average distance between cells is taken to be much less than R_c. Consider a spherical region of radius R_m within the medium that contains n cells. The potential at a point outside this sphere will comprise the summation of the induced dipole fields of the n cells and a directional component of the applied field. From Box 6.1 the potential at a distance r from the centre of our imagined spherical volume of suspended cells is of the form:

$$\phi_1 = \left(\frac{nA}{r^2} - Br\right) E\cos\theta, \text{ where } A = \frac{\varepsilon_c - \varepsilon_m}{\varepsilon_c + 2\varepsilon_m} R_c^3$$

The volume fraction v_c of suspended cells is given by:

$$v_c = \frac{nR_c^3}{R_m^3}$$

so that

$$\phi_1 = \left(v_c R_m^2 \frac{\varepsilon_c - \varepsilon_m}{\varepsilon_c + 2\varepsilon_m}\frac{1}{r^2} - Br\right) E\cos\theta$$

Replacing our imagined sphere containing n cells with an equal spherical volume of material of homogeneous permittivity ε_{eff}, the new potential is:

$$\phi_2 = \left(R_m^3 \frac{\varepsilon_{eff} - \varepsilon_m}{\varepsilon_{eff} + 2\varepsilon_m}\frac{1}{r^2} - Br\right) E\cos\theta$$

We now wish our imagined homogeneous sphere of permittivity ε_{eff} to be equivalent in dielectric properties to the sphere containing n cells (i.e., we want $\phi_2 = \phi_1$). This requires the following equality:

$$\frac{\varepsilon_{eff} - \varepsilon_m}{\varepsilon_{eff} + 2\varepsilon_m} = v_c \frac{\varepsilon_c - \varepsilon_m}{\varepsilon_c + 2\varepsilon_m} \tag{9.2}$$

approximation into the denominator of the left-hand side of equation (9.3) we obtain the relationship:

$$\varepsilon^*_{eff} = \varepsilon^*_m\left(1 + 3v_c \frac{\varepsilon^*_c - \varepsilon^*_m}{\varepsilon^*_c + 2\varepsilon^*_m}\right) \tag{9.5}$$

This equation, generally known as the Maxwell–Wagner equation, predicts that as the volume concentration v_c of the cells is increased from zero (when $\varepsilon^*_{eff} = \varepsilon^*_m$) the measured value of the suspension's permittivity increases linearly with increasing cell concentration. This is in fact observed experimentally, so that Equation (9.5) can be employed for very dilute cell suspensions. In this respect we should note that the results shown in Figure 9.8 correspond to a 43.8% volume-based concentration of cells. This is not a dilute cell suspension. It corresponds, for example, to the situation of a suspension of cells of diameter 10 μm having their centres spaced just 13 μm apart. The cells will experience each other's induced dipole field that locally distorts the applied uniform field. This violates the assumption made in Box 9.1 that the cells are *not* influenced by such distortions of the applied field. With a 4% volume concentration the cells will on average be spaced three cell diameters apart. In this case the error using Equation (9.5) is probably 'tolerable'.

Relationships of the form of Equation (7.12) are obtained on separating the real and imaginary terms in Equation (9.5):

$$\varepsilon'_{eff} = \varepsilon'_\infty + \frac{(\varepsilon'_s - \varepsilon'_\infty)}{1 + i\omega\tau}; \quad \sigma'_{eff} = \sigma'_\infty - \frac{(\sigma'_\infty - \sigma'_s)}{1 + i\omega\tau} \tag{9.6}$$

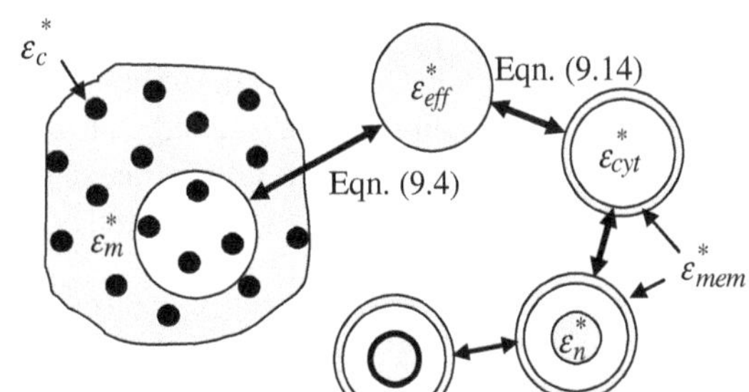

Figure 9.9 A schematic to show how through Equation (9.4) the effective 'smeared-out' complex permittivity ε^*_{eff} of a cell can be obtained from measurement of the complex permittivity of a suspension of such cells. Alternatively, Equation (8.32) can be used to predict the complex permittivity of a compound dielectric composed of known particles dispersed in a known medium. Through a repeat procedure used to derive Equation (9.14) the dielectric properties of a multicomponent model of a cell can be simplified to that of a homogeneous sphere of effective permittivity ε^*_{eff}, or *vice versa*. Thus, through Equations (9.4) and (9.14) the dielectric properties of the components of a cell can be deduced from the dielectric properties of a suspension of the cells.

with

$$\tau = \varepsilon_o \frac{\varepsilon'_c + 2\varepsilon'_m}{\sigma'_c + 2\sigma'_m} \tag{9.7}$$

$$(\varepsilon'_s - \varepsilon'_\infty) = \Delta\varepsilon'_{eff} = 9\,v_c \frac{(\varepsilon'_c\sigma'_m - \varepsilon'_m\sigma'_m)^2}{(\varepsilon'_c + 2\varepsilon'_m)(\sigma'_c + 2\sigma'_m)^2} \tag{9.8}$$

$$(\sigma'_\infty - \sigma'_s) = \Delta\sigma'_{eff} = \frac{1}{\tau}\Delta\varepsilon'_{eff} \text{ (see Equation 7.18)} \tag{9.9}$$

The approximation that, for a sufficiently large observation scale, a heterogeneous compound material can be considered as a homogeneous one is the basis for what is called the *effective medium theory*. It has been employed with various degrees of success to formulate theories that describe the ferroelectric, magnetic, optical and piezoelectric properties of heterogeneous materials. Of particular interest to us, of course, are those that are directed towards describing the dielectric properties of mixtures of high particle concentration [e.g., 48–53].

A particularly successful extension of Equation (9.5) for highly concentrated suspensions is that formulated by Hanai [54]. His ingenious procedure involves incrementing, step by step, the volume fraction v_c so that an increment Δv_c increases the permittivity of the mixture from ε^*_{mix} to $(\varepsilon^*_{mix} + \Delta\varepsilon^*_{mix})$. At the same time the permittivity of the medium is assumed to increase from ε^*_m to ε^*_{mix}. The volume fraction v_c is also replaced by the factor $\Delta\hat{v}_c/(1-\hat{v}_c)$, where $\hat{v}_c$ is the new volume fraction. Placing these substitutions into Equation (8.32) and equating ε^*_{mix} to ε^*_{eff} gives the relationship:

$$\Delta\varepsilon^*_{mix} = \frac{3\varepsilon^*_{mix}\Delta\hat{v}_c(\varepsilon^*_c - \varepsilon^*_{mix})}{(1-\hat{v}_c)(2\varepsilon^*_{mix} + \varepsilon^*_c)} \tag{9.10}$$

The act of incrementally increasing the volume fraction continues until a final value for v_c is reached. The summation of these incremental actions is given after rearranging Equation (9.10) and by the following integral relationships:

$$\int_{\varepsilon^*_m}^{\varepsilon^*_{mix}} \frac{(2\varepsilon^*_{mix} + \varepsilon^*_c)}{3\varepsilon^*_{mix}(\varepsilon^*_{mix} - \varepsilon^*_c)} d\varepsilon^*_{\text{mix}} = \int_0^{v_c} -\frac{1}{1-\hat{v}_c} d\hat{v}_c \tag{9.11}$$

The right-hand definite integral is straightforward and equal to $\ln(1 - v_c)$, but the left-hand function is discontinuous (heading off to infinity) at $\varepsilon^*_{mix} = 0$ and at $\varepsilon^*_{mix} = \varepsilon^*_c$. However, ε^*_{mix} is always greater than zero and less than ε^*_c. Within the limits of the integration this function is thus continuous and its integral can be obtained:

$$\int_{\varepsilon^*_m}^{\varepsilon^*_{mix}} \frac{(2\varepsilon^*_{mix} + \varepsilon^*_c)}{3\varepsilon^*_{mix}(\varepsilon^*_{mix} - \varepsilon^*_c)} d\varepsilon^*_{\text{mix}} = \frac{1}{3}\ln\left(\frac{\varepsilon^*_m}{\varepsilon^*_{mix}}\right) + \ln\left(\frac{\varepsilon^*_{mix} - \varepsilon^*_c}{\varepsilon^*_m - \varepsilon^*_c}\right)$$

From Equation (9.11) we then have the following relationship:

$$\frac{\varepsilon^*_{mix} - \varepsilon^*_c}{\varepsilon^*_m - \varepsilon^*_c}\left(\frac{\varepsilon^*_m}{\varepsilon^*_{mix}}\right)^{1/3} = 1 - \hat{v}_c \tag{9.12}$$

This is generally known as the Hanai mixture equation and is the relationship commonly employed to analyse dielectric measurements performed on cell suspensions. As far as the author can find, Hanai's mixture equation has not been taken further to give the dielectric increments and relaxation time of the forms of Equations (9.6) to (9.9). To obtain the value of ε^*_c from the known value of the volume fraction v_c of the cell suspension and the measured values for ε^*_{mix} and ε^*_m, requires the aid of a computer program to find the complex roots of the polynomial equation that results from cubing both sides of Equation (9.12):

$$(\varepsilon^*_{mix})^3 - 3\varepsilon^*_c(\varepsilon^*_{mix})^2 + \left(3(\varepsilon^*_c)^2 + \frac{[(v_c-1)(\varepsilon^*_m - \varepsilon^*_c)]^3}{\varepsilon^{*3}_m}\right)\varepsilon^*_{mix} - (\varepsilon^*_c)^3 = 0$$

Employing the expression $\varepsilon^* = (\varepsilon' - i\sigma'/\varepsilon_o\omega)$ for the complex permittivity, values of ε' and σ' can be calculated. This method was used to obtain the membrane capacitance of erythrocytes from the results shown in Figure 9.9 [23]. A value of 6.5–7.0 mF/m^2 was obtained using Equation (9.8), compared with 7.0–7.5 mF/m^2 obtained from a direct patch-clamp measurement. Analysis of the data using the Maxwell–Wagner Equation (9.5) gave a membrane capacitance value of 8.0 mF/m^2 [23].

These consistent values for the membrane capacitance of the human erythrocyte can also be compared with the value of 8.1 mF/m^2 obtained by Fricke for the dog erythrocyte [55]. He employed an equivalent electrical circuit for a cell suspension of the configuration shown in Figure 9.1(b) and assumed that, at low frequencies, the impedance of the cell membrane greatly exceeded that of the cell interior. He obtained the membrane capacitance value using the relationship:

$$C = \alpha A\left(1 - \frac{r_1}{r}\right)C_o = C_{100}\left(1 - \frac{r_1}{r}\right) \tag{9.13}$$

in which C is the measured specific capacity of the cell suspension; C_o the static capacity per sq. cm of the cell membrane, r and r_1 the specific resistances respectively of the cell suspension and suspending liquid, 2α the major axis of the spheroid (cell). The factor A is a constant equivalent to the depolarizing factor of Equations (7.33) and (7.34). Fricke defined the specific electric capacity of a cell suspension as the capacity that, combined in parallel with a certain resistance, electrically balances 1 cm. cube of the suspension [56]. The parameter C_{100} is the specific capacity of a suspension with a concentration of 100% and its theoretical value was determined based on measurements of C and r for various dilutions (with serum, $r_1 = 75.8\,\Omega$) of whole blood. Based on $2\alpha = 7.2$ µm and an axial ratio of 4 : 1 for the discoid erythrocyte, A = 1.28. The value for C_{100} was found to be 372 picofarads (±2%) per 1 cm^3, so that, based on the relationship $C_o = (C_{100}/\alpha A)$ obtained from Equation (9.13), the membrane capacitance was calculated to be 8.1 mF/m^2. Assuming a value of 3 for the relative permittivity of the lipid membrane material, a membrane thickness of 33 nm was derived [55]. This was the first indication of the molecular dimension of the membrane of an intact and physiologically viable cell.

From numerous chemical, mechanical, optical, X-ray and electron microscopy studies that have been performed over the decades since Fricke's pioneering work, a value for the cell membrane thickness of 7.5~10 nm has emerged. A light-microscope technique, in which the membrane material of a single intact human erythrocyte is extracted into a long cylindrical strand, gave a membrane thickness of 7.8 nm [57]. Justification is therefore required as to why a membrane thickness of 7 nm is commonly assumed in the analysis of the dielectric properties of cells (e.g., see Tables 9.2 and 9.3). A cell membrane is not a homogeneous structure. It consists of three main regions connected 'electrically' in series. For example around the outer surface of an erythrocyte there is a 'fuzzy' coat of thickness ~100 nm, known as the glycocalyx, consisting of hydrophilic oligosaccharides, glycoproteins and glycolipids. Ions can freely migrate through this structure and so is not the resistive interface responsible for the Maxwell–Wagner β-dispersion. The interface of relevance is revealed after the cell is 'fixed' with glutaraldehyde and osmium tetroxide treatment. A transmission electron micrograph of a section through the membrane of a 'fixed' cell shows two dense lines that correspond to the inner and outer polar layers of the lipid head groups. The clear region between these lines is the hydrophobic portion of the lipid bilayer, depicted in Box 8.1. Coster and Smith [58] considered this molecular organization in black lipid membranes (BLMs) and concluded that the capacitance associated with the polar head region was ~300 mF/m^2. Because this capacitance is 'connected' in series with that (~8 mF/m^2) of the hydrophobic lipid bilayer, the total effective value of the membrane capacitance is given by:

$$\frac{1}{C_{mem}} = \frac{1}{C_{lipid}} + \frac{1}{C_{polar\,heads}} \approx \frac{1}{8} + \frac{2}{300}\ (\text{mF/m}^2)^{-1}$$

From this relationship we can deduce that the outer and inner membrane layers containing the polar head groups contribute no more than ~5% of the total membrane capacitance. It is therefore reasonable to take the 'electrical' thickness of a cell membrane to be at the lower end of the range of observed physico-chemical or 'mechanical' thickness values. A nominal membrane thickness of ~7 nm is reasonable.

Example 9.2 Deriving the Volume Fraction of Cells in a Suspension using Hanai's Mixture Equation
Human erythrocytes are suspended in a hypotonic solution of 50% phosphate buffered saline of conductivity 8 mS/cm. Dielectric measurements reveal the existence of a large dispersion with lower and upper frequency limits of 300 kHz and 30 MHz respectively. The conductivity of the suspension was measured to be 3.8 mS/cm at 10 kHz. Use Equation (9.12) to derive the volume fraction of cells in the suspension.

Solution 9.2 When suspended in 50% PBS erythrocytes have a spherical shape and Hanai's mixture equation can be used for cell suspensions with volume fractions up to ~60%. At 10 kHz, a frequency well below the range occupied by (what can be assumed to be) the β-dispersion, the relationship $\varepsilon^* = (\varepsilon' - i\sigma'/\varepsilon_o\omega)$ approximates to $\varepsilon^* = -i\sigma'/\varepsilon_o\omega$, so that the Hanai mixture equation given by Equation (9.12) simplifies to:

$$\frac{\sigma'_c - \sigma'_{mix}}{\sigma'_c - \sigma'_m}\left(\frac{\sigma'_m}{\sigma'_{mix}}\right)^{\frac{1}{3}} = 1 - \widehat{v}_c$$

Assuming that the cells have intact, high-resistance, plasma membranes then $\sigma'_m >> \sigma'_c$ and to a good approximation the left-hand side of this equation is equal to $(\sigma'_{mix}/\sigma'_m)^{2/3}$, so that

$$\widehat{v}_c = 1 - \left(\frac{\sigma'_{mix}}{\sigma'_m}\right)^{\frac{2}{3}}$$

For $\sigma'_{mix} = 3.8$ mS/cm and $\sigma'_m = 8$ mS/cm, the volume fraction $\widehat{v}_c = 0.39$.

9.4.1 Single-Shell Model of a Cell

Considering the range of measurement and analysis techniques employed, it is remarkable that the derived membrane capacitance for mammalian erythrocyte is consistently determined as 7.3 ± 0.8 mF/m^2. There is also the experimental fact that erythrocytes do not always retain their discoid shape in some suspension media, becoming spherical in moderately hypotonic (~70% PBS) solutions, for example. Takashima *et al.* [23] did not employ Equation (9.13) derived by Fricke to calculate the membrane capacitance. So how did they achieve this from their derivation of ε^*_c? How were the dielectric parameters for the cytoplasm, nuclear envelope and nucleoplasm cited in Table 9.3 derived? A clue to this is provided by considering the case n = 1 for the procedure outlined in Box 9.1. This can represent a single sphere of radius R_1 located concentrically inside a slightly larger spherical cell of radius R_2. If we let the inner sphere have dielectric properties similar to that of the cytosol of a cell, with the material in the 'shell' between the inner and outer sphere gap having a permittivity and conductivity similar to that of a cell membrane, we have an elementary model for a spherical cell such as an erythrocyte, which does not possess a nucleus or internal organelles. Proceeding as in Box

9.1 and noting that, for n = 1, the volume fraction is given by $v_c = (R_1/R_2)^3$, the following relationship is derived:

$$\varepsilon^*_{cell} = \frac{(2\varepsilon^*_{mem} + \varepsilon^*_{cyt})R_2^3 + 2(\varepsilon^*_{cyt} - \varepsilon^*_{mem})R_1^3}{(2\varepsilon^*_{mem} + \varepsilon^*_{cyt})R_2^3 - (\varepsilon^*_{cyt} - \varepsilon^*_{mem})R_1^3}\varepsilon^*_{mem} \quad (9.14)$$

where the suffices *mem* and *cyt* represent the membrane and cytoplasm phases, R_2 is the cell radius and the membrane thickness d is given by $d = R_2 - R_1$. Our system of two concentric spheres, representing a single shell model for a cell, is therefore equivalent to a homogeneous single sphere whose complex permittivity is expressed in terms of the cell radius and the permittivities of the cytoplasm and membrane. The permittivity ε^*_m of the surrounding medium does not appear in Equation (9.14) and so our result is completely general. Altering the suspending medium will not influence the theoretical result, but in an experiment this may cause osmotic stress to the cell so that some of the parameters such as R_1 and ε^*_m may change. The procedure used to derive Equation (9.14) can now be used to derive the equivalent homogeneous permittivity of a sphere of permittivity ε^*_m that is inserted into another sphere and so on. As shown in Figure 9.9 we can reach the stage of modelling a cell consisting of a nucleus with its nuclear membrane, a cytoplasm containing organelles and an outer cytoplasmic membrane. Furthermore, through a combination of Equations (9.4) and (9.14) the dielectric properties of the components of a cell can be deduced from the dielectric properties of a suspension of the cells.

The cell membrane thickness d is exquisitely small, so that, in Equation (9.4), we can make the following approximations: $R_1 \gg d$; $R_1 \approx R_2 = R$; terms such as d/R and $d^3 \approx 0$. With these approximations Equation (9.4) simplifies to:

$$\varepsilon^*_{cell} = \varepsilon^*_{cyt} + \frac{2d}{R}\varepsilon^*_{mem} \quad (9.15)$$

Expressing the complex permittivity as $\varepsilon^* = (\varepsilon' - i\sigma'/\varepsilon_o\omega)$ and $\sigma^* = i\omega\varepsilon_o\varepsilon^*$, the effective permittivity and conductivity of the cell can be given in terms of the cell radius R, with the cytoplasm and membrane dielectric properties, as:

$$\varepsilon^*_{cell} = \left(\varepsilon'_{cyt} + \frac{2d}{R}\varepsilon'_{mem}\right) - i\frac{1}{\varepsilon_o\omega}\left(\sigma'_{cyt} + \frac{2d}{R}\sigma'_{mem}\right) \quad (9.16a)$$

$$\sigma^*_{cell} = \left(\sigma'_{cyt} + \frac{2d}{R}\sigma'_{mem}\right) + i\varepsilon_o\omega\left(\varepsilon'_{cyt} + \frac{2d}{R}\varepsilon'_{mem}\right) \quad (9.16b)$$

where

$$\varepsilon'_{cell} = \varepsilon'_{cyt} + \frac{2d}{R}\varepsilon'_{mem} \quad \text{and} \quad \sigma'_{cell} = \sigma'_{cyt} + \frac{2d}{R}\sigma'_{mem}$$

The real and imaginary components of this equation can be substituted into Equations (9.6) to (9.9) to derive expressions for the dielectric increment and relaxation time of the interfacial polarization. For example, the relaxation time is given by:

$$\tau = \varepsilon_o \frac{\varepsilon'_{cyt} + 2\varepsilon'_m + \frac{2d}{R}\varepsilon'_{mem}}{\sigma'_{cyt} + 2\sigma'_m + \frac{2d}{R}\sigma'_{mem}} \quad (9.17)$$

To check the validity of this result, on removal of the membrane the relaxation time should return to the value given by Equation (9.7). With the membrane removed (i.e., $d = 0$) parameters ε'_{cyt} and σ'_{cyt} revert back to the permittivity and conductivity of the cell (ε'_c and σ'_c) and Equation (9.17) does indeed become Equation (9.7).

Miles and Robertson [59] employed the single-shell geometry shown in Figure 9.10(a) to describe the dielectric behaviour of a compound material consisting of colloidal particles each surrounded by an electrical double layer. Thus, implicit in their model is the assumption that the particles carry a net surface charge density and so attract counter-ions to their surfaces. The particles are modelled as spheres of radius R surrounded by a thin ionic double layer, represented as a concentric conducting shell of thickness d (~1 nm) with $d \ll R$. It is also assumed that the particles are poor conductors, so that

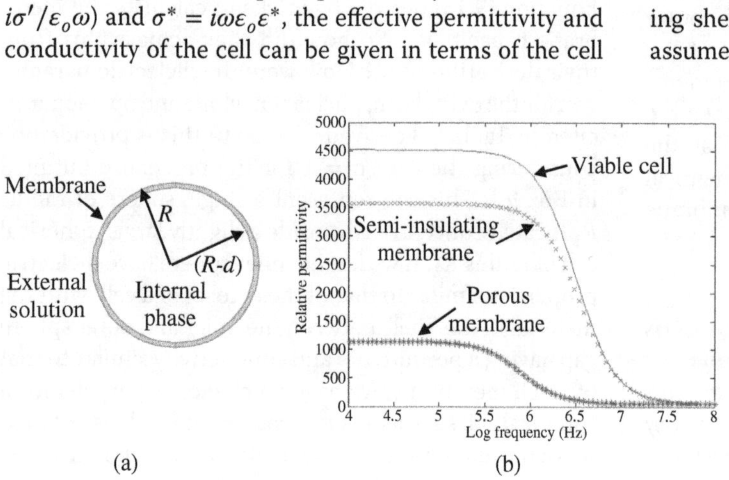

Figure 9.10 (a) Single-shell model of a cell of radius R, consisting of a plasma membrane of thickness d and an internal phase. (b) Results obtained using the MATLAB program in Box 9.2 to show how the interfacial polarization (β-dispersion) of a cell depends on the integrity of the plasma membrane.

the conductivity of the ionic shell is large compared with that of the particle core. Following a mathematical treatment similar to that described in Box 6.1 and adopted in Box 9.1, Miles and Robertson derived an equation of the same form as Equation (9.14) to describe the effective dielectric properties of the compound material, but with the dielectric parameters expressed as complex conductivity terms rather than complex permittivities [59]. Miles and Robertson concluded that the spread of relaxation times, predicted by Wagner [47] to describe the dielectric behaviour of a compound dielectric formed of spheres having a range of specific conductivities, may also result from the colloidal particles having a spread of diameters.

9.4.1.1 Maxwell–Wagner Interfacial Polarization (β-Dispersion)

The Maxwell–Wagner interfacial polarization of a cell can be modelled using the MATLAB program given in Box 9.2. An example to evaluate how interfacial polarization is influenced by the cell membrane is given in Figure 9.10. With an intact membrane exhibiting a high resistance to passive ion flow, the modeled permittivity shows a large dielectric dispersion, known as the β-dispersion, of the form shown in Figure 9.8. The magnitude of this dispersion reduces significantly as the high resistance value of the membrane is degraded and ions in the cytoplasm are allowed to diffuse down their concentration gradient from the cytoplasm to the outside medium. The characteristic frequency of the dispersion, corresponding to the point of inflection of the dielectric decrement, is also observed to decrease. This corresponds to an increase of the relaxation time τ, in agreement with the form of Equation (9.17) which predicts that the value of τ should decrease as the values for σ'_{mem} and σ'_{cyt} increase.

The MATLAB program in Box 9.2 can also be used to investigate how the β-dispersion exhibited by a suspension of cells is influenced by the size of the cells and the magnitude of their membrane specific capacitance. The influence of cell size, whilst maintaining a constant membrane specific capacitance and specific resistance, is shown in Figure 9.11. This reveals that the magnitude of the interfacial polarization increases with increasing

Box 9.2 MATLAB Program for the Single-Shell Model of a Cell

This program, which extends the one given in Box 6.9, treats the case of a particle surrounded by a thin layer that mimics the membrane of a spherical cell. When used together lines 19 and 23 model a viable cell. Changing line 23 for 24 degrades the resistance of the membrane, whilst introducing line 20 with line 24 models the stage where ions have diffused out of the cell to decrease the cytoplasm conductivity.

```
1 %  SSMC.m
2 % Internal phase assigned conductivity kc1
    and permittivity kp1.
3 % Membrane assigned capacitance C (F/m^2) &
    conductance G (S/m^2).
4 % From C & G derive membrane conductivity
    kc2 and permittivity kp2.
6 % Table 9.2 used as a guide to assign
    dielectric values.
7 % ki (i=1,2) complex permittivity of
    cytoplasm, membrane.
8 % -------------------------------------------
9 % p0: permittivity of free space (8.854 x
    10^-12 F/m)
10 p0=8.854e-12;
11 % Membrane thickness d; cell radius a2
12 d=7e-9;
13 a2=5e-6;
14 a1=a2-d;
15 f=logspace(3,9,120);
16 zeroline=f-f;
17 w=2*pi*f;
18 % Cytoplasm conductivity and permittivity
19 kc1=0.8;
20 % kc1=0.1;
21 kp1=50*p0;
22 % Membrane conductivity and permittivity
23 G=0.2;
24 % G=2e4;
25 kc2=G*d;
26 C=8e-3;
27 kp2=C*d;
28 % Calculate Effective Permittivity -
    using Equation (9.14)
29 k1=kp1-i*kc1 ./w;
30 k2=kp2-i*kc2 ./w;
31 am1=a1^3;
32 am2=a2^3;
33 keff2=k2 .*(am2*(k1+2*k2)-2*am1*(k2-
k1)) ./(am2*(k1+2*k2)+... am1*(k2-k1));
34 rm=real(keff2);
35 plot(log10(f),rm/p0,'-', log10(f),
    zeroline,'-');
36 xlabel('Log Frequency (Hz)')
37 ylabel('Relative Permittivity')
38 hold on
```

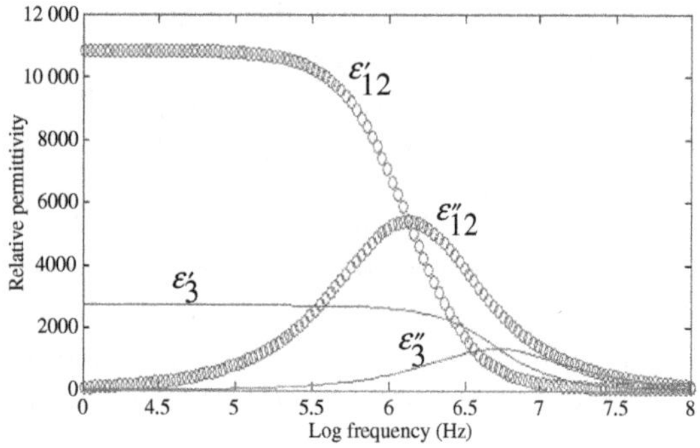

Figure 9.11 Plots of the real (ε') and imaginary (ε'') components of the β-dispersion exhibited by a cell as modelled using the MATLAB program in Box 9.2. The subscripts 3 and 12 refer to a cell radius of 3 μm and 12 μm, respectively. The magnitude of the dispersion increases, whereas the characteristic frequency decreases as the radius of the cell increases.

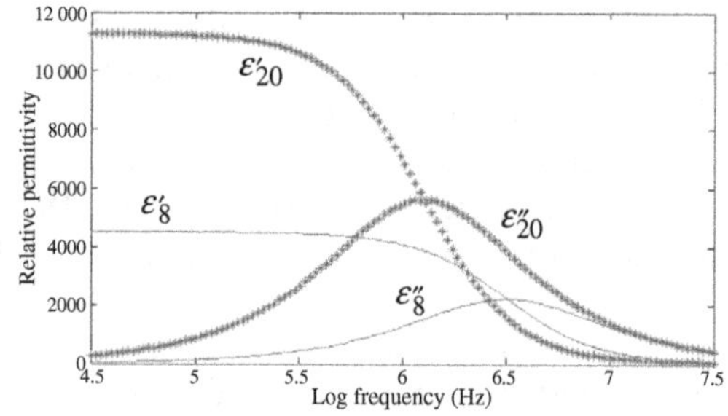

Figure 9.12 The real (ε') and imaginary (ε'') components of the β-dispersion exhibited by a cell of radius 5 μm, modelled using the program in Box 9.2. The subscripts 8 and 20 refer to membrane capacitance values of 8 and 20 mF/m^2, respectively. The magnitude of the dispersion increases, whereas the characteristic frequency decreases, as the membrane capacitance increases.

radius of the cell. This result is consistent with an effective increase of the total surface charge storing capacity the cell as its surface area increases with cell diameter. It is also evident in Figure 9.11 that the characteristic frequency of the dispersion decreases as the radius of the cell increases. This can also be related to a geometrical effect. The charges accumulating on the outer cell surface are ions from the surrounding electrolyte, while counter charges from the internal electrolyte also accumulate on the inner cell membrane surface. At each half-cycle of the applied AC electric field these accumulated ions reverse polarity with respect to each other. The characteristic time for this charge displacement increases as the distance between opposite sides of the cell increases, so that the characteristic displacement frequency (being proportional to the reciprocal of the displacement time) decreases, as shown in Figure 9.11. Another way of viewing this is that the length (hence effective resistance) of the ionic conducting path around the cell and between opposing internal membrane surfaces increases as the diameter of the cell increases. This, combined with the increase of the area for charge accumulation raises the effective RC time constant for charge displacement. The influence of a change of membrane capacitance, for a cell of constant diameter, is shown in Figure 9.12.

9.4.1.2 Factors Influencing Membrane Capacitance Value

A cell's plasma membrane acts as an electrical charge storing element because it has the same basic architecture as a capacitor in an electrical circuit. It consists of a thin dielectric sandwiched between two conductors, namely the outer electrolyte and the cytoplasm. For a spherical cell this architecture is the same as that depicted in Figure 3.21 for applying Gauss's Law to calculate the field produced in a dielectric situated between two spherical conducting shells. The solution to this exercise is given by Equation (3.43):

$$E = \frac{Q}{4\pi\varepsilon_o\varepsilon_r r^2} \quad \text{for the condition } R > r > (R-d)$$

We have assigned the radii of the outer and inner surfaces of the cell membrane to be R and $(R - d)$, respectively, where d is the membrane thickness. From Equation (3.8) and Table 3.2 we see that this is the same expression for E created by a point charge Q, as employed in Example 4.4. We can therefore apply Equation (4.23) to obtain an expression for the potential V of the inner conductor (the cytoplasm) at $r = (R - d)$ with respect to that of the outer conductor (the external electrolyte) at $r = R$:

$$V = V_{R-d} - V_R = \frac{Q}{4\pi\varepsilon_o\varepsilon_r}\left(\frac{1}{(R-d)} - \frac{1}{R}\right) = \frac{Q}{4\pi\varepsilon_o\varepsilon_r}\frac{d}{R(R-d)}$$

The capacitance of this system is given by

$$C = \frac{Q}{V} = 4\pi\varepsilon_o\varepsilon_r = \frac{R(R-d)}{d} \tag{9.18}$$

The quantity $4\pi R(R - d)$ is the geometric mean of the areas $4\pi R^2$ and $4\pi\ (R - d)^2$. Assigning this area as A, we can rewrite Equation (9.18) as

$$C_m = \frac{A\varepsilon_o\varepsilon_r}{d} \tag{9.19}$$

For the case of a mammalian cell, where we always have the situation that $R \gg d$, the membrane capacitance can therefore be calculated using the same formula as that for a capacitor composed of a dielectric sandwiched between two parallel plate electrodes distance d apart and each of area A. As indicated in Table 9.2, if nominal values of ~5 and ~7 nm are adopted for ε_r and d, respectively, a specific membrane capacitance

(C_m/A) of ~6 mF/m^2 is obtained for a smooth cell membrane. Values in the range 6 ~ 8 mF/m^2 are observed for erythrocytes, which have relatively smooth membranes [e.g., 23, 55]. However, from Table 9.2 it is clear that values significantly larger than this are also observed. The only parameter in Equation (9.19) capable of sufficient adjustment to produce large values for C_m is the effective area of the membrane surface. This can be quantified by introducing a membrane topography parameter ϕ that represents the ratio of the actual membrane area of the cell to the membrane area ($4\pi R^2$) that would form a perfectly smooth and spherical covering of the cytoplasm. A perfectly smooth membrane corresponds to ϕ having a value of 1.0. The area A in Equation (9.19) is thus replaced with the factor ϕA_o, where ϕ is proportional to the 'roughness' of the membrane surface:

$$C_m = \frac{\phi A_o \varepsilon_o \varepsilon_r}{d} \tag{9.20}$$

Tests of this concept have been performed by subjecting cells to changes of osmotic pressure. In section 8.3 it is described how cells are 'happy' in an isotonic medium, equivalent to a ~145 mM KCl solution, which has an osmolarity of ~290 mOsm. Cells suspended in a medium with a tonicity larger than this value are in a hypertonic medium. The number of 'particles' per unit volume of water outside the cell is larger than that in the cytoplasm. Alternatively, we can say that the concentration (also termed *activity*) of water inside the cell is larger than that of the suspending medium. These two solutions are separated by a water-permeable membrane and when confronted with this situation water will always flow in the direction that will dilute the more concentrated solution. When placed in a hypertonic solution, water will thus flow out of the cell. This causes the cell to shrink and the membrane to 'wrinkle'. The value of the membrane topography parameter ϕ in Equation (9.20) increases. Conversely, cells suspended in a hypotonic medium will swell as water flows into the cell. To accommodate this increase in cell surface area the membrane will stretch and smooth out membrane folds or reduce the number of microvilli by utilizing their membrane material. This should have the effect of reducing the value of ϕ. Irimajiri *et al.* [60] studied the dielectric properties of basophil leukaemia cells and found that the specific membrane capacitance value was strongly correlated with the number density of microvilli on the membrane surface. Sukhorukov *et al.* [61] performed electrorotation studies on three cultured murine cell lines (myeloma (SP2); hybridoma (G8); L-cells (fibroblasts)) suspended in media of osmotic strengths ranging from 60 mOsm to 330 mOsm. Under isotonic conditions (280 mOsm) the myeloma and hybridoma cells exhibited C_m values of 10.1 and 10.9 mS/m^2, respectively, whilst the fibroblasts exhibited a value of 21.8 mS/m^2. For all three cell types the C_m value decreased with decreasing osmolality. For example, for the G8 cells the C_m value decreased from 12.7 mS/m^2 at 320 mOsm to reach a plateau of 7.7 ~ 8.0 mS/m^2 below 150 mOsm, representing a fall of ϕ from ~1.6 to ~1.0. Electrorotation studies were also performed by Wang *et al.* [62] for murine erythroleukaemia (DS19) cells during their erythropoietic differentiation, over which time their C_m value fell from 17.4 to 15.3 mS/m^2 at 300 mOsm. Scanning and transmission electron microscopy revealed that this fall in C_m correlated with a reduction in the density of complex surface features, including microvilli. When the osmolality was decreased from 450 to 210 mOsm, the mean C_m of undifferentiated DS19 cells changed from 20.5 to 15.8 mS/m^2 (a decrease of 22.9%) while that for differentiated cells changed from 17.2 to 14.7 mS/m^2 (a decrease of 14.5%). This demonstrated that cells exposed to the differentiation treatment had an enhanced mechanical resilience compared with their untreated counterparts, evidencing the early stages of the development of the membrane skeleton, which becomes fully developed in mature erythrocytes. Asami [63] performed a three-dimensional finite element method (FEM) simulation of the dielectric properties of a cell model that emulates the morphological changes of microvilli during osmotic perturbation. In this simulation a microvillous cell is modeled as a sphere with cylindrical projections. He confirmed that the dielectric properties of cells are strongly influenced by such surface morphology features.

The value of the membrane capacitance could also be influenced by changes in the values of ε_r and d in Equation (9.20) arising, for example, from differences in the protein content of the membrane. Gentet *et al.* [64] performed patch-clamp measurements of C_m on embryonic kidney cells (HEK-293) before and after transfection to increase the content (expression) of glycine receptors and other membrane proteins. Values of 11.1 ± 0.8 mS/m^2 and 10.5 ± 0.9 mS/m^2 were obtained for the untransfected and transfected cells, respectively. Thus, despite the large increase in the membrane protein content of transfected HEK-293 cells, there was no detectable change in the C_m value. Stoneman *et al.* [65] studied the dielectric properties of yeast cells (*Saccharomyces cerevisiae*) that had been genetically engineered to overexpress a G protein-coupled receptor (Ste2p protein) in the membrane of these yeast cells. These proteins were fused with the green fluorescent protein (GFP) to enable determination the extent of this expression by fluorescence microscopy. The double-shell model of a cell was used to analyse the data. The control cells exhibited a C_m value of 17.5 ± 1.6 mS/m^2, while for the cells with an increased protein content in their membrane the C_m

value was 16.3 ± 1.1 mS/m^2. Again, this demonstrates that the protein content of a membrane plays a minor part in determining its specific capacitance. Instead, it may be assumed that the C_m value mostly reflects the properties of the hydrophobic layer of the membrane, which is populated by the hydrocarbon tails of the phospholipids and hydrophobic segments of integral membrane proteins. Of relevance to this are the highly efficient (98%) dielectrophoretic separation of co-cultured C2C12 myoblasts and GFP-expressing MRC-5 fibroblasts reported by Muratore *et al.* [66]. Because of their similar size and shape this high purity separation was attributed to differences in the membrane capacitance values of the two cell types. Raman spectra obtained for their membranes indicated that the fibroblast membranes contained a smaller proportion of saturated lipids than those of the myoblasts. It was concluded that, apart from the membrane topography parameter ϕ, membrane thickness and molecular composition may thus also contribute to the specific membrane capacitance of a cell.

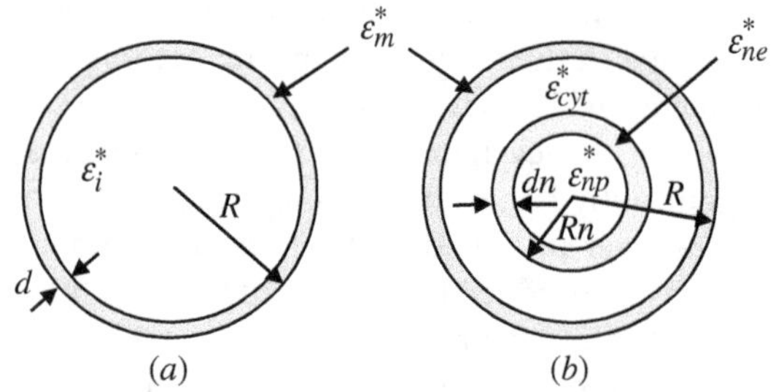

Figure 9.13 (a) Single-shell model of a shell. (b) Double-shell model. The number of shells is denoted by the number of membranes, not the number of represented cell components. The subscripts for the complex permittivities refer to the following compartments: *m*, plasma membrane; *i*, inner phase of cell; *cyt*, cytoplasm; *ne*, nuclear envelope; *np*, nucleoplasm. The morphological parameters are: *R*, outer cell radius; *Rn*, outer radius of nucleus; *d*, plasma membrane thickness; *dn*, thickness of nuclear envelope. The double-membrane structure of the nuclear envelope is modelled as a single homogeneous shell. To account for the double membrane of the nuclear envelope requires a triple-shell model of the cell.

9.4.2 Double-Shell Model of a Cell

Irimajiri *et al.* [67] measured the permittivity and conductivity of suspensions of cultured rat basophil leukaemia cells over the frequency range 10 kHz–500 MHz. Two dielectric dispersions were observed: a low-frequency dispersion with a characteristic frequency of several hundred kHz and $\Delta\varepsilon = 10^3 \sim 10^4$; a high-frequency dispersion with a characteristic frequency ~20 MHz and $\Delta\varepsilon = 10 \sim 100$. An analysis of the data was attempted using the single-shell model of a cell and the Hanai mixture theory expressed by Equation 9.12. A reasonable fit of the experimental data to theory was obtained for frequencies below ~1 MHz, yielding a value of 14 mF/m^2 for the plasma membrane capacitance. However, the single-shell model failed in its ability to simulate the experimental data above 1 MHz. It was concluded that the single-shell model has limited ability to describe the dielectric properties of cells, which may contain subcellular organelles delineated with their own membranes, such as nuclei, mitochondria and secretory granules [60]. To account for the discrepancy between experiment and theory, it was proposed that the double-shell model proposed by Irimajiri *et al.* [68, 69] should be used rather than the single-shell model. The number of shells implied by the terminology corresponds to the number of membranes represented in the model. Thus, the single-shell model takes into the account the plasma membrane for a cell without a nucleus (namely a normal erythrocyte), whilst the double-shell model includes the nuclear envelope (treated as a single homogeneous membrane) [69] or pentaphase systems consisting of shell spheres [70]. The double-shell model has been applied to analyse of changes of the dielectric properties of lymphocyte membranes induced by GM1 and GM3 glycolipids [71], as well as the dielectric properties of normal and malignant leukocytes [72]. Other systems to have been studied in this way include organelles such as mitochondria [73], yeast cells [74, 75], *E. coli* [76] and plant protoplasts [77]. Schematics of the single-shell and double-shell models for a spherical cell are shown in Figure 9.13. Asami *et al.* [78] have developed these models and theory to describe nonspherical (i.e., ellipsoidal) particles. In Box 9.3 the MATLAB program given in Box 9.2, for the case of the single-shell model, is extended to encompass the double-shell model.

An application of the double-shell model is demonstrated in Figure 9.14 to show how the form of the β-dispersion is sensitive to the relative size of the nucleus. The two relaxation processes observed by Irimajiri *et al.* [67] for leukaemia cells become more evident as the nucleus-cytoplasm volumetric ratio increases. This shows why nucleated cells cannot be analysed accurately using the single-shell model.

The MATLAB program in Box 9.3 assumes that the various components of the cell, including the nucleoplasm, are homogenous mediums whose dielectric properties do not vary as a function of frequency. This is of no major consequence if the cytoplasm, nuclear envelope or nucleoplasm contain polarizable entities with relaxation times longer than that exhibited by the β-dispersion. In this frequency range the cell interior is largely shielded from the applied electric field. At the high-frequency

Box 9.3 MATLAB Program for the Double-Shell Model of a Cell

This program models the dielectric properties of a cell in terms of the nucleus-cytoplasm volume ratio. Dielectric values are assigned according to Table 9.3.

```
% DSMC.m
% Nuclear envelope represented as a single
  homogeneous membrane
% Nucleoplasm(radius a1): conductivity kc1,
  permittivity kp1
% Envelope: conductivity kc2; permittivity
  kp2; thickness d2.
% Cytoplasm (radius a3): conductivity kc3
  and permittivity kp3.
% Membrane: conductivity kc4; permittivity
  kp4; thickness d4.
% ki (i=1,2,3,4) complex permittivity of
  each compartment.
% ------------------------------------------------
%
  clear;
% Establish relative size of nucleus wrt
  total cell diameter
% Cytoplasm inner radius a3
  a3=5.0e-6;
% Internal radius of nucleus a1
% Define fractional volume 'v' of cell
  occupied by nucleus
  v=0.8;
  a1=(v*a3^3)^0.333;
% Thickness of nuclear envelope d2
  d2=40e-9;
  a2=a1+d2;
% Cytoplasm membrane thickness d4
  d4=7e-9;
  a4=a3+d4;
  p0=8.854e-12;
  f=logspace(4.01, 9.0,100);
  zeroline=f-f;
  w=2*pi*f;
% Nucleoplasm conductivity kc1 and
  permittivity kp1
  kc1=1.35;
  kp1=52*p0;
% Nuclear envelope conductivity kc2 and
  permittivity kp2
 kc2=1.1e-3;
 kp2=28*p0;
% Cytoplasm conductivity kc3 and
  permittivity kp3
 kc3=0.8;
 kp3=50*p0;
% Plasma membrane conductivity kc4,
  permittivity kp4
 Gm=0.2;
 kc4=Gm*d4;
 Cm=8e-3;
 kp4=Cm*d4;

 k1=kp1-i*kc1 ./w;
 k3=kp3-i*kc3 ./w;
 k2=kp2-i*kc2 ./w;
 k4=kp4-i*kc4 ./w;

 am1=a1^3;
 am2=a2^3;
 am3=a3^3;
 am4=a4^3;

 keff2=k2 .*(am2*(k1+2*k2)-2*am1*(k2-k1))
 ./(am2*(k1+2*k2)+... am1*(k2-k1));

 keff3=k3 .*(am3*(keff2+2*k3)-2*am2*
 (k3-keff2)) ./(am3*(keff2+2*k3)+...
  am2*(k3-keff2));

 keff4=k4 .*(am4*(keff3+2*k4)-2*am3*
 (k4-keff3)) ./(am4*(keff3+2*k4)+...
  am3*(k4-keff3));

 rm=real(keff4);
 im=imag(keff4);

  plot(log10(f), rm/p0,'+', log10(f),
  zeroline,'-');
% plot(log10(f), -im/p0,'+', log10(f),
  zeroline,'-');

  hold on

 xlabel('Log Frequency (Hz)')
 ylabel('Relative Permittivity')

hold on
```

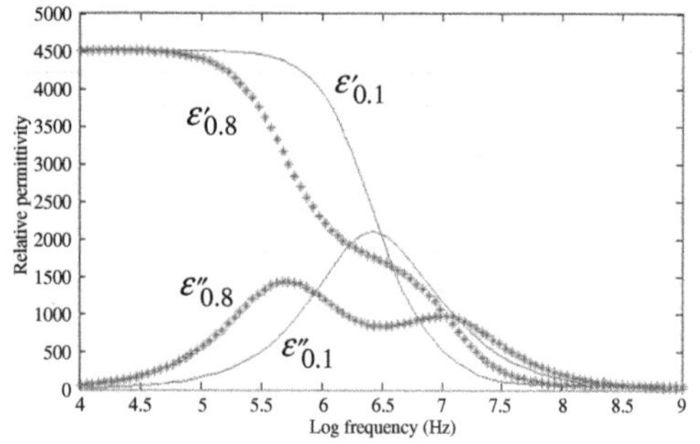

Figure 9.14 Plots of the real (ε') and imaginary (ε'') components of the β-dispersion exhibited by a nucleated cell, obtained using the double-shell MATLAB program in Box 9.3. The subscripts 0.1 and 0.8 refer to nucleus-cytoplasm volume ratios of 0.1 and 0.8, respectively. The additional component of the dispersion associated with the nucleus becomes more apparent as its relative size increases.

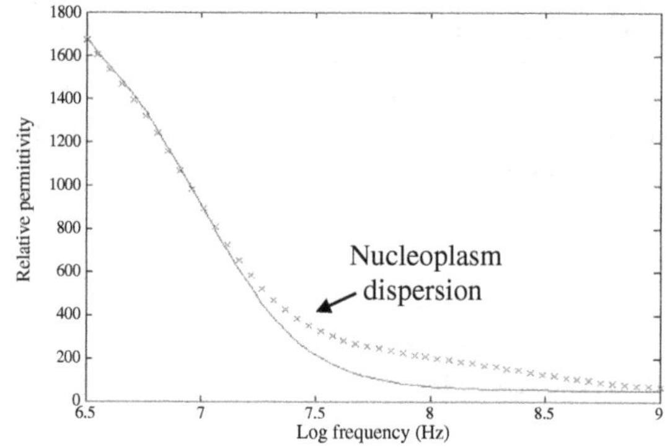

Figure 9.15 Plot of the real (ε') component of the β-dispersion exhibited by a nucleated cell, with (dotted curve) and without (solid line) a dielectric relaxation occurring in the nucleoplasm. See main text for details.

limit of the β-dispersion, however, the field penetrates into the cell. Although there is no evidence (as yet) that the nucleoplasm exhibits dipolar relaxations at high frequencies, their potential influence can be considered by replacing line 30 in the MATLAB program of Box 9.3 with the following ones:

```
delta=350*p0;   % Magnitude of dielectric
                  dispersion
kp1h=50*p0;     % Permittivity at high-
                  frequency limit of
                  dispersion
fc=3e8;         % Characteristic frequency
                  of dispersion
kp1=kp1h + delta ./(1 + (f ./fc).^2); %
                  Equation (7.36)
```

These extra lines of the program incorporate Equation (7.36)

$$\varepsilon'(\omega) = \varepsilon_\infty + \frac{(\varepsilon_s - \varepsilon_\infty)}{1 + f^2/f_c^2}$$

with 'delta' defining the magnitude $\Delta\varepsilon = (\varepsilon_s - \varepsilon_\infty)$ of the dispersion, 'kp1h' the permittivity ε_∞ at the high-frequency limit of the dispersion, f_c its characteristic frequency. An example of executing this program, for the case of a nucleus occupying a fractional volume $v = 0.6$ of the cytoplasm and $f_c = 300$ MHz, is shown in Figure 9.15. Compared to the overall magnitude of the β-dispersion, the effect of an extra dispersion arising, for example, from relaxations induced in the DNA-protein complexes in the nucleoplasm is small. It could, however, influence the value of the high-frequency DEP crossover effect to be described in the next chapter.

9.4.2.1 Modelling the Nuclear Envelope

In the double-shell model the nuclear envelope is represented as a single, homogeneous, shell. In fact, it is composed of two lipid bilayers, the inner nuclear membrane and the outer nuclear membrane, separated by a fluid-filled region known as the perinuclear space of thickness 30 ± 10 nm. The outer nuclear membrane is physically linked to the endoplasmic reticulum, which consists of a network of membranes that form sac-like structures known as cisternae and held together by the cytoskeleton. Their function is to synthesize and export proteins and lipids. The effective area of the outer nuclear membrane is therefore much larger than that required to surround the nucleus and is the reason why it exhibits anomalous specific membrane capacitances ranging from 1000 to 5000 mF/m^2 [30, 31]. The inner nuclear membrane encloses the nucleoplasm and is connected to the outer membrane by nuclear pores. These pores are composed of proteins that form a hollow tube that penetrating each nuclear membrane, through which small particles can pass by passive diffusion. The fluid in the perinuclear space can thus be considered to have a similar conductivity value to that of the electrolyte in the cytoplasm. The capacitances of the inner nuclear membrane and outer nuclear membranes (C_{inm} and C_{onm}, respectively) are thus electrically connected in series. Their combined value to give the nuclear envelope capacitance C_{ne} is thus given by the following formula:

$$\frac{1}{C_{ne}} = \frac{1}{C_{inm}} + \frac{1}{C_{onm}}, \quad \text{so that} \quad C_{ne} = \frac{C_{inm}C_{onm}}{C_{inm} + C_{onm}}$$

There is no experimental evidence to suggest that the capacitance of the inner nuclear membrane has an anomalously high specific capacitance, so the assumption can be made that $C_{onm} \gg C_{inm}$. In this case, based on the above formula, $C_{ne} \approx C_{inm}$. Therefore, the anomalous specific capacitance of the outer nuclear membrane does not form part of the numerical data to be entered into the double-shell model of a cell. In their double-shell model of mouse lymphocytes, Asami *et al.* [34] assumed that the capacitances of the inner nuclear membrane and outer nuclear membrane were the same and each equal to 2 C_{ne}. The best fit value to the experimental dielectric data gave $C_{ne} = 6.2\ \mathrm{mF/m^2}$, in reasonable agreement with the value of 8 $\mathrm{mF/m^2}$ reported by Irimajiri *et al.* [69].

Asami *et al.* [34] also found that the best fit to the experimental data produced a relatively high conductance value of 15 $\mathrm{S/cm^2}$ for the nuclear envelope. This is to be expected because of the large number of electrolyte-filled nuclear pores that penetrate the inner and outer nuclear membranes. Analysis of the dielectric data, obtained over a frequency range from 0.1 to 250 MHz, was performed using both a double-shell model and a single-shell model of the nucleus of the lymphocyte. No difference between the two models and only one dispersion was found [34]. It was therefore concluded that the nuclear envelope can be represented as a single shell. This is why the MATLAB program in Box 9.3 treats a nucleated cell in terms of the double-shell model. The two nuclear membranes and the perinuclear space that comprise the nuclear envelope is treated as a single, homogeneous, shell. The plasma membrane is represented by the second shell.

9.5 Effect of Cell Surface Charge on Maxwell–Wagner Relaxation

As discussed in Chapter 2, cells carry a net surface charge that can be characterized by the electrophoretic mobility of a cell [79]. X-ray standing waves can also be used to investigate the extent to which counterions condense on a charged membrane surface or form a diffuse layer with a near exponential decay of concentration with distance from the surface [49]. Although it was implicit in their model that their particles under consideration carried a net surface charge density so as to attract counterions, the theory adopted by Miles and Robertson [59] is similar to that described in Box 6.1 for the solution of Laplace's equation. However, for a particle carrying a net surface charge, rather than working with Equation (3.31) we should in principle be seeking a solution to Equation (3.30), namely Poisson's equation:

$$\nabla^2 V = -\frac{\rho_T}{\varepsilon_o \varepsilon'_m}$$

where ρ_T is the net charge density carried by the cell. Attention should also be given to the dynamics of the diffuse layer of counterions in the electrical double layer when an external AC field is applied.

As a first approximation we can treat a cell as a homogeneous solid particle, with no inner charged membrane surface. In this case the internal potential of the cell can be obtained by solving Laplace's equation and we need only solve Poisson's equation for the external potential. This is the approach of Bonincontro *et al.* [80] who consider the case of spherical particles of net negative surface charge σ_o suspended in an electrolyte containing monovalent positive and negative ions. The ions in the diffuse region of the electrical double layer formed around the particle are assumed to be subjected to thermal agitation, as well as being under the influence of the field produced by the fixed charges on the particle's surface. Near the charged surface the positive ions (acting as counterions) in the double layer will outnumber the negative ions in a distribution profile dictated by the Poisson–Boltzmann equation described in Chapter 12. On application of an external electric field this distribution of ions is perturbed by what we can term as electromigration, together with the diffusion of counterions and co-ions down induced concentration gradients. Taking these various factors into account, Bonincontro *et al.* [80] considered the spatial distribution of the charge density to be of the form:

$$\rho_o(r) + \rho(r,\theta) = \frac{A}{\kappa r}\exp(-Kr) + \frac{B}{\gamma r}\left(1 + \frac{1}{\gamma r}\right) \exp(-\gamma r)\cos\theta \tag{9.21}$$

where A and B are constants to be determined and κ is equal to the reciprocal of the Debye screening radius. The concept of the Debye screening radius is described in Chapter 12, but in brief we can understand its significance by saying that for distances less than the Debye radius the distribution of ions in the solution is strongly influenced by their electrostatic interactions with the fixed surface charges. For larger distances the counterions already attracted to the charged surface act as an ionic screen, so that electrostatic interactions of the electrolyte ions with the charged surface are weaker. At a distance of around $10/\kappa$ the ions in the bulk solution are not perturbed by the charged cell. The parameter γ in Equation (9.21) is given by $\gamma^2 = \kappa^2 + i\omega/D$, where D is the ion diffusion coefficient. At low frequencies, as $\omega \to 0$, γ approaches a value equal to the reciprocal of the Debye screening radius. With increasing frequency the value of γ is increased (i.e., the effective Debye screening radius is decreased) by an incoming flux of counterions that is out of phase with the applied field. As described in section 7.4, a phase lag between a polarization process and the applied field leads to energy absorption and a dielectric dispersion. This dispersion is the manifestation of the Maxwell–Wagner interfacial polarization.

The potentials ϕ_o and ϕ_i outside and inside the charged cells, respectively, have to satisfy Poisson's equation and Laplace's equation, respectively, as follows:

$$\nabla^2 \phi_o = -\frac{1}{\varepsilon_o \varepsilon'_m}[\rho_o(r) + \rho(r,\theta)];\ \nabla^2 \phi_i = 0$$

Bonincontro *et al.* [80] determined the potentials ϕ_o and ϕ_I on applying the boundary conditions described in Chapter 5 at the surface of the cell ($r = R$), namely:

$$\phi_i(r,\theta) = \phi_i(r,\theta) \quad \text{and} \quad \varepsilon'_c \frac{\partial \phi_i}{\partial r} - \varepsilon'_m \frac{\partial \phi_o}{\partial r} = \sigma_o \quad (9.22)$$

Based on their solution for ϕ_o, Bonincontro *et al.* derive the following expression for the induced dipole moment of a charged sphere (cell):

$$\mathrm{p} = 4\pi\varepsilon_o\varepsilon'_m R^3 \left[\varepsilon'_c\left(1 - \frac{\kappa^2R^2}{\gamma^2R^2 + 2\gamma R + 2}\right) - \varepsilon'_m\right] \left[2\varepsilon'_m + \varepsilon'_c\left(1 - \frac{\kappa^2R^2}{\gamma^2R^2 + 2\gamma R + 2}\right)\right]^{-1}$$

Comparing this result with the form of the Clausius–Mossotti factor given in Equation (6.4) it is apparent that this induced dipole moment is equivalent to that which would be induced in a spherical particle of effective permittivity given by:

$$\varepsilon'_{ceff} = \varepsilon'_c\left(1 - \frac{\kappa^2R^2}{\gamma^2R^2 + 2\gamma R + 2}\right) \quad (9.23)$$

On modelling the implications of this result, Bonincontro *et al.* conclude that only a small and in some cases a negligible, correction to the characteristic parameters of the Maxwell–Wagner interfacial polarization is required to account for cell surface charge. An interesting aspect of the result expressed by Equation (9.23) is that it does not include the net negative surface charge σ_o on the cell. Neither the factor κ nor γ depend on the net charge of the cell. The Debye screening radius $1/\kappa$ depends on the concentration and charges of ions in the solution, the permittivity of the solution and temperature. The diffusion constant of the ions depends on their charge, mobility and the temperature. Another interesting feature of Equation (9.21) is that the existence of an electrical double layer around a charged cell is predicted to reduce the magnitude of its induced dipole moment. This is in agreement with expectations. The Maxwell–Wagner interfacial polarization, as manifested by the appearance of the β-dispersion, occurs at frequencies where the effective polarizability of the cell is less than that of the surrounding medium. According to the scheme shown in Figure 6.12 and reproduced in Figure 9.16, this corresponds to an induced dipole moment aligned against the direction of an applied electric field. In the diffuse region of the double layer around a cell that carries a fixed negative charge, positively charged counterions dominate in number over negative ions. As shown (in an exaggerated form) in Figure 9.16 we can expect the distribution profile of the counterions to be distorted on application of an external field. For the instantaneous field direction shown, positive counterions from the bulk electrolyte will be driven the left-hand side of the cell, to then move by conduction and diffusion within the double layer to the right-hand side. At low frequencies the conductive and diffusive counterion fluxes will be in phase with the applied field, but with increasing frequency the diffusive flux will not be in phase. Field-driven negative ions approaching the right-hand side of the cell will be confronted by coulombic repulsion from the fixed surface charge in attempting to conduct and diffuse within the double layer to the left-hand side of the cell. This produces a nonsymmetrical distribution of counterions and a dipole moment aligned with the field. The net result is a reduction of the total induced moment compared to that exhibited by an uncharged cell. For the case of an uncharged cell, the Maxwell–Wagner polarization arises from the accumulation of field driven ions at the boundary between two dielectric phases, namely the electrolyte and the cell membrane. When fixed charges are added to the membrane, the accumulated ions are envisaged to form inside the diffuse region of the electrical double layer. This provides a surface conduction path for the accumulated counterions to redistribute themselves.

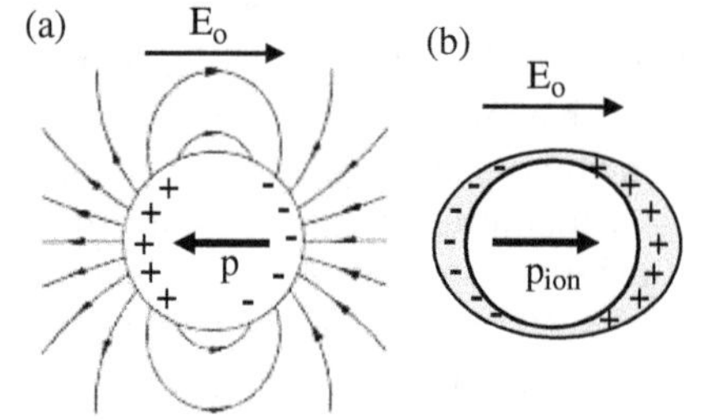

Figure 9.16 (a) The induced dipole moment p of a viable cell at low frequencies is aligned against the direction of the applied field E_o. (b) A cell with a net charge on its surface is surrounded by an electrical double layer. An applied field distorts the diffuse part of the double layer to produce an 'ionic' dipole moment aligned with the field.

The result given by Equation (9.23) has been criticized by Garcia *et al.* [81] who point out that in the limit as $\gamma R \rightarrow \infty$ the counterions have effectively all condensed onto the surface of the cell (radius R) and we should expect the external potential to have the same form as for an uncharged spherical particle. In their view this limiting situation corresponds to either the ionic diffusion coefficient tending to zero with the medium conductivity remaining constant, or the radius of the cell tending to infinity. In either case the result expressed by Equation (9.21) does not have the correct limiting value as $\gamma R \rightarrow \infty$.

Analytical solutions of the equations obtained by Garcia *et al.* [81] are not straightforward, but numerical calculations show that the characteristic frequency of the interfacial dispersion is always higher than that corresponding to the Maxwell–Wagner approximation

$$\omega_{MW} = \frac{2\sigma_m}{\varepsilon_o(\varepsilon_c + 2\varepsilon_m)}$$

by an amount that depends only on the ratio of the permittivities of the cell and the surrounding solution, together with the value of γR, but does not depend on the conductivity of the solution. The effect of the charge on a spherical particle can be neglected only when $\kappa R > 100$. For example, for $\kappa R = 1$, corresponding to a particle of radius 10^{-8} m suspended in an ionic solution for which $\kappa = 10^8\ \mathrm{m}^{-1}$, the dielectric increment is five times lower and the relaxation frequency three times greater than the values predicted by the Maxwell–Wagner approximation for dilute suspensions [81]. For a cell of radius 10 μm suspended in a 10 mM ionic solution ($\kappa \approx 3 \times 10^8\ \mathrm{m}^{-1}$) then $\kappa R > 1000$ and according to Garcia *et al.* the influence on the β-dispersion and hence also on the form of the Clausius–Mossoti factor, is negligible. Aspects of the theoretical approach adopted by Garcia *et al.* [81] should be mentioned. They determine the potential outside the spherical particle using the Poisson equation:

$$\nabla^2\phi_m(r,t) = -\frac{e}{\varepsilon_o\varepsilon_m}[n^+(r,t) - n^-(r,t)]$$

where $n^{\pm}(r,t)$ represent the spatial and temporal densities of monovalent ions of charge $\pm e$ in the diffuse region of the electrical double layer. Note that the angular component of the potential involving $\cos\theta$, considered by Boninicontro *et al.* [80] in Equation (9.21), is not included in the treatment by Garcia *et al.* For the second boundary condition given in Equation (9.21) the negative surface charge σ_o is not included, as a consequence 'of the fact that no true surface charge is considered' [81]. This is an interesting, almost philosophical, point that appears to introduce an approximation of a key boundary condition required for a solution of Poisson's equation. Garcia *et al.* [81] derive the following expression for the induced dipole moment of a charged sphere:

$$p = 4\pi\varepsilon_o\varepsilon'_m R^3 \frac{\varepsilon'_c - \left(\varepsilon'_{gm} + \varepsilon'_c R\right)}{\varepsilon'_c + 2\left(\varepsilon'_{gm} + \varepsilon'_c R\right)}$$

where $\varepsilon'_{gm} = \varepsilon'_m - i\sigma'_m/\omega\varepsilon_o$. This result implies that the induced dipole moment of a charged cell surrounded by an electrical double layer in an ionic solution of permittivity ε'_m and conductivity σ'_m is equivalent to that which would be induced in an uncharged spherical particle when suspended in a medium of effective permittivity:

$$\varepsilon'_{meff} = \varepsilon'_{gm} + \varepsilon'_c R \tag{9.24}$$

Bonincontro *et al.* [80] and Garcia *et al.* [81] thus present us with two quite different forms of a result. Cesare Cametti's group [80] endows an effective permittivity to the charged particle plus its electrical double layer, whilst Constantino Grosse and colleagues [81] focus attention only on the properties of the ionic phase surrounding the particle. The Maxwell–Wagner interfacial polarization involves the interface between the particle and its surrounding medium, leading to both viewpoints being valid.

Both results predict that the effect of surface charge is to reduce the magnitude of the induced dipole moment, as well as concluding that the influence on the magnitude of the interfacial polarization and its relaxation time are likely to be negligible for the case of cell suspensions. However, clarification is required regarding the dielectric properties of smaller charged particles such as proteins, viruses and bacteria. The form of Equation (9.24) predicts a linear proportionality between particle size and the influence of charge on the particle, which does not readily equate to the requirement that $\kappa R > 100$ for such influence to be neglected. This author is unaware of any published, analytical, theory for Maxwell–Wagner interfacial polarization that treats the case of charged particles with an electrical double layer that has a thickness comparable to that of the particle's size.

Grosse and colleagues refined their theory in subsequent publications [82–84], which can be summarized as indicating that the dielectric behaviour of suspensions of charged particles depends little on the ionic composition of the electrolyte or on the fixed charge density of the particle's surface. The particle is considered to be surrounded by a thin layer of counterions (which neutralize it and hence avoids having to solve Poisson's equation) and by a diffused non-neutral cloud of ions. Furthermore, the thin layer of counterions behaves electrically as an electrical conductor, in electrical contact with the counterions from the electrolyte but insulated from the co-ions. When an external electric field is applied, an ion cloud begins to build up around the layer of counterions and finally attains a finite equilibrium value. For a negatively charged particle the counterion layer consists of positive ions and bears a surface current in the direction of the applied field. This transfers positive ions from one side of the particle to the other. Negative ions cannot be conducted in this fashion so that their concentration increases. As a simple picture, the positive ions behave as if the particle was a perfect conductor, while for the negative ions the particle appears as a perfect insulator. The resulting polarization is thus due essentially to this asymmetry of the system and for highly charged particles in low-conductivity electrolytes the relaxation mechanism is rather insensitive to the detailed distribution of ions in the double layer and even to the value of the surface charge of the particles. The magnitude $\Delta\varepsilon_r$ of the corresponding dielectric dispersion is given by:

$$\Delta\varepsilon_r = \frac{9}{16}\frac{(\kappa R)^2(\kappa R+2)}{(\kappa R+1)}\widehat{v}_c\varepsilon_r \tag{9.25}$$

where R is the particle radius and, as in Equation (9.21), κ is the reciprocal of the Debye screening radius [84]. For a 10% concentration of cells of radius 5 μm suspended in an aqueous 10 mM ionic solution $\kappa R \approx 1500$ and Equation (9.25) predicts a value for $\Delta\varepsilon_r \approx 10^7$. This is certainly large enough to be detected experimentally. However, the characteristic frequency f_c of the polarization mechanism is given approximately by

$$f_c = \frac{D}{2\pi R^2} \tag{9.26}$$

where R is the particle radius and D is the diffusion coefficient of ions in the bulk electrolyte [84]. A typical value of D for small ions in an aqueous medium is 2×10^{-9} m^2/s, so that for a cell of radius 5 μm the characteristic frequency can be expected to have a value of around 10 Hz. Unless special care is taken to either avoid or to compensate for electrode polarization effects, this ionic polarization effect will be difficult to detect for mammalian cell suspensions. An example of where the high-frequency tail of this dispersion may have been detected for erythrocytes is shown in Figure 9.8, where it is designated as the α-dispersion.

Cametti and colleagues have extended their theoretical approach to analyse the electrical polarizability for the single-shell spherical model of a cell in the presence of a layer of localized, partially bounded, charges at the two cell membrane interfaces [85]. Allowing these partially bounded charges freedom to conduct over the surface of the cell produces a further dielectric relaxation, besides the one due to the usual Maxwell–Wagner effect. This extra dielectric relaxation is in fact the α-dispersion to be described for the case of bacteria in the next section. This is then extended for the case of shelled ellipsoidal cells with a layer of localized charges at the outer and inner membrane interfaces [86], as well as cylindrical and toroidal particles [87]. Cametti *et al.* also apply their theoretical framework to analyse the alterations observed by dielectric spectroscopy measurements of the passive electrical parameters of the human erythrocyte cell membrane induced by the presence of glucose in the extracellular medium [88]. This analysis includes both the low-frequency α-dispersion and the β-dispersion. A marked increase of the membrane permittivity is induced by a glucose concentration of 20 mM, but with no significant changes of the electrical properties of the cytosol [88].

9.6 Dielectric Properties of Bacteria

Living organisms can be classified on the basis of their cell structure, namely the *eukaryotes* and the *prokaryotes*. The eukaryotes (which include animals, plants and fungi) possess a distinct envelope-enclosed nucleus containing their DNA. Bacteria are prokaryotes – they do not have a distinct nuclear compartment to contain their DNA. Their genetic material is located in the cytoplasm inside an irregular shaped body called the nucleoid, which contains a single chromosome and associated proteins and RNA. Bacteria are typically spherical or rod shaped and of no more than a few micrometres in diameter or length. They often have a protective cell wall around a single cytoplasmic compartment that is contained within one or two plasma membranes. For many years molecular biologists used a particular strain (K-12) of *Escherichia coli* (*E. coli*) as the model bacteria to study, which provided most of our understanding of how cells replicate their DNA and decode its 'instructions' for the synthesis of specific proteins or RNA. The genome of *E. coli* contains 4.6 million nucleotide pairs, represented in texts on molecular biology as a circle because the DNA double helix of prokaryotes forms a single closed loop. Some genes are transcribed from the origin of replication along one strand of the double helix to the terminus of replication in the loop, while other genes are transcribed in the opposite direction along the other strand of bases.

Bacteria are differentiated into two large groups, Gram-positive and Gram-negative, based on a staining procedure developed by the Danish bacteriologist Hans Christian Gram. This process works because Gram-positive bacteria generally have a single plasma membrane surrounded by a thick peptidoglycan, whereas Gram-negative bacteria generally possess a thin layer of peptidoglycan between two membranes. Gram staining involves applying crystal violet to a heat-fixed smear of a bacterial culture, followed by the addition of iodine. A rapid decolouring procedure is then employed using alcohol or acetone. After this decolouring procedure a Gram-negative cell has lost its outer membrane and loses its purple colour, whereas the Gram-positive cell remains purple. A counterstain, using safranin or basic fuchsin, is applied last to give decoloured Gram-negative bacteria a pink or red colour. This staining procedure works because Gram-positive bacteria possess a thick cell wall containing many layers of peptidoglycan and teichoic acids. The crystal violet-iodine complex becomes trapped within the peptidoglycan multilayers. Gram-negative bacteria, on the other hand, have a relatively thin cell wall consisting of a few layers of peptidoglycan surrounded by a second lipid membrane containing lipopolysaccharides and lipoproteins. These differences in structure can produce differences in antibiotic susceptibility. As a general rule Gram-negative bacteria are more pathogenic because the lipopolysacharide in their outer membrane breaks down into an endotoxin, which increases the severity of inflammation. The human body does not contain peptidoglycan and humans produce an enzyme called lysozyme that attacks the open peptidoglycan layer of Gram-positive bacteria. Gram-positive bacteria are also more susceptible to antibiotics such as penicillin, which inhibit a step in the synthesis of peptidoglycan.

Some types of bacteria, mainly of the *Bacillus* and *Clostridium* genus, form into an endospore when confronted with a hostile environment. This is a dehydrated form of a bacterium, consisting of a core that contains only the basic requirements for bacterial growth. It is surrounded by the exosporium, a tough, hydrophobic and highly impermeable multilayer coating consisting mainly of proteins, lipids and carbohydrates. This protects the spore from extremes of temperatures, pH and pressure, as well as harmful chemicals and radiation. A quite amazing example of this survival strategy was the recovery and revival of a *Bacillus* spore from 25 ~ 40-million year old amber [89].

The first study of the dielectric properties of bacteria appears to be that reported by Fricke *et al.* [90] for suspensions of *E. coli* (5×10^{11} cells/ml) suspended in 0.25% NaCl solution. Measurements were taken over the frequency range 50 Hz to 150 MHz for suspensions of *E. coli* (5×10^{11} cells/ml) suspended in 0.25% NaCl solution. A dielectric relaxation of the form of the β-dispersion was observed, centred around 6 MHz and with a decrement $\Delta\varepsilon'_r \approx 500$. This was taken as evidence that *E. coli* possesses a low-conductance surface membrane of approximately the same thickness and permittivity as those found for cells (erythrocytes, leucocytes and yeast) in previous studies. A membrane capacitance value of 7 mF/m^2 was determined and a large dispersion below 100 kHz was also observed.

In a dozen or so papers [e.g., 91–95], Edwin Carstensen and collaborators extended the findings of Fricke *et al.* [90] to show that the low-frequency dispersion (the α-dispersion) exhibited by *E. coli* and *M. lysodeikticus* can be explained in terms of their conducting cell wall [91, 92]. At low conductivities of the environment, the conductivity of the cell appears to be dominated by the counterions of the fixed charge of the cell wall. At higher conductivities the ions from the environment invade the cell wall, causing an increase in the effective conductivity of the cell so that it takes on values roughly proportional to that of the environment. It was possible to rule out ion leakage from the cytoplasm as an important contribution to the population of ions in the wall [92]. That the cell wall's conductivity and concentration of fixed charge are directly related was later confirmed by direct observations of isolated bacterial cell walls [93]. The numbers and sign of fixed charge sites in the wall structure are of the correct order of magnitude to explain the wall conductivity in terms of mobile counterions. The dominant effect of the cell wall on the α-dispersion exhibited by bacteria was later confirmed [94] by performing dielectric measurements on protoplasts of *Micrococcus lysodeikticus* (obtained by removing its cell wall). Removal of the cell wall decreased the low-frequency permittivity by two orders of magnitude. The upper limit of the effective, homogeneous, conductivity for the protoplast was 1 mS/m as compared with 45 mS/m for the intact cell. Thus, on removal of the cell wall the α-dispersion disappeared, but the β-dispersion due to the plasma membrane was still evident. A significant finding was that ion exchange resins exhibit low-frequency dielectric properties similar to that observed for bacteria [95]. Such resins comprise porous particles with a uniform volume distribution of fixed charges. Schwarz [96] had already proposed a mechanism to explain the low-frequency dielectric dispersion in solid colloidal particles. This mechanism involves movement of ions along the charged surface of the particles and relates the magnitude of the low-frequency dielectric constant to the surface charge density of the particle. Ion exchange resins therefore provided a nearly ideal model system to test the conclusions reached by Carstensen *et al.* in their dielectric measurements, as well as the theory proposed by Schwarz.

Carstensen's group also studied the dielectric properties of an endospore [97]. Dormant spores of *Bacillus cereus* were found to have extremely low conductivities at 50 MHz, indicating they contained very low concentrations of mobile ions, both within the core and its surrounding structure. Activation, germination and outgrowth of the endospore were all accompanied by increases in conductivity of the cells and their suspending medium, indicating that intracellular electrolytes had leaked from the spores. The 50 MHz permittivity values of spores were also consistent with normal states for cell water. This observed rise in cell water content increased during successive stages of development from dormant spore to vegetative bacillus [97].

Asami *et al.* [98–100] have also contributed greatly to our understanding of the dielectric properties of *E. coli*. Dielectric measurements of *E. coli* cell suspensions were carried out over a frequency range from 10 kHz to 100 MHz [98]. The β-dispersions were analysed using as a cell model an ellipsoid covered with two confocal shells, representing the plasma membrane and the cell wall. The cells were determined to have lengths in the range 1.2–5.2 μm and diameter 0.68 μm, with cell wall and plasma membrane thickness assumed to be 20 nm and 5 nm, respectively. When suspended in a 10 mM NaCl solution the β-dispersion exhibited a characteristic frequency of 1.1 MHz, with bounding low- and high-frequency relative permittivity values of 460 and 71, respectively. The characteristic frequency increased to 1.6 MHz as the NaCl concentration was increased to 50 mM, but values derived for the dielectric parameters of the *E. coli* components did not change. Based on an assumed plasma membrane conductivity of 5×10^{-8} S/m and a cell wall relative permittivity of 60, relative permittivity values of 11.1 ± 0.3 and 63 ± 3 were determined for the plasma membrane and protoplasm, respectively. This gives the plasma membrane capacitance as 19.6 ± 0.5 mF/m^2, based on the relationship $C_m = \varepsilon_m \varepsilon_o / d_m$. Conductivity values of 0.82 ± 0.14 S/m and 0.2 ± 0.01 S/m were derived for the cell wall and protoplasm, respectively.

Bai *et al.* [99] improved the double-shell model of *E. coli* to that of a three-shell spheroidal model, where the three shells correspond to the outer membrane, the periplasmic space and the inner membrane, respectively. A curve-fitting procedure was developed to analyse the β-dispersion, so that the dielectric properties of the outer membrane and periplasmic space could be evaluated quantitatively for the first time. For suspending medium conductivities less than 10^{-4} S/m the relative permittivity of the outer membrane was determined to be 11 ± 1, increasing to 34 for a medium conductivity of 10^{-3} S/m, corresponding to specific membrane capacitances of 13–15 mF/m^2 and 43 mF/m^2, respectively. These relatively high capacitances probably relate to the composition and structure of the outer membrane being different to the plasma membranes of mammalian cells. The outer 'leaflet' of the bacteria membrane consists of lipopolysaccharide, while the inner 'leaflet' consists of phospholipids. In addition, the outer membrane is much more permeable to sugar and ions than the inner membrane and is not regarded as a barrier for ions but as a filter to exclude large molecules. The outer membrane can therefore be expected to be more hydrophilic and have a higher polarizability than mammalian plasma membranes. The specific capacitance of the inner membrane was found to be comparable to that for mammalian cells, namely 7 mF/m^2. Finally, the conductivity of the periplasmic space was estimated to be 2.7 ± 0.5 S/m, some ten times higher than the conductivity of the outer medium. This large value could result from the peptidoglycan that fills this space acting as an ion-exchange resin, which absorbs ions from the surrounding medium. The relative permittivity and conductivity of the cytoplasm was determined

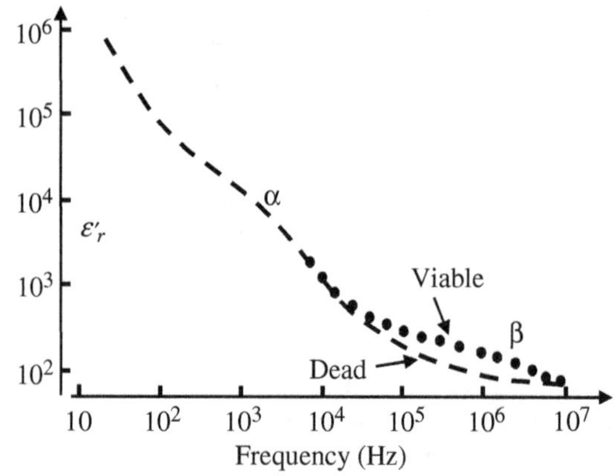

Figure 9.17 Frequency dependence of the relative permittivity for viable and heat-treated (dead) *E. coli* cells suspended in 1 mM NaCl solution. The β-dispersion exhibited by the viable cells is not observed for those that have been heat treated. Both the viable and dead cells exhibit the low-frequency α-dispersion.

to be 100 and 0.22 ± 0.02 S/m, respectively. This large value for the permittivity could be due to the low accuracy in its determination, or the presence of dispersed proteins, granules and DNA, which give to the cytoplasm the high permittivity properties of a microemulsion.

Asami [100] investigated the effect on the α- and β-dispersion of killing *E. coli* cells by heating at 60 °C for 30 min. The result is shown in Figure 9.17, where it is seen that the β-dispersion, due to interfacial polarization, appearing above 100 kHz disappeared completely, whereas the α-dispersion observed below 100 kHz was hardly modified at all. This indicates that the plasma membranes of dead bacteria no longer act as an imperpermeable barrier to ions and also that the mechanism giving rise to the α-dispersion is not related to Maxwell–Wagner interfacial polarization. This also disproves the analytical result reported by Prodan *et al.* [101] that, for a given cell concentration and geometry, the α-dispersion correlates with the magnitude of the plasma membrane potential. Asami [100] also investigated the influence of the ionic concentration of the suspending medium on the magnitude of the α-dispersion, finding that its magnitude increased as the NaCl concentration was increased from 0.1 mM, reaching a maximum of around a threefold increase at 2 mM and then decreasing. The magnitude of the α-dispersion was also found by Asami [100] to decrease as the pH of the suspending medium (for heat-treated cells) decreased from 10 to 4 and to disappear at a pH of ~3.5 that corresponds to the isoelectric point reported for many bacteria and isolated cell walls [102]. The pH dependence of the α-dispersion on pH is also similar to the acid-base titration curves of *E. coli* suspensions [103], suggesting that the α-dispersion is proportional to the density of net charges on the cell surface and / or within the cell wall. Although the simplified model of Grosse and Foster [83] does not predict that the α-dispersion depends on the magnitude of the surface charge, certain aspects of their model do agree with the findings of Assami [100]. For example, Equations (9.25) and (9.27) give reasonable predictions of the magnitude and characteristic frequency of the low-frequency dispersion found for bacteria of micron-scale dimension, as well as the effects expected on changing the ionic concentrations of relatively weak ionic solutions.

Finally, it is reasonable to suggest that aspects of the counterion polarization theory developed to describe the α-dispersion exhibited by DNA solutions (see Chapter 8, section 8.5.1) are of relevance to the mechanism responsible for the α-dispersion exhibited by bacteria. In particular, we can predict that the relaxation time τ of the α-dispersion is proportional to the square of a *characteristic* length L (if not the radius R) of the bacteria cell wall. From Equation (8.29) the induced dipole moment ($\mathrm{m} = \alpha E$) associated with field-induced displacements of n counterions of valency Z at a fixed temperature is given by:

$$\mathrm{m} \propto \alpha E \propto Z^2 q^2 n L^2 E \propto Zqn \left\langle ZqL^2 \right\rangle E$$

where the term $\langle ZqL^2 \rangle E$ can be considered to be the mean displacement of the counterions. The relaxation time is calculated by dividing this mean displacement by the average velocity v of the fluctuating counterions:

$$\tau = \frac{ZqL^2E}{\mu ZqE} = \frac{L^2}{\mu}$$

where μ is the mobility of the counterions. This predicts that the relaxation time is proportional to the square of a characteristic length associated with the surface of the cell wall. We can also adopt the counterion polarization theory developed by Takashima [2, pp. 204–209] for a spherical particle of radius R, which gives:

$$\tau = \frac{R^2}{2\mu kT}$$

Noting that $\mu kT = D$, the ionic diffusion coefficient and that the characteristic frequency of the dispersion is given by $f_c = (1/2\pi\tau)$, this result is the same (apart from a factor of $\frac{1}{2}$) as Equation (9.26) derived by Grosse and Foster [83].

9.7 Summary

The dielectric (hence dielectrophoretic) behaviour of a mammalian cell is primarily controlled by the *passive* electrical properties of its plasma membrane, namely its *specific* conductance and *specific* capacitance. For a dielectric material held between two plane-parallel electrodes of area A separated by a distance d, its conductance G_m and capacitance C_m are defined by the relationships:

$$G_m = \frac{A\sigma}{d}; \qquad C_m = \frac{A\varepsilon_o\varepsilon_r}{d}$$

The plasma membrane takes the form of a spheroidal envelope, rather than a flat sheet of material. However, because the membrane's thickness (d) is so small compared with the size of a cell, these two formulae can also be used to describe the *specific* conductance (G_m/A) and *specific* capacitance (C_m/A) of the plasma membrane of an intact cell. The conductivity σ is the proportionality factor between the induced electric current density and the applied electric field ($J = \sigma E$). It is a measure of the ease with which delocalized charge carriers (ions) migrate across the membrane under the influence of an applied electric field. The intact plasma membrane of a viable cell behaves essentially as an electrical insulator, so that the value of σ (and hence G_m/A) is practically zero in comparison to the conductivity of a physiological electrolyte such as the cytoplasm. The factor ε_o is the dielectric permittivity of free space and has the value 8.854×10^{-12} Farad/m, whilst ε_r is the permittivity of the material *relative* to that of free space and is a measure of the extent to which localized charge distributions can be distorted or polarized under the influence of the field. The internal lipid-protein structure of the plasma membrane has a very low polarizability, with an ε_r value only slightly larger than that of a nonpolar insulating plastic such as polypropylene. High-performance capacitors used in electronic circuits are constructed by metalizing the surfaces of thin sheets ($2 \sim 10$ μm) of polypropylene. The plasma membrane of a cell is thus constructed in the same form as an electronic capacitor device, namely as a thin dielectric sandwiched between two electrical conductors (the cytoplasm and the external physiological medium). The equivalent electrical circuit of the plasma membrane can thus be represented, as depicted in Figures 9.1 and 9.7, as a parallel combination of a resistance ($R_m = 1/G_m$) and capacitance C_m. This provides a tool for deriving values of G_m and C_m from measurements of the electrical impedance of a suspension of cells, performed for a range of frequencies of an applied electric field. Erythrocytes exhibit a specific membrane capacitance of $\sim$8 mF/m^2. This is at least 1000 times larger than the specific capacitance exhibited by high-performance capacitors composed of polypropylene and reflects the very thin nature of the membrane. In fact, the experimental determination more than 90 years ago of the membrane capacitance of erythrocytes provided the first direct evidence of how exquisitely thin a cell membrane might be [55,56].

As for the case of a capacitor in an electric circuit, the magnitude of the membrane capacitance is a measure of the electrical charge that can accumulate at the boundaries between its dielectric and conducting 'electrodes'. For the case of cells suspended in an aqueous electrolyte, this effect can be analysed in terms of the properties of a heterogeneous dielectric as described in Chapter 7. The field-induced accumulation of charges at the boundaries between different constituents of a heterogeneous dielectric is known as Maxwell–Wagner interfacial polarization. For suspensions of viable cells this is manifested as the so-called β-dispersion, typically observed in the 50 kHz–50 MHz frequency range. An example is shown in Figure 9.8 for a suspension of erythrocytes. We can understand the frequency profiles of the conductivity and permittivity of the cell suspension in terms of the parallel combination of G_m and C_m. The values of G_m and C_m remain fixed as a function of frequency, but the equivalent resistance (the *reactance*) of C_m is equal to $1/\omega C_m$. At low frequencies the capacitive reactance is very high, so that field-driven ions accumulate at the membrane surface because it represents a barrier to current flow (G_m has a very low value). This accumulation of charges at the membrane surface equates to a high value of the permittivity (ε_r) of the cell suspension. At low frequencies the cells act as insulating spheroids, so that the conductivity (σ) of the cell suspension appears to be much lower than that of the suspending medium by an amount proportional to the cell concentration. With increasing frequency the reactance of C_m decreases, so that this component of the membrane's equivalent circuit starts to provide a path for the passive flow of ions, effectively bypassing G_m. This is accompanied by an increase of σ, whilst a reduction of charge accumulation leads to a decrease of the effective value for ε_r. At a sufficiently high frequency the membrane reactance ($1/\omega C_m$) becomes negligibly small, the applied electric field penetrates into the cytoplasm and there is no field-driven charge accumulation at the membrane surface. The value of ε_r falls to less than that of the suspending medium (by an amount related to the content of lipids and proteins in each cell) and σ increases to a value that depends on the relative magnitudes of the medium and cytoplasm conductivities.

It follows that the existence of the β-dispersion depends on the plasma membrane acting as a high resistance to passive ionic conduction. If this property is impaired, through cell death or the cytotoxic action of a chemical agent, for example, the β-dispersion disappears. This was first observed more than 100 years ago by Höber, when he found that the low-frequency conductivity of compacted erythrocytes, after lysis of their plasma membrane by saponin, increased to the value observed at MHz frequencies [8,9]. A more recent demonstration, shown in Figure 9.17, was obtained by Asami when he found that the β-dispersion exhibited by bacteria disappears after impairment of their plasma membrane by heat treatment [100]. Experimental studies and applications of dielectrophoresis for cellular systems are mostly performed in the frequency range where the β-dispersion exists. This enables dielectrophoresis to fully exploit the sensitivity of the β-dispersion, in terms of its magnitude and characteristic frequency, to changes in the physico-chemical properties of the plasma membrane or other cell components as a result of cell death (with subtle differences exhibited if this results from necrosis or apoptosis), cell differentiation, medium osmolarity, or the effects of chemical agents, for example. The physico-chemical characteristics that can be monitored by dielectrophoresis, or used to selectively isolate or enrich cell subpopulations, include changes of: cell size and / or shape; membrane surface area arising, for example, from the appearance of microvilli or blebs; membrane conductivity associated with a degradation of membrane structure; cytoplasm conductivity associated with leakage of ions or osmosis; nucleus-cytoplasm volume ratio. Thus, although cell biologists might view electrical circuits such as those shown in Figure 9.1 with suspicion regarding their possible cytological relevance, these examples might reassure them that the end product yields information of some relevance to their work.

A useful way to analyse the dielectric properties of cell suspensions and to derive the conductivity and permittivity values of the various components of a cell, is by means of the so-called multishell model [67–69]. The *single-shell* model of a cell treats the cytoplasm and its contents (organelles, reticulum and nucleus) as a homogeneous medium surrounded by the plasma membrane. As shown in Figures 9.10–9.12, this can be employed to investigate how the β-dispersion depends on cell size, membrane conductance, membrane capacitance and possible leakage of ions from the cytoplasm, for example. At the high-frequency tail of the β-dispersion the applied electric field penetrates into the cytoplasm, so that the presence of the nucleus may influence the dielectric (and hence dielectrophoretic) behaviour of a cell. This can be analysed by adding an extra shell, to represent the nuclear envelope, in the form of the *double-shell* model of a cell. This procedure is outlined in Figure 9.13 and examples of its application to investigate the influence of the nucleus-cytoplasm volume ratio or possible dielectric relaxations in the nucleoplasm are given in Figures 9.14 and 9.15, respectively. The double-shell model can be extended to that of the *three-shell* model to incorporate the outer membrane, the periplasmic space and the inner membrane of Gram negative bacteria, such as *Escherichia coli* [99].

Mammalian cells and bacteria carry a net negative charge on their outer surface. Theoretical analyses [80, 81] indicate that the presence of this surface charge may have only a small, even negligible, influence on the characteristics of the β-dispersion. However, mobile counterions that are attracted to this surface charge are considered [83] to be responsible for a large low-frequency dispersion, known as the α-dispersion, which has been observed [90–93] for suspensions of bacteria cells. Certain aspects of this polarization mechanism are similar to the field-induced fluctuating counterion model described in Chapter 8 for the α-dispersion exhibited by DNA solutions. In particular, the relaxation time of the dispersion is predicted to be proportional to the square of a characteristic dimension (e.g., the radius) of the cell. This means that the α-dispersion occurs in a frequency range that can be masked by electrode polarization effects and is lower than that normally used for dielectrophoresis experiments. The finding [95] that bacteria exhibit dielectric properties similar to that of ion exchange resins is of relevance to understanding and predicting the dielectrophoretic properties of bacteria.

9.8 References

1 Cole, K. S. (1968) *Membranes, Ions and Impulses*, University of California Press, Berkeley, CA.

2 Takashima, S. (1989) *Electrical Properties of Biopolymers and Membranes*, Adam Hilger, Bristol.

3 Hille, B. (2001) *Ionic Channels of Excitable Membranes*, 3rd edn. Sinauer Associates, Inc., Sunderland, MA.

4 Ohm, G. S. (1827) *Die galvanische Kette, mathematisch bearbeitet*, T. H. Riemann, Berlin.

5 Planck, M. (1890) Über die Potentialdifferenz zwischen zwei verdünnten Lösungen binärar Elektrolyte. *Ann. Physik u. Chem.* **40**, 561–576.

6 Nernst, W. (1899) Zur Theorie der elektrischen Reizung. *Nachrichten von der Königl. Gesellschaft der Wissenschaften zu Göttingen* 104–108.

7 Bernstein, J. (1902) Untersuchungen zur Thermodynamik der bioelektrischen Ströme. *Arch. Ges. Physiol.* **92**, 521–562.

8 Höber, R. (1910) Eine Methode, die elektrische Leitfähigkeit im Innern von Zellen zu messen. *Pflügers Archiv Physiologie des Menschen und der Tiere* **133**, 237–253.

9 Höber, R. (1912) Ein zweites Verfähren die elektrische Leitfähigkeit im Innern von Zellen zu messen. *Pflügers Archiv Physiologie des Menschen und der Tiere* **148**, 189–221.

10 Höber, R. (1913 Messungen der inneren Leitfähigkeit von Zellen. *Pflügers Archiv Physiologie des Menschen und der Tiere* **150**, 15–45.

11 Pethig, R. and Schmueser, I. (2012) Marking 100 years since Rudolf Höber's discovery of the insulating envelope surrounding cells and the β-dispersion exhibited by tissue. *J. Electr. Bioimp.* **3**, 74–79.

12 Schwan, H. P. (1957) Electrical properties of tissue and cell suspensions, in *Advances in Biological and Medical Physics*, Vol. V (eds J. H. Lawrence and C. A. Tobias). Academic Press, New York, NY, p. 162.

13 Fricke, H. (1925) The electrical capacity of suspensions with special reference to blood. *J. Gen. Physiol.* **9**(2), 137–152.

14 Fricke, H. and Morse, S. (1925) The electrical resistance and capacity of blood for frequencies between 800 and $4\frac{1}{2}$ million cycles. *J. Gen. Physiol.* **9**(2), 153–167.

15 Fricke, H. and Curtis, H. J. (1935) The electric impedance of hemolyzed suspensions of mammalian erythrocytes. *J. Gen. Physiol.* **18**, 821–836.

16 Philippson, M. (1920) Sur la résistance électrique des cellules et des tissus. *Compt. Rend. Soc. Biol.* **83**, 1399–1402.

17 Philippson, M. (1921) Les lois de la résistance électrique des tissus vivants. *Bull. Acad. Roy. Belgique* 7(Ser. 5), 387–403.

18 Lollike, K. and Lindau, M. (1999) Membrane capacitance techniques to monitor granule exocytosis in neutrophils. *J. Immunol. Methods* **232**(1), 111–120.

19 Scepek, S., Coorssen, J. R. and Lindau, M. (1998) Fusion pore expansion in horse eosinophils is modulated by Ca^{2+} and protein kinase C via distinct mechanisms. *EMBO J.* **17**(15), 4340–4345.

20 Fernandez, J. M., Neher, E. and Gomperts, B. D. (1984) Capacitance measurements reveal stepwise fusion events in degranulating mast cells. *Nature* **312**(5993), 453–455.

21 Henkel, A. W., Meiri, H., Horstmann, H. *et al.* (2000) Rhythmic opening and closing of vesicles during constitutive exo- and endocytosis in chromaffin cells. *EMBO J.* **19**(1), 84–93.

22 Neher, E. and Sakmann, B. (1992) The patch clamp technique. *Sci. Am.* **266**(1), 44–51.

23 Takashima, S., Asami, K. and Takahashi, Y. (1988) Frequency domain studies of impedance characteristics of biological cells using micropipette technique. *Biophys. J.* **54**, 995–1000.

24 Asami, K., Takahashi, Y. and Takashima, S. (1990) Frequency domain analysis of membrane capacitance of cultured cells (HeLa and myeloma) using the micropipette technique. *Biophys. J.* **58**(1), 143–148.

25 Cole, K. S. (1937) Electric impedance of marine egg membranes. *Trans. Faraday Soc.* **33**(8), 966–972.

26 Cole, K. S. and Spencer, J. M. (1938) Electric impedance of fertilized arbacia egg suspensions. *J. Gen. Physiol.* **21**(5), 583–590.

27 Cole, K. S. and Curtis, H. J. (1938) Electric impedance of single marine eggs. *J. Gen. Physiol.* **21**(5), 591–599.

28 Cole, K. S. (1941) Impedance of single cells. *Tabulae Biologicae (Cellula)* **19**(2), 24–27.

29 Blinks, L. R. (1930) The direct current resistance of Valonia. *J. Gen. Physiol.* **13**(3), 361–378.

30 Loewenstein, W. R. and Kanno, Y. (1962) Some electrical properties of the membrane of a cell nucleus. *Nature* **195**(4840), 462–464.

31 Loewenstein, W. R. and Kanno, Y. (1963) Some electrical properties of a nuclear membrane examined with a microelectrode. *J. Gen. Physiol.* **46**(6), 1123–1140.

32 Pethig, R. and Kell, D. B. (1987) The passive electrical properties of biological systems: their significance in physiology, biophysics and biotechnology. *Phys. Med. Biol.* **32**(8), 933–970.

33 Reynolds, C. R. and Tedeschi, H. (1984) Permeability properties of mammalian cell nuclei in living cells and *in vitro*. *J. Cell. Sci.* **70**(1), 197–207.

34 Asami, K., Takahashi, Y. and Takashima, S. (1989) Dielectric properties of mouse lymphocytes and erythrocytes. *Biochim. Biophys. Acta* **1010**(1), 49–55.

35 Bustamante, J. O. (2006) Current concepts in nuclear pore electrophysiology. *Can. J. Physiol. Pharm.* **84**, 347–365.

36 Pauly, H. (1959) Electrical conductance and dielectric constant of the interior of erythrocytes. *Nature* **183**(4657), 333–334.

37 Schwan, H. P. (1959) Alternating current spectroscopy of biological substances. *Proc. Inst. Radio Eng.* **47**(11), 1841–1855.

38 Hodgkin, A. L. and Rushton, W. A. H. (1946) The electrical constants of a crustacean nerve fibre. *Proc. Roy. Soc. London, Series B, Biol. Sci.* **133**(873), 444–479.

39 Curtis, H. J. and K. S. Cole, K. S. (1938) Transverse electric impedance of the squid giant axon. *J. Gen. Physiol.* **21**(6), 757–765.

40 Cole, K. S. (1941) Rectification and inductance in the squid giant axon. *J. Gen. Physiol.* **25**(1), 29–51.

41 Cole, K. S. and Baker, R. F. (1941) Longitudinal impedance of the squid giant axon. *J. Gen. Physiol.* **24**(3), 771–778.

42 Cole, K. S. (1949) Dynamic electrical characteristics of the squid axon membrane. *Archives des Sciences Physiologiques* **3**, 253–258.

43 Hodgkin, A. L. and Huxley, A. F. (1952) A quantitative description of membrane current and its application to conduction and excitation in nerve. *J. Physiol.* **117**(4), 500–544.

44 Takashima, S. (1979) Admittance change of squid axon during action potentials. Change in capacitance component due to sodium currents. *Biophys. J.* **26**(1), 133–142.

45 Grimnes, S. and Martinsen, O. G. (2015) *Bioimpedance and Bioelectricity Basics*, 3rd edn. Academic Press, Amsterdam, Ch. 8.

46 Maxwell, J. C. (1873) *A Treatise on Electricity and Magnetism*, Vol. 1. 1st edn. Clarendon Press, Oxford, Art. 314, p. 440.

47 Wagner, K. W. (1914) Erklärung der dielektrischen Nachwirkungsvorgänge auf Grund Maxwellscher Vorstellungen. *Archiv f. Electrotechnik* **2**(9), 371–387.

48 Slivinsky, G. G., Hymel, W. C., Bauer, J. and Morrison, D. R. (1997) Cellular electrophoretic mobility data: A first approach to a database. *Electrophoresis* **18**, 1109–1119.

49 Bedzyk, M. J., Bommarito, G. M., Caffrey, M. and Penner, T. L. (1990) Diffuse-double layer at a membrane-aqueous interface measured with X-ray standing waves. *Science* **248**, 52–56.

50 Polder, P. and Van Santen, J. H. (1946) The effective permittivity of mixtures of solids. *Physica* **12**, 257–265.

51 Böttcher, C. J. F. and Bordewijk, P. (1978) *Theory of Electric Polarization*, Vol. II, Elsevier, New York, NY, pp. 476–491.

52 Aspnes, D. E. (1982) Local-field effects and effective medium theory – a microscopic perspective. *Am. J. Phys.* **50**(8), 704–709.

53 Norris, A. N., Sheng, P. and Callegan, A. J. (1985) Effective-medium theories for two-phase dielectric media. *J. Appl. Phys.* **57**, 1990–1996.

54 Hanai, T. (1960) Theory of the dielectric dispersion due to the interfacial polarization and its application to emulsion. *Kolloid Z.* **171**, 23–31.

55 Fricke, H. (1925) The electric capacity of suspensions of red corpuscles of a dog. *Phys. Rev.* **26**(5), 682–687.

56 Fricke, H. (1925) A mathematical treatment of the electric conductivity and capacity of disperse systems. II. The capacity of a suspension of conducting spheroids surrounded by a non-conducting membrane for a current of low frequency. *Phys. Rev.* **26**(5), 678–681.

57 Hochmuth, R. M., Evans, C. A., Wiles, H. C. and McCown, J. T. (1983) Mechanical measurement of red cell membrane thickness. *Science* **220**(4592), 101–102.

58 Coster, H. G. L. and Smith, J. R. (1974) The molecular organization of bimolecular lipid-membranes. A study of the low frequency Maxwell–Wagner impedance dispersion. *Biochim. Biophys. Acta* **373**(2), 151–164.

59 Miles, J. B. and Robertson, H. P. (1932) The dielectric behavior of colloidal particles with an electric double-layer. *Phys. Rev.* **40**(4), 583–591.

60 Irimajiri, A., Asami, K., Ichinowatari, T. and Kinoshita, Y. (1987) Passive electrical properties of the membrane and cytoplasm of cultured rat basophil leukaemia cells: II. Effects of osmotic perturbation. *Biochim. Biophys. Acta* **896**(2), 214–223.

61 Sukhorukov, V. I., Arnold, W. M. and Zimmermann, U. (1993) Hypotonically induced changes in the plasma membrane of cultured mammalian cells. *J. Membr. Biol.* **132**, 27–40.

62 Wang, X-B., Huang, Y., Gascoyne, P. R. C. *et al.* (1994) Changes in Friend murine erythroleukaemia cell membranes during induced differentiation determined by electrorotation. *Biochim. Biophys. Acta* **1193**, 330–344.

63 Asami, A. (2011) Dielectric properties of microvillous cells simulated by the three-dimensional finite-element method. *Bioelectrochemistry* **81**, 28–33.

64 Gentet, L. J., Stuart, G. J. and Clements, J. D. (2000) Direct measurement of specific membrane capacitance in neurons. *Biophys. J.* **79**, 314–320.

65 Stoneman, M., Chaturvedi, A., Jansma, D. B. *et al.* (2007) Protein influence on the plasma membrane dielectric properties: In vivo study utilizing dielectric spectroscopy and fluorescence microscopy. *Bioelectrochem.* **70**, 542–550.

66 Muratore, M., Srsen, V., Waterfall, M. *et al.* (2012) Biomarker-free dielectrophoretic sorting of differentiating myoblast multipotent progenitor cells and their membrane analysis by Raman spectroscopy. *Biomicrofluidics* **6**, 034113.

67 Irimajiri, A., Asami, K., Ichinowatari, T. and Kinoshita, Y. (1987) Passive electrical properties of the membrane and cytoplasm of cultured rat basophil leukaemia cells: I. Dielectric behaviour of cell suspensions in 0.01-500 MHz and its simulation with a single-shell model. *Biochim. Biophys. Acta* **896**(2), 203–213.

68 Irimajiri, A., Hanai, T. and Inouye, A. (1979) Dielectric theory of multi-stratified shell-model with its application to a lymphoma cell. *J. Theor. Biol.* **78**(2), 251–269.

69 Irimajiri, A., Doida, Y., Hanai, T. and Inouye, A. (1978) Passive electrical properties of cultured marine lymphoblasts (L5178Y) with reference to its cytoplasmic membrane, nuclear-envelope and intracellular phases. *J. Membrane Biol.* **38**(3), 209–232.

70 Hanai, T., Zhang, H. Z., Sekine, K. *et al.* (1988) The number of interfaces and the associated dielectric relaxations in heterogeneous systems. *Ferroelectrics* **86**, 191–204.

71 Cametti, C., De Luca, F., D'Ilario, A. *et al.* (1992) Alteration of the passive electrical properties of lymphocyte membranes induced by GM1 and GM3 glycolipids. *Biochim. Biophys. Acta* **1111**(2), 197–203.

72 Polevaya, Y., Ermolina, I., Schlesinger, M. *et al.* (1999) Time domain dielectric spectroscopy study of human cells: II Normal and malignant white blood cells. *Biochim. Biophys. Acta* **1419**(2), 257–271.

73 Asami, K. and Irimajiri, A. (1984) Dielectric analysis of mitochondria isolated from rat liver II. Intact mitochondria as simulated by a double-shell model. *Biochim. Biophys. Acta* **778**(3), 570–578.

74 Asami, K. and Yonezawa, T. Biophys. J. (1996) Dielectric behavior of wild-type yeast and vacuole-deficient mutant over a frequency range of 10 kHz to 10 GHz. *Biophys. J.* **71**(4), 2192–2200.

75 Raicu, V., Raicu, G. and Turcu, G. (1996) Dielectric properties of yeast cells as simulated by the two-shell model. *Biochim. Biophys. Acta* **1274**, 143–148.

76 Asami, K., Hanai, T. and Koizumi, N. (1980) Dielectric analysis of *Escherichia-Coli* suspensions in the light of the theory of interfacial polarization. *Biophys. J.* **31**(2), 215–228.

77 Asami, K. and Yamaguchi, T. (1992) Dielectric spectroscopy of plant protoplasts. *Biophys. J.* **63**(6), 1493–1499.

78 Asami, K., Hanai, T. and Koizumi, N. (1980) Dielectric approach to suspensions of ellipsoidal particles covered with a shell in particular reference to biological cells. *Jap. J. Appl. Phys.* **19**(2), 359–365.

79 Slivinsky, G. G., Hymel, W. C., Bauer, J. and Morrison, D. R. (1997) Cellular electrophoretic mobility data: A first approach to a database. *Electrophoresis* **18**, 1109–1119.

80 Bonincontro, A., Cametti, C. and Di Biaso, A. (1980) Effect of volume ion polarisations on Maxwell–Wagner dielectric dispersions. *J. Phys. D: Appl. Phys.* **13**, 1529–1535.

81 Garcia, A., Grosse, C. and Brito, P. (1985) On the effect of volume charge distribution on the Maxwell–Wagner relaxation. *J. Phys. D: Appl. Phys.* **18**, 739–745.

82 Grosse, C. and Barchin, R. (1986) On the permittivity of a suspension of charged spherical particles in an electrolyte. *J. Phys. D: Appl. Phys.* **19**, 1113–1127.

83 Grosse, C. and Foster, K. R. (1987) Permittivity of a suspension of charged spherical particles in electrolyte solution, *J. Phys. Chem.* **91**, 3073–3076.

84 Grosse, C. (1988) Permittivity of a suspension of charged spherical particles in electrolyte solution. 2.

Influence of the surface conductivity and asymmetry of the electrolyte on the low- and high-frequency relaxations. *J. Phys. Chem.* **92**(13), 3905–3910.

85 Di Biasio, A. and Cametti, C. (2010) Polarizability of spherical biological cells in the presence of localized surface charge distributions at the membrane interfaces. *Phys. Rev. E* **82**, 021917.

86 Di Biasio, A., Ambrosone, L. and Cametti, C. (2010) Dielectric properties of biological cells in the dipolar approximation for the single-shell ellipsoidal model: The effect of localized surface charge distributions at the membrane interface. *Phys. Rev. E.* **82**, 041916.

87 Di Biasio, A., Ambrosone, L. and Cametti, C. (2013) Electrical polarizability of differently shaped dielectric objects in the presence of localized interfacial charge distribution: a unifying scenario. *J. Phys. D.: Appl. Phys.* **46**, 055305.

88 Di Biasio, A. and Cametti, C. (2009) D-glucose-induced alterations in the electrical parameters of human erythrocyte cell membrane. *Bioelectrochemistry* **77**, 151–157.

89 Cano, R. J. and Borucki, M. K. (1995) Revival and identification of bacterial spores in 25- to 40-million-year-old Dominican amber. *Science* **268** (5213), 1060–1064.

90 Fricke, H., Schwan, H. P., Li, K. and Bryson, V. (1956) A dielectric study of the low-conductance surface membrane in *E. coli*. *Nature* **177**(4499), 134–135.

91 Carstensen, E. L., Cox, H. A., Mercer, W. B. and Natale, L. A. (1965) Passive electrical properties of microorganisms. I. Conductivity of *Escherichia coli* and *Micrococcus lysodeikticus*. *Biophys. J.* **5**, 289–300.

92 Carstensen, E. L. (1967) Passive electrical properties of microorganisms. II. Resistance of the bacterial membrane. *Biophys. J.* **7**, 493–503.

93 Carstensen, E. L. and Marquis, R. E. (1968) Passive electrical properties of microorganisms. III. Conductivity of isolated bacterial cell walls. *Biophys. J.* **8**, 536–548.

94 Einolf, C W. and Carstensen, E. L. (1969) Passive electrical properties of microorganisms. IV. Studies of the protoplasts of *Micrococcus lysodeikticus*. *Biophys. J.* **9**, 634–643.

95 Einolf, C W. and Carstensen, E. L. (1971) Low-frequency dielectric dispersion in suspensions of ion-exchange resins. *J. Phys. Chem.* **75**(8), 1091–1099.

96 Schwarz, G. (1962) A theory of the low-frequency dielectric dispersion of colloidal particles in electrolyte solution. *J. Phys. Chem.* **66**(12), 2636–2642.

97 Carstensen, E. L., Marquis, R. E. and Gerhardt, P. (1971) Dielectric study of the physical state of electrolytes and water within *Bacillus cereus* spores. *J. Bacteriol.* **107**, 106–113.

98 Asami, K., Hanai, T. and Koizumi, N. (1980) Dielectric analysis of *Escherichia coli* suspensions in the light of the theory of interfacial polarization. *Biophys. J.* **31**, 215–228.

99 Bai, W., Zhao, K. S. and Asami, K. (2006) Dielectric properties of *E. coli* cell as simulated by the three-shell spheroidal model. *Biophys. Chem.* **122**, 136–142.

100 Asami, K. (2014) Low-frequency dielectric dispersion of bacterial suspensions. *Coll. Surf. B: Biointerfaces* **119**, 1–5.

101 Prodan, E., Prodan, C. and Miller, J. H. (2008) The dielectric response of spherical live cells in solution: An analytical solution. *Biophys. J.* **95**, 4174–4182.

102 Harden, V. P. and Harris, J. O. (1952) The isoelectric point of bacterial cells. *J. Bacteriol.* **65**, 198–202.

103 Martinez, R. E., Smith, D. S., Kulczycki, E. and Ferris, F. G. (2002) Determination of intrinsic bacterial surface acidity constants using a Donnan shell model and a continuous pK_a distribution method. *J. Colloid Interface Sci.* **253**, 130–139.

10

Dielectrophoresis: Theoretical and Practical Considerations

10.1 Introduction

Herb Pohl describes his book [1, Preface] as a 'survey of the principles, aims, and practical effects of nonuniform fields on matter'. Although it remains a valuable entry point into the subject of dielectrophoresis (DEP) we can expect that key aspects may need to be clarified or updated. In this chapter, two main areas are considered, namely refinements of the basic theory and the impact of microfabrication and microfluidic technologies. As outlined by Pohl: 'Provided certain simplifying assumptions are made, the theory for the force exerted upon a body suspended in a fluid medium when in a nonuniform field can be expressed in relatively simple terms' [1, p. 15]. These simplifying assumptions are reviewed in this chapter, along with the work that has been achieved to refine them and to understand their implications more completely. The electrodes available to Pohl and others for their experiments largely took the form of metal wires, pins or foil, with applied voltage potentials above 1 kV sometimes required to produce an observable DEP effect. The quantification of the DEP collection of cells at an electrode was often facilitated by time-lapse photography of the growth of pearl-chain whiskers. This particular aspect of the subject has witnessed many changes that are ongoing and reflect advances being made in miniaturization techniques and the quest to exploit DEP in new devices for biomedical, clinical, drug discovery, environmental and food safety applications.

Cells suspended in media of relatively low conductivity ($\leq$ 200 mS/m) can exhibit a transition from negative DEP to positive DEP, as the applied field frequency is increased to where the high resistance to passive ion flow of the plasma membrane begins to be short circuited by its capacitive reactance. This transition takes place at the DEP crossover frequency (f_{xo1}) and typically occurs in the readily accessible frequency range from ~10 kHz to ~1 MHz. If the typical value of f_{xo1} and cell size is known for a particular cell type, it can be characterized in terms of its membrane conductance and capacitance. These dielectric parameters can in turn provide details of cell viability and surface morphology, as well as enabling protocols to be devised for separating different cell types from mixtures of cells (based on their different f_{xo1} values). The value of f_{xo1} is determined by the interfacial polarization taking place at the interface between the plasma membrane and the suspending medium – it is insensitive to the dielectric properties of the cell interior. As the field frequency is increased beyond f_{xo1} the plasma membrane becomes essentially 'invisible' to the applied electric field, so that the cell can be regarded in dielectric terms as a ball of cytoplasm in direct contact with the outside medium. If the permittivity of the cytoplasm is less than that of the surrounding medium, the interfacial polarization at the medium-cytosol interface can result in a second DEP crossover (f_{xo2}) occurring at a frequency, typically >200 MHz for a viable cell, that depends on the effective conductivity of the cell interior. The frequency range for f_{xo2} is not as readily accessible as for f_{xo1} and so has been relatively unexplored. However, some progress has been made in opening up what could provide the means for simultaneously characterizing and monitoring the physico-chemical properties of the plasma membrane and cell interior.

10.2 Inherent Approximations in the DEP Force Equation

10.2.1 Field Gradient across a Particle

In Chapter 2, Box 2.4, the following equation is derived for the dielectrophoretic (DEP) force acting on a spherical particle, of dipole moment p and radius R, subjected to a field gradient ∇E:

$$\mathrm{F}_{DEP} = (\mathrm{p} \cdot \nabla)\mathrm{E} \tag{10.1}$$

A net electrostatic force is exerted on the particle because the two charges ($+q$ and $-q$) forming the dipole moment ($p = qd$) are located a vector distance d apart and so experience different field values. As described in Box 2.4, the effective field gradient acting across the dipole

Dielectrophoresis: Theory, Methodology and Biological Applications, First Edition. Ronald Pethig.

is obtained by performing a Taylor series expansion of $\mathrm{E}(r + d)$ about r and taking the effective length of the dipole moment p as $d = 2R$. This expansion is incorporated into the following expression for the field gradient across the particle:

$$\frac{(\mathrm{E}(r+d) - \mathrm{E}(r))}{2R} = \frac{\partial \mathrm{E}}{\partial r} + R\frac{\partial^2 \mathrm{E}}{\partial r^2} \cdots + \frac{(2R)^{n-1}}{n!} \cdot \frac{\partial^n \mathrm{E}}{\partial r^n} + \cdots \quad (10.2)$$

Equation (10.1) is obtained by ignoring the terms beyond $\partial \mathrm{E}/\partial r$. In other words it is assumed that the sphere's diameter ($2R$) is much smaller than the scale of the field's nonuniformity. The validity of this assumption can be tested for the particular case of the spherical electrode geometry shown in Figure 2.10, which together with a coaxial geometry and the isomotive (polynomial) design [1 (pp. 50–55), 2] is amenable to analytical calculation of the generated field and field gradient. Computer-aided numerical methods are required to calculate the fields and field gradients generated by other electrode geometries commonly used in DEP devices [3, 4]. For the spherical electrode design of Figure 2.10 the potential $V(r)$ at a point r, when a voltage V is applied to the inner electrode and the outer one is grounded, is given by (see Example 2.7):

$$V(r) = \frac{Vr_1(r_2 - r)}{r(r_2 - r_1)}$$

The field at point r is given by:

$$\mathrm{E} = -\nabla V = \frac{AV}{r^2}\mathrm{r}_o$$

where $A = \frac{r_1 r_2}{(r_2 - r_1)}$ and r_o is the unit radial vector.

On differentiation we obtain

$$\frac{\partial \mathrm{E}}{\partial r} = -2\frac{AV}{r^3}\mathrm{r}_o$$

The second and third differentials are given by:

$$\frac{\partial^2 \mathrm{E}}{\partial r^2} = 6\frac{AV}{r^4}\mathrm{r}_o = -\frac{3}{r}\frac{\partial \mathrm{E}}{\partial r}; \quad \frac{\partial^3 \mathrm{E}}{\partial r^3} = -24\frac{AV}{r^5}\mathrm{r}_o = \frac{12}{r^2}\frac{\partial \mathrm{E}}{\partial r}$$

For terms up to the third differential, Equation (10.2) can thus be written as:

$$\frac{(\mathrm{E}(r+d) - \mathrm{E}(r))}{2R} = \frac{\partial \mathrm{E}}{\partial r}\left[1 - \frac{3R}{r} + \frac{8R^2}{r^2}\right] \quad (10.3)$$

For the spherical electrode geometry shown in Figure 2.10 and the dimensions ($r = 190$ μm, $R = 5$ μm) specified in Examples 2.7 and 2.8, the factor $3R/r = 0.079$ and $8R^2/r^2 = 0.006$. Ignoring the terms beyond $\partial \mathrm{E}/\partial r$ in Equation (10.1) leads to an overestimation by 7.3% of the field gradient, and thus also to the value of the DEP force. The error in assuming that the sphere's diameter is much smaller than the scale of the field's nonuniformity becomes less as the value of R is reduced. In fact, Equation (10.1) should strictly describe the DEP force acting on an infinitesimally small dipole, where in the limit as $d \to 0$ the value of the dipole moment remains constant and finite in magnitude. This procedure is adopted in formulating Equation (5.34) to describe the potential of a point dipole in an external field.

10.2.2 Macroscopic Clausius–Mossotti Factor

Chapter 6 shows that the dielectrics literature describes and uses two forms of the Clausius–Mossotti factor, namely the *microscopic* (molecular) and *macroscopic* form. The macroscopic form is the one adopted in the DEP literature for describing the polarization of a macroscopic particle (such as a mammalian cell, bacteria, oil droplet or gas bubble, for example) situated in a medium whose dielectric properties differ from those of the particle. The *molecular* version(s), for which we have adopted the term *Clausius–Mossotti–Lorentz relation*, might be applicable to a small bioparticle such as a protein or RNA molecule. Such bioparticles could exhibit anisotropic polarizability and possess a permanent, conformation dependent, dipole moment, which interacts electrostatically with polar molecules in the medium.

As described in Chapter 6 the derivation of the macroscopic Clausius–Mossotti factor assumes that the particle is uncharged and composed of an isotropic and homogeneous perfect dielectric material of relative permittivity ε_p, embedded in a homogeneous dielectric medium of relative permittivity ε_m. A *uniform* electric field E is assumed to have already been established in the medium, which for distances far from the particle is not distorted by the field of the particle's induced dipole moment. From the particle's perspective the surrounding medium should thus appear to be of infinite extent. It should not be located near another polarizable particle or a boundary where an image potential such as that shown in Figure 5.18 is created.

The task is to deduce the form of particle polarization whose field, when superposed onto E, produces a resultant potential ϕ, which satisfies the electrostatic conditions described in Chapter 5, section 5.6, which are summarized as:

1. On either side of the sphere's surface the normal component of the gradient of ϕ changes such that $\varepsilon(\partial\phi/\partial n)$ remains constant.
2. ϕ is continuous across this boundary defined by the sphere's surface.
3. In all of the space, ϕ satisfies Laplace's equation ($\nabla^2\phi = 0$).

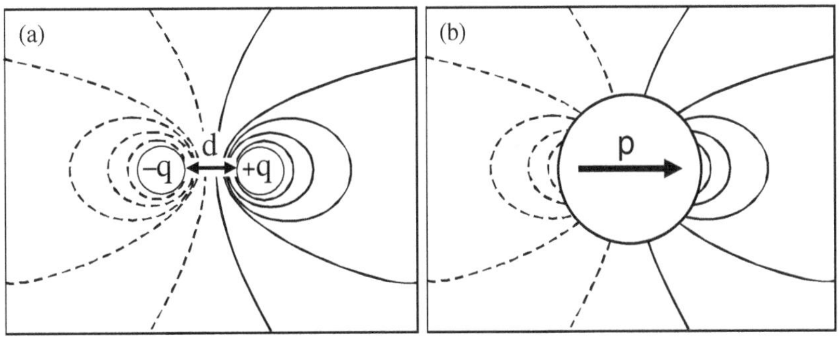

Figure 10.1 (a) The lines of electric potential associated with a dipole of moment qd. (b) In the dipole approximation for the DEP force, the potential generated outside an uncharged dielectric sphere polarized by a *uniform* imposed field E, is identical to that produced by an induced dipole moment p. The resultant potential, when this dipole potential is superposed onto that of the original field E, must satisfy the electrostatic boundary conditions at the surface of the sphere as described in Chapter 5, section 5.6.

For the field E directed along the positive z-axis, we also have the condition:

4. At distances far beyond the sphere, the external potential is given by $\phi_o = -\mathrm{E}z$.

If the sphere consists of an isotropic and homogeneous dielectric, it will be homogeneously polarized by the external field E to create an internal field $\mathrm{E_i}$ symmetric about the x-axis:

$$\phi_i = -\mathrm{E}_i z = \mathrm{E}_i r\cos\theta \quad \text{(spherical coordinates, } z = r\cos\theta) \tag{10.4}$$

Boundary conditions 3 and 4 are satisfied by assuming an external potential of the form:

$$\begin{aligned}\phi_o &= -\mathrm{E}z + \mathrm{E}\frac{k}{4\pi\varepsilon_o\varepsilon_m r^2}\cdot\frac{z}{r}\\ &= -\mathrm{E}\cos\theta\left(r - \frac{k}{4\pi\varepsilon_o\varepsilon_m r^2}\right)\end{aligned} \tag{10.5}$$

The particle thus assumes the form of an induced dipole of effective moment $\mathrm{p} = k\mathrm{E}$ located at the sphere's centre. The potential profile generated by this dipole is depicted in Figure 10.1. As a result of the spherical geometry and the perfect (i.e., lossless) nature of the dielectric, the induced moment field is aligned with and superimposed on the original external field E. Boundary conditions 1 and 2 require the following equalities:

$$\phi_i = \phi_o \quad \text{and} \quad \varepsilon_p\frac{\partial\phi_i}{\partial r} = \varepsilon_m\frac{\partial\phi_o}{\partial r}, \text{when } r = \mathrm{R} \tag{10.6}$$

These conditions are satisfied by the following relationships:

$$\mathrm{E}_i = \mathrm{E}\left(1 - \frac{k}{4\pi\varepsilon_o\varepsilon_m R^3}\right)$$

and

$$\varepsilon_p\mathrm{E}_i = \varepsilon_o\varepsilon_m\mathrm{E}\left(1 + \frac{2k}{4\pi\varepsilon_o\varepsilon_m R^3}\right)$$

Substituting $\mathrm{E_i}$ from the first relationship into the second, we can solve for k and derive the sphere's induced dipole moment as:

$$\mathrm{p} = k\mathrm{E} = 4\pi\varepsilon_o\varepsilon_m R^3\left(\frac{\varepsilon_p - \varepsilon_m}{\varepsilon_p + 2\varepsilon_m}\right)\mathrm{E} \tag{10.7}$$

Equation (10.4) describes the potential profile generated by a dipole, as depicted in Figure 10.1. However, we should note from Example 5.7 that this profile applies only to the potential at a radial distance $r \gg R$. For distances closer to the particle than this the potential has contributions from higher order multipoles. The function enclosed in brackets in Equation (10.7) can thus be considered as being *approximately* proportional to the polarizability per unit volume of the particle. This function is known as the macroscopic Clausius–Mossotti factor.

10.2.3 The DEP Force Equation

On substituting the value of the dipole moment p from Equation (10.7) into Equation (10.1), the DEP force acting on a *spherical* particle in a DC nonuniform field is given by:

$$\mathrm{F}_{DEP} = 4\pi R^3\varepsilon_o\varepsilon_m[CM](\mathrm{E}\cdot\nabla)\mathrm{E} \tag{10.8}$$

where $[CM]$ is the Clausius–Mossotti factor. As described in Box 2.5, the vector transformation of $(\mathbf{E}\cdot\nabla)\mathbf{E}$ leads to the identity:

$$2(\mathrm{E}\cdot\nabla)\mathrm{E} = \nabla\mathrm{E}^2$$

Equation (10.7) can thus be written as:

$$\mathrm{F}_{DEP} = 2\pi R^3\varepsilon_o\varepsilon_m[CM]\nabla\mathrm{E}^2 \tag{10.9}$$

In this expression for the DEP force, the AC field is assumed to be sinusoidal and of magnitude given by its *root mean square* (rms) value, as defined for the voltage waveform in Figure 2.13.

We can now appreciate that this formulation of the DEP force equation, known as the *dipole approximation*, involves the following deficiencies:

- Derivation of the Clausius–Mossotti factor assumes that the imposed electric field is *uniform*, rather than *nonuniform*.
- The assumption is made that the particle's diameter is much smaller than the scale of the nonuniformity of the applied electric field.
- The particle's induced polarization is assumed to take the form of a simple dipole moment, ignoring contributions from higher order multipoles such as quadrupoles.
- The particle is composed of a *homogeneous* dielectric, which exhibits *no conductive losses* and carries *no net charge*.
- The particle exists in a dielectric medium of *infinite* extent, implying that the field in the vicinity of the particle is not perturbed by the presence of a boundary such as a metal or dielectric surface, or another polarizable particle, for example.

Before reviewing how these deficiencies can be addressed, the positive aspects of Equation (10.9) should be emphasized. This equation correctly informs us that:

The DEP force is zero if the field is uniform (i.e., for $\nabla E = 0$).

- The dependence on R^3 shows that the DEP force is ponderomotive. With all other factors remaining constant, the larger the particle volume the greater is the DEP force acting on it.
- The direction of the induced dipole moment can be with or against the applied field, depending on whether CM has a positive or negative value, respectively. A positive CM value corresponds to the particle's permittivity being greater than that of the surrounding medium. In this case, the particle is directed by a DEP force along a path leading to a field maximum, which at every point follows the steepest field gradient (see Figure 5.20). A field maximum can only exist at an electrode edge, never in free space (see Figure 5.11). The DEP force therefore directs a particle to an electrode. The opposite situation, where the particle's permittivity is less than that of the medium, results in a DEP force that directs the particle away from an electrode towards a field *minimum*.
- The DEP force depends on the square of the applied field magnitude, indicating that DEP can be observed either using a direct current (DC) or alternating (AC) electric field. Reversing the polarity of the voltage applied to an electrode (e.g., from $+V$ to $-V$) does not reverse the direction of the DEP force, because $(-V)^2$ is equivalent to $(+V)^2$. In an AC field of constant periodicity, T the time-averaged DEP force acting on a spherical particle is obtained by integrating Equation (10.9) over a complete cycle as follows:

$$\langle \mathrm{F}_{DEP} \rangle = \frac{4\pi R^3 \varepsilon_o \varepsilon_m}{T} \int_0^T [CM^*](\mathrm{E}_{AC} \cdot \nabla)\mathrm{E}_{AC} dt \tag{10.10}$$

where CM^* is the complex Clausius–Mossotti factor introduced in Chapter 6. The integration involved is equivalent to that solved in Box 2.6 to derive the RMS voltage.

With reference to the concepts described in Chapter 4, especially Equation (4.38), we can also interpret Equation (10.9) in terms of electrostatic energy, with the factor $\varepsilon_m R^3 (CM)\mathrm{E}^2$ being proportional to the energy required to withdraw a sphere of radius R from a field E into a region where there is no field (assuming that the medium is isotropic and linear, such that ε_m may be a function of position but not of the field E). A positive value for CM indicates that work will be required to withdraw the particle from the highest field region, whereas for a negative CM value work is required to push the particle from a low to a high field region. As discussed in section 4.4.2 in terms of the energy of a lossless dielectric particle in a field, these two situations describe positive and negative DEP, respectively.

From Equation (10.3) we can also appreciate that electrode geometry is an important design factor with respect to the magnitude of the factor (E.∇)E, which can be obtained for a specified applied voltage. There are advantages, including the ability to use low-cost AC voltage generators and avoiding electrolysis effects at DC or low frequencies, in choosing the applied voltage to be as low as possible. The field E and its gradient ∇E have dimensions of V/m and V/m^2, respectively, so that their scalar product (E.∇)E has dimensions of V^2/m^3. A constant value for (E.∇)E can thus be maintained on combining, for example, a hundredfold reduction of the applied voltage with a reduction by a factor of 1000 of the scale of the electrode geometry. It also follows that, although the DEP force is directly proportional to the volume of the target particle, it is particularly well suited to applications in micro- and nanoscaled technologies. As demonstrated in Examples 2.7 and 2.8, a value for (E.∇)E of ~10^{13} V^2/m^3 is achieved using spherical electrodes of radii 150 μm and 750 μm with an applied voltage of 5 V. This induces a significant DEP force of magnitude ~5×10^{-12} N on a biological cell of diameter 10 μm. By *significant*, we mean that this DEP force is considerably larger by a factor of ~25 than the sedimentation force of ~2×10^{-13} N acting on the cell (see Example 2.3) and nearly 10 000 times greater than the thermal (Brownian) randomizing force ($kT/2R$) of

~4 × 10^{-16} N. If the diameter of the target biological particle is reduced to that of a virus (i.e., 10 ~ 100 nm) the magnitude of the DEP force acting on it would still be ~25 times greater than the sedimentation force, because that too scales with particle volume (see Example 2.3). However, the thermal randomizing force would now be only 10 ~ 100 times smaller than the DEP force. Rather than compensating for this by increasing the magnitude of the applied voltage, the DEP force acting on the virus can be increased by (say) a factor of 4.6 ($100^{1/3}$) on reducing the radii of the spherical electrodes to 1.5 μm and 7.5 μm. In planning DEP experiments for small particles it is in fact preferable to explore how electrode geometries other than the simple spherical design of Figure 2.9 can produce larger values for the factor (E.∇)E.

10.3 Refinements of the DEP Force Equation

10.3.1 The Induced Dipole Approximation

Equation (10.9) is based on the assumption that the field's nonuniformity is large enough to create a significant DEP force on the particle, but that the field does not vary so strongly throughout the particle as to induce within it a polarization not properly described by equation (10.7). A uniform polarization of the particle is assumed and, as shown in Figure 10.1, is represented as an induced dipole moment located at the particle centre. In reality the response of a dielectric to an imposed electric field involves the polarization of its constituent atoms and molecules. These are distributed as a continuous function of position and their polarizations will be sensitive to the value of the local electric field. A nonuniform field will result in an inhomogeneous polarization of the particle.

The treatment of the polarization of a particle in a nonuniform field can be refined by incorporating the 'theorem' given in Chapter 5, which states that: *the potential generated outside an uncharged sphere, by an arbitrary distribution of charges within it, is identical to the potential of a system of multipoles located at its centre.*

A multipole, p_n, is classified according to the scheme: n = 0 for a point charge; n = 1 a dipole; n = 2 a quadrupole and so on. They can be distributed along an axis of symmetry in a particle, as shown in Figure 10.2, or take a more general form, as shown in Figure 5.15 for the quadrupole and octupole. For the linear quadrupole shown in Figure 10.2, located on and aligned with the field E along the x-axis, it is straightforward to derive the x-directed DEP force as:

$$F_{DEP_{n=2}} = q\,[E(x+d) - 2E(x) + E(x-d)] = qd^2 \cdot \frac{\partial^2 E}{\partial x^2}$$

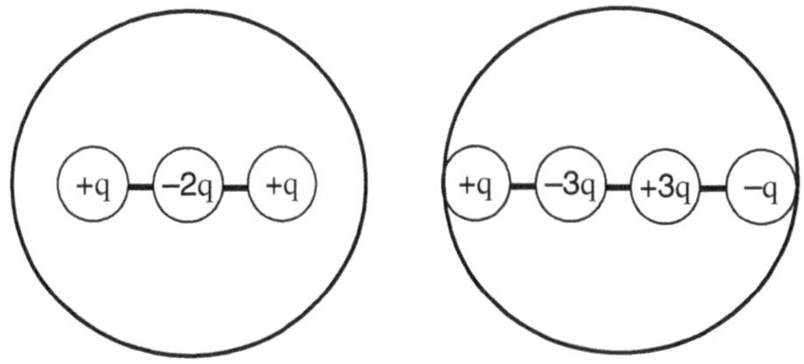

Figure 10.2 Examples of axial multipoles constructed from evenly spaced point charges. Examples of an axial quadrupole and octupole are given in Figure 5.15.

For the octupole (n = 3) shown in Figure 5.15 we have:

$$F_{DEP_{n=3}} = q\left[E\left(z+\frac{3d}{2}\right) - 3E\left(z+\frac{d}{2}\right) + 3E\left(z-\frac{d}{2}\right) - E\left(z-\frac{3d}{2}\right)\right] = qd^3 \cdot \frac{\partial^3 E}{\partial x^3}$$

From this example we find that only nth-order terms in the Taylor series survive these calculations, so that for the general nth-order linear multipole we have:

$$F_{DEP_n} = qd^n \cdot \frac{\partial^n E}{\partial x^n}$$

The more general multipoles are treated by Washizu and Jones [5]. Green and Jones [6] provide a method for determining the linear multipoles (up to the ninth order) for a range of particle shapes other than spheres (ellipsoids, truncated cylinders and an approximation of an erythrocyte).

For the situation where the particle is subjected to an arbitrary nonuniform field we are required to derive a resultant potential ϕ, which satisfies the standard boundary conditions as well as Laplace's equation ($\nabla^2\phi = 0$). Equation (10.5) is thus replaced to give a more general expression for the potential ϕ_o generated outside the sphere by a system of multipoles located within it:

$$\phi_o = -Ez + \sum_{n=0}^{\infty} \phi_n$$

The partial potential ϕ_n of the general multipole p_n of nth order is given by:

$$\phi_n = \frac{1}{4\pi\varepsilon_o\varepsilon_m} p_n \frac{Y_n}{r^{n+1}}$$

in which $p_n = n!qd^n$. Inspection of Box 5.5 and Equation (5.29) shows that Y_n involves a Legendre polynomial function determined by the spherical coordinates defining the multipole geometry. Thus, the potentials of higher order moments fall off more rapidly with distance than for a dipole (r^{-4} versus r^{-2} for a quadrupole, for example) and so their relative importance increases with increasing field nonuniformity.

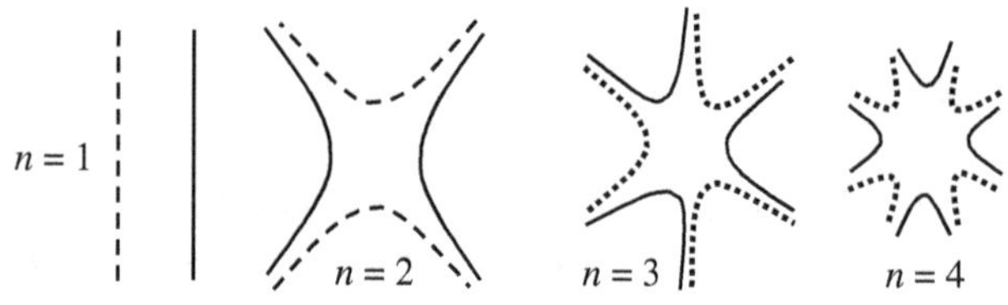

Figure 10.3 Boundaries of *polynomial* electrodes, where the full and dotted lines denote electrodes of opposite polarity. The number of electrodes of each polarity is equal to n. The electrodes are designed to produce potentials defined by a polynomial that obeys Laplace's equation. (Based on Huang and Pethig [2].)

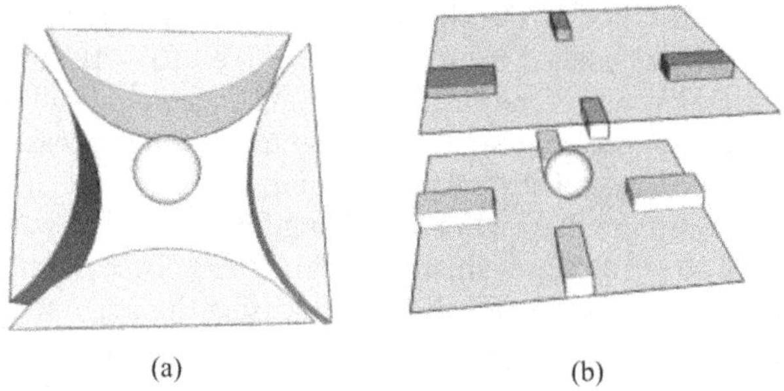

Figure 10.4 (a) A spherical particle levitated and trapped in the centre of planar quadrupole 'polynomial' electrodes [2]. (b) A particle trapped in a field cage produced between two planar quadrupole electrodes [10]. Zero field is generated along the axis of symmetry in these two electrode assemblies, so no dipole moment can be induced in the particles. Higher order moments are induced and account for the DEP forces [7, 10].

The first report of a DEP force calculation to include the higher order components of an arbitrary field and the induced multipole moments is that of Washizu [7], who illustrated the importance of this for the quadrupole 'polynomial' electrode design shown in Figure 10.3. An outline of his procedure is given in Box 10.1. As shown schematically in Figure 10.4, particles can be directed into the central location and slightly levitated above the plane of these electrodes by a negative DEP effect [2]. At this central location the factor (E.∇E) is zero and so no DEP force can be exerted. Washizu [7] demonstrated that, in this situation and especially for distances within one particle radius from the centre, the induced quadrupole moment should be included in the DEP force calculation. However, for distances far from the centre the dipole approximation gives accurate results. This condition is readily satisfied by taking DEP measurements on cells located nearer to the quadrupole electrode boundaries than to the central location. The importance of higher order moments was also found to apply to circumferentially periodic electrode structures used to achieve passive DEP levitation of particles, where pronounced size-dependent effects not anticipated by the

Box 10.1 The Dielectrophoretic Force in an Arbitrary Field

Based on the procedure described in Example 5.4 and Box 5.5 for solving Laplace's equation using Legendre polynomials and the spherical (polar) coordinate system shown below ($x = r\sin\theta\cos\phi$; $y = r\sin\theta\sin\phi$; $z = r\cos\theta$) Washizu [7] gives the following expression for an arbitrary externally applied potential:

$$\phi_{ext} = \sum_{n=0}^{N}\sum_{m=0}^{n} r^n P_n^m(\cos\theta)\left[A_{n,m}^c \cos m\phi + A_{n,m}^s \sin m\phi\right]$$

in which P_n^m is the associated Legendre function, N is the order of the potential and A is a constant. The induced surface charge σ_i on a spherical particle of complex permittivity ε_p^* and radius R, subjected to this potential in an external medium, is found by solving Laplace's equation with the standard boundary conditions at $r = R$:

$$\sigma_i = -\sum_{n=0}^{N}\sum_{m=0}^{n} \varepsilon_o K_n^* R^{n-1} P_n^m(\cos\theta) \times \left[A_{n,m}^c \cos m\phi + A_{n,m}^s \sin m\phi\right]$$

where

$$K_n^* = \frac{n(2n+1)(\varepsilon_p^* - \varepsilon_m^*)}{(n+1)\varepsilon_m^* + 2\varepsilon_p^*}$$

The DEP force is then given by the following surface integral:

$$F_{DEP} = \frac{\varepsilon_m}{\varepsilon_o}\int_s \sigma_i E dS$$

As shown by Washizu [7], because σ_i and E are expressed as sums of spherical harmonics, their orthogonality (see Box 5.5) can be used in this integration. Only those terms with the same n and m yield a nonzero integral. Furthermore, the DEP force appears only when the external potential contains n-th and (n+1)-th components simultaneously. This represents a logical extension of Equation (10.1), where the surface charge produced by a uniform field (i.e., first-order potential) interacts with a constant field gradient (second-order potential). For the case of the polynomial electrode design shown in Figure 10.3, the dipole approximation for the DEP force breaks down if the particle's displacement from the centre of symmetry of the electrodes is less than the particle radius. In this situation higher order field effects should be taken into account when comparing experimental to theoretical values of the DEP force.

conventional dipole-based theory were observed [8, 9]. Higher order moments are responsible for the levitation force achieved by such electrode structures because the electric field is zero along the central axis, resulting in the lack of an induced dipole moment. Schnelle *et al.* [10] investigated the situation for DEP field cages formed by a sandwich structure of two planar quadrupole electrodes, also shown in Figure 10.4 and concluded that quadrupole moment forces contribute ~5% of the total DEP force for particles larger than about a quarter of the electrode spacing. For particles smaller in diameter than about a tenth of the electrode spacing, the error arising from a DEP force calculation that ignores the quadrupole contribution is typically smaller than 1%.

For a spherical dielectric particle of radius R with complex permittivity ε_p^*, suspended in a medium of permittivity ε_m^* and subjected to a sinusoidal steady-state electric field vector E, the effective nth-order multipolar moment is a tensor phasor of the form [9, 11]:

$$\mathrm{p}_n = \frac{4\pi\varepsilon_m R^{2n+1}}{(2n-1)!!} CM_{(n)} (\nabla)^{n-1}\mathrm{E}$$

$CM_{(n)}$ is the general multipole form of the Clausius–Mossotti factor, given by:

$$CM_{(n)} = \frac{\varepsilon_p^* - \varepsilon_m^*}{n\varepsilon_p^* + (n+1)\varepsilon_m^*}$$

where the asterisk denotes that the complex form of the permittivity is considered. As shown in the next section this accommodates the presence of dielectric losses in the particle and surrounding medium. Following the form of Equation (10.10) the general, time-average, expression for the DEP force acting on the nth-order multipole is given by:

$$\langle \mathrm{F}_{DEP_n} \rangle = \mathrm{Re}\left[\frac{p_n[\cdot]^n(\nabla)^n\mathrm{E}}{n!}\right] \quad (10.11)$$

where $[\cdot]^n$ represents n dot products performed on the dyadic tensors and $(\nabla)^n$ represents n vector ∇ operations [11]. The parameter 'Re' denotes that the real part of the complex function is taken (see Equation 10.22). This multipolar analysis was also applied by Jones and Washizu [11] to traveling wave dielectrophoresis. An important conclusion was that the various multipolar DEP forces show similar frequency dependencies, especially in terms of the value of the so-called DEP crossover frequency that marks the point where the DEP force reverses polarity. This is important because, as described in the next chapter, DEP-based cell separation protocols rely on an interpretation of DEP crossover measurements. Washizu [12] later extended the multipole moment analysis of the DEP force to the case of spherical particles being manipulated by focused laser beams (i.e., optical tweezers).

10.3.2 Consideration of Electrical Energy Loss

Bioparticles usually exhibit dielectric losses associated with relaxations of permanent or induced dipoles, as well as ohmic conduction loss due to the displacement of mobile ions associated with their structure. The medium in which particles are suspended during DEP experiments may also be a buffered electrolyte that exhibits ionic conduction and dipole relaxation losses. We have noted that the factor $\varepsilon_m R^3(CM)\mathrm{E}^2$ in Equation (10.9) is proportional to the energy required to withdraw a sphere of radius R from a field E into a region where there is no field. However, this equation employs the CM factor obtained from a solution of the boundary conditions described in Box 6.1, for an ideal dielectric sphere introduced into an ideal dielectric medium. The existence of electrical energy losses, manifested as the dissipation of heat, negates the principle of energy conservation. The energy variation approach used to derive Equation (4.38) is therefore no longer valid and will tend to underestimate the magnitude of the DEP force. The *effective* dipole moment p used in Equation (10.1) is also a point dipole, composed of free charges, which when exchanged with the particle, produces the same dipole field as that derived in Box 6.1. We can ask to what extent the neglect of electrical losses, together with the use of the effective moment method, influences the validity of Equation (10.9).

As described in Chapters 6 and 7, AC dielectric and conduction losses are encompassed as either a complex permittivity ε^* or complex conductivity σ^*. It is of value to rehearse these concepts. The total current density in a dielectric is given by:

$$\mathrm{J}_T = \mathrm{J}_C + \partial \mathrm{D}/\partial t$$

where J_C is the conduction current given by Ohm's Law ($\mathrm{J}_C = \sigma\mathrm{E}$) and D is the electrical displacement ($\mathrm{D} = \varepsilon\mathrm{E}$). Assuming that all of the fields have a sinusoidal time variation given by the exponential $e^{i\omega t}$ (see ahead to Box 10.3) then

$$\mathrm{J}_T = \mathrm{J}_C + \mathrm{i}\omega\mathrm{D} = (\sigma + \mathrm{i}\omega\varepsilon)\mathrm{E} \quad (10.12)$$

where i is the imaginary vector ($i = \sqrt{-1}$) and ω is the angular frequency ($\omega = 2\pi f$) of the applied sinusoidal electric field. This shows that the displacement current leads the conduction current by a phase angle of $\pi/2$ radians. Ohm's Law can therefore be written as $\mathrm{J}_C = \sigma^*\mathrm{E}$, where σ^* is the complex conductivity given by:

$$\sigma^* = \sigma' + \sigma'' \quad (10.13)$$

The real component σ' is the conductivity σ used in the 'standard' form of Ohm's Law, while the imaginary component σ'' is equal to $\omega\varepsilon$, with ε the absolute value of the permittivity ($\varepsilon = \varepsilon_0\varepsilon_r$). The form of the total current given by Equation (10.12) can be modified as follows:

$$\begin{aligned} J_T &= (\sigma + i\omega\varepsilon)E = i\omega(\sigma/i\omega + \varepsilon)E \\ &= i\omega(\varepsilon - i\sigma/\omega)E = i\omega\varepsilon^* E \end{aligned}$$

where ε^* is the complex permittivity given by:

$$\varepsilon^* = \varepsilon' - \varepsilon''$$

The complex relative permittivity is thus given by:

$$\varepsilon_r^* = \frac{\varepsilon^*}{\varepsilon_o} = \varepsilon_r' - i\varepsilon_r''$$

The real component ε_r' of the complex relative permittivity is identical to the parameter ε_r used to define capacitance ($C = A\varepsilon_o\varepsilon_r/d$), whilst the imaginary component can be separated into two parts, namely the DC conductivity and the high-frequency part ε_r'', to give:

$$\varepsilon_r^* = \varepsilon_r' - i\left(\varepsilon_r'' + \frac{\sigma_{dc}}{\omega\varepsilon_o}\right) \tag{10.14}$$

As described in Chapter 6, these complex permittivity and conductivity parameters can be inserted into two equivalent forms of the Clausius–Mossotti factor:

$$CM^* = \left(\frac{\varepsilon_p^* - \varepsilon_m^*}{\varepsilon_p^* + 2\varepsilon_m^*}\right)$$

or

$$CM^* = \left(\frac{\sigma_p^* - \sigma_m^*}{\sigma_p^* + 2\sigma_m^*}\right)$$

As shown in Box 10.1, Washizu [7] adopted complex permittivity quantities and by doing so included electric losses in his multipole refinement of the DEP force equation. However, this theory leads to Equation (10.11) in which the time average of the DEP force is determined by the *real* part of the induced moment and hence also by the real part of the *CM* factor. (An explanation as to why the real part of *CM* is involved is deferred until Box 10.4.) The dielectric loss parameter ε'' and DC conductivity σ_{dc} in Equation (10.14) are therefore not taken into account. The question thus arises as to the extent that this omission affects the validity of Equation (10.11). Obviously, as long as the condition $\varepsilon'' \ll \varepsilon'$ holds and the ohmic conductivity is small, then any error involved is also small. Fortunately, DEP experiments on cell suspensions are often performed using low conductivity aqueous media and at frequencies well below the dielectric loss peak for water. The ohmic conductivity of viable cells is also small. The extent to which this can be properly quantified, however, requires further study.

The most rigorous method to derive a field-induced force employs the Maxwell stress tensor. In this method the DEP force is found by integrating the stress tensor T over the surface of the test particle:

$$F_{DEP} = \int_S (T \cdot n)dA \tag{10.15}$$

in which n is the unit vector normal to the particle surface element dA. The method uses the concept of electromagnetic momentum density, commonly known as the Poynting vector [13], which remains constant in its mathematical form irrespective of whether or not the medium involved is loss free. The method has been used by Mognaschi and Savini [14] to calculate the forces and torques acting in nonhomogeneous lossy dielectric systems, under the influence of nonuniform fields and without recourse to the effective dipole approximation. Sauer and Schlögl [15, 16] also employed an integration of the Maxwell stress tensor over the surface of a spherical particle to derive an expression for the DEP force similar to that of Equation (10.11). The complication involved in taking into account the nonuniform nature of the applied electric field meant that the analysis was restricted to the case of a very small gradient of the field. Later work by Wang *et al.* [17], which was not restricted to small field gradients, led to the important conclusion that the Maxwell stress tensor method gives the same expression for the DEP force as the effective dipole and multipole methods. We can conclude from this that application of the effective dipole method does not result in an erroneous result for the DEP force. However, the effective dipole method is best suited when dealing with particles of well defined geometry, such as a sphere or ellipsoid. An advantage of the Maxwell stress tensor method is that for arbitrary particle shapes, together with knowledge of how the electric field is distributed over its surface, the DEP force can be calculated directly by summing the scalar product (T.n) in Equation (10.15) over the whole particle surface [17].

It is useful to have these two forms for *CM* given above, because the total induced current in the particle can be considered to comprise two components – a component in phase with the applied electric field arising from field-induced movements of free charges, as well as an out-of-phase component associated with relaxations of dipoles. At low frequencies ($\omega \to 0$) the current is dominated by the conduction of charges and its magnitude is determined by the electrical conductivity of the particle. If the applied electric field changes with time, the current will change in step with this. For a particle that behaves as an ideal conductor (or resistor) there is no phase difference between the field and the current. However, at high frequencies ($\omega \to \infty$) the dielectric displacement current dominates, so that the particle behaves as an ideal dielectric that has no leakage conductance. As shown in Box 10.2, for a particle that behaves as the dielectric of an ideal capacitor, the current *leads* the applied field by a phase angle of 90° (equivalent to a lead of $\pi/2$ radians). This appears to negate the relationship between cause and effect. Surely, the current can flow only *after* the voltage is applied. However, the situation shown in Box 10.2, where the sine and cosine waveforms are assumed to be fully formed at time zero, is misleading. A mathematician will inform us that we should assume the waveforms were established a very long time (theoretically an infinitely long time) before our chosen reference of time zero. The cosine function is given by $\cos(x) = \sin(x + \pi/2)$ and so is a sine wave, but with an advanced phase shift of $\pi/2$ radians. The current waveform shown in Box 10.2 thus has an automatic 'head start' of $\pi/2$ radians on the voltage waveform. This situation is a consequence of the current acting to oppose changes of the capacitor voltage. An analogy is the interest rate set by banking authorities to control an undesired rate of monitory inflation. We can say that a cyclical change of inflation leads the cyclical change of the interest rate. In a capacitor the cosine current function leads the voltage sine function.

The opposition to current flow in an ideal conductor is called its resistance, arising from thermally induced, randomizing, deviations from the electric field-forced paths that should be taken by the mobile charges. The relationship between the applied voltage and the current is given by Ohm's Law $V = IR$, where V and I are either the DC or RMS values (see Box 2.6). The resistance R is measured in units of ohms. The current does not oppose voltage changes, so that there is no phase difference between the current and applied voltage. At any instant in time the resistive current in an ideal conductor, subjected to a sinusoidal voltage waveform, is given by:

$$i_R(t) = \frac{v(t)}{R} = \frac{1}{R} V_{pk} \sin(\omega t) \tag{10.16}$$

The conduction current stays in phase with the applied voltage and its magnitude does not change as a function of the voltage frequency. From Box 10.2 we know that a capacitor opposes a voltage change, by inducing either a charging or discharging current that is directly proportional to the rate of change (frequency) of the applied voltage. For a constant DC voltage the current is zero. In Chapter 2 we refer to this current as the *displacement* current. From Box 10.2 the instantaneous current $i_c(t)$ at any time is given by:

$$\begin{aligned} i_c(t) &= \omega C V_{pk} \cos(\omega t) \\ &= I_{pk} \sin(\omega t + \pi/2) \quad (I_{pk} = \omega C V_{pk}) \end{aligned} \tag{10.17}$$

Box 10.2 AC Capacitance

Figures 2.1 and 3.1 depict the charging of a capacitance C when a DC voltage source is connected across it. The charging current i_c, depicted in Figure 9.7, is initially large and decreases to zero as the voltage difference v_c across the capacitor approaches that of the applied voltage. At any instant in time the current is given by $i_c = dq_c/dt$, where $q_c = v_c C$ is the charge stored on each capacitor plate. If a sinusoidal voltage of the form shown in Figure 2.13 is applied, the charging current at any instant is given by:

$$i_c(t) = \frac{dq_c}{dt} = C\frac{dv_c}{dt} = C\frac{dV_{pk}\sin(\omega t)}{dt} = \omega C V_{pk}\cos(\omega t)$$

The current is thus directly proportional to the frequency of the applied voltage. The current waveform, together with that of the applied voltage, is shown below:

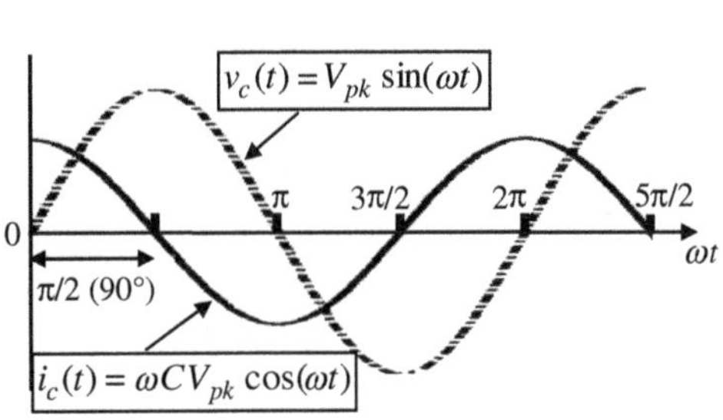

When the applied voltage completes its first quarter cycle (at $\pi/2$ radians) the charging rate (dv_c/dt) and current i_c are both zero. Although the applied voltage is still positive for the next quarter-cycle, current i_c flows in the opposite (negative) sense in an effort to maintain a constant voltage. At the half cycle (π radians) point the rate of change of the applied voltage is at a maximum in the negative sense and i_c attains its peak negative value. At this point the capacitor plates carry equal but opposite charges, so that no net voltage appears between them. During the second half cycle the applied voltage proceeds to its negative peak value at $3\pi/2$ radians, the capacitor is fully charged in the opposite sense to that at $\pi/2$ radians and the current flow is again zero. As the applied voltage head backs to zero to complete a full cycle at 2π radians, the fully charged capacitor attempts to maintain the voltage across its plates by discharging. The charging / discharging process then begins its next full cycle.

We conclude that a capacitor opposes a change in voltage and this is manifested as the current waveform *leading* the applied voltage waveform by $\pi/2$ radians (90°). The magnitude of the current is directly proportional to the frequency of the voltage waveform, being zero under steady-state DC conditions.

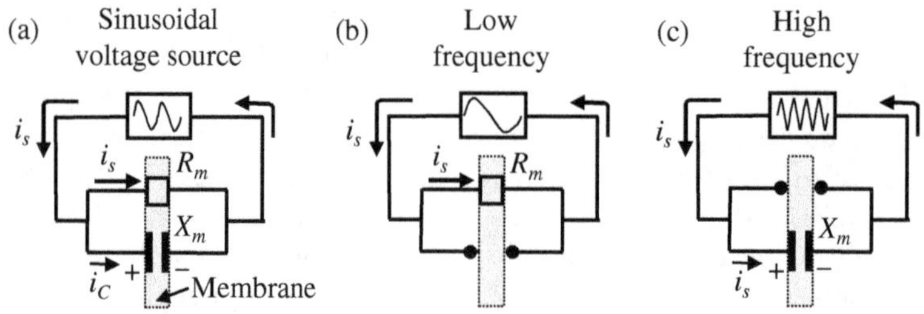

Figure 10.5 (a) As depicted in Figure 9.7 the equivalent circuit for a cell membrane consists of the passive membrane resistance R_m in parallel with the membrane capacitance C_m. (b) At very low frequencies C_m appears as an electrical open circuit so that all of the signal current must pass through R_m. The electrical impedance of a cell is dominated by its conductive rather than capacitive elements. (c) At a sufficiently high frequency the membrane reactance X_m falls to a low value and short circuits the membrane's high resistance. The permittivity parameters of the cell's constituents now dominate over its conductive properties.

The relationship between the peak voltage across a dielectric in a capacitor and the peak induced (displacement) current takes the same form as Ohm's Law, namely $V_{pk} = I_{pk}X_c$. The symbol X_c is referred to as the *reactance* of the capacitor. From Equation (10.16) we deduce that $X_c = 1/\omega C$.

In Figure 9.7 the observed electrical properties of a mammalian cell membrane can be interpreted as an equivalent circuit composed of the membrane resistance in parallel with the membrane capacitance. Equation (10.16) informs us that, at low frequencies (as $\omega \rightarrow 0$), the current induced in the capacitor will tend to zero, so that the capacitor appears as an electrical open circuit of infinite resistance. Although the specific resistance of the membrane of a viable mammalian cell is large, considered to be at least 100 000 times greater than that of the cytoplasm (see Table 9.3), it is certainly not as large as that presented by an open circuit. As depicted in Figure 10.5 the current drawn from the applied voltage source will thus flow predominantly through the resistive component of the membrane. If instead of being applied directly across the membrane, as depicted in Figure 10.5, the voltage is applied across the whole cell (as it would be when suspended in an electrolyte for a DEP experiment) the current drawn from the voltage source will skirt around the cell, as depicted in Figure 2.5. With increasing frequency of the applied voltage signal, the effective resistance (reactance) of the capacitive element of the membrane gets smaller, in direct proportion to the reciprocal value of the frequency. At a sufficiently high frequency, the capacitive element of the membrane will electrically short out the membrane resistance, as depicted in Figure 10.5. If the voltage is applied across the whole cell, when suspended in an electrolyte of lower conductivity than the cytoplasm, the current drawn from the voltage source will flow across the membrane and into the cytoplasm, as depicted in Figure 2.5.

The situations shown in Figure 10.5 represent the limiting low-frequency and high-frequency cases. At low frequencies the current drawn from the supply is to a good approximation in phase with the applied voltage, whilst for much higher frequencies the current leads the voltage by nearly $\pi/2$ radians. At intermediate frequencies, the current is distributed between the resistive and capacitive components of the membrane and so will lead the voltage by less than $\pi/2$ radians. A helpful way to depict this is the *phasor* representation of the voltage and current waveforms. As the name suggests, a phasor is a vector that quantifies the magnitude and phase of a waveform. It is a complex number (i.e., has a real and imaginary component), which represents a sinusoidal waveform whose peak magnitude, angular frequency and phase angle remain constant with time. As shown in Figure 10.6 a phasor takes the form of a vector of length

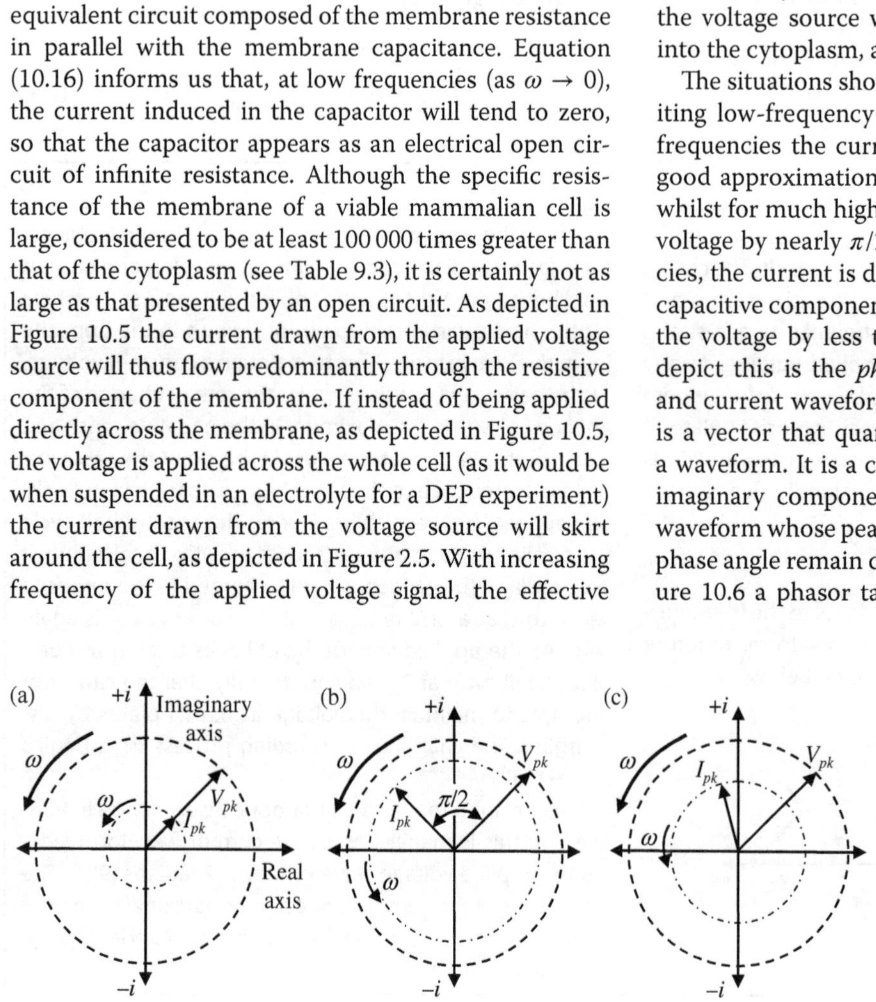

Figure 10.6 (a) At very low frequencies the voltage and current phasors, associated with conduction through the membrane shown in Figure 10.5, are in phase. (b) At very high frequencies the current phasor leads the voltage phasor by 90° ($\pi/2$ radians). (c) At an intermediate frequency the current phasor leads the voltage phasor by less than ($\pi/2$ radians).

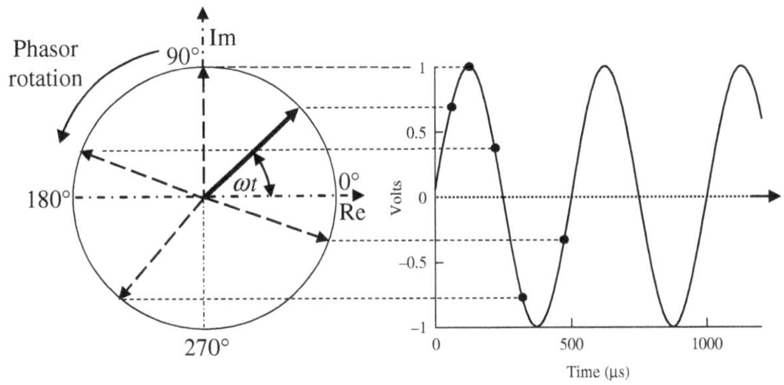

Figure 10.7 A voltage phasor is shown rotating in an anticlockwise direction at ω radians per second (in this 4π thousand radians/s, corresponding to 2 kHz). At each angular position the location of its tip, equal to $V_{pk}\sin(\omega t)$, plots out the sinewave as a function of time.

equal to the peak amplitude of the sinusoid and by convention rotates in the anticlockwise direction at an angular frequency of ω radians per second. The voltage and current phasors for the limiting low- and high-frequency cases of Figure 10.5 are shown in Figure 10.6, together with a situation between these two limits.

As shown in Figure 10.7 for the case of a voltage phasor, the projected height of its tip on the imaginary (Im) axis maps out a voltage sinewave as a function of time. A horizontal orientation of the phasor along the real axis corresponds to an angular position (ωt) of $n\pi$ ($n = 0, 1, 2, \cdots$) radians, whilst an orientation along the imaginary axis corresponds to angular positions of $n\pi/2$ ($n = 1, 3, 5, \cdots$). At any instant a projection of the phasor tip's height onto the imaginary axis is equal to $V_{pk}\sin(\omega t)$, which is the function sketched out in Figure 10.7. The corresponding projected heights along the real (Re) axis map out the function $V_{pk}\cos(\omega t)$. This illustrates in graphical form the result derived in Box 10.3, known as Euler's formula, for the phasor representation of a sinusoidal waveform:

$$V(t) = V_{pk}(\cos(\omega t) + i\sin(\omega t)) = V_{pk}e^{i\omega t} \tag{10.18}$$

which also provides a formal mathematical definition of the sine and cosine functions:

$$\cos x = \mathrm{Re}[e^{ix}] = \frac{e^{ix} + e^{-ix}}{2}$$

$$\sin x = \mathrm{Im}[e^{ix}] = \frac{e^{ix} - e^{-ix}}{2i}$$

Box 10.3 Expressing a Sinusoidal Voltage as a Complex Exponential Function

In Chapter 6 the value of representing relative permittivity as a complex number was introduced. A complex number takes the form $z = x + iy$, with x the real and y the imaginary value of z. The horizontal and vertical components of the phasor shown in Figure 10.7 are $x = V_{pk}\cos\omega t$ and $y = V_{pk}\sin\omega t$, so that $V(t)$ may be expressed as:

$$V(t) = V_{pk}(\cos\omega t + i\sin\omega t)$$

The Maclaurin series for the cosine and sine functions are:

$$\cos\omega t = 1 - \frac{(\omega t)^2}{2!} + \frac{(\omega t)^4}{4!} + \cdots \cdot + \frac{(-1)^{n-1}}{(2n-2)!}(\omega t)^{2n-2} + \cdots$$

$$\sin\omega t = 0 - \frac{(\omega t)^3}{3!} + \frac{(\omega t)^5}{5!} + \cdots \cdot + \frac{(-1)^{n-1}}{(2n-1)!}(\omega t)^{2n-1} + \cdots$$

Substituting these series into the complex expression for $V(t)$ we obtain:

$$V(t) = V_{pk}(\cos\omega t + i\sin\omega t)$$

$$= V_{pk}\left[1 + i\omega t - \frac{(\omega t)^2}{2!} - \frac{i(\omega t)^3}{3!} + \frac{(\omega t)^4}{4!} + \frac{i(\omega t)^5}{5!} + \cdots\right]$$

Substituting into this equation the following identities $i^2 = -1$; $i^3 = -i$; $i^4 = 1$; $i^5 = i$; etc.:

$$V(t) = V_{pk}\left[1 + i\omega t + \frac{(i\omega t)^2}{2!} + \frac{(i\omega t)^3}{3!} + \frac{(i\omega t)^4}{4!} + \frac{(i\omega t)^5}{5!} + \cdots\right]$$

Within the brackets is the Maclaurin series for the exponential function $e^{i\omega t}$, so that $V(t)$ can be expressed in the compact (Euler's formula) format:

$$V(t) = V_{pk}e^{i\omega t} = V_{pk}(\cos\omega t + i\sin\omega t)$$

For a stationary AC field, where the phase remains constant as a function of location, the field is given by $E(t) = -\nabla V(t)$. The expression for the field waveform in Equation (6.39) is obtained as follows:

$$E(t) = \mathrm{Im}\left[E_o e^{i\omega t}\right] = \mathrm{Im}\left[E_o(\cos\omega t + i\sin\omega t)\right] = E_o\sin\omega t$$

where Im[..] indicates the imaginary part of the complex function inside the brackets.

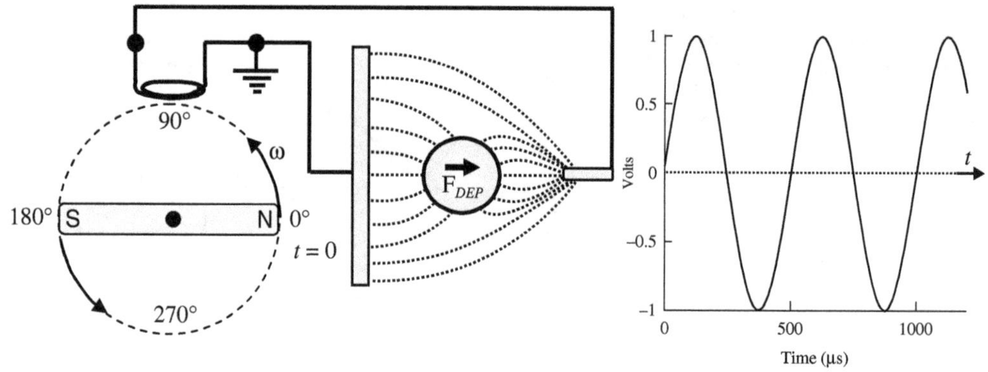

Figure 10.8 The rotating bar-magnet generator shown in Figure 2.12 is used to create a nonuniform AC field between DEP electrodes. At reference time zero, with the north pole at the 0° position, the applied voltage is zero. As the north pole approaches and passes the 90° position the voltage at the pin electrode attains its maximum *positive* value with respect to the grounded plate electrode. As the south pole passes the 90° position the maximum *negative* voltage is generated at the pin electrode. A time-averaged positive DEP force acts on the spherical particle.

A schematic is given in Figure 10.8 of how the rotating bar magnet shown in Figure 2.12 represents the generation of a voltage phasor, which can be applied to two electrodes to produce a nonuniform electric field for a DEP experiment.

In Figure 10.7 the voltage phasor is shown 'frozen in time' at an angle of 45° ($\pi/4$ radians). If another phasor begins to rotate at the same rate, starting at the zero-angle position, its sinusoidal waveform will *lag* behind the voltage phasor by $\pi/4$ radians. Alternatively we can say that the voltage phasor will *lead* the second one by $\pi/4$ radians. The following mathematical relationship describes the magnitude and phase angle of a phasor:

$$\begin{aligned} V(t) &= V_{pk}\,(\cos(\omega t \pm \phi) + i\sin(\omega t \pm \phi)) \\ &= V_{pk} e^{i(\omega t \pm \phi)} \end{aligned} \tag{10.19}$$

A positive value ($+\pi > \phi > 0$) indicates a *leading* phase angle, whereas a negative value $-\pi < \phi < 0$ indicates that the signal *lags* the reference signal. A zero value for ϕ signifies that the signal is in phase with the reference, taken to be $V_{pk}\sin(\omega t)$ for example.

From Equation (10.21) in Box 10.4 the time-dependent DEP force is given by:

$$\mathrm{F}_{DEP}(t) = \mathrm{Re}\left[(\mathrm{p}^* \cdot \nabla)\,\mathrm{E}_{\mathrm{pk}} e^{i\omega t}\right] \tag{10.23}$$

where p* is the particle's induced (complex) dipole moment, given by:

$$\mathrm{p}(t) = v\alpha^* \mathrm{E}_{\mathrm{pk}} e^{i\omega t}$$

with v the volume of the particle. The parameter α is the moment per unit volume, induced by a field of unit magnitude, as given in Equation (6.4). From Equation (10.17) the time-averaged DEP force, over a large number of periodic cycles of the field, is given by:

$$\begin{aligned} \langle F_{DEP} \rangle &= \frac{v}{T}\int_0^T \mathrm{Re}\left[(\alpha^* \mathrm{E}_{\mathrm{pk}} e^{i\omega t} \cdot \nabla)\mathrm{E}_{\mathrm{pk}} e^{i\omega t}\right] dt \\ &= \frac{v}{2T}\int_0^T \mathrm{Re}\left[\alpha^* \nabla(\mathrm{E}_{\mathrm{pk}}^2 e^{2i\omega t})\right] dt \end{aligned} \tag{10.24}$$

in which $T = 2\pi/\omega$ and the vector relationship $2(\mathrm{E}\cdot\nabla)\mathrm{E} = \nabla \mathrm{E}^2$ is employed. The integration step in this equation can be performed as follows:

$$\begin{aligned} \int_0^T \mathrm{Re}\left[\alpha^* \nabla\left(\mathrm{E}_{\mathrm{pk}}^2 e^{2i\omega t}\right)\right] dt &= \int_0^{2\pi/\omega} \mathrm{Re}[\alpha^*] \cos^2 \omega t\, dt \\ &= \mathrm{Re}[\alpha^*]\left[\frac{t}{2} + \frac{\sin 4\omega t}{4\omega}\right]_0^{2\pi/\omega} = \frac{\pi}{\omega}\mathrm{Re}[\alpha^*] \end{aligned}$$

Using this result in Equation (10.24) the time-averaged DEP force is thus given by:

$$\langle F_{DEP} \rangle = \frac{v}{2T}\frac{\pi}{\omega}\mathrm{Re}[\alpha^*] = \frac{v}{4}\mathrm{Re}[\alpha^*]\nabla \mathrm{E}_{pk}^2 \tag{10.25}$$

For a spherical particle, $v = (4\pi R^3)/3$ and α is derived in Box 6.1 as:

$$\alpha = 3\varepsilon_o \varepsilon_m \left(\frac{\varepsilon_p - \varepsilon_m}{\varepsilon_p + 2\varepsilon_m}\right)$$

From Equation (10.25) the time-averaged DEP force acting on a spherical particle of radius R in a nonuniform time-varying, sinusoidal, field is thus:

$$\langle \mathrm{F}_{DEP} \rangle = \pi \varepsilon_o \varepsilon_m R^3 \mathrm{Re}\left[\frac{\varepsilon_p^* - \varepsilon_m^*}{\varepsilon_p^* + 2\varepsilon_m^*}\right] \nabla \mathrm{E}_{\mathrm{pk}}^2 \tag{10.26}$$

This result for the AC dielectrophoretic force follows naturally from our definition of the *magnitude* of a sinusoidal voltage or field in terms of its *peak* amplitude. It

Box 10.4 The DEP Force is related to the Real Part of [(p·∇)E]

The DEP force acting on a particle is given according to Equation (10.1) as the scalar (*dot*) product of the particle's dipole moment p and the field gradient ∇E. The scalar product of any two vectors A and B is defined as the magnitude of B multiplied by the component of A parallel to B. Thus, as shown in the following diagram:

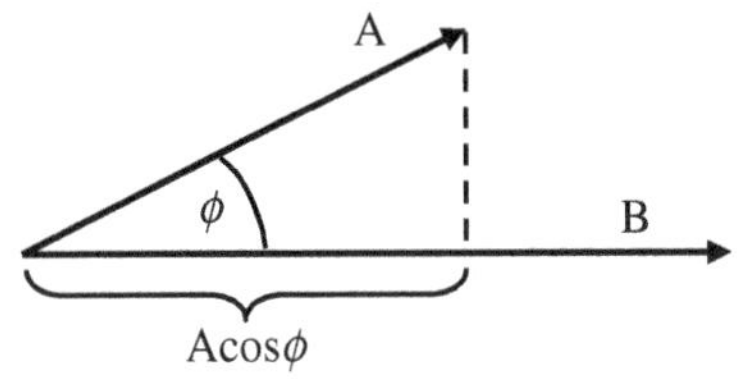

$$A \cdot B = (|A| \cos \phi)\, |B| = |A|\, |B| \cos \phi$$

If A and B are phasors of the same angular frequency, then

$$A(t) = |A|\, e^{j(\omega t + \phi)} \quad \text{and} \quad B(t) = |B|\, e^{j\omega t}$$

A formal way to multiply together two complex phasors is to use the conjugate of one of them (A and B are conjugates if they have the same real parts and equal and opposite imaginary parts). Thus, the product of A(t) and B(t) is given by:

$$\begin{aligned} A(t)\bar{B}(t) &= |A|\, e^{j(\omega t+\phi)}\, |B|\, e^{-i\omega t} = |A||B|e^{i\phi} \\ &= |A|\, |B|\, [\cos\phi + i \sin\phi] \end{aligned}$$

Comparing this with the above expression for A · B it follows that:

$$A(t) \cdot B(t) = \text{Re}\left[|A||B|e^{i\phi}\right] \quad (10.20)$$

where Re signifies the real part of [...]. From inspection of Equations (10.1) and (10.20) the time-dependent DEP force is thus given by:

$$F_{DEP}(t) = \text{Re}\,[(p \cdot \nabla)E] \quad (10.21)$$

The DEP force is thus related to the real part of the Clausius–Mossotti factor. The product of two phasors is also given by its time-averaged value (e.g., [18]):

$$\langle A(t) \cdot B(t) \rangle = \frac{1}{2}\text{Re}\left[A(t) \cdot \bar{B}(t)\right]$$

so that from inspection of Equation (10.1):

$$\langle F_{DEP}(t) \rangle = \frac{1}{2}\text{Re}\left[(p \cdot \nabla)\bar{E}\right] \quad (10.22)$$

is the preferred form adopted by Morgan and Green [4], for example. However, most of the DEP literature, including relevant texts [e.g., 1, 3, 19, 20], adopt the *root mean square* value of the field, derived in Box 2.6 as $E_{rms} = E_{pk}/\sqrt{2}$. Equation (10.26) is thus more usually presented as:

$$\langle F_{DEP} \rangle = 2\pi\varepsilon_o\varepsilon_m R^3 \text{Re}\left[\frac{\varepsilon_p^* - \varepsilon_m^*}{\varepsilon_p^* + 2\varepsilon_m^*}\right] \nabla E_{rms}^2 \quad (10.27)$$

with E_{rms} usually given simply as E. A perceived advantage of the form given by Equation (10.27) is that it also gives the DEP force in a DC field and is compatible with the use of a digital multimeter (usually calibrated to read rms values) to monitor the voltage signal applied to electrodes during a DEP cell separation procedure, for example. However, if an oscilloscope is used to monitor the applied signal (the past normal practice in the author's labs) then the peak magnitude of the signal is readily determined and Equation (10.26) is the more useful. It is also helpful during an electrorotation experiment to monitor the quadrature-phased signals applied to the electrodes by means of an oscilloscope. As shown in section 10.3.4 the magnitude of the rotating field is determined by the peak voltage of the applied signals.

From Washizu's result [7] the time-averaged DEP force acting on a spherical particle, taking into account the dipole and quadrupole terms is given by:

$$\begin{aligned} \langle F_{DEP} \rangle = {} & 2\pi\varepsilon_m R^3 \text{Re}\left[\frac{\varepsilon_p^* - \varepsilon_m^*}{\varepsilon_p^* + 2\varepsilon_m^*}\right] \nabla E^2 \\ & + \frac{2}{3}\pi\varepsilon_m R^5 \text{Re}\left[\frac{\varepsilon_p^* - \varepsilon_m^*}{2\varepsilon_p^* + 3\varepsilon_m^*}\right] \nabla \cdot \nabla E^2 \end{aligned} \quad (10.28)$$

For particles with diameters less than around one-tenth of the interelectrode spacing, the first (dipole force) term in Equation (10.28) dominates the overall DEP force. For larger particles than this the second (quadrupole) term becomes important. For the dipole case we can use the expression given by Equation (6.50) for Re[CM^*]:

$$\begin{aligned} \text{Re}[CM^*] = \Bigg[& \left(\frac{\omega^2\tau^2}{1+\omega^2\tau^2}\right)\left(\frac{\varepsilon_p - \varepsilon_m}{\varepsilon_p + 2\varepsilon_m}\right) \\ & + \left(\frac{1}{1+\omega^2\tau^2}\right)\left(\frac{\sigma_p - \sigma_m}{\sigma_p + 2\sigma_m}\right)\Bigg] \end{aligned} \quad (10.29)$$

where τ is the relaxation time for the interfacial charging given by Equation (6.49), reproduced here for convenience:

$$\tau = \varepsilon_o \frac{\varepsilon_p + 2\varepsilon_m}{\sigma_p + 2\sigma_m} \quad (10.30)$$

At low frequencies, where $\omega\tau \ll 1$ (a signal frequency typically below 10 kHz):

$$\mathrm{Re}[CM^*] \approx \left(\frac{\sigma_p - \sigma_m}{\sigma_p + 2\sigma_m}\right) \tag{10.31a}$$

For the case $\omega\tau \gg 1$ (typically above 500 MHz):

$$\mathrm{Re}[CM^*] \approx \left(\frac{\varepsilon_p - \varepsilon_m}{\varepsilon_p + 2\varepsilon_m}\right) \tag{10.31b}$$

At low frequencies the DEP force thus depends on the conductive properties of the particle and suspending medium, whilst at high frequencies the permittivity values are important. Figures 6.5 shows how the function Re[CM^*] given by Equation (10.29) varies with frequency for the situation where the conductivity of a homogeneous particle (σ_p) exceeds that (σ_m) of the surrounding medium (σ_m), with the medium permittivity (ε_m) being larger than that (ε_p) of the particle. We find that Re[CM^*] and hence the DEP force, makes the transition from a positive to negative value as the frequency increases. For a homogeneous metallic particle we have the situation $\sigma_p \gg \sigma_m$ and $\varepsilon_p \gg \varepsilon_m$. In this case, from Equations (10.31), Re[CM^*] remains positive on sweeping from a low to a high frequency, with a value close to the maximum theoretical value of +1.0. A system possessing electrical properties in strong contrast to that of a suspended metallic particle would be an ideal dielectric particle suspended in an aqueous electrolyte (i.e., $\sigma_p \ll \sigma_m$; $\varepsilon_p \ll \varepsilon_m$). In this case Re[$CM^*$] will remain negative as a function of frequency, with a value close to the theoretical minimum of −0.5. From this we can appreciate that the dynamic range of $-0.5 \leq \mathrm{Re}[CM^*] \leq +1.0$ is not large, implying that it will be difficult to distinguish between particles having subtle differences in their electrical properties. However, inspection of Figures 6.5 and 6.6 show that apart from a difference in the polarity of Re[CM^*], hence also of the DEP force, a significant variation can occur of the frequency at which Re[CM^*] and the DEP force change polarity. This effect is commonly referred to as the DEP crossover (DEP$_{\mathrm{XO}}$) and the frequency ($f_{xo} = \omega_{xo}/2\pi$) at which it occurs can be calculated by setting Re[CM^*] equal to zero in Equation (10.29). On substituting for the relaxation time τ given by Equation (10.30) we obtain the following relationship for the crossover radian frequency:

$$\omega_{xo}^2 = \frac{1}{\varepsilon_o^2}\frac{(\sigma_m - \sigma_p)(\sigma_p + 2\sigma_m)}{(\varepsilon_p - \varepsilon_m)(\varepsilon_p + 2\varepsilon_m)} \tag{10.32}$$

This equation is quadratic in form and so in principle has two roots – in other words two possible values for the crossover frequency can exist. We can test the validity of this conclusion for the case of a lymphocyte suspended in an aqueous medium. As described in Chapter 9, Asami *et al.* [21] adopted the multishell model of a cell to analyse

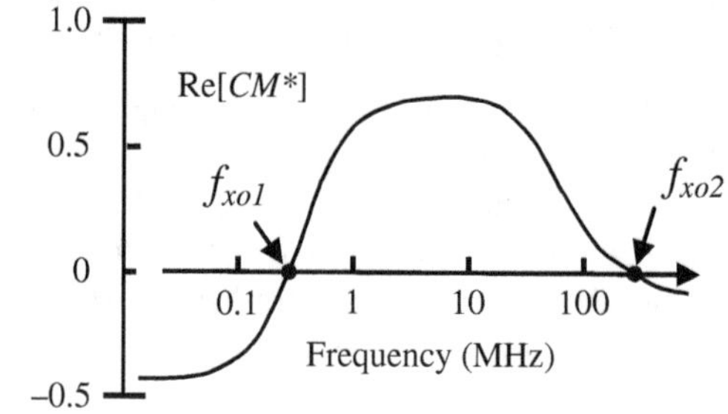

Figure 10.9 The frequency variation of Re[CM^*] derived from the dielectric data obtained by Asami *et al.* [21] for lymphocytes. The DEP force acting on the cell is proportional to Re[CM^*] and so follows the same frequency profile. Two DEP crossover points exist, at f_{xo1} and f_{xo2}.

impedance measurements performed on suspensions of viable lymphocytes, the results of which are presented in Table 9.3. Using these dielectric parameters to derive the effective values for σ_p and ε_p as a function of frequency, on inserting them into Equation (10.29) the frequency variation of Re[CM^*] obtained for a viable lymphocyte is shown in Figure 10.9. Two crossover frequencies, f_{xo1} and f_{xo2}, are shown located at ~280 kHz and ~300 MHz, respectively.

Experimentally relevant solutions (i.e., real rather than imaginary ones obtained from taking the square root of a negative number) for ω_{xo} in Equation (10.32) only occur where the following condition is satisfied:

$$\frac{(\sigma_m - \sigma_p)}{(\varepsilon_p - \varepsilon_m)} > 0 \tag{10.33}$$

The effective values for σ_p and ε_p obtained from the data of Asami *et al.* [21] for viable lymphocytes are shown in Figure 10.10. For cells suspended in an aqueous electrolyte of conductivity 30 mS/m and relative permittivity 79, it is apparent from this figure that the necessary condition defined by Equation (10.33) is *not* met in the frequency range between ~2 MHz to ~100 MHz. Below

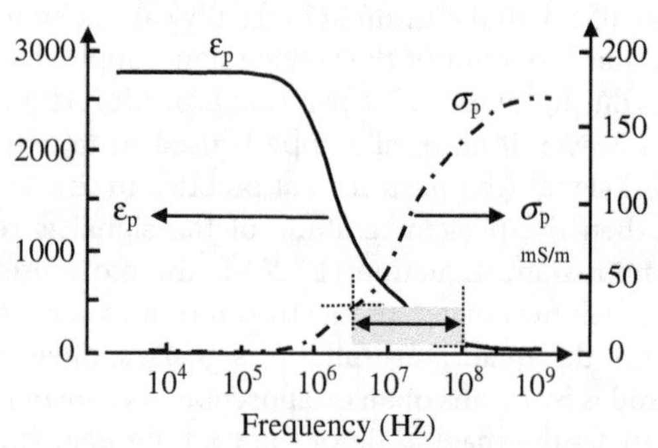

Figure 10.10 Effective values for the permittivity (ε_p) and conductivity (σ_p) of a lymphocyte as a function of frequency, derived from the dielectric data of Asami *et al.* [21].

2 MHz the factors $(\sigma_m - \sigma_p)$ and $(\varepsilon_p - \varepsilon_m)$ in Equation (10.33) are both positive, whilst above 100 MHz they are both negative, to give a positive value for their ratios. Between 2 MHz and 100 MHz, $(\varepsilon_p - \varepsilon_m)$ is positive but $(\sigma_m - \sigma_p)$ is negative, so that the solution for ω_{xo} involves the square root of a negative number. The crossover frequencies shown in Figure (10.9) do not lie within the frequency range yielding a nonreal value for ω_{xo}.

From the single-shell model of a cell described in Chapter 9, the effective complex permittivity of a cell is given by Equation (9.14) as:

$$\varepsilon^*_{cell} = \frac{(2\varepsilon^*_{mem} + \varepsilon^*_{cyt})R_2^3 + 2(\varepsilon^*_{cyt} - \varepsilon^*_{mem})R_1^3}{(2\varepsilon^*_{mem} + \varepsilon^*_{cyt})R_2^3 - (\varepsilon^*_{cyt} - \varepsilon^*_{mem})R_1^3}\varepsilon^*_{mem}$$

where the suffices *mem* and *cyt* represent the membrane and cytoplasm phases, R_2 is the cell radius and the membrane thickness d is given by $d = R_2 - R_1$. This equation can be rearranged to the form:

$$\varepsilon^*_{cell} = \varepsilon^*_{mem}\frac{\left(\frac{R_2}{R_1}\right)^3 + 2\left(\frac{(\varepsilon^*_{cyt} - \varepsilon^*_{mem})}{(\varepsilon^*_{cyt} + 2\varepsilon^*_{mem})}\right)}{\left(\frac{R_2}{R_1}\right)^3 - \left(\frac{(\varepsilon^*_{cyt} - \varepsilon^*_{mem})}{(\varepsilon^*_{cyt} + 2\varepsilon^*_{mem})}\right)} \tag{10.34}$$

For frequencies well below 1 MHz, the values for the various dielectric parameters of the cell are such that [22]:

$$\left(\frac{(\varepsilon^*_{cyt} - \varepsilon^*_{mem})}{(\varepsilon^*_{cyt} + 2\varepsilon^*_{mem})}\right) \cong 1$$

Redefining the cell radius to be R, with the assumption that this radius is far greater than the membrane thickness d, Equation (10.34) simplifies to:

$$\varepsilon^*_{cell} = \varepsilon^*_{mem}\left(\frac{\left(\frac{1}{1 - d/R}\right)^3 + 2}{\left(\frac{1}{1 - d/R}\right)^3 - 1}\right)$$

$$\cong \varepsilon^*_{mem}\left(\frac{R + d}{d}\right) \cong \varepsilon^*_{mem}\left(\frac{R}{d}\right) \tag{10.35}$$

It is evident from Figure 10.10 that for frequencies well below the region of the interfacial polarization, where the low conductivity of the membrane dominates the effective conductivity of the cell (σ_p), the effective permittivity (ε_p) of the cell greatly exceeds that of the suspending medium (ε_m). The value for σ_p is also much less than the conductivity of the medium (σ_m). In this case, to a good approximation, Equation (10.32) reduces to:

$$\omega^2_{xo1} \cong \frac{1}{\varepsilon_o^2}\frac{2\sigma_m^2}{\varepsilon_p^2}$$

The DEP crossover frequency f_{xo1} can thus be given as:

$$f_{xo1} = \frac{\omega_{xo1}}{2\pi} \approx \sqrt{2}\frac{\sigma_m}{2\pi\varepsilon_o\varepsilon_p} \tag{10.36}$$

For the special situation where the particle is a viable cell with an intact membrane, suspended in a medium of conductivity less than that of its cytoplasm, we can assume that $\sigma_p \ll \sigma_m$ and $\varepsilon_p \gg \varepsilon_m$. From the above relationship and Equations (10.30) the relaxation time τ can be approximated as:

$$\tau^2 \approx \frac{2}{\omega^2_{xo1}}$$

Substituting this relationship into Equation (10.29), together with the conditions $\sigma_p \ll \sigma_m$, $\varepsilon_p \gg \varepsilon_m$, we obtain as the approximate frequency dependence of the Clausius–Mossotti factor in the frequency range close to the DEP crossover frequency:

$$\text{Re}[CM(f)] \approx \frac{f^2 - f^2_{xo1}}{f^2 + 2f^2_{xo1}}$$

From Equation (10.27) we therefore have the following approximation for the time-averaged DEP force of a spherical, viable, cell:

$$\langle F_{DEP}\rangle \approx 2\pi\varepsilon_o\varepsilon_m R^3\left[\frac{f^2 - f^2_{xo1}}{f^2 + 2f^2_{xo1}}\right]\nabla E^2_{rms} \tag{10.37}$$

Given in this form the DEP force equation illustrates the significance of the crossover frequency f_{xo1}, as well as the practical interpretation that can be given of the Clausius–Mossotti factor. For a relatively narrow frequency range that encompasses f_{xo1}, if the operating frequency f is less than f_{xo1} the DEP force effect acting on a viable cell is negative. The cell will be repelled from a metal electrode. As the frequency f is increased beyond the value for f_{xo1}, the DEP force effect makes the transition from negative to positive DEP and the cell is attracted to the electrode. Equation (10.37) represents the key guideline as to how DEP-based cell separation protocols exploit the different values for f_{xo1} exhibited by different cell types in a cell mixture.

The specific membrane capacitance C_{mem} can be defined as $\text{Re}[\varepsilon^*_{mem}]/d$ (units of F m^{-2}). Noting that $\varepsilon^*_p \equiv \varepsilon^*_{cell}$, from Equation (10.35):

$$C_{mem} = \frac{\varepsilon_o\varepsilon_p}{R} \tag{10.38}$$

Substituting this relationship for $\varepsilon_o\varepsilon_p$ into Equation (10.36) we obtain the following expression for the crossover frequency f_{xo1}:

$$f_{xo1} \approx \frac{1}{\sqrt{2}}\frac{\sigma_m}{\pi R C_{mem}} \tag{10.39}$$

In a DEP experiment, the conductivity of the medium (σ_m) and the radius of a cell can be measured directly. A determination of the crossover frequency f_{xo1} therefore provides a method for determining the capacitance of the plasma membrane, provided that the high resistance value of the membrane has not been degraded due to structural damage or the onset of cell death, for example. Equation (10.38) can be compared with Equation (9.19) obtained from treating the membrane as a dielectric slab, of relative permittivity ε_r, sandwiched between two parallel plate electrodes distance d apart and each of area A:

$$C_{mem} = \frac{\varepsilon_o \varepsilon_r}{d} \text{ per unit area } A \qquad (10.40)$$

From Equations (10.38) and (10.40) we note that the relationship between the macroscopic, *effective*, permittivity (ε_p) value of the *whole* cell and the microscopic relative permittivity (ε_r) of the membrane is given by $(\varepsilon_p/\varepsilon_r) = (R/d)$. Equation (10.40) was derived in Chapter 9 by adopting the same geometrical scheme as that given in Figure 3.21 when applying Gauss's Law to calculate the field produced in a dielectric situated between two spherical conducting shells. The conducting shells are unable to sustain an electric field, so that the result given by Equation (10.40) assumes that *all* the applied electric potential is dropped across the membrane. The approximations used to derive Equation (10.36) also employ this assumption because the conductivity of the membrane is taken to be negligible. The voltage dropped across a cell membrane on application of an external AC electric field of magnitude E is given by the following equation [23]:

$$V_{mem}(\omega, \theta) = \frac{3}{2}\frac{RE\cos\theta}{(1 + i\omega\tau)} \qquad (10.41)$$

where R is the cell radius and θ is the polar angle with respect to the field direction as shown in Figure 2.7. V_{mem} is thus a complex quantity, with a real and imaginary component, the latter having either a positive or negative value (depending on whether the voltage leads or lags the applied field, respectively). The field E_{mem} across the membrane is given by $|V_{mem}|/d$, where d is the membrane thickness:

$$E_{mem}(\omega, \theta) = \frac{3}{2}\frac{(R/d)E\cos\theta}{(1 + \omega^2\tau^2)^{1/2}} \qquad (10.42)$$

This function is plotted in Figure (10.11) and superimposed on the frequency variation of Re$\{CM^*\}$ shown in Figure 10.9 for lymphocytes. The value for the interfacial relaxation time τ is calculated using Equation (6.49) and the dielectric data derived for lymphocytes as given in Table 9.3. At low frequencies ($\omega\tau < 1$) the field E_{mem} across the membrane can exceed the applied field by a factor of 10^3 or greater, depending on the size of the cell. As the frequency is increased and approaches the DEP crossover frequency, it can be seen from Figure 10.11 that the magnitude of E_{mem} begins to decrease, indicating that the total voltage drop across the cell is shared between the membrane and cell interior. The applied field begins to penetrate into the cell interior and the extent of this increases with increasing frequency. The conductivity of the cell membrane can no longer be neglected.

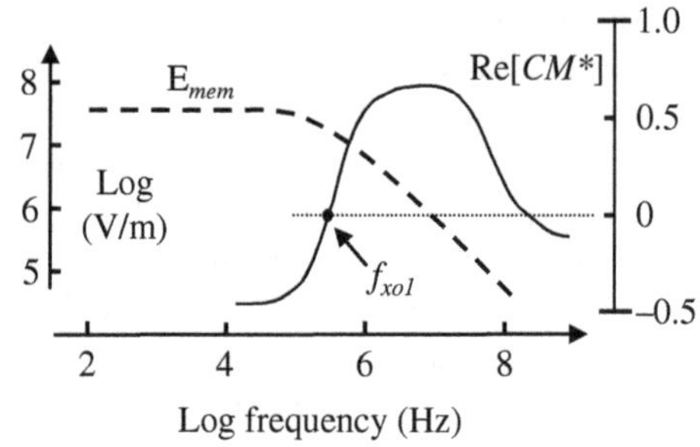

Figure 10.11 A plot of the transmembrane field E_{mem}, derived from Equation (10.41), is shown superimposed on the plot of Re[CM*] given in Figure 10.9 for lymphocytes. At the DEP crossover frequency f_{xo1} there is a significant penetration of the external field across the membrane and into the cell interior.

An improvement of the approximate nature of equation (10.37) can be made by including the membrane's specific conductance G_{mem} [24]:

$$f_{xo1} = \frac{1}{\sqrt{2}}\frac{\sigma_m}{\pi RC_m}\sqrt{1 - \frac{RG_{mem}}{2\sigma_m} - 2\left(\frac{RG_{mem}}{2\sigma_m}\right)^2}$$

For values of G_{mem} less than around 600 S/m^2, this relationship can be simplified to the form [25]:

$$f_{xo1} = \frac{\sqrt{2}}{2\pi RC_{mem}}\left(\sigma_m - \frac{G_{mem}}{4}\right) \qquad (10.43)$$

The value of the G_{mem} should include the surface conductance (K_{ms}) of the cell membrane [25]. This contribution to the total effective conductance of the membrane is described in section 10.3.6 of this chapter. Values for the membrane capacitance for a particular cell type can be obtained to a reasonable level of accuracy (but less so for the membrane conductance) by determining their average radii (R) and f_{xo1} value at different values of the medium conductivity (σ_m). From Equation (10.43) we expect a plot of (f_{xo1} R) against σ_m to produce a straight line ($y = mx + c$, see ahead to Figure 10.26b) of slope m given by:

$$m = \frac{\sqrt{2}}{2\pi C_{mem}}$$

and an intercept c given by:

$$c = -\frac{\sqrt{2}}{8\pi C_{mem}}RG_{mem}$$

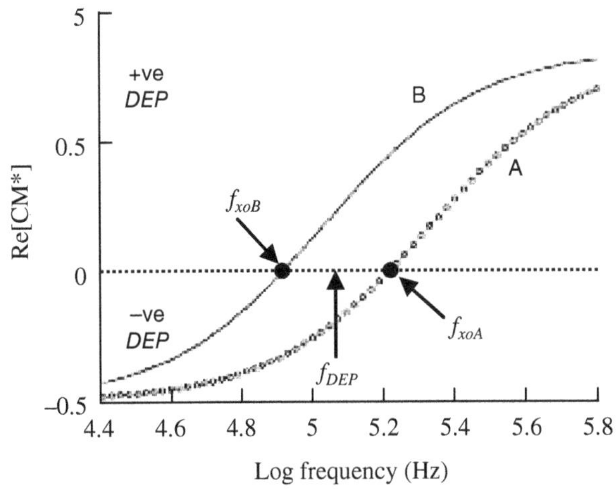

Figure 10.12 Plots of Re[CM^*] for two cells of the same radius, suspended in a medium of conductivity 40 mS/m. Cells A and B have membrane capacitances of 8 mF/m^2 and 16 mF/m^2, respectively. As expected from Equation (10.39) cell A with the smaller C_{mem} value exhibits a higher DEP crossover frequency f_{xoA} than cell B. On applying a field of frequency f_{DEP} cell B should experience a positive DEP force, which attracts it to an electrode, whilst cell A should be pushed away from the electrodes and into the surrounding medium by a negative DEP force.

As shown in Table 9.2, if nominal values of ~5 and ~7 nm are adopted in Equation (10.40) for ε_r and d, respectively, a specific membrane capacitance (C_{mem}/A) of ~6 mF/m^2 is obtained for an ideal smooth cell membrane. Values in the range 6~8 mF/m^2 are observed for erythrocytes, which have relatively smooth membranes, but values significantly larger than this are observed for other cell types. The only parameter in Equation (10.40) capable of sufficient adjustment to produce large values for C_{mem} is the value we take for the *effective* area of the membrane surface. As discussed in Chapter 9, this can be quantified by introducing a membrane topography parameter ϕ. This parameter represents the ratio of the actual membrane area of the cell to that which would form a perfectly smooth covering of the cytoplasm. A perfectly smooth membrane corresponds to ϕ having a value of 1.0. A cell whose membrane becomes 'roughened', as a result of the appearance of membrane folds, blebs or microvilli, for example, should therefore exhibit an increased membrane capacitance in direct proportion to the increase of ϕ. With all other factors remaining constant (e.g., cell radius, medium conductivity) then from Equation (10.38) this should result in a lowering of the DEP crossover frequency. This effect is shown schematically in Figure 10.12 for two cells of the same radius, suspended in the same medium conductivity, but with one cell having a surface roughness that effectively doubles the membrane capacitance (from 8 mF/m^2 to 16 mF/m^2). This example demonstrates an important application of dielectrophoresis. By setting the frequency of the applied nonuniform field to a value between the DEP crossover frequencies exhibited by the two cells, they experience opposite DEP forces. The cell exhibiting the lower f_{xo1} value experiences a positive DEP force that attracts it towards the nearest electrode, possibly trapping it there. The other cell experiences a negative DEP force that repels it from the electrodes and into the bulk of the fluid medium. The two cell types are thus physically separated from one another and completely so if the fluid flows over the electrodes and is collected downstream.

For Figure 10.12 it was specified that the radii and the suspending medium for the two cells were the same. The reason for this is clear from Equation (10.39), where the value for f_{xo1} is shown to be directly proportional to the medium conductivity and inversely proportional to the cell radius. The value for f_{xo1} is related to the frequency dependence of Re[CM^*] and so depends on the relaxation time given by Equation (10.30). This equation does not contain the cell radius as an implicit factor. So why should f_{xo1} for an isolated cell be a function of its radius? For a particle that consists of a homogeneous material then, in theory, the relaxation time τ for interfacial polarization should not vary as a function of particle size. However, no particle is strictly homogeneous because its surface is likely to have physico-chemical properties that differ from its bulk. The single-shell model of a cell described in Chapter 9 can be modified to accommodate the general case of a particle of radius R, having a surface layer of thickness d with dielectric properties that differ from its bulk. In this case the relationships given in Equation (9.17) can be modified as follows:

$$\varepsilon'_p = \varepsilon'_{bulk} + \frac{2d}{R}\varepsilon'_{surface} \text{ and } \sigma'_p = \sigma'_{bulk} + \frac{2d}{R}\sigma'_{surface}$$

When these relationships are inserted into Equation (10.30) it is evident that for some situations the relaxation time for interfacial polarization can be sensitive to particle size. From Equations (9.16) we also have the following general relationship:

$$\varepsilon^*_p = \left(\varepsilon'_{bulk} + \frac{2d}{R}\varepsilon'_{surface}\right) - i\frac{1}{\varepsilon_o\omega}\left(\sigma'_{bulk} + \frac{2d}{R}\sigma'_{surface}\right)$$

The dependence of the DEP force on particle size as a function of frequency is obtained by inserting this expression for the complex permittivity into Equation (10. 27). For the particular case of a cell enclosed by a membrane, then from Equation (10.30) the time constant can be given as [26,27]:

$$\tau = RC_{mem}\left(\frac{\sigma_{cyt} + 2\sigma_m}{2\sigma_{cyt}\sigma_m + RG_{mem}(\sigma_{cyt} + 2\sigma_m)}\right) \quad (10.44)$$

This relationship is derived by employing the limiting low-frequency (DC) approximations:

$$\sigma_p = \sigma_{pDC} \equiv \sigma_{mem}\frac{R}{d} = RG_{mem};\ \varepsilon_p = \varepsilon_{pDC}$$

$$\equiv \varepsilon_{mem}\frac{R}{d} = RC_{mem}$$

Adopting these DC approximations can lead to significant errors when analysing DEP and electrorotation data, especially for frequencies and medium conductivities above 100 kHz and 1 mS/m, respectively. A refinement of the theories has been described by Lei *et al.* [28] to enable more accurate extraction of the dielectric parameters of cells from such data.

The influence of cell radius on the values for f_{xo1} and f_{xo2} is shown in Figure 10.13. This modelling was achieved using the modifications, described in Box 10.5, to the MATLAB program given in Box 9.3. An increase of the cell radius results in a shift of f_{xo1} to a lower frequency, but the DEP crossover frequency (f_{xo2}) at ~300 MHz

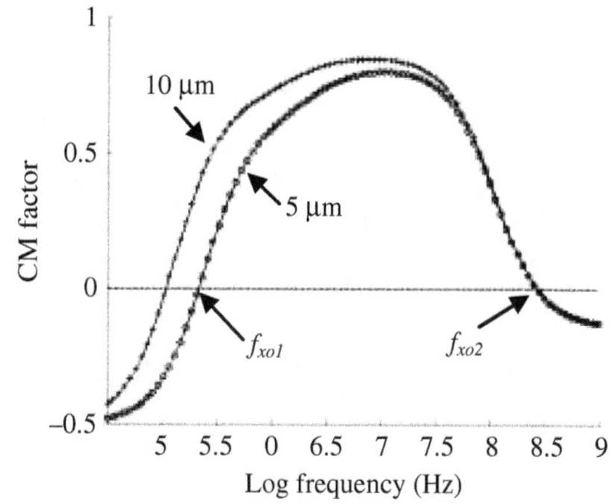

Figure 10.13 Re[CM^*] modelled for a nucleated cell as a function of cell radius (with all other factors fixed). The value for f_{xo1} decreases as the radius increases but f_{xo2} remains constant.

Box 10.5 MATLAB Program to Model the DEP Characteristics of a Cell

An extension is given from line 43 of the program given in Box 9.3 to include the conductivity and permittivity of the suspending medium. This enables Re[CM^*] and Im[CM^*] to be modelled for a cell containing a nucleus. Dielectric values for the various components of the cell are guided by the data given in Table 9.3.

```
42
43  % Suspending medium conductivity kc5
    and permittivity kp5
44  kc5=40e-3;
45  kp5=79*p0;
46
47  k1=kp1-i*kc1 ./w;
48  k3=kp3-i*kc3 ./w;
49  k2=kp2-i*kc2 ./w;
50  k4=kp4-i*kc4 ./w;
51  k5=kp5-i*kc5 ./w;
52
53  am1=a1^3;
54  am2=a2^3;
55  am3=a3^3;
56  am4=a4^3;
57
58  keff2=k2 .*(am2*(k1+2*k2)-2*am1*
    (k2-k1)) ./(am2*(k1+2*k2)+...
    am1*(k2-k1));
59
60  keff3=k3 .*(am3*(keff2+2*k3)-2*am2*
    (k3-keff2)) ./(am3*(keff2+2*k3)
    +am2*(k3-keff2));
61
62  keff4=k4 .*(am4*(keff3+2*k4)-2*am3*
    (k4-keff3)) ./(am4*(keff3+2*k4)+
    am3*(k4-keff3));
63
64  m=(keff4-k5) ./(3*(k5+D*(keff4-k5)));
65
66  rm=real(m);
67  im=imag(m);
68
69  plot(log10(f), rm,'o', log10(f),
    zeroline,'-');
70  % plot(log10(f), im,'+', log10(f),
    zeroline,'-');
71
72   hold on
73
74  xlabel('Log Frequency (Hz)')
75  ylabel('CM Factor')
76
77 hold on
```

remains unchanged. This reinforces our view that the DEP force acting on a cell at the lower frequencies is strongly influenced by the presence and integrity of the plasma membrane. At high frequencies the membrane capacitance electrically shorts out the large resistance presented by the membrane (as shown in Figure 10.5), so that the cell appears electrically more like a homogeneous sphere with a single DEP crossover frequency whose value is governed by permittivity parameters rather than conductive properties. This same effect can be demonstrated, as shown in Figure 10.14, by modelling Re[CM^*] for steadily increasing values of the plasma membrane conductance. A cell with a membrane that has lost its resistance to passive ion flow exhibits positive DEP in the frequency range where viable cells usually exhibit negative DEP. This demonstrates an important feature of DEP, namely its ability to not only distinguish between viable and dead cells, but also to provide a simple method for separating them. Figure 10.14 in fact presents a simplified scenario for what might be observed during a DEP experiment as a cell progresses from the viable to apoptotic state, for example. The size of the cell will probably change and this will alter the finer details of the DEP response near f_{xo1}. If, as is usually the case, the cell is suspended in an aqueous medium of relatively low conductivity, ions will also leak from the cytoplasm. As we will see in section 10.5, this will lead to a lowering of the value for f_{xo2}. However, the final result, that the dead cell will exhibit positive DEP instead of negative DEP will be the same.

Finally, it should be noted that the analyses of the DEP force given here are not restricted to a simple sinusoidal

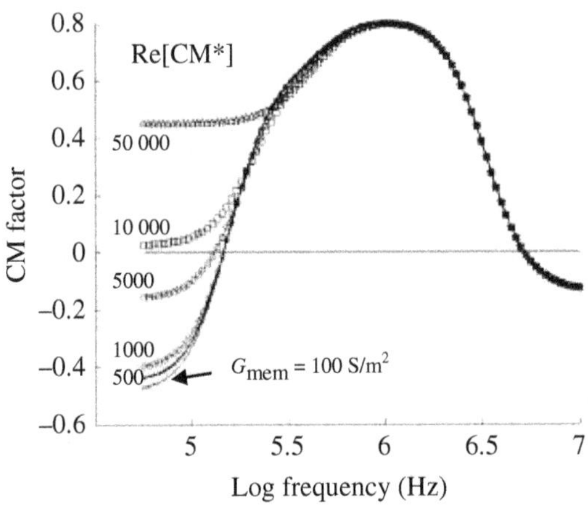

Figure 10.14 Plots of Re[CM^*] are shown as the plasma membrane conductance G_{mem} of a nucleated cell increases. This can mimic the effect of degradation of the membrane as a result of cell damage and death. With an applied field set at a frequency below the usual value of f_{xo1} observed for a viable cell, the nonviable cells will be attracted to the electrodes by positive DEP, whilst the viable cells will be forced into the bulk fluid by negative DEP.

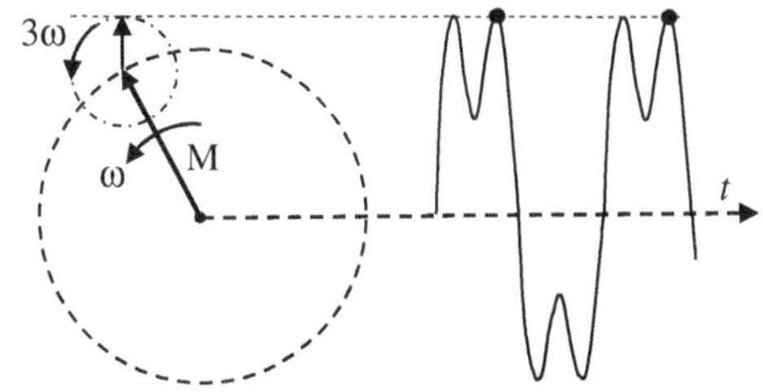

Figure 10.15 An application of Equation (10.34) with two terms of the Fourier series. The waveform is created by adding a second phasor, of angular frequency 3ω and magnitude M/3, to the primary phasor of frequency ω and magnitude M. A good approximation to a square wave is produced by adding phasors of frequencies 5ω, 7ω, 9ω and magnitudes M/5, M/7, M/9, respectively.

field. DEP experiments can be performed using voltage signals in the form of square waves or multiple superimposed frequencies, for example. In such cases the applied field can be expressed as a summation (a Fourier series [29]) of the form:

$$E(t) = \sum_{n=1}^{N} E_{pk} e^{i\omega nt} \text{ (where integer } N \geq 1) \quad (10.45)$$

An example of the summation of two phasors is shown schematically in Figure 10.15. By adding the appropriate number of harmonically related frequencies and appropriate amplitudes to the primary phasor it is possible to generate a wide range of periodic waveforms (e.g., square, saw tooth). The periodic time T to complete one cycle of such waveforms is given by $T = 2\pi/\omega$. From the *Principle of Superposition of Forces* described in Chapter 3 the total DEP force acting on a particle is given by the summation of the forces produced by each component of the harmonic fields that make up the total electric field waveform applied to the particle. The mathematics described here to calculate the time-averaged DEP force for the case of a single applied sinusoidal field can therefore be applied to more complicated waveforms.

10.3.3 Assumption of a Quasi-Static, Stationary AC Field

We have assumed that the applied AC field is of the form:

$$E(t) = \text{Re}[E_{pk} e^{i\omega t}] \quad (10.46)$$

The magnitude of the sinusoidal waveform is thus taken to be a function of time, but to be independent of the location between the electrodes. A relationship of the form:

$$E(t, x) = \text{Re}[E_{pk}(x) e^{i\omega t}]$$

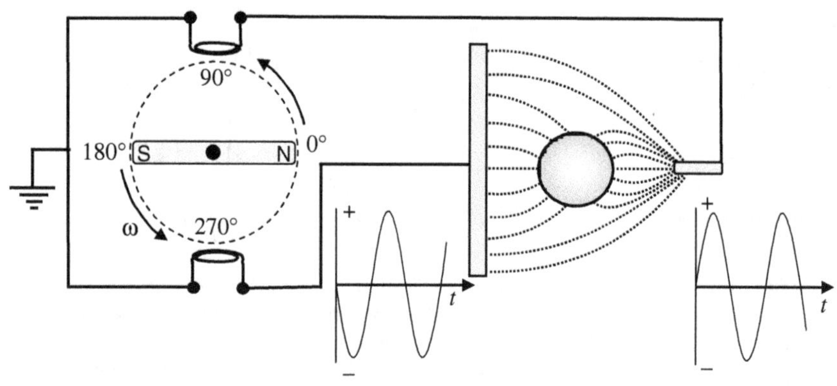

Figure 10.16 The peak voltage applied to the DEP electrodes shown in Figure 10.8 is doubled by applying a voltage $V_{pk}\sin(\omega t)$ to the pin electrode and $-V_{pk}\sin(\omega t)$ to the plate electrode. There is no phase difference between these two applied AC signals. By doubling the peak voltage the DEP force acting on the particle is quadrupled.

indicates that the waveform attains its peak magnitude at different times t according to its location x. This is equivalent to the sinewave shown in Figure 10.7 exhibiting a phase shift as a function of location – in other words the field exhibits a phase gradient. It is also implicit in Equation (10.46) that the field vector is real and not a complex vector having an imaginary component at right angles to the assumed direction (real component) of the field. To what extent can we ignore complications arising from a field phase gradient and an imaginary component of the field? Before answering this we will first review the nature of an AC electric field.

An AC electric field can be launched into free space by generating a regularly repeating electrical breakdown of the air gap between two point electrodes, for example. The electric field generated close to the electrodes is termed the *near* field and for our purposes we will assume that it propagates into free space as a sinewave. After one or two complete oscillations (wavelengths) of this near field an orthogonal magnetic field is generated. Likewise, a magnetic field launched into space from an induction coil, for example, induces the appearance of an electric field. These two fields travel together as an electromagnetic (EM) wave at the speed of light. An EM wave has a wavelength λ given by the relationship $\lambda = c/f$, where c is the speed of light ($\sim 3 \times 10^8$ m/s) and f is the frequency. Thus, even at a frequency as high as 1 GHz the wavelength (~30 cm) of the E-field is much larger than the distance between the electrodes used in dielectrophoresis. At 1 MHz the wavelength is more than a million times larger than the characteristic interelectrode spacing (typically ~200 μm or less) of a DEP electrode array.

In a DEP experiment the electric field, at any instant in time, can therefore be considered as the *near* field with no accompanying magnetic field component. The field can also be considered to be *quasistatic*, because at (say) 1 MHz the spatial scale (~200 μm) divided by the temporal scale (10^{-6} s) equates to ~200 m/s and is insignificant compared to the velocity ($\sim 3 \times 10^8$ m/s) at which the field is transmitted in free space. For a 'static' or 'stationary' wave the phase of the applied voltage remains constant in time and space with respect to a stationary reference waveform. It is therefore safe to assume that the nonuniform E-field generated in the schematic of Figure 10.8 can be considered to be quasistatic with no phase gradient. Figure 10.16 illustrates how the peak (and rms) voltage signal to the electrodes can be doubled by using two voltage generators. Because the DEP force is proportional to E^2, this provides a simple way to quadruple the DEP force when using signal generators of limited output magnitude. In practice, of course, modern voltage signal generators employ digital electronics and not rotating bar magnets (although this was the method used to generate AC voltages in the early studies of the dielectric properties of cells, as for example those performed by Rudolf Höber in 1910, as described in Chapter 9). The rotating magnet scheme has been used simply to illustrate the generation of a sinusoidal voltage signal and the concept of a phasor.

To what extent can we assume that the field vector is real, without an imaginary component? For the majority of reported DEP experiments, the electrodes used to generate the electric field make contact with the fluid medium in which the test particles are suspended. The possibility that electrode polarization effects might influence the observed DEP responses must therefore be considered. As described by Schwan [30] electrode polarization can be represented as an equivalent circuit composed of a capacitance in series with the resistance of the bulk solution. The capacitance is associated with the electrical double layer at the electrode-solution interface. At low frequencies the reactance of the double layer ($1/\omega C$) can greatly exceed the resistance of the bulk fluid and so the applied voltage difference between the electrodes acts predominantly across the electrical double layer. Even though the voltage signal applied to the electrodes might remain constant, the field actually experienced by the test particles can thus vary significantly as a function of frequency in the range where electrode polarization occurs. The extent of this can be determined and correction made to the magnitude of the DEP

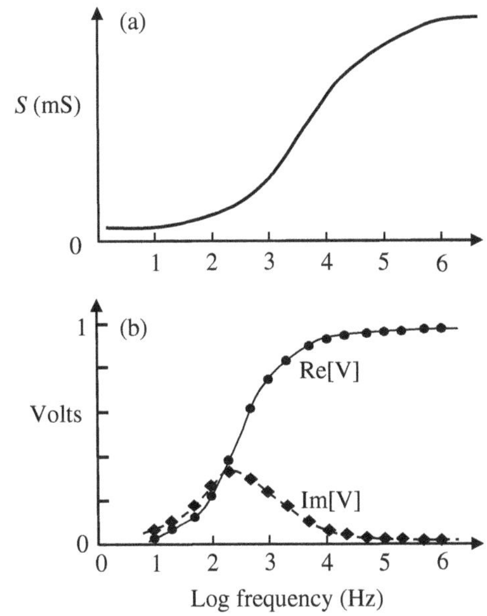

Figure 10.17 (a) The variation in apparent conductance *S* of a DEP cell suspension medium as a function of frequency (based on Burt *et al.* [31]). (b) The real and imaginary components of the normalized voltage drop across a solution above microelectrodes, as a function of frequency (based on González *et al.* [32]).

effect, by measuring the apparent conductance of the suspension medium in the DEP chamber as a function of frequency. An example of such a measurement is shown in Figure 10.17(a), based on the procedure described by Burt *et al.* [28] to correct for the effect of electrode polarization. It can be seen that electrode polarization can extend up to a frequency as high as 100 kHz and significant correction for it may be required below 10 kHz. However, relying solely on data such as that shown in Figure 10.17(a) to correct for the magnitude of the field in the medium is not sufficient. It does not take into account the fact that the effective capacitance of the double layer introduces a phase difference between the resultant current in the medium and the applied field. This is equivalent to the voltage drop across the medium and hence the field, having a real and imaginary component. This was analysed for the first time for practical DEP electrodes by González *et al.* [32]. An example of their analysis is shown in Figure 10.17(b). In the frequency range where electrode polarization occurs and particularly below ~10 kHz, it cannot be assumed that the field E in Equation 10.27, for example, is a purely real quantity – it will be complex and have an imaginary component. As shown in Box 10.6 this leads to the introduction of an imaginary component of the DEP force. As we will see in the next section, the imaginary component of the field exerts a rotational torque on a test particle.

10.3.4 Rotating Fields and Electrorotation

Consider the scheme shown in Figure 10.18 where four symmetrically arranged electrodes are energized by a sequence of sinusoidal voltages of equal magnitude and phased 90° ($\pi/2$ radians) apart. A practical electronic circuit that can be 'home built' for producing quadrature-phased signals is shown in Figure 10.19. Alternatively, commercial generators are available that provide quadrature phased waveforms of different types.

The orthogonal field vectors E_x and E_y, shown in Figure 10.18, are given mathematically as:

$$E_x = E_{pk}\sin(\omega t)\hat{x};$$
$$E_y = E_{pk}\sin(\omega t + \pi/2)\hat{y} = E_{pk}\cos(\omega t)\hat{y}$$

Figure 10.18 Quadrature-phased AC voltages of equal magnitudes are shown applied to four electrodes.

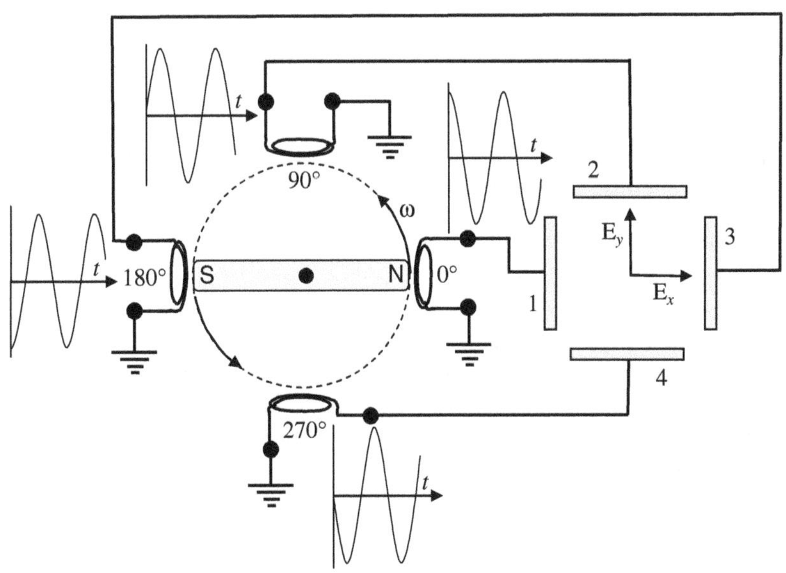

Box 10.6 The DEP Force arising from a Gradient of Field Phase

From Equation (3.10) and Box 10.4 an electric field can be expressed in Cartesian co-ordinates, in terms of its magnitude, frequency ω and phase ϕ, as:

$$\begin{aligned} E &= E_x(t)\hat{i} + E_y(t)\hat{j} + E_z(t)\hat{k} \\ &= \sum_{n=x,y,z} E_{n0}(x,y,z)\cos(\omega t + \phi_n(x,y,z))\hat{n}_n \end{aligned}$$

where $\hat{n}$ is a unit vector. From Equation (10.1) the DEP force is then given by:

$$\begin{aligned} F_{DEP}(x,y,x,t) &= F_{DEPx}(t)\hat{i} + F_{DEPy}(t)\hat{j} + F_{DEPz}(t)\hat{k} \\ &= \sum_{n=x,y,z}\left(p_x(t)\frac{\partial E_x(t)}{\partial n} + p_y(t)\frac{\partial E_y(t)}{\partial n} + p_z(t)\frac{\partial E_z(t)}{\partial n}\right) \end{aligned} \quad (10.47)$$

Referring to Box 10.4 this equation can be expanded in terms of its real and imaginary components. For example, for the first term of the summation in Equation (10.47):

$$\begin{aligned} p_x(t)\frac{\partial E_x(t)}{\partial n} &= 4\pi\varepsilon_o\varepsilon_m R^3\,(\mathrm{Re}[CM]\cos(\omega t + \phi_x) \\ &\quad -\mathrm{Im}[CM]\sin(\omega t + \phi_x)) \\ &\quad \times\left(\frac{\partial E_{x0}}{\partial n}\cos(\omega t + \phi_x) - \frac{\partial \phi_x}{\partial n}\sin(\omega t + \phi_x)\right) \end{aligned}$$

The time-averaged DEP force for this component is given by:

$$\begin{aligned} &\left\langle p_x(t)\frac{\partial E_x(t)}{\partial n}\right\rangle \\ &= 2\pi\varepsilon_o\varepsilon_m R^3\left(\mathrm{Re}[CM]E_{x0}\frac{\partial E_{x0}}{\partial n} + \mathrm{Im}[CM]E_{x0}^2\frac{\partial \phi_x}{\partial n}\right) \end{aligned}$$

The second and third terms of the summation in Equation (10.47) have similar time averages, so that the time-average DEP force for the situation of a complex nonuniform field is given by:

$$\begin{aligned} &\langle F_{DEP}(t)\rangle \\ &= 2\pi\varepsilon_o\varepsilon_m R^3\left(\mathrm{Re}[CM]\nabla E_{rms}^2 + \mathrm{Im}[CM]\sum E^2\nabla\phi\right) \end{aligned} \quad (10.48)$$

where

$$\sum E^2\nabla\phi = (E_{x0}^2\nabla\phi_x + E_{y0}^2\nabla\phi_y + E_{z0}^2\nabla\phi_z)$$

In section 10.3.5 we find that the term involving Im[CM] in Equation (10.48) introduces a rotational torque to the total DEP force.

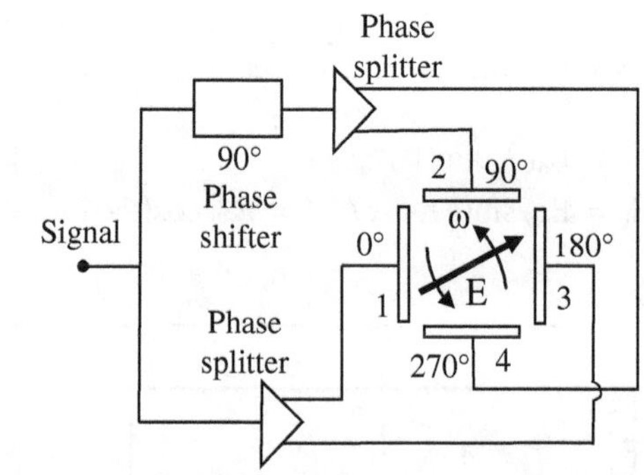

Figure 10.19 Schematic of an electronic circuit, which can be used to produce the quadrature phased sinusoidal signals shown in Figure 10.18. A rotating electric field is created between the electrodes.

where $\hat{x}$ and $\hat{y}$ are unit vectors along the x- and y-axes, respectively and ω is the radian frequency of the applied voltage signals. The amplitude of the field E produced between the electrodes is in principle given by the vector sum of these two vectors:

$$E = E_{pk}\sin(\omega t)\hat{x} + E_{pk}\cos(\omega t)\hat{y} \quad (10.49)$$

so that

$$E^2 = E_x^2 + E_y^2 = E_{pk}^2(\sin^2\omega t + \cos^2\omega t) = E_{pk}^2$$

to give $E = E_{pk}$. The angle θ shown in Figure 10.20 is given by:

$$\tan\theta = \frac{E_{pk}\sin\omega t}{E_{pk}\cos\omega t} = \tan\omega t$$

so that $\theta = \omega t$. The resultant field E is thus constant in magnitude, but rotates with an angular frequency ω. A rotating field can be generated using the four electrodes with quadrature phases, as shown in Figure 10.19, or

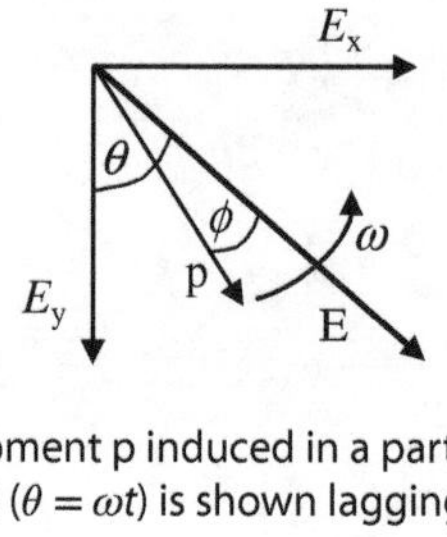

Figure 10.20 The moment p induced in a particle exposed to a rotating electric field ($\theta = \omega t$) is shown lagging the field by ϕ degrees.

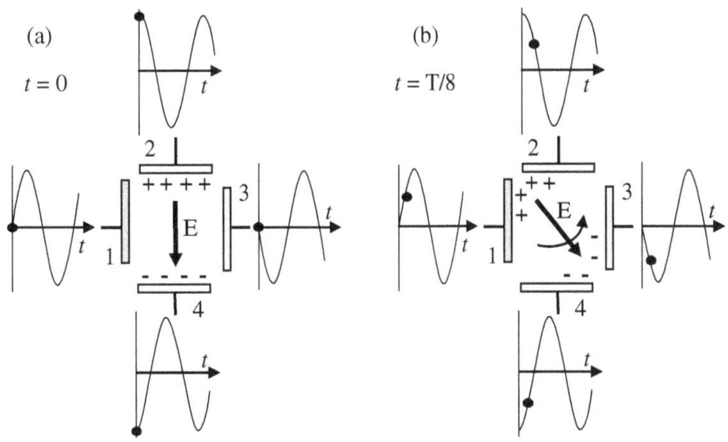

Figure 10.21 The charges on the electrified electrodes of a quadrupole phase system are shown at the initiation and completion of one-eighth of a phasor cycle. (a) At time t = 0 the field vector generated is directed from electrode 2 (90°) to 4 (270°). (b) At $t = T/8$, corresponding to a phase advance of $\pi/4$ radians (45°) the field vector rotates in the anticlockwise direction (i.e., against the clockwise sense of increasing phase angle applied to the electrodes). After a further time lapse of T/8, the field vector will be directed from electrode 1 (0°) to 3 (180°).

using n electrodes with sinusoidal signals that are phase shifted from each other by $2\pi/n$ radians. The direction of rotation of the field vector shown in Figure 10.19 is against the sense of the increasing phase angle applied to the electrodes (i.e., anticlockwise for a clockwise application of increasing voltage phase). This may appear to be counterintuitive but is explained in Figure 10.21.

For regions away fringe fields in the gaps between adjacent electrodes shown in Figures 10.19 and 10.21, the rotating field E is uniform. A uniform field does not induce a DEP force, but a particle in a rotating field experiences a rotational torque. In Figure 10.20 the induced moment p is shown lagging the rotating field E by ϕ degrees, but it can also lead the field. From Chapter 2 (see Figure 2.8) we know that an out-of-phase dipole moment experiences a torque that aligns it with a stationary field. In a rotating field the moment will experience a torque causing it to rotate constantly in the direction that should align it with the field. If the moment lags the field it will continually attempt to 'catch up' with the field under the action of a positive rotational torque. The polarized particle will rotate in the same direction as the field. If the moment leads the field it will continually try to 'backtrack', so that under the action of a negative rotational torque the particle rotates in the opposite sense to the field. These two situations are depicted in Figure 10.22. We see that a positive rotational torque is created when the induced charges on the particle are of opposite sign to that of the adjacent electrified electrode(s). This situation arises when the rate of electrical response of the particle to a change of field direction is less rapid than that of its surrounding medium. This is equivalent to the effective charge density relaxation time ($\varepsilon_o\varepsilon_p/\sigma_p$) of the particle being longer than the charge density relaxation time ($\varepsilon_o\varepsilon_m/\sigma_m$) of the medium. When the relaxation time of the particle is less than that of the medium, the induced charges on the particle are of the same sign as that of the adjacent electrode(s). This leads to a negative torque that causes the particle to rotate in the opposite sense to that of the field.

As shown in Figure 10.22, the particle rotates about an axis that passes through its centre of symmetry and is normal to the x-y plane in which the field and induced moment rotate. The torque T acting about this central point is given by the product of the moment p (a vector) and the vector component of the field E that acts

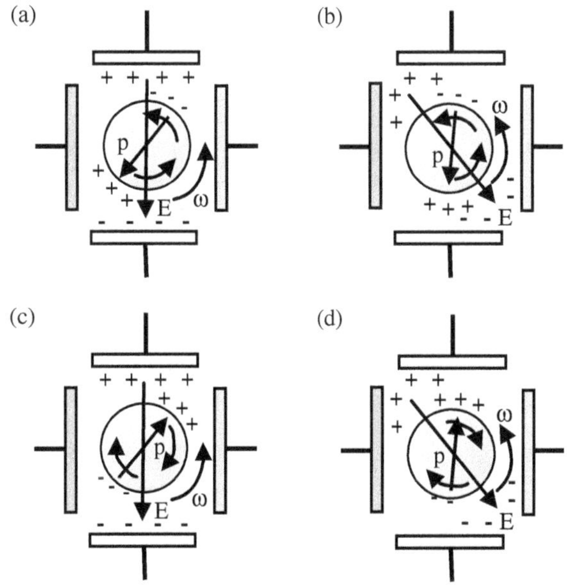

Figure 10.22 (a) The distributions of charges are shown for electrodes and a particle whose induced moment p *lags* a rotating field vector E. The moment p and the charges rotate synchronously with the field E. (b) As the field rotates the attractive forces between the charge distributions leads to the particle rotating in the same sense as the field. (c) The electrode and particle charge distributions are shown where the induced moment p *leads* the rotating field. (d) The repulsion between the charge distributions leads to the particle rotating in the opposite sense to that of the field.

Box 10.7 Electrorotational Torque is related to the Imaginary Part of [p × E]

From Equation (10.50) the torque T is given by a vector (cross) product of the form:

$$T = p \times E$$

Vectors p and E have components $[p_x, p_y, p_z]$ and $[E_x, E_y, E_z]$, respectively, with T also being a vector $[T_x, T_y, T_z]$. The magnitude (length) of T is given by:

$$|T| = |p||E| \sin \phi$$

where ϕ is the angle between p and E and is equivalent to the area of the shaded parallelogram shown below. The direction of T is perpendicular to both p and E, such that p, E and T (in this order) form a right-handed triple of vectors as shown below. In the right-handed Cartesian coordinate system (see below) the components of T are given by:

$$T_x = p_yE_z - p_zE_y;\ T_y = p_zE_x - p_xE_z;\ T_z = p_xE_y - p_yE_x$$

For p and E both located in the x-y plane, then E_z and p_z are zero and $T_x = T_y = 0$. In phasor notation a right, circularly polarized, field vector is given by:

$$E = E(\hat{x} - i\hat{y})e^{i\omega t}$$

where $\hat{x}$ and $\hat{y}$ are unit vectors $(\hat{i}x, \hat{j}y)$. For a particle of volume v and *isotropic* complex polarizability α^*, so that $p_x = p_y = v\alpha^* Ee^{i(\omega t+\phi)}$, we obtain the result:

$$T_z = -i(2v\alpha^* E^2 e^{i\phi})$$

The real components of T_z have cancelled, to give a purely imaginary value. From the formal definition of a vector product [18] the time-averaged torque is given by:

$$\langle T(t)\rangle = \frac{1}{2}[p \times \bar{E}] = -v\text{Im}[\alpha^*]E^2\hat{z} \quad (10.51)$$

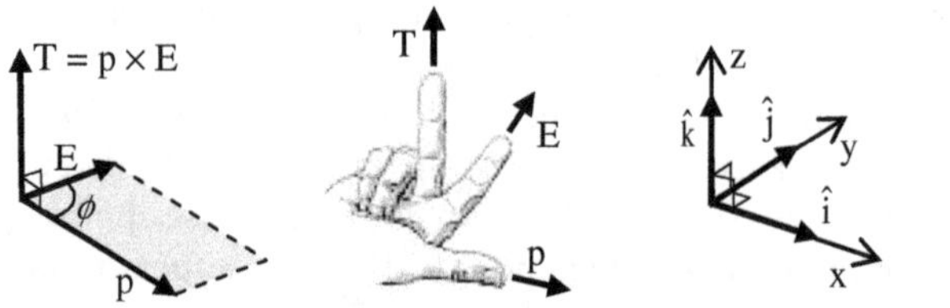

(a) Vector product. (b) Right-handed triple of vectors and coordinates.

orthogonally to and in the same plane as the induced moment. From Figure 2.8 this product given by:

$$T = pE\sin\phi$$

where ϕ is the angle between the vectors p and E. This in fact is the result of the so-called vector (cross) product of p and E expressed by:

$$T = p \times E = pE \sin \phi \quad (10.50)$$

As given by Equation (10.51) in Box 10.7 the time-averaged torque acting on a particle is:

$$\langle T(t)\rangle = -v\text{Im}[\alpha^*]E^2\hat{z}$$

where $\hat{z}$ is the unit vector $(\hat{k}z)$ along the z-axis in a right-handed Cartesian coordinate system (see Box 10.7). This torque causes the particle to rotate about the z-axis that passes through the centre of the particle and is directed at right angles to the xy-plane, which contains the rotating field and the induced dipole moment. For a spherical particle of radius R, $v = (4\pi R^3)/3$ and from Box 6.1:

$$\alpha^* = 3\varepsilon_o\varepsilon_m\left(\frac{\varepsilon_p^* - \varepsilon_m^*}{\varepsilon_p^* + 2\varepsilon_m^*}\right)$$

The function inside the brackets is the complex form of the CM-factor. The time-averaged torque exerted on a spherical particle of radius R, suspended in a medium of absolute permittivity $\varepsilon_o\varepsilon_m$ and subjected to a uniform rotating field E in the xy-plane is:

$$\langle T(t)\rangle = -4\pi\varepsilon_o\varepsilon_m R^3\text{Im}[CM^*]E^2\hat{z} \quad (10.52)$$

Thus, unlike the DEP force, which depends on the *real* component of the Clausius–Mossotti factor, the electrorotational torque depends on the *imaginary* component. The minus sign in Equation (10.52) indicates that the direction of rotation of the particle is opposite to the direction of rotation of the field when the imaginary component of the CM factor has a positive value. This is termed *contrafield* electrorotation. *Cofield* rotation occurs when the imaginary component of the CM factor has a *negative* value.

The frequency dependence of $\text{Im}[CM^*]$ is given in Box 6.8 by the relationship:

$$\text{Im}[CM^*] = \frac{3\omega(\varepsilon_p\sigma_m - \varepsilon_m\sigma_p)}{(\sigma_p + 2\sigma_m)^2 + \omega^2(\varepsilon_p + 2\varepsilon_m)^2} \quad (10.53)$$

$\text{Im}[CM^*]$ tends to a value of zero for very low and very high frequencies and has a peak value at the radian frequency where $\omega = 1/\tau$. Substituting this frequency into Equation (10.53), with τ given by Equation (10.30), the peak value for $\text{Im}[CM^*]$ is:

$$\text{Im}[CM^*]_{pk} = \frac{3}{2}\frac{\varepsilon_p\sigma_m - \varepsilon_m\sigma_p}{(\sigma_p + 2\sigma_m)(\varepsilon_p + 2\varepsilon_m)} \tag{10.54}$$

For the situation $\varepsilon_p \gg \varepsilon_m$ and with $\sigma_m \gg \sigma_p$, $\text{Im}[CM^*]$ has a peak value of 0.75. For $\varepsilon_m \gg \varepsilon_p$, with $\sigma_p \gg \sigma_m$, $\text{Im}[CM^*]_{pk} = -0.75$. From Equation (10.52) the torque exerted on a spherical particle is given by:

$$\langle \text{T}(t)\rangle = -12\pi\varepsilon_o\varepsilon_m R^3 \times \left[\frac{\omega(\varepsilon_p\sigma_m - \varepsilon_m\sigma_p)}{(\sigma_p + 2\sigma_m)^2 + \omega^2(\varepsilon_p + 2\varepsilon_m)^2}\right]\text{E}^2 \tag{10.55}$$

From this equation we find that contrafield rotation occurs for $\varepsilon_p\sigma_m > \varepsilon_m\sigma_p$. This condition for contrafield rotation is equivalent to having $(\varepsilon_o\varepsilon_p/\sigma_p) > (\varepsilon_o\varepsilon_m/\sigma_m)$ – in other words the charge density relaxation time of the particle exceeds that for the suspending medium. This agrees with the conclusion drawn from inspection of Figure 10.22. Cofield rotation occurs for the condition $\varepsilon_p\sigma_m < \varepsilon_m\sigma_p$, where the charge density relaxation time of the suspending medium exceeds that for the particle.

As the particle rotates, it experiences a resistive viscous torque (T_η) from the surrounding fluid, given by:

$$\text{T}_\eta = -8\pi\eta R^3\Omega_o$$

where Ω_o is the constant angular velocity (radians per second) of the particle and η is the dynamic viscosity of the surrounding fluid [33]. The fluid is assumed to be stationary, with the angular velocity low enough for the fluid streamlines around the particle to be of a laminar rather than turbulent nature. At the steady state the electrorotational torque is exactly balanced by the viscous torque, so that

$$\text{T}_\eta + \langle \text{T}(t)\rangle = 0$$

to give:

$$\Omega_o = \frac{\langle \text{T}(t)\rangle}{8\pi\eta R^3}$$

From Equation (10.52)

$$\Omega_o = -\frac{\varepsilon_o\varepsilon_m}{2\eta}\text{Im}[CM^*]\text{E}^2 \tag{10.56}$$

Thus, although from Equation (10.53) the variation of the rotation rate with frequency is determined by the frequency dependence of $\text{Im}[CM^*]$, the rotation is not

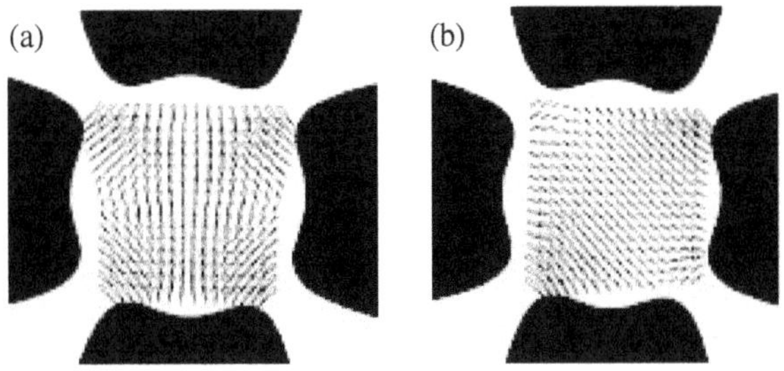

Figure 10.23 (a) Quadrupole electrodes of the 'bone' design create a rotating field of constant magnitude over a significant area between the electrodes. (b) A map of the field vectors following a clockwise advance by 30° of the signal phases applied to the electrodes. (Based on Dalton *et al.* [36].)

synchronous with the rate of rotation of the applied field. The magnitude of the rotation depends on the square of the applied field, which is determined by the peak voltage of the quadrature phases signals applied to the electrodes. A scaling factor should be applied, whose value depends on the geometry (e.g., straight, circular, elliptic) of the electrodes and the specific location of the particle within the space bounded by the electrodes [34, 35]. A design that produces a rotating field that is uniform over a good working area between the electrodes is the so-called 'bone' design [36] shown in Figure 10.23.

Equation (10.56) informs us that particle size does not influence the steady-state rotation rate. Unlike DEP, electrorotation (ROT) is not a ponderomotive effect. This is a consequence of the fact that Ω_o is attained when the viscous torque exactly balances the rotational torque. Because these torques depend on particle size, when the steady-state condition is reached the particle size cancels 'out of the equation'. A comparable situation occurs in dielectrophoresis regarding the stable levitation above an electrode plane acquired by a particle that experiences both a negative buoyancy force and a negative DEP force. Negative buoyancy acts to bring the particle down to the electrode plane and this force is proportional to the particle volume and the difference in density between the particle and the suspending fluid. The negative DEP force pushing against this sedimentation force is proportional to the particle volume. Equilibrium is attained when the negative buoyancy is counterbalanced by the DEP force, at which point the particle volume cancels 'out of the equation'. It should be noted, though, that the electrical frequency at which electrorotation attains its peak magnitude can depend on particle size. If the dipole moment of a particle undergoing electrorotation arises principally from interfacial polarization, the peak rotation depicted in Figure 10.24 is determined by the value of $\text{Im}[CM^*]_{pk}$, given by Equation (10.53) and corresponds to the condition $\omega\tau = 1$. For a heterogeneous particle whose structure can be modelled in the form of multiple

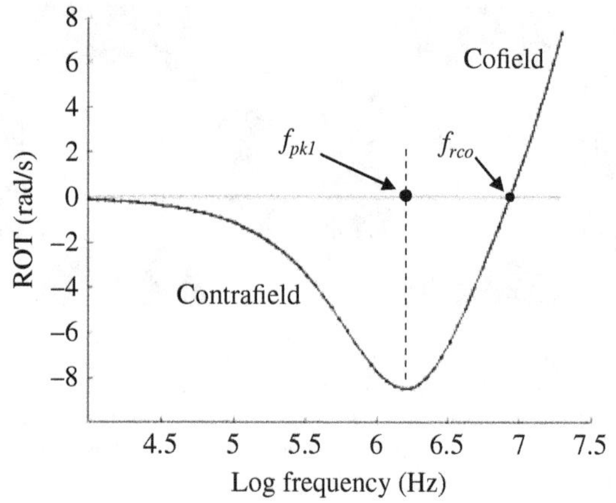

Figure 10.24 Electrorotation (ROT) response for a viable lymphocyte suspended in a medium of conductivity 40 mS/m, modelled using the MATLAB program given in Box 10.5, with line 70 modified according to details given in the main text. Below 1 MHz the sense of rotation is contrafield, with a peak rotation rate of −8.5 radians s^{-1} (−1.35 revs s^{-1}) at f_{pk1}. A transition to cofield rotation occurs at the crossover frequency f_{rco} (which does not coincide with the DEP crossover at f_{xo1}).

shells, the value of τ is given by Equation (10.44). Thus, as for the DEP crossover frequency f_{xo1}, the frequency f_{pk1} at which the peak electrorotation occurs can depend on particle size. Further details of the interrelationship between dielectrophoresis and electrorotation are presented in Box 10.8.

The following important relationship, which relates the frequency dependence of the electrorotation torque T to that of the DEP force, is derived in Box 10.8:

$$T(\omega_o) = -\frac{2\omega_o}{\pi k}\int_0^{\infty}\frac{F_{DEP}(\omega)}{\omega^2-\omega_o^2}d\omega$$

where k is a frequency independent parameter given by:

$$k = \frac{\left|\nabla E_{DEP}^2\right|}{(E_{ROT}^2)}$$

From this the following interrelationship between DEP and ROT can be derived [22, 39]:

- in a frequency range where the DEP force acting on a particle increases with frequency $\partial F_{DEP}/\partial\omega > 0$) the rotational torque $T(\omega)$ is negative, to give contrafield rotation;
- where the DEP force decreases with frequency ($\partial F_{DEP}/\partial\omega < 0$) the rotational torque $T(\omega)$ is positive, to give co-field rotation;
- at frequencies where $\partial F_{DEP}/\partial\omega = 0$ the rotational torque is zero.

Box 10.8 Relationship between Dielectrophoresis and Electrorotation

We have treated the induced dipole moment of a particle as a linear response of the applied electric field ($p = v\alpha E$). Thus, using the terminology of electronic signal processing, the temporal variation of the moment $p(t)$ can be expressed in terms of the applied field $E(t)$ and the impulse response $h(t)$ by means of the convolution integral [37]:

$$p(t) = \int_0^{\infty} h(\lambda)E(t-\lambda)d\lambda$$

Taking the Fourier transform of both sides of this equation:

$$p(\omega) = E(\omega)H(\omega)$$

where $H(\omega)$ is the Fourier response function of the induced moment, which referring to Equation (10.26), is given by:

$$H(\omega) = 4\pi\varepsilon_m R^3\left(\frac{\varepsilon_p^* - \varepsilon_m^*}{\varepsilon_p^* + 2\varepsilon_m^*}\right)$$

For any practical (therefore causal) linear system, the Kramers–Krönig theorem relates the real and imaginary components of the system's Fourier response [38]. This should also be valid for our response function $H(\omega)$. From Equations (10.26) and (10.52) the DEP force F_{DEP} and ROT torque T are determined by the real and imaginary components of the *CM* factor, respectively. Thus, invoking the Kramers–Krönig theorem, we have:

$$F_{DEP}(\omega_o) - F_{\infty} = \frac{2k}{\pi}\int_0^{\infty}\frac{T(\omega)\omega}{\omega^2-\omega_o^2}d\omega$$

and

$$T(\omega_o) = -\frac{2\omega_o}{\pi k}\int_0^{\infty}\frac{F_{DEP}(\omega)}{\omega^2-\omega_o^2}d\omega$$

where F_{∞} is the DEP force at the high frequency limit, reflecting the instantaneous response of the dipole moment to the external field. The quantity k is a frequency independent, positive, factor related to the DEP and ROT applied fields by:

$$k = \frac{|\nabla E_{DEP}^2|}{(E_{ROT}^2)}$$

This relationship indicates that through the Kramers–Krönig theorem we can obtain the DEP frequency response of a particle from its ROT spectrum and *vice versa*.

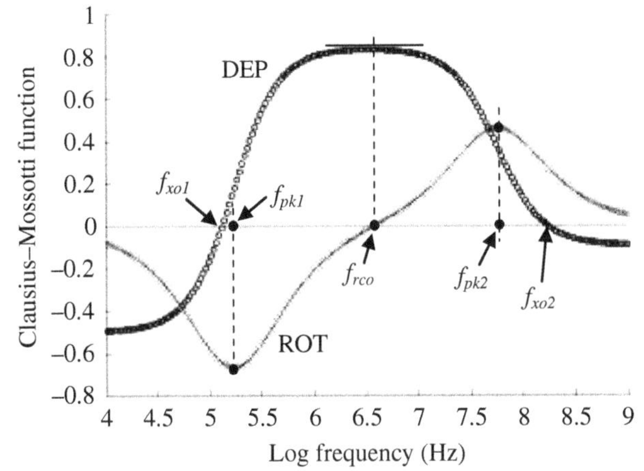

Figure 10.25 Dielectrophoresis (DEP) and electrorotation (ROT) responses modelled for a lymphocyte suspended in a medium of conductivity 40 mS/m, using the MATLAB program given in Box 10.5 and modified according to details given in the main text. The low- and high-frequency DEP crossover points (f_{xo1}, f_{xo2}) are indicated, together with the contrafield and co-field peaks of rotation at f_{pk1} and f_{pk2}, respectively. A transition from contra- to cofield ROT occurs at f_{rco}.

These relationships can be seen in Figure 10.25 for the DEP and ROT response modelled for a lymphocyte, using a version of the MATLAB program given in Box 10.5 in which line 70 has been modified to read:

```
70  plot(log10(f), -55.3*im,'·', log10(f),
    zeroline,'-');
```

As indicated in Figure 10.25, when viable lymphocytes are suspended in a medium of conductivity 40 mS/m they can be expected to exhibit negative DEP at frequencies below 100 kHz. The magnitude of this negative DEP force decreases and tends towards a positive value with increasing frequency (i.e., $\partial F_{DEP}/\partial\omega > 0$). If the electrorotation response were to be measured at the same time as the DEP force, it would be negative (i.e., contrafield). As the frequency is increased to the point of inflection of curvature of the DEP frequency response, corresponding to the midpoint between its limiting low- and high-frequency values, the electrorotation rate exhibits its peak negative value at f_{pk1}. The MATLAB program in Box 10.5 is based on the multishell model of a cell. In this model, if the cell possesses a viable membrane its low-frequency dielectric properties are dominated by the high-resistance (low conductivity) of the membrane. The cell appears as an insulating particle suspended in a conducting medium and so exhibits negative DEP determined by the Maxwell–Wagner interfacial polarization taking place at the interface between the cell membrane and the suspending medium. At frequencies in the MHz range, the capacitance of the membrane shorts out the high resistance of the membrane and the field penetrates the cell interior. The membrane appears invisible to the field, so that the corresponding interface for the Maxwell–Wagner polarization is now that between the outside suspending medium and a sphere composed of the cytoplasm and its internal contents. If, as is usually the case, the effective permittivity of the cytoplasm and its contents is less than that of the suspending medium, the DEP response will decrease towards a negative value (or at the very least a less positive one) so that $\partial F_{DEP}/\partial\omega < 0$. In this case, as shown in Figure 10.25, the cell should now exhibit co-field electrorotation with a peak rate at the frequency (f_{pk2}) corresponding to the point of inflection of the high-frequency DEP characteristic. The author is not aware of a reported determination of f_{pk2} for a viable cell. As also shown in Figure 10.25, where the DEP response attains a plateau (i.e., $\partial F_{DEP}/\partial\omega = 0$) the rotational torque is zero. This occurs at very low and very high frequencies, as well as at the frequency where there is a transition from $\partial F_{DEP}/\partial\omega > 0$ to $\partial F_{DEP}/\partial\omega < 0$. At this frequency value (f_{rco}) the electrorotation makes the transition from contra- to co-field rotation.

Example 10.1 Estimation of the Peak Angular Velocity of Electrorotation

Quadrature phased sinusoidal voltages of peak magnitude 5 V are applied for an electrorotation (ROT) experiment to the electrodes shown in Figure 10.19. Frequencies between 10 kHz to 30 MHz are to be investigated, within which range Maxwell–Wagner polarization of a test spherical particle is known to occur. The distance between opposite electrode pairs is 400 μm. Estimate the peak angular velocity expected for the electrorotation (ROT) response of the particle when located in the central region between the electrodes.

Solution 10.1 From Equation (10.56) the maximum (peak) magnitude of the rotation is given by:

$$\Omega_{pk} = -\frac{\varepsilon_o\varepsilon_m}{2\eta}\mathrm{Im}[CM^*]_{\mathrm{pk}}E^2 \tag{10.57}$$

From Equation 10.49) the magnitude of the rotating field is determined by the peak value of the applied voltage signal to each electrode. The peak voltage is 5 V and the gap between opposing electrodes is 400 μm. Neglecting the fact that a correction factor of ~0.9 may be required to obtain the actual field profile [35], we can estimate the field as $5\,\mathrm{V}/(4 \times 10^{-4}\,\mathrm{m}) = 1.25 \times 10^4\,\mathrm{V/m}$ (N/C). For an aqueous medium we have $\varepsilon_m \approx 80$ and $\eta \approx 1\,\mathrm{mPa\,s}$, so that from Equation (10.57) the predicted maximum rotation rate is:

$$|\Omega_{pk}| \approx \left(\frac{(8.854\times10^{-12}\ \mathrm{C^2N^{-1}m^{-2}})\times 80}{2\times10^{-3}\ \mathrm{Nm^{-2}s}}\right)$$
$$\times(1.25\times10^{4}\ \mathrm{NC^{-1}})^2\mathrm{Im}[CM^*]_{pk}\mathrm{s}^{-1}$$
$$\approx 55.3\ \mathrm{Im}[CM^*]_{pk}\mathrm{s}^{-1}$$

In this calculation, the dimensional units (F/m) for the electric constant ε_o are treated as equivalent to $\mathrm{C^2N^{-1}m^{-2}}$, which can be verified from Appendix D. From Equation (10.54) we have deduced that values for $\mathrm{Im}[CM^*]$ lie within the bounds $\pm$ 0.75. The maximum electrorotation rate expected is thus $\pm(0.75 \times 55.3) = \pm41.5$ radians per second, equivalent to 6.6 revs / second in either the co-field or contrafield direction.

Rotation rates for frequencies either side of the peak value can, to a fair degree of accuracy, be determined by eye and a stopwatch (e.g., measuring the time taken for 10 revolutions). However, the most important parameter to determine in a ROT experiment is the critical frequency f_{pk1} at which the peak rotation rate occurs. If the rotation rate is too high, determination of f_{pk1} can be achieved by reducing the amplitude of the quadrature voltage signals. An alternative, quite ingenious technique involving two contrarotating fields of equal amplitude is described by Arnold and Zimmermann [40], who performed pioneering studies of cell electrorotation [41,42]. Other early studies of cell electrorotation were reported by Glaser *et al.* [43].

The frequency f_{pk1} at which the maximum contrafield rotation occurs is when $\omega\tau = 1$, with τ given by Equation (10.30), so that:

$$f_{pk1} = \frac{1}{2\pi\tau} = \frac{1}{2\pi}\frac{(\sigma_p + 2\sigma_m)}{\varepsilon_o(\varepsilon_p + 2\varepsilon_m)}$$
$$= \frac{\sigma_m}{\pi\varepsilon_o(\varepsilon_p + 2\varepsilon_m)} + \frac{\sigma_p}{2\pi\varepsilon_o(\varepsilon_p + 2\varepsilon_m)} \tag{10.58}$$

Thus, if ROT experiments are performed as a function of the medium conductivity (σ_m), whilst maintaining isotonic conditions for the cells (see Example 8.3), a plot of f_{pk1} versus σ_m should thus produce a straight line with a slope given by:

$$\mathrm{Slope} = \frac{1}{\pi\varepsilon_o(\varepsilon_p + 2\varepsilon_m)} \tag{10.59}$$

If the cells (or particles) are suspended in an aqueous medium the value for ε_m will be known (~79) and so the value for permittivity of the particle can be obtained from the slope value. If cells are being investigated and their radii R have been measured, then the membrane capacitance C_{mem} can be determined with the aid of Equation (10.38). The intercept of the straight line on the conductivity axis is given by setting f_{pk1} in Equation (10.58) to zero, to give

$$\mathrm{Intercept} = -\frac{\sigma_p}{2} \tag{10.60}$$

This relationship enables a determination of the particle conductivity. Furthermore, as described in the derivation of Equation (10.44), an estimate of the membrane conductance G_{mem} of a cell can then be made using the low-frequency approximation that $G_{mem} = \sigma_p/R$. It is usually the case that the uncertainty in the value obtained for G_{mem} is much larger than that obtained for the membrane capacitance C_{mem} [28]. One reason for this is that the transmembrane conductance G_{mem}, which for a viable cell is expected to have a value smaller than ~10 S $\mathrm{m^{-2}}$, can be masked by the surface conductance of the cell [25]. This is discussed further in section 10.3.6.

A schematic of the plot of f_{pk1} versus σ_m is shown in Figure 10.26, together with a plot of DEP data corresponding to Equation (10.43).

10.3.5 Traveling Fields and Traveling Wave Dielectrophoresis (TWD)

The rotational torque generated by the scheme shown in Figure 10.19 can be converted to produce a translational force on a particle by 'opening up' the electrodes and 'laying them down' to form a repeating linear array of electrodes as depicted in Figure 10.27. A particle situated above these electrodes will experience an electric

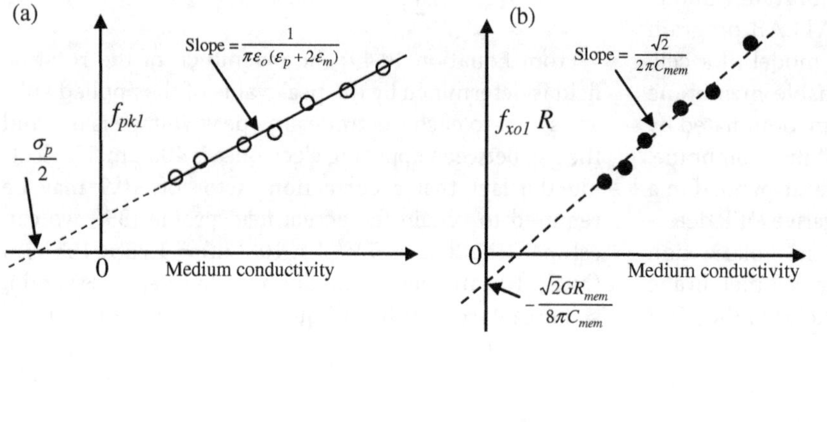

Figure 10.26 (a) A plot of the frequency (f_{pk1}) at which the peak contrafield rotation occurs in a ROT experiment on viable cells, as a function of the suspending medium conductivity. (b) A plot of the product of the DEP crossover frequency (f_{xo1}) and cell radius (R) as a function of the medium conductivity. The expressions for the slopes and intercepts of the linear plots are given by Equations 10.43 and 10.58.

field having a gradient of both field magnitude and phase. This is the situation analysed in Box 10.6, which gives the resultant time-averaged DEP force acting on a spherical particle of radius R as:

$$\langle F_{DEP} \rangle = 2\pi\varepsilon_m R^3\{\mathrm{Re}(CM^*)\nabla E^2 + \mathrm{Im}(CM^*)\Sigma E^2 \nabla\phi\} \tag{10.61}$$

where $\Sigma E^2\nabla\phi$ represents a summation of the magnitude and phase ϕ of each field component in Cartesian coordinates. The first term on the right relates to the in-phase component of the induced dipole moment and represents the dipole approximation of the DEP force, as given by Equation (10.27). This force directs the particle towards either strong field regions at the electrodes or weak field regions above the electrodes, depending on whether Re[CM^*] is positive or negative, respectively. The second term relates to the out-of-phase component of the induced moment and produces a force that acts to move the particle in a direction either with or against the direction of the phase gradient, depending on the magnitudes and sign of Im[CM^*] and the phase gradient factor $\Sigma E^2\nabla\phi$ [44]. The field travels towards the smaller phase regions as indicated in Figure 10.27. A positive value for Im[CM^*] produces a force that directs the particle towards regions of larger phases of the field component along the electrode array – in other words *against* the direction of the traveling field. For the case where Im[CM^*] < 0, the particle is directed towards regions of smaller phase and so with the direction of the traveling field. If, as is usually the case, quadrature-phase voltages are applied to the interdigitated electrodes shown in Figure 10.27, the distance between every second electrode ($2d_1 + 2d_2$) corresponds to a one-half wavelength of the traveling field and is thus equal to the distance of maximum phase difference on the electrode track.

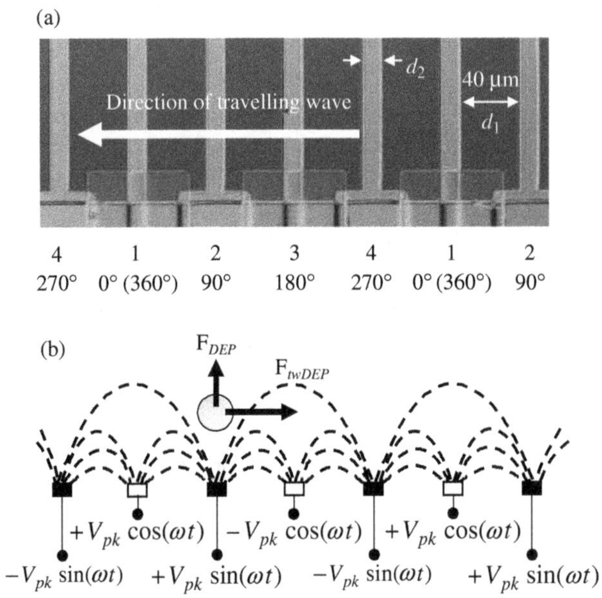

Figure 10.27 (a) View looking down on an interdigitated electrode array designed for travelling wave dielectrophoresis. The direction of the travelling field, as for the sense of rotation of a rotating field, is against to the direction of increasing phase angle applied to the electrodes. (b) Side view showing the sequence of quadrature-phased signals at an instant in time. A particle is shown levitated above the electrodes, under the action of a negative DEP effect and propelled by a twDEP force against the direction of the travelling wave at frequencies where In[CM^*] > 0.

As depicted in Figure 10.27, a particle suspended above the electrodes is not capable of translational motion if it is brought down to and trapped on the electrodes by a positive DEP force. If the particle is levitated above the electrodes by a negative DEP force, it experiences a translational force provided the magnitude of Im[CM^*] is large enough. This restricts traveling wave dielectrophoresis (TWD) to the frequency range shown in Figure 10.28. To simplify the discussion of this, the particular situation has been chosen of a particle made from a poorly conducting polymer suspended in an aqueous electrolyte. This system therefore possesses a single interface and so will exhibit a single Maxwell–Wagner interfacial dispersion. The particle can be expected to exhibit a DEP response that follows the frequency-dependent form of Re[CM^*] shown in Figure 10.28. A negative DEP force will act to levitate the particle for frequencies up to the DEP crossover point at around 300 MHz, but TWD will only commence at ~200 kHz where Im[CM^*] is large enough to produce a significant translational force.

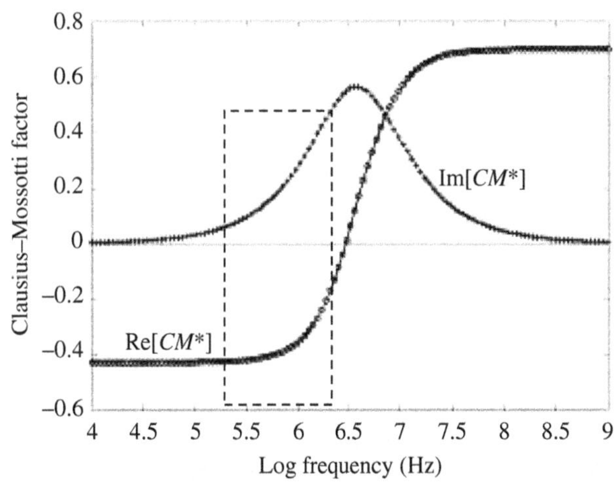

Figure 10.28 Plots of the real and imaginary components, Re[CM^*] and Im[CM^*], respectively, of the Clausius–Mossotti factor for a homogeneous spherical particle. Travelling wave DEP is restricted to the frequency range bounded by the rectangular area, where a negative DEP force levitates the particle and Im[CM^*] is of sufficient magnitude to induce translational motion.

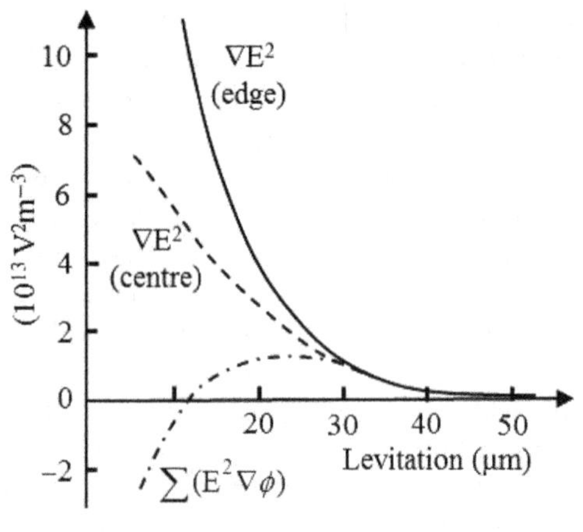

Figure 10.29 Variations of the TWD force factors ∇E^2 and $\Sigma(E^2\nabla\phi)$ as a function of levitation height above TWD electrodes, with $(d_1 + d_2)/2 = 10$ μm and an applied quadrature-phase voltage signal of $1\,V_{rms}$ [45].

Above ~200 MHz the value for Re[CM^*] is either not large enough to levitate the particle or has a positive value, which leads to its immobilization on the electrodes by positive DEP. As shown in Figure 10.28 the imaginary component of the CM factor is positive in this frequency range extending from 200 kHz to 200 MHz. The particle will move along the electrode track in a direction opposite to that of the traveling field.

The values for ∇E^2 and $\Sigma E^2 \nabla\phi$ in Equation (10.61) vary as a function of height above the electrode plane. An example is shown in Figure 10.29. This example indicates a general trend, in that as a particle approaches the electrode plane a strong DEP force directs it towards the edge rather than the top of an electrode, whilst the TWD force reverses in polarity. The same forms of variations of ∇E^2 and $\Sigma E^2 \nabla\phi$ have been derived using the charge density method [46], Green's theorem [47] finite and Fourier series analysis [48]. These various methods of analysis indicate that TWD takes the form of simple translational motion for particles levitated higher than $d = (d_1 + d_2)/2$ above the electrode plane, where as indicated in Figure 10.27 the distance between electrodes is d_1 and d_2' is the electrode width. When particles approach the electrodes, they can exhibit circular and spinning motions as well as sudden reversals in their direction of travel. Such effects were noted in earlier studies of TWD and designated as the FUN (fundamentally unstable) regime [49]. The term 'FUN' was adopted to reflect the amusement caused in viewing these effects down a microscope.

For levitation heights greater than d, the time-averaged TWD velocity is given by the following expression [47]:

$$v_{TWD} = -\frac{\varepsilon_o \varepsilon_m R^2 V^2}{3\eta} \mathrm{Im}(m) \Sigma E^2 \nabla\phi \tag{10.62}$$

in which η is the viscosity of the suspending electrolyte and V the applied (rms) voltage. For an aqueous electrolyte we have $\varepsilon_o \varepsilon_m \sim 7 \times 10^{-10}$ F m^{-1} and $\eta \sim 10^{-3}$ kg m^{-1}.s^{-1}. For the case of a lymphocyte ($r \sim 5$ μm) levitated to a height of 25 μm on applying a 1 V_{rms} signal and assuming Im(m) ~0.6, then from Figure (10.29) and Equation (10.62) we obtain a contrafield TWD velocity of ~22 μm.s^{-1}. Particle velocity values of this order are typically observed in TWD experiments. As discussed by Wang *et al.* [47] the TWD velocity increases with increasing applied voltage, but not as a voltage squared dependency as suggested by Equation (10.62). With increasing voltage the increased DEP force causes an increase of particle levitation height and as shown in Figure 10.29 this leads to a smaller value for $\Sigma E^2 \nabla\phi$ for levitation heights greater than 15 μm. If a particle is assumed to be way above the electrode plane, where the field and DEP force decay exponentially as a function of levitation height, the following voltage-independent expression can be derived for the TWD velocity [48]:

$$v_{TWD} = \frac{2}{9} \frac{r^2 \Delta\rho g}{\eta} \frac{\mathrm{Im}(m)}{\mathrm{Re}(m)} \tag{10.63}$$

$\Delta\rho$ is the difference between the specific densities of the particle and suspending electrolyte and g is the gravitational acceleration factor (9.81 m.s^{-2}). For lymphocytes suspended in an aqueous electrolyte $\Delta\rho$ has a value ~0.04 kg.dm^{-3}. For Re(CM^*) = −0.4 and Im(CM^*) = 0.6, Equation (10.63) predicts a contrafield TWD velocity of ~3.3 μm.s^{-1} for lymphocytes. In most work reported in the literature, TWD velocities are considerably larger than this value and are voltage dependent. This indicates that most reported experiments have operated under conditions where Equation (10.62) rather than Equation (10.63) is pertinent.

If a DEP or second TWD signal is added to the first TWD signal, the resulting electrostatic potential at the electrodes is a superposition of the separate voltage functions. Provided that nonlinear effects are absent, such as those arising from electrode polarization, high power dissipation, or a frequency variation of ε_m, for example, there is a linear superposition of the effects caused by these two signals [45]. The total resultant DEP and TWD force acting on a cell will be the vector sum of the individual forces produced by each signal acting separately. This can be used to physically separate and isolate cells of different types or stages of cell cycle, for example, by inducing them to move in opposite directions along the TWD electrode track [45].

Travelling waves dielectrophoresis electrodes have mostly been designed to operate as linear 'tracks', but spiral designs have also been used [47, 50]. The spiral design shown in Figure 10.30 combines traveling wave DEP and

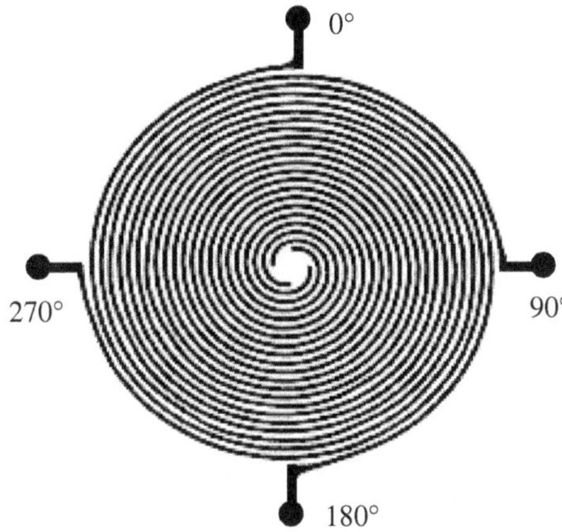

Figure 10.30 A combined traveling wave (TWD) and electrorotation (ROT) electrode design. On energizing the four spiral electrode elements with sinusoidal voltages of the indicated phase relationships, a travelling field propagates radially from the centre to the outer periphery of the electrode array. Particles directed to the centre by TWD are exposed to an anticlockwise rotating field [50].

electrorotation to concentrate and assay the viability of microorganisms [50].

10.3.6 Particle Inhomogeneity, Net Charge and Surface Conductance

Most particles, especially biological ones, are not homogeneous. As shown in Chapter 9, bacteria and cells can be modelled to take account of their heterogeneous structures using the so-called *multishell* model. The Clausius–Mossotti function for a multishell particle and hence the DEP force that acts on it, can be obtained by evaluating *effective* values for the relative complex permittivity and conductivity values of both the particle and its suspending medium. This has been described in some detail in section 9.4 of Chapter 9.

Particles generally carry a net charge associated with the presence of charged polymers and / or ionizable acidic and basic groups on their surfaces. Early studies [51, 52] demonstrated that the low frequency DEP response for erythrocytes and erythroleukemia cells was affected following neuraminidase treatment to reduce their cell membrane charge by 50~60%. The fixed charge on the surface of a particle can influence its DEP behaviour through electrophoresis and counterion relaxations and conduction in the double layer that forms around all charged particles when they are suspended in aqueous media. Evidence for an electrophoretic contribution to the low frequency (<10 Hz) DEP response of a suspension of bacteria was observed by Burt *et al.* [31], but inertial constraints will render such contributions to be insignificant at higher frequencies, even for very small particles. Counterion relaxation processes, involving both ionic diffusion and ionic conduction around particle surfaces, will however contribute to the total polarizability and influence the DEP response for frequencies up to ~1 MHz. A comprehensive treatment of electrical double layer polarizations is given by Lyklema [53]. In summary, as described in Chapter 12, the electrical double layer can be considered to consist of two classes of counterions – those strongly attracted to the fixed charges on the particle surface (termed the Stern layer) and those more loosely associated in the diffuse outer layer where the electrical potential of the charged particle approaches that of the surrounding bulk medium with increasing distance from the particle surface. The 'physical' boundary between these two populations of counterions can be considered to be the so-called 'slip plane' at the zeta potential, where a charged particle plus its Stern layer separates from the outer diffuse population of counter ions during electrophoresis. Fluid flow occurs beyond the slip plane, but beneath the fluid layer next to the charged surface can be considered to be 'stagnant'.

The influence of the electrical double layer on DEP behaviour is particularly important for nanoparticles, where the thickness of this layer can approach, or even exceed, the particle's 'physical' diameter ($2R$). Field-induced mobility of counterions in the double layer give rise to a surface conductance K_s whose influence on the overall polarizability can exceed that of the particle's bulk conductivity (σ_{bulk}). The total conductivity of a particle can be described as the sum of its bulk and surface conductivity [54]:

$$\sigma_p = \sigma_{bulk} + \frac{2K_s}{R} \qquad (10.64)$$

Values for K_s were determined from measurements of the electrorotation response of different types of latex particle of radii R in the range 2.5–5 μm [54]. Depending on particle type and pretreatment, the range of observed surface conductivities was 0.2 to 2.1 nS and found to be independent of the conductivity of the suspending medium (0.2 – 1.6 S/m). Significantly larger values have been reported [55] for 20-mer single-stranded DNA (7.9 ± 1.9 nS), 40-mer double-stranded DNA (5.3 ± 0.7 nS) and yellow fluorescent protein (21.5 ± 1.6 nS). There are two main contributions to the value of K_s, namely that due to field-induced charge movement in the Stern layer (K_{Stern}) and charge movement in the diffuse part of the electrical double layer (K_{diff}). Both of these contributions are given by the product of the equivalent surface charge density and ion mobility in the two layers. For the Stern layer, the surface charge density is defined by the zeta potential (see Box 2.3 and Figure 12.21). The total of K_s in Equation (10.64) is thus given by $K_s = K_{Stern} + K_{diff}$.

Referring to the expression for the low-frequency DEP crossover frequency for a cell, given by Equation (10.43) and repeated here for convenience:

$$f_{xo1} = \frac{\sqrt{2}}{2\pi RC_{mem}} \left(\sigma_m - \frac{G_{mem}}{4} \right)$$

the value of the membrane G_{mem} should include the membrane surface conductance (K_{ms}). Adopting the low-frequency approximation $G_{mem} = \sigma_p/R$:

$$G_{mem} = G_{bulkmem} + 2K_{ms}/R^2.$$

For a cell of radius 5 μm with $K_{ms} = 2$ nS, then $2K_{ms}/R^2 = 160$ S/m^2. The value of K_{ms} obtained for pancreatic β cells from electrorotation measurements is 179 ± 61 S/m^2 [25]. This suggests that the membrane surface conductance may well be greater than the bulk conductance of the membrane. Furthermore, the relationship given by Equation (10.64) indicates that the influence of K_s may dominate over the bulk conductivity for particles of diameter less than 1 μm. As discussed in Chapter 12, the surface conductivity in the diffuse part of the double layer consists of two parts, namely a contribution caused by the movement of charges with respect to the liquid, plus a contribution caused by liquid flow beyond the shear plane, which gives rise to an additional mobility of the charges. Beneath the shear plane, in the 'stagnant' fluid layer next to the charged surface, only electrically induced movement of the counterions can contribute to K_{ms}. Lyklema and Minor [56] have shown that for nonpenetrable surfaces, lateral mobilities of monovalent ions in the Stern layer are not much lower than those in the bulk fluid. Under certain conditions conduction beneath the slip plane, in the stagnant layer, may be of the same order of magnitude as that beyond it. Conduction in the diffuse layer and in the Stern layer could therefore contribute separately to the overall magnitude of K_s and thus to the DEP behaviour of nanoparticles [57,58].

A comprehensive study of the AC and DC electrokinetic properties of latex nanoparticles, as a function of suspending medium conductivity and viscosity, has been reported by Ermolina and Morgan [59] and a theoretical modelling of the DEP force that takes into account the influence of the electrical double layer has been presented by Zhou *et al.* [60]. Analysis of the normal and tangential ionic currents that occur around and at the surface of a particle, when its diameter approaches and becomes smaller than the width of its own electrical double layer, indicates that a capacitance effect contributes to the total polarizability of the particle and exceeds the influence of the surface conductance *Ks* [61, 62]. The DEP crossover frequency for nanocolloids has also been shown to be inversely proportional to the *RC* time constant of the diffuse layer component of the electrical double layer [62]. This is considered to offer a sensitive method for detecting the hybridization of target molecules onto functionalized nanocolloid probes, based on the change of the surface conductance of these probes and the corresponding change of their DEP crossover frequency [62]. Hoffman *et al.* [63, 64] have demonstrated that electrical double layer effects influence the way nanoparticles aggregate under the influence of a DEP force.

10.3.7 Presence of Perturbing Particles or Boundaries

The derivation of the macroscopic Clausius–Mossotti factor assumes that a particle is embedded in a homogeneous dielectric medium, which from the particle's perspective should appear to be of infinite extent. The particle is thus assumed to be far removed from the influence of the dipole field of another polarizable particle, or from a boundary where image potentials such as those shown in section 5.4.1 of Chapter 5 are created. However, it is not uncommon when viewing the DEP response of cells through a microscope to see that within a few seconds of applying the field neighbouring cells move together to form 'pearl chains' of two or more connected cells. Pearl chaining also occurs in a uniform field (see Figure 10.31) so that we can assume the electrostatic interaction between the particles arises from a 'micro-DEP' effect, whereby neighbouring polarized particles respond to the local distortions of the field caused by their dipole fields. We can also assume that this mutual dipole-dipole attraction overcomes electrostatic repulsion between the negatively charged surfaces of the cells. Pearl chaining is particularly evident if the average spatial distance between

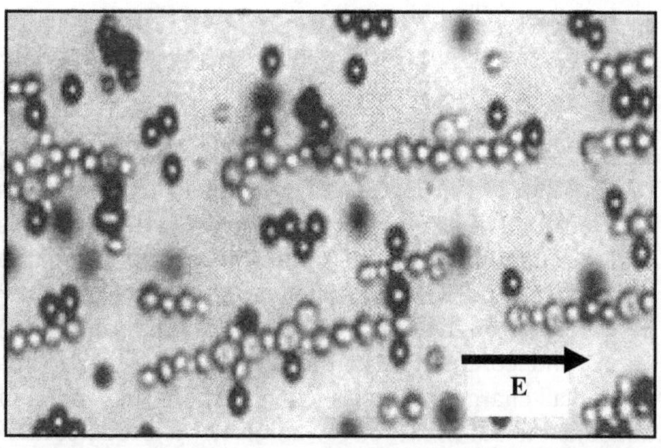

Figure 10.31 Longitudinal (parallel to the field) and transverse (perpendicular to the field) chaining of a mixture of yeast cells (white) and polystyrene beads (dark) subjected to a 100 kHz uniform field of 1 kV/m. (Lee, R. S. and Pethig, R., unpublished.)

cells in a stationary fluid is seen to be less than 5∼6 particle diameters (i.e., a concentration greater than ∼10^7/ml for cells of diameter ∼10 μm). The DEP behaviour of a cell can be expected to change after incorporation into a doublet, triplet or higher order cell aggregate, as well as when it approaches a metal or dielectric boundary.

The influence of a neighbouring particle can be appreciated from the following simple example, based on a model described by Stoy [65]. Instead of a single sphere inserted into an imposed uniform field E within an infinite medium, we consider the case of two identical spheres placed distance D apart along an axis colinear with the field E. The centre of sphere 1 is placed at the origin of the spherical coordinate system employed to derive Equation (10.5). The second sphere thus experiences a net field E_2 composed of the applied field E and the dipole field created by particle 1:

$$E_2 = E\left(1 + \frac{2(CM)}{D^3}\right)$$

This in turn gives rise to an increased induced moment created by particle 2, which sphere 1 now reacts to in the form of an increase of its own induced dipole moment:

$$p_1 = 4\pi\varepsilon_o\varepsilon_m(CM)E\left[1 + \frac{2(CM)}{D^3}\left(1 + \frac{2(CM)}{D^3}\right)\right]$$

Sphere 2 in turn senses this extra dipole field created by the induced dipole of sphere 1 and so its own dipole field changes accordingly. If this procedure is repeated *ad infinitum* the interactive dipole moments p_1 and p_2 can be obtained in closed form according to the expression [65]:

$$p_1 = p_2 = 4\pi\varepsilon_o\varepsilon_m(CM)E\left(\frac{1}{1 - 2(CM)/D^3}\right)$$

Each polarized sphere distorts the local field of the other, so that they experience a mutual attractive DEP force given by:

$$F_{DEP2} = -F_{DEP1} = -\frac{6p_1 \cdot p_2}{4\pi\varepsilon_o\varepsilon_m D^4}$$

The major contribution to this mutual DEP force will be that of an assumed dipole-dipole interaction only when the spheres are separated by more than a distance equal to that of their radii. As the spheres approach more closely than a distance equal to that of their radii, the mutual DEP force between them depends not just on induced dipole-dipole interactions, but of all the multipoles induced in sphere 1 and sphere 2. Stoy [66] extended his model to compute the DEP force between identical touching spheres in a parallel field, to include up to 2800 interactions of linear multipoles.

Early computer-aided studies (Monte Carlo simulations) of how dipole-dipole interactions can lead to particle chain structures considered the particles to be hard spheres with permanent dipole moments [67, 68]. The first attempt to investigate the situation of relevance to DEP studies was made by Sancho *et al.* [69], who treated the spheres as nonpolar so that chain formation resulted from *induced* dipole-dipole interactions driven by an external polarizing field. The spheres were also treated as lossy dielectric particles immersed in a conductive medium. The existence of a threshold of field intensity, necessary for chain structures to appear, was predicted – together with a frequency dependence of the process of chain formation that reflected the presence of Maxwell–Wagner polarization. This work was later refined to take account of the interaction between induced multipoles [70]. A multiple image approximation has also been employed to study the polarization spectra of a pair of touching cells and colloidal particles by Huang *et al.* [71]. These studies were restricted to plain spherical particles, aligned with the direction of the electric field. The practical calculation of the interaction requires truncation of expansions at some finite number of terms and is affected by convergence problems when particles have high permittivity or they are closely spaced.

An important application of DEP is the selective separation of target cells from a cell mixture. For suspensions of the same type of particle, the alignment of pearl chaining is always seen to be parallel with the applied field. However, Griffin and Ferris observed transverse pearl chains in mixtures of polystyrene spheres with either erythrocytes or unicellular algae [72]. An example of this phenomenon is shown in Figure 10.31 for a mixture of polystyrene spheres with yeast cells.

An explanation for the transverse orientation of pearl chains shown in Figure 10.31 has been provided by Giner *et al.* [73]. For simplicity we restrict ourselves to a suspension of particles, half of type 1 and half of type 2, subjected to a uniform electric field E_z directed along the z-axis. This field induces a dipole moment for each particle type given by

$$p_{1,2} = \nu_{1,2}\alpha^*_{1,2}E_z \tag{10.65}$$

where $\nu_{1,2}$ and $\alpha*_{1,2}$ are the particle volumes and complex polarizabilities, respectively. From Box 6.1:

$$\mathrm{Re}\left[\alpha^*_{1,2}\right] = 3\varepsilon_o\varepsilon_m\left(\frac{\varepsilon_{1,2} - \varepsilon_m}{\varepsilon_{1,2} + 2\varepsilon_m}\right) \tag{10.66}$$

From Equations (4.41) and (10.65) the time average of the interaction energy (per unit volume) between a particle of type 1 and type 2 is given by:

$$\langle U_{1,2}\rangle = -\frac{1}{2}\alpha_1 E_z^2 = -\frac{1}{2}\mathrm{Re}[p_1 \cdot E_2] \tag{10.67}$$

where E_2 is the dipole field created by particle type 2, given by Equation (5.42). For the case where the two particles are located at r_i and r_j situated on a plane parallel to the z-axis:

$$E_2 = \frac{p_2}{4\pi\varepsilon_o\varepsilon_m}\frac{1}{R_{ij}^3}(3\cos^2\theta_{ij} - 1)$$

where $R_{ij} = |r_i - r_j|$ and θ_{ij} is the angle between R_{ij} and E_z. From Equation (10.67) and the above expression for E_2 the time average of the potential energy of the interacting particles is:

$$\left\langle U_{1,2}^{i,j}\right\rangle = \frac{1}{4\pi\varepsilon_o\varepsilon_m R_{ij}^3}\mathrm{Re}[\bar{\alpha}_1^*\alpha_2^*](1 - 3\cos^2\theta_{ij})E_z^2 \tag{10.68}$$

where $\bar{\alpha}_1^*$ represents the complex conjugate of α_1^*. Once the dielectric properties of the interacting particles and their suspending medium are chosen, whether this interaction energy is positive or negative will depend only on the frequency of the applied field E_z. If the particles are of the same type then $Re[\bar{\alpha}_1^*\alpha_2^*] = |\alpha_1^2|$ or $|\alpha_2^2|$ and will always have a positive value, irrespective of whether a particle is more or less polarizable than the surrounding medium. In this case the particles can minimize their potential energy by adjusting their positions so that the factor $(1 - 3\cos^2\theta_{ij})$ achieves its maximum negative value, corresponding to when they are touching (R_{ij} a minimum) and $\theta_{ij} = 0°$. This corresponds to a doublet pearl chain aligned with the field direction along the z-axis. For particles of a different type, field frequencies can exist where the factor $\mathrm{Re}[\bar{\alpha}_1^*\alpha_2^*]$ is negative. In this case to minimize their potential energy the particles will endeavour to rearrange themselves such that they touch with $\theta_{ij} = 90°$, in other words forming a transverse doublet chain. This argument can be extended to understand the formation of the triplet and higher order transverse chains shown in Figure 10.31. Examples of how the factor $\mathrm{Re}[\bar{\alpha}_1^*\alpha_2^*]$ varies as a function of frequency is shown in Figure 10.32 for a mixture of yeast cells and polystyrene beads, using literature values for the dielectric properties of yeast and modelling the polystyrene beads as spheres surrounded by a mobile ion double layer and appropriate surface conductivity value. Between 20 kHz and 300 MHz the negative value exhibited by $\mathrm{Re}[\bar{\alpha}_1^*\alpha_2^*]$ for the interaction of a polystyrene bead with a yeast cell indicates that they form transverse chains aligned at right angles to the applied field direction. This effect can be seen in Figure 10.31.

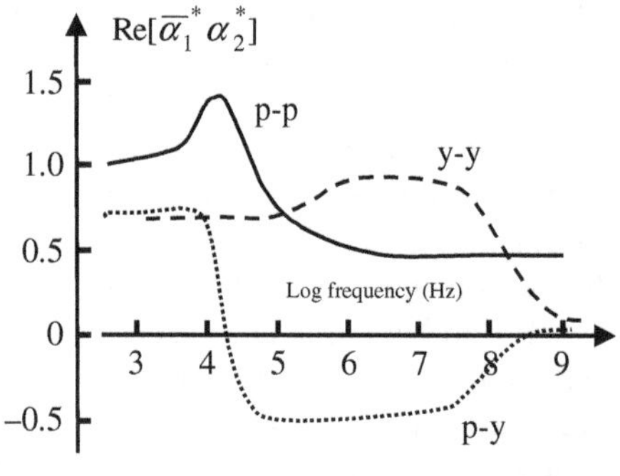

Figure 10.32 Values of the factor Re $[\bar{\alpha}_1^*\alpha_2^*]$ in Equation (10.68) for the interaction of polystyrene-polystyrene (p-p) beads, polystyrene beads with yeast cells (p-y) and pairs of yeast cells (y-y) as a function of frequency. Over the whole frequency range the interaction of the same particle types results in longitudinal chaining. From 20 kHz to 300 MHz the negative value for Re $[\bar{\alpha}_1^*\alpha_2^*]$ indicates that the interaction of a polystyrene bead with a yeast cell leads to transverse chaining. (Based on Giner *et al.* [73].)

In the Monte Carlo simulation performed by Giner *et al.* [73] an initial two-dimensional configuration was established that mimics a random distribution of particles settled on a plane parallel to the applied field. The surface concentration was equivalent to a volume concentration of ~10^7 particles/ml. Avoiding a situation where particles overlap, each particle was then moved at random on the plane and its energy was computed according to Equation (10.68). A new configuration was allowed if either the particle energy was lower than the initial configuration, or the exponential of minus the difference of these energies relative to kT (with $T = 300$ K) was greater than a certain number picked at random between zero and unity. The simulation was supposed to be a part of an indefinite system, so periodic boundary conditions were imposed and long-range interactions calculated. The simulation was performed for a constant field amplitude 13 kV/m over the frequency range 10 Hz to 10 MHz.

To quantify the Monte Carlo simulation results and compare them with experiments on mixtures of yeast cells and polystyrene beads, the connectivity of the particles was defined as the average number of particles (cells or polystyrene beads) contacting each other. Two particles were considered to be in contact when the distance between the centre coordinates of two particles was less than $2.1r$, r being the radius of the yeast cell or polystyrene bead. If two particles satisfied this criterion and a line joining the centres of the two particles made a small angle (±7°) with respect to the field direction, then the contact was defined as *longitudinal*. A *transverse* contact was defined for connected particles with this angle lying within ±7° from a line perpendicular to the field direction. The longitudinal and transverse connectivity obtained from the Monte Carlo simulation compared well with experimental results [73], an example of which is shown in Figure 10.31. Mutual interactions between polarized particles also couples their electrorotation response [74] and produces an effect whereby two neighbouring rotating cells revolve around each other as they slowly come together to form a rotating doublet [75].

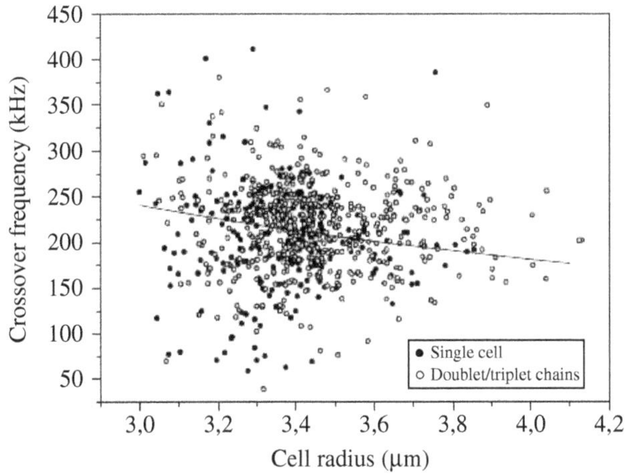

Figure 10.33 Experimental DEP crossover frequencies (f_{xo1}) for single T-cells and doublet / triplet chains [75]. The solid line represents the theoretically modelled f_{xo1} versus radius for a single cell of membrane capacitance 6 mF/m^2. The scattering of data shows the variability of the morphology and dielectric characteristics of the cell population. (Reproduced with permission.)

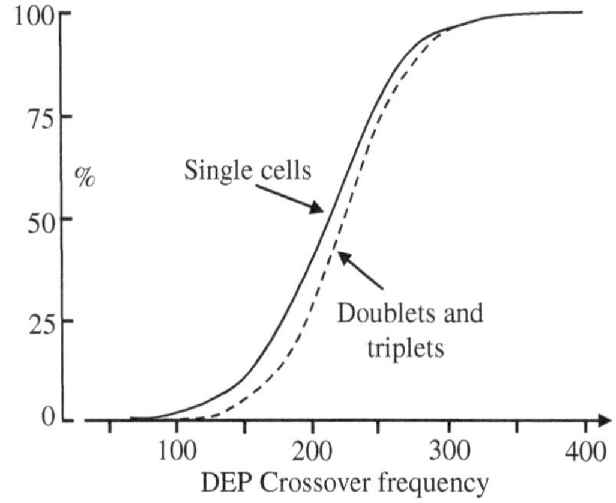

Figure 10.34 Best fit sigmoidal curves to the cumulative percentage distributions of the experimental f_{xo1} values shown in Figure 10.32 for single T-cells and doublet / triplet chains. (Based on Sancho *et al.* [75].)

Sancho *et al.* [75] also investigated both experimentally and theoretically the effect that pearl chaining of cells has on their DEP cross-over frequency (f_{xo1}). Human T-lymphocytes (~10^6 cells/ml) were suspended in an isotonic medium of conductivity 40 mS/m. From observations of more than 500 cells it was found that single cells exhibited a mean f_{xo1} of ~210 kHz, whereas for those that formed doublets, triplets and higher order pearl chains (65% of the total cell population) the value for f_{xo1} shifted on average 10.7 kHz to a higher frequency. The corresponding distributions of the f_{xo1} values and cell radii are shown in Figure 10.33. The largest shift in f_{xo1} from the single cell value occurred for doublets and triplets, with the addition thereafter of extra cells to the chain producing a much smaller shift of the f_{xo1} value. As shown in Figure 10.34, the theoretically derived shift of f_{xo1} agreed qualitatively in sign and order of magnitude with experimental observations [75]. Although pearl chaining results in a relatively small change of the DEP crossover frequency, it is an important consideration when designing efficient DEP-based cell separation protocols.

Camarda *et al.* [76] performed a three-dimensional Monte Carlo simulation to determine the equilibration of a 3D suspension of 1920 interacting cells suspended in a defined volume (1600 × 1600 × 1500 μm^3) of static liquid, subjected to a DEP force generated by polynomial electrodes. The purpose of this study was to compare this simulation with an actual experiment performed using the same cell type; cell concentration (5 × 10^5 cells/ml); aqueous medium conductivity (30 mS/m); polynomial electrode geometry (similar to Figure 2.15) and applied voltage signal (1 MHz, 8 $V_{pk\text{-}pk}$). No mention is made of the mass density of the cells compared to that of the suspending medium, so that the buoyancy of the cells does not appear to have been factored into the simulation. The cells were chosen to be a breast cancer cell line (MDA-MB-231) whose average radius (6.2 μm) and dielectric properties had been evaluated by Gascoyne *et al.* [77]. The Monte Carlo simulation consisted of 2 × 10^8 iterations, equivalent to 180 seconds in real experiment time, which allowed sufficient time for the cells to attain equilibrium positions. The local field at each simulation step was taken to be the sum of the field generated by the electrodes and the contributions of the induced dipole fields of the cells. The configuration potential energy of the cells was taken to be the sum of the potential energy given by Equation (4.41), whose minimization leads to DEP capture of the cells at the electrode edges and the interaction energy given by Equation (10.68) whose minimization drives pearl chain formation. Excellent agreement was found between the final spatial distribution of cells at the electrode edges obtained by the Monte Carlo simulation with that found experimentally. At a frequency of 1 MHz the cells are expected to collect at the electrode edges under the influence positive DEP. In both the simulation and experiment this was observed, with single cells and small pearl chains distributed at the electrodes in a manner similar to that shown in Figure 2.15(a). However, away from the electrode edges the cells appear to be distributed randomly. This resulted from the fact that the distance between opposing electrode edges was 780 μm, much less than the distance 128 μm reported in reference [2], so that the field and field gradient would have been low and zero at the centre. The evidence for pearl chaining at the electrodes was taken to indicate that the effects of particle-particle interactions play a crucial role, even for concentrations as low as 10^5 cells/ml. Taking into account the fact that the large distance between the electrode reduced the volume for effective cell trapping region, Camarda *et al.* [76] concluded that to avoid dipole-dipole interaction the concentration of cells should be below 3×10^4 cells/ml! This represents an average distance of 203 μm between cell centres (i.e., greater than 16 cancer cell diameters).

The three-dimensional Monte Carlo method developed by Camarda *et al.* [76] is undoubtedly a powerful one, but their interesting conclusion regarding a maximum cell concentration of 3×10^4 cells/ml to avoid dipole-dipole interactions appears too conservative and is worthy of further study regarding its relevance to the development of protocols for DEP-based cell separation protocols. For example, this proposed limit can be compared to the concentration of $\sim 10^9$ bacteria/ml [78], $\sim 4 \times 10^6$ myoblasts/ml [79], $\sim 4 \times 10^6$ mononuclear blood cells/ml [80, 81] and $\sim 10^7$ mononuclear blood cells/ml [82] reported for various cell enrichment or separation protocols using various forms of fluid-flow assisted DEP devices. These various cell concentrations probably represent the upper limit for efficient DEP processing and will depend on such factors as the geometry and dimensions of the electrodes and the fluidic chamber; the magnitude of the applied voltage signal; the fluid flow rate; the difference in dielectric properties of the target cells and those from which they are to be separated. Because DEP is a ponderomotive effect, the occurrence of pearl chaining may in fact be beneficial in increasing the rate at which unwanted cells are either attracted or repelled from the electrodes. However, an undesired effect would be cell-cell interactions that lead to the entrapment of target cells by the unwanted cells. A fine line will divide these two situations. Of great value would be 3D Monte Carlo simulations of the kinetic evolution of the equivalent experiment where rare target cells, to be collected by positive DEP, are spiked into a much larger concentration of cells that should exhibit negative DEP. The chamber height chosen by Camarda *et al.* [76] is about ten times higher than the range at which a DEP force on a cell can be effective. In their simulation, single cells may have been attracted to the electrodes by positive DEP, whilst the dipole-dipole interactions that formed pearl chains may only have been a significant factor in determining the way the cells built up at the electrodes. In other words, were the cells attracted to cells already trapped at the electrode? This is commonly observed in DEP experiments and would be an important aspect to clarify. It is also clear that when fluid flow is incorporated into a DEP cell sorting protocol, large cell concentrations can be considered when sample volume size and processing time is important. Finally, would it be pushing the wish list too far to wonder if the normal biological distribution of cell size and dielectric properties, such as that shown in Figure 10.33, together with cell buoyancy (sedimentation) and surface charge, can be incorporated into the powerful Monte Carlo method described by Camarda *et al.* [76]?

The effect of the perturbing influence of boundaries has not received as much attention as that given to particle-particle interactions. A polarized particle located near a metal or dielectric wall will induce image charges in the wall material to create an attractive or repulsive force, respectively. Lo and Lei [83–85] have employed the theory of images described in Chapter 5 to derive DEP force and torque expressions for a sphere of radius R located a distance h from a wall. The ratio of the wall perturbing force to the DEP force was found to be of the order $(L/h)(R/2h)^3$. L is the length scale of the electric field, taken to be the distance between opposing electrodes for DEP and electrorotation experiments. For L = 100 μm and a particle radius of 5 μm, the particle centre would need to be more than 12 μm from a boundary for the wall effect to be negligible (i.e., less than 5% of the primary DEP force or electrorotation torque). This condition can readily be satisfied for electrorotation experiments, but this is not the case when particles are being attracted to an electrode by positive DEP. For traveling-wave DEP, L is the distance between every second electrode, corresponding to a one-half wavelength when quadrature-phase voltages are used and thus equal to the distance of maximum phase difference on the electrode track [45]. A typical value for L is 20 μm. A 5 μm radius particle would need to be levitated more than 8 μm above the electrode plane for the wall effect to be negligible. Levitation heights above 25 μm are common and so this condition is readily met in most applications and studies of traveling wave DEP.

Camarda *et al.* [86] have described the existence of 'anomalous' regions on the electrode and interelectrode surfaces of a planar array of interdigitated electrodes (i.e., the geometry depicted in Figure 10.27) where the effective DEP force does not depend directly on the complex permittivity of the particles and suspending medium (i.e., is independent of the *CM* factor). As shown in Figure 5.21, the image force resulting from a positive *CM* factor is expected to enhance the attractive DEP force at an electrode surface. Likewise, a negative *CM* factor that drives a particle from an electrode to an insulating surface should give rise to a repulsive force with its dipole image. However, based on both analytical models and finite element simulations, Camarda *et al.* [86] have identified situations where negative DEP at an electrode edge can result in a positive DEP force of attraction towards the centre of the electrode strip! This effect appears to be related to the effective values of the field E and the factor ∇E^2, as averaged over the particle volume and determined by solving Poisson's equation (which is scale invariant). If the particle radius R is less than a critical value given by $0.19W_{el}$, where W_{el} is the electrode width, the average value of E exceeds ∇E^2 at the surface so that the image dipole force dominates and leads to a cohesive rather than repulsive force. Thus, for the case of $W_{el} = 50$ μm, all cells with radius smaller than 9.5 μm will be attracted to the centre of the electrodes, rather than being repelled by negative DEP. Camarda *et al.* [86] suggest that the separation efficiency of field-flow assisted DEP separation of rare cells (e.g., cancer cells) from blood could be improved by using an interdigitated design of electrode width 14 μm and gaps of 50 μm. It would be of interest to investigate to what extent this anomalous dielectrophoretic effect compares with that described in early studies using castellated interdigitated electrodes [39] and later determined to be caused by electro-osmotic fluid flow [3 (pp. 65–69), 4 (pp. 144–145), 32].

10.4 Electrodes: Fabrication, Materials and Modelling

10.4.1 Metal Electrodes

In his early DEP experiments Pohl [87, 88] employed metal wires and thin sheet electrodes to produce nonuniform fields. For example, he used a 10 mm diameter tungsten wire as the central electrode and a band of tinfoil as the outer concentric electrode to remove by positive DEP the carbon black filler from solutions of polyvinylchloride. DC or AC voltage differences of up to 11 kV were applied across the electrodes [1, pp. 122–367]. To quantify DEP collection of particles, the electrodes were observed through a microscope and particle collection (often in the form of pearl chain growth) was photographed over several minutes [1, pp. 361–380]. The use of high voltages often resulted in fluid motion, arising from thermal effects, which perturbed the DEP-induced motions of the particles. In more recent times this problem has been reduced in a DEP microfluidic device in which 100 μm diameter platinum wires, in a pin-plate configuration, span the entire depth of the DEP chamber and are energized with 15 $V_{pk\text{-}pk}$ voltages [89].

The fact that the parameter $(E.\nabla)E$ in Equation (10.8) has units of V^2m^{-3} provided the clue that, by miniaturizing the electrodes and thus being able to use much smaller applied voltages, thermal and electrolysis effects could be avoided. The application of photolithography and metal vapour deposition for fabricating microelectrodes was described in 1979 [90] and modern developments of this are now well documented [e.g., 91, 92]. The interdigitated, castellated, geometry shown in Figure 10.35 was chosen because it provided a large value for $(E.\nabla)E$, using modest values of applied voltage and enabled both positive and negative DEP behaviour to be observed [93]. The first sets of electrodes were fabricated by sputtering onto a glass substrate a seed layer of chrome, followed by a 1 μm layer of sputtered copper, which was then etched by a standard photolithographic method and finally covered with a 0.2 μm layer of chemically deposited gold. To avoid the cytotoxic effects of copper, this microfabrication procedure was modified to sputter gold directly onto the chrome seed layer, to give a final electrode thickness of ~70 nm. The characteristic dimension defining the planar geometry typically ranged from 10 – 120 μm, chosen to be 5~10 times the diameter of the particles to be manipulated by DEP. These dimensions were refined using a finite-difference computational method to solve Laplace's equation for defined electrode voltage potentials [94]. Field and potential energy profiles were later obtained using the charge density method with the aid of a VAX computer and FORTRAN (VAX/VMS) operation system [95]. Commercial software packages are now available for such modelling to be performed on a personal rather than mainframe computer. An example is shown in Figure 10.36 using COMSOL Multiphysics® finite element analysis software. Reproduction in grayscale of the original

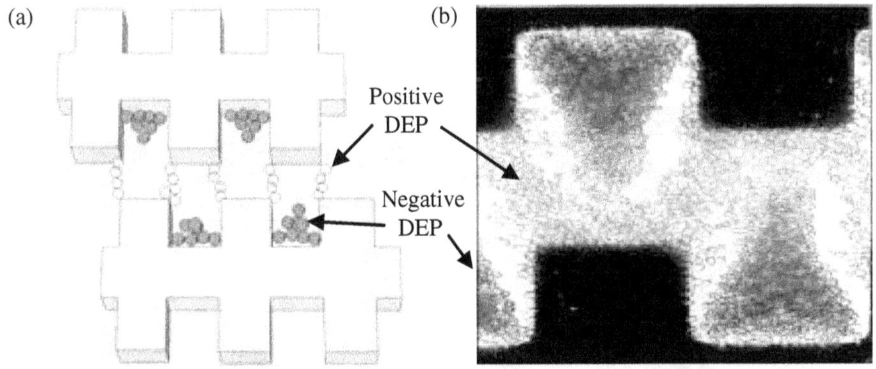

Figure 10.35 (a) The interdigitated, castellated, microelectrode design for facilitating both positive and negative DEP collection of cells [39]. (b) Dead cells (stained) held by negative DEP are weakly trapped and can readily be removed by flowing fluid over the electrodes, whereas viable cells held by positive DEP remain trapped [93].

Figure 10.36 COMSOL Multiphysics® model of an array of interdigitated electrodes. The original colour scale more clearly revealed the voltage potential profiles along each electrode element (unpublished).

colour-scale output does not adequately demonstrate the value of such modelling, which in this case was performed to analyse the voltage drop along the whole length, as well as along individual finger elements, of an array of interdigitated electrodes. Along with the development of user-friendly software packages, the technical challenge in producing microelectrodes has also been eased. Rapid and low cost microfabrication methods to produce electrode arrays of the interdigitated, castellated, geometry have been described by Rajaraman *et al.* [96].

As depicted in Figure 10.35, the interdigitated, castellated, electrode geometry can be used to observe both positive and negative DEP of cells across an array of microelectrodes simultaneously [39]. Cells experiencing negative DEP are directed to regions of local field minima in the 'bay' regions of the castellations. When located in such a potential energy well, the cells are in fact elevated slightly above the electrode plane in the fluid medium. This has been exploited in flow-through devices for the DEP separation and isolation of different cell types in mixtures (e.g., [93]). By aligning the castellations, as shown in Figure 10.37(a), particles elevated into a flowing fluid are directed into well defined flow paths [97]. This effect has been exploited by Yasukawa *et al.* [98] in a modified design, shown in Figure 10.37(b), to separate particles according to their size. By coupling acoustic waves into an interdigitated microelectrode system, particles can first be preconcentrated before focusing them into flow channels and to precise locations using DEP forces [99].

Another electrode design, the quadrupole 'polynomial' electrode system shown in Figures 10.3 and 10.23, was chosen to provide defined analytical expressions for the spatial variation of the factor ∇E^2 [2]. The design of polynomial electrodes is based on the assumption that the electrical potential at any point created by an electrode system of interest is defined by a polynomial that obeys Laplace's equation. By substituting this polynomial into Laplace's equation the corresponding equipotentials can therefore be determined and these in turn can be used to define the electrode boundaries. The general design comprises $2n$ electrodes, but the most common one employs $n = 2$, namely the quadrupole. Electrode separations from 5 μm to 500 μm (as measured between opposing electrodes across the centre) have been extensively used for both DEP and electrorotation experiments. The quadrupole design has found wide application. For example, a quadrupole electrode array has been used by Gagnon *et al.* [100] in a device to monitor the DEP crossover frequencies (f_{xo1}) of oligonucleotide-functionalized silica nanoparticles as they participate in DNA-DNA hybridization reactions. Kuo and Hsieh [101] describe a method for performing single-bead-based biochemical assays on a quadrupole DEP microfluidic chip, whilst Voldman *et al.* [102, 103] have designed extruded quadrupole electrodes, with an asymmetric trapezoidal geometry, which are electrically switchable and can be scaled up to form a dynamic array cytometer. The ability of quadrupole electrodes to form negative DEP traps is useful when operating at high values of the fluid medium conductivity. For example, they have been developed as single-cell trapping devices for fluids having conductivities (1.25 S/m) typical of physiological fluids and culture growth conditions [104]. A 'zipper' electrode design has been described by Hoettges *et al.* [105] in the form of an array of interlocking, approximately circular, electrode pads. This design exploits field-induced electrohydrodynamic fluid flow to direct particles towards the electrode pads and then DEP forces to trap them. The effective capture volume for particles is increased, so there is also

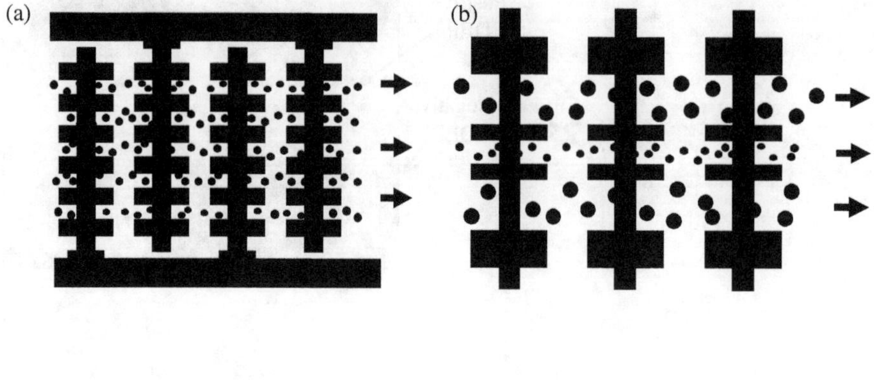

Figure 10.37 (a) Cells focused into narrow bands of fluid flow by negative DEP using interdigitated, castellated, electrodes oriented at right angles to the fluid flow [97]. (b) A modified geometry of the electrodes designed to separate particles into separate fluid flow lines according to their size [98].

the advantage that the particles are deposited at the centres of relatively large electrodes and can potentially be detected by surface plasmon resonance or evanescent light scattering. Hoettges has also reviewed the types of electrodes commonly used for DEP studies of cells [106].

The consequence of a particle experiencing positive DEP is that it should always tend to move against a local field gradient that leads to a local maximum of the field. For the particle this represents it quest to search for a local potential energy minimum. Field maxima always occur at the edges of an electrode, so that is where particles will collect under the action of positive DEP [95]. For the early efforts to quantify such DEP collection, the electrodes were observed through a microscope and recorded as a sequence of photographs. More sophisticated methods have been developed, including the monitoring of changes in the optical scattering of light beams through particle suspensions [107]; computerized image analysis of particle motion [108, 109]; changes of the impedance of the electrodes [110, 111]; fluorescence detection for the case of submicron particles [112, 113]. An array of zipper DEP electrodes has been fabricated onto a quartz crystal to provide an extra force to drive particles towards the crystal surface [114]. Particles loading onto the crystal can be detected very sensitively as a shift in resonant frequency of the crystal and this DEP-aided device was found to perform up to five times faster than other quartz crystal microbalance surface loading techniques described in the literature. Surface enhanced Raman scattering has been used as an on-chip detection method in an integrated DEP for the continuous filtering, trapping and sorting of bacteria [115].

10.4.1.1 Materials

A significant development has also been the introduction of new materials and methods for fabricating microelectrode DEP devices. The early ones (e.g., [94]) consisted of two glass slides held apart by a thin gasket, with microelectrodes deposited on one or both inner faces, sealed together with inlet and outlet fluid ports using epoxy resin. The microelectrode fabrication required access to photolithography and metallic vapour deposition facilities in a clean room. The introduction of the silicone polymer poly(dimethylsiloxane) (PDMS) has enabled fast and inexpensive fabrication of microfluidic devices by 'soft lithography' under normal benchtop conditions [116, 117]. Apart from being robust, flexible, biocompatible and of low thermal conductivity, PDMS is also ideal for producing devices that require more than one material in fabrication, since it can seal to a variety of materials. 3D devices are easily fabricated by aligning and sealing different layers of PDMS containing channels, reservoirs, valves and electrode-bearing substrates. Valuable protocols for using PDMS in the fabrication of microfluidic devices has been provided by Friend and Yeo [118]. A reconfigurable microfluidic chip system has been developed by Dalton and Kaler [119], in which PDMS microchannels are reversibly bonded onto the chip, allowing them to be readily removed for cleaning, changing of analytes or to accommodate different microfluidic channel and electrode geometries. Using this technique, rapid prototyping of both microelectrode designs and microfluidic systems can be performed by most research groups (PDMS is commercially available as a two-part self-curing material supplied as liquids). Using photo-patternable silicones, two levels of metal deposited electrodes can be sandwiched into a microchannel layer, without an extra etching step being required to make electrical contact between the fluid in the channel and electrodes on the upper and lower layers [120]. Patterned microfluidic networks can also be created using dry film resist in either a cleanroom or basic laboratory conditions. The resist can be double bonded at relatively low temperatures without the use of extra adhesives, so that complex devices can be fabricated with active elements on two substrate layers [121]. Aspect ratios of more than two can be achieved for free standing structures such as channels and pillars. The dry resist is inexpensive, fluid sealable, biocompatible and can be processed on almost any substrate with any dimension, ranging from a single chip to complete silicon wafer.

Electron beam lithography and thin film techniques have been used to fabricate vertical microelectrodes, with electrode gaps down to 0.2 μm, made by the superposition of niobium, titanium and gold layers [122]. The niobium layer improves the mechanical strength and ensures a good resistance of the structure to galvanic corrosion. Excimer laser ablation, capable of high-resolution patterning over large areas, has also been used to fabricate DEP devices. Examples of this are glass-based chips, combining traveling wave DEP and electrorotation, to concentrate and assay the viability of microorganisms [50], as well as structures incorporating DEP, traveling wave DEP and electrorotation for selectively isolating and then characterizing cells, microorganisms and other particles [123]. Figure 10.38 shows a traveling wave DEP junction, fabricated by excimer laser ablation and designed to either bring together or separate different particles types [123]. Other examples of combining the complementary techniques of DEP, traveling wave DEP and electrorotation include their combination onto a single, PC-controlled, printed circuit board, which was tested by manipulating tumour cells [124]. All three AC electrokinetic techniques have also been incorporated onto a single, integrated, silicon chip (3 × 6 mm) using conventional microfabrication and tested using human malignant cells to demonstrate the ability to perform as a programmable microsystem [125]. Electrode arrays have also been produced by the excimer laser ablation of indium tin oxide (ITO) electrodes [126, 127, 128].

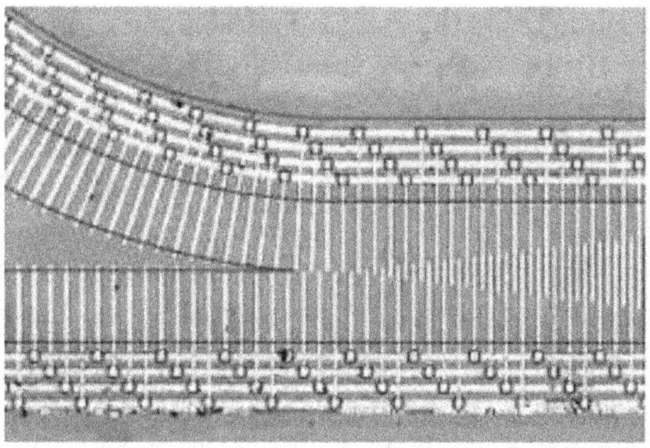

Figure 10.38 Part of a travelling wave DEP (TWD) device that was fabricated using excimer laser ablation [123]. The section shown here acts as a TWD junction to either bring together (from left to right) two types of particle, or to separate (from right to left) particle types according to their dielectric properties.

Although the optical transparency of an ITO film is high, making this electrode material particularly suitable for observing DEP particle manipulation with transmission microscopy, it can be micromachined using an excimer laser beam because UV radiation is strongly absorbed by ITO.

The development of DEP technology has always been driven by both curiosity and the search for potential practical applications. The separation of target cells from cell mixtures or biofluids for diagnosis, drug screening assays, therapeutic applications or for further analysis are obvious applications. Examples of how DEP technology has evolved to address such opportunities include the 'funnel' electrode design shown in Figure 10.39 and developed by Fuhr *et al.* [129,130]. A positive DEP force,

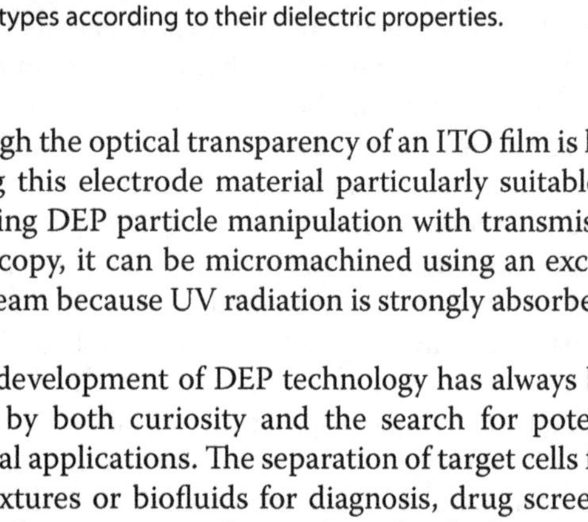

Figure 10.39 (a) DEP funnel electrode design for focussing and concentrating particles in a flowing aqueous suspension (based on Fiedler *et al.* [129]). (b) 'Herringbone' electrodes located at the top and bottom of a DEP chamber. Cells experiencing positive DEP remain in two outer fluid streams, whilst those experiencing negative DEP are deflected into the centre output fluid port (based on Muratore *et al.* [79]).

directed normal to the angled electrodes, guides particles to a small exit gap, at which point particle concentration can be considerably enhanced over the starting concentration introduced into the device. Angled electrodes are employed in an integrated DEP chip design by Cheng *et al.* [131] for the continuous filtering and sorting of bioparticles and also used to great effect in a continuous DEP size-based particle sorter [132] and a multitarget DEP activated cell sorter [133]. A simple arrangement for aligning cells in fluidic channels, consisting of two face-to-face strip electrodes mounted on the top and bottom of a microchannel, has been described by Schnelle *et al.* [134]. In this device, particles exhibiting negative DEP are brought by fluid flow to an energized electrode pair and as a result of experiencing repulsion forces from both electrodes are lifted into the central stream of the fluid flow. This basic concept has been refined by Demierre *et al.* [135] who fabricated a microfluidic device based on an arrangement of lateral metal electrodes and a patterned insulator. This device combines the concept of insulator-based 'electrodeless' DEP with multiple frequencies to achieve focusing and continuous separation of dielectric particles flowing through a channel. The opposition of two DEP-force fields, operating at different frequencies, defines a position of equilibrium for the dielectric particles placed in these fields. As well as being able to direct particles into specific fluidic streams or ports, the selective trapping of them is also important. A simple method for producing microwell DEP traps has been described by Fatoyinbo *et al.* [136] and involves drilling holes through a laminate consisting of 20 aluminium layers and 19 epoxy layers. Bocchi *et al.* [137] describe a similar method, which involves drilling holes through a polyimide substrate containing copper-gold or aluminium metal layers that form three annular electrodes within the well. A channel under the device provides the means for fluid flow into the microwells by capillary action. An array of traps designed for single-particle patterning and capable of holding cells in position against the force produced by practical fluid flow rates, has been described by Rosenthal and Voldman [138]. Thomas *et al.* [139] describe a particle trap, consisting of a metal ring electrode and a surrounding ground plane, to create a closed electric field cage for particles experiencing negative DEP. The simplification that each trap requires just one electrical connection allows for the fabrication of a large array of single particle trapping centres, capable of holding the particles in a flowing fluid. The operation of the device was demonstrated by trapping single latex spheres and HeLa cells against a moving fluid. Cells can be maintained on the chip for further culture, or released from the chip by fluid flow. Kang *et al.* [140] describe a microfluidic device for separating particles based on a hybrid design of a PDMS insulating hurdle and a pair of embedded metal electrodes to generate DEP forces. DEP forces can be used to increase the rate at which particles drift towards a sensing element. An example of this is described by Pham *et al.* [141], who theoretically modelled and fabricated electrodes, in the form of pyramidal shaped trenches, to enhance the transport of bioparticles such as cells, proteins and DNA towards reactive surfaces. Particles experiencing either positive or negative DEP can be attracted towards separate collection regions in this pyramidal design. Buyong *et al.* have described the design and use of tapered aluminium microelectrodes for improvement of DEP particle manipulation [142].

10.4.1.2 Theoretical Modelling

A detailed analysis of the factor $(E.\nabla)E$ is often required for the design of the electrodes in a DEP device. A common problem when attempting this arises from the nature of the boundary conditions. The applied electric potential is chosen for the surface of metal electrodes (the Dirichet condition) and in the rest of the space to be analysed the Neumann condition is used to specify that the normal derivative of electric potential at a boundary surface is zero. This can limit the number of simple geometries able to be analysed accurately. Numerical methods have been adopted to derive approximate solutions, based on Green's theorem [143] and Fourier series [48]. However, as discussed by Green *et al.* [144] these methods can lead to inaccurate results. A closed-form solution with the exact boundary conditions for conventional DEP electrodes has been presented by Chang *et al.* [145], but is difficult to apply to traveling wave DEP, for example. Analytical solutions of the electric potential for planar electrodes that are relevant to both normal DEP and traveling wave DEP applications have been presented by Sun *et al.* [146] and Alazzam *et al.* [147]. Song and Bennett [148] have described a new semianalytical approach to the modelling of the DEP force generated by planar parallel electrodes as well as those forming 3D arrays of the form developed by Chen *et al.* [149]. The unknown coefficients of the Fourier series derived for the electric potential equation were determined by training a linear neural network, using appropriate data that satisfy both the Dirichet and Neumann conditions.

Voldman *et al.* [102, 103] have developed simulation tools to model the performance of quadrupole DEP particle traps. This simulation takes as inputs the electric-field data and other experimental parameters and computes the total force acting on a particle everywhere in space. From this it can be determined if the total force on the particle in the trap stably goes to zero at some location. Such locations are called *holding points* and represent where the particle will be held in a trap. By varying the applied flow rate for a given experimental condition, this modelling environment can determine when the *holding points* cease to exist and therefore the strength of the DEP particle trap. Schnelle *et al.* [150] conducted a comprehensive analysis and experimental investigation of the forces acting on dielectric particles and living cells exposed to alternating and rotating fields generated by three-dimensional multielectrode microsystems. This analysis included a description of numerical procedures for calculating the electric field distribution and negative DEP forces for electrodes of any shape and dielectric particles of complex structure, produced by high-frequency AC or rotating electric fields up to 400 MHz. Various multielectrode systems were tested for their ability to move and assemble microparticles or living cells without contact with the electrodes. Park and Beskok [151] have provided a simple, but valuable, theoretical model that considers the relative magnitudes of DEP, electrophoresis, AC-electro-osmosis and Brownian motion forces acting on microparticles. This theoretical model provides quantitative descriptions of ac electrokinetic transport, for a given target species and suspending medium conductivity, over a wide spectrum of electric field amplitude and frequency. Experimental validations of the model were conducted using interdigitated microelectrodes for polystyrene and gold particles, as well as *Clostridium sporogenes* bacterial spores [151].

10.4.2 Insulator-Based Dielectrophoresis (iDEP)

Before describing the development of DEP devices in which metal electrodes are replaced with electrical insulators, a brief explanation of why this could be useful will be helpful. We have seen that electrode-based DEP devices have been applied successfully to the physico-chemical characterization of inanimate particles and bioparticles and especially for the selective isolation or separation of cells from cell mixtures. However, the frequency range over which cell separation protocols mostly operate is from around 10 kHz to 50 MHz. It is not sensible to operate electrode-based DEP devices at lower frequencies because of the disturbing effects of electrode polarization, electrolysis and electro-osmosis. For mixtures of viable cell types of roughly the same size, efficient cell separation is based on exploiting the difference in membrane capacitance of the target cell from other cell types, which manifests itself as a difference in the DEP crossover frequency f_{xo1} given by Equation (10.43). This equation informs us that cell separation can also be based on finding the appropriate operating frequency that exploits the difference in the sizes of the cell types. If the target cells to be removed are nonviable cells, then differences in the membrane conductance is exploited. If we wish to extend the basis of DEP cell separation to include differences in cell surface charge, we

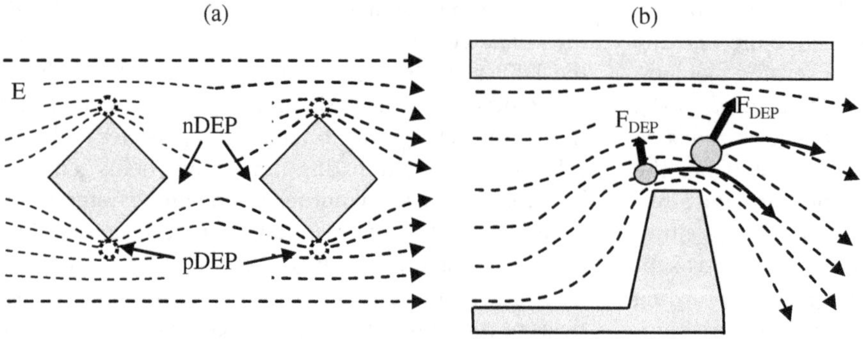

Figure 10.40 (a) The distortion, caused by two diamond-shaped insulator posts, of a uniform electric field in a conducting fluid can create regions where particles collect by either positive or negative DEP. (b) Two particles of the same polarizability are shown driven along field lines by electro-osmosis towards an insulating obstacle in a fluidic channel. The larger particle is diverted further away from the field lines because it experiences the larger negative DEP force.

need to operate below ~1 kHz and avoid the situation where metal electrodes make contact with the particle suspension. Removing the metal electrodes and replacing them with insulating structures becomes an option. For example, external electrodes can be used to generate a large DC electric field across a channel containing an electrically conducting fluid. Distortions of this field can be created by placing insulating constrictions inside the channel, as first demonstrated by Washizu *et al.* [152,153] in devices that bring together two cell types before electrofusing them into a hybrid cell. If instead of a constriction, insulator posts are placed in a channel containing a conducting fluid, an imposed electric field is distorted according to the basic scheme shown in Figure 6.12 for the case of a particle less polarizable than its surrounding medium. Localized regions in the conducting medium around such posts can be created, which represent potential energy minima to which particles may be attracted by DEP. This is illustrated in Figure 10.40(a) for the case of two diamond-shaped insulating posts located in a conducting medium. For a fixed shape and size of the post, the depths of these potential energy minima will depend on the magnitude of the applied field. Washizu *et al.* [152, 153] generated the fields using metal electrodes embedded into fluidic channels and they also used mechanical pumps to produce fluid flow. If the electrodes are placed outside the chamber and energized with a DC voltage, particle and fluid flow can be generated by electrophoresis and electroosmosis, without the need for mechanical pumps. In this situation an insulating constriction, such as that depicted in Figure 10.40(b), with an imposed DC field strong enough to induce a significant DEP force, can be used to deflect particles of different size (or conductivity) into different flow streams. The electrokinetic forces of electrophoresis and electroosmosis are linearly dependent on the DC field strength, whereas the DEP force depends on the square of this field strength. We can therefore expect some form of critical field value at which DEP will become effective.

The first approach to the development of iDEP devices that incorporate electrokinetic fluid flow was that of Chou *et al.* [154, 155] who described the fabrication on quartz wafers of DEP traps that can operate at frequencies far lower than is practicable using metal electrodes. An array of insulator constrictions, of the form depicted in Figure 10.41, was fabricated using reactive ion etching techniques. After pretreating the internal walls and surfaces of the fluidic channel with oxygen plasma, so as to make them hydrophilic and wettable, the influence of fluid flow caused by electro-osmosis was minimized by adding polyacrylamide to the buffer solution. The concentration and patterning of both single-strand and double-strand DNA was observed at the insulating constrictions. The DNA was driven through the array of constrictions by electrophoresis, generating the required field using external gold electrodes and a high voltage amplifier with 1 kHz bandwidth and an output of ±1 kV. The DEP trapping force on the DNA was found to increase with increasing applied voltage and frequency, up to the maximum assessable frequency of 1 kHz [154]. This frequency dependent polarizability (i.e., CM factor) of the DNA was attributed to it exhibiting a dielectric relaxation associated with counterion fluctuations, as described in Chapter 8. Chou and Zenhausern [155] later described how this device could efficiently separate a mixture of *E. coli* in 2% whole blood in a 100 mm salt solution. With a 100 $V_{pk\text{-}pk}$ applied signal at 2 kHz, *E. coli*

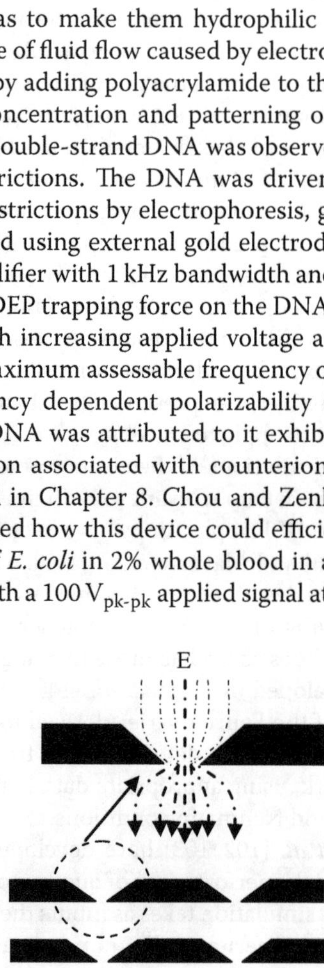

Figure 10.41 An electrodeless DEP trapping device consisting of an array of constrictions (1 μm wide and 1.26 μm deep) etched in quartz. In a conducting solution an applied electric field E is 'squeezed' to produce high field regions for the positive DEP collection of molecular species such as DNA (based on Chou *et al.* [154]).

cells were trapped by positive DEP at the restrictions, whilst the blood cells were repelled from them by negative DEP.

Following the work of Chou *et al.* [154], Cummings and Singh [156] described what they termed as insulator-based dielectrophoresis (iDEP). They investigated the DC electrokinetic behaviour of fluorescent latex nanoparticles suspended in a series of 7 μm deep channels containing 1 mM phosphate solution and uniformly patterned arrays of insulating glass posts. These posts were spaced 63 μm apart and fabricated on glass wafers using standard photolithography and wet etch techniques. Different channels contained posts of different shapes (round, square, diamond) oriented at different angles to an applied DC field. The diamond and square posts had sides of 36 μm, whilst the circular posts were 36 μm in diameter. Under normal conditions, in the absence of insulator posts, the particles should flow along the DC field lines with electrokinetic behaviour that combines the effects of electrophoresis and electro-osmosis. However, as shown in Figure 10.40 the introduction of insulator posts introduces another electrokinetic factor – namely dielectrophoresis. Very interesting regimes of electrokinetically driven particle transport were observed by Cummings and Singh [156]. For relatively low field strengths (~250 V/cm), dielectrophoresis was found to be small compared to the electrokinetic flow. At moderate applied electric fields (800 V/cm) dielectrophoresis overwhelmed diffusion and electrostatic repulsion among the particles, but not electrokinetic flow. This resulted in streams of concentrated and rarefied particles. At a critical field of ~1 kV/cm, dielectrophoresis dominated over the other transport mechanisms, resulting in the trapping of particles at field maxima. For the case of diamond-shaped posts the trapping sites corresponded to the regions of field maxima shown in Figure 10.40 and could be coherently reinforced depending on the patterning, shape and orientation of the posts. This effect was termed as *streaming* dielectrophoresis. Simple mathematical models and continuum simulations, based on ideal electrokinetic flow and dielectrophoresis, for various DEP particle concentrators and sorters were described by Cummings and Singh [156]. The perceived advantages of using insulating posts rather than metal electrodes were considered to include: simpler and cheaper fabrication (by stamping, etching or moulding); lower susceptibility to fouling; no electrochemical alteration of the fluid or particles, manifested for example by the generation of gas bubbles.

The iDEP device described by Cummings and Singh [156] was later shown to be capable of selectively separating and concentrating in a continuous manner two species of live bacteria [157]. Four species of bacteria were studied, namely the Gram-negative *Escherichia coli* and the Gram-positive *Bacillus subtilis, B. cereus* and *B. megaterium*. All four types of bacteria exhibited negative DEP. *E. coli* was found to become trapped at the weakest applied electric field, while the *Bacillus* species were trapped at different characteristic threshold fields. The order of trapping, from the lowest to the highest electric field required, was *B. megaterium* < *B. subtilis,* < *B. cereus*. This same order of trapping was observed at two different values of the solution conductivity (2.25 and 10.4 mS/m). This trend was not in agreement with the relative hydrodynamic diameters of the bacteria, which was determined from light scattering to be: *B. subtilis* (5.65 ± 1.23 μm) > *B. cereus* (4.01 ± 0.66 μm) > *B. megaterium* (3.15 ± 0.86 μm). The hydrodynamic diameter of *E. coli* was determined as 1.09 ± 0.32 μm. Because DEP is a ponderomotive effect we would expect the order of trapping to be that *B. subtilis* would be trapped at the lowest threshold field and *E. coli* at the largest field. Their different shapes and the fact that *E. coli* and *B. cereus* possess flagella was probably a major contributing factor for the order of trapping. The observed order of trapping means that mixtures of the bacteria can be separated by choosing the appropriate magnitude of the applied field. At stronger applied electric fields, two different species of bacteria in the microchannel were dielectrophoretically trapped into two spatially distinct bands. In another paper by the same team [158], using the same device, live and dead *E. coli* were concentrated and selectively released by applying stepped DC voltages. The dead cells were observed to have significantly lower dielectrophoretic mobility than live cells, whereas the electrokinetic mobilities (electrophoretic plus electro-osmotic) of live and dead cells were indistinguishable. The electrodes were two platinum wires placed in the inlet and outlet reservoirs, spaced 10.2 mm apart, producing mean electric fields of up to 2000 V/cm across the insulators (i.e., applied voltages up to 2 kV). In a separate set of experiments [159] the threshold field to trap *B. subtilis* spores, suspended in distilled water at a concentration of ~6 × 10^8 cells/ml, was found to be larger than those of vegetative *B. subtilis* cells (~500 versus ~1500 V/cm). This difference allowed the iDEP device to separate vegetative cells from spores. The threshold field (~2 kV/cm) for trapping Tobacco Mosaic Virus (TMV) was observed to be larger than those required to trap bacterial cells or spores. Additionally, TMV was selectively concentrated against a background of 200 nm polystyrene particles, demonstrating the ability of iDEP to separate particles having similar sizes but different dielectric properties. Cummings and his colleagues [160] later demonstrated the potential of insulator-based dielectrophoresis (iDEP) for particle filtration and concentration by describing devices that incorporated three-dimensional ridgelike glass structures in micrometre-sized fluid channels. They were created using a two-level etching process of an insulating substrate, which produced channels of depth 50 μm, width ~200 μm, length ~1 cm and an insulator-based ridge of height 50 μm in this channel. Conceptually, this ridge design can be employed to trap particles selectively in desired locations of a microchannel, so that they can be detected directly at these locations or held in position for chemical reactions (e.g., labelling) or washing steps. Alternatively, particles can be released after a desired period of concentration or trapping for later batch processing. A continuous processing mode was demonstrated whereby a mixture of bacteria (*B. subtilis*, 6×10^7 cells/ml) and polystyrene nanoparticles (4.5×10^8 per ml) were separated. With channel volumes $\sim 10^{-4}$ ml, this represents the processing per batch of $5 \sim 50 \times 10^4$ bacteria plus nanoparticles at a rate dictated by the electrophoretic flow velocity.

Baylon-Cardiel *et al.* [161] have shown that a simple set of equations can effectively model the performance of iDEP devices. The devices studied comprised a microchannel of length 10.22 mm, width 2 mm wide and depth 10 μm, with an array of 8 columns × 4 rows of cylindrical insulating posts 470 μm in diameter and arranged 510 μm centre to centre. The total volume of their chamber was thus ~0.2 μl, of which ~28% was occupied by the glass posts. By employing a relationship involving electrokinetic and dielectrophoretic velocities, the location and magnitude of the regions of DEP trapping along an array of insulating cylindrical structures can be predicted. The computational simulations correlated accurately with their experimental observations. With 500 V DC applied across the channel the maximum electric field strength (2.5 kV/cm) occurred in the narrow region between posts, coinciding with the location of the maximum value for ∇E^2 ($7 \times 10^{14}\ V^2/m^3$). This value can be compared to that of $-7.1 \times 10^{12}\ V^2/m^3$ calculated for the metal electrode based DEP device described in Example 2.7 and of similar dimensional scale (but with an applied voltage of 5 V and not 500 V). However, in order to trap DNA much larger values of ∇E^2 must have been generated by the constriction channels described by Chou *et al.* [154]. Another factor of relevance to the performance of DEP-based particle sorting devices is the sample throughput, which for a set concentration of suspended particles depends on the volumetric flow rate. For an open channel, with no obstructions such as posts, this quantity can be obtained by the product of the cross-sectional area of the channel and the velocity of the fluid. The fluid in the channel described by Baylon-Cardiel *et al.* [161] was essentially driven by electroosmosis, with a maximum electrosmotic mobility of $\sim 2 \times 10^{-4}\ cm^2V^{-1}s^{-1}$. For an applied potential difference of 500 V along the channel, this produces an electrokinetic velocity of $\sim 0.1\ cm\ s^{-1}$. The volumetric rate of fluid processed through their channel was thus $1\ cm \times (2 \times 10^{-3}\ cm) \times 0.1\ cm\ s^{-1} = 2 \times 10^{-4}\ ml\ s^{-1}$. To account for the cross-sectional area occupied by the posts, this estimate reduces to $\sim 1.4 \times 10^{-4}\ ml\ s^{-1}$, which is rather modest compared to the rates of $10^{-2}\ ml\ s^{-1}$ readily obtained using electrode-based DEP and mechanical pumps [e.g., 162]. Another important factor in microfluidic systems, especially when using high applied voltages and electric fields, is the generation of Joule heating. The electrical power W generated per unit volume is given by

$$W = \sigma_m E^2 \text{ (watts per cubic metre)}$$

where σ_m is the electrical conductivity of the fluid medium. With an applied voltage of 500 V, the field generated in the narrow fluidic region ($\sigma = 10$ mS/m) between the insulating posts was determined as 2.5×10^5 V/m [161]. This gives $W = 6.25 \times 10^8\ W/m^3$, which is generated in a very small volume ($\sim 40\ \mu m \times 80\ \mu m \times 10\ \mu m = 3.2 \times 10^{-14}\ m^3$) to give a power dissipation of ~20 μW within each active DEP region. There are 52 such micro-DEP regions in the chip design described by Baylon-Cardiel *et al.* [161], to give a total power dissipation of ~1 mW. The generation of this amount of power over time in a slowly moving liquid of very small volume ~0.15 μl could potentially result in a significant increase of temperature within the iDEP device. An evaluation of this potential problem requires solving the following energy balance equation [163, 164]:

$$\rho_m C_p \left(\frac{\partial T}{\partial t} + \text{v} \cdot \text{grad}T \right) = \text{div}(k\,\text{grad}T) + \sigma E^2 \tag{10.69}$$

where ρ_m, C_p, v, k is the mass density, specific heat, velocity and thermal conductivity, respectively, of the fluid medium. Based on an order-of-magnitude analysis of Equation (10.69) by Ramos *et al.* [164] for a typical electrode-based DEP device, the incremental temperature rise can be estimated from the relationship:

$$\text{grad}T \approx \frac{\sigma_m V^2}{k} \tag{10.70}$$

where V is the rms or DC voltage applied across the fluid to produce the imposed field E. In the iDEP system described by Baylon-Cardiel *et al.* [161], three fluid stream paths can be identified across which there is an applied potential difference of 500 V. These fluid streams have conductivities of 10 mS/m and for an aqueous electrolyte $k \approx 0.6\ Wm^{-1}K^{-1}$, so that Equation (10.70) predicts for these regions of fluid an incremental temperature rise of $\sim 4 \times 10^3$ °C! Although this probably represents a gross overestimation of what can happen in practice, it does draw attention in a dramatic way to the inherent thermal problems that could be a feature of iDEP devices. Equation (10.70) does however offer clear guidelines for minimizing possible Joule heating – namely: the electrical conductivity σ_m of the fluid should be as low as possible; the main channel along which the field is applied should be as short as possible (to minimize the magnitude of the applied voltage V).

Following these early pioneering studies [154–161] numerous other laboratories explored different designs and applications of iDEP devices. These included the use of various types of insulating material to distort the field (e.g., oil droplets [165], pairs of oil menisci [166], PDMS [167, 168], hot embossed thermoplastic [169]); post shapes other than circular, square and diamond (e.g., rectangular [167], triangular [170], saw tooth [171, 172]; different geometries of the fluid channel, including serpentine [173, 174]. Numerical simulations, validated by experiments with polystyrene particles, were reported, which related an average trapping condition and the average lateral-to-longitudinal force ratio experienced by particles [175]. These were used to determine improved geometrical parameters (e.g., shape, length and width) and arrangement (e.g., lateral and longitudinal spacing) of the insulator posts. The use of DC-biased AC fields has also been investigated [169, 176–178]. Hawkins *et al.* [169] describe the operation of a continuous-flow iDEP 'spectrometer' for sorting particles according to size. In this device a 50 V DC voltage is applied to induce a combined electrophoretic and electroosmotic fluid flow along a channel of high aspect ratio (250 : 1, width to depth). A thermoplastic ridge across the channel is curved in such a way that the angle between the ridge and the direction of the electrokinetic flow varies continuously across the channel. A 1 kHz AC signal is superimposed on the DC voltage to induce a DEP force on the particles as they flow over the ridge and results in larger particles being deflected in the fluid stream to a greater extent than smaller particles. The extent of this deflection was determined by the magnitude of the 1 kHz signal (up to 250 V_{rms}) and was demonstrated for the case of a mixture of 2- and 3-micron diameter polystyrene spheres. At the other end of the range of applied DC voltage and AC frequency is a system that employs asymmetric insulating posts and a square DC-biased AC signal, with a positive peak voltage of +500 V, a negative peak voltage of −700 V and a frequency of 0.4 Hz [178]. This signal profile had the effect of producing a net movement in the upstream direction for the smaller particles in a mixture, whilst preferentially trapping the larger particles in the downstream direction. This effect was validated for a mixture of yeast cells (6.3 μm diameter) and 2 μm polystyrene beads.

The first study of Joule heating effects on electroosmotic flow in a typical iDEP device appears to be that of Sridharan *et al.* [179], for a constriction microchannel under DC-biased AC voltages. The fluid was found to reach a high temperature (e.g., 325 K) within the constriction. Depending on the magnitude of the DC voltage, a pair of thermally induced counter-rotating fluid circulations was also found to occur at either the downstream end alone or each end of the channel constriction. A transient, 3D, full-scale numerical model to study Joule heating and its effects on the coupled transport of charge, heat and fluid in an iDEP device with a rectangular constriction microchannel was also developed by Kale *et al.* [180]. This model was validated by comparing the simulation results with experimentally obtained fluid flow patterns. A significant difference was identified in the time scales of the electric, temperature and flow fields in iDEP microdevices. Predictions could also be made of the locations of electrothermal flow circulations in different halves of the channel at the upstream and downstream of the constriction. Other studies have drawn attention to the importance of considering Joule heating effects when designing iDEP systems for manipulating bioparticles [181, 182]. In some of these systems temperature variations above 50 °C can be obtained, in some cases reaching 100 °C at the end of an insulator post array and presenting a threat to the viability of cells trapped in that location.

A significant advance in the building of iDEP devices that reduce Joule heating effects is that of Braff *et al.* [183], who employed micromilling to construct devices with three-dimensional features, which exhibit very large constriction ratios. These three-dimensional iDEP devices allow for the trapping of microparticles at average electric fields one order of magnitude lower than two-dimensional designs with the same footprint. For example, *E. coli* samples were observed to be trapped near constriction regions at potentials as low as 50 V/cm, whilst *B. cereus* was hardly trapped at all for applied potentials in excess of 80 V. These three-dimensional iDEP devices therefore provide a clear differentiation between these bacteria species while keeping temperature variation within the channel to the order of 1 °C or less [183].

Finally, note should be made of what has sometimes been referred to as a sister technique to iDEP, namely, *contactless* DEP (cDEP), which generates the DEP field gradient without having electrodes in contact with the sample fluid. Shafiee *et al.* [184, 185] described a proof-of-principle method for this where metal electrodes are located in highly conductive solutions, which are isolated from the main fluidic channel by a thin PDMS membrane. This membrane should be thin enough to provide good capacitive coupling of the electrode field into the main channel, but not so thin as to lack structural integrity. A membrane thickness of ~20 μm appears to be suitable. A perceived advantage of this method is that electrode electrolysis and fouling is avoided, but a disadvantage over bare metal electrodes is that higher applied voltages are required to generate sufficiently large values for ∇E^2. The electrodes are connected to the secondary side of a transformer, the primary of which can be connected to the output of a wideband power amplifier. The resonant frequency at which this circuit operates is highly dependent on the load impedance connected to the secondary side of the transformer. Two high-voltage power supplies were fabricated, with resonant frequencies tested in the range 85 kHz – 500 kHz, with an output voltage that could be adjusted from approximately 20 V_{rms} to 500 V_{rms} and capacitively coupled to a fluidic channel bounded by the PDMS membrane. Viable human leukaemia monocytes were observed to be influenced by positive DEP and trapped when flowing past the shaped PDMS channel, under the influence of a capacitively coupled 500 kHz, 40 V_{rms} signal, whilst the passage of dead cells was not influenced by a DEP force. Reported trapping (live / dead cell) efficiencies of 89.6% at a flow rate of 0.02 mL/h and 44.8% (±14.2) at 0.8 mL/h, were reported. Cell damage, in the form of lysis, was seen at all frequencies when a voltage of 50 V_{rms} was applied [185]. An interesting application of this method was found to be the rapid mixing of 0.5 μm diameter beads, indicating that the method can be extended to the mixing of low diffusivity biological samples such as cells, which can be a challenging problem in laminar flows at small scales [186]. Čemažar *et al.* [187] advanced the design of contactless DEP to include insulating pillars with diameters of the same order as the cells to be trapped. The viability and trapping efficiency of a mouse ovarian surface epithelial cell line was tested for a design consisting of 68 664 pillars of diameter 20 μm located in four parallel fluidic chambers of height 50 μm. Electrode channels running parallel to the fluidic channels were filled with tenfold concentrated PBS solution, with immersed electrodes being energized by the combination of a function generator and high-voltage amplifier, so that the applied field was directed at right angles to the direction of externally pumped fluid flow. The cancer cells were suspended at 2.5×10^6 cells/ml in a buffer of conductivity ~12 mS/m and processed at a flow rate of 2.2 ml/h with an applied 30 kHz signal of 350 V_{rms}. The DEP crossover frequency for these cells was stated to be 11.9 kHz and the *CM* factor as equal to 0.5 at the operating frequency of 30 kHz [187]. Subpopulations of viable cells were collected at the pillars, to be released later for possible off-chip analysis. To what extent the characteristics of the collected cells were determined by *CM*-factor value or cell size was not discussed by the authors.

10.4.2.1 Niche Applications of iDEP

An objective assessment of the capabilities of iDEP would suggest that it will not compete well against electrode-based and contactless DEP devices for applications that require high volumetric processing of samples, in combination with fine tuning of the frequency of the applied field to achieve sufficient dielectric discrimination between target and background cells. Particle and fluid flow rates achieved by electroosmosis and electrophoresis tend to be much less than that achieved using pressure-driven flow and it is more straightforward to design and operate AC voltage generators, with tuning capabilities in the kHz–MHz range, than is the case for the combination of high voltage DC and AC sources. However, a consistent finding with the iDEP devices described in the literature is that they have the capability to efficiently sort micron- and submicron-sized particles on the basis of size, using device structures that are cheaper and simpler to construct than a typical electrode-based DEP device. A good example of this is a device described by LaLonde *et al.* [188], which is capable of efficiently and selectively trapping 2 μm-sized particles against a background of 100 000 smaller (1 μm) particles. This discrimination increased to a concentration ratio of 1 : 1 000 000 when the smaller particle was reduced to a diameter of 500 nm. An iDEP device would also be the method of choice for DEP sorting of bacteria based on size and surface charge [e.g., 159, 183]. These relative advantages, especially for situations where only low volumes of pure product are required, offer niche areas of application for iDEP.

One such niche application of iDEP in nanocrystallography has been identified by Abdallah *et al.* [189]. To obtain high-resolution diffraction patterns from crystals, a well ordered crystal is necessary so that the diffracted signal is void of crystal lattice imperfections. Consequently, crystals in the sub-500 nm size regime are desired for improved shape transforms, crystal phasing uniformity and compatibility with the beam diameter and sample injection of current state-of-the-art free electron lasers employed for nanocrystallography. Variations in crystal size and shape lead to large amounts of single-crystal diffraction data, with several hundred thousand images needed for one data set. A monodispersed sample of nanocrystals with a narrow size distribution could therefore reduce the amount of data required by an order of magnitude. Abdallah *et al.* [189] were able to sort by iDEP, without further sample treatment, crystals from batch crystallization broths of the huge membrane protein complex photosystem I (PSI). A high degree of monodispersity and crystallinity in the ~100 nm size range was obtained. To avoid Joule heating effects and sample destruction, the iDEP device was restricted to a length of 5 mm, with a single inlet channel (100 μm width, 12 μm depth) leading to five outlet channels. Applied potentials of no more than 55 V were used and a low conductivity buffer of 1.5 mS/m was employed.

Nanocrystals of ∼100 nm in size were isolated from a bulk solution containing a broad crystal size range. Even when multiple experiments were performed to provide a large volume of sorted sample, the process was reproducible and resulted in a large volume (∼300 μL) of fractionated nanocrystals (∼60–300 nm). This volume is in the range typically required for nanocrystallography experiments.

The potential for iDEP to provide a simple on-chip procedure for concentrating DNA from such samples as cytoplasm, saliva and blood has been highlighted by Li *et al.* [190], who designed a simple chip, with microchannels separated by nanoslits, capable of trapping small fragments (2k bp) of DNA in both high (< 1 S/m) and low conductivity (< 5 mS/m) media. This was achieved for a low field strength of 10 V/cm, so that any significant Joule heating was avoided. iDEP has also been demonstrated to offer a new method for the isolation and separation of subcellular organelles [191]. The dielectrophoretic properties of isolated Fischer 344 (F344) rat semimembranous muscle mitochondria and C57BL/6 mouse hepatic mitochondria in low conductivity buffer (0.025–0.030 S/m) at physiological pH (7.2–7.4) were studied using polydimethylsiloxane (PDMS) microfluidic devices and AC signals up to 50 kHz. This established that the mitochondria exhibited negative DEP. The DC potential required to trap the mitochondria was found to be generally above 200 V, when applied over a channel length of 1 cm and to be only weakly dependent on the signal frequency. A separation scheme using DC potentials less than 100 V was demonstrated to perform a size-based iDEP sorting of mitochondria. Samples of isolated mitochondria with heterogeneous sizes, 150 nm to 2 μm diameters, were successfully separated into submicron fractions, indicating the ability to isolate mitochondria into populations based on their size.

10.4.3 Liquid Electrodes

The concept of a *liquid* electrode, as first described and named by Demierre *et al.* [192, 193] is shown in Figure 10.42. The electrodes producing the field are in fact made of metal, but are located at the base of narrow access channels leading up to the main fluid channel in which the DEP manipulation of particles takes place. The field lines from an electrode to the main channel are constricted to follow parallel paths along the access channel, so that at the boundary interface to the main channel the emerging field lines create an almost equipotential surface, as if this boundary was in fact metallic rather than a fluid surface. In any theoretical analysis or modelling of the field produced in the main channel, the electrical resistance of the access channel has to be taken into account. The perceived advantage of a liquid electrode is

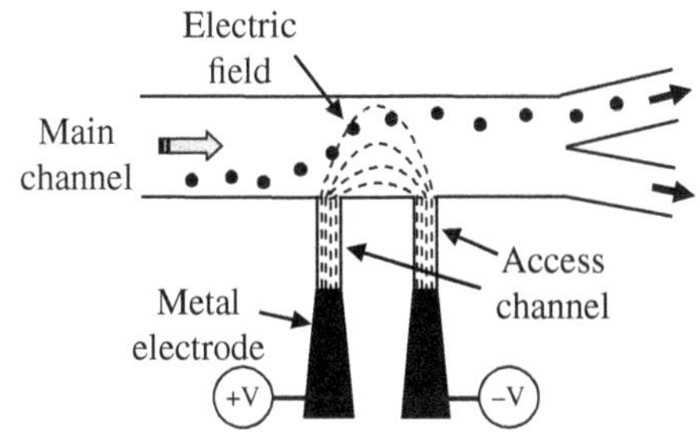

Figure 10.42 Particles are shown deflected across a main fluid channel by negative DEP. The nonuniform field is created by energized metal electrodes located in the narrow access fluid channels. As the field lines emerge from an access channel they form an equipotential surface at the boundary with the main channel, to form a *liquid electrode* [193].

that it avoids electrode electrolysis effects (e.g., bubbles and electrochemical byproducts) appearing in the main channel, as well as fouling of the electrode surfaces by cell fragments or the like. A potential problem could be the blocking of an electrode access channel by cellular debris or even whole cells, which could alter the effective electrical resistance to the main channel.

Demierre and Braschler *et al.* [194, 195] incorporated arrays of liquid electrodes along the side walls of a main flow channel, with a low frequency voltage signal applied on both sides and a high frequency signal superimposed on one side only of the channel. Particles with different dielectric properties experienced different DEP forces and were continuously focused to different streamlines in the flow channel. A particle's dielectric response could be related to its position in the downstream channel. This effect was demonstrated by separating a mixed yeast cell population into pure fractions of viable and nonviable cells with nearly 100% efficiency and also to enrich erythrocytes infected with a major pathogen in cattle, namely *Babesia bovis*. This result also confirmed the working hypothesis that infection with *B. bovis* causes significant changes in the dielectric properties of erythrocytes [195–197].

10.4.4 Carbon Electrodes

Pyrolysed, glasslike, carbon electrodes are widely used in analytical electrochemical techniques because of their excellent mechanical and inert chemical properties in both aqueous and nonaqueous electrolytes. They exhibit a wider electrochemical window of ∼4.4 V between oxidation and reduction potentials than that exhibited (∼2.8 V) by platinum or gold electrodes, for example and can be microfabricated into three-dimensional structures such as wires, plates, ribbons and pillars [198–200]. Carbon electrodes thus share the advantage with insulator-based electrodes that sample electrolysis can be minimized, but have the added advantage of being able to generate suitable fields for DEP with applied voltages in the range of tens of volts, rather than the hundreds or thousands of volts typically required for iDEP devices. This also means that although they have much lower electrical conductivities than metal electrodes, carbon electrodes can be used over the same broad frequency range and not be limited to the lower kHz range available to IDEP devices.

The first indication that carbon-electrode DEP devices could be employed in future analytical applications such as DNA preconcentration and fractionation was obtained by Martinez-Duarte *et al.* [201]. The DEP response of λ-DNA (48.5 kbp) under various frequencies and flow conditions necessary for retention of λ-DNA were studied. When suspended in phosphate buffer at pH 8.1 and a conductivity of 19 mS/m, the DNA samples exhibited positive DEP from 5 kHz to 75 kHz, with a crossover to negative DEP occurring above 75 kHz. With a potential of 14 $V_{pk\text{-}pk}$ applied between two rows of carbon posts, an analysis of the field generated gave a maximum value for ∇E^2 of $7 \times 10^{15}\ V^2/m^3$. Elitas *et al.* [202] also employed three-dimensional carbon-electrode arrays to demonstrate for the first time that antibiotic-treated mycobacterial subpopulations can be both enriched and recovered for downstream analysis using a DEP system. Their efforts were focused on optimizing the method to allow separation of *Mycobacterium smegmatis* cells based on only small changes in their intrinsic properties, such as membrane integrity, rather than differences in size, shape, or volume. Following antibiotic treatment, intact and damaged cells were separated via DEP; intact cells were then washed and recovered from the device. Enrichment was confirmed by using flow cytometry and purification exceeding 99% was achieved, with recovery of up to 3×10^4 cells of interest from the DEP device for further analysis.

10.4.5 DEP 'Tweezers'

The ability to isolate and accurately position single cells in three dimensions is becoming increasingly important in many areas of cell biology and tissue engineering. The simplest electromanipulation device, utilizing DEP, would consist of a single wire. Schnelle *et al.* [203] demonstrated that yeast cells could be attracted to a single gold wire, capacitively coupled to a counter electrode. Maintaining a consistent DEP force using capacitive coupling to a moveable single wire is difficult, as is the selective trapping of a single cell from a collection of cells. In a variation of this single wire design, Lee *et al.* [204] demonstrated that yeast cells and erythrocytes can be manipulated in three dimensions using an electrode arrangement comprising an electrochemically sharpened tungsten wire positioned above a planar electrode. However, the DEP force generated at the wire tip was very sensitive to its height above the planar electrode and this compromised spatial accuracy for cell capture and release. The fact that a captured cell is held by a single point of contact can limit the extent to which it can resist fluid drag forces and hence limit the speed at which it can be translocated through a fluid medium. The first reported design of a twin-electrode DEP tweezer for manipulating cells was that by Matsue *et al.* [205]. The tweezer took the form of a microring+ring electrode assembly formed by the vacuum deposition of platinum and gold films onto the outer and inner surfaces, respectively, of a 6~10 μm diameter capillary tip. Single myeloma cells could be trapped by positive DEP, but this was found to result in cell damage arising from electroporation of the cell membrane [205]. With a later design, in the form of a dual microdisk comprising two platinum-rhodium electrodes of diameter ~2 μm, single chlorella cells were captured by positive DEP and released by negative DEP at a new location, without harming the cell [206]. Hunt and Westervelt [207] have also demonstrated that a DEP tweezer, fabricated by vacuum evaporating electrically isolated Ti-Au films onto two sides of a sharpened glass rod, can trap yeast cells for several hours without harming them (the trapped cells were observed to bud and form daughter cells).

Menachery *et al.* [208] describe the design, theoretical modelling and testing of a DEP 'tweezer' for picking out and relocating single target cells. The device took the form of two electrochemically etched gold wires insulated from each other except for a short region near the electrode tips, which formed the working ends of the DEP tweezer. It was constructed using facilities available in most electrophysiology laboratories, without the requirement of sophisticated microfabrication technology. A three-dimensional modelling of the tweezer using COMSOL Multiphysics software, revealed that the field parameter ∇E^2 had a peak value of $\sim 10^{18}\ V^2/m^3$ at the tips of the gold wires, reducing rapidly to less than $10^{11}\ V^2/m^3$ for distances greater than 20 μm from the electrodes, with an applied voltage of 1 V_{pk}. The practical use of the tweezer was demonstrated by isolating schwannoma cells expressing a membrane channel protein of interest, tagged to the green fluorescent protein (GFP), from nontransfected cells. A schematic of the procedure is shown in Figure 10.43. While traditional methods for production of stably transfected cell lines involve transfection of cells and selection by drug susceptibility, this new procedure has the potential to simplify and shorten the process.

An improvement of the earlier tweezer design [208] was described later by Graham *et al.* [209]. It consists of a single, commercially available, microelectrode that

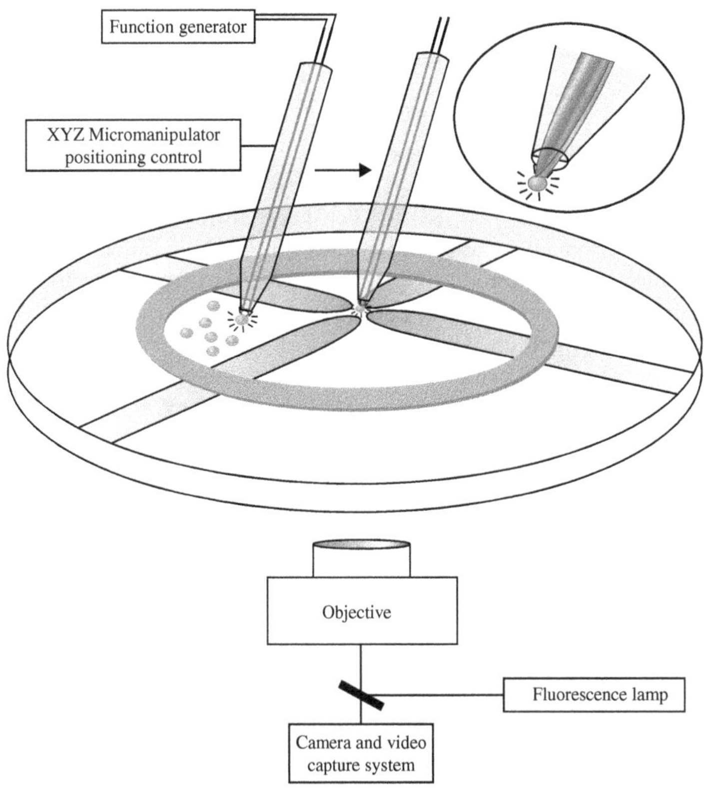

Figure 10.43 Schematic of a DEP tweezer, energized with a voltage function generator, involving an inverted fluorescence microscope equipped with a micromanipulator and video camera. A target cell, tagged with a green fluorescent protein, is shown selectively picked and placed within an electrode array for electrokinetic studies (Menachery *et al.* [208], reproduced with permission).

is capacitively coupled to a grounded thin metal ring below a cell culture dish, negating the need for a direct ground in the liquid medium. Although this design generates larger DEP forces (e.g., $\nabla E^2 \sim 10^{11}\ V^2/m^3$ at a radial distance of ~40 μm from the electrode tip) than the previous design [208], cell damage is avoided due to a thin porous metal oxide coating on the microelectrode tip. The overall usefulness of this single DEP electrode in spatial manipulation of living cells was demonstrated using CHO and HeLa cell clonal cell lines. Cells labelled with calcein-AM and nonloaded cells were actively separated from the mixed population using the single DEP electrode with an applied voltage of 1 V_{rms} and frequency settings of 10 MHz and 50 kHz for positive DEP (cell pickup) and negative DEP (cell release), respectively. An example of how the cells could be separated into two distinct populations of loaded calcein-AM and nonloaded cells is shown in Figure 10.44. The use of the single electrode DEP was also tested on smaller diameter organisms. Fresh-water green algae, *Eremosphaera viridis* and growing lily pollen tubes were used for their pellucid unicellular bodies, which contain several chloroplasts and rapidly trafficking organelles, respectively. Both *Eremosphaera* and lily pollen tubes thrive in low conductivity media, making them amenable to DEP studies. A single DEP electrode was positioned near either a single *Eremosphaera* cell or a growing lily pollen tube. On application of a 100 kHz signal, the *Eremosphaera* were attracted to the electrode by positive DEP, indicating that the electric field penetrated the outer membrane and into the cell interior. The intracellular chloroplasts were also attracted towards the electrode tip. In the growing pollen tube, a flowing stream of organelles was repelled by negative DEP, greatly slowing the rate of cytoplasmic streaming within the growing tube. Both types of intracellular organelles exhibited strong redistributions in response to the DEP force. Upon removal of the voltage signal, the intracellular organelles evenly redistributed themselves

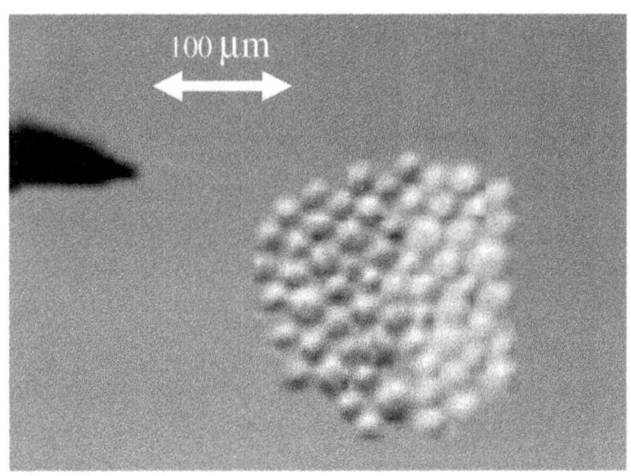

Figure 10.44 A mixed population of Calcein-AM loaded and nonloaded CHO cells separated into two distinct populations using a single electrode DEP 'tweezer'. (Graham *et al.* [209], unpublished image.)

in the *Eremosphaera* after approximately 1 h, whilst cytoplasmic streaming recommenced within the lily pollen tube after seconds.

10.4.6 Isodielectric Cell Separation

Vahey and Voldman [210, 211] have introduced a new type of DEP-based cell separation method called isodielectric separation (IDS). This falls into the *equilibrium gradient* category of particle separation methods, wherein a medium with some type of spatial physico-chemical nonuniformity is combined with a force field to focus particles to equilibrium positions related to their intrinsic physico-chemical properties. A well exploited example of this is isoelectric focusing, where amphoteric molecules are forced via electrophoresis to the location in a pH gradient where their net charge vanishes at their isoelectric point. Cells can be sorted using this method, but the net surface charge of a cell does not provide a particularly specific indicator of its phenotype. Cell and particle separations in biology and biotechnology predominantly rely on techniques that fall within the nonequilibrium category of separation. These methods exploit labels or tags to sort cells based on their extrinsic properties, exemplified by fluorescence-activated cell sorting (FACS) and magnetic-activated cell sorting (MACS). In IDS the forcing mechanism is DEP, which directs cells to the point in a conductivity gradient where the net induced polarizability is zero.

Markx *et al.* [212] describe a cell separation procedure that uses a conductivity gradient of the fluid suspending medium. This was tested on pure suspensions and mixtures of bacteria, which were first immobilized by positive DEP at 10 kHz onto electrodes in a fluidic chamber. A conductivity gradient was then established along the chamber as a function of time. The bacteria were released from the electrodes according to their own dielectric properties and as a function of flow rate and the local medium conductivity. For example, at a flow rate of 0.44 ml/minute and with the conductivity gradient established between values of 0.4 mS/m and 307 mS/m, 90~100% of *E. coli, B. subtilis* and *M. luteus* in a mixture of ~5 × 10^7 cells/ml were released at medium conductivites of ~2, 100 and 120 mS/m. respectively. Lee *et al.* [213] have also described how yeast cells, trapped at 4 MHz by positive DEP at a microelectrode array, were released within an isotonic permittivity gradient. The upper and lower permittivity values in the gradient were established using 0.8 M glycylglycine ($\varepsilon_r = 136.2$) and 0.6 M glucose ($\varepsilon_r = 76.7$) solutions. However, these methods [212, 213] employed *time-varying* gradients of conductivity or permittivity in the same direction as the fluid flow – they do not represent *equilibrium gradient* categories of particle separation. They do, however,

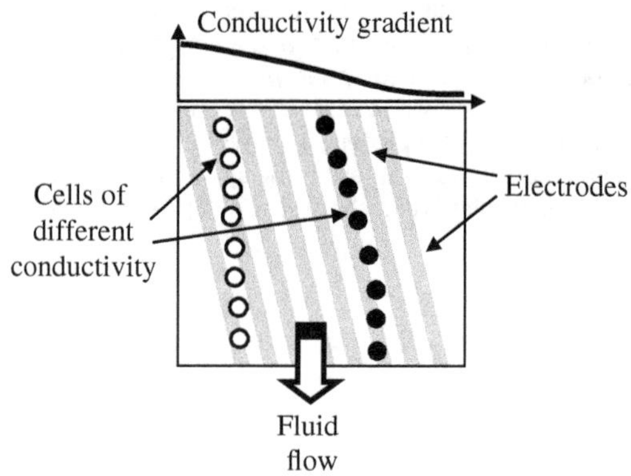

Figure 10.45 The isodielectric method of cell separation involves using a diffusive mixer to establish a conductivity gradient across a diagonal array of electrodes. A combination of the DEP force and hydrodynamic drag carries the cells across the width of the fluidic chamber, in the direction of decreasing conductivity, until the iosdielectric point is reached. (Based on Vahey and Voldman [210].)

demonstrate the principle of operation of the isodielectric method. The trick performed by Vahey and Voldman [210] is to create a conductivity gradient normal to the fluid flow and to angle the electrodes so as to also generate a component of the DEP force normal to the flow.

The principle of operation of the isodielectric method of cell separation is shown in Figure 10.45. A monotonic gradient in electrical conductivity across the width of a microfluidic channel is established by injecting one solution of relatively high conductivity containing the cell mixture and a second solution of relatively low conductivity through a diffusive mixer. This mixer generates a smooth monotonic conductivity profile that flows directly into a channel containing a diagonal array of electrodes. These electrodes guide the cells in the direction of decreasing medium conductivity, until the DEP force becomes sufficiently small that it is overwhelmed by hydrodynamic drag. The cells then continue downstream for collection at different sampling ports, segregated according to their dielectric properties. The IDS device must operate in an advection-dominated transport regime to ensure that the conductivity gradient is preserved over the channel length. This requirement is quantified by the dimensionless Peclet number (*Pe*), which gives the ratio of advection and diffusion. This number is given by $Pe = UL/D$, where U denotes the characteristic velocity of the fluid, L denotes the relevant length scale and D is the diffusivity of the chemical species of interest (see Chapter 12, section 12.6.1). As the diffusion coefficient of the species increases, the relative significance of advection decreases (Vahey and Voldman use the term *convection*, but this is commonly restricted to where heat is being transferred). The two length scales of importance are the channel width (w) from which the time scale for diffusion is obtained ($\sim w^2/D$) and the channel length (l), which enters through the advective time constant ($\sim l/U$). This suggests that transport in the device is best governed in terms of the modified Peclet number (w^2U/Dl), which defines for a given separation a critical Pe value below which cell separation is no longer possible. Vahey and Voldman [210] investigated the relationship between transport (as determined by Pe) and the conductivity profile using a two-dimensional model (2-D rather than 3-D because of the rapid equilibration of the conductivity over the depth of the shallow fluidic channel). Since the separation is based upon a mapping of the effective conductivity of a particle to the position along the channel width where it matches the solution conductivity, there is a direct correspondence between the ranges of conductivities preserved in the device and the maximum range of conductivities that can be simultaneously resolved. For example, if a sample contains particles varying in conductivity by a factor of 5, the Pe number must be maintained above ~18 to accommodate the full range, independent of other parameters and operating conditions [210]. The parameters that determine Pe also influence the forces throughout the system and in turn the maximum flow rate (and Pe) at which the device can operate. Specifically, a particle will pass over the electrode barrier when the axial component of the drag and dielectrophoretic forces balance. This occurs when the real part of the complex Clausius–Mossotti factor given by:

$$|\mathrm{Re}[CM]| = f\frac{\eta \sin(\theta) h^2 U}{\varepsilon_o \varepsilon_m R V_o^2} \approx f\frac{\eta D h^2}{\varepsilon_m R V_o^2 w} Pe$$

where η denotes the fluid viscosity, θ ($\approx w/l$) is the angle of the electrode with respect to the axis of the channel, h is the channel height, U is the mean fluid velocity, V_o is the amplitude of the applied voltage and ε_m is the relative permittivity of the fluid. The function f (≈ 250) is a dimensionless number, which depends upon the geometric ratio of the channel height and electrode spacing.

Vahey and Voldman [210] also considered the constraints arising from electrohydrodynamic flows. The conductivity gradient will modify the local electric field, whilst at the same time the electric field can perturb the fluid through both Joule heating and induced electrokinetic fluid flow. Predictions of the electric field intensities that can be used in the isodielectric device were obtained by considering the influence of induced electroosmosis (ICEO) and electrohydrodynamics (EHD) driven by polarization of both the imposed conductivity gradient (intrinsic EHD) as well as gradients in conductivity and permittivity induced by Joule heating (thermal EHD). The strong frequency dependence of ICEO indicates that it will be negligible when the frequency of the applied voltage signal is greater than ~10 kHz. From a scaling analysis [210] the DEP force and intrinsic EHD both scale as E^2, while thermal EHD scales as E^4. At sufficiently low electric fields, thermal EHD will be less than intrinsic EHD. The similar physics underlying intrinsic EHD and the DEP force on a particle, however, make decoupling these two phenomena by tuning the electric field intensity impossible. A simple scaling argument reveals that the relative magnitude of the DEP and induced drag forces is $\sim R^2w/h^3$, implying that the smallest particle that can be separated has a radius of the order $\sqrt{(h^3/w)}$. Although this was shown [210] by both numerical simulation and experiments to overpredict the minimum particle size by approximately three times, it provides reasonable guidelines for sizing an isodielectric device.

The isodielectric separator was tested using polystyrene beads of different diameters, as well as live and dead yeast cells (*S. cerevisiae*). Beads of diameter 1.60, 1.75 and 1.90 μm, respectively, when suspended in a medium of conductivity 33 mS/m exhibited DEP crossover frequencies of 180, 120 and 40 kHz, respectively [210]. From Equation (10.64) we expect the effective conductivity of these beads to vary as a function of both their surface conductance K_s and radius R:

$$\sigma_p = \sigma_{bulk} + \frac{2K_s}{R}$$

It was verified that the separation of the beads was based upon differences in the equilibrium positions of the beads in a conductivity gradient, rather than differences in bead size. Additionally, the requirement for the conductivity gradient and the variation of the separation with frequency together precluded the possibility that the separation was due to any other nonelectrical parameter. To further verify that the separations were based upon surface conductance as opposed to other properties that differ between the beads, complementary experiments were performed in which beads with comparable sizes but different surface coatings were sorted. For carboxyl-modified beads with comparable surface charge densities, it was found that the smaller beads separated at the higher end of the conductivity gradient, whilst small unmodified polystyrene beads exhibited lower effective conductivities than their larger and more highly charged counterparts. These experiments demonstrated that IDS may be used to separate particles according to their electrical properties, even in the presence of competing differences in the sizes of the particles. The separation of live and dead yeast cells was also demonstrated, with or without the use of heat treatment to deliberately kill the cells. Further separation experiments were later performed on polystyrene beads and yeast cells, as well as BA/F3 murine pro B cells [211]. After passing through the isodielectric separator (operating at 6 $V_{pk\text{-}pk}$, 10 MHz and a flow rate of 4 μL/min) the murine cells were found to have no significant difference in viability before and after the cells were passed through the device, with viability >85% in both cases. The lack of adverse effects on cell viability is consistent with the brief residence time of around 16 s of the cells in the device, together with operation in the MHz frequency range where the imposed transmembrane voltage would have been relatively small [214]. From knowledge of the point along the conductivity gradient at which the DEP force succumbed to the fluid drag force, the electrical properties of the beads and cells could be determined, including how these electrical properties varied with the electrical conductivity of the suspending medium [211]. By far the most exciting application of the isodielectric separator has been its application to identify for yeast cells those genes whose deletion changes their DEP phenotype [214]. This is described and discussed in detail in Chapter 11 (section 11.2.2.2).

10.5 The Second (High-Frequency) DEP Crossover Frequency (f_{xo2})

The DEP response, as represented by Re[CM^*], in Figure 10.13 shows the existence of two DEP crossover points, at f_{xo1} and f_{xo2}. The physical reason for the two DEP crossovers can be explained in terms of the relaxation time τ for the interfacial polarization given by Equation (10.30) and repeated here for convenience:

$$\tau = \varepsilon_o \frac{\varepsilon_p + 2\varepsilon_m}{\sigma_p + 2\sigma_m}$$

At low frequencies the dielectric parameters ε_p and σ_p refer to the effective permittivity and conductivity of the cell and is dominated by the presence and dielectric properties (especially the high resistance) of the plasma membrane. At high frequencies, much higher than f_{xo1}, the effective resistance of the plasma membrane is zero, having been shortcircuited by the capacitive reactance of the membrane. The applied electric field penetrates the membrane and effectively 'sees' the cell as a bag of cytosol and its contents. We now have a mismatch between the conductivity of the outside medium and the conductivity of the cell interior – which gives rise to a second relaxation time τ_2 given by:

$$\tau_2 = \varepsilon_o \frac{\varepsilon_{\text{int}} + 2\varepsilon_m}{\sigma_{\text{int}} + 2\sigma_m} \tag{10.71}$$

where ε_p and σ_p refer to the effective permittivity and conductivity of the cell interior. At high frequencies we can therefore rewrite Equation (10.32) to give the value for f_{xo2} as:

$$f_{xo2}^2 = \frac{1}{4\pi^2}\frac{1}{\varepsilon_o^2}\frac{(\sigma_m - \sigma_{\text{int}})(\sigma_{\text{int}} + 2\sigma_m)}{(\varepsilon_{\text{int}} - \varepsilon_m)(\varepsilon_{\text{int}} + 2\varepsilon_m)} \tag{10.72}$$

which is a result previously reported by Gimsa *et al.* [215]. This quadratic equation in principle has two roots, real (rather than imaginary) ones existing when the following condition is obeyed:

$$\frac{(\sigma_m - \sigma_{\text{int}})}{(\varepsilon_{\text{int}} - \varepsilon_m)} > 0 \tag{10.73}$$

For an aqueous suspending medium, where $\varepsilon_m \approx 79$, it is usually found from analysis of dielectric spectroscopy measurements on cells that $\varepsilon_m > \varepsilon_{int}$ (e.g., see Tables 9.2 and 9.3). To satisfy the condition given by Equation (10.73) requires that $\sigma_{int} > \sigma_m$, which will usually be the case when using media of conductivity less than around 200 mS/m. In fact, in a typical DEP cell-separation protocol (e.g., 79–82) the conductivity (σ_m) of the suspending medium is some 20-times less than that of the cell interior (σ_{int}). In this situation Equation (10.72) can to a good approximation be simplified to give:

$$f_{xo2} = \frac{\sigma_{\text{int}}}{2\pi}\frac{1}{\varepsilon_o}\sqrt{\frac{1}{(\varepsilon_{\text{int}} - \varepsilon_m)(\varepsilon_{\text{int}} + 2\varepsilon_m)}} \tag{10.74}$$

This indicates that the value for f_{xo2} is largely controlled by the conductivity of the cell interior and to a lesser extent by its permittivity. This conclusion was previously obtained, through different theoretical routes, by Broche *et al.* [216] and Chung *et al.* [217]. The relative contributions that σ_{int} and ε_{int} have on the value of f_{xo2} are shown in Figure 10.46, obtained using the single-shell model for a cell. It can be seen that f_{xo2} is particularly sensitive to the cytoplasm conductivity, but not so sensitive to changes in of its permittivity. The DEP crossover at f_{xo1} is not influenced by the dielectric properties of the cytoplasm. Figures 10.13 and 10.14 further illustrate how the values of f_{xo1} and f_{xo2} differ in their sensitivity to changes in cell size and plasma membrane conductance. These various sensitivities to changes of cell dielectric parameters are summarized in Table 10.1.

Chung *et al.* [217] described the circuitry and electrical load modelling of an interdigitated microelectrode array for performing DEP measurements on cells up to a frequency of 400 MHz. Particular efforts were made to avoid the electrical resonance effects reported by Gimsa *et al.* [218] in their DEP and electrorotation measurements on erythrocytes at 80 MHz. The first systematic determination of the high-frequency DEP crossover (f_{xo2}) exhibited by mammalian cells (murine myeloma) could then be reported, an example of which is shown in Figure 10.47(a). The value of f_{xo2} was found to be sensitive to

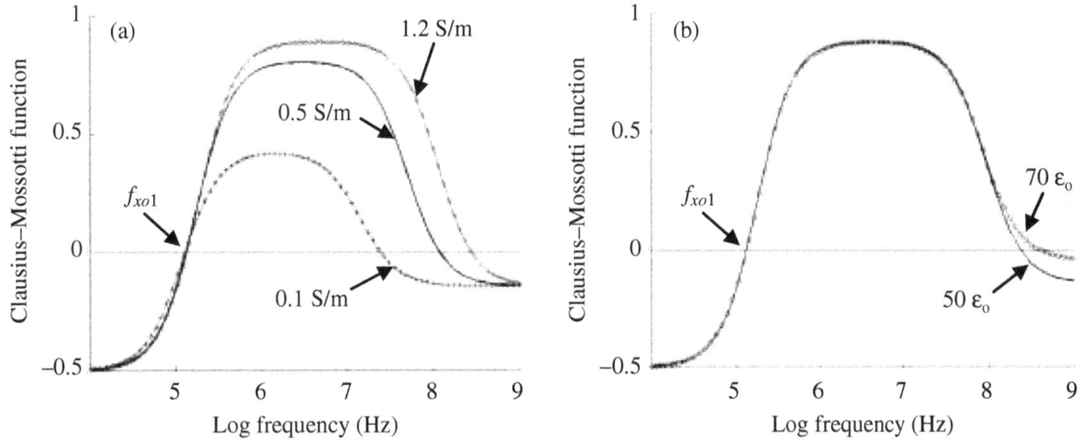

Figure 10.46 Single-shell modelling of a cell to show (a) that the high-frequency DEP cross-over at f_{xo2} is particularly sensitive to the value of the cytoplasm conductivity, but as indicated in (b) is not so sensitive to changes in the permittivity of the cytoplasm. The DEP crossover at f_{xo1} is not influenced by the dielectric properties of the cytoplasm. This model is for a cell of fixed diameter 10 μm and plasma membrane capacitance of 10 mF/m^2.

Table 10.1 Relative sensitivities of the DEP crossover frequency values of f_{xo1} and f_{xo2} to changes of various cell parameters. In summary, f_{xo1} is sensitive to changes in cell size and the dielectric properties and morphology of the plasma membrane. The value of f_{xo2} is insensitive to these changes but is highly sensitive to a reduction of conductivity associated with ion leakage from the cytoplasm.

	Sensitivity	
Parameter	f_{xo1}	f_{xo2}
Cell diameter	High	None
Ion leakage from cytoplasm	None	High
Cytoplasm permittivity	None	Moderate
Membrane capacitance	High	None
Membrane passive conductance of ions	High	None

both the osmolarity and temperature of the cell suspending medium [217]. A rapid decrease in the mean value for f_{xo2} was consistently observed and found to be a function of the medium temperature. For example, at 37 °C the value of f_{xo2} typically dropped from ~200 MHz to ~90 MHz over a period of 2 h, whereas at 10 °C the fall was from ~200 MHz to ~120 MHz over 6 h. During this period the value observed for f_{xo1} did not change. An Arrhenius plot, of the rate of fall of f_{xo2} versus reciprocal absolute temperature, produced a straight line giving an activation energy of ~44 kJ/mol for the underlying process leading to this temporal behaviour of f_{xo2}. This is close to the activation energy of ~41 kJ/mol obtained for water self-diffusion through lipid bi-layer membranes [219], which is a coincidence worth further investigation. As shown in Figure 10.47(b), erythrocytes were found to have a less wide distribution and a lower mean value of f_{xo2} values compared to the myeloma cells, but still exhibited a shift to lower frequencies with time. In later experiments [220] ratiometric flow cytometry of murine myeloma cells loaded with a potassium-sensitive fluorophore revealed temporal trends in the intracellular concentration of potassium ions that were similar to the observed time course of the f_{xo2} values. This indicated that f_{xo2} did scale with intracellular conductivity and that changes to the intracellular permittivity were minimal by comparison.

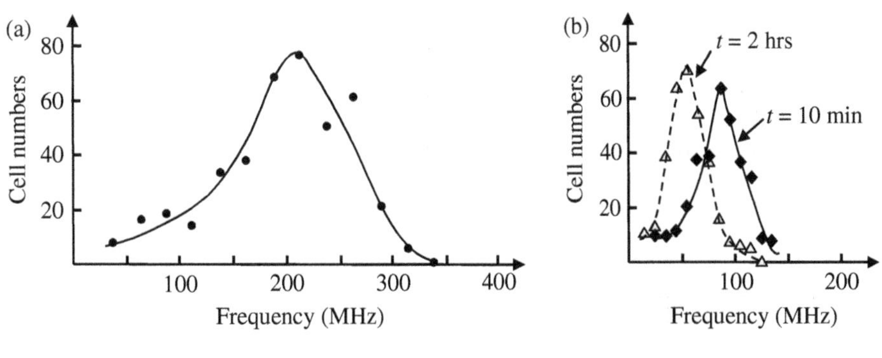

Figure 10.47 (a) Distribution of f_{xo2} values for murine myeloma cells (n = 418), 10 minutes after their suspension in a medium of conductivity 33 mS/m. (b) Distribution of f_{xo2} values for human erythrocytes (n = 328) 10 minutes after their suspension and 2 hours later (n = 315) in a medium of conductivity 33 mS/m. (Chung [220], to be published.)

10.6 Summary

The basic DEP force equation:

$$F_{DEP} = (p \cdot \nabla)E$$

as derived by Pohl in his book [1, pp. 15–17] and referred to as the *ponderomotive* force (die ponderomotorischen Kräfte) in text books on electricity way before that time (e.g., [221]), requires some simplifying assumptions for it to be presented in this straightforward form. Two significant assumptions are that the particle is much smaller than the scale of the nonuniformity ∇E of the field and that the particle's induced polarization is equivalent to that of a simple dipole moment p. The basic DEP force equation is in fact only strictly valid for the case of a point dipole, corresponding to a vanishingly small particle radius. As indicated by the simple assessment made in section 10.2.1 we can expect that for cell-sized particles this particular assumption typically leads to an overestimate of the DEP force's magnitude by a factor of 5% to 10%. For most, if not all, situations this is of no major consequence. The assumption that the particle's polarization takes the form of a simple dipole moment is more interesting and for some situations quite significant. Expressed as an induced dipole moment p, the polarization of a spherical particle of radius R is given by the Equation (10.7) and as derived in the early literature [e.g., 203]:

$$p = 4\pi\varepsilon_o\varepsilon_m R^3 \left(\frac{\varepsilon_p - \varepsilon_m}{\varepsilon_p + 2\varepsilon_m}\right) E$$

where E is the mean value of the local field, assumed to be *uniform*, before insertion of the particle. The parameter ε_o is the permittivity of free space, whilst ε_p, ε_m refer to the relative permittivity of the particle and surrounding medium, respectively. This induced dipole produces a field with lines of electric potential as shown in Figure 10.1. For the case of a uniform field E the particle is uniformly polarized and the term in brackets in the above equation represents the polarizability per unit volume of the particle – otherwise known as the Clausius–Mossotti factor *CM*. But for a DEP force to act on the particle it must be subjected to a nonuniform field and so it will not be uniformly polarized. Using the procedure outlined in Box 10.1, Washizu [7] took account of this by introducing into the DEP force equation induced moments of higher order, known as multipoles. Jones and Washizu [9, 11] derived the followed expression for the nth-order multipolar moment:

$$p_n = \frac{4\pi\varepsilon_m R^{2n+1}}{(2n-1)!!} CM_{(n)} (\nabla)^{n-1} E$$

where $CM_{(n)}$ is the general multipole form of the Clausius–Mossotti factor, given by:

$$CM_{(n)} = \frac{\varepsilon_p^* - \varepsilon_m^*}{n\varepsilon_p^* + (n+1)\varepsilon_m^*}$$

In this expression the dipole moment corresponds to $n = 1$, with $n = 2$ the quadrupole; $n = 3$ the octupole and so on. Consideration of multipoles beyond the quadrupole are normally not required. The inclusion of the asterisks denotes that the complex form ε^* of the permittivity is considered. This accommodates the presence of dielectric losses in the particle and surrounding medium, addressing another simplification in the basic DEP force equation, namely that it does not take into account such energy losses. Washizu [7] demonstrated that contributions from the quadrupole moment should be included for particles located near the centre of the 'polynomial' electrode design shown in Figure 10.3. At the central location the factor (E.∇E) is zero and so no DEP force is created. At locations within one particle radius from the centre, the induced quadrupole moment contributes significantly to the overall DEP force and should be included. For distances greater than a particle diameter from the centre, the quadrupole contribution is insignificant, so that the dipole approximation that gives the basic DEP equation is sufficiently accurate. Higher order moments are also important for the circumferentially periodic electrode structures used to achieve passive DEP levitation of particles, where pronounced size-dependent effects not anticipated by the standard DEP theory have been observed by Washizu, Jones and Kaler [8, 9]. Higher order moments are responsible for the levitation force achieved by such electrode structures because the electric field is zero along the central axis of such electrodes. As a general rule, higher order moment effects can be ignored where the particle diameter is less than around one-tenth of the interelectrode spacing. For example, Schnelle *et al.* [10] investigated the situation for DEP field cages formed by a sandwich structure of two planar quadrupole electrodes (see Figure 10.4) and concluded that quadrupole moment forces contribute ~5% of the total DEP force for particles larger than about a quarter of the electrode spacing. For particles smaller in diameter than about a tenth of the electrode spacing, the error arising from a DEP force calculation that ignores the quadrupole contribution is typically smaller than 1%.

A factor to be considered with the higher order induced moments is that their potentials fall off more rapidly with distance r as their order increases. For example, the dipole potential falls off as $1/r^2$, whereas the quadrupole potential falls off as $1/r^4$. This is of relevance regarding another significant approximation made in the derivation of the basic DEP force equation, namely that the polarized particle is assumed to be far removed from the perturbing influence of another polarized particle, or from the surface of an electrode or a boundary wall of the fluidic chamber, for example. The effects of higher order moments can therefore become significant in the situation where cells approach each other to form pearl chains, such as those shown in Figure 10.31, or where particles experience the image forces shown in Figure 5.21 when they are close to an electrode surface or chamber wall.

A multipole analysis was also applied by Jones and Washizu [11] to traveling wave dielectrophoresis. An important conclusion was that the various multipolar DEP forces show similar frequency dependencies. This is significant in the context of Equation (10.37):

$$\langle F_{DEP} \rangle \approx 2\pi\varepsilon_o\varepsilon_m R^3 \left[\frac{f^2 - f_{xo1}^2}{f^2 + 2f_{xo1}^2}\right] \nabla E_{rms}^2$$

which demonstrates the practical significance of the Clausius–Mossotti factor in the DEP separation of different cell types in a cell mixture. Multipole effects could slightly influence the magnitude of the DEP force, but should not influence the value of the DEP crossover frequency f_{xo1} and thus the protocol adopted for the DEP sorting of cell mixtures. In most applications of DEP the most important experimental factor is identification of the appropriate operating frequency f, taking into account the value (or range of values) of f_{xo1} for the target cell, with other considerations such as the rate of fluid flow. Accurate knowledge, to within ±10%, of the DEP force is of no real practical consequence. The basic DEP force equation, formulated using the equivalent dipole moment approximation and the assumption that the particle is much smaller than the scale of the field nonuniformity, serves most purposes very well. These approximations are also valid for the electrorotation and traveling wave dielectrophoresis phenomena not considered by Pohl. However, the approximation that the polarized particle is exposed to an external field of infinite extent, unperturbed by other fields, can lead to inaccuracies when the particle is in close vicinity to a metal surface or dielectric boundary. This includes the anomalous DEP effects identified by Camarda *et al.* [86].

Pohl [87,88] sometimes employed metal wires and thin sheet electrodes to produce nonuniform fields, which required the application of DC or AC voltage potentials of up to 11 kV. Joule heating effects often perturbed the DEP-induced motions of the particles. The parameter (E.∇)E in Equation (10.8) has units of V^2m^{-3}, providing the clue that by miniaturizing the electrodes much smaller applied voltages would result in the same magnitude of DEP force, but would avoid undue thermal and electrolysis effects [94]. The impact that microfabrication technologies have made on the DEP field has been summarized in this chapter. A particularly good review has also been given by Martinez-Duarte [199]. Until comparatively recently, DEP devices that use metal-based microelectrodes have tended to dominate the field, but they are perceived to have disadvantages related to their cost and complication of fabrication. In this author's view, more valid objections relate to the fact that frequencies below ~5 kHz are not attainable because of problems associated with fluid motion induced by electroosmosis and electrochemical / electrolysis effects that can lead to the generation of gas bubbles. DEP 'tweezers', capable of selecting an individual cell from amongst other cells, picking it up and then releasing it at a new location for further investigations or manipulation (e.g., patch-clamp or electrorotation studies) can find useful applications in many areas of cell biology and tissue engineering. They operate at frequencies above 5 kHz and can be fabricated quite simply and at low cost, in the form of vacuum deposited metal electrodes at a capillary or sharpened glass tip [205, 207]. An even simpler design uses two electrochemically etched gold wires insulated from each other, except for a short region near the electrode tips [208]. The simplest electromanipulation device, utilizing DEP, consists of a single wire and can be realized using low cost and commercially available microelectrodes commonly found in electrophysiology laboratories [209]. Apart from being able to manipulate individual cells (e.g., see Figure 10.44) this form of device has also been demonstrated to alter the location of intracellular organelles, such as chloroplasts in algae or pollen tubes [209]. A relatively simple metal electrode structure is also used in an innovative isodielectric cell separation device that incorporates a fluidic conductivity gradient [210, 211]. The electrodes guide the cells in the direction of decreasing medium conductivity, until the DEP force becomes comparable to the viscous drag force of a fluid flowing over the electrodes. The cells then move with the fluid flow (as shown in Figure 10.45) for collection at different sampling ports, segregated according to their dielectric properties (and not on cell size, for example).

Masuda, Washizu and Nanba [152, 153] demonstrated that insulating structures placed in a microfluidic channel could distort an imposed electric field sufficiently enough to generate a DEP force on cells. This concept of insulator-based (iDEP) devices has been significantly advanced by the pioneering works of Chou *et al.* [154, 155], Cummings, Singh, Simmons and Lapizco-Encinas, for example [e.g., 156–161]. Interesting and exploitable balances between electrophoretic, electroosmotic and dielectrophoretic forces have been demonstrated and niche applications such as the efficient sorting of rare (larger sized) particles from a million-fold or greater number of smaller particles has been demonstrated by LaLonde *et al.* [188], together with the concentrating of protein nanocrystals [189], DNA [190] and mitochondria [191]. The advantages that iDEP devices have over electrode-based DEP devices include their easier construction and ability to operate at DC or low frequencies. Downsides are that iDEP devices have generally been operated at much larger voltage potentials and field strengths, which can create undesirable Joule heating effects, coupled with the fact that the perceived advantage of using electroosmotic fluid flow rather than pressure-driven pumping can lead to much lower volumetric particle throughput. This means that heat-sensitive particles such as cells can be exposed to high fields for longer periods in an iDEP device than is generally the case with an electrode-based device. Three-dimensional structures, which lead to increased volumetric throughput, are certainly possible in iDEP devices [e.g., 187] and can probably be addressed. Liquid electrodes, where conducting narrow fluid channels electrically connect metal electrodes to a main channel have also been described [192–197] and merit further study. Finally, a recent innovation has been the use of three-dimensional carbon electrodes in DEP devices [201,202]. The indications are that such electrodes will compete very well with metal ones. Although they are less conductive and so require higher applied voltage potentials, carbon electrodes are electrochemically less active than metal ones, can operate at frequencies below 1 kHz and are readily fabricated into three-dimensional structures [200].

10.7 References

1 Pohl, H. A. (1978) *Dielectrophoresis: The Behaviour of Neutral Matter in Nonuniform Electric Fields*, Cambridge University Press, Cambridge.

2 Huang, Y. and Pethig, R. (1991) Electrode design for negative dielectrophoresis. *Meas. Sci. Technol.* **2**, 1142–1146.

3 Hughes, M. P. (2003) *Nanoelectromechanics in Engineering and Biology*, CRC Press, Boca Raton, FL, Ch. 10.

4 Morgan, H. and Green, N. G. (2003) *AC Electrokinetics: Colloids and Nanoparticles*, Research Studies Press Ltd, Baldock, Ch. 10.

5 Washizu, M. and Jones, T. B. (1994) Multipolar dielectrophoretic force calculation. *J. Electrostat.* **33**, 187–198.

6 Green, N. G. and Jones, T. B. (2007) Numerical determination of the effective moments of non-spherical particles. *J. Phys. D: Appl. Phys.* **40**, 78–85.

7 Washizu, M. (1992) Precise calculation of dielectrophoretic force in arbitrary field. *J. Electrostat.* **29**, 177–188.

8 Washizu, M., Jones, T. B. and Kaler, K. V. I. S. (1993) Higher-order dielectrophoretic effects: levitation at a field null. *Biochim Biophys. Acta* **1158**, 40–46.

9 Jones, T. B. and Washizu, M. (1994) Equilibria and dynamics of DEP-levitated particles: multipolar theory. *J. Electrostat.* **33**, 199–212.

10 Schnelle, T., Müller, T., Fiedler, S. and Fuhr, G. (1999) The influence of higher moments on particle behaviour in dielectrophoretic field cages. *J. Electrostat.* **46**, 13–28.

11 Jones, T. B. and Washizu, M. (1996) Multipolar dielectrophoretic and electrorotation theory. *J. Electrostat.* **37**, 121–134.

12 Washizu, M. (2004) Equivalent multipole-moment theory for dielectrophoresis and electrorotation in electromagnetic field. *J. Electrostat.* **62**, 15–33.

13 Ida, N. (2000) *Engineering Electromagnetics*, Vol. 1. Springer Verlag, New York, NY, p. 762.

14 Mognashi, E. R. and Savini, A. (1983) The action of a non-uniform electric field upon lossy dielectric systems – ponderomotive force on a dielectric sphere in the field of a point charge. *J. Phys. D: Appl. Phys.* **16**, 1533–1541.

15 Sauer, F. A. (1985) Interaction forces between microscopic particles in an external electromagnetic field, in *Interactions between Electromagnetic Fields and Cells* (eds A. Chiabrera, C. Nicolini and H. P. Schwan). Plenum Press, New York, NY, pp. 181–202.

16 Sauer, F. A. and Schlögl, R. W. (1985) Torques exerted on cylinders and spheres by external electromagnetic fields, in *Interactions between Electromagnetic Fields and Cells* (eds A. Chiabrera, C. Nicolini and H. P. Schwan). Plenum Press, New York, NY, pp. 203–251.

17 Wang, X., Wang, X.-B. and Gascoyne, P. R. C. (1997) General expressions for dielectrophoretic force and electrorotational torque derived using the Maxwell stress tensor method. *J. Electrostatics* **39**, 277–295.

18 Ramo, S., Whinnery, J. R. and Van Duzer, T. (1994) *Fields and Waves in Communication Electronics*, 3rd edn. John Wiley & Sons, Inc., New York, NY, Appendix 4, pp. 825–827.

19 Jones, T. B. (1995) *Electromechanics of Particles*, Cambridge University Press, Cambridge.

20 Chang, H.-C. and Yeo, L. Y. (2010) *Electrokinetically Driven Microfluidics and Nanofluidics*, Cambridge University Press, Cambridge.

21 Asami, K., Takahashi, Y. and Takashima, S. (1989) Dielectric properties of mouse lymphocytes and erythrocytes. *Biochim. Biophys. Acta* **1010**(1), 49–55.
22 Wang, X.-B., Huang, Y., Hölzel, R., Burt, J. P. H. and Pethig, R. (1993) Theoretical and experimental investigations of the interdependence of the dielectric, dielectrophoretic and electrorotational behaviour of colloidal particles. *J. Phys. D: Appl. Phys.* **26**, 312–322.
23 Grosse, C. and Schwan, H. P. (1992) Cellular membrane potentials induced by alternating fields. *Biophys. J.* **63**, 1632–1642.
24 Huang, Y., Wang, X.-B., Becker, F. F. and Gascoyne, P. R. C. (1996) Membrane changes associated with the temperature-sensitive P85$^{gag\text{-}mos}$-dependent transformation of rat kidney cells as determined by dielectrophoresis and electrorotation. *Biochim. Biophys. Acta* **1282**, 76–84.
25 Pethig, R., Jakubek, L., Sanger, R. H. *et al.* (2005) Electrokinetic measurements of membrane capacitance and conductance for pancreatic β-cells. *IEE Proc. Nanobiotechnol.* **152**(6), 189–193.
26 Schwan, H. P. (1957) Electrical properties of tissue and cell suspensions, in *Advances in Biological and Medical Physics* (eds J. H. Lawrence and C. A. Tobias). Vol. V, Academic Press, New York, NY, pp. 147–209.
27 Pauly, von H. and Schwan, H. P. (1959) Über die Impedanz einer Suspension von kugelförmigen Teilchen mit einer Schale, *Z. Naturforsch.* **14b**, 125–131
28 Lei, U., Sun, P.-H. and Pethig, R. (2011) Refinement of the theory for extracting cell dielectric properties from dielectrophoresis and electrorotation experiments. *Biomicrofluidics* **5**, 044109.
29 Remmert, R. (1991) *Theory of Complex Functions*, Springer-Verlag, New York, NY, Ch. 12, pp. 361–376.
30 Schwan, H. P. (1966) Alternating current electrode polarization. *Biophysik.* **3**, 181–201.
31 Burt, J. P. H., Al-Ameen, T. A. K. and Pethig, R. (1989) An optical dielectrophoresis spectrometer for low-frequency measurements on colloidal suspensions. *J. Phys. E: Sci. Instrum.* **22**, 952–957.
32 González, A., Ramos, A., Green, N. G., Castellanos, A. and Morgan, H. (2000) Fluid flow induced by non-uniform fields in electrolytes on microelectrodes. II. A linear double-layer analysis. *Phys. Rev. E.* **61**(4), 4019–4028.
33 Lei, U., Yang, C. Y. and Wu, K. C. (2006) Viscous torque on a sphere under arbitrary rotation. *Appl. Phys. Lett.* **89**, 181908.
34 Hölzel, R. (1993) Electric field calculation for electrorotation electrodes. *J. Phys. D: Appl. Phys.* **26**, 2112–2116.
35 Hughes, M. P., Wang, X.-B., Becker, F. F. *et al.* (1994) Computer-aided analyses of electric fields used in electrorotation studies. *J. Phys. D: Appl. Phys.* **27**, 1564–1570.
36 Dalton, C., Goater, A. D., Drysdale, J. and Pethig, R. (2001) Parasite viability by electrorotation, Colloid. *Surfaces A.* **195**, 263–268.
37 McGillem, C. D. and Cooper, G. R. (1984) *Continuous and Discrete Signal and System Analysis*, Holt, Reinhart & Winston, New York, NY.
38 Bunget, I. and Popescu, M. (1984) *Physics of Solid Dielectrics* (Materials Science Monographs, 19). Elsevier, Amsterdam, pp. 291–296.
39 Pethig, R., Huang, Y., Wang, X.-B. and Burt, J. P. H. (1992) Positive and negative dielectrophoretic collection of colloidal particles using interdigitated castellated electrodes. *J. Phys. D: Appl. Phys.* **24**, 881–888.
40 Arnold, W. M. and Zimmermann, U. (1988) Electro-rotation: Development of a technique for dielectric measurements on individual cells and particles. *J. Electrostatics* **21**, 151–191.
41 Arnold, W. M. and Zimmermann, U. (1982) Rotation of an isolated cell in a rotating electric field. *Naturwiss.* **69**, 297.
42 Arnold, W. M. and Zimmermann, U. (1982) Rotating-field induced rotation and measurement of the membrane capacitance of single mesophyll cells of *Avea sativa. Z. Naturforsch.* **37c**, 908–915.
43 Glaser, R., Fuhr, G. and Gimsa, J. (1983) Rotation of erythrocytes, plant cells, and protoplasts in an outside rotating electric field. *Stud. Biophys.* **96**, 11–20.
44 Wang, X.-B., Huang, Y., Becker, F. F. and Gascoyne, P. R. C. (1994) A unified theory of dielectrophoresis and travelling wave dielectrophoresis. *J. Phys. D: Appl. Phys.* **27**, 1571–1574.
45 Pethig, R., Talary, M. S. and Lee, R. S. (2003) Cell separation and characterization using traveling wave dielectrophoresis enhanced by signal superposition. *IEEE Eng. Med. Biol. Mag.* (Nov./Dec.), 43–50.
46 Hughes, M. P., Pethig, R. and Wang, X.-B. (1996) Dielectrophoretic forces on particles in traveling electric fields. *J. Phys. D: Appl. Phys.* **29**, 474–482.
47 Wang, X.-B., Huang, Y., Wang, X. *et al.* (1997) Dielectrophoretic manipulation of cells in spiral electrodes. *Biophys. J.* **72**, 1887–1899.
48 Morgan, H., Izquierdo, A. G., Bakewell, D. *et al.* (2001) The dielectrophoretic and traveling wave forces generated by interdigitated electrode arrays: analytical solution using Fourier series. *J. Phys. D: Appl. Phys.* **34**, 1553–1561.
49 Huang, Y., Tame, J. A. and Pethig, R. (1993) Electrokinetic behaviour of colloidal particles in traveling electric fields: studies using yeast cells. *J. Phys. D: Appl. Phys.* **26**, 1528–1535.

50 Goater, A. D., Burt, J. P. H. and Pethig, R. (1997) Combined traveling wave dielectrophoresis and electrorotation device: Applied to the concentration and viability determination of Cryptosporidium. *J. Phys. D.: Appl. Phys.* **30**, L65–L70.

51 Burt, J. P. H., Pethig, R., Gascoyne, P. R. C. and Becker, F. F. (1990) Dielectrophoretic characterization of Friend murine erythroleukaemic cells as a measure of induced differentiation. *Biochim. Biophys. Acta* **1034**, 93–101.

52 Gascoyne, P. R. C., Pethig, R., Burt, J. P. H. and Becker, F. F. (1993) Membrane changes accompanying the induced differentiation of Friend murine erythroleukemia cells studied by dielectrophoresis. *Biochim. Biophys. Acta* **1149**, 119–126.

53 Lyklema, J. (1995) *Fundamentals of Interface and Colloid Science*, Vol. 2: Solid-Liquid Interfaces, Academic Press, San Diego, CA.

54 Arnold, W. M., Schwan, H. P. and Zimmermann, U. (1987) Surface conductance and other properties of latex particles measured by electrorotation. *J. Phys. Chem.* **91**, 5093–5098.

55 Clarke, R. W., Piper, J. D., Ying, L. and Klenerman, D. (2007) Surface conductivity of biological macromolecules measured by nanopipette dielectrophoresis. *Phys. Rev. Lett.* **98**, 198102.

56 Lyklema, J. and Minor, M. (1998) On surface conduction and its role in electrokinetics. *Colloids Surf. A* **140**, 33–41.

57 Hughes, M. P., Morgan, H. and Flynn, M. F. (1999) The dielectrophoretic behavior of submicron latex spheres: influence of surface conductance. *J. Colloid Interface Sci.* **220**, 454–457.

58 Hughes, M. P. and Green, N. G. (2002) The influence of Stern layer conductance on the dielectrophoretic behavior of latex nanospheres. *J. Colloid Interface Sci.* **250**, 266–268.

59 Ermolina, I. and Morgan, H. (2005) The electrokinetic properties of latex particles: comparison of electrophoresis and dielectrophoresis. *J. Colloid Interface Sci.* **285**, 419–428.

60 Zhou, H., Preston, M. A., Tilton, R. D. and White, L. R. (2005) Calculation of the electric polarizability of a charged spherical dielectric particle by the theory of colloid electrokinetics. *J. Colloid Interface Sci.* **285**, 845–856.

61 Basuray, S. and Chang, H. C. (2007) Induced dipoles and dielectrophoresis of nanocolloids in electrolytes. *Phys. Rev. E.* **75**(6), 06051.

62 Basuray, S. and Chang, H. C. (2010) Designing a sensitive and quantifiable nanocolloid assay with dielectrophoretic crossover frequencies. *Biomicrofluidics* **4**(1), 013205.

63 Hoffman, P. D. and Zhu, Y. X. (2008) Double-layer effects on low frequency dielectrophoresis-induced colloidal assembly. *Appl. Phys. Lett.* **92:** 224103.

64 Hoffman, P. D., Sarangapani, P. S. and Zhu, Y. X. (2008) Dielectrophoresis and AC–induced assembly in binary colloidal suspensions. *Langmuir* **24**, 12164–12171.

65 Stoy, D. S. (1994) Interactive dipole model for two sphere system. *J. Electrostatics* **33**, 385–392.

66 Stoy, D. S. (1995) Force on two touching dielectric spheres in a parallel field. *J. Electrostatics* **35**, 297–308.

67 Wei, D. and Patey, G. N. (1992) Orientational order in simple dipolar liquids – computer-simulation of a ferroelectrical nematic phase. *Phys. Rev. Lett.* **68**, 2043–2045.

68 Levesque, D. and Weis, J. J. (1994) Orientational and structural order in strongly interacting dipolar hard-spheres. *Phys. Rev. E* **49**, 5131–5140.

69 Sancho, M., Giner, V. and Martinez, G. (1997) Monte Carlo simulation of dielectrophoretic particle chain formation. *Phys. Rev. E* **55**, 544–548.

70 Sancho, M., Martinez, G. and Llamas, M. (1988) Multipole interaction between dielectric particles. *J. Electrostatics* **21**, 135–144.

71 Huang, J. P., Karttunen, M., Yu, K. W. *et al.* (2004) Electrokinetic behaviour of two touching inhomogeneous biological cells and colloidal particles: Effects of multipolar interactions. *Phys. Rev. E* **69**, 051402.

72 Griffin, J. L. and Ferris, C. D. (1970) Pearl chain formation across radio frequency fields. *Nature* **226**, 152–154.

73 Giner, V., Sancho, M., Lee, R. S. *et al.* (1999) Transverse dipolar chaining in binary suspensions induced by rf fields. *J. Phys. D: Appl. Phys.* **32**, 1182–1186.

74 Simpson, G. J., Wilson, C. F., Gericke, K.-H. and Zare, R. N. (2002) Coupled electrorotation: Two proximate microspheres in registry with an AC electric field. *Chem. Phys. Chem.* **3**, 416–423.

75 Sancho, M., Martinez, G., Munoz, S. *et al.* (2010) Interaction between cells in dielectrophoresis and electrorotation experiments. *Biomicrofluidics* **4**, 022802.

76 Camarda, M., Fisicaro, G., Anzalone, R. *et al.* (2014) Theoretical and experimental study of the role of cell-cell dipole interaction in dielectrophoretic devices: application to polynomial electrodes. *BioMed. Eng. OnLine* **13**, 71–81.

77 Gascoyne, P. R. C., Shim, S., Noshari. J. *et al.* (2013) Correlations between the dielectric properties and exterior morphology of cells revealed by dielectrophoretic field-flow fractionation. *Electrophoresis* **34**, 1042–1050.

78 Hu, X., Bessette, P. H., Qian, J. *et al.* (2005) Marker-specific sorting of rare cells using dielectrophoresis. *PNAS* **102**(44), 15757–15761.

79 Muratore, M., Srsen, V., Waterfall, M. *et al.* (2012) Biomarker-free dielectrophoretic sorting of differentiating myoblast multipotent progenitor cells and their membrane analysis by Raman spectroscopy. *Biomicrofluidics* **6**, 034113.

80 Gascoyne, P. R. C., Noshari, J., Anderson, T. J. and Becker, F. F. (2009) Isolation of rare cells from cell mixtures by dielectrophoresis. *Electrophoresis* **30**, 1388–1398.

81 Shim, S., Stemke-Hale, K., Tsimeridou, A. M. *et al.* (2013) Antibody-independent isolation of circulating tumor cells by continuous-flow dielectrophoresis. *Biomicrofluidics* **7**, 011807.

82 Gupta, V., Jafferji, I., Garza, M. *et al.* (2012) ApoStream^{TM}, a new dielectrophoretic device for antibody independent isolation and recovery of viable cancer cells from blood. *Biomicrofluidics* **6**, 024133.

83 Lo, Y. J. and Lei, U. (2009) Quasistatic force and torque on a spherical particle under generalized dielectrophoresis in the vicinity of walls. *Appl. Phys. Let.* **95**, 253701.

84 Lo, Y. J. and Lei, U. (2010) Experimental validation of the theory of wall effect on dielectrophoresis in the vicinity of walls. *Appl. Phys. Let.* **97**, 093702.

85 Lei, U. and Lo, Y. J. (2011) Review of the theory of generalised dielectrophoresis. *IET Nanobiotechnol.* **5**, 86–106.

86 Camarda, M., Scalese, S. and La Magna, A. (2015) Analysis of the role of the particle-wall interaction on the separation efficiencies of field-flow fractionation dielectrophoretic devices. *Electrophoresis* **36**, 1396–1404.

87 Pohl, H. A. (1951) The motion and precipitation of suspensoids in divergent electric fields. *J. Appl. Phys.* **22**, 869–871.

88 Pohl, H. A. (1951) Some effects of nonuniform fields on dielectrics. *J. Appl. Phys.* **29**, 1182–1188.

89 Minerick, A. R., Zhou, R., Takhistov, P. and Chang, H.-C. (2003) Manipulation and characterization of red blood cells with alternating current fields in microdevices. *Electrophoresis* **24**: 3703–3717.

90 Pacansky, J. and Lyerla, J. R. (1979) Photochemical decomposition mechanisms for AZ-type photoresists. *IBM Res. Dev.* **23**, 42–55.

91 Fransilla, S. (2010) *Introduction to Microfabrication*, 2nd edn, John Wiley & Sons, Chichester.

92 Madou, M. J. (2011) *Fundamentals of Microfabrication and Nanotechnology*, 3rd edn, Vol. 2, CRC Press, Boca Raton, FL.

93 Markx, G. H.; Talary, M. S. and Pethig, R. (1994) Separation of viable and nonviable yeast using dielectrophoresis. *J. Biotechnology* **32**, 29–37.

94 Price, J. A. R., Burt, J. P. H. and Pethig, R. (1988) Applications of a new optical technique for measuring the dielectrophoretic behaviour of micro-organisms. *Biochim. Biophys. Acta* **964**, 221–230.

95 Wang, X.-B., Huang, Y., Burt, J. P. H. *et al.* (1993) Selective dielectrophoretic confinement of bioparticles in potential energy wells. *J. Phys. D: Appl. Phys.* **26**, 1278–1285.

96 Rajaraman, S., Noh, H.-S., Hesketh, P. J. and Gottfried, D. S. (2006) Rapid, low cost microfabrication technologies toward realization of devices for dielectrophoretic manipulation of particles and nanowires. *Sensors Actuators B: Chemical* **114**, 392–401.

97 Becker, F. F., Wang, X.-B., Huang, Y. *et al.* (1995) Separation of human breast cancer cells from blood by differential dielectric affinity. *Proc. Natl. Acad. Sci. USA* **92**, 860–864.

98 Yasukawa, T., Suzuki, M., Shiku, H. and Matsue, T. (2009) Control of the microparticle position in the channel based on dielectrophoresis. *Sensors Actuators B: Chemical* **142**, 400–403.

99 Ravula, S. K., Branch, D. W., James, C. D. *et al.* (2008) A microfluidic system combining acoustic and dielectrophoretic particle preconcentration and focusing. *Sensors Actuators B: Chemical* **130**, 645–652.

100 Gagnon, Z., Senapati, S., Gordon, J. and Chang, H.-C. (2008) Dielectrophoretic detection and quantification of hybridized DNA molecules on nano-genetic particles. *Electrophoresis* **29**, 4808–4812.

101 Kuo, Z. T. and Hsieh, W. H. (2009) Single-bead-based consecutive biochemical assays using a dielectrophoretic microfluidic platform. *Sensors Actuators B: Chemical* **141**, 293–300.

102 Voldman, J., Gray, M. L., Toner, M. and Schmidt, M. A. (2002) A microfabrication-based dynamic array cytometer. *Anal. Chem.* **74**, 3984–3990.

103 Voldman, J., Toner, M., Gray, M. L. and Schmidt, M. A. (2003) Design and analysis of extruded quadrupolar dielectrophoretic traps. *J. Electrostatics* **57**, 69–90, 2003.

104 Jang, L.-S., Huang, P.-H. and Lan, K.-C. (2009) Single-cell trapping utilizing negative dielectrophoretic quadrupole and microwell electrodes. *Biosensors Bioelectr.* **24**, 3637–3644.

105 Hoettges, K. F., Hughes, M. P., Cotton, A. *et al.* (2003) Optimizing particle collection or enhanced surface-based biosensors. *IEEE Eng. Med. Biol. Mag.* **22**(6), 68–74.

106 Hoettges, K. F. (2010) Dielectrophoresis as a cell characterization tool, in *Microengineering in*

Biotechnology (eds M. P. Hughes and K. F. Hoettges). Humana Press, New York, NY, pp. 183–198.

107 Talary, M. S. and Pethig, R. (1994) Optical technique for measuring the positive and negative dielectrophoretic behaviour of cells and colloidal suspensions. *IEE Proc. Sci. Meas. Technol.* **141**, 395–399.

108 Gascoyne, P. R. C., Noshari, J., Becker, F. F. and Pethig, R. (1994) Use of dielectrophoretic collection spectra for characterising differences between normal and cancerous cells. *IEEE Trans. Ind. Appl.* **30**, 829–834.

109 Pethig, R., Bressler, V., Carswell-Crumpton, C. *et al.* (2002) Dielectrophoresis studies of the activation of human T lymphocytes using a newly developed cell profiling system. *Electrophoresis* **23**, 2057–2063.

110 Milner, K. R., Brown, A. P., Betts, W. B. *et al.* (1998) Analysis of biological particles using dielectrophoresis and impedance measurement. *Biomed. Sci. Instrum.* **34**, 157–162.

111 Suehiro, J., Yatsunami, R., Hamada, R. and Hara, M. (1999) Quantitative estimation of biological cell concentration suspended in aqueous medium by using dielectrophoretic impedance measurement method. *J. Phys. D: Appl. Phys.* **32**, 2814–2820.

112 Bakewell, D. J. and Morgan, H. (2004) Quantifying dielectrophoretic collections of sub-micron particles on microelectrodes. *Meas. Sci. Technol.* **15**, 254–266.

113 Giraud, G., Pethig, R., Schulze, H. *et al.* (2010) Dielectrophoretic manipulation of ribosomal RNA. *Biomicrofluidics* **5**, 024116.

114 Fatoyinbo, H. O., Hoettges, K. F., Reddy, S. M. and Hughes, M. P. (2007) An integrated dielectrophoretic quartz crystal microbalance (DEP-QCM) device for rapid biosensing applications. *Biosensors Bioelectr.* **23**, 225–232.

115 Cheng, I.-F., Chang, H.-C., Hou, D. and Chang, H.-C. (2007) An integrated dielectrophoretic chip for continuous bioparticle filtering, focussing, sorting, trapping and detecting. *Biomicrofluidics* **1**, 021503.

116 McDonald, J. C., Duffy, D. C., Anderson, J. R. *et al.* (2000) Fabrication of microfluidic systems in poly(dimethysiloxane). *Electrophoresis* **21**, 27–40.

117 Wolfe, D. B., Qin, D. and Whitesides, G. M. (2010) Rapid prototyping of microstructures by soft lithography for biotechnology, in *Microengineering in Biotechnology* (eds. M. P. Hughes and K. F. Hoettges). Humana Press, New York, NY, pp. 81–107.

118 Friend, J. and Yeo, L. (2010) Fabrication of microfluidic devices using polydimethylsiloxane. *Biomicrofluidics* **4**, 026502.

119 Dalton, C. and Kaler, K. V. I. S. (2007) A cost effective, re-configurable electrokinetic microfluidic chip platform. *Sensors Actuators B: Chemical* **123**, 628–635.

120 Honegger, T., Berton, K., Pinedo-Rivera, T. and Peyrade, D. (2009) Design and realization of a microfluidic system for dielectrophoretic colloidal handling. *Microelectronic Eng.* **86**, 1401–1403.

121 Vulto, P., Glade, N., Altomare, L. *et al.* (2005) Microfluidic channel fabrication in dry film resist for production and prototyping of hybrid chips. *Lab Chip* **5**, 158–162.

122 Leoni, R., Castellano, M. G., Gerardino, A. *et al.* (1996) Sub-micron niobium electrodes for dielectrophoresis applications. *Microelectronic Eng.* **30**, 555–558.

123 Pethig, R., Burt, J. P. H., Parton, A. *et al.* (1998) Development of biofactory-on-a-chip technology using Excimer laser micromachining. *J. Micromech. Microeng.* **8**, 57–63.

124 Altomare, L., Borgatti, M., Medoro, G. *et al.* (2003) Levitation and movement of human tumor cells using a printed circuit board device based on software-controlled dielectrophoresis. *Biotechnol. Bioeng.* **82**, 474–479.

125 Cen, E. G., Dalton, C., Li, Y. *et al.* (2004) A combined dielectrophoresis, traveling wave dielectrophoresis and electrorotation microchip for the manipulation and characterization of human malignant cells. *J. Methods Microbiol.* **58**, 387–401.

126 Suehiro, J. and Pethig, R. (1998) The dielectrophoretic movement and positioning of a biological cell using a three-dimensional grid electrode system. *J. Phys. D: Appl. Phys.* **31**, 3298–3305.

127 Burt, J. P. H., Goater, A. D., Menachery, A. *et al.* (2007) Development of microtitre plates for electrokinetic assays. *J. Micromech. Microeng.* **17**, 250–257.

128 Pethig, R., Menachery, A., Heart, E. *et al.* (2008) Dielectrophoretic assembly of insulinoma cells and fluorescent nanosensors into three-dimensional pseudo-islet constructs. *IET Nanobiotechnol.* **2**, 31–38.

129 Fiedler, S., Shirley, S. G., Schnelle, T. and Fuhr, G. (1998) Dielectrophoretic sorting of particles and cells in a microsystem. *Anal. Chem.* **70**, 1909–1915.

130 Müller, T., Gradl, G., Howitz, S. *et al.* (1999) A 3-D microelectrode system for handling and caging single cells and particles. *Biosensors and Bioelectronics* **14**(3), 247–256.

131 Cheng, I.-F., Chang, H.-C., Hou, D. and Chang, H.-C. (2007) An integrated dielectrophoretic chip for continuous bioparticle filtering, focussing, sorting, trapping and detecting. *Biomicrofluidics* **1**, 021503.

132 Kralj, J. G., Lis, M. T. W., Schmidt, M. A. and Jensen, K. F. (2006) Continuous dielectrophoretic size-based particle sorting. *Anal. Chem.* **78**, 5019–5025.

133 Kim, U., Qian, J., Kenrick, S. A. *et al.* (2008) Multitarget dielectrophoresis activated cell sorter. *Anal. Chem.* **80**, 8656–8661.

134 Schnelle, T., Muller, T., Gradl, G. *et al.* (1999) Paired microelectrode system: dielectrophoretic particle sorting and force calibration. *J. Electrostatics* **47**, 121–132.

135 Demierre, N., Braschler, T., Muller, R. and Renaud, P. (2008) Focusing and continuous separation of cells in a microfluidic device using lateral dielectrophoresis. *Sensors and Actuators B: Chemical* **132**, 388–396.

136 Fatoyinbo, H. O., Kamchis, D., Whattingham, R. *et al.* (2005) A high-throughput 3-D composite dielectrophoretic separator. *IEEE Trans. Biomed. Eng.* **52**, 1347–1349.

137 Bocchi, M., Lombardini, M., Faenza, A. *et al.* (2009) Dielectrophoretic trapping in microwells for manipulation of single cells and small aggregates of particles. *Biosensors Bioelectr.* **24**, 1177–1183.

138 Rosenthal, A. and Voldman, J. (2005) Dielectrophoretic traps for single-particle patterning. *Biophys. J.* **88**, 2193–2205.

139 Thomas, R. S., Morgan, H. and Green, N. G. (2009) Negative DEP traps for single cell immobilisation. *Lab on a Chip* **9**, 1534–1540.

140 Kang, Y., Cetin, B., Wu, Z. and Li, D. (2009) Continuous particle separation with localized AC-dielectrophoresis using embedded electrodes and an insulating hurdle. *Electrochimica Acta* **54**, 1715–1720.

141 Pham, P., Texier, I., Larrea, A. S. *et al.* (2007) Numerical design of a 3-D microsystem for bioparticle dielectrophoresis: The pyramidal microdevice. *J. Electrostatics* **65**, 511–520.

142 Buyong, M. R., Larki, F., Faiz, M. S. *et al.* (2015) A tapered aluminium microelectrode array for improvement of dielectrophoresis-based particle manipulation. *Sensors* **15**, 10973–10990.

143 Wang, X. J., Wang, X. B., Becker, F. F. and Gascoyne, P. R. C. (1996) A theoretical model of electric field analysis for dielectrophoretic electrode arrays using Green's theorem. *J. Phys. D: Appl. Phys.* **29**, 1649–1660.

144 Green, N. G., Ramos, A. and Morgan, H. (2002) Numerical solution of the dielectrophoretic and traveling wave forces for interdigitated electrode arrays using the finite element method. *J. Electrostat.* **56**, 235–254.

145 Chang, D. H., Loire, S. and Mezic, I. (2002) Closed-form solutions in the electric field analysis for dielectrophoretic and traveling wave inter-digitated electrode arrays. *J. Phys. D: Appl. Phys.* **36**, 3073–3078.

146 Sun, T., Morgan, H. and Green, N. G. (2007) Analytical solutions of ac electrokinetics in interdigitated electrode arrays: electric field, dielectrophoretic and traveling wave dielectrophoretic forces. *Phys. Rev. E.* **76**, 1–18.

147 Alazzam, A., Roman, D., Nerguizian, V. *et al.* (2010) Analytical formulation of electric field and dielectrophoretic force for moving dielectrophoresis using Fourier series. *Microfluid Nanofluid* **9**, 1115–1124.

148 Song, H.-G. and Bennett, D. J. (2010) A semi-analytical approach using artificial neural network for dielectrophoresis generated by parallel electrodes. *J. Electrostatics* **68**, 49–56.

149 Chen, D. F., Du, H. and Li, W. H. (2006) A 3D paired microelectrode array for accumulation and separation of microparticles. *J. Micromech. Microeng.* **16**, 1162–1169.

150 Schnelle, T., Hagedorn, R., Fuhr, G. *et al.* (1993) Three-dimensional electric field traps for manipulation of cells – calculation and experimental verification. *Biochim. Biophys. Acta* **1157**, 127–140.

151 Park, S. and Beskok, A. (2008) Alternating current electrokinetic motion of colloidal particles on interdigitated microelectrodes. *Anal. Chem.* **80**, 2832–2841.

152 Masuda, S., Washizu, M. and Nanba, T. (1989) Novel-method of cell-fusion in field constriction area in fluid integrated circuit. *IEEE Trans. Ind. Appl.* **25**, 732–737.

153 Washizu, M., Nanba, T. and Masuda, S. (1990) Handling biological cells using a fluid integrated circuit. *IEEE Trans. Ind. Appl.* **26**, 352–358.

154 Chou, C.-F., Tegenfeldt, J. O., Bakajin, O. *et al.* (2002) Electrodeless dielectrophoresis of single- and double-stranded DNA. *Biophys. J.* **83**, 2170–2179.

155 Chou, C.-F. and Zenhausern, F. (2003) Electrodeless dielectrophoresis for micro total analysis systems. *IEEE Eng. Med. Biol. Mag.* **22**, 62–67, 2003.

156 Cummings, E. B. and Singh, A. K. (2003) Dielectrophoresis in microchips containing arrays of insulating posts: theoretical and experimental results. *Anal. Chem.* **75**, 4724–4731.

157 Lapizco-Encinas, B. H., Simmons, B. A., Cummings, E. B. and Fintschenko, Y. (2004) Insulator-based dielectrophoresis for the selective concentration and separation of live bacteria in water. *Electrophoresis* **25**, 1695–1704.

158 Lapizco-Encinas, B. H., Simmons, B. A., Cummings, E. B. and Fintschenko, Y. (2004) Dielectrophoretic concentration and separation of live and dead bacteria in an array of insulators. *Anal. Chem.* **76**, 1571–1579.

159 Lapizco-Encinas, B. H., Davalos, R. V., Simmons, B. A. *et al.* (2005) An insulator-based (electrodeless) dielectrophoretic concentrator for microbes in water. *J. Microbiol. Meth.* **62**, 317–326.

160 Barrett, L. M., Skulan, A. J., Singh, A. K. *et al.* (2005) Dielectrophoretic manipulation of particles and cells

using insulating ridges in faceted prism microchannels. *Anal. Chem.* **77**, 6798–6804.

161 Baylon-Cardiel, J. L., Lapizco-Encinas, B. H., Reyes-Betanzo, C. *et al.* (2009) Prediction of trapping zones in an insulator-based dielectrophoretic device. *Lab Chip* **9**, 2896–2901.

162 Markx, G. H. and Pethig, R. (1995) Dielectrophoretic separation of cells: continuous separation. *Biotechnol. Bioeng.* **45**, 337–343.

163 Landau, L. D. and Lifshitz, E. M. (1959) *Fluid Mechanics*, Pergamon, Oxford, Ch. 5, p. 184.

164 Ramos, A., Morgan, H., Green, N. G. and Castellanos, A. (1998) Ac electrokinetics: a review of forces in microelectrode structures. *J. Phys. D: Appl. Phys.* **31**, 2338–2353.

165 Barbulovic-Nad, I., Xuan, X. C., Lee, J. S. H. and Li, D. (2006) DC-dielectrophoretic separation of microparticles using an oil droplet obstacle. *Lab Chip* **6**, 274–279.

166 Thwar, P. K., Linderman, J. J. and Burns, M. A. (2007) Electrodeless direct current dielectrophoresis using reconfigurable field-shaping oil barriers. *Electrophoresis* 4572–4581.

167 Srivastava, S. K., Baylon-Cardiel, J. L., Lapizco-Encinas, B. H. and Minerick, A. R. (2011) A continuous DC-insulator dielectrophoretic sorter of microparticles. *J. Chromatogr. A* **1218**, 1780–1789.

168 Lewpiriyawong, N., Yang, C. and Lam, Y. C. (2008) Continuous separation and sorting of microparticles by size using AC dielectrophoresis in a PDMS microfluidic device with 3-D conducting PDMS composite electrodes. *Electrophoresis* **31**, 2622–2631.

169 Hawkins, B. G., Smith, A. E., Syed, Y. A. and Kirby, B. J. (2007) Continuous-flow particle separation by 3D insulative dielectrophoresis using coherently shaped DC-biased, AC electric fields. *Anal. Chem.* **79**, 7291–7300.

170 Kang, Y., Li, D., Kalams, S. A. and Eid, J. E. (2008) DC-Dielectrophoretic separation of biological cells by size. *Biomed. Microdev.* **10**, 243–249.

171 Chen, K. P., Pacheco, J. R., Hayes, M. A. and Staton, S. J. R. (2009) Insulator-based dielectrophoretic separation of small particles in a sawtooth channel. *Electrophoresis* **30**, 1441–1448.

172 Pysher, M. D. and Hayes, M. A. (2007) Electrophoretic and dielectrophoretic field gradient technique for separating bioparticles. *Anal. Chem.* **79**, 4552–4557.

173 Zhu, J., Tzeng, T.-R. J., Hu, G. and Xuan, X. (2009) DC dielectrophoretic focusing of particles in a serpentine microchannel. *Microfluid. Nanofluid.* **7**, 751–756.

174 Church, C., Zhu, J. and Xuan, X. (2011) Negative dielectrophoresis-based particle separation by size in a serpentine microchannel. *Electrophoresis* **32**, 527–531.

175 Saucedo-Espinosa, M. A. and Lapizco-Encinas, B. H. (2015) Design of insulator-based dielectrophoretic devices: Effect of insulator posts characteristics. *J. Chromatogr., A* **1422**, 325–333.

176 Zhu, J. and Xuan, X. (2009) Dielectrophoretic focusing of particles in a microchannel constriction using DC-biased AC electric fields. *Electrophoresis* **30**, 2668–2675.

177 Lewpiriyawong, N., Yang, C. and Lam, Y. C. (2012) Electrokinetically driven concentration of particles and cells by dielectrophoresis with DC-offset AC electric field. *Microfluid. Nanofluid.* **12**, 723–733.

178 Saucedo-Espinosa, M. A., LaLonde, A., Gencoglu, A. *et al.* (2016) Dielectrophoretic manipulation of particle mixtures employing asymmetric insulating posts. *Electrophoresis* **37**, 282–290.

179 Sridharan, S., Zhu, J., Hu, G. and Xuan, X. (2011) Joule heating effects on electroosmotic flow in insulator-based dielectrophoresis. *Electrophoresis* **32**, 2274–2281.

180 Kale, A., Patel, S., Hu, G. and Xuan, X. (2013) Numerical modeling of Joule heating effects in insulator-based dielectrophoresis microdevices. *Electrophoresis* **34**, 674–683.

181 Gallo-Villanueva, R. C., Sano, M. B., Lapizco-Encinas, B. H. and Davalos, R. V. (2014) Joule heating effects on particle immobilization in insulator-based dielectrophoretic devices. *Electrophoresis* **35**, 352–361.

182 LaLonde, A., Romero-Creel, M. F. and Lapizco-Encinas, B. H. (2015) Assessment of cell viability after manipulation with insulator-based dielectrophoresis. *Electrophoresis* **36**, 1479–1484.

183 Braff, W. A., Pignier, A. and Buie, C. R. (2012) High sensitivity three-dimensional insulator-based dielectrophoresis. *Lab Chip* **12**, 1327–1331.

184 Shafiee, H., Caldwell, J. L., Sano, M. B., Davalos, R. V. (2009) Contactless dielectrophoresis: a new technique for cell manipulation. *Biomed. Microdev.* **11**, 997–1006.

185 Shafiee, H., Sano, M. B., Henslee, E. A. *et al.* (2010) Selective isolation of live / dead cells using contactless dielectrophoresis (cDEP). *Lab Chip* **10**, 438–445.

186 Salanzadeh, A., Shafiee, H., Davalos, R. V. and Stremier, M. A. (2011) Microfluidic mixing using contactless dielectrophoresis. *Electrophoresis* **32**, 2569–2578.

187 Čemažar, J., Douglas, T. A., Schmelz, E. M. and Davalos, R. V. (2016) Enhanced contactless dielectrophoresis enrichment and isolation platform via cell-scale microstructures. *Biomicrofluidics* **10**, 014109.

188 LaLonde, A., Romero-Creel, M. F., Saucedo-Espinosa, M. A. and Lapizco-Encinas, B. H. (2015) Isolation and enrichment of low abundant particles with insulator-based dielectrophoresis. *Biomicrofluidics* **9**, 064113.

189 Abdallah, B. G., Chao, T.-C., Kupitz, C. *et al.* (2013) Dielectric sorting of membrane protein nanocrystals. *ACS Nano* **7**, 9129–9137.

190 Li, S., Ye, Z., Hui, Y. S. *et al.* (2015) On-chip DNA preconcentration in different media conductivities by electrodeless dielectrophoresis. *Biomicrofluidics* **9**, 054115.

191 Luo, J., Abdallah, B. G., Wolken, G. G. *et al.* (2014) Insulator-based dielectrophoresis of mitochondria. *Biomicrofluidics* **8**, 021801.

192 Demierre, N., Braschler, T., Linderholm, P. *et al.* (2007) Characterization and optimization of liquid electrodes for lateral dielectrophoresis. *Lab Chip* **7**, 355–365.

193 Tornay, R., Braschler, T., Demierre, N. *et al.* (2008) Dielectrophoresis-based particle exchanger for the manipulation and surface functionalization of particles. *Lab Chip* **8**, 267–273.

194 Demierre, N., Braschler, T., Muller, R. and Renaud, P. (2008) Focusing and continuous separation of cells in a microfluidic device using lateral dielectrophoresis. *Sens. Actuator. B* **132**, 388–396.

195 Braschler, T., Demierre, N., Nascimento, E. *et al.* (2008) Continuous separation of cells by balanced dielectrophoretic forces at multiple frequencies. *Lab Chip* **8**, 280–286.

196 Nascimento, E., Nogueira, N., Silva, T. *et al.* (2008) Dielectrophoretic sorting on a microfabricated flow cytometer: Label free separation of *Babesia bovis* infected erythrocytes. *Bioelectrochem.* **73**, 123–128.

197 Valero, A., Braschler, T., Demierre, N. and Renaud, P. (2010) A miniaturised continuous dielectrophoretic cell sorter and its applications. *Biomicrofluidics* **4**, 022807.

198 Wang, C., Jia, G., Taherabadi, L. H. and Madou, M. J. (2005) A novel method for the fabrication of high-aspect ratio C-MEMS structures. *J. Microelectromech. Syst.* **14**, 348–358.

199 Martinez-Duarte, R. (2012) Microfabrication technologies in dielectrophoresis applications – a review. *Electrophoresis* **33**, 3110–3132.

200 Martinez-Duarte, R. (2012) SU-8 photolithography as a toolbox for carbon MEMS. *Micromachines* **5**, 766–782.

201 Martinez-Duarte, R., Camacho-Alanis, F., Renaud, P. and Ros, A. (2013) Dielectrophoresis of lambda-DNA using 3D carbon electrodes. *Electrophoresis* **34**, 1113–1122.

202 Elitas, M., Martinez-Duarte, R., Dhar, N. *et al.* (2014) Dielectrophoresis-based purification of antibiotic-treated bacterial subpopulations. *Lab Chip* **14**, 1850–1857.

203 Schnelle, T., Müller, T., Hagedorn, R. *et al.* (1999) Single micro electrode dielectrophoretic tweezers for manipulation of suspended cells and particles. *Biochim. Biophys. Acta* **1428**, 99–105.

204 Lee, K., Kwon, S. G., Kim, S. H. and Kwak, Y. K. (2007) Dielectrophoretic tweezers using sharp probe electrode. *Sensor. Actuators A* **136**, 154–160.

205 Matsue, T., Matsumoto, N., Koike, S. and Uchida, I. (1993) Microring-ring electrode for manipulation of a single cell. *Biochim. Biophys. Acta* **1157**, 332–335.

206 Ogata, S., Yasukawa, T. and Matsue, T. (2001) Dielectrophoretic manipulation of a single chlorella cell with dual-microdisk electrode. *Bioelectrochem.* **54**, 33–37.

207 Hunt, T. P. and Westervelt, R. M. (2006) Dielectrophoresis tweezers for single cell manipulation. *Biomed. Microdevices* **8**, 227–230.

208 Menachery, A., Graham, D., Messerli, S. M. *et al.* (2010) Dielectrophoretic tweezer for isolating and manipulating target cells. *IET Nanobiotechnol.* **5**, 1–7.

209 Graham, D., Messerli, M. A. and Pethig, R. (2012) Spatial manipulation of cells and organeles using single electrode dielectrophoresis. *BioTechniques* **52**, 39–43.

210 Vahey, M. D. and Voldman, J. (2008) An equilibrium method for continuous-flow cell sorting using dielectrophoresis. *Anal. Chem.* **80**, 3135–3143.

211 Vahey, M. D. and Voldman, J. (2009) High-throughput cell and particle characterisation using isodielectric separation. *Anal. Chem.* **81**, 2446–2455.

212 Markx, G. H., Dyda, P. A. and Pethig, R. (1996) Dielectrophoretic separation of bacteria using a conductivity gradient. *J. Biotechnol.* **51**, 175–180.

213 Lee, R. S., Arnold, W. M. and Pethig, R. (2004) Control of cell elution from a dielectrophoresis micro-array using a permittivity gradient. Annual Report – Conference on Electrical Insulation and Dielectric Phenomena, pp. 352–355.

214 Vahey, M. D., Pesudo, L. Q., Svensson, J. P. *et al.* (2013) Microfluidic genome-wide profiling of intrinsic electrical properties in *Saccharomyces cerevisiae*. *Lab Chip* **13**, 2754–2763.

215 Gimsa, J., Marszalek, P., Loewe, U. and Tsong, T. Y. (1991) Dielectrophoresis and electrorotation of neurospora slime and murine myeloma cells. *Biophys. J.* **60**, 749–760.

216 Broche, L. M., Labeed, F. H. and Hughes, M. P. (2005) Extraction of dielectric properties of multiple populations from dielectrophoretic collection spectrum data. *Phys. Med. Biol.* **50**, 2267–2274.

217 Chung, C., Waterfall, M., Pells, S. *et al.* (2011) Dielectrophoretic characterisation of mammalian cells above 100 MHz. *J. Electrical Bioimpedance* **2**, 64–71.

218 Gimsa, J., Müller, T., Schnelle, T. and Fuhr, G. (1996) Dielectric spectroscopy of single human erythrocytes at physiological ionic strength. *Biophys. J.* **71**, 495–506.

219 Khakimova, A. M., Rudakova, M. A., Doroginitskií, M. M. and Filippov, A. V. (2008) An NMR study of the temperature dependence of the coefficient of water self-diffusion through lipid bilayer membranes. *Biofizika* **53**, 271–280.

220 Chung, C. (2015) Dielectrophoretic Investigations of Internal Cell Properties. Ph.D thesis, University of Edinburgh.

221 Abraham, M. and Föppl, A. (1907) *Theorie der Elektrizität*, Teubner, Leipzig, pp. 167–185.

11

Dielectrophoretic Studies of Bioparticles

11.1 Introduction

The first demonstration that dielectrophoresis (DEP) could find useful applications in the biomedical sciences was the report in 1966 by Pohl and Hawk [1] that heat-killed yeast cells (*Saccharomyces cerevisiae*) could be separated from living yeast. Herb Pohl thought this was probably the first time a purely physical technique had been used to both distinguish and separate live and dead cells at the same time. There is no evidence in the literature to suggest otherwise. Twelve years later, in his seminal book on dielectrophoresis, Pohl could describe in some detail the DEP characterization of yeast cells and several types of bacteria, as well as preliminary results for blood platelets, chloroplasts, erythrocytes, green algae and mitochondria. He also provided preliminary results for the use of DEP to form masses of living cells, as well as the continuous DEP separation of some cell types from cell mixtures. The preliminary results were either described in the theses of Pohl's MSc students (I. L. Hawk [1967]; K. L. Wiley [1970]; C. S. Chen [1972]; J. E. Rhoads [1973]) or formed part of work in progress with Kaler for later publication [2]. This work with Kaler describes the continuous DEP separation of yeast from green algae.

The objective of this chapter is to outline, with references to carefully chosen publications, the progress that has been achieved over the past 50 years in the various avenues of DEP described by Pohl. At the time of Pohl's book (1978), there were 16 reports on biological DEP to be found in the published literature, all but four of which had originated from his laboratory in Oklahoma. A search in November 2015, using the Web of Science Core Collection and other databases, revealed that well over 4500 publications on the theory and applications of DEP have been published since 1978!

Of Pohl's many achievements, the demonstration that DEP is able to separate dead cells from live ones is of particular importance. It is the first topic in this chapter.

11.2 DEP Characterization and Separation of Live and Dead Cells

To employ an expression that comes from basketball in the United States, distinguishing and selectively separating dead cells (having degraded plasma membranes) from live cells is a 'slam-dunk' application for DEP. Before describing the reasons for this and giving examples from the literature, it is of value to consider the ways in which a cell can die.

11.2.1 Types of Cell Death

There are three main processes leading to cell death, namely necrosis, apoptosis and autophagy [3–13]. With review titles such as: 'A blast from the past' [6]; 'Eat me or die' [7], 'Can't live without them, can live with them' [9], 'Cell death in the neighbourhood' [10] and 'Means to an end' [11], the subject appears to attract those with a special, if not morbid, sense of humour. Schematics of necrosis and apoptosis are shown in Figure 11.1. A comprehensive review of the dielectric techniques, including dielectric spectroscopy and DEP, which have been used to investigate cell viability is given by Patel and Markx [14].

Necrosis is the premature death of a cell due to acute stress or injury, caused for example by exposure to toxins, change of pH, heat or radiation, nutrient or oxygen deprivation, hypertonic or hypotonic conditions, viral or parasite infection. Many of these can also elicit programmed cell death (apoptosis). These external assaults lead to unregulated self-digestion (autolysis) of cell components, performed by its own enzymes, which degrade proteins. This results in loss of integrity of the plasma membrane and the membranes of internal organelles, producing uncontrolled release of cell death products into the external environment. This commonly leads to an inflammatory response in tissues. Unless the cells are

Dielectrophoresis: Theory, Methodology and Biological Applications, First Edition. Ronald Pethig.

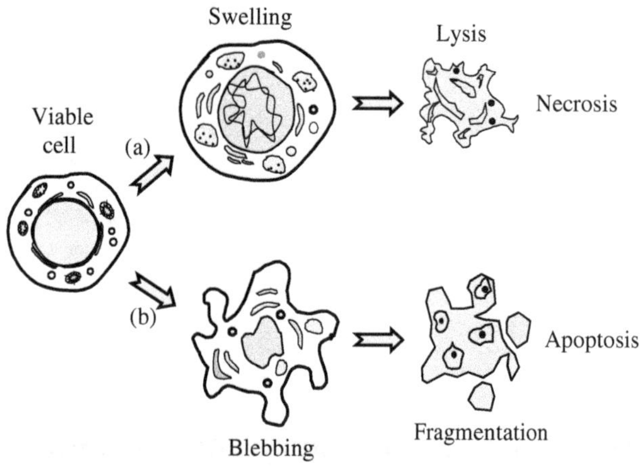

Figure 11.1 Two of the ways that a viable cell may die. (a) Necrosis: this is initiated by cell trauma. The plasma membrane loses its integrity, the cell's organelles swell, leading to cell lysis and rupture with leakage of its contents. (b) Apoptosis: a biochemically regulated process – the cell shrinks, blebs appear on the plasma membrane, leading to disassembly and fragmentation of the cell's contents.

deliberately exposed to a chemical that can induce apoptosis, necrosis is probably the most common form of cell death to occur in a DEP experiment. It is important to remember that necrosis is a 'downhill' process while apoptosis requires energy, so that in DEP experiments necrosis is likely to occur when cells are not at physiological temperature, for example.

Programmed cell death is the process mediated by an intracellular sequence of biochemical events, the best known form of which is apoptosis. This leads to characteristic changes in the morphology (shape, structure, size) of a cell and finally to its destruction. One of the first changes is the degrading of the cytoskeleton and its decoupling from the plasma membrane. This gives the membrane sufficient flexibility to bulge and form what are known as blebs. The normally asymmetrical transmembrane distribution of phospholipids also reorganizes in such a way that phosphatidylserine, normally localized exclusively in the cell's inner membrane leaflet, redistributes to the outer membrane leaflet. This can be identified by the ability of the cells to bind Annexin-V [5]. (Annexin-V is a Ca^{2+}-dependent anticoagulant protein that has high affinity for negatively charged phosphatidylserine and when conjugated with a fluorochrome can be used as a marker to identify apoptosis.) The cell nucleus also fragments, so that complexes of DNA and protein dissociate from the chromosomes and are released and distributed within the cytoplasm. The various components of the fragmented cell are partitioned into small packages known as apoptotic bodies, before being engulfed by phagocytes. Phagocytosis prevents the contents of the cell leaking out, which could cause inflammatory sequelae and include death of additional cells. One of the pathways to apoptosis is the so-called *extrinsic pathway* of apoptosis, mediated by death-receptor ligands on the cell surface. An alternative route, the *intrinsic* pathway, is activated by multiple stimuli and is characterized by regulatory proteins being released from the intermembrane space of mitochondria located in the cytoplasm. Both pathways (which are interconnected) activate enzymes, called caspases, which degrade other proteins. Controlled apoptosis endows advantages to an organism. An often cited example of this is the removal of cells that initially bind together the fingers and toes of a developing human embryo. Defective apoptotic processes can lead to disease. For example, excessive apoptosis can result in the breakdown of tissue, whereas the opposite situation can produce uncontrolled cell proliferation – as in cancer.

Another pathway that can lead to cell death, not shown in Figure 11.1, is a regulated process known as autophagy [12]. Autophagy is responsible for the lysosome-mediated degradation of damaged proteins and organelles. This process starts with the formation of a double-membrane bound vesicle in the cytoplasm, known as an autophagosome. This vesicle fuses with a lysosome to become an autolysosome and acts to 'eat up' damaged organelles in the cytoplasm.

11.2.2 Yeast Cells

11.2.2.1 Structure and Life Cycle

Yeast cells are unicellular eukaryotes (they possess a nucleus) and are classified as members of the fungus kingdom. They number around 1500 species, although when referring to yeast most people have in mind the species *Saccharomyces cerevisiae* – the name given by Franz Meyen in 1837 to an organism discovered in malt used for beer brewing. By fermentation, *S. cerevisiae* converts carbohydrates to carbon dioxide and alcohol, which for at least two thousand years has been exploited in baking and wine / beer production. As the fruit fly or squid axon is to genetics or electrophysiology, respectively, so is *S. cerevisiae* to the study of dielectrophoresis. It is readily available as bakers' or brewers' yeast; can be readily cultured by engineers or physicists lacking cell biology training; harvested at various stages of its life cycle; is resilient enough to withstand abuse in the hands of an engineer (e.g., suspending in distilled water, exposing to high electric fields); nonviable (dead) cells are easily identified by a simple dye treatment using methylene blue. The first test bioparticle of choice for newcomers to DEP has been *S. cerevisiae*. They (the cells) are typically round to ovoid in shape and 5~10 μm in diameter, thus lending themselves amenable to theoretical modelling and observation of their DEP response

Figure 11.2 Yeast cells collecting as pearl chains by positive DEP at the rounded tip of a platinum wire [15].

using standard microscopy. A photomicrograph [15] of viable yeast cells collecting by positive DEP at the rounded tip of a platinum wire is shown in Figure 11.2.

As for all fungi, yeast may have asexual and sexual life cycles. The most common form of vegetative yeast growth is asexual reproduction by budding – where a daughter cell in the form of a small bud appears on the parent cell. The nucleus of the parent cell splits into a daughter nucleus and migrates into the daughter cell. The bud continues to grow until it forms a new cell on separating from the parent cell. A daughter cell is generally smaller than its parent. A photomicrograph of a dividing yeast cell is shown in Figure 11.3. Another species of yeast, also used in brewing and baking, which has been studied by DEP, is *Schizosaccharomyces pombe*. This is also known as fission yeast, which, as its name implies, reproduces by fission instead of budding. It is a rod-shaped cell, 3~4 μm in diameter and 7~14 μm in length, which grows from its tips and then divides to create two identical daughter cells (~7 μm long). *S. pombe* is an

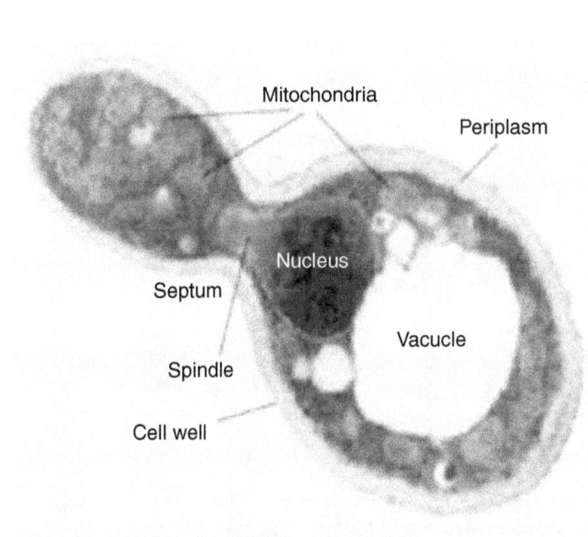

Figure 11.3 Photomicrograph of a dividing yeast cell [16].

important organism in the study of cellular responses to DNA damage and the process of DNA replication.

The cytosol of the dividing yeast cell shown in Figure 11.3 is surrounded by a cell envelope. As viewed from outside the cell and progressing into its interior, the cell envelope consists of an external cell wall, the periplasmic space and then the plasma membrane. The cell wall is a rigid structure about 100–200 nm thick and constitutes about 25% of the total dry mass of the cell. This wall prevents the cell from overexpansion when water enters its interior, as will occur when suspended (for a DEP experiment) in an aqueous medium of low ionic strength. The composition of the cell wall can vary according to growth conditions, but is essentially composed of just four types of macromolecule, namely glycoproteins, chitin and two types of β-glucans. Chitin is a tough material formed of a long-chain polymer of glucosamine, constituting the shell of a crab and the exoskeleton of a beetle, for example. It is the reason why yeast cells are so robust. The periplasmic space is narrow, 3.5~4.5 nm in width, containing mainly secreted proteins such as invertase and phosphatase, which catabolize substrates that are unable to cross the plasma membrane into the cytosol. The plasma membrane of *S. cerevisiae* is about 7.5 nm thick, which like mammalian cells consists of a lipid bilayer containing proteins that act as cytoskeletal anchors and as enzymes for ion channel transport, cell wall synthesis, osmotic control and signal transduction, for example. The cytoplasm contains soluble macromolecules such as proteins and glycogen, as well as ribosomes, lipid particles and mitochondria. The nucleus has a diameter of ~1.5 μm, surrounded by an inner and outer nuclear membrane. Nuclear pore complexes, 50~100 nm in diameter, form channels through these membranes for the exchange of components between the nucleus and cytosol. At cell division, motor proteins (kinesins) move the mitotic spindle attached to the nucleus into the narrow neck between the parent and daughter cell, to segregate accurately the duplicated chromosomes. As can be seen in Figure 11.3 a dominant feature of a yeast cell is the vacuole, which may occupy up to 30% of the total cell volume. This organelle has a single membrane and amongst other functions acts as a store of simple amino acids and metal cations (K^+, Mg^{2+}, Ca^{2+}). They are dynamic structures, which may exist in a cell as a single large compartment or as several smaller ones. Whilst the cytosol is maintained at ~pH 7.2, an enzyme (ATPase) in the vacuole membrane maintains the vacuolar pH at 5.0. This is the optimum pH for the functioning of more than 40 different hydrolytic enzymes stored within the vacuole, so that another of its roles is to degrade unwanted proteins and the products from autophagy.

For at least 100 years in the brewing industry, methylene blue has been used as the standard vital stain for yeast. This dye takes on a blue colour when dissolved in oxygenated water (i.e., in its oxidized form) and can readily pass through the cell wall and plasma membrane of either a live or dead yeast cell. In a living yeast cell, active enzymes called dehydrogenases are involved in the breaking down of glucose (glycolysis) and in the Krebs cycle, to produce ATP. These enzymes are capable of transferring their captured hydrogen atoms to methylene blue, thereby chemically reducing it. In its reduced state methylene blue is colourless. Methylene blue that enters into a dead cell should not be reduced and will retain its colour. Methylene violet can also be used in this way as a vital stain. However, the method is not perfect. Some dead cells could have enough active enzymes remaining in them to reduce the dye; in a live cell the dye may bind to a protein other than a hydrogenase so that it cannot be reduced. The method can also fail to distinguishing live and dead cells that have formed a clump of cells. A significant disadvantage is that the method does not depend on the integrity or otherwise of the plasma membrane. For these reasons trypan blue is favoured. Trypan blue, as its name implies, is also a blue dye and stains dead cells blue. Unlike the situation for methylene blue and methylene violet, the staining action depends on the plasma membrane losing its integrity. An intact and viable plasma membrane represents an impermeable barrier to trypan blue – it can only pass across a damaged membrane. It does though have one disadvantage. Although it has become the gold standard for nonfluorescent vital dyes, it is cytotoxic. Cells exposed to trypan blue for too long will die and stain blue.

11.2.2.2 Live and Dead Yeast

In their DEP experiments with yeast cells, Pohl and Hawk [1] used a pin-plate electrode arrangement, with a carefully rounded 0.66 mm stainless steel wire facing a flat steel plate positioned 1 mm away. Yeast cells (not stated, but presumably *S. cerevisiae*) were killed by heat treatment at 60–70 °C for 3 min and simultaneously stained with crystal violet. These dead cells were mixed with live ones in water. After successive suspending and washing of the cells in high-resistivity water, the final conductivity of the suspending medium fell to a value in the range from 2.8 to 10 mS/m. This conductivity is equivalent to that of 20–35 mM KCl. On application of a 2.55 MHz, 30 V_{rms}, signal to the electrodes, the live cells clustered at the pin electrode by positive DEP within a period of 15 to 30 s, whilst the stained dead cells tended to remain unperturbed in the suspension medium. Viable cells that had been exposed to the DEP field for several minutes were removed and cultured. They grew again on an agar medium, showing that the cells had not been damaged irreversibly by exposure to the low ionic strength medium and the high-voltage AC signal. Further studies by Pohl and Crane [17, 18], Mason and Townsley [19] confirmed these results and added new details. When using a cell separator with coaxial cylindrical electrodes, the rate of collection (yield) of cells was determined in Pohl's lab by obtaining the slope of a plot of the total length of pearl chains (such as those depicted in Figure 11.2) as a function of the square-root of the length of time (up to ~5 min) the electric field had been applied. This approach is credited by Pohl to an original theoretical analysis by John A. Wheeler [20] and is expanded further in Box 11.1. This analysis predicts that for a coaxial cylindrical geometry a plot of cell yield versus $\sqrt{t}$ is proportional to the product of the initial cell concentration and the square root of the Clausius–Mossotti factor. A linear dependence on $\sqrt{t}$ was observed for cell concentrations up to $\sim 10^7$ cells/ml, beyond which concentration the dependence was superlinear [22, p. 364]. The superlinearity probably arose from the formation of cell doublets and higher order pearl chaining before the cells reached the electrode. This would have the effect of increasing the effective size of the cells and the DEP force acting on them. As shown in Box 11.1, for the case of spherical electrode geometry, a linear relationship is expected for a plot of the cell yield versus $t^{5/2}$. Mason and Townsley [19] determined their cell collection rates by optical density measurements of the effluent from the sample port, compared with that of the sample before introduction into a DEP particle separator. They used a cell separator of the coaxial cylindrical design shown in Figure 11.4, which is based on that originally described by Pohl [21].

A basic understanding of why live and dead yeast cells exhibit a significant difference in their DEP response is obtained from the modelled DEP frequency-dependent characteristics shown in Figure 11.5, derived using the MATLAB program detailed in Box 11.2. This program employs the double-shell model of a cell, as described in

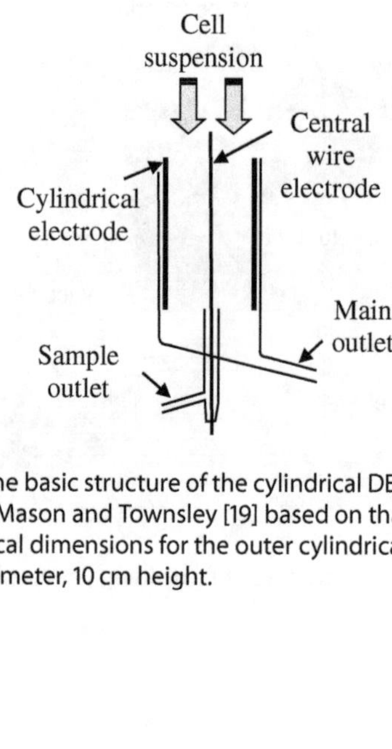

Figure 11.4 The basic structure of the cylindrical DEP cell separator described by Mason and Townsley [19] based on that described by Pohl [21]. Typical dimensions for the outer cylindrical electrode are 2 cm inner diameter, 10 cm height.

Box 11.1 Particle Yields for Cylindrical and Spherical Electrode Geometries

In the absence of fluid flow or other disturbing influences, the mass yield W by positive DEP of particles at an electrode depends on the particle mass concentration c and the volume swept in (V_{si}) over time t. For cylindrical and spherical geometries:

$$W_{cyl} = c\,V_{si} = ch(\pi r_t^2);\; W_{sph} = c\,V_{si} = 4c(\pi r_t^3)/3 \quad (11.1)$$

where c is the mass concentration of particles, h is the height of a cylindrical separator and r_t is the radius swept clean in time t. The value of r_t will depend on the velocity v imparted on the particles, which is given by the balance of the DEP force and viscous drag force. From Stokes' Law given by Equation 2.4 and the DEP force given by Equation (2.19), for a suspension of spherical particles of mean radius R and suspending medium viscosity η, this balance is given by:

$$2\pi R^3\varepsilon_o\varepsilon_m[CM]\partial^2E/\partial r^2 = 6\pi R\eta v$$

to give:

$$v = \frac{\varepsilon_o\varepsilon_m R^2[CM]}{3\eta}\frac{\partial^2 E}{\partial r^2}$$

Values for $\partial^2E/\partial r^2$ are obtained by differentiating the expressions for ∇E given in Table 3.2 for the cylindrical electrode and spherical electrode geometries:

$$\text{Cylindrical: } \partial^2E/\partial r^2 = \lambda/(\pi\varepsilon_o\varepsilon_m r^3).$$
$$\text{Spherical: } \partial^2E/\partial r^2 = 3Q/(2\pi\,\varepsilon_o\varepsilon_m r^4)$$

where λ and Q are the charge densities on the inner electrode of the cylindrical and spherical electrodes, respectively. The corresponding particle velocities are thus:

$$v_{cyl} = R^2\lambda[CM]/(3\pi\eta r^3);\; v_{sph} = R^2Q[CM]/(2\pi\eta\, r^4)$$

A time interval dt can be given as: $dt = dr/(\partial r/\partial t) = dr/v$. For the two geometries:

$$\text{Cylindrical: } dt = (3\pi\eta\, r^3)dr/(R^2\lambda[CM]) = Kr^3dr\,;$$
$$\text{Spherical: } dt = Kr^4dr$$

where K is of fixed magnitude, proportional to $[CM]^{-1}$, for a particular experiment. For the two geometries we assume that the inner electrode radius $r_i \ll r$, so that

$$t_{cyl} = K\int_r^{r_i} r^3dr = \frac{K}{4}\left[r^4 - r_i\right] \approx \frac{K}{4}r^4;$$
$$t_{sph} = K\int_r^{r_i} r^4dr \approx \frac{K}{5}r^5$$

From Equation (11.1) the corresponding collected yields over time t are:

$$W_{cyl} = ch\left(\pi r_t^2\right) = ch(2\pi/\sqrt{K})t^{1/2};$$
$$W_{sph} = 4c(\pi r_t^3)/3 = 4c(2\pi/\sqrt{K})t^{5/2}$$

A plot of cell yield versus $t^{1/2}$ or $t^{5/2}$, for the cylindrical and spherical geometry, respectively, should be linear with a slope proportional to the cell concentration, the cell radius and the square root of the Clausius–Mossotti factor [CM].

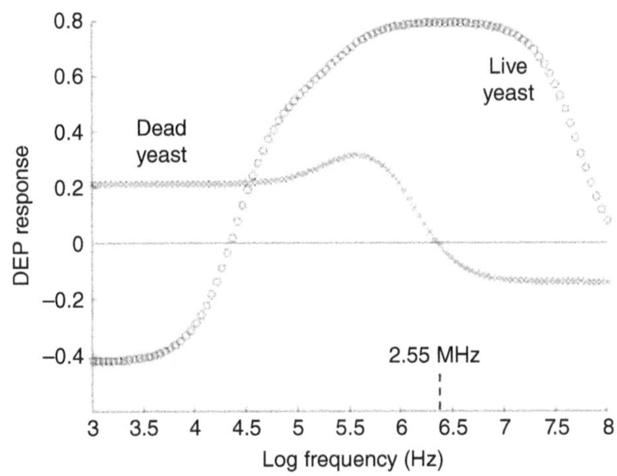

Figure 11.5 Relative DEP responses (as determined by CM factor evaluation using the MATLAB program in Box 11.2) of viable and dead yeast suspended in a medium of conductivity 3 mS/m. The responses at 2.55 MHz agree qualitatively with those reported at this conductivity and frequency by Pohl and Hawk [1]. At 10 kHz dead cells are collected at an electrode whilst live ones are repelled. The opposite result occurs at 10 MHz.

Chapter 9. It is basically the same as the program given in Box 9.3, but modified to take into account the existence of the cell wall. The presence of the vacuole and nucleus is incorporated into the effective dielectric properties of the cytosol. This simplification results in only a small difference in modelled results, compared to those that would be obtained if the vacuole and nucleus were included [23, 24].

In early analyses of the dielectric properties of yeast, it was assumed that the wall conductivity is almost the same as that of the external suspending medium [25, 26]. However, this did not take account of the work of Carstensen *et al.* [27] on the dielectric properties of the bacterium *M. lysodeikticus*, regarding counterion conduction associated with fixed charges (e.g., ionized groups on glycoproteins) in the cell wall. The following relationship for the cell wall conductivity σ_w was established [27]:

$$\sigma_w \approx c_{fw}\mu_w\left[1 + \left(2\frac{c_m}{c_{fw}}\right)^2\right]^{1/2} \quad (11.2)$$

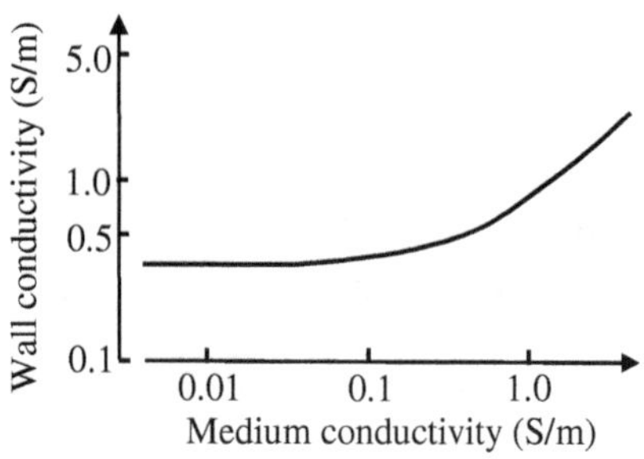

Figure 11.6 The conductivity of an isolated cell wall of *Micrococcus luteus* as a function of the conductivity of the suspending medium (from Carstensen *et al.* [27]).

where c_{fw} is the fixed charge concentration in the cell wall, μ_w is the mobility of the counterions in the cell wall and c_m is the ionic concentration of the suspending medium. For high values of the medium's ionic concentration, from Equation (11.2) we have:

$$\sigma_w \approx 2c_m\mu_w$$

The conductivity of the medium is given by $\sigma_m = c_m\mu_m$ where μ_m is the mobility of the ions in the bulk solution. Thus, for high values of the medium conductivity:

$$\sigma_w \approx 2\sigma_m\frac{\mu_w}{\mu_m}$$

so that the cell wall conductivity varies in direct proportion to the medium conductivity and the ratio of the ion mobilities in the wall and bulk electrolyte. Asami and Yonezawa refer to this ratio as the effective porosity of the cell wall [23]. As the medium conductivity tends to lower values, from Equation (11.2) we expect the cell wall conductivity to reach a constant value determined by the concentration of fixed charge density in the wall and the mobility of the counterions. This form of relationship for the cell wall conductivity as a function of medium conductivity is shown in Figure 11.6 for the case of the bacteria *Micrococcus lysodeikticus* [27]. For the case of yeast cells, for a suspending medium conductivity of 2.8 mS/m the ratio (σ_w/σ_m) of the cell wall and medium conductivity was determined as $\sigma_w/\sigma_m = 0.24$, tending to a constant value of 0.1 with increasing medium conductivity [23]. The capacitance of the plasma membrane was also determined to be ~6.5 mF/m^2 [23]. This low value indicates that the membrane surface is relatively free of membrane folds or blebs, possibly reflecting the fact that under hypotonic conditions it is pressed hard against the inner cell wall. The relationship between cell wall and medium conductivity [23], together with the relatively small membrane capacitance value, is taken into account in the MATLAB program of Box 11.2. It is also assumed in the program that the major effect of the heat treatment to kill the yeast is to degrade the structure of the plasma membrane, resulting in a 500-fold reduction of its resistance to passive ion flow and the leakage of ions from the cytosol into the surrounding medium. The dielectric properties of the more robust cell wall are assumed to remain unaltered. Even with these arbitrary assumptions, it is interesting to see that the DEP responses shown in Figure 11.5 are in good agreement with those reported [1, 17–19]. At a frequency of 2.55 MHz live yeast cells exhibit strong positive DEP, whilst the dead cells experience no or little DEP force. Mason and Townsley [19] found good collection of dead cells (by positive DEP) at 10 kHz, whereas at 100 kHz the rate of collection of dead cells was less than that of live cells. These results can also be predicted by the DEP responses shown in Figure 11.5. In Figure 15.7 of his book [22, p. 367] Pohl provides a plot of the collection by positive DEP of live yeast cells, as a function of the conductivity of the suspending medium and the frequency of the applied voltage signal. The same conductivity values used by Pohl are employed in the program detailed in Box 11.2 to provide the plots shown in Figure 11.7. The form of the positive DEP responses (Pohl was unable to evaluate negative DEP effects) shown in Figure 11.7 closely mirror those described by Pohl.

Mason and Townsley [19] also investigated the possibility that DEP could separate cells, which differ less markedly than being dead or alive. It was hoped that cells different in their respective biochemical makeup would also differ in their DEP responses. They cultured yeast under three different nutrient conditions. It was expected that cells cultured under conditions of a balanced medium might have a higher RNA:protein ratio than cells cultured under conditions of nitrogen starvation. Likewise, cells grown in a rich medium might have higher nucleic acid content per cell than cells cultured in a minimal medium. Although cells grown under different culture media could be separated, it was apparent that a complex function of both cell size and polarizability was involved. This situation, also revealed through the theoretical treatment in Box 11.1, remains an important consideration in devising protocols for the DEP separation of different cell types.

The formation of cell pearl chains through dielectrophoresis can be used as the first step in a procedure to hybridize cells through electrofusion, with early tests of this being performed with protoplasts of *S. cerevisiae* [28]. The yeast cells first have their cell wall removed, usually by enzyme digestion, to produce protoplasts. They are then brought together by DEP into an electrode gap, using a nonuniform field of 33~670 kV/m at 1–2 MHz, over a period of 2–10 min [28, 29]. Membrane fusion is then induced by subjecting the cells to a high voltage pulse, of typical magnitude ~1500 kV/m and duration 7~40 μs. Förster and Emeis [29] tested the viability of protoplasts from *S. cerevisiae* and *S. diastaticus* that

Box 11.2 MATLAB Program for Double-Shell Model of a Yeast Cell

```
% YEAST.m
% Cytosol: conductivity kc1; permittivity kp1; radius a1.
% Plasma membrane: conductivity kc2; permittivity kp2.
% Cell wall: conductivity kc3, permittivity kp3, radius a3.
% Suspending medium: conductivity kc4; permittivity kp4.
% Plasma membrane thickness d2; Cell Wall thickness d3.
% ----------------------------------------
clear;
d2=7.5e-9;
 d3=1.5e-7;
 a1=4.0e-6;
 a2=a1+d2;
 a3=a2+d3;
 p0=8.854e-12;
 f=logspace(3, 8,100);
 zeroline=f-f;
 w=2*pi*f;
% Cytosol conductivity and permittivity.
% Live Cell  : kc1=0.5 S/m; kp1=50.
% Dead Cell  : kc1=50 mS/m; kp1=50.
 kc1=0.5;
%kc1=5e-2;
 kp1=50*p0;
% Plasma membrane conductivity kc2 and capacitance Cm.
% Live Cell: kc2=5 exp-7 S/m; Cm=7 mS/m^2.
% Dead Cell: kc2=1 exp-5 S/m; Cm=7 mS/m^2.
  kc2=5e-7;
% kc2=1e-5;
 Cm=7e-3;
 kp2=Cm*d2;
% Cell Wall conductivity kc3 and permittivity kp3.
% Viable and dead cell: kc3=0.24*kc4; kp3=60.
 kc3=0.24*kc4;
 kp3=60*p0;
% Suspending medium conductivity kc4 and permittivity kp4.
 kc4=2.8e-3;
 kp4=79*p0;
 k1=kp1-i*kc1 ./w;
 k2=kp2-i*kc2 ./w;
 k3=kp3-i*kc3 ./w;
 k4=kp4-i*kc4 ./w;
 am1=a1^3;
 am2=a2^3;
 am3=a3^3;
 keff2=k2 .*(am2*(k1+2*k2)-2*am1*(k2-k1))
 ./(am2*(k1+2*k2)+...am1*(k2-k1));
 keff3=k3 .*(am3*(keff2+2*k3)-2*am2*(k3- keff2))
 ./(am3*(keff2+2*k3)+keff2)) ./(am3*(keff2+2*k3)
 +am2*(k3-keff2));
 m=(keff3-k4) ./(keff3+2*k4);
 rm=real(m);
 plot(log10(f), rm,'o', log10(f),zeroline,'-');
 xlabel('Log Frequency (Hz)')
 ylabel('DEP Response')
 hold on
```

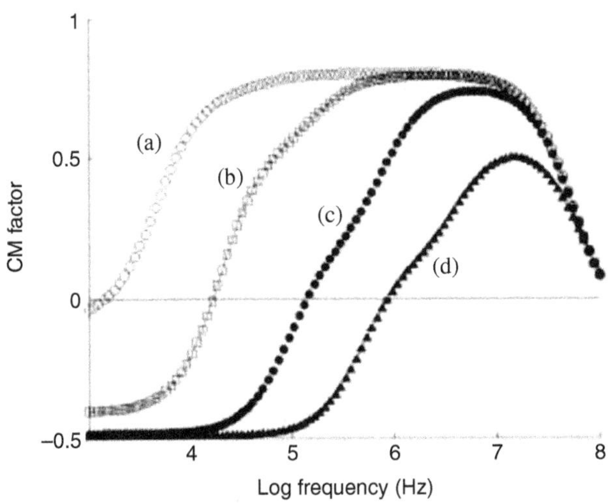

Figure 11.7 Relative DEP response (as determined by *CM*-factor evaluation using the MATLAB program in Box 11.2) of viable yeast as a function of frequency and medium conductivity. (a) 0.32 mS/m; (b) 2.12 mS/m; (c) 15.6 mS/m; (d) 91 mS/m.

had been electrofused over this range of electrical treatments. The regeneration rates of yeast protoplasts collected under the conditions employed for the DEP and electrofusion steps did not differ from those of protoplasts that had been maintained under the same experimental conditions, but had not been subjected to the electric field treatments. This was an important fact to establish, because a common first question by biologists introduced to DEP is to enquire about the cell damage caused by exposure to the electric field.

The first detailed differences in the DEP behaviour of live and dead yeast cells appear to be those reported by Huang *et al.* [30]. The best fit curves to this data (which also included electrorotation responses) are shown in Figure 11.8. The relative permittivity values of the cell wall (60), plasma membrane (6) and cytosol (50) were assumed to remain unchanged after heat treatment (75 °C for 5 min), as is the case for the modeled DEP responses shown in Figure 11.5. The best-fit conductivity values of the cell wall, plasma membrane and cytosol were determined to be 14 mS/m; 2.5×10^{-7} S/m and

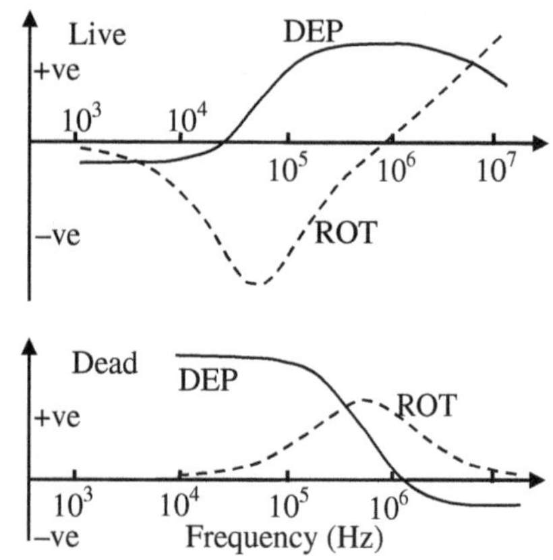

Figure 11.8 Dielectrophoretic (DEP) and electrorotation (ROT) spectra obtained for live *S. cerevisiae* and heat-treated (dead) cells. The curves are the best fits to experimental data reported by Huang *et al.* [30].

0.2 S/m, respectively [30]. After heat treatment, these values changed to 1.5 mS/m; 1.6×10^{-4} S/m and 7 mS/m, respectively. The large reduction in cytosol conductivity after cell death indicates that ions in the cytosol were able to leak across the damaged plasma membrane, down their osmotic gradients, into the suspending medium (of conductivity 0.5 mS/m). The difference in electrokinetic behaviour of the live and dead yeast cells is particularly evident when the DEP and electrorotation results (ROT) are displayed as a function of frequency in the geometrical form of an Argand plot, as shown in Figure 11.9. (Sight of this geometric distinction between life and death still induces a sense of awe in the author.) The semicircular form of these Argand plots is a manifestation of the Kramers–Krönig relations described in Chapter 7 (section 7.3.1) and outlined further in Box 10.8. The underlying dielectric phenomena that give rise to their semicircular form are also described by Wang *et al.* [31].

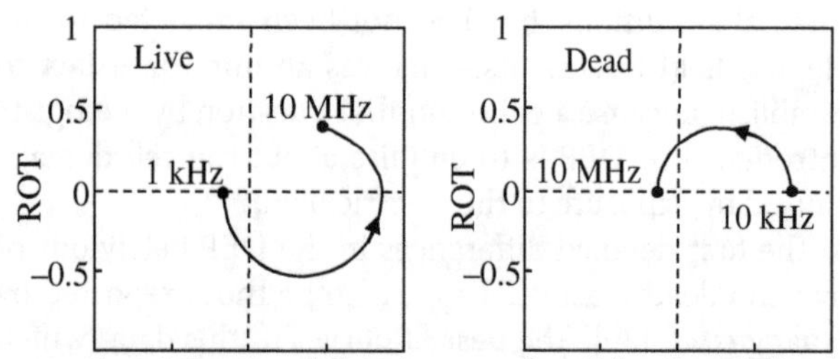

Figure 11.9 Geometrical representations of the variations of the normalized DEP velocity and electrorotation (ROT) rate as a function of frequency for live and dead (heat-treated) *S. cerevisiae* cells. The normalizations were such that the velocities were directly equal to the DEP force and ROT torque factors Re[*CM*] and –Im[*CM*] of Equations (10.27) and (10.52), respectively. (Based on Wang *et al.* [31].)

Markx *et al.* [32, 33] demonstrated that known mixtures of live and dead *S. cerevisiae* cells could be separated with good efficiency by DEP using castellated, interdigitated, microelectrodes. Through measurement of cell viability by staining with methylene blue and plate counts, for an initial cell concentration of $\sim 1.4 \times 10^7$ cells/ml containing 60% nonviable cells, the DEP separated nonviable fraction contained 3% viable cells, with the viable fraction 8% dead cells. From direct microscopic observations of the DEP effect on methylene blue-treated suspensions, this 'contamination' was found to occur because nonviable cells were sterically hindered and even trapped by the viable cells. This effect was reduced significantly on tenfold dilution of the initial suspension. Improved efficiency of separation was also obtained by passing the cells through two or more stages of DEP separation. Patel *et al.* [34] reported that the dielectric capacitance of suspensions and the DEP behaviour of dying and dead yeast cells were both strongly dependent on the method used to induce cell death. Methods (e.g., heat treatment) that cause denaturation of proteins and directly affect the membrane permeability and consequently the membrane and internal conductivities, resulted in large changes in DEP behaviour, whereas methods that affected the cell interior but had little effect on the cell membrane resulted in small or no changes in the dielectric properties of the cells. Methods deemed to damage the plasma membrane were heat treatment and solubilization with iso-octonal, whilst lethal treatments that did not damage the membrane were glutaraldehyde to stabilize (fix by crosslinking) the membrane and the multipurpose disinfectant, Virkon. These findings were taken to imply that, depending on the method by which cell death is induced, DEP will not always be able to separate viable from nonviable cells. The difference in the DEP behaviour of viable and nonviable cells, as demonstrated in Figures 11.8 and 11.9 and also discussed by Patel and Markx [13], is caused by the Maxwell–Wagner interfacial polarization arising from the plasma membrane acting as a high resistance to passive ion flow across it. The fact that an environmental stress on the cell, which causes minimal damage to the plasma membrane and leads to no or minimal change in its DEP behaviour, is on its own not surprising. (Although it should be mentioned that fixing a cell with glutaraldehyde typically results in cell shrinkage and membrane 'ruffling', which is an effect usually distinguishable by DEP.) However, if this stress leads to lethal damage to the cell interior and nucleus, such as protein synthesis inhibition or DNA fragmentation, it should be expected that amongst secondary effects there would be damage to the membrane. For example, it is known that cells undergoing apoptosis maintain their plasma membrane integrity for several hours after initiating nuclear damage [14]. Membrane permeabilization, resulting in a large increase in membrane conductance and ion leakage is considered a relatively late-stage event of apoptosis. It may simply be the case that a toxicant that primarily targets the plasma membrane results in a *more rapid* change of a cell's DEP characteristic than a toxicant that primarily attacks enzyme activity or nucleic acid. Patel *et al.* [34] have raised important questions regarding the efficacy of DEP to distinguish between live and dead cells, which certainly require further study and understanding. For example, it would be of value to repeat their work using trypan blue as the vital stain. As discussed in the previous section, this dye can only pass across a damaged membrane, whereas methylene blue can pass through both intact and damaged membranes. The use of Virkon, which is widely used as a disinfectant, could also be explored further. It is composed of a surfactant (sodium dodecybenzene-sulphonate), which disrupts the lipid structure of a membrane, as well as an oxidizing agent (potassium peroxymonosulphate). This oxidizing activity could interfere with the efficacy of methylene blue as a vital stain (remaining blue even in a viable cell with active enzymes that might otherwise have reduced it).

Figure 11.8 indicates that, for an appropriate value of the suspending medium conductivity, whether or not viable or dead yeast cells exhibit positive or negative DEP depends on the frequency of the applied field. At low frequencies (below ~10 kHz) viable and dead cells exhibit negative and positive DEP, respectively. At 10 MHz this behaviour is reversed. Based on an analysis of the potential energy surfaces generated by microelectrodes of interdigitated castellated geometry, as well as from experimental findings, it was concluded that particles trapped in potential energy wells under the action of negative DEP can more easily be removed from an interdigitated electrode array (e.g., by fluid flow or gravitational forces) than those trapped under positive DEP [35]. This is of relevance to the protocol to be adopted for the separation of viable and dead cells. By operating at a low frequency the dead cells can be trapped at the electrodes, whilst the live ones are carried away over the electrodes and eluted from the DEP chamber. Trapping the viable cells at a high frequency (and eluting the dead cells) would expose them to a high field gradient, increasing the risk of damage or stress-induced physiological changes. As given by Equation (10.62) the direction and rate of movement of a cell exposed to a traveling electric field depends on the polarity and magnitude of Im[Re]. This has been verified using mixtures of live and dead yeast cells [36, 37]. In their studies of the electrokinetic behaviour of yeast cells in travelling electric fields, Huang *et al.* [36] reported that within a relatively narrow range of frequencies (400 kHz to 1 MHz) the cells exhibited a random range of spinning, circular and zig-zag motions along the channel between the tips of opposing electrodes. This was designated as the FUN regime of electrokinetic activity, ostensibly as the acronym for 'fundamentally unstable' but really as an expression of how amusing it was to observe a cell executing these random motions. However, unless controlled, this effect can compromise the main advantage of travelling-wave DEP in being able to selectively transport cells along a microchannel without having to pump the fluid itself. Through both numerical simulations of the forces involved and experiments with yeast cells, Nudurupati *et al.* [38] have provided considerable insights into what controls the FUN regime and of ways to minimize its effect.

An advantage of using DEP to sort cells is that it is a label-free method, not requiring add-on optical or magnetic techniques to interrogate fluorescent probes or to capture magnetically labeled cells, for example. This has led to investigations of the possibility that DEP can be used in point-of-care diagnostics. An excellent example of this is the combination of a DEP cell separator and cell counter on a single chip, as described by Mernier *et al.* [39]. In this device living and dead yeast cells were separated by DEP and counted using the coulter method, to determine the percentage of living and dead cells for viability studies of cell samples. The authors suggest that this device could further be used, for sorting and counting blood cells, in applications such as diagnosis of insufficient cell concentrations, identification of cell deficiencies or bacterial contamination. A novel microfluidic approach, termed reservoir-based dielectrophoresis (rDEP) to separate cells by viability has been described by Patel *et al.* [40]. Its effectiveness was demonstrated by selectively trapping dead yeast cells at a microchannel junction within a reservoir containing the cells, whilst continuously separating them from live cells. This has the significant advantage of eliminating 'dead spaces' within fluidic channels and as envisaged by the authors can be readily integrated with other components into lab-on-a-chip devices for applications to biomedical diagnostics and therapeutics. Budding yeast cells offer a versatile model of eukaryotic cells for cytological studies because their genome is well understood and links can be drawn to higher eukaryotes. Tang *et al.* [41] have demonstrated that the DEP responses of budding and nonbudding yeast cells are so different that they can efficiently be separated and immobilized onto microelectrodes at desired densities down to the level of a single cell. Once immobilized the budding yeasts can be converted to sphereoplasts or protoplasts using the enzyme lyticase to lyse their cell walls [41].

An exciting development has been the identification by Vahey *et al.* [42] of genes whose deletions change the electrical conductivity of a yeast cell. This work used the genetically bar-coded yeast deletion library and high-throughput sequencing for quantifying strain abundance [43]. The *Saccharomyces cerevisiae* sequencing project (1998–2002) resulted in the yeast deletion collection, also known as the yeast knockout set, which comprises 21 000 mutant strains and represents the only complete collection for any organism [44]. Using their iso-dielectrophoretic separation method [45] described in Chapter 10, Vahey *et al.* [42] characterized the DEP properties of $\sim 10^7$ cells that had been pooled from approximately 5000 strains of *S. cerevisiae*. Through its barcoded DNA the strain type of each yeast cell was identified and matched to its DEP characteristic at 300 kHz (to probe the electrical properties of the cell envelope) and at 10 MHz (to probe the cell interior). A determination was made of the mean and variance in conductivity for each strain, as a measure of how electrically distinguishable they were. 419 strains were identified whose dielectric properties differed on average from the composite pool. An interesting correlation was found between strains exhibiting altered dielectric properties and those exhibiting defects in fitness (i.e., growth under various stresses). For example, strains with increased cell envelope conductivity had a high incidence of ionic sensitivity (more than tenfold higher than expectation) while strains with decreased cytoplasmic conductivity were enriched for sensitivity to nutritional limitations (more than fivefold higher representation than across the genome). Overall, ~30% of strains with altered electrical properties exhibited a fitness defect, with the largest overlap occurring for strains with decreased cyoplasmic conductivity. Strains with decreased effective conductivity were also found to be significantly enriched for shapes that could be described as round, small and dumped, whereas strains with increased effective conductivity were enriched for ellipsoidal morphologies. However, the data also revealed that morphological changes were neither necessary nor sufficient to change the electrical properties of a cell, pointing to a substantial set of mutations that alter electrical properties through changes that are more subtle than those associated with visibly different morphology (such as having different depolarization factors related to an ellipsoidal rather than spherical shape). The important question was also addressed as to whether changes in the electrical properties of the cell envelope or intracellular space would result from the deletion of proteins localizing to these compartments, or whether the connection between gene deletion and electrical change is more complex. It was found that a significantly enriched fraction of genes whose deletion changes cell envelope conductivity code for proteins that localize to the cytoplasm and nucleus. This suggests that downstream functional consequences of a protein's absence determine electrical differences, rather than the direct physical consequences of the protein's absence. So, to summarize, by sorting the deletion collection into fractions with different electrical characteristics and counting the relative abundance of each strain within the different fractions, Vahey *et al.* [42] were able to generate for the first time a genome-wide mapping between genotype and DEP phenotype. This mapping revealed that dielectric properties are largely independent of, and thus complimentary to, other phenotypic data including fitness and morphology. This enables the ability to identify specific processes and pathways whose perturbations can be detected through changes in DEP properties. It also demonstrates the feasibility of performing whole-genome screens based on intrinsic properties – a methodology that can be translated to other devices (including the hybrid-DEP devices described in Chapter 1) that sort cells based upon physical properties other than their dielectric ones.

11.2.3 Yeast as a Model Cell for DEP Studies

Since the publication of Pohl's book [22] in 1978, around 300 publications have described the use of yeast as the model particle in various aspects of DEP. The extent of such studies can be appreciated from the following examples of this work.

The kinetics of pearl-chain formation of yeast cells were studied by Schmidt *et al.* [46] for application as a parameter for measuring the DEP of cells, whilst Venkatesh and Markx [47] performed a systematic study of the parameters that influence the height of cell aggregates formed by positive DEP. Interdigitated electrodes (of optimum characteristic size, which depended on cell type and size) with oppositely placed castellations gave higher aggregate heights than interdigitated parallel electrodes. Following simple rules, such as an optimal frequency of 1 MHz; as low as possible conductivity of the suspending medium and fluid flow rate through the chamber; it was shown possible to create aggregate cell heights of over 150 μm for all the three cell types (bacteria, yeasts and mammalian cells) studied, using voltages of only 20 Vpk-pk. In studies directed towards developing electrorheological fluids, Kadaksham *et al.* [48] studied the induced transient clustering of suspended viable yeast cells in a microelectrode device on the sudden application of an AC field gradient. Two distinct regimes of positive DEP were identified. Although for both regimes the cells eventually clustered at the electrodes edges, their transient behavior and their final arrangement were quite different. When the frequency was much smaller than the crossover frequency, yeast cells near the electrodes quickly rearranged in well defined chains and then moved toward the electrode edges, remaining aligned as elongated chains at their final location. However, when the frequency was close to the crossover frequency, the cells moved individually towards the regions of collection and simply agglomerated along the electrode edges. Analysis of these effects showed that in the first regime both the DEP force and the mutual DEP force arising from electrostatic particle-particle interactions were important. In the second regime only the DEP force dominated [48]. Different methods have been described to quantitatively monitor the DEP behaviour of cells and other colloidal particles. A dual beam optical spectrometer was developed by Talary and Pethig [49] having the advantage over previous DEP experiments of having the ability to characterize both positive and negative DEP. This provided a simple method for deriving the practical conditions required for the selective manipulation and separation of cells using DEP forces. Measurements were reported for viable and nonviable yeast cells, as a function of the conductivity of their suspending medium and also of the magnitude and frequency of the applied AC voltage.

Yeast cells have been employed as the test bioparticle for developing methods to control the translational movement of cells, an interesting example being the use of two AC fields exhibiting antiparallel field gradients that can be operated at frequencies ranging from 10 Hz to 1 GHz [50]. Prasad *et al.* [51] describe an efficient real-time, multiple-cell tracking platform coupled with DEP to quantify the dynamics of cell motion and obtain cell viability information. The use of DEP to pattern and immobilize cells in a hydrogel [52], a microfluidic cell-culture chip for trapping, cultivating and releasing selected individual cells [53] and studies of the effects of cell viability from exposure to DEP fields [54] have all employed yeast cells. Other examples include the application of DEP to improve the sensitivity of PCR [55–57]; programmable microfluidic chips to trap and move cells and droplets with DEP [58] or as continuous cell separators [59]; novel electrode structures, based on textile technology, for large scale DEP cell separations [60]; traveling wave DEP to concentrate suspended particles in solution at a subset of electrodes, with impedance sensing to determine the particle concentrations [61].

An interesting example of insulator-based DEP is that of Suehiro *et al.* [62], who demonstrated the continuous separation and recovery of biological cells suspended in water using a DEP filter. The filter consisted of a chamber containing about one million glass beads of diameter 220 μm packed between two parallel stainless steel electrodes. The total volume of beads and free space in the chamber was 4.2 and 3.8 ml, respectively, whilst each electrode had a small hole that served as inlet and outlet ports for fluid flow. Without a voltage applied to the electrodes, viable yeast cells in suspension flowed freely through the gaps between the beads. However, at a flow rate of 30 ml/hr and an applied electrode potential of ~50 V_{pk-pk} the cells were trapped by positive DEP, preferentially at the lateral surfaces of two adjacent beads. On removing the applied potential the cells were released into the flowing fluid. In an experiment where viable and nonviable yeast cells were mixed at the same density of 10^6 cells/ml in 100 ml of an aqueous electrolyte of 0.2 mS/m, the DEP filter was operated in a circulation-flow mode (the outlet fluid returning to the inlet). The density of viable cells was decreased from 10^6 cells to 10^2 cells/ml in a period of 5 h at a fluid flow rate of 60 ml/hr. The density of nonviable yeast cells decreased to ~10^5 cells/ml. The yeast *Pichia pastoris* is a popular eukaryotic host used in studies of heterologous protein expression. Strains are available, for example, which are capable of overexpressing various membrane proteins. Due to their small size and cell wall, however, *P. pastoris* cells are not suitable for direct electrophysiological studies such as those shown in Figures 9.2 and 9.3. To overcome these limitations Terpitz *et al.* [63] produced giant protoplasts of *P. pastoris* by means of multicell electrofusion. The protocol for the electrofusion process was refined by thorough analysis of the dielectric properties of parental protoplasts (2–4 μm diameter) by means of DEP and electrorotation. Stable multinucleated protoplasts of a *P. pastoris* strain expressing channelrhodopsin-2 (ChR2), with diameters of up to 35 μm, were produced. These giant protoplasts were suitable for electrophysiological measurements, as proved by whole-cell patch clamp recordings of light-induced ChR2-mediated currents, which was not possible with parental protoplasts. Finally, Tang *et al.* [64] have developed a novel DEP-based microfluidic platform for interfacing nonadherent cells with high-resolution scanning electron microscopy (SEM) in the low vacuum mode. This system enables rapid immobilization and dehydration of cells without deposition of chemical residues over their surface, also enabling on-chip chemical stimulation and fixation of immobilized cells with minimum dislodgement. The system was tested by comparing the morphological changes of nonbudding and budding yeast cells following lyticase treatment.

11.2.4 Bacteria

Basic details of the structure, morphologies, pathogenecity and their division into Gram-positive and Gram-negative types are given in Chapter 9 and Appendix L. In brief, all bacteria are prokaryotes – they do not have a well defined nucleus surrounded by a nuclear envelope. Genetic information is stored in the form of either a bacterial chromosome or plasmid. The chromosome contains genes essential for cellular functions, is located in what is referred to as the nucleoid

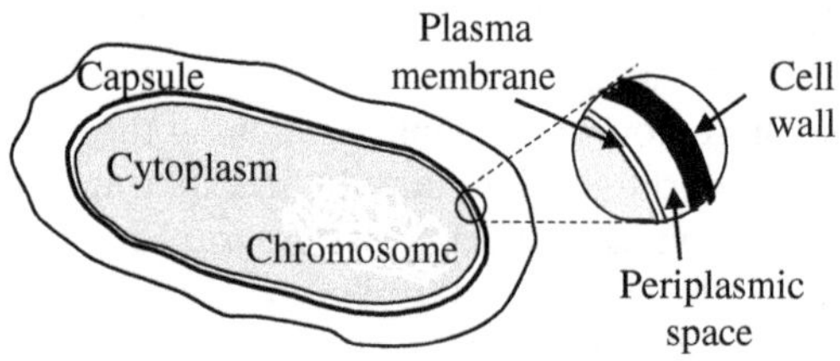

Figure 11.10 Basic structure of a Gram-positive bacteria (e.g., *B. megaterium*). Gram-negative bacteria (e.g., *E. coli*) have a wider periplasmic space, an additional (outer) bilayer lipid membrane and a thinner cell wall. The outer capsule is a polysaccharide layer found most commonly among Gram-negative bacteria but also for some Gram-positive bacteria.

region (not surrounded by nuclear envelope) and consists of single, double-stranded, circular DNA. Plasmids also consist of a single, double-stranded, circular form of DNA, but are much smaller than a chromosome – they number from one to several and are located in the cytosol. Unlike a chromosome they do not contain information essential for growth and metabolism, but have genes giving resistance to drugs and heavy metals. Of particular relevance to DEP studies is the fact that many types of bacteria possess a single flagellum or several flagella. By either pushing or pulling the cell through a liquid medium these long filamentous appendages enable the cell to move to different parts of their environment, helping them to find new resources for survival. Such locomotion will obviously compete against a DEP force.

The basic structure of a Gram-positive bacterium, without flagella, is shown in Figure 11.10. It typically has a cell wall of thickness ~20 nm, composed mainly (~90%) of peptidoglycan and ~10% teichoic acid. Peptidoglycan, also called murein, is made from polysaccharide chains cross-linked by unusual peptides containing D-amino acids. Gram-positive bacteria are also characterized by having no, or a small, periplasmic space and a single lipid bilayer membrane – the inner plasma membrane. Gram-negative bacteria are characterized by having a thinner cell wall (that does not contain teichoic acid) and a larger periplasmic space, as well as a *second* lipid bilayer membrane at the outside surface of the cell wall. This outer membrane contains protein channels called porins, which allow the passage of small hydrophilic molecules across it, as well as lipopolysaccharide molecules that extend into the external medium. Because of their thick cell wall Gram-positive bacteria are stained purple by crystal violet, whereas Gram-negative bacteria with their much thinner wall do not (but can be counterstained pink by safranin). Through microscope inspection, we can readily familiarize ourselves with some of the characteristic shapes of bacteria and the way they group or cluster together after cell division. Common examples include: rod-shaped, as for the bacillus genus of Gram-positive bacteria (e.g., *Bacillus cereus*) and the Gram-negative *Escherichia coli*; spherical (coccus) such as the Staphylococcus and Streptococcus genus; or spiral-shaped such as the Spirochaetes. Other forms include square-shaped (Arcula) and star-shaped (Stella). *E. coli* often group in pairs, Streptococcus in chains and Staphylococcus in clusters.

The vast majority of bacteria bear a net negative charge at normal pH values, associated with ionizable groups of the polysaccharide plus teichoic acid cell wall content of Gram-positive bacteria and the lipopolysaccharide of Gram-negative bacteria. The plasma membrane also carries a net negative charge. These charges are often stabilized by the presence of divalent counterions, such as Ca^{2+} and Mg^{2+}. As a result of this electrical characteristic, many antibacterial agents are cationic – requiring only a strong positive charge together with a hydrophobic region in their molecular structure to interact with the cell surface and integrate into the inner plasma membrane [65]. Inhibitors of peptidoglycan biosynthesis also act as antibiotics – producing loss of cell wall integrity and cell lysis. Specific antibacterial compounds that contain a β-lactam structure (e.g., penicillin) interfere with the synthesis of the cell wall for both Gram-positive and Gram-negative bacteria, by competing for the binding sites of the D-amino acids that crosslink the polysaccharide chains. Mammalian cells lack a cell wall structure, so that this class of antibiotic selectively targets bacteria without harming the cells of the animal or human host.

11.2.4.1 Live and Dead Bacteria

The first report of the difference between the DEP characteristics of a viable and nonviable bacterium is that given by Chen in his M.Sc thesis [66] and later discussed by Pohl [22] who supervised this work. The DEP collection yield for *Bacillus cereus* was found to be remarkably sensitive to the age of the colony. For a suspension medium of conductivity ~1 mS/m, the yield peak occurred at a frequency that advanced from ~100 kHz to ~300 kHz. For an 11-hours old culture that had been briefly extracted with chloroform, there was a marked shift of the peak yield to a lower frequency. This was attributed [22, p. 417] to a lessening of the cytoplasmic ionic concentration as the membrane is destroyed. The influence of chloroform on membrane structure (and its anesthetic effect) in fact remains an active research topic. There is strong evidence to indicate that chloroform loosens cholesterol-containing bilayers, thereby changing their lateral lipid organization [67]. Zhou *et al.* [68] performed electrorotation measurements on polystyrene beads on whose surfaces biofilms of *Klebsiella pneumoniae* had been grown. This bacterium is Gram-negative, rod-shaped, nonmotile and naturally occurs in soil. The presence of a biofilm coating was found to alter the electrototation properties of a bare bead and, in turn, these properties were altered by adding a biocide (Cosmocil™) to the surrounding aqueous medium. Cosmocil is a cationic polymeric biguanide, commonly used as a surface and topical antimicrobial, which displaces divalent cations from the walls and membranes of both Gram-positive and Gram-negative bacteria [65]. A single-shell model was used to analyse the dielectric properties of the biofilm and thus the overall effective dielectric properties of a *K. pneumoniae* bacterium before and after biocide treatment [68]. The dielectric properties of its various compartments were then obtained by finding the best fit curve to the data points, employing a triple-shell model of a cell so as to take account of the bacterium's cytosol, plasma membrane, cell wall and outer membrane. The presence of the periplasmic space was not included in the model – its dielectric properties were in effect merged with that of the cell wall. The dielectric properties and physical characteristics determined for live and biocide-treated *K. pneumoniae* by Zhou *et al.* [68] are presented in Table 11.1. The most significant changes resulting from biocide treatment were found to be a thousandfold decrease of the plasma membrane's effective resistivity, with a 22-fold decrease of the cytosol conductivity. This indicates that damage was caused to the plasma membrane, resulting in leakage of ions from the cytosol. Interestingly, no change was indicated to have occurred to the effective resistivity of the outer lipid membrane, implying that the biocide acted mainly on the plasma

Table 11.1 Dielectric properties of the compartments of viable and biocide-treated *Klebsiella pneumoniae* derived from electrorotation measurements on biofilm-covered beads [68].

Compartment	No treatment	Biocide treatment
Cell Wall		
Conductivity	5.5 ± 0.5 mS/m	0.17 ± 0.03 mS/m
Permittivity	$(60 \pm 15)\ \varepsilon_o$	$(60 \pm 10)\ \varepsilon_o$
Outer Membrane		
Conductivity	0.1 ± 0.05 μS/m	0.1 ± 0.09 μS/m
Permittivity	$(8 \pm 0.5)\ \varepsilon_o$	$(8 \pm 1)\ \varepsilon_o$
Plasma Membrane		
Conductivity	0.6 ± 0.3 μS/m	0.6 ± 0.1 mS/m
Permittivity	$(6 \pm 2)\ \varepsilon_o$	$(6 \pm 2)\ \varepsilon_o$
Cytoplasm		
Conductivity	0.44 ± 0.1 S/m	20 ± 4 mS/m
Permittivity	$(60 \pm 20)\ \varepsilon_o$	$(60 \pm 10)\ \varepsilon_o$

Notes: The suspending medium conductivity values were 0.4 mS/m and 1.3 mS/m for measurements performed on the untreated and biocide-treated samples, respectively. The following dimensions were also obtained from a sensitivity analysis of the best fit data: cell-wall thickness 20 ± 2.5 nm; membrane thickness 7 ± 1 nm; bacteria diameter 0.5 ± 0.01 μm [68].

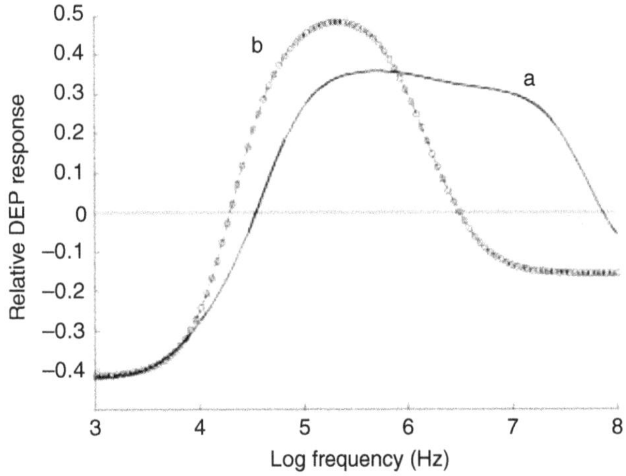

Figure 11.11 DEP responses for (a) viable and (b) biocide-treated *Klebsiella pneumoniae*, derived from the data given in Table 11.1 and a three-shell version of the MATLAB program given in Box 11.2. These responses are modelled for a suspending medium conductivity of 0.4 mS/m.

membrane. After biocide treatment there was a 32-fold decrease of the cell wall conductivity. This implies that in replacing the divalent cations in the cell wall, the cationic biocide depleted the concentration of mobile counterions that contributed to the cell wall's conductivity. The dielectric data presented in Table 11.1 can be used to derive the DEP characteristics of a *K. pneumoniae* bacterium using a modified (three-shell) form of the MATLAB program given in Box 11.2. The result is shown in Figure 11.11. The effect of biocide-treatment is clearly evident – mainly manifested as a replacement of the positive DEP response at high frequencies by negative DEP. This result implies that selective DEP separation of a mixture of viable and nonviable *K. pneumoniae* can be achieved at a frequency of ~10 MHz, for an aqueous suspending medium conductivity of 0.4 mS/m.

In response to the risk to public healthcare posed by increasing numbers of antibiotic-resistant strains of bacteria (e.g., methicillin-resistant *S. aureus* (MRSA), vancomycin-resistant *S. aureus* (VRSA), ciprofloxacin-resistant *E. coli*) there is the demand for rapid antibiotic testing. The common test, which can take from 16–24 h or even several days to complete, involves streaking bacteria onto a culture dish in which antibiotic impregnated disks are placed. Another test, known as the broth dilution method, involves incubating the bacteria in soy broth in tubes containing various dilutions of the antibiotic and examining the changes of suspension turbidity. The use of DEP as a more rapid test of antibiotic efficacy or antibacterial resistance is thus worth exploring.

Hoettges *et al.* [69] have demonstrated that the action of polymyxin B as an antibiotic against *E. coli* can be detected as a significant change in this bacterium's

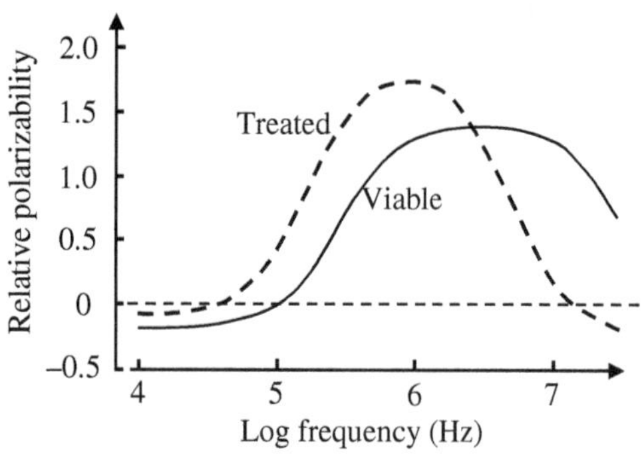

Figure 11.12 Best fit lines (using a single-shell model of a cell) of the relative DEP response of *E. coli* before and after 4 h treatment with 20 μg/ml polymyxin B (based on Hoettges *et al.* [69]).

DEP characteristics as soon as 1 h after isolating a treated culture from nutrient broth. Polymyxin is a cationic antimicrobial that acts in a similar manner as that described above for cosmocil [65]. For their DEP experiments, control and treated samples were suspended in an iso-osmotic medium, consisting of 280 mM mannitol adjusted to a conductivity of 5 mS/m by adding phosphate buffered saline. An example of their findings is shown in Figure 11.12. To better understand the significant change of the DEP characteristic as a function of exposure time to the antibiotic, a single-shell model of a spherical cell was manually fitted onto the experimental data [69]. As the authors state, this model will provide qualitative rather than absolute values for the dielectric properties of the components of *E. coli* (no account is taken of the cell wall and outer membrane). From their model Hoettges *et al.* concluded that even 1 h after exposure to polymyxin B the cytoplasm conductivity dropped by nearly two-thirds from 0.35 S/m to 0.13 S/m and then down to 0.05 mS/m after 4 h. This mirrors, but more dramatically so, the fall in cytoplasm conductivity following biocide treatment of *K. pneumoniae* (see Table 11.1). The membrane conductance was deduced to increase from 7.75 kS/m^2 to 17.8 kS/m^2 after 2 h treatment. This appears to be too modest a fall in conductance to account for the dramatic leakage of ions that must have occurred from the cytosol across the membrane. Qualitatively, the trends shown in Figures 11.11 and 11.12 are similar regarding the change in DEP response following drug exposure. Hoettges *et al.* [69] also deduced that the plasma membrane capacitance nearly doubled from a value of 13.3 mF/m^2 for untreated *E. coli* to 24.3 mF/m^2 after 1 h of treatment, remaining constant at this value even after 4 h. This suggests that the membrane shrivelled as the ions leaked from the cytosol into the external medium, followed by water driven down its concentration gradient. The fact that the membrane capacitance remained constant in the studies of *K. pneumoniae* [68] may reflect that they were suspended in a hypotonic medium where there would be a constant osmotic pressure for water to enter the cytosol, pressing the plasma membrane hard against the cell wall and preventing the formation of blebs (which would increase the membrane capacitance).

Chung *et al.* [70] have reported a significant lowering within 1 h of the DEP crossover frequency exhibited by *E. coli* exposed to 32 μg/ml of the β-lactam antibiotic cephalexin. This antibiotic inhibits cell division by binding to proteins responsible for the polymerization of the peptidoglycan in the cell wall, causing elongation of the *E. coli* cell [71]. The change in the DEP crossover frequency was thus considered to primarily result from this elongation of the cell. Chung *et al.* [72] extended their studies to include the reduction of the DEP crossover frequency caused by cefazolin treatment of *E. coli* and *K. pneumoniae*. The bacteria became filamentous due to the inhibition of cell wall synthesis and cell division, with cell lysis occurring for the higher antibiotic dose. The crossover frequency decreased from ~2 MHz down to the hundreds of kHz range within 2 h, whilst cell lengths extended to more than 10 μm, even up to 20–30 μm for the higher drug doses. The drug resistant strains did not behave in this way. Tests on control cells, untreated and treated cells could be carried out simultaneously using eight sets of quadrupole electrodes, fabricated on indium tin oxide glass slides. The minimum inhibitory concentration determined using the DEP-based method was consistent with the results of the broth dilution method. An analysis was also performed of possible damage to the cells due to electrothermal effects near the microelectrodes. According to the applied voltage of 10 V_{pk-pk}, the frequency range of 100 kHz – 10 MHz, the medium conductivity of 0.3 S/m and the thermal conductivity of the aqueous medium (1.0 W/mK), the temperature rise was estimated to be no more than ~7.5 °C. Moreover, each DEP examination of the bacteria did not exceed 3 min. Heat damage to the cells should not have occurred under these conditions. It is of interest and value to consider to what extent the lowering of the DEP crossover frequency depended on elongation of the bacteria caused by the cefazolin treatment. The Clausius–Mossotti factor is given by:

$$CM = \frac{\varepsilon_p^* - \varepsilon_m^*}{3(\varepsilon_m^* + (\varepsilon_p^* - \varepsilon_m^*)A)}$$

where A is the depolarization factor described in Chapter 7. From Figure 7.11, $A = 1/3$ for a sphere, so that:

$$CM = \frac{\varepsilon_p^* - \varepsilon_m^*}{\varepsilon_p^* + 2\varepsilon_m^*}$$

in agreement with Equation 6.1 and its form in the expression for the DEP force acting on a spherical particle given by Equation (10.26). Chung *et al.* [72] describe elongation of the bacteria to filaments of rough dimension 1 μm and lengths exceeding 10 μm. From Figure 7.11 this corresponds to $A = 0.5$. The MATLAB program in Box 11.2 can be modified to incorporate values for the factor A obtained from Figure 7.11 (or by inserting Equation (7.34) to approximate a rod-shaped bacterium as a prolate spheroid):

```
A=0.333;
% A=0.5;
m=(keff3-k4) ./(3*(k4+A*(keff3-k4)));
```

This change in the program produced the plots shown in Figure 11.13. The change from a sphere-shaped bacterium to one resembling a rod does lower the DEP crossover frequency, but not to the extent reported by Chung *et al.* [72]. Also shown in Figure 11.13 is the result of assuming that along with elongation of the bacterium there was also damage to the plasma membrane, resulting in a lowering of its resistivity. It is likely that this combination did occur for their bacteria. An increase of the cross-sectional area normal to the long axis of the bacterium would also result in a lowering of the DEP crossover frequency, but there is no evidence for this.

Jones *et al.* [73] have exploited the fact that their insulator-based DEP separator, described in Chapter 10 (references 171 and172) that uses a field gradient and

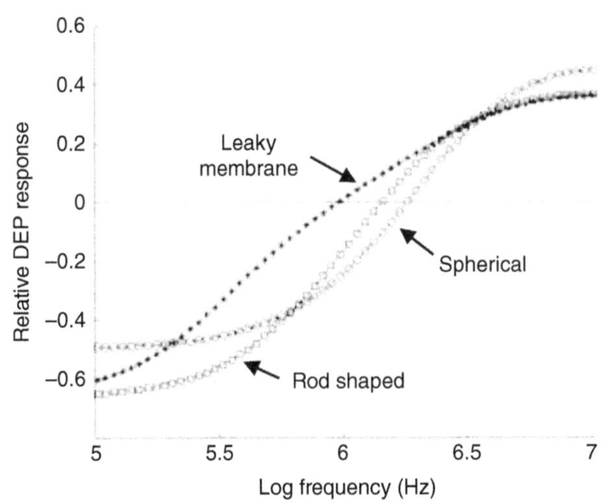

Figure 11.13 From this modelling of the DEP behaviour of a bacterium (using a modification of the program in Box 11.2) the DEP crossover frequency is expected to fall if, due to the effects of an antibiotic, it changes from a spherical to rod shape. This behaviour is enhanced if the plasma membrane is also damaged so as to increase its electrical conductance and the leakage of ions across it.

sawtooth channel, can be used to separate particles on the basis of a combination of their electrophoretic and DEP characteristics. They achieved resolution of gentamicin resistant and susceptible strains of *Staphylococcus epidermidis.* This demonstrated that the presence of antibiotic resistance enzymes (or secondary effects) produces a sufficient degree of electrophysical difference to allow separation of these strains. The differentiating factor was found to be the ratio of electrophoretic to dielectrophoretic mobilities. This factor was $4.6 \pm 0.6 \times 10^9$ V/m^2 for the resistant strain, versus $9.2 \pm 0.4 \times 10^9$ V/m^2 for the susceptible strain. This difference was sufficient to produce clear and easily discerned differentiation of the two *S. epidermidis* strains.

Elitas *et al.* [74] have addressed the problem of the persistence of bacteria during antibiotic therapy. This is of particular importance in refractory mycobacterial infections such as leprosy and tuberculosis. Persistence is characterized by the phenotypic tolerance of a subpopulation of baxcterial cells to antibiotics. Characterization of these 'persister' cells is often difficult due to the transient, nonheritable nature of the phenotype and due to the presence of contaminating material from nonpersisting cells, which usually comprise the larger fraction. In their study Elitas *et al.* [74] used 3D carbon-electrode arrays to purify intact cells from cultures of *Mycobacterium smegmatis* treated with isoniazid, a frontline antituberculosis drug. The rate of persistence against this drug is relatively high compared to other antituberculosis drugs and drug combinations. Intact persister cells were differentiated and separated from damaged cells by differential staining with propidium iodide and flow cytometry. Most of the 'omics-based' approaches for downstream characterization of purified bacterial populations, such as transcriptomic and proteomic analysis, require a sample containing at least $10^5 \sim 10^6$ targeted cells. Elitas *et al.* [74] were able to recover up to 3×10^4 intact cells, with up to 99% purity, following their DEP assay protocol. Serial assays, or preferably improvement of the DEP device's performance in terms of throughput, could provide enough material to perform downstream analysis.

As emphasized by Chung *et al.* [72], for microfluidic applications of DEP it is important to evaluate the limiting conditions that can be used in terms of applied voltage potentials and flow rates that allow damage-free cell manipulation. Donato *et al.* [75] have investigated the physiological impact on *E. coli* resulting from their DEP capture in a PDMS microfluidic channel. Cell suspensions (5×10^6 cells/ml) at fluid-flow velocities between 10^{-2} and 2.6×10^{-2} m/s were trapped by quadrupole, titanium-tungsten, electrodes. Trapping by positive DEP was performed at 15 MHz and a medium conductivity of 30 mS/m. This medium conductivity was chosen as it is low enough to allow strong pDEP trapping, but high enough to avoid ionic losses by the cells. Trapping by negative DEP was performed at 1 MHz and a medium conductivity of 0.35 S/m (chosen to avoid electrolytic corrosion of the electrodes observed at higher medium conductivities). The metabolic viability of cells during DEP trapping was studied by first performing a negative viability control with an *E. coli* population previously submitted to an ethanol shock for 30 min at 65 °C. These conditions effectively killed the cells, as confirmed using the trypan blue stain test and they were not trapped when subjected to the nDEP and pDEP experimental conditions. A second, positive test was performed by trapping cells under flow conditions using pDEP, testing for viability using two dyes to discriminate cells on the basis of the state of their membranes. Cells with compromised membrane were stained red with ethidium bromide, whereas cells with intact membranes were stained green. The trapped cells were green, suggesting that cells trapped by DEP were alive. The impact of DEP immobilization on cell viability was further studied using genetically modified *E. coli* cells to express green fluorescence protein (GFP). By monitoring the expression of the GFP reporter, Donato *et al.* [75] were thus also able to confirm cell viability by means of a metabolic outcome. This work demonstrated that *E. coli* cells trapped in a microchannel, using either pDEP or nDEP under conditions of fluid flow or nonflow, remained viable. However, beyond a certain electrical field magnitude (1.3×10^6 V/m) and 15 minutes of pDEP exposure time, the viability of the cells was shown to be compromised. (A field magnitude of ($\sim 1.3 \times 10^6$ V/m corresponds to a peak applied voltage of 130 V across a 100 μm electrode gap.) Meanwhile, after 10 min of pDEP exposure time in the same conditions, the cell viability was confirmed. These results provided the following important guidance [75] for the development of a number of lab-on-chip applications: (i) the observed effect of cell concentration can potentially be used to increase the sensitivity of integrated biosensors and the speed of miniaturized bioanalysis; (ii) the metabolic viability of trapped cells suggests that genetically modified trapped cells can be used as live biosensors in which the timing and amplitude of the expression of a marker protein can be correlated with a complex fluidic stimulation; and (iii) the robust definition of trapping conditions, in particular, the time of exposure to the electric field and the observation that dead cells are not trapped indicates that complex spatial and chemical manipulation of cells can be performed in microfluidic devices without compromising cell viability.

An electrical impedance method for monitoring the DEP collection of bacteria at an interdigitated chrome electrode array has been developed by Suehiro *et al.* [76, 77]. As the cells collect at the edges of the electrodes by positive DEP and span the interelectrode gap as pearl chains, the electrical impedance between the electrodes changes. If the cells are suspended in a low-conductivity medium, both the conductive and capacitive components of this impedance will increase. Higher cell concentrations will exhibit a faster development of the pearl chains. By monitoring the temporal change of the impedance a quantitative evaluation of the cell number density is possible. For example, it was demonstrated that an *E. Coli* suspension of 10^5 cells/ml could be accurately assayed in about 10 min [76]. In later studies [77] selective DEP-impedance inspection of viable *E. coli* from a mix of viable and nonviable bacteria was performed. Nonviable *E. coli* were prepared by two different sterilization methods, namely heat treatment (80 °C for 15 min) and by 10 s UV irradiation (254 nm, 2.5 mW/cm^2 power density). Cell viability was determined by incubating control and sterilized bacteria on agar plates for 48 h. It was found that heat-treated bacteria exhibited a considerable change in their DEP and dielectric parameters, whilst UV-based sterilization hardly affected those properties. For example, DEP collection observations were made at two different electric field frequencies, 100 kHz and 1 MHz, using interdigitated, castellated, electrodes. At 100 kHz, both viable and nonviable *E. coli* were trapped at the castellated edges due to positive DEP, whereas at 1 MHz only viable bacteria were trapped and nonviable cells were not collected by positive DEP. These observations suggested that the positive DEP force exerted on the nonviable heat-treated bacteria was negligibly small at 1 MHz. In contrast to the heat sterilization case, both viable and UV-treated (nonviable) *E. coli* collected under positive DEP at both 100 kHz and 1 MHz. The threshold for impedance detection of bacteria is reduced to 10^4 cells/ml in less than 5 min in a technique described by Zhou *et al.* [78]. The method uses the high polarizability and DEP mobility of single-walled carbon nanotubes (SWNT). Concentrated SWNT solutions are mixed with the test sample and a low frequency (<100 kHz) field is applied by a microelectrode array, to enhance bulk absorption of the bacteria by the SWNTs via dipole–dipole interactions and to then drive the SWNT+bacteria aggregates to the microelectrodes by positive DEP. The SWNTs and absorbed bacteria assemble rapidly (<5 min) into conducting linear aggregates between the electrodes. Measured AC impedance spectra by the same trapping electrodes and fields show a detection threshold of 10^4 bacteria/ml with this pathogen trapping and concentration technique [78].

Amako *et al.* [79] investigated the DEP properties of *E. coli* as a function of heat stress. The cells were collected by positive DEP, at 100 kHz, between 10 pairs of interdigitated chromium electrodes. The rate of cell capture was determined from the temporal change in the electrical impedance measured across the electrode gaps

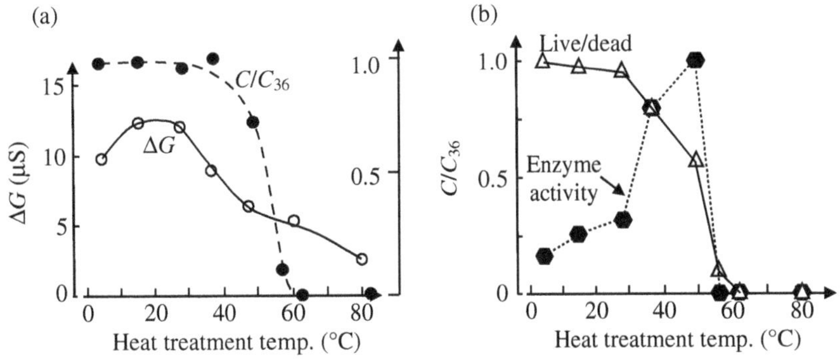

Figure 11.14 (a) Temporal change ΔG of the conductance of *E.coli* collected at DEP electrodes as a function of the temperature of heat treatment. Also shown is an indicator of growth ability in terms of colony count C relative to that at 36 °C. (b) Normalized values of cell viability and respiration enzyme activity as a function of the heat treatment temperature (based on Amako *et al.* [79]).

where the cells collected. This was determined for various temperatures and time lengths of heat treatment in a water bath. Control samples were maintained at 4 °C, whilst the test samples were heated at temperatures ranging from 15–80 °C for up to 15 min before determining their DEP collection rate. The cell suspensions (~3 × 10^7 cells/ml) were pumped over the electrodes at a flow rate of 60 ml/hr. The conductivity of the cell suspension medium was not specified, but presumably the medium was high-resistivity water so as to maximize the sensitivity of conductance change measurements. The conductance component of the measured impedance was found to change more significantly than the capacitance component. In parallel with these electrical measurements the following biological methods were employed to verify the effect of heat treatment on the *E. coli*:

- Enzyme activity associated with respiratory activity of the mitochondrial electron transport chain was evaluated by staining the cells with cyanoditolyl tetrazolium chloride (CTC) and observing their fluorescence intensity. Under the influence of this enzyme activity CTC is reduced to red fluorescent CTC formazan, which accumulates inside the cells. Thus, an observed increase in fluorescence intensity corresponded to enzyme activity.
- Viability was determined by dispensing control and heat-treated cells into sterilized Petri dishes, then monitoring bacterial growth from a colony count determined by the pour plate method.
- Observation of surface morphology was achieved by fixation with glutaraldehyde and osmium tetroxide, followed by gold coating and then observation under a scanning electron microscope.
- The viability and membrane state of the cells was determined using a membrane permeable, green DNA stain, together with a propidium iodide red stain (which is non-membrane-permeable). Viable cells with undamaged plasma membranes thus exhibited green DNA fluorescence, whilst those with a damaged membrane assumed a red fluorescence.

The relationship between the temporal change ΔG in the measured conductance and the heat treatment temperature is shown in Figure 11.14(a). It can be seen that ΔG increased with increasing heat treatment temperature in the range 4°–20 °C, decreasing thereafter until at 80 °C the collection rate was very low. Also included in this figure is the variation in observed growth ability (relative to the control at 36 °C). Growth ability was maintained between 4° and 38 °C, but dropped dramatically thereafter. Bacteria are capable of recovering from heat stress damage, depending on the level of this damage and their growth medium. However, no such recovery occurred for treatment temperatures above 60 °C, presumably because metabolic functions were lost due to denaturation of proteins and damage to the plasma membrane and entire cell surface. The relationship between heat treatment and the live / dead stain results are shown in Figure 11.14(b), together with the determined enzyme activity. The viability data mirrored to some extent the growth ability trend, whilst the enzyme activity (normalized to results for 5 min heat exposure) exhibited a peak value at a heat treatment temperature of ~50 °C. This appearance of a peak in activity can partly be explained by the fact that above this heat stress level membrane-bound enzymes are damaged, whereas cold shock induces a ribosomal-associated protein that unwinds double-stranded RNA so that protein synthesis is inhibited [80]. The SEM observations on the fixed cells found that with heat treatment temperatures above 47 °C numerous indentations and folds appeared on their surface, whilst above 80 °C shrinkage of the bacterial body was added to these surface changes. Amako *et al.* [79] concluded that their combined DEP and impedance results provided an integrated estimate, including bacterial metabolic conditions, degree of damage and viable-to-dead ratio. Their stated future objectives were to further clarify the correlation between variations of bacterial metabolic activity and DEP properties under different heat treatment conditions and to carry out metabolism monitoring experiments with yeasts and lactobacilli of relevance to fermentation processes.

11.2.4.2 DEP and Electrorotation Studies of Flagella

Some bacteria, for example strains of *E.coli* and *salmonella*, are motile. They possess flagella that act as rotary molecular engines, driven by ions moving inwards across the plasma membrane. This generates the thrust enabling the cells to swim. The ion flux across the membrane is powered by an electrochemical gradient, either a proton-motive force or sodium-motive force in motors driven by protons or sodium ions, respectively. Their rotation speeds are remarkable, ranging up to a few hundred Hz in cells driven by proton-motors and 1000 Hz for those driven by sodium motors. Typical speeds up to 50 μm/s are achieved by motile bacterium, with some species moving at over 500 μm/s [81].

Washizu *et al.* [82] investigated the external force-velocity characteristics of swimming *Salmonella typhimurium* using DEP, as well as the torque-speed characteristics of their flagella motor by electrorotation. Electrostatic orientation of the bacteria parallel to an applied electric field was established using concentric electrodes of spherical geometry. A two-level DEP force was applied at 1 MHz by switching several times between a low and high applied voltage potential, while gradually letting a bacterium under observation go outward from the inner electrode. This continued until the critical radius from the inner electrode was finally exceeded and bacterium could not be pulled back by positive DEP. By repeating this procedure the force-velocity relationship for a given position could be measured several times. After accomplishing this for live bacteria, the whole area between the electrodes was irradiated by a UV beam for several minutes until the bacteria lost motility. The force-velocity relationships for dead bacteria were then obtained. Because dead bacteria lacked motility they were unable to 'fight against' and move outwards from the inner electrode under conditions of an applied positive DEP force. An example of the force-velocity characteristic obtained for live and dead bacterium, of roughly similar size, is shown in Figure 11.15. Rotating fields generated by six electrodes were used to apply an

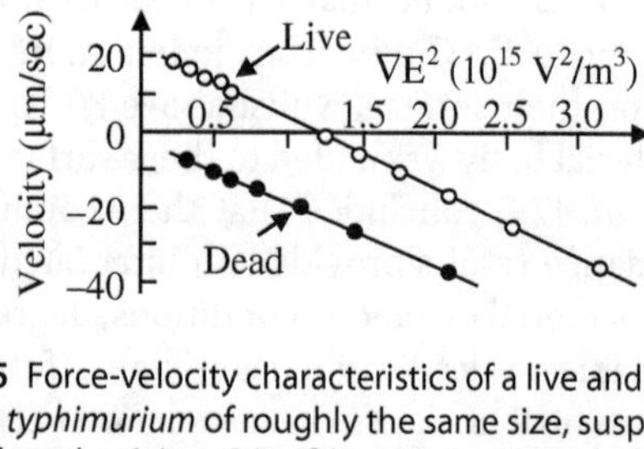

Figure 11.15 Force-velocity characteristics of a live and dead *Salmonella typhimurium* of roughly the same size, suspended in a medium of conductivity ~3.5 mS/m. Using a 1 MHz, bi-level applied potential, the DEP force was applied in the direction to reverse the bacterium from a positive (outward) velocity and back towards the inner of two spherically concentric electrodes. Dead cells were unable to oppose a positive DEP force (based on Washizu *et al.* [82]).

external torque to tethered bacterium. By attaching one of the flagella to aminosilane or mercaptosilane-coated glass, with the main body of the bacterium and its other flagella remaining free, the motor of the attached flagellum rotated the whole cell. The torque-to-speed characteristic was obtained by changing the magnitude of the applied rotating field. Over their operating angular velocity of 0–100 Hz, Washizu *et al.* [82] found that the molecular motor generates approximately constant torque, regardless of its sense of rotation. This result was confirmed by Berry and Berg [83] in electrorotation studies of motile *E. coli*. The torque generated by the flagellar motor was estimated using an analysis that explicitly considered the angular dependence of both the viscous drag coefficient of the cell and the torque produced by electrorotation. In agreement with Washizu *et al.* [82], the motor torque was found to vary approximately linearly with speed up to over 100 Hz in either sense of rotation. This places constraints on mechanisms for torque generation in which rates of proton transfer for backward rotation are limiting. For example, it rules out those models of the mechanism of flagellar rotation that predict a barrier to backward rotation. Barriers to backward rotation are predicted by models where rotation is tightly coupled to the flux of ions through the motor and where the rate of transit of ions against their electrochemical gradient is strictly limited. At a kinetic rather than mechanical level, the torque-speed curve can be understood in terms of the torque dependence of the rate-limiting step in the torque-generating cycle. Interpreted in the context of a simple three-state kinetic model, this suggests that the rate-limiting step in the torque-generating cycle is a power stroke in which motor rotation and dissipation of the energy available from proton transit occur synchronously [83]. Hughes and Morgan [84] employed the polynomial, quadrupole, electrode design shown in Figure 10.4(a) to exert a negative DEP force on motile *S. typhimurium* suspended in brain / heart infusion medium. The interelectrode gap at the centre was 25 μm. For 1 MHz, 1 V_{pk} applied signals, no collection of the bacteria was observed between the electrodes. Presumably the magnitude of the DEP force was insufficient to hold the bacteria within the confines of the electrodes. At an applied potential of 2 V_{pk} the DEP force was sufficient for some bacteria to become trapped, with some of them occasionally swimming away and escaping along the upper electrode surfaces. This only happened when a bacterium was moving straight toward the electrode edge. At higher voltages this did not happen; the bacteria in the electrode vicinity were forced into an approximately circular pattern. At a signal of 10 V_{pk} the bacteria were observed to levitate above the centre of the electrodes. Adopting a double-shell model for the bacterium as a spherical particle with a

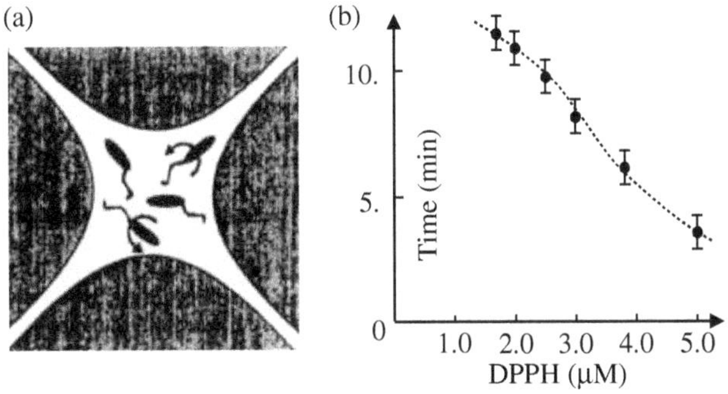

Figure 11.16 (a) *Euglena gracilis* suspended between polynomial electrodes. Half of the algae are shown swimming against the electrorotational torque (b) Plot of the time period after exposure to DPPH when half the algae rotate with the field [85].

cytoplasm ($\varepsilon_r = 60$, $\sigma = 0.19$ S/m), membrane ($\varepsilon_r = 10$, $\sigma = 5 \times 10^{-8}$ S/m) and cell wall ($\varepsilon_r = 60$, $\sigma = 0.9$ S/m) with a measured suspending medium conductivity of 1.58 S/m, the Clausius–Mossotti factor at 1 MHz was calculated to have a value of −0.488. The average value of the field gradient factor ∇E^2 for the threshold of containment of the bacteria was determined to be 1.4×10^{15} V^2/m^3. For an assumed bacteria radius of 0.5 μm, from the expression for the DEP force acting on a spherical particle given by Equation (10.26), Hughes and Morgan [84] estimated the flagellar motor force to be ~0.37 pN.

Electrorotation can also be employed to evaluate the relative effectiveness of, or resistance to, the exposure of toxic agents to motile microorganisms. An example of this is shown in Figure 11.16 in a study of how the motility of a common algae (*Euglena gracilis*) is affected by exposure to the well known free radical compound diphenyl-picryl hydrazyl (DPPH). This chemical causes oxidative damage to membranes and has been used in assays to characterize new antioxidants. These microorganisms were taken straight from a pond. After simple filtration to remove debris they were resuspended in fresh water and retained within the bounds of quadrupole polynomial electrodes of an electrorotation chamber. The length of time taken for half of the organisms in any one experiment to succumb to the rotational torque and exhibit a steady rotation rather than random motility was determined [85]. This was repeated for various concentrations of added DPPH. This procedure could, if required, be adopted in the form of a handheld device for detection of toxic chemicals in the environment, using as test microorganisms those that are readily available for collection from natural sources.

11.2.5 Mammalian Cell Apoptosis

The first detailed study of the correlation between the effect of a cytotoxic agent and its influence on the DEP characteristics of a cell appears to be that by Ratanachoo *et al.* [86]. They studied the responses to toxicants of HL-60 cells, which is a cultured human leukemia cell line. The responses were determined as temporal changes of the DEP crossover frequency (f_{xo1}) and changes of surface morphology as observed by scanning electron microscopy (SEM). From measurements of f_{xo1} and the cell radius, values of the specific membrane capacitance (C_{mem}) were determined using Equation (10.39), as well as from the procedure outlined in Figure 10.26(b) and Equation (10.43). Four toxicants were chosen because of their different mechanisms of cytotoxic action, namely: free radical attack on the plasma membrane (paraquat); simultaneous membrane and nucleic acid attack (styrene oxide); nucleic acid alkylation (N-nitroso-N-methylurea); protein synthesis inhibition (puromycin). Exposure to all four toxicants resulted in a decrease with time of the value for C_{mem}, while the specific membrane conductance (G_{mem}) increased. A decrease of C_{mem} implies a reduction of surface morphological features, such as membrane folds, microvilli and blebs. This was confirmed by SEM inspection of the cell surfaces. For a given dose concentration, the rate of reduction of C_{mem} with time was more rapid for the agents that had a direct action on the membrane than to agents for which membrane alterations were secondary. For example, responses to paraquat and styrene oxide, which directly damaged the cell membrane, could be detected 15 min after exposure, while those for puromycin and N-nitroso-N-methylurea, which acted on intracellular targets, could be detected after 30 min. As described in section 11.2.2.2, studies of this nature were performed some six years later by Patel *et al.* [34] for different methods of inducing the death of yeast cells. Whereas Patel *et al.* concluded that DEP will not always be able to separate viable yeast cells from those killed by toxicants that act primarily on the cytosol rather than the plasma membrane, the work of Ratanachoo *et al.* [86] imply otherwize. Toxicants that kill cells by acting on the cytosol do change the dielectric properties of the plasma membrane, which can potentially be exploited by DEP to separate the dead cells from the live ones. However, in terms of a temporal response, a DEP-based sensor will be more sensitive to detecting toxic agents that have a direct action on the plasma membrane than to agents for which membrane alterations are secondary effects.

Wang *et al.* [87] also studied how the DEP characteristics of HL-60 cells changed following exposure to genistein. This chemical agent is a topoisomerase and tyrosine protein kinase inhibitor that induces apoptosis at all phases of the HL-60 cell cycle. After adding genistein at 100 μg/ml to the cell cultures, cell viability remained above 90% for up to 8 h, as determined using the trypan blue dye-exclusion method. The cell DNA content was also quantified by permeabilizing the cells and treating them with a fluorescent dye (ethidium bromide) followed by flow cytometry. Because of DNA fragmentation, cells undergoing apoptosis have a diminished susceptibility to DNA staining. The nuclei of apoptotic cells contain less DNA than nuclei of healthy cells in the (G_0/G_1) phase of the cell cycle, resulting in a sub-G_1 peak in the fluorescence histogram. At two hours post-treatment there was no significant evidence of this effect, but the sub-G_1 population increased thereafter from ~22% of the total cells at four hours, to ~26% at six hours, ~45% at eight hours and ~52% at ten hours [87]. A decrease in forward light scattering, beginning at two hours post-treatment, was also observed but with no alteration of side scattering, indicating a reduction in cell size. A significant translocation of phosphatidylserine from the inner plasma membrane leaflet to the outer leaflet occurs for apoptotic cells and this can be detected by staining with the phosphatidylserine-binding protein Annexin-V. The number of Annexin-V positive cells one hour post-treatment was found to be similar to that of untreated control cells and only became different two hours after treatment. Annexin-V positive cells represented ~5.7% of the whole cell population, at both zero and one hour, increasing to 22.3% at two hours, 27.2% at three hours, 47.9% at four hours, 56.7% at five hours and 62.4% at six hours post-treatment [87]. In summary, for genistein-treated HL-60 cells it took two hours to detect significant changes of cell size, quantitative staining of DNA and Annexin-V staining. In sharp contrast to this, values for the DEP crossover frequency (f_{xo1}) were observed to increase within minutes of treatment. At a suspending medium conductivity of 56 mS/m, the crossover frequencies for control cells remained bracketed between around 80 to 130 kHz for four hours after treatment, whereas for the cells undergoing apoptosis the values were between ~130 and ~200 kHz. This corresponded to the apparent specific cell membrane capacitance of the cells falling from an initial value of 17.6 ± 0.9 to $13.1 \pm 0.8\,\text{mF/m}^2$ after 2 h and then down to $9.1 \pm 0.5\,\text{mF/m}^2$ 4 h after genistein treatment. The differences in the temporal responses of the DEP crossover frequency, against those of DNA content and Annexin-V staining, are shown in Figure 11.17. These results show that detection of the DEP crossover-frequency f_{xo1} was the most sensitive method, especially at the early time points post genistein treatment.

In the analysis of their DEP data, Wang *et al.* [87] assumed the single-shell model of a cell and calculated the membrane capacitance according to Equation (10.43), namely:

$$f_{xo1} = \frac{\sqrt{2}}{2\pi R C_{mem}}\left(\sigma_m - \frac{G_{mem}}{4}\right) \tag{11.3}$$

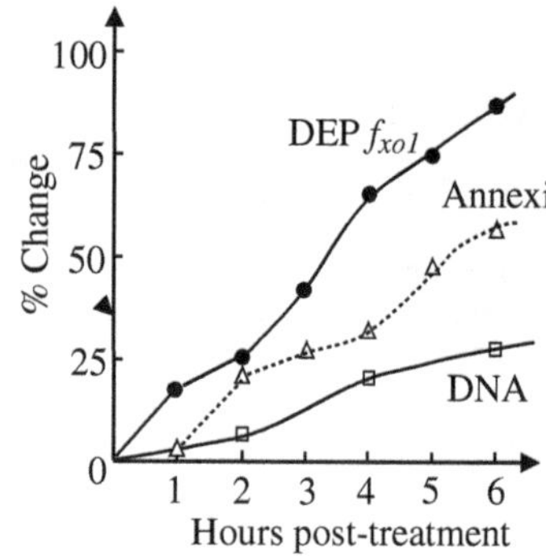

Figure 11.17 Based on the studies of Wang *et al.* [87], these plotted data show that the temporal change of the DEP crossover frequency f_{x01}, especially for short times after treatment of HL-60 cells with genistein, provides a more sensitive detection of apoptosis than the standard DNA content and Annexin V staining methods.

During the first two hours of genistein treatment, the medium conductivity σ_m was held constant and the values of the cell radius R and membrane conductance G_{mem} were not found to change. This indicates that for the first two hours post-treatment, the observed increase of f_{xo1} resulted from a decrease of the membrane capacitance C_{mem} caused by a general smoothing of the membrane. Thereafter, R decreased and f_{xo1} continued to increase, suggesting that two distinct morphological responses were occurring during the early stages of apoptosis. It was found that treatment by the broad spectrum caspase inhibitor *N*-benzyloxycarbony-Val-Ala-Asp(*O*-methyl)-fluoromethyketone (zVAD-fmk) did not prevent these early DEP-detectable cell membrane responses, suggesting that the caspase system was not involved [87]. The membrane conductance G_{mem} did not alter during the first four hours post genistein treatment, but increased significantly and progressively thereafter. Finally, as the barrier function of the plasma membrane failed and the cells became necrotic, the value of G_{mem} increased by many orders of magnitude.

Results in broad agreement with those observed for HL-60 cells were obtained by Pethig and Talary [88] for Jurkat cells undergoing induced apoptosis using etoposide. A dose of 50 μM etoposide was used because after six hours of exposure about half the cells remained viable and the other half exhibited early stages of apoptosis comparable in time scales to those studied by Wang *et al.* [87]. Annexin-V positive Jurkat cells (i.e., apoptotic cells) constituted around 4% of the untreated control cell population, whilst after two hours exposure to etoposide 14% of the cells were Annexin-V positive. The corresponding populations at four and six hours of exposure were 38% and 52%, respectively. A progressive decrease in *forward* light scattering was observed at four and six hours of etoposide exposure, which, as for the HL-60 cells, was

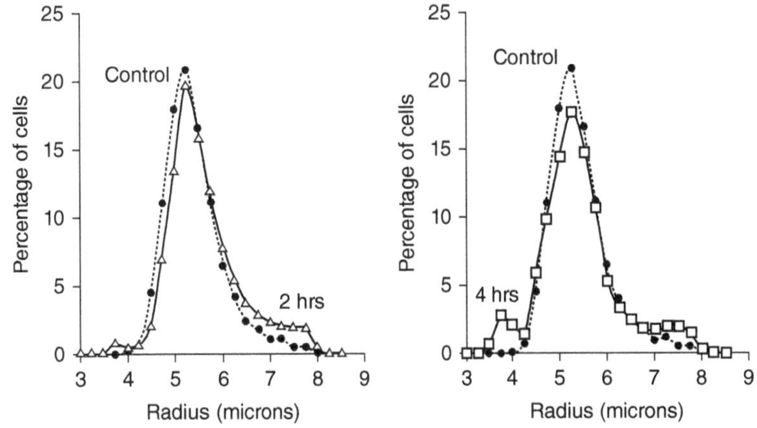

Figure 11.18 The distributions of radii for the untreated (control) cells (n = 256) and those (n = 283; 253) exposed to etoposide for two hours and four hours, respectively. At two hours post-treatment cells of both reduced and increased size appear. At four hours a larger population of apoptotic cells, characterized by their size reduction, is evident. This population increases further as time progresses after treatment (Pethig and Talary [88], reproduced with permission).

indicative of an overall reduction in the average cell diameter. No significant change in *side* scatter intensity was detected over the first four hours of etoposide exposure, but a slight increase was detected at six hours, indicative of a small increase in cell granularity. The results reported by Wang *et al.* [87] were obtained from the study of tens of cells, whereas the later work on Jurkat cells used a cell 'physiometry' profiling technique capable of determining at the same time the size and DEP crossover frequency of hundreds of cells. Advantages of this capability are demonstrated in Figures 11.18 and 11.19.

Figure 11.18 shows the distributions of size of the untreated Jurkat cells and those for cells at two hours and four hours post-treatment. Although the light scattering measurements indicated that there was very little change in the average value of cell size at a post-treatment time of two hours, it is evident from Figure 11.18 that exposure to etoposide resulted in changes in distributions of the Jurkat cell radii, with the appearance of more

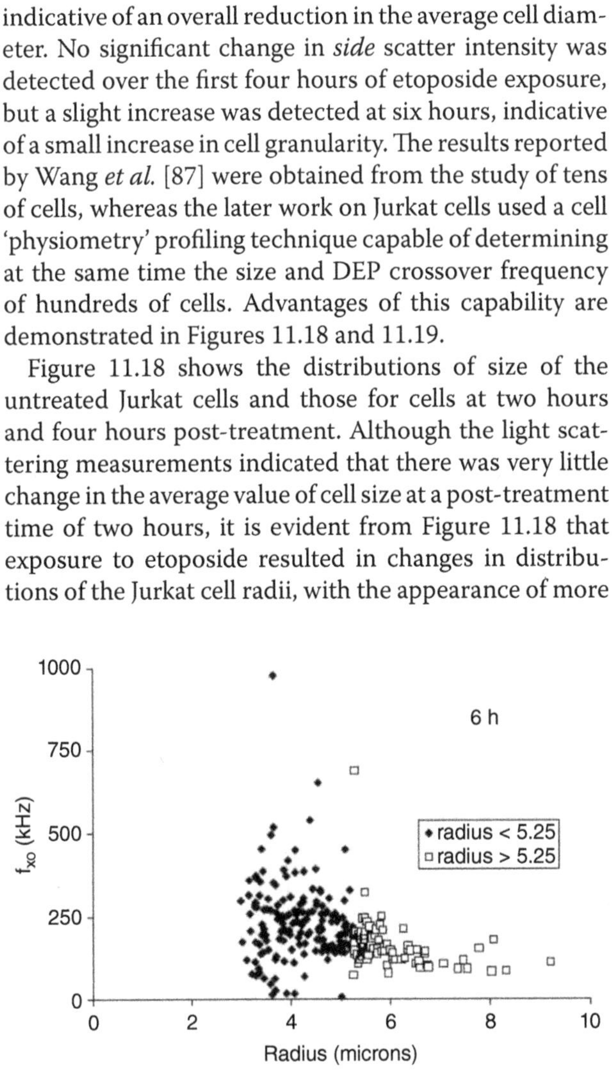

Figure 11.19 A scatter plot of individual cell size and DEP crossover frequency f_{xo1} for Jurkat T-cells (n = 217) exposed to etoposide for 6 hours. Cells of radius less than 5.25 μm are arbitrarily designated as 'small cells'. The highest f_{xo1} values (> 250 kHz) are predominantly exhibited by the smallest cells (R < 5.25 μm). (Pethig and Talary [88], reproduced with permission.)

cells having radii larger than 7.2 μm and a new population of presumptive necrotic cells having radii less than 4.3 μm. These changes accompanied a small reduction in the number of cells of radii close to the average of 5.3 μm. These trends continued for cells exposed to etoposide for six hours, with a reduction of the number of cells of radii between 4.5 and 6.5 μm and the appearance of an increasing number of small cells (R < 4.5 μm) being particularly evident [88]. The relationship between the size of a Jurkat cell and its DEP crossover frequency f_{xo1} is shown in Figure 11.19 for 217 cells 6 h post-treatment. Inspection of this data reveals that the cells exhibiting the highest f_{xo1} values (> 250 kHz) are predominantly exhibited by the smallest cells, with radii less than 5.25 μm.

The DEP crossover frequency (f_{xo1}) values for the control and induced-apoptotic HL-60 cells (n = 20) suspended in a 56 mS/m solution were found to fall into two separable, relatively narrow, frequency bands (80∼130 kHz) and (130∼200 kHz), respectively [87]. This situation was not observed for the Jurkat cells. The control cells (n = 526) suspended in a 40 mS/m solution exhibited f_{xo1} values ranging from 50 to 250 kHz As apoptosis progressed over six hours the upper value for f_{xo1} progressively increased and extended beyond 500 kHz [88]. Also, unlike the situation for HL-60 cells, which were reported to exhibit a change in f_{xo1} values within minutes of exposure to the apoptosis-inducing agent [87], the Jurkat cells exhibited a relatively small initial change. With increasing exposure, larger numbers of cells exhibited f_{xo1} values above 250 kHz and, as shown in Figure 11.20(a), extended beyond 400 kHz for some cells exposed to etoposide for six hours. With the aid of Equation (11.3) and the simultaneous determinations of the f_{xo1} value and radius of each cell, this corresponded to a reduction in plasma membrane capacitance from 13.34 (± 2.88) to 10.49 (± 4.00) mF/m^2, reflecting (as for the HL-60 cells) a general smoothing of the membrane through loss of microvilli, for example. The distribution of the membrane capacitances of the control

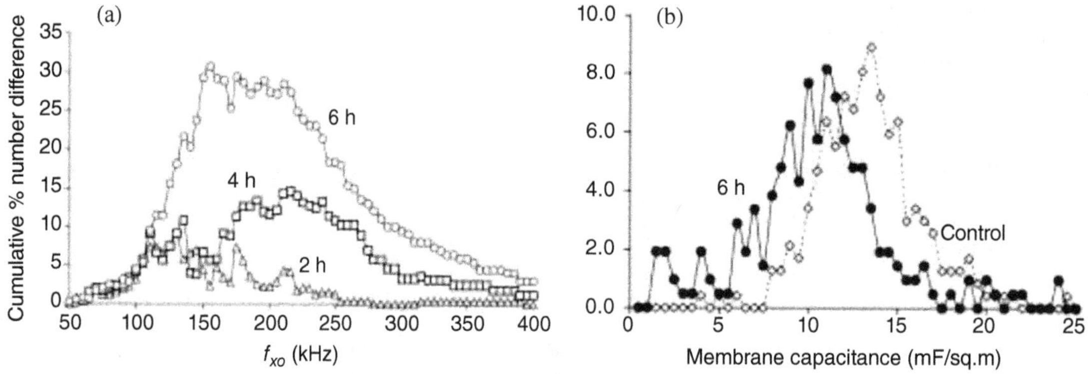

Figure 11.20 (a) Relative changes, using untreated cells as the reference, of the distribution of DEP crossover frequency values (f_{xo1}) exhibited by Jurkat T-cells with time after treatment with etoposide. (b) The distribution of plasma membrane capacitance values for control (n = 526) and treated cells (n = 217) after six hours of etoposide exposure. Treated cells exhibit a definite trend towards lower membrane capacitance values than the untreated (control) cells (reproduced with permission: Pethig and Talary [88]).

Jurkat cells and those treated for six hours are shown in Figure 11.20(b).

An interesting conclusion, not described in the previous work on HL-60 cells [87], is that the observed [88] reduction in the plasma membrane capacitance of the Jurkat cells after two hours of exposure resulted mainly from the cells exhibiting an increase of their average size. Thereafter, as exposure to etoposide increased the reduction of membrane capacitance was mainly associated with the appearance of smaller cells. Wang *et al.* [87] also reported that putatively necrotic cells exhibited f_{xo1} values in the megahertz range, with some above 10 MHz. As shown in Figure (11.19) just one cell in this category was found for the Jurkat cells, with an f_{xo1} value near 1 MHz. Mulhall *et al.* [89] investigated the change in DEP behaviour of Jurkat cells after exposure to the apoptosis-inducing agent staurosporine. They used an optical method to investigate the DEP-induced motions of cells contained within microwells, each well capable of holding ~0.3 μl of a cell suspension [90, 91]. After incubating with 0.5 μM staurosporine, Jurket cells at a concentration of ~5×10^6 cells/ml were placed in each well (i.e., ~1500 cells). Cell viability was measured using the trypan blue assay. Cell radii were obtained by capturing images of 100 cells per experiment on a haemocytometer and measuring the cell diameter using image analysis software. The cell radii showed a reduction from 5.7 μm to 4.5 μm during the first four hours of treatment, after which they remained stable. The rate of change of optical absorbance of the cell suspension was converted into a DEP response of the form of those shown in Figures 11.11 and analysed using the single-shell model of a cell. The average plasma membrane capacitance of the cell population was determined to remain relatively stable, beginning at 9.1 mF/m^2, varying within ±12% of this value throughout the experiment and ending at 8.0 mF/m^2 after 24 h of treatment [89]. The observed reduction (9.1 down to 8.0 mF/m^2) in the average membrane capacitance of staurosporine-treated Jurkat cells [89] was thus in broad agreement with the reduction (13.34 down to 10.49 mF/m^2) observed for etoposide-treated Jurkat cells [88].

Lv *et al.* [92] followed changes in the value of f_{xo1} of NB4 cells after their treatment with cytosine arabinoside (Ara-C). NB4 is a cancer cell line originally established from the bone marrow of a patient with acute promyelocytic leukaemia, whilst Ara-C is the most widely used antimetabolite for inducing remission of acute leukemia. Ara-C is a potent killer of dividing cells through its action in inducing the mechanisms of cellular apoptosis. The DEP electrodes were fabricated by patterning non-closed ring gold electrodes (width 20 μm, inner diameter 200 μm) onto a glass wafer using standard photolithography. A 200 μL Eppendorf tube was cut and assembled on the chip to construct the DEP measurement chamber. The proportion of Annexin-V positive (i.e., apoptotic cells) at two hours post-treatment was not significantly different compared to control cells. Among the entire cell population, the proportion of apoptotic cells increased from 4.51% ± 0.44% at two hours, to 7.19% ± 0.26% at four hours, 12.42% ± 0.68% at six hours and 21.68% ± 0.45% at 12 h. Values of f_{xo1} for around 20 cells were measured over a time period of less than 20 min for each experiment and at 2 h post-treatment displayed distinct differences with those obtained for the untreated control cells. The f_{xo1} values were reported to continually increase over 12 h post-treatment, from 96 ± 4.73 to 354 ± 6.11 kHz, whilst at the same time the cell diameter constantly decreased from 16.30 ± 0.35 to 13.81± 0.62 μm over 12 h. (Deliberate selections of cell size and / or f_{xo1} values will have been made by the experimenters, because the reported standard deviations are much smaller than that expected, based on biological variation alone. For example, inspection of Figure 1A of the paper [92] shows NB4 cells of diameters ranging from 13 to 23 μm.) Together with Equation (11.3) and the single-shell model of a cell, the (presumably selected) values of f_{xo1} and cell radius were used to determine the plasma membrane capacitance C_{mem}. This was reported to decrease from 9.42 to 7.63 mF/m^2 in the first two hours following treatment with Ara-C, finally falling to a very low value of 3.03 ± 0.05 mF/m^2 after 12 h. A measurement of the high-frequency DEP crossover f_{xo2} (see Figure 10.25) was also conducted and observed to fall from an initial value of 301 ± 7.09 to 165 ± 7.78 MHz over the 12 h post-treatment. (This, again, is a surprisingly small spread of values. For example, a typical value of f_{xo2} for SP2/O cells was found to be 190 ± 60 MHz [93].) A fall in the value of f_{xo2} signifies a drop in the cytoplasmic conductivity σ_{cyt} [93, 94], so that a decrease of σ_{cyt} from 0.217 to 0.190 S/m within 2 h of Ara-C treatment was determined. This lower value was maintained for a short period of time before decreasing further [92].

To better understand the molecular mechanism underlying apoptosis using DEP monitoring, Lv *et al.* [92] examined mRNA changes in NB4 cells after Ara-C treatment over a 12-hour time course using a human whole genome oligo array. A total of nine, 37, 42 and 117 genes displayed altered expression levels at two, four, six and 12 h, respectively. Gene ontology analysis revealed that differentially expressed genes could be grouped into five main categories, namely apoptosis; cell division and proliferation; cell morphogenesis; ion transport [92]. Additionally, cluster analysis confirmed the expression level of some known apoptosis-related genes changed after Ara-C treatment, including the upregulated proapoptosis genes and the downregulated antiapoptosis genes. The expression level of nearly all genes involved in cell division and proliferation was downregulated. Notably, the expression level of some genes involved in cell morphogenesis or ion transport changed as early as two hours post-Ara-C treatment, which may correlate with the measured alteration in membrane capacitance or alteration in cytoplasmic conductivity. For example, their studies indicated that expression levels of KIF20A and CENPE, which are related to the synthesis of cytoskeletal proteins such as tubulin (and hence to membrane morphology and its capacitance) are downregulated, while expression levels of P2RX4 and KCTD9, which are related to ion transport are upregulated as early as two hours post-Ara-C treatment. Thus, a decrease in intracellular potassium concentration induces a loss of cytoplasmic conductivity, even though the intracellular calcium and sodium levels increase. As remarked by Lv *et al.* [92] the changes in cell membrane capacitance and cytoplasmic conductivity make apoptotic cells amenable for characterization using DEP analysis. They also suggested that this could be used to help physicians detect apoptosis earlier and that as DEP technology develops, individually tailored and more personalized patient treatment will be possible.

Nikoloic-Jaric *et al.* [95] describe a DEP cytometer that incorporates a differential coplanar electrode array that allows independent detection and actuation of single cells within a short section (~300 μm) of a microfluidic channel. A first electrode pair detects an electrical impedance signal (P_1) as each cell passes through the channel. The next pair of electrodes is electrically energized to produce a DEP force on the passing cell, which either elevates the cell higher into the fluid flow stream by negative DEP, or reduces its elevation by attracting it towards the electrode pair by positive DEP. The cell then passes over the second set of passive electrodes where a second impedance signal is (P_1) detected. A change in altitude smaller than 0.25 μm of a cell flowing between the two impedance detection sites can be detected. By analysing the experimental signatures of cells of known dielectric properties a simple connection can be made between the Clausius–Mossotti factor and the amount of vertical cell deflection during DEP actuation. To quantify the changes in impedance signals, a force index ϕ was used:

$$\phi = \frac{P_2 - P_1}{P_2 + P_1}$$

Positive or negative ϕ is associated with pDEP or nDEP, respectively, with $\phi = 0$ corresponding to no DEP actuation close to the DEP crossover frequency. The magnitude of ϕ is related to the strength of the DEP force that caused the altitude change. A linear correlation between the two exists for small values of ϕ. Theoretically, for a large population of cells with a range of different polarizabilities, values of ϕ are expected to follow a sigmoid function that saturates at the two extreme values (examples of such curves are shown by Pethig and Talary [88]). In practice, neither of these limits could be reached: the lower limit ($\phi = -1$) was unattainable due to the limits of equipment sensitivity and signal-to-noise issues, while the higher limit ($\phi = +1$) lay beyond the cutoff imposed by how close the attracted cells could approach the electrodes [95]. The DEP cytometer was validated using Chinese hamster ovary (CHO) cells that were followed in their rapid transition from a healthy viable to an early apoptotic state. The CHO cells were suspended in a medium of conductivity 0.17 S/m and exhibited a DEP crossover frequency for the healthy cell of ~0.5 MHz. An operating frequency of 6 MHz was thus chosen in order to monitor both negative and positive DEP through a frequency range of 0.1 to 6 MHz. The nonviable (apoptotic) cells exhibited negative DEP throughout this frequency range, indicating that their crossover frequencies had advanced above 6 MHz. The DEP cytometer cell viability estimates closely matched an Annexin-V assay on the same population of cells. Fluid flow rates were limited to the range 5~10 nl/s, so that for cell concentrations of ~10^6 cells/ml only 500~1000 cells were typically analysed. However, this was sufficient to determine the fraction of cells involved in early apoptosis. As pointed out by Nikoloic-Jaric *et al.* [95], even at a low throughput rate of one event per second, it would take only 10–15 min to determine the fraction of cells involved in early apoptosis in a given dilute sample. By comparison, the standard fluorescent flow cytometer assay for detection of early apoptosis, namely the Annexin-V assay, requires between 2000 and 10 000 cells (typically from dilutions of an initial concentration of around 10^6 cells/ml). These cells then have to be incubated for at least 15 min before the measurements can be made. This illustrates the advantages of the DEP cytometer in terms of its sensitivity and ability to quantify changes in the electronic signature produced by single cells.

An important requirement for a rapid and sensitive detection of apoptosis is in the large-scale production of glycoproteins from mammalian cell culture processes, driven by the application of these molecules as biopharmaceuticals for unmet medical needs [96, 97]. Large volumetric production and maintenance of a constant product quality are essential requirements. Indicators of early events leading to apoptosis are highly desirable in such cell bioprocesses because these initial stages may be reversible by appropriate intervention, such as nutrient feeding. Mid-stages of apoptosis indicate the end of protein production from the cells, but late stages may cause harm as lysed cells release proteinases and glycosidases into the medium with a potential to degrade the glycoprotein product. As a further demonstration of the usefulness of the DEP cytometer described by Nikoloic-Jaric *et al.* [95], Braasch *et al.* [98] investigated five different and independent methods of monitoring cell density and / or cell viability of CHO cells grown in a batch culture in a controlled bioreactor. These methods included: a particle counter; trypan blue exclusion; an *in situ* bulk capacitance probe; an offline fluorescent flow cytometer and a prototype DEP cytometer. These various techniques gave similar values during the exponential growth phase of the cells. However, beyond the exponential growth phase the viability measurements diverged. Fluorescent flow cytometry with a range of fluorescent markers was used to investigate this divergence and to establish the progress of cell apoptosis. The cell density estimates by the intermediate stage apoptosis assay agreed with those obtained by the bulk capacitance probe and the early stage apoptosis assay viability measurements correlated well with those of the DEP cytometer. The trypan blue assay showed higher estimates of viable cell density and viability compared to the capacitance probe or the DEP cytometer. The DEP cytometer identified at least two populations of cells, each with a distinct dielectric polarizability, one population being associated with viable (nonapoptotic) cells and the other with apoptotic cells. From the end of the exponential through the stationary and cell decline stages there was a gradual shift of cell count from the viable into the apoptotic population. However, the two populations maintained their individual dielectric properties throughout this shift. This led to the conclusion that changes in bulk dielectric properties of cultures might be better modeled as shifts in cells between different dielectric subpopulations, rather than assuming a homogeneous dielectric population. This demonstrates that bulk dielectric probes are sensitive to the early apoptotic changes in cells and that DEP cytometry in particular offers potential applications as a low-cost, label-free, electronic monitor of physiological changes in cells.

Differences in the way various apoptosis-inducing agents act on different stages of the cell cycle for different cell types can be expected. This therefore remains an interesting area for future study – one in which a DEP cell profiling technique could play an important role. Of relevance to this is the interesting fact that in a study of the DEP spectra of K562 (chronic myeloid leukemia) cell suspensions [99, 100] a reduction of average cell size was observed as expected, but the membrane capacitance was deduced to *increase* from 9.7 to 14.9 mF/m^2 (rather than to decrease in value as for other studies [87, 88, 92] after eight hours incubation with the apoptosis-inducing agent staurosporine. The cytoplasmic conductivity of the K562 cells was also deduced to *increase* from 0.28 to 0.45 S/m [100], a result again contrasting sharply with conclusions derived for HL-60 cells [87], Jurkat cells [88] and NB4 cells [92]. Could it be that there was a greater population of necrotic K562 cells than realized? K562 cells are also known to be refractory to induction of apoptosis by topoisomerase II-targeting agents, thus leading to the suggested alternative [100] that the morphological differences caused by apoptosis in HL-60, Jurkat and NB4 cells differ from those in K562 cells. This is worthy of further study, along with studies of whether DEP can distinguish between apoptotic and necrotic cells.

11.3 Mammalian Cells

11.3.1 Blood Cells

In 1986 Tsoneva *et al.* [101] determined the DEP velocities of human erythrocytes in the field produced between two concentric cylindrical metal electrodes, of inner and outer radii 0.24 mm and 1 mm, respectively. The AC frequency was fixed at 2 MHz and the voltage was varied up to 19 V_{rms}. Two formulations for the suspending medium were used: (i) 90% of a 2.1% glycine solution mixed with 10% of a 5.5% glucose solution; (ii) 5.4% sorbitol solution. In both mediums the DEP velocities were found to be proportional to the square of the applied voltage and inversely proportional to the cube of the distance from the symmetry axis of the coaxial cylindrical electrodes (as predicted by Equation (2.15) in Box 2.4) and the expression for (E.∇)E deduced from Table 3.2. Tsoneva *et al.* [101] also reported the interesting observation that the DEP coefficient of proportionality depended on the human donor of the erythrocytes, further commenting that this might be used for diagnostic purposes. These words were endorsed (the echo taking 22 years in its journey) by Srivastava *et al.* [102] who quantified the DEP responses of different positive blood types (A+, B+, AB+ and O+) at 1 MHz in a pin-plate electrode arrangement. The whole blood samples were diluted 60-fold using isotonic, physiological strength, phosphate buffered saline (PBS). Thus, although not cited, we can assume a conductivity of ~1.5 S/m for the cell suspending medium. From Equations (10.36) and (10.39) and assuming a radius of 7 μm and membrane capacitance of 8 mF/m^2 for an erythrocyte, the DEP crossover (f_{x01}) frequency can be estimated as ~5 MHz. The DEP characteristics were thus obtained at an operating frequency (1 MHz) close to but below f_{x01}. From Equation (10.37) a negative DEP response is expected, in agreement with that reported by Srivastava *et al.* [102]. At an operating frequency close to f_{x01} the DEP response can also be expected to be sensitive to subtle differences in the membrane capacitance of the erythrocytes. Cells of O+ type exhibited a relatively attenuated DEP response and could be distinguished with greater than 95% confidence from all the other three blood types. The practical significance of this is that O-type blood can be universally used in blood transfusions. AB+ cell responses differed from A+ and B+ blood types, a result considered to arise because AB+ erythrocytes express both the A and B glycoforms on their membrane. Srivastava *et al.* [102] suggested that from these findings the DEP of untreated erythrocytes beyond simple dilution could be used in portable blood typing devices. In later work Leonard and Minerick [103] compared the lower (f_{x01}) and upper (f_{x02}) crossover frequencies of erythrocytes with known ABO-Rh antigen expression against the same erythrocytes modified to remove sugar units from the A, B and O antigens. The presence of the transmembrane Rhesus factor (i.e., positive blood types) increased the value of f_{x02} by ~9 MHz in the 70–80 MHz range. The Rhesus factor is a transmembrane protein that likely functions as an ion channel, thus affecting the membrane permeability, supporting the view that the value of f_{x02} is associated with cytosolic properties [103]. The lower f_{x02} values observed in the 40–45 MHz range implied that DEP could be exploited to distinguish ABO blood types. Erythrocytes with both A and B antigens displayed unique concurrent negative and positive DEP signatures over a wide frequency range. There was also good reproducibility between donors, suggesting the potential for determining the ABO-Rh blood type of an unknown sample using multistep DEP comparisons. Srivastava *et al.* [104] have also investigated the potential of being able to separate erythrocytes, based on their ABO-Rh blood group, through their deflection across the inlet and outlet fluidic channel of a DC insulator-based DEP device. Under optimized conductivity (in the range 52 mS/m – 0.91 S/m) and field conditions (1.71–6.85 kV/m), A+ blood could be discerned with >99% confidence from all other blood types. B− exhibited a 99.4% discernability in this test. An unexpected finding was that similar channel fractions were observed with B+, A−, O+ and AB− cells, even though they represent dissimilar antigen expressions.

The different patterns of cell collection at castellated interdigitated electrodes (pearl chaining or triangular depending on positive or negative DEP, respectively) observed by Wang *et al.* [35] for erythrocytes was confirmed by Xu *et al.* [105] – with the added observation that at 2 MHz and in a medium of conductivity 23.5 mS/m a fraction of the cells collected by positive DEP could be released with gentle fluid flow. This implied that subpopulations of the cells exhibited different dielectric properties. Aceti *et al.* [106] measured the electrical conductivity of trophozoite-infected erythrocytes obtained from a patient suffering from cerebral malaria, with a very high level of parasitaemia (27%), before and after starting chemotherapy with quinine hydrochloride. Before starting therapy, the membrane conductivity of infected red blood cells was significantly higher than that of normal erythrocytes, which suggests that the ionic transport and permeation processes occurring in the membrane were enhanced by the presence of the parasite, possibly due to structural disorder induced by *Plasmodium falciparum* in the phospholipid bilayer. After initiation of antimalarial chemotherapy, a marked reduction of electrical conductivity of the parasitized cell, to values below those of normal cells, was observed. This reduction appeared to be related to the number of infected erythrocytes, since there was a direct correlation between the parasite density and the conductivity value, which reached normal levels after the clearance of parasitaemia. To demonstrate that quinine alone was not responsible for any possible alteration in the membrane structure and function, the *in vitro* effect of this drug on the conductivity of normal erythrocytes was investigated. The membrane conductivity was found to be ~1.1×10^{-4} S/m, in good agreement with the value estimated in normal conditions. The markedly reduced conductivity following quinine therapy suggests that the drug can significantly decrease ion transport across the infected membrane. Thus, it appears that quinine makes the cellular environment unfavourable for *P. falciparum* replication, by blocking the transport of substances physiologically relevant for the rapidly growing parasite. The drug does not kill the parasite directly, but its toxic effect may be a consequence of the erythrocyte's membrane modification. Furthermore, the membrane conductivity of *P. fulciparum* infected erythrocytes, after the initiation of chemotherapy, became similar to that of effete and senescent red cells, which are susceptible to filtration by the spleen. Therefore, it seems that quinine has an 'ageing' effect on parasitized erythrocytes, at least with respect to the electrical properties of the membrane. These cells may be removed from the circulation because they are susceptible to the culling process within the spleen. The sudden increase of splenic filtration observed in patients with acute *falciparum* malaria after starting quinine therapy had been thought to be the result of a direct effect of the drug on splenic function. The study by Aceti *et al.* [106] indicated that it may be a consequence of alterations of the infected erythrocyte membrane induced by quinine treatment. From measurements of the crossover frequency (f_{x01}) values of 11.8 mF/m^2 and 271 S/m^2 were derived for the membrane capacitance and membrane conductivity, respectively, of normal erythrocytes [107]. The value for f_{x01} depended linearly with the value of the medium conductivity, in accordance with Equation (10.39). However, for the parasitized cells the value for f_{x01} exhibited a superlinear dependence on medium conductivity above 20 mS/m. The parasitized cells exhibited a membrane capacitance of ~9 mF/m^2, with a membrane conductance of 1130 S/m^2 that increased steadily as the medium conductivity was increased above 20 mS/m [107]. This increase was considered to result from the known presence of parasite-associated membrane pores in parasitized cells. The large differences between the f_{x01} values of normal and parasitized cells was considered to provide straightforward sorting of these cell types by DEP, with a particular application of providing presample enrichment of clinical blood samples for PCR analysis [108, 109].

Piacentini *et al.* [110] have described a microfluidic device that is capable of separating platelets to a purity of 98.8% from other blood cells, with less than 2% cell loss. The device combines hydrodynamic focusing in combination with DEP field flow fractionation, using the so-called 'liquid electrodes' design described in Chapter 10, section 10.4.3. Whole blood was centrifuged for five minutes at 1000 rpm and the sample on the top of the tube (containing mostly platelets) was remixed with the sample at the bottom at the tube (containing erythrocytes and leucocytes) was diluted to obtain a concentration of 1~2×10^8 cells and platelets per ml. The suspending medium was phosphate buffer saline, diluted in sucrose solution to give a conductivity of 55 mS/m, while keeping an osmolarity of 300 mOsm/l. The addition of 1% w/v bovine serum albumin reduced the adhesion of cells on the microchannel walls. At this conductivity both the platelets and other blood cells exhibited the required magnitude of negative DEP at 100 kHz using a modest applied voltage of 10 V_{pk-pk}. Pethig *et al.* [111] validated a DEP cell profiler by following changes in the dielectric properties of human T lymphocytes (Jurkat E6-1 cells) after their mitogenic stimulation with phorbol myristate acetate (PMA) and ionomycin. B lymphocytes are activated to commence cell division (mitosis) when they encounter an antigen matching their specific immunoglobulin, whereas T cells undergo mitosis when stimulated by mitogens to produce small lymphocytes that produce lymphokines for bolstering the host organism's immunity. The cell profiler automatically applied a sequence of AC voltages to the microelectrodes and images of the DEP-induced motions of the cells were captured at 30 frames per second. The location of each cell was continuously tracked and a value for the diameter of each cell was obtained from the average obtained from the images collected over time, to an estimated accuracy of ±0.25 μm. The velocities of the cells were computed to an accuracy of ±0.1 μm/s or better and these values were normalized with respect to the variation of the electric field strength and gradient between the electrode edges. The data for each cell was fitted to the idealized DEP response curve, corresponding to the single shell model of a cell, for the purpose of computing the cell's crossover frequency (f_{xo1}). With a suspending medium conductivity of 40 mS/m, membrane capacitance values in the range 9.0–16.9 mF/m^2 were obtained for the control Jurkat cells, whilst the corresponding range for activated cells was 7.0–13.8 mF/m^2. This reduction of the mean membrane capacitance of the cells was the main dielectric effect found to occur 24 h following PMA and ionomycin treatment, indicating a reduction in the complexity of the membrane topography that was related to the observed reduction in the percentage of S phase cells [111]. These results are consistent with the findings of electrorotation studies of human T lymphocytes [112, 113]. Mitogenic stimulation of resting T lymphocytes resulted in an increase of membrane capacitance as the cells moved from their normally resting G0 phase into the cell division cycle. Huang *et al.* [112] found the largest increase (77%) in membrane capacitance to occur as the cells progressed from the G1 phase through the S phase of the cell cycle.

11.3.2 Cancer Cells

Simply stated, cancer cells can multiply rapidly and proliferate independently of both internal and external signals that normally restrain cell growth. The DNA of their genes that regulate normal cell division is altered or damaged in some way, either by multiple mutations that accumulate over time or through exposure to toxic chemicals, radiation and other environmental sources. When a cell transforms into a cancer cell certain changes of its behaviour occur, such as loss of contact inhibition and anchorage independence. With loss of contact inhibition the dividing cells are not restricted by cell-cell contact and can grow over one another. With anchorage independence the cells, instead of having to attach to a solid surface before they are able to divide, can form colonies in a viscous fluid or soft agar gel. The number of cancer cells growing in a culture dish is commonly determined qualitatively by estimating their *confluence* – a term referring to the proportion of the surface area of the dish substrate covered by the cells. Thus, 80% confluence (a typical value for cell removal) corresponds to ~80% of the dish surface being covered by the growing cells. Adherent cells in culture are typically detached and dispersed into solution for DEP study using a combination of trypsin and ethylenediamine-tetra-acetic acid (EDTA). Trypsin is a proteolytic enzyme that breaks down proteins, whilst EDTA chelates divalent ions (such as Ca^{2+}), so that acting together they weaken cell-cell and cell-substrate interactions. The extent to which this procedure might modify the cell's DEP properties has not been quantified (to the author's knowledge), but it is generally considered to have a minimal effect.

Burt *et al.* [114] investigated the DEP properties of three lines (DS19, R1 and DR10) of Friend murine erythroleukaemic (MEL) cells. These cells, derived from a B-cell lymphoma, are rapidly dividing mouse cell lines that are maintained in suspension cultures and grow as loose clumps at a density between 10^5 and 10^6 cells/ml. They should be split roughly tenfold every two days or so to maintain this concentration. MEL cells are transformed precursors of erythrocytes that have not differentiated beyond the stage of colony-forming cells. They are well established as a model for the study of cell differentiation because of their susceptibility to a variety of chemical agents that can, in some clones, induce varying degrees of differentiation up to an advanced step of erythroid differentiation, which has much in common with the late steps of erythropoiesis and the normal nuclear extrusion process (a fully formed erythrocyte does not have a nucleus). Cell line DS19 progresses beyond the colony-forming stage to the stage of terminal differentiation in response to treatment with hexamethylene bisacetamide (HMBA), evidenced by haemoglobin production and loss of the ability of the cells to grow in soft agar. The R1 cell line, cloned from DS19, is characterized by its inability to terminally differentiate and produce haemoglobin following treatment with HMBA or dimethylsulphoxide (DMSO). R1 is thus classified as being noninducible, whilst DR10 is characterized by a differential response to HMBA and DMSO. Whereas DR10 responds to HMBA treatment by producing haemoglobin and losing its ability to grow in soft agar, it is resistant to DMSO, which is a strong inducing agent for DS19. Treatment of the three cell lines with HMBA was already known to decrease their average cell diameters by ~12% and their plasma membrane surface charge densities by ~14%. The DEP characterization [114] of the MEL cells used an optical technique capable of performing measurements in the frequency range 1 Hz to 4 MHz [115]. To investigate the influence of surface charge, human erythrocytes were also investigated before and after treatment with neuraminidase (an enzyme that cleaves sialic acid from cell surfaces). This investigation revealed that a reduction of cell surface charge significantly reduced the magnitude of the DEP response for erythrocytes in the frequency range 1–30 Hz, but had a negligible effect at higher frequencies [114]. HMBA-treatment *reduced* the low-frequency (1–200 Hz) DEP response of the inducible DS19 cells as well as the noninducible R1 cells, reflecting a loss of surface charge for both cell types. However, HMBA-treatment *increased* the DEP response of the DS19 cells in the frequency range 200 Hz–10 kHz, but had no effect on the R1 cells. This result was mirrored in the results obtained for the DR10 cells, where HMBA-treatment *increased* the DEP response in the frequency range 200 Hz–10 kHz, but was not altered by exposure to DMSO. An increase in DEP response in this frequency range implied an increase of effective cell conductivity. These results were taken to confirm that DEP can distinguish between cell surface charge and cell membrane polarizability parameters.

The DEP characteristics of clone DS19 of the MEL cell line, together with the effect of treatment with the inducing agent HMBA, were later studied in more detail by Gascoyne *et al.* [116]. In this work a method was devised in which the DEP properties of 200–300 cells could be observed simultaneously on an interdigitated, castellated, electrode array and the results quantified using computerized image analysis. This electrode design (see Figure 10.35) had been shown to collect cells in aggregation patterns that can clearly be interpreted in terms of positive and negative DEP [117]. Software was used either to measure the averaged behaviour of all the cells, to distinguish cell subpopulations by identifying their morphological differences or to track individual cells of interest. This method thus retained the ability to discriminate individual cells and cell subpopulations, while permitting observation of sufficient numbers to

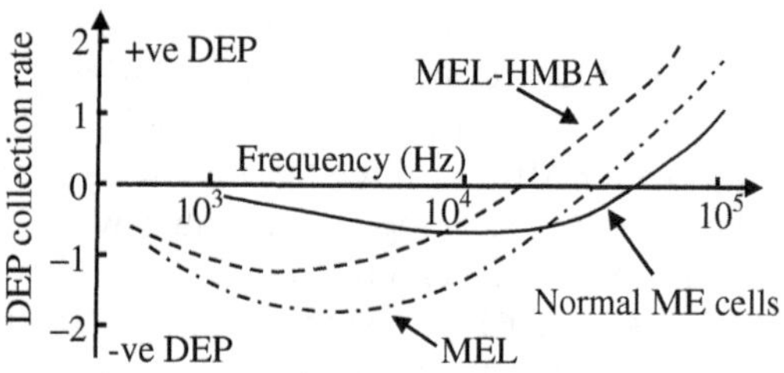

Figure 11.21 DEP collection spectra for normal murine erythrocytes (ME); murine erythroleukemia (MEL) cells; and MEL cells treated with the inducing agent hexamethylene bisacetamide (HMBA). The conductivity of the isotonic cell suspension (10^7 cells/ml) was 2 mS/m (based on Gascoyne *et al.* [116]).

allow meaningful statistical analyses. After growth and harvesting, the cells were washed three times in 320 mM sucrose solution containing 2 mg/ml dextrose, finally being suspended at 10^7 cells/ml in the same solution. The cell suspension was adjusted to a conductivity of 2 mS/m by titration with NaC1, kept on ice until DEP characterization, which was typically completed within 45 min. The frequency-dependencies of the rates at which normal murine erythrocytes, untreated MEL cells and those treated with HMBA, collected by either positive or negative DEP at the castellated electrodes are shown in Figure 11.21.

From Figure 11.21 it can be seen that at a frequency of 22 kHz the HMBA-treated leukemic cells displayed positive DEP collection, whilst untreated leukemic cells exhibited negative DEP. It follows that a mixture of these different cell types could in principle be separated to different regions of the electrode array at this frequency. Likewise, a mixture of leukemic and healthy mouse erythrocytes should be capable of such separation at 30 kHz. To demonstrate that this prediction could be realized in practice, leukemic and normal erythrocytes were mixed, injected onto the electrode array and subjected to a 30 kHz field [118]. Good local separation of the leukemic and normal cells occurred within one minute, the different cell types being readily distinguished by their different sizes. Normal erythrocytes are smaller than erythroleukaemic cells, so that a simple size discrimination algorithm allowed the image of all cells to be split into separate images containing only erythrocytes or only leukemic cells. Because the normal erythrocytes were so much smaller than erythroleukemia cells (less than $\frac{1}{4}$ of the area) cells were correctly identified by the image processor to >99% accuracy.

The image processing also allowed quantitative measurements to be made of the DEP-induced motions of the cells, so that any changes could be determined of the membrane dielectric properties of the MEL clones during the process of cell differentiation induced by HMBA or DMSO [118]. The results of this investigation are shown in Table 11.2. From this Table, it is clear that HMBA induced at least a sixfold decrease in cell membrane conductivity and a decrease of approximately 30% in membrane permittivity in DS19 cells. DMSO was found to induce similar effects. The membrane changes developed over four days of treatment with HMBA or DMSO and paralleled morphological alterations in the cells and the expression of haemoglobin as the cells differentiated [119]. Table 11.2 also shows that the DEP properties of the noninducible R1 exhibited almost no response following exposure to these agents. Decreases in membrane conductance and permittivity, which occurred for DS19 but not for R1 following treatment by HMBA or DMSO, therefore appeared to be correlated with cell differentiation – as judged by the expression of haemoglobin and morphological changes that included a reduction of cell volume and a shrinkage of the cell nucleus during erythropoiesis.

The earlier investigations [116, 117] had achieved DEP-assisted separations of different cell types at a localized level on the microelectrode array. For most practical applications of cell separation, however, the objective is to physically isolate and collect target cells from a cell mixture for further study. Becker *et al.* [120] described a DEP device with interdigitated, castellated, electrodes that they referred to as an 'electroaffinity' column. A demonstration of this technology was given through the practical example of removing leukemic cells from human blood. The cell mixture consisted of blood cells (collected by venipuncture from healthy volunteers and diluted with 90 parts Ca^{2+}/Mg^{2+}-free PBS containing 5 mM hemisodium EDTA) mixed in a ratio of 3 : 2 with HL-60 leukaemia cells that had been cultured under standard conditions and harvested by centrifugation. The mixed cells were then washed twice

Table 11.2 Dielectric properties of the plasma membranes of responsive (DS19) and nonresponsive (R1) murine erythroleukaemic (MEL) cells [118].

	DS19 Cells		R1 Cells	
Membrane property	**No treatment**	**HMBA treated**	**No treatment**	**HMBA treated**
Conductivity (S/m)	5.4×10^{-7}	$<1 \times 10^{-7}$	7.7×10^{-7}	5.1×10^{-7}
Permittivity (F/m)	$6.39\ \varepsilon_o$	$4.49\ \varepsilon_o$	$6.86\ \varepsilon_o$	$6.67\ \varepsilon_o$

in isotonic (8.5%) sucrose containing 3 mg/ml dextrose and resuspended at a final concentration of 2×10^7 malignant cells and 3×10^7 normal blood cells/ml in this same solution. The suspension conductivity was adjusted to 10 mS/m by addition of hemisodium EDTA to a final concentration of approximately 0.7 mM. The DEP separation chamber was operated under conditions where the DEP force for one cell type was strong enough to prevail against the horizontal fluid drag force and the hydrodynamic lift force, causing this cell type to be retained on the electrode tips while other cell types were eluted by the combined influences of the hydrodynamic forces. Thus, the DEP chamber was considered to operate as an electroaffinity column having affinity for only one cell type. Electrorotation measurements were also performed on the erythrocytes and HL-60 cells. Based on the values of the cellular dielectric parameters derived (e.g., membrane capacitances of 15 and 9 mF/m^2 for the HL-60 and erythrocytes, respectively) from an analysis of this electrorotation data, it was calculated that at 80 kHz the HL-60 and erythrocytes would exhibit Clausius–Mossotti factors of +0.6 and −0.1, respectively. The DEP force acting on the HL-60 cells would thus be strongly positive and exceed the horizontal drag force by a factor of three at 80 kHz, whilst erythrocytes would be repelled from the electrodes by a negative DEP force. This prediction, that the HL-60 cells would be retained in the chamber and the erythrocytes eluted, was observed in practice [120]. The HL-60 cancer cells were retained at the tips of the electrode castellations, while the normal blood cells were eluted with the fluid flow from the chamber, when the applied voltage signal was repetitively swept between 20 and 80 kHz. This procedure loosened the packing density of the collected cells and helped to repel any blood cells that had become entrapped with the cancer cells. After 20 min the cancer cells were retained at a purity of ~80%. It was also noted that because the cells had been sorted at a rate of $\sim 10^3$/s, thus faster than conventional fluorescent activated and other refined cell sorters, it should be feasible to increase this rate by at least two orders of magnitude using a larger device, making the technology applicable to practical large-scale cell sorting problems. Furthermore, the technique is noninvasive to the cells and does not depend on tagging or labelling the cell surface with potentially disruptive antibodies.

The knowledge gained from these studies [114, 116, 118, 120] to investigate the DEP characteristics of cancer and normal blood cells have been directed towards the development of DEP-based technologies for detecting and isolating circulating cancer cells (CTCs) from the peripheral blood of cancer patients. The potential impacts and technical challenges of achieving this aim are significant. The prognosis and treatment of various cancers (e.g., breast, prostate, ovarian, colon) is aided by knowledge of the concentration of CTCs. At any time during therapy the detection of elevated CTCs provides an accurate indication of subsequent rapid disease progression and mortality for a patient. However, CTC concentrations are extremely low compared to the background count of normal blood cells. For example, detection of more than five CTCs per 7.5 ml peripheral blood is indicative of a worsening outcome for breast cancer patients [121]. From the composition of human blood detailed in Appendix K, the challenge for a DEP-device is thus equivalent to isolating from (say) 15 ml of whole blood at least ten cancer cells from amongst $\sim 8 \times 10^{10}$ erythrocytes and other blood cells! If, and not an insignificant 'if', the vast majority of the erythrocytes can be removed by lysis or density-gradient separation without loss of any of the cancer cells, the problem is reduced to isolating ten cancer cells from $\sim 10^8$ peripheral blood mononuclear cells (leucocytes) of an original 15 ml blood sample. This simple analysis provides an idea of the scale of the challenge in terms of cell density and fluid volume throughput for 100% efficiency and purity of target cancer cell capture. Progress has been made through the use of larger electrode arrays and DEP field-flow-fractionation (DEP-FFF), together with an evaluation of how processing time, fluid flow rate and cell concentration influences the efficiency of target cancer cell separation. In initial tests using the batch-mode form of DEP-FFF, mixtures of peripheral blood mononuclear cells (PBMNs) and three cultured cancer cell lines, with a ratio of 1 : 1000 for cancer-to-PBMN cell content, the efficiency of cancer cell recovery was 92% for a chamber cell load of 10^5 cells, falling to 10% efficiency at a cell loading of 2.25×10^7 cells [122]. After isolation, the cancer cells were successfully returned to growth in culture to demonstrate functional integrity and viability. Shim *et al.* [123] describe the continuous-flow form of DEP-FFF that can process larger sample volumes at processing rates of 10^6 PBMN cells / minute. The peripheral blood mononuclear cell fraction of a clinical specimen is slowly injected, deionized by diffusion and then subjected to a balance of DEP, sedimentation and hydrodynamic lift forces. These forces cause cancer cells to be transported close to the floor of the chamber, while blood cells are carried about three cell diameters above them. The cancer cells are isolated by skimming them from the bottom of the chamber while the blood cells flow to waste. To illustrate operation of the technology, the isolation of circulating colon cancer cells from clinical specimens was achieved, with the tumour origin of these cells verified by molecular analysis [123]. A commercial form of this instrumentation is described by Gupta *et al.* [124]. To investigate the potential applicability of this technology to different cancer types, Shim *et al.* [125] measured the DEP and specific density properties of the NCI-60 panel of cancer cell types (54 types in total), whilst comparing these properties to those exhibited by the subpopulations (e.g., lymphocytes, monocytes, basophils, neutrophils) of normal peripheral blood cells. Apart from one of the five breast cancers and one of the nine nonsmall cell lung cancers, all of the NCI-60 cell types had densities in the range 1048 to 1068 kg/m^3. It was also found that the NCI-60 cell types, regardless of tissue of origin, exhibited DEP properties that facilitate their isolation from blood. Cell types derived from solid tumours that grew in adherent cultures exhibited DEP crossover frequencies (f_{xo1} = 20~65 kHz) in a 30 mS/m medium that were strikingly different from those of peripheral nucleated blood cells (80~220 kHz), while leukemia-derived lines that grew in nonadherent cultures exhibited f_{xo1} (55~110 kHz) values that approached or overlapped those of peripheral blood cell types (monocytes and CD34+ haemopoietic cells). A cautionary remark may be appropriate here regarding the DEP properties of the NCI-60 panel of cancer cells that are cultured *in vitro*. The important DEP data obtained by Shim *et al.* [125] may have to be taken as useful rather than precise guidelines for devising DEP separation protocols, because the DEP properties of the cultured cell lines may not exactly mirror the corresponding properties exhibited by cancer cells that have metastasized into peripheral blood.

Other approaches to the DEP separation of cancer cells include that by An *et al.* [126] who used fan-shaped, asymmetric, electrodes to direct different cell types into two outlet fluid channels. Malignant human breast cancer epithelial cells were successfully separated from healthy breast cells of similar size, based on subtle differences in their dielectric properties. Yang *et al.* [127] separated colorectal cancer cells from human embryonic cells and *E. coli* using opposing wedge-shaped electrodes in a fluidic channel. Moon *et al.* [128] combined a multi-orifice fractionation in series with slanted, interdigitated, DEP electrodes to separate human breast cancer cells from a spiked blood cell sample. A 162-fold enrichment of the cancer cells was obtained for a sample flow rate of 126 ml/minute, whilst the erythrocytes and leukocytes were removed with separation efficiencies of 99.24% and 94.23%, respectively. In a series of publications Salmanzadeh and co-workers used contactless-DEP to isolate prostate tumour initiating cells from noninitiating cells [129] and to investigate the DEP responses of progressive stages of mouse ovarian epithelial cells [130, 131]. The crossover frequencies (f_{xo1}) were determined for the four different stages of malignancy of the epithelial cells, finding that the plasma membrane capacitance advanced from a value of 15.39 ± 1.54 mF/m^2 for a nonmalignant benign stage to 26.42 ± 1.22 mF/m^2 for the most aggressive stage [131]. These differences were postulated to be the result of morphological variations due to changes in the cytoskeleton structure, specifically the decrease of the level of actin filaments in the cytoskeleton structure of the transformed epithelial cells [131]. Finally, Mulhall *et al.* [132] employed their DEP-microwell system to show that the cytoplasmic conductivity and plasma membrane capacitance of normal, precancerous and cancerous oral keratinocytes are distinct. The mean membrane capacitance values of dysplastic (DOK) and cancerous (H357 and H157) oral keratinocytes were higher than those for primary normal oral keratinocytes (HOK).

11.3.3 Stem Cells

In brief, stem cells are unspecialized cells that can differentiate to one or more specialized types of cell. Throughout its life a stem cell can replicate indefinitely through mitosis to generate more stem cells or *progenitor* cells that become highly specialized (e.g., a blood cell, muscle cell, neuron) and maintain the normal turnover of regenerative organs, such as blood, skin, or intestinal tissues. All of the cells in blood are derived from haematopoietoc stem cells, which reside in bone marrow, in a process known as haematopoiesis. Unlike stem cells, progenitor cells can divide a limited number of times only – they are said to exhibit oligopotency in that they can differentiate into only a few cell types. Examples are progenitor vascular cells that can become smooth muscle cells or endothelial cells, as well as a progenitor lymphoid cell that can give rise to a B or T blood cell (but not an erythrocyte, which is derived from a myeloid progenitor). Another term commonly used by stem cell biologists is *precurser* cell, which implies that it can differentiate into only one cell type. Adult stem cells (also known as *somatic* stem cells) and progenitor cells can be found in various tissues of young or adult animals and humans and act to either replenish or repair tissues. Embryonic stem cells can be isolated from the inner cell mass of blastocysts – in a developing embryo (or laboratory) they can differentiate into all the specialized cells of the body. For the purpose of stem cell therapy, human mesenchymal stem cells (hMSCs) have attracted particular attention because they have three important properties – they have a high differentiation capacity (being able to produce bone cells, cartilage cells, muscle cells and fat cells, for example) high growth (trophic) activity, plus the ability to self-renew.

The propensity of pluripotent stem cells (pSCs) to spontaneously differentiate and their unpredictability to commit to specific lineages opens up the opportunity for DEP to contribute to stem cell research and therapy [133]. For example, it could provide the means to act as a sensitive and noninvasive method to monitor and separate stem cell populations. Whether derived from embryos of varying stages of development, or induced by expression of nuclear factors in somatic cells, the isolation and renewal of pSCs are significantly challenged by the absence of noninvasive methods to discriminate and specifically promote the growth of this cell type, either from limiting quantities of tissue (e.g., embryos) or amidst competing unreprogrammed somatic cells during induction protocols. This affects the efficiency of their isolation and interferes with the ability to achieve clonal cell lines. This is especially the case when co-culturing these cells with other cell types (e.g., feeders) that support self-renewal and / or differentiation. Alternatively, in a therapeutic context, DEP may provide a simple and noninvasive way to positively or negatively select for target or contaminating cell types prior to transplantation. There are comparatively few publications that describe the DEP characterization or manipulate of stem cells and the following citations hopefully cover most of them up to early 2016.

The CD34+ antigen is a marker that has been used to identify haematopoietic precurser cells and is progressively lost as the cells differentiate. It is thought to have a role in early haematopoiesis by mediating the attachment of stem cells to the bone marrow extracellular matrix or directly to connective tissue (stromal) cells. Talary *et al.* [134] demonstrated that enrichment of CD34+ cells that contain the stem cell subpopulations in bone marrow and peripheral blood can be achieved using DEP. The cell samples were prepared by taking 0.5 ml and 0.25 ml of peripheral blood from stem cell harvests and bone marrow, respectively and making the solution up to 20 ml with phosphate buffered saline. The erythrocytes were removed in a lymphocyte density gradient separator. The remaining leucocytes were washed and resuspended in 320 mM sucrose solution containing 3 mg/ml glucose (conductivity 1 mS/m) and labelled with a fluorescent antibody against the CD34+ antigen. The cell suspensions were then pumped through a DEP chamber containing interdigitated, castellated, microelectrodes and subjected to a batch-mode form of DEP-FFF. At 500 kHz the majority of cells exhibited positive DEP and were retained in the chamber after flushing with the cell suspension medium. Cell fractions exhibiting negative DEP and eluted from the chamber were obtained by lowering the voltage in steps from 500 kHz down to 1 kHz. To test the efficacy of the DEP force the cell suspensions were passed through the DEP chamber, with and without an applied voltage signal, so that the relative concentrations of the CD34+ stem cells in the bone marrow and peripheral blood could be determined using standard FACS analysis. The sample fraction collected from the eluate at 5 kHz displayed the presence of a 4.97% population of CD34+ cells, representing a 5.9-fold enrichment of stem cells [134]. In later work Stephens *et al.* [135] increased the number of applied voltage frequencies and also demonstrated that the collected CD34+ stem cells grew when plated in colony assay cultures. This showed that the collected stem cells remained normal, viable and capable of colony formation when cultured for two weeks. The quantity of colonies formed correlated with the percentage of CD34+ cells in each fraction collected. The number of myeloid and erythroid colonies produced varied with each fraction, with the highest concentration in the 10 kHz fraction. Between samples collected from different donor patients there was a spread of frequencies in which enrichment occurred, but the peak enrichment was consistently obtained at 5 kHz.

Vykoukal *et al.* [136] later used a DEP-FFF device to enrich putative stem cells from adipose tissue. This device incorporated an electrode structure fabricated using a flex-circuit strip of gold microelectrodes on polyimide substrate (to anticipate scalable and low-cost volume manufacturing) later employed in the devices for isolating circulating tumour cells from peripheral blood [123–125]. Adipose tissue was manually minced and collagenase-digested; centrifuged to remove adipocytes and liquid fat; the resultant cellular fraction filtered; the erythrocyte quantity reduced by standard density gradient centrifugation. The adipose-derived cells were then labeled with various fluorescent antibodies, washed and suspended in iso-osmotic buffer (9.5% sucrose, 0.3% dextrose, conductivity adjusted to 30 mS/m with PBS) and resuspended at a concentration of 2×10^6 cells/ml. The frequency of the applied AC electric field was decreased linearly from 200 kHz to 60 kHz over 40 min, the eluted fractions being analysed by flow cytometry to track the elution by negative DEP of the various cell populations from the chamber. The elution profiles as a function of voltage frequency were similar in distribution to those obtained for CD34+ cells [134, 135]. Independent DEP-FFF runs yielded nearly identical elution profiles for cells labeled with FITC-conjugated antibodies against two putative stem cell markers, NG2 and nestin. Furthermore, the elution peaks for the putative stem cells were shifted from the peak containing the CD45+ leukocytes. Overall, the relatively rare (<2% in the starting mixture) NG2-positive cells were enriched up to fourteenfold. (The clinical use of adipose stem cells, isolated from fat by DEP, for the treatment of hand astrophy has been reported [137]. However, at the time of reading this paper it carried the editor's note that concerns have been raised over the reliability of the data presented!)

Flanagan *et al.* [138] found that populations of mouse neural stem / precursor cells (NSPCs), differentiated neurons and differentiated astrocytes exhibited different DEP-frequency profiles. This not only gave an indication that the different stem cell types had different dielectric parameters, but also that they exhibited different levels of heterogeneity. By isolating NSPCs from developmental ages at which they were more likely to generate neurons, or astrocytes, they also obtained evidence to support the very interesting conclusion that a shift in DEP-frequency profile reflected their fate bias *before* detectable marker expression in these cells. Measurements of the DEP crossover frequency (f_{xo1}) and cell radius were not made in this study, but these aspects were addressed in later work [139]. NSPCs of similar size and morphology but different in their ability to form neurons and astrocytes were investigated. From the single-shell analysis of the DEP data, it was concluded that plasma membrane capacitance, but not membrane conductance, is a specific and dynamic indicator of NSPC fate potential. Furthermore, NSPCs that differ in fate potential have distinct f_{xo1} values, suggesting that DEP may be used to isolate undifferentiated NSPCs based on their propensity to form either neurons or glia. It was further hypothesized by Nourse *et al.* [140] that inherent electrophysiological properties, as expressed by their DEP properties, are sufficient to define neorogenic progenitors (NPs) and astrogenic progenitors (APs) and that this could be tested by determining whether isolation of cells solely by these properties specifically separates NPs and APs. Nourse *et al.* found NPs and APs could be enriched in distinct fractions after separation by DEP. A single round of DEP isolation provided greater NP enrichment than sorting with PSA-NCAM, which is considered an NP marker. Additionally, cell surface N-linked glycosylation was found to significantly affect cell fate-specific electrophysiological (DEP) properties, providing a molecular basis for the cell membrane characteristics. Inherent plasma membrane biophysical properties that influence their DEP properties were thus thought sufficient to define progenitor cells of differing fate potential in the neural lineage [140]. Furthermore, these properties can be used to specifically isolate these cells and are linked to patterns of glycosylation on the cell surface.

Assessment of the DEP crossover frequency (f_{xo1}) values, cell diameter and plasma membrane capacitance (C_m) values have been reported by Velugotla *et al.* [141] for a group of human embryonic stem cell (hESC) lines. Undifferentiated hESC lines (H1, H9, RCM1, RH1) and a transgenic subclone of H1 (T8) exhibited similar f_{xo1} and C_m values ($14 \sim 20\,\text{mF/m}^2$) that did not allow for their clear discrimination by DEP. However, the differentiation of H1 and H9 to a mesenchymal stem cell-like phenotype resulted in significant increases in their mean C_m values to $41{\sim}49\,\text{mF/m}^2$ in both lines ($p < 0.0001$). BMP4-induced differentiation of RCM1 to a trophoblast cell-like phenotype also resulted in a distinct and significant increase in mean C_m value to $28\,\text{mF/m}^2$ ($p < 0.0001$). The progressive transition to a higher membrane capacitance was also evident after each passage of cell culture as H9 cells transitioned to a mesenchymal stem cell-like state induced by growth on a substrate of hyaluronan. It was concluded that these findings confirm the existence of distinctive parameters between undifferentiated and differentiating cells on which future application of DEP in the context of hESC manufacturing can be based.

Adams *et al.* [142] characterized the DEP frequency-responses of native human mesenchymal stem cells (hMSCs) using quadrupole electrodes (see Figure 10.4) for electrical frequencies in the range 10 kHz to 35 MHz. Measurements were performed for two dextrose buffer solutions of conductivities 30 mS/m and 100 mS/m. The data obtained for the native cells was compared with those obtained for hMSCs that had been morphologically standardized to a spheroidal shape using an elastinlike polypeptide-polyethyleneimine copolymer. Computer-aided image analysis was employed to identify the DEP crossover frequency (f_{xo1}) by interpolation. Estimated values for f_{xo1} of 0.62 MHz and 1.3 MHz were obtained for the native cells at conductivities of 30 mS/m and 100 mS/m, respectively. With all other factors (e.g., cell size and shape, plasma membrane capacitance) remaining constant, then according to 'standard' DEP theory given by Equation (10.39) the value of f_{xo1} should vary in direct proportion to the medium conductivity. Based on this assumption the 100 mS/m value of f_{xo1} should have been ~2 MHz (rather than 1.3 MHz), implying that the product RC_m in Equation (10.39) increased when the medium conductivity was raised from 30 mS/m to 100 mS/m. Adams *et al.* [142] obtained best fit values a plasma membrane permittivity factor ε_{mem} of 0.79 and 1.1 at 30 mS/m and 100 mS/m, respectively, indicating that indeed the membrane capacitance had increased. This permittivity factor is given by the relationship [141]:

$$\varepsilon_{mem} = \frac{C_{mem} d}{4\pi R^2 \varepsilon_o} \tag{11.4}$$

and so mirrors Equation (11.3) but with C_{mem} representing the total capacitance (farads) of the cell, rather than the specific capacitance (F/m^2). The corresponding values derived [142] for C_{mem} are 2.2 pF and 4.5 pF for medium conductivities of 30 mS/m and 100 mS/m, respectively. Adams *et al.* [142] do not cite the sizes of their cells, but assuming a membrane thickness d of 7 nm (see Table 9.2) and using the 30 mS/m values $\varepsilon_{mem} = 0.79$, $C_{mem} = 2.2\,\text{pF}$, from Equation (11.4) the mean cell radius R is calculated to be 13.2 μm. This is a cell size twice that of the mesenchymal-like stem cells studied by Velugotla *et al.* [141]. The specific membrane capacitance (assuming a spherical shape) corresponding to a total cell capacitance of 2.2 pF can thus be estimated as $(2.2 \times 10^{-12})/(4\pi R^2) = 1\,\text{mF/m}^2$. This is a very low value, well below that of $\sim 5\,\text{mF/m}^2$ estimated for a spherical cell with a perfectly smooth membrane [143]. A value of $\varepsilon_{mem} = 0.050$ was obtained for the copolymer-treated hMSCs of spherical shape, which leads to an even lower specific membrane capacitance. Taken as they stand, the DEP-frequency characteristics obtained by Adams *et al.* [142] are of value and interest. Images provided of the cells undergoing DEP between the quadrupole electrodes indicate a high cell packing density, with many cells in contact. So, as a cautionary note, it is possible that cell-cell interactions may have influenced the DEP behaviour of the cells, as well as the determination of their sizes.

Muratore *et al.* [144] have demonstrated that DEP can be used to discriminate cells between stages of differentiation in the C2C12 myoblast multipotent mouse model. Terminally differentiated myotubes were separated from C2C12 myoblasts to better than 96% purity, a result validated by flow cytometry and Western blotting. To determine the extent to which cell membrane capacitance, rather than cell size, determined the DEP response of a cell, C2C12 myoblasts were co-cultured with GFP-expressing MRC-5 fibroblasts of comparable size distributions (mean diameter ~10 μm). A DEP sorting efficiency greater than 98% was achieved for these two cell types, a result concluded to arise from the fibroblasts possessing a larger membrane capacitance (~11.5 mF/m^2) than the myoblasts (~7.3 mF/m^2). The finding by Raman spectroscopy that the fibroblast membranes contained a smaller proportion of saturated lipids than those of the myoblasts was taken to imply that changes in membrane chemistry, as well as surface morphology, should be taken into account when considering changes of the plasma membrane capacitance. In further work [145] analysis of the cell size and 'smoothness' by light microscopy and scanning electron microscopy, respectively, showed that there were differences between the cell types. C2C12 myoblasts and induced myotubes mainly differed in size, whereas fibroblasts and C2C12 were found to have very different plasma membrane microvilli arrangements, with fibroblasts having apparently many more microvilli. This difference in the cell surface could have been responsible for the observed changes in membrane capacitance and crossover frequency. No statistical differences were found between the stage of the cell cycle before and after DEP sorting. If there had been a change in size during stages of cell cycle, it was considered insufficient to account for the high-efficiency of cell separations reported earlier [144].

11.3.4 Neurons

The first investigations of the DEP behaviour of isolated neurons appear to be those reported by Heida *et al.* [146]. Cortical neurons isolated from rat fetuses were suspended at 10^6 neurons/ml in a medium of conductivity 1.6 S/m. 20 μl of this cell suspension was pipetted into a 4 mm ring placed around the centre of quadrupole electrodes that had a gap of 100 μm between opposing electrodes. The number of neurons 'trapped' by negative DEP within the central region of the quadrupole microelectrode structure was determined for two different amplitudes (3 V and 5 V) and six different frequencies in the range from 1 MHz to 18 MHz. This yield of trapped cells was taken to give a measure of the DEP force as a function of the amplitude and frequency of the applied field. The experiments were time lapse recorded by taking an image every 10 s for a time period of 30 min. However, a contradictory trend was considered to be found for the yield of trapped neurons for the two voltage amplitudes as a function of frequency. Increasing the frequency resulted in a decrease in the number of cells trapped at an amplitude of 3 V, whereas with an amplitude of 5 V an increase in the number of cells trapped was observed as the frequency was increased. Some mechanism was considered to be present that inverted the behaviour of the cells and / or the medium when the voltage amplitude was increased. It was suggested that their calculation of the Clausius–Mossotti factor might not give 'a good representation of reality due to the incomplete knowledge of the electric properties of neurons'. On the basis of the single-shell model of a cell and a calculation of the complex form of the Clausius–Mossotti factor given by Equation (6.1), the neurons were predicted [146] to exhibit negative DEP throughout the frequency range investigated. The following parameter values were assumed for the neuron: radius 5 μm; cytoplasm conductivity 0.75 S/m; cytoplasm relative permittivity 80; membrane capacitance 1.8 mF/m^2. With this very low value of membrane capacitance, Equation (10.39) predicts the DEP crossover frequency (f_{xo1}) to be 40 MHz. However, assigning a more realistic membrane capacitance value of (say) 18 mF/m^2 gives $f_{xo1} = 4$ MHz. Heida *et al.* [146] were probably operating just below f_{xo1} where the negative DEP force would be weak and also decrease with a small increase of frequency. Coupled with thermally induced fluid flow effects in their high conductivity medium, this may have influenced the observed motions of the cells. In fact, in later work where measurements were made between 10 kHz and 50 MHz, Heida *et al.* [147] concluded that the amplitude-dependent frequency behaviour and unexpected DEP-spectra peaks above 1 MHz were induced by heating of the medium, especially for larger voltage amplitudes and higher frequencies. Thermal effects are commonly observed in DEP studies, but not so common is such honest reporting of them. Heida *et al.* [148] also investigated the viability of postnatal cortical rat cells that were trapped within quadrupole electrodes by DEP using a 14 MHz, 3 V voltage signal. Morphological characteristics as well as the ratio of the number of outgrowing to nonoutgrowing cortical cells were used to compare the viability of trapped cells to that of cells in the same environment but without the imposed nonuniform electric field. The morphological characteristics included the area of the cell, representing adhesive properties and the number and length of the outgrowing processes, as a measure for functional recovery. It was concluded that no difference could be detected in the viable state of DEP-trapped postnatal cortical rat cells and those not exposed to the electric field.

To facilitate the *in vitro* extracellular recording of the electrical activity of individual neurons and as the first stage in forming a neuronal network, Prasad *et al.* [149] separated neurons from glial cells and positioned them into DEP traps in a 4 × 4 microelectrode array. To estimate the final locations of the cells over the electrode array, the electric field distribution was determined using three-dimensional finite element modelling. Dissociated neurons and glial cells from hippocampi were suspended at 2500 cells/ml into DEP separation media suitable for achieving long term cell viability. This medium comprised minimum essential medium, 10% fetal bovine serum and 5% PBS, with a conductivity of 248 mS/m and pH of 7.4. Positive DEP forces were used to trap the cells over the electrodes. Electrical signals of 8 V_{pk-pk}, 4.6 MHz were employed for the neurons and 152 kHz, 2 V_{pk-pk} for the glial cells.

For the purposes of recording spontaneous and evoked electrical potential signals from isolated neurons located in a multielectrode array, Zhou *et al.* [150] also describe the DEP separation of embryonic mouse hippocampal neurons from glial cells. Glial cells serve important supporting roles for neuronal networks and are always present in dissociated neural cultures. For their specific application, the goal was to actively and exclusively recruit neurons over the stimulation and recording sites. Each multielectrode array consisted of 16 electrodes of diameter 25 μm. A mathematical simulation and analysis was performed, for a range of suspending medium conductivities, to anticipate the DEP frequency at which the neurons and glial cells could be separated. These simulations revealed that positive DEP was not possible in pure cell media (conductivity 1.1 S/M) and so sucrose was added to reduce its conductivity. A 20% cell media composition (212 mS/m) resulted in a measured crossover frequency (using quadrupole electrodes) of 750 kHz for the neurons and 500 kHz for the glial cells. The dielectric parameters employed in the single-shell modelling of the DEP behaviour of neurons and glial cells are shown in Table 11.3. The simulations agreed well with the experimental observations, especially with respect to predicting the DEP crossover frequency. It is of interest to note that the dielectric parameters for the glial cells required significant modification from those previously reported in the literature, especially regarding the effective conductivity of the cytoplasm and the capacitance of the plasma membrane. From their patch-clamp measurements of cortical pyramidal neurons, spinal cord neurons and hippocampal neurons, Gentet *et al.* [151] obtained a membrane capacitance value of 9 mF/m^2 for each class of neuron. The value required to accurately model their observed DEP experiments is 12 mF/m^2. This can be reconciled by the fact that a patch-clamp determination corresponds to that of a small selected area of the membrane, whilst a DEP experiment is sensitive to the average capacitance over the whole cell area and to the contributions made by the presence of morphological features such s membrane folds and blebs.

Table 11.3 Dielectric parameters employed by Zhou *et al.* [150] in their single-shell model of the measured DEP characteristics of hippocampal neurons and glial cells.

Parameter	Neurons	Glial cells
Radius	4 μm	6 μm
Cytoplasm conductivity (S/m)	0.65	0.35
Cytoplasm permittivity (F/m)	$80\varepsilon_o$	$80\varepsilon_o$
Membrane capacitance (mF/m^2)	8.0	12.0

11.3.5 Spermatozoa

Fuhr *et al.* [152] have described a method for bringing individual spermatozoa to a defined position for characterization followed by sampling with capillaries. The sperm cells were suspended in standard *in vitro* fertilization (IVF) medium, in either a planar four microelectrode 'field funnel', or within a three-dimensional cage created by an octopole electrode system. In these electrode systems, rapidly swimming spermatozoa could be trapped under the action of a negative DEP force at MHz frequencies for just a few seconds, but some spermatozoa stopped moving if exposed to field strengths exceeding 50 kV/m. However, in stripwise and interdigitated structures with electrode gaps of less than 40 μm, rapidly swimming sperm cells could be positioned quite accurately in front of a so-called 'break electrode' by a combination of electric field trapping and field induced fluid streaming.

Boon and Marcos [153] have theoretically explored the use of DEP to sort and select spermatozoa for male and female gender preselection in artificial fertilization techniques. Techniques already developed include flow cytometry where X- and Y-spermatozoa can be differentiated by virtue of their difference in chromosome composition. It is also possible to separate the X- and Y-spermatozoa into two different layers, when then are passed through human serum albumin solution. Theoretical modeling was performed of the variation in the time-averaged velocity of spermatozoa when located in a field gradient [152]. It was found that the velocity can be increased or decreased significantly. More notably, the X- and Y-spermatozoa respond to the DEP force to different extents because of their size variations. Therefore, a high-gradient electric field should result in a notable difference between the swimming velocities of the X- and Y-spermatozoa. It was suggested that the medical field could use this technique to increase the chances of selecting spermatozoa containing chromosomes for a particular gender, thereby advancing gender selection in artificial fertilization techniques.

11.3.6 Bio-Mechanical Properties

Although the ways in which a cell responds to an externally applied physical force may depend on its type and physiological state, it is the cell's cytoskeleton that mainly determines the extent of this response. The cytoskeleton spans almost the entire cell and is built of protein filaments of three main types, namely actin, intermediate filaments and microtubules. Depending on their organization and density these filaments determine the shape and the rigidity of the cell and are also actively involved in cellular processes such as intracellular transport of organelles and cytokinesis. It is intimately connected to the global physiological status of the cell, unlike molecular markers, which are linked to a particular biochemical process. The screening of cells based on both their expression of molecular markers and mechanical response to an external force may have useful application in various areas of biomedical research, drug discovery and diagnostics. The possibility that DEP forces can be employed in this manner is worthy of investigation.

Early studies to demonstrate that DEP forces, generated within a microelectrode device, are capable of characterizing the mechanical properties of cells were described by MacQueen *et al.* [154]. They used the 'fringing' fields between adjacent planar electrodes in a microelectrode array, which penetrated into a cell suspension dispensed as droplets over the electrodes. Electrical stresses (i.e., DEP forces) generated by the planar microelectrodes were used to trap and stretch cells, while cell deformation was observed using optical microscopy. The strain and relaxation of two distinct cell types were investigated, namely Chinese hamster ovary (CHO) cells, which are adherent epithelial cells; and U937 human promonocytes, which are nonadherent. The mechanical properties of these cell types are relevant to their use in suspension cultures since CHO cells are used in large-scale bioreactors for the production of recombinant proteins and U937 cells are used to study differentiation along the monocyte–macrophage pathway. To demonstrate the potential use of electrodeformation for the mechanical characterization of these two cell types, strain and relaxation data were fitted with a three-parameter 'standard linear solid' model of visco-elasticity, as well as with a two-parameter power-law method. The CHO cells were found to be approximately twice as stiff as U937 human promonocytes. The CHO cells also displayed an elastic behaviour with recovery of initial shape, while U937 strain data exhibited plastic deformation. Chen *et al.* [155] performed similar experiments to characterize the mechanical properties of SiHa and ME180 cells (two cervical cancer cell lines). The cells were placed between two microelectrodes fabricated on ITO coated glass slides, with rectangular voltage waveforms applied at one of three frequencies (500 kHz, 1 MHz and 5 MHz) and voltage magnitudes between 4 V and 24 V. Cell deformation was recorded using computer-aided image analysis. Numerical simulations were performed to model cell electro deformation based on the Maxwell stress tensor formulation. Because exact electrical properties of SiHa and ME180 cells were not known, these simulations explored the theoretical effects on the cell's electro-deformed behaviour across the wide range of cell dielectric parameters reported in the literature. The ranges of cell electrical properties tested were as follows: membrane relative permittivity 10, 20 and 30; cytoplasm relative permittivity 40, 80 and 120; cytoplasm conductivity 0.1, 0.4 and 0.7 S/m. By comparing the measured morphological changes with those obtained from numerical simulations, Chen *et al.* [155] were able to quantify Young's modulus of SiHa cells (601 ± 183 Pa) and ME180 cells (1463 ± 649 Pa). These values were consistent with Young's modulus values (SiHa: 400 ± 290 Pa and ME180: 1070 ± 580 Pa) obtained from their own determinations using conventional micropipette aspiration.

Guido *et al.* [156] stretched cells through the application of DEP forces. The cells tested were of two types, namely human breast adenocarcinoma cells (MCF-7) and human nontumorigenic epithelial cells (MCF-10A) derived from benign breast tissue with fibrocystic disease. For the DEP stretching experiments the cells were suspended in an isotonic, commercially available aqueous polymer solution that effectively reduced undesirable nonspecific adhesion of the cells to the microchip surfaces. The electrical conductivity of this solution was adjusted to 5 mS/m, without altering its isotonicity, by mixing with 0.3 M inositol solution and phosphate-buffered saline. The DEP force was produced using a planar arrangement of two microelectrodes, each of width 250 μm and separated by a gap of 20 μm, fabricated by structuring indium tin oxide-plated glass slides using pulsed laser ablation. Fluidic access was achieved by mounting a silicone trough onto the processed slides, forming a sample volume of roughly 200 μl. The cell stretching experiments were performed by first applying a square waveform signal (15 MHz) at 2 V_{rms} to the electrodes. The cells were trapped by positive DEP at the electrodes – but this voltage was too low to induce any visible deformation. Visible stress was generated by increasing the voltage to 6 V for 60 s and the relaxation process was recorded for another 60 s. The DEP force acting on the cells was calculated to be ~56 Pa. The response curves indicated the strain response of the noncancerous (MCF-10A) cells was approximately two and a half times stronger than that of the cancerous (MCF-7) cells derived from a human adenocarcinoma. The noncancerous cells appeared considerably softer than the related cells from the cancer cell line. The mean volume of MCF-7 cells (1920 μm^3) was 47.4% larger than that of the MCF-10A cells (1303 μm^3). Therefore, the MCF-7 cells experienced the larger DEP force, but were stretched less. This indicated that the observed difference in strain was not due to the difference in cell size. The differences in the deformation response between the cancerous and noncancerous cell types were considered to be caused by the structural architecture of their cytoskeletons. To discover which part was mainly responsible, Guido *et al.* [154] performed stretching experiments in the presence of cytoskeleton-active toxins. Both cell types were treated with either latrunculin A or colchicine, which are inhibitors of actin and microtubule polymerization, respectively. After the latrunculin A treatment, both MCF-7 and MCF-10A cells appeared considerably softer (i.e., the strain increased. Although the strain of MCF-7 cells increased by approximately 110% and that of MCF-10A cells by 65%, the MCF-7 cells still remained stiffer than the MCF-10A cells. In contrast, the colchicine treatment resulted in a softening of both cell types, such that the responses of MCF-7 and MCF-10A cells to stretching became indistinguishable. Their interpretation of these results was that differences in microtubule structures between the two cell types were primarily responsible for the different deformation responses of the two cell types.

Artificial tissues or organs for implant as a repair or replacement involves the initial seeding of appropriate cells onto a biocompatible scaffold. The outcome of this procedure relies on successful *in vitro* cell proliferation and differentiation and in turn this requires that the cultured cells adhere well to the scaffold. An improved understanding of cell adhesion will not only benefit tissue engineering, but also aspects of cell migration, tumour growth and metastasis. Lin *et al.* [157] used DEP forces to determine the adhesion characteristics of bovine endothelial cells that had been seeded onto polydimethylsiloxane (PDMS) and polylactide (PLA) substrates. The cultured cells on their substrate were removed from the incubator, the culture medium was replaced with the selected DEP medium and the cell culturing substrate was overturned and placed on top of DEP electrodes. The electrodes took the form of two parallel strips, one with a flat surface and the other with triangular- or rectangular-shaped features for producing the nonuniform field. The spatial characteristics of the electric field were simulated using commercial finite element analysis software (COMSOL). The closest distance apart of the electrodes was 40 μm, chosen because the initial diameter of the endothelial cells was 10~15 μm, increasing to 35~40 μm after 24 h of seeding. Voltage potential differences at 200 kHz were applied, scanning from 400 mV to 50 V, whilst examining the deformation and further detachment of cells due to the effects of DEP. (The term 'electrophoresis' rather than 'dielectrophoresis' is employed in parts of this paper, but DEP is implied throughout.) Contact angle determinations of hydrophilicity indicated that PLA is more hydrophilic than PDMS, suggesting that the endothelial cells should have better adhesion on PLA than on PDMA. However, the cell detachment results did not fully agree with this inference. An applied DEP force was not able to detach the adhered cells from a PDMS substrate when the seeding time was longer than 4 h. However, the DEP force produced by a 6 V applied potential was enough to lift endothelial cells cultured on the PLA substrate after four and six hours of culture. When the culture time was increased to 8 h, the cells apparently stretched out and a higher voltage was required to lift and move them. These results indicated that the adherence of bovine endothelial cell to PLA was more stable after 8 h of seeding. It was concluded that other features of the substrate were considered to be more crucial for the adhesion of the cells than its hydrophilicity.

Haque *et al.* [158] describe a combined application of a DEP force and a single-beam optical tweezer for determining the stretching stiffness of single human erythrocytes. A particular advantage of this approach is the lack of mechanical contact with the cells. When cells are trapped in a medium of conductivity 124 mS/m at pH 7.3 by the combination of DEP and laser-tweezer forces, the two gradient forces keep balance with each other and deform the trapped cells. From this the elastic shear modulus of the cell membrane can be determined. The DEP force was applied between electrodes of triangular shape at 10 MHz, a frequency high enough to avoid conductivity losses, but below the second DEP crossover frequency (f_{xo2}). With assumed dielectric parameters for the erythrocytes (cytoplasm conductivity 0.312 S/m, membrane thickness 8 nm, medium and cytosol relative permittivities of 80 and 60, respectively) the real part of the Clausius–Mossotti factor was calculated to be +0.8. By measuring the geometrical parameters of single healthy human erythrocytes as a function of the applied voltage, the elastic modulus of the erythrocytes was determined to be 1.80 ± 0.5 μN/m. This result is within the range (1~10 μN/m) reported from dynamic membrane fluctuation measurements and also for values of the shear modulus of the membrane and the underlying spectrin skeleton (1.4–2.5 μN/m) reported by independent laser tweezers experiments using silica microbeads attached to the cell membrane. Given its sensitivity and the lack of mechanical contact with the cells, Haque *et al.* [158] considered the method an easy-to-use tool for determining the mechano-elastic properties of living cells. It could prove useful in distinguishing between healthy and diseased cells, or monitoring the effects of aging, both of which are crucial for the preservation of blood for medical use and forensic applications.

Finally, Zhang *et al.* [159] investigated the stretching by a DEP force of human acute promyelocytic leukemia (NB4) cells before and after treatment with all-trans retinoic acid, a drug used in the treatment of acute promyelocytic leukaemia. The cells were trapped between parallel electrodes, spaced 20 μm apart, patterned onto an ITO layer in a PDMS microfluidic chip. A low voltage was used to first immobilize the cells at an electrode edge, followed by an increase of the voltage to stretch the cell. Elongation of the cell was observed using an optical microscope and the results showed that the untreated and treated cells were both deformed by the induced DEP force. After 180 s of applied DEP force, the strain of the untreated NB4 cells was recorded to be ~0.08, compared to that of ~0.21 for the treated cells. This indicated that a decrease in the stiffness of the cell occurred after drug treatment. The elastic modulus of the cell was also evaluated and the modulus changed from 140 Pa to 41 Pa after drug treatment. Zhang *et al.* [159] point out that the DEP force acting on the cells is correlated to unique properties, such as cell size and the dielectric properties of the cells and the medium and so it is difficult to perform force calibration with other specimens. They employed the standard linear solid model to estimate the modulus of the cell and the cell was assumed to be a viscoelastic solid body under the DEP stretching force. This model is considered to accurately estimate the biomechanical properties of leukocytes, for example [160]. Nevertheless, other reports have also indicated that the liquid droplet model can better characterize the plasma membrane and the cytoplasmic viscosity of leukocytes under persistent tension [e.g., 161]. Although several models have been proposed and examined, these models are mostly derived based on the assumptions or experimental conditions for a particular cell deformation technique, which may not be applicable to other cell deformation techniques. Zhang *et al.* [159] suggest that further investigation should be conducted to examine the elastic modulus for the same cell type with different techniques in order to obtain a more comprehensive data for comparison. They also provide a useful table of the biomechanical properties of different cell types derived using different experimental methods (micropipette, atomic force microscopy, optical indentation, optical stretching, magnetic cytometry) including DEP [154, 155].

11.4 Bacteria

Representative citations of the literature on the DEP properties of bacteria are given in section 11.2.4 of this chapter. The following citations describe works that were not specifically directed towards identifying the differences between live and dead bacteria.

Markx *et al.* [162] investigated the DEP behaviour of various bacteria when suspended between polynomial, quadrupole, electrodes at various values of the medium conductivity, over the frequency range 1 kHz to 10 MHz. The DEP-frequency responses for all the bacteria types investigated exhibited a 'plateau' commencing at frequency between 10 ~ 100 kHz, which extended up to ~10 MHz where the electric field was assumed to begin to penetrate into cytosol. In this plateau range of frequencies the conductivity parameters (rather than the permittivity parameters) of the cell wall and membrane(s) were assumed to dominate the complex expression for the Clausius–Mossotti factor. The effective conductivities derived for the bacteria are given in Table 11.4 at a frequency that marked the lower end of the DEP-frequency plateau.

Inspection of Table 11.4 indicates that the effective particle conductivities of the various bacteria that were investigated covered a broad range. There is a slight tendency among the 14 species investigated for the Gram positive bacteria to exhibit the larger conductivity, but many more bacterial types need to be investigated before such a trend can be confirmed. However, in many cases the separation of many bacteria types from each other by DEP should be possible. The essential requirements are a suspending medium of conductivity value between the effective conductivities of the two species, with an applied voltage frequency intermediate between those used to determine the effective particle conductivity. An example of this is shown in Figure 11.22, for the DEP separation of *E. coli* (experiencing negative DEP and 'caged' at the centre of a polynomial electrode system) from *M. lysodeikticus* collected at the electrode edges by positive DEP. The suspending medium was 280 mM mannitol adjusted to a conductivity of 55 mS/m with NaCl and a 4 V_{pk-pk} voltage signal at 100 kHz was employed. Separations of different bacteria types were

Table 11.4 Effective conductivity of some Gram-positive and Gram-negative bacteria derived from measurements at a frequency where the dielectric properties of the cell wall and membrane(s) are considered to dominate their DEP responses (Markx *et al.* [162]). Yeast (*S. cerevisiae*) is added for comparison.

Species	Gram stain	Conductivity (mS/m)	Frequency (kHz)
Acinetobacter calcoaceticus	Negative	< 2.0	25
Erwinia carotovore	Negative	2.0 ± 0.9	10
Lactobacillus brevis	Positive	15.3 ± 1.8	25
Pseudomonus putida	Negative	19.5 ± 1.4	10
Enterococcus faecalis	Positive	23.0 ± 2.1	10
Escherichia coli	Negative	41.2 ± 2.5	100
Lactococcus lactis	Positive	41.6 ± 4.6	25
Agrobacterium tumefaciens	Negative	45.2 ± 5.0	100
Klebsiela rubiacearum	Negative	51.3 ± 10.2	10
Bacillus megaterium	Positive	88.8 ± 3.9	10
Bacillus subtilis	Positive	93.5 ± 9.6	10
Rhodobacter sphaeroides	Negative	96.7 ± 5.3	15
Pediococcus damnosus	Positive	101.8 ± 6.8	10
Micrococcus lysodeikticus	Positive	155.7 ± 11.6	15
(*Saccharomyces cerevisiae*)		1.6	10

also [performed using other electrode designs, including interdigitated, castellated, microelectrode arrays [162].

A bacterial enumeration system is described by Brown *et al.* [163] that is based on the fact that the amount of particles collected by positive DEP on an array of microelectrodes is influenced by the number of those particles passing near the electrodes in a given time period.

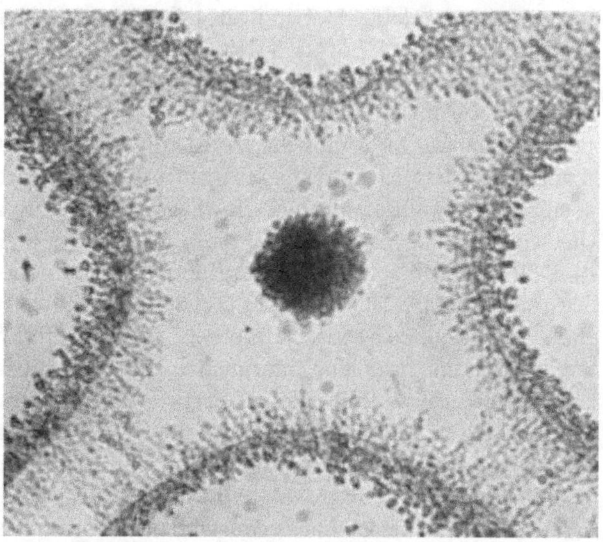

Figure 11.22 Separation of *E. coli* (collected in the centre by negative DEP) from *Micrococcus luteus* that are collecting at the edges of quadrupole electrodes (unpublished image from Markx *et al.* [162]).

Increasing the cell concentration in a particle suspension should result in increased levels of collection at electrode surfaces. Consequently, the relationship between DEP collection and concentration determined for a range of microbiological particles should be useful in the rapid evaluation of a comparable suspension of unknown particle concentration. Such relationships were observed between suspension concentrations and the extent of DEP collection for polystyrene latex beads, pure bacterial samples, as well as mixtures of bacterial species including *Escherichia coli, Serratia marcescens, Pseudomonas aeruginosa* and *Bacillus subtilis.* A similar relationship was used for polystyrene latex as a calibration line to enable the concentration of particles in a suspension to be determined according to the level of DEP collection. The particle concentration of an unknown test sample was found to lie within the predicted concentration range determined on the basis of DEP collection. In addition, the predicted limits were found only to deviate between −6.2 and +6.9% from the mean particle concentration.

Castellarnau *et al.* [164] report on an experimental method based on dielectrophoretic analysis to identify changes in four *E. coli* isogenic strains that differed exclusively in one mutant allele. The DEP properties of wild-type cells were compared to those of *hns, hha* and *hha hns* mutant derivatives. The *hns* and *hha* genes code respectively for the global regulators Hha and H-NS. The Hha and H-NS proteins modulate gene expression in *E. coli* and other Gram negative bacteria. Mutations in either *hha* or *hns* genes result in a pleiotropic phenotype. A two-shell prolate ellipsoidal model was used to fit the experimental data, obtained from DEP measurements and to study the differences in the dielectric properties of the bacterial strains. The experimental results indicated that the mutant genotype can be predicted from the DEP analysis of the corresponding cultures, opening the way to the development of microdevices for specific bacteria identification.

Following on from their previous descriptions of a DEP-based bacteria detection system [76, 77] Suehiro *et al.* [165] have described a selective detection method for specific bacteria by using a DEP impedance measurement method in conjunction with an antigen–antibody reaction. Antibody molecules were immobilized on the electrodes, so that the bacteria were attracted to the electrode gap under the action of positive DEP and finally brought into contact with the glass surface to be bound with the immobilized antibody. By appropriately adjusting the balance between the DEP force and the drag force caused by liquid flow, the target bacteria (*E. coli*) could be selectively retained on the chip surface, so avoiding undesired nonspecific binding. The retained bacteria were electrically detected by their impedance system. It was also confirmed that the proposed method was able

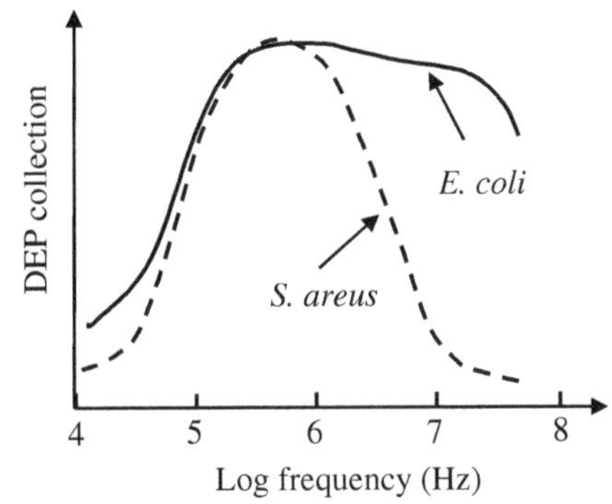

Figure 11.23 DEP collection spectra for Gram negative *Escherichia coli* and Gram positive *Staphylococcus aureus* (based on Sanchis *et al.* [166]).

to selectively detect the target bacteria from a mixed suspension of different bacteria types.

Automated measurements of DEP collection spectra of *E. coli* and *Staphylococcus aureus* suspensions were used by Sanchis *et al.* [166] to characterize the dielectric properties of these two quite different bacteria. *E. coli*, a Gram negative bacteria, was modeled as a three-shell rodlike particle with length 3.33 μm and width 0.82 μm, composed of cytoplasm, a plasma membrane of thickness 8 nm, a wall of thickness 15 nm and an outer membrane of thickness 8 nm. *S. aureusis* (Gram positive) was modelled as a spherical particle of radius 0.6 μm with two shells – a plasma membrane of thickness 8 nm and a wall of thickness 20 nm. The results of the modelling matched closely their experimentally obtained DEP-frequency spectra and these results are shown in Table 11.5. The effect of the extra outer membrane for the Gram negative *E. coli* in giving rise to the high-frequency 'shoulder' is clearly seen in the schematics of the DEP collection spectra shown in Figure 11.23, mirroring the profile shown for the Gram negative *K. pneumoniae* shown in Figure 11.11. In Table 11.5 the dielectric parameters obtained by Sanchis *et al.* are compared to those previously reported in the literature. Close agreement is found with the results of Suehiro *et al.* [77]. but not so closely with those of Johari *et al.* [167] – particularly regarding the derived values for the permittivity of the plasma membrane and the conductivity of the cell wall for the closely related species *S. aureusis* ($\varepsilon_r = 4.5$, $\sigma_w = 0.3$ S/m [166]) and *S. epidermidis* ($\varepsilon_r = 16$, $\sigma_w = 0.01$ S/m [167]), respectively.

11.5 Other Cell Types (Plant, Algae, Oocytes, Oocysts) and Worms

Kaler *et al.* [168–170] developed a dual-frequency excitation technique that uses feedback-control of a DEP force to levitate particles and to investigate both positive and negative DEP characteristics in the frequency range from ~1 Hz to ~ 50 MHz. Rapeseed (canola) and tobacco plant protoplasts, obtained from 6 to 8 weeks old plant tissue subjected to enzymatic digestion were investigated using this method. The upper and intermediate DEP force frequency spectra obtained were generally consistent with the conventional understanding of DEP, but below a certain frequency the protoplasts exhibited an anomalous positive DEP response. Examples of this for a single tobacco protoplast [170], the DS19 clone of Friend murine erythroleukaemic cells [114] and a suspension of *M. lysodeikticus protoplasts* [171] are shown in Figure 11.24. The low-frequency DEP response is considered anomalous because the Clausius–Mossotti factor is not permitted to decrease and then increase again as the frequency of the electric field is changed [172]. The frequency at which this anomalous crossover occurred was found to be: linearly related to suspension conductivity; virtually independent of the suspension pH; inversely proportional to the square of the cell radius. This behaviour had already been observed for mammalian cells and bacteria using conventional DEP studies, having been attributed to surface charge effects (e.g. [114, 117, 171]). Examination of the complex Clausius–Mossotti

Table 11.5 Dielectric parameters determined by Sanchis *et al.* [166] for *E. coli* and *Staphylococcus aureus* by fitting DEP collection spectra to compartmented spherical or rodlike particles. Their results are compared with values reported by Suehiro *et al.* [77] and Johari *et al.* [167].

	E. coli				***S. aureus***			
	Suehiro [77]		Sanchis [166]		Johari [167]		Sanchis [166]	
Compartment	$\varepsilon/\varepsilon_o$	σ (S/m)	$\varepsilon/\varepsilon_o$	σ (S/m)	$\varepsilon/\varepsilon_o$	σ (S/m)	$\varepsilon/\varepsilon_o$	σ (S/m)
Cytoplasm	60	0.1	70 ± 10	0.07± 0.01	–	0.2	70±10	0.8 ± 0.05
Plasma membrane	10	5E-8	10 ± 5	<5E-6	4.5	5E-8	16 ± 0.5	<1.5E-6
Cell wall	60	0.5	60	0.5 ± 0.01	–	0.3	60 ± 5	0.01
Outer membrane	–	–	10 ± 1	(2 ± 1)E-6	–	–	–	–
Suspending medium	80	2E-4	80	2.7E-3	80	1E-3	80	2.5E-3

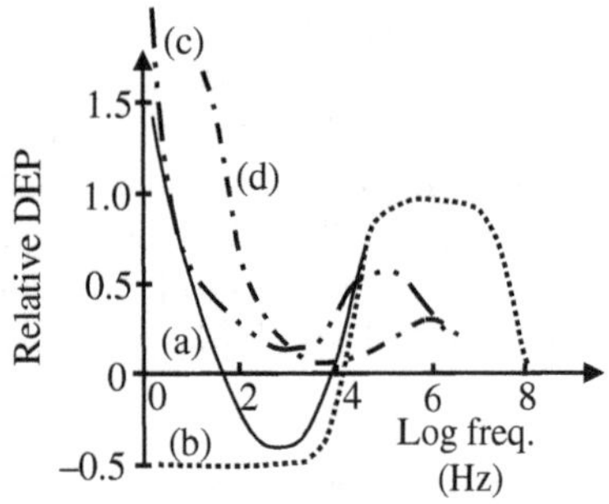

Figure 11.24 (a) DEP spectrum of a single tobacco plant protoplast. The anomalous positive DEP response occurs below ~ 1 kHz [170]. (b) Predicted DEP response of the tobacco protoplast based on the standard multishell model of a cell [170]. (c) DEP spectrum of erythroleukaemia cells [114]. (d) DEP spectrum of *Micrococcus luteus* protoplasts [171].

polarization coefficient revealed that the observed positive DEP response could not be accounted for in terms of Maxwell–Wagner polarization associated with a conventional multishell model for the protoplast [170]. The failure of straightforward enhancements to the protoplast model in explaining the low frequency behaviour was considered to indicate the presence of an electrophoretic contribution to the net observable force on the particle and hence upon the net particle charge [173]. A hysteresis effect was also observed in the low-frequency response of levitated cells. During experiments on tobacco protoplasts the direction of the frequency scan was reversed, with the result that the Clausius–Mossotti factor appeared to be double valued for each value of frequency. This effect was analysed analytically by Paddison *et al.* [174] by introducing a nonlinear dielectric displacement vector at the interface of the cell surface and the surrounding medium. A possible cause of the hysteresis effect was suggested in terms of the motion and relaxation of transmembrane proteins under the influence of the local electrical field at the cell surface [174].

The electrokinetic characteristics of single cells derived from Neurospora slime were studied by Gimsa *et al.* [175] using a technique that enabled simultaneous measurements of their DEP and electrorotation properties in the same fluidic chamber. The membrane conductivity and capacitance of these cells was determined to be 500 S/m2 and 8.1 mF/m^2, respectively.

Based on its biochemical conversion sustainability, social and ecological costs, the use of the lipid content of microalgae cells as a biofuel is becoming increasingly important [176, 177]. Marine microalgal species are highly efficient in transferring solar energy into energy-rich compounds and their cultivation has a low requirement in arable land and freshwater. When microalgae are cultured under suboptimal conditions, as for example under high intensity light, low temperature and limited carbon in their nutrients, they accumulate neutral lipids in their cytoplasm as an energy store. Deng *et al.* [178] have investigated the influence of the solution conductivity on the DEP crossover frequency of microalgal cells (*Chlorella vulgaris*) with different lipid contents. This species has a single spherical cell, with a diameter in the range 2–10 μm and no flagella. The conductivity of the suspending medium was adjusted using different concentrations of KH_2PO_4 buffer solution. Algal cells with 11 wt% and 45 wt% lipid content were obtained by varying the light exposure and carbon source during their culture. To distinguish the cells of different lipid content, *Nile Red* fluorescence dye was used to label only the microalgae with 45 wt% lipid content. For algal cells with 11 wt% lipid, on increasing the suspension medium conductivity from 140 to 295 mS/m, the value for the DEP crossover frequency presumably (f_{xo1}) increased from ~2 to ~10 MHz. Cells with 45 wt% lipid content exhibited negative DEP at frequencies below 20 MHz when the solution conductivity was within the range 206–295 mS/m. However, positive DEP was observed when the solution conductivity was lowered to 140 mS/m. Successful local separation by DEP of a mixture of algal cells of these two different lipid contents was achieved using an array of parallel gold microelectrodes, a medium conductivity of 295 mS/m and frequency of 20 MHz [178].

Hadady *et al.* [179] monitored the DEP response of the microalgae *Chlamydomonas reinhartii* in the frequency range 20–80 MHz and different cell suspension media over time. This species of microalgae is a green single cell, about 10 μm in diameter with a cell wall, two flagella and an *eyespot* that senses light. It is widely studied for its potential in producing biopharmaceuticals, biofuel and hydrogen. The rationale for studying the DEP response at high frequencies (i.e., above the crossover frequency f_{xo1}) was the expectation that the lipid content in the cytosol would influence the DEP behaviour and especially the value of the second DEP crossover at f_{xo2}. *C. reinhardtii* was cultured in regular medium and under nitrogen-depleted conditions, respectively, in order to produce populations of cells with low and high lipid content, respectively. Relative lipid content was estimated with a fluorescent dye by calculating the area-weighted intensity average of fluorescent images. The microalgal cells grown in nitrogen-deficient medium exhibited an increase in fluorescent intensity average that correlated with increasing lipid concentrations. The value of f_{xo2} of the nitrogen-deficient (i.e., lipid-rich) cells decreased from ~55 MHz at day 5 to ~40 MHz at day 16. In contrast, the cells grown in a nitrogen-rich medium did not exhibit fluorescence when stained, indicating negligible lipid accumulation. The value of f_{xo2} also shifted to lower frequencies over time, but to a lesser extent than the nitrogen-deficient cells (from ~60 MHz at day 5 to

~52 MHz at day 16). Since f_{xo2} is primarily a function of the dielectric properties of the cytoplasm, it was inferred that the cytoplasm of the cells cultured in both the nitrogen-rich and nitrogen-poor media changed during the cultivation period. The cells cultured in the nitrogen-deficient medium were observed to have an f_{xo2} value approximately 13 MHz lower than the nitrogen-rich cells at the end of the experiment, due to the difference in their lipid content. The high-lipid content cells were calculated to have a cytoplasm conductivity of 95 mS/m, compared to 227 mS/m for the high-lipid cells. The conductivity of the suspending medium for these DEP experiments was 10.6 mS/m. To demonstrate the ability to separate microalgal cells on the basis of lipid content, Hadady *et al.* [179] mixed portions of the nitrogen-rich and nitrogen-poor cultures together in the DEP medium. The high frequency DEP response was observed by suspending the cells above chromium-gold sputtered electrodes patterned using standard photolithography into a needle pattern with 100 μm spacing. As expected, at frequencies lower than f_{xo2} for both cell types, the cells exhibited positive DEP. Likewise, at frequencies above f_{xo2} for both cells, they exhibited negative DEP. At intermediate frequencies (i.e., frequencies above the nitrogen-deficient cell crossover and below the nitrogen-rich cell crossover) the two cell types could be spatially separated. To quantify the efficiency of the separation, a region of interest was defined around the electrode. Using UV excitation, high- and low-lipid cells were distinguished by their fluorescence intensity. Image analysis was used to quantify the amount of each type of cell in the region of interest and the ratio of low- to high-lipid cells was used as a figure of merit for the separation efficiency. The separation efficiency exhibited a peak at a figure of merit ~15 for an applied voltage frequency of 41 MHz.

It is of value to note the experimental differences in the separations of the lipid-rich and lipid-poor cells reported by Deng *et al.*, [178] and Hadady *et al.* [179]. Deng *et al.* were not able to generate the high frequencies required to reach the f_{xo2} value reported by Hadady *et al.* They also operated with a medium conductivity (206–295 mS/m) that probably lay between that of the cytoplasmic conductivity of the two cell types. The lipid-rich cells would thus have exhibited negative DEP throughout the whole frequency range, whereas the lipid-poor cells would have exhibited positive DEP at their operating frequency of 20 MHz, which would have been below f_{xo2}. On the other hand, Hadady *et al.* [179] used a suspending medium conductivity that was lower than that of the cytoplasmic conductivity of the lipid-rich cells. Their cell separation depended on the difference between the f_{xo2} values for the two cell types.

The transplantation of insulin secreting islets of Langerhans into the liver circulation of patients affected by type I diabetes offers benefits for their quality of life and may decrease the number of hypoglycaemic episodes [180]. Islets are composed of pancreas cellular aggregates consisting of different cell types, each of them producing particular hormones. The vast majority (~75%) of these cell types are the insulin-producing cells – the beta-type cells. Membrane capacitance and conductance values have been determined for primary β-cells and INS-1 insulinoma cells using DEP and electrorotation with the polynomial and 'bone' design electrodes shown in Figures 10.4 and 10.23 [181]. These measurements were performed in isotonic media of conductivity ranging from 11.5 to 101.4 mS/m and analysed using data plots of the form shown in Figure 10.26. The membrane capacitance value of 12.57 (± 1.46) mF/m^2, obtained for β-cells and the values from 9.96 (± 1.89) to 10.65 (± 2.1) mF/m^2, obtained for INS-1 cells, fall within the range expected for mammalian cells (e.g., see Tables 9.1–9.3). The electrorotation results for the INS-1 cells lead to a value of 36 (± 22) S/m^2 for the membrane conductance associated with ion channels, if values in the range 2–3 nS are assumed for the membrane surface conductance parameter K_s given in Equation (10.64). This K_s value falls within the range reported for INS cells obtained using the whole-cell patch-clamp technique. However, the total 'effective' membrane conductance value of 601 (± 182) S/m^2 obtained for the INS-1 cells by DEP is significantly larger (by a factor of around three) than the values obtained by electrorotation. This could have resulted from an increased membrane surface conductance, or increased passive conduction of ions through membrane pores, induced by the larger electric field stresses experienced by cells in the DEP experiments.

Negative DEP forces, generated by 10 kHz, 4 V_{pk-pk}, voltages applied to micromachined, transparent, indium tin oxide electrodes, were used to condense suspensions of β-cells and insulinoma cells (BETA-TC-6 and INS-1) into a 10 × 10 array of three-dimensional cell constructs [182]. The suspending medium conductivity was 55 mS/m and the nonuniform field was generated between two ITO-coated glass slides placed one above the other. In the upper slide an array of 200 μm diameter holes had been produced by laser ablation of the ITO. Some of the cell constructs, measuring ~150 μm in diameter, 120 μm in height and containing around 1000 cells, were of the same size and cell density as a typical islet of Langerhans. With the DEP force maintained, these engineered cell constructs were able to withstand mechanical shock and fluid flow forces. Reproducibility of the process required knowledge of cellular dielectric properties, in terms of membrane capacitance and membrane conductance, which were obtained from previous results [181] and additional electrorotation measurements. Fluorescent nanosensors were incorporated into these 'pseudo

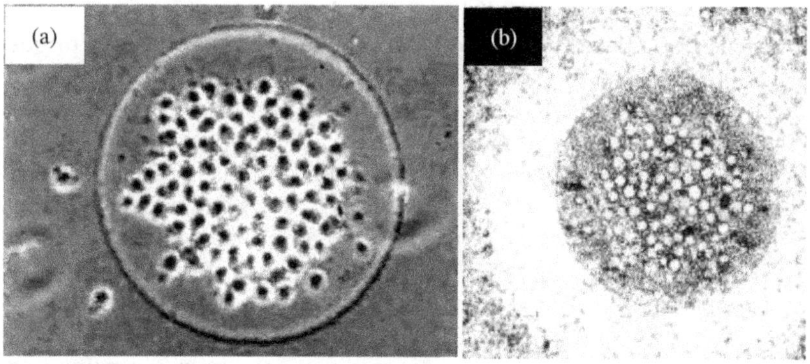

Figure 11.25 (a) BETA-TC-6 insulinoma cells (~10^8 cells/ml) directed by negative DEP into a densely packed assembly within the perimeter of a 100 μm laser-machined hole on an ITO-coated glass slide. (b) Confocal microscope image, taken at a plane 60 μm below the top of an assembly of INS-1 insulinoma cells in which are embedded *PEBBLE* nanosensors for monitoring pH and oxygen levels (from Pethig *et al.* [182], reproduced with permission).

islets' as probes of cellular oxygen and pH levels as a function of glucose concentration. Examples are given in Figure 11.25. The footprint of the 10 × 10 array of cell constructs was compatible with that of a 1536 microtitre plate and thus amenable to optical interrogation using automated plate reading equipment.

The purification of islets from fragments of pancreatic exocrine tissue is a critical stage in the transplantation of insulin secreting islets. Burgarella *et al.* [183] have evaluated the potential of DEP as a contactless method for the isolation of islets. Using the dielectric parameters reported for β-cells by Pethig *et al.* [181], Burgarella *et al.* performed numerical simulations to optimize the exact shape and size of a quadrupole microelectrode configuration suitable for DEP measurements on islets obtained from the pancreas of rats. A Langerhans islet was modeled as an approximately spherical aggregate of hexaogonal close-packed, 10 μm diameter, β-cells with an overall diameter in the range between 50 μm and 250 μm. Using finite element modeling software the effective permittivity ε_p and conductivity σ_p of the islet was determined by considering it as a single body composed of different elements, namely cells and intercellular spaces. The computation was performed by integration over the volume taken up by the aggregate. The real part of the Clausius–Mossotti (*CM*) factor was calculated to have a near constant *negative* value in the frequency range 200 kHz–1 MHz for a suspending medium conductivity of 130 mS/m. The real component of the *CM* value at 1 MHz was −0.2388 (with an imaginary component of +0.0016). This prediction of a negative DEP force at 1 MHz was confirmed by experiments for islets suspended in a serum medium of conductivity 130 mS/m, using an applied voltage of 50 V_{pk-pk} with a gap of 0.5 mm between the quadrupole electrode tips. This was taken as confirmation that Langerhans islets can be manipulated without physical contact by DEP.

Oocytes are diploid cells that develop by meiosis, a process where a single cell divides twice to produce four cells containing half the chromosomes of the parent cell. They are the sex cells – eggs in females and sperm in males. A zygote is a cell that is formed when an egg and sperm combine after successful fertilization of the egg. Oocytes are retrieved from cattle ovaries as a source for embryo *in vitro* production. Influenced by their opinion that the selection of developmentally competent oocytes and zygotes based on their morphology is more often influenced by personal judgments and lacks universal standards, Dessie *et al.* [184] investigated the rate of development and mRNA level of DEP-separated oocytes and zygotes to validate if this offered a noninvasive option for their selection. Experiments were conducted to: (i) evaluate the DEP behaviour of oocytes and zygotes obtained from cows; (ii) investigate the relationship between the DEP mobility of metaphase II (MII) oocytes and zygotes and their developmental competence; (iii) evaluate the size of the blastocyst derived from DEP-separated zygotes; (iv) investigate the DEP behaviour of MII oocytes and zygotes in relation to the mRNA expression to generate candidate genes related to the developmental competence. A single cell at a time was positioned midway between two electrodes of a DEP chamber, in which the suspending medium had a conductivity of 8 mS/m. Depending on the time taken to reach an electrode under the influence of a positive DEP force, the oocytes and zygotes were classified into four DEP mobility categories, namely: very fast, fast, slow and very slow. The low standard error mean revealed the repeatability of measurements of a single cell in all the four DEP categories, with the DEP mobility directly relating to the effective dielectric polarizabiity of a cell. A correlation analysis performed between the diameter of the cell and its DEP mobility revealed there was no such correlation for the oocytes, but for the zygotes there was a tendency (but not statistically significant) to show a negative correlation. At 6 and 7 days of postinsemination, the blastocyst rate of the 'very slow' oocytes was significantly lower than for the other groups. Similarly, for zygotes, the blastocyst rate at 7 days postinsemination was higher in the 'very fast' category when compared with the 'slow' and 'very slow' categories. The mRNA level was analysed for the 'very fast' and 'very slow' oocytes and zygotes, using the bovine cDNA microarray. The result showed that 36 and 42 transcripts were differentially regulated between the 'very fast' and 'very slow' DEP mobility categories of oocytes and zygotes, respectively. It was concluded [184] that DEP-separated oocytes and zygotes exhibited difference in the rate of blastocyst development accompanied by differences in transcriptional abundances, in direct proportion to their effective dielectric polarizabilities.

The fact that travelling wave DEP (TWD) and electrotation can be induced using similar electrical signals implies that they should be capable of integration into a single microelectrode structure. Goater *et al.* [185] described such a device consisting of four spiral electrode elements (see Figure 10.30) and demonstrated that it can be used to concentrate and determine the viability of relatively dilute concentrations of microorganisms. The model microorganism used in this study was the oocyst of *Cryptosporidium parvum*. The presence, even at very low concentrations, of these oocysts in drinking water has led to outbreaks of human infection (cryptosporidiosis), which occurs as self-limiting diarrhoea in healthy adults but may lead to death in infants and immunocompromised people. Because the oocysts do not multiply outside their hosts they cannot be cultured *in vitro*. Current diagnostic techniques are based on the filtration of large quantities of water followed by fluorescence microscopy and can be inaccurate. *C. parvum* oocysts of human origin were washed twice in ultra-pure water, centrifuged and then suspended at various diluted concentrations in phosphate buffered saline solutions. When the four spiral electrode elements were energized with sinusoidal voltages of appropriate phase differences, a travelling electric field was generated that traveled radially from the centre towards the periphery of the spiral array. A rotating electric field was also generated at the same time in the central free-space region between the ends of each spiral electrode element. In the frequency range where a particle exhibited induced TWD motion in the direction of propagation of the travelling field (i.e., was nonviable) it also exhibited co-field electrorotation. Viable oocysts exhibited the opposite behaviour. The TWD and electrorotation behaviour of the *C. parvum* oocysts observed on the spiral electrode structure correlated very well with morphological and fluorogenic vital dye examinations [185]. The device used in these investigations had an effective sample capture area of around 0.1 cm^2, for samples of 10^4 oocysts/ml suspended in a sample volume of 1.45 μl, corresponding to an average 55 ± 1 oocysts being suspended within the field of influence of the spiral microelectrodes. The laser machining procedure for producing the electrodes is capable of high-resolution patterning over large areas, to accuracies of 1 μm over 20 cm. In principle at least, an electrode array capable of handling sample volumes 10/ml is possible.

Finally, the first demonstration that the DEP trapping and manipulation of a whole animal can be achieved was reported by Chuang *et al.* [186] for the case of the nematode *Caenorhabditis elegans*. This transparent roundworm, of length ~1 mm, was the first multicellular organism to have its whole genome sequenced and to have its complete neuronal 'wiring diagram' (connectome) described. It has a relatively short life cycle of 2.5 days and remains a much investigated animal for biomedical research. In many studies, it is necessary to selectively sort the worms and immobilize them for observations, by manually picking up an individual and gluing it to a surface. They can also be immobilized in microfluidic channels and their location controlled pneumatically. The ability to manipulate *C. elegans* with a noncontact force such as DEP could be useful. Chuang *et al.* [186] studied the effect of a nonuniform electric field on *C. elegans* as a function of field intensity and frequency. They identified a range of electric field intensities and frequencies that trap them without apparent adverse effect on their viability. It has been known for some time that *C. elegans* exhibits electrotaxis and crawls or swims towards the negative electrode in a DC electric field [e.g., 187]. At frequencies from 1 Hz to ~3 kHz the worms tend to localize in a small region, probably due to frequent changes in the direction of electrotaxis. Chuang *et al.* investigated the effect of relatively high frequency electric fields up to 800 kHz. In contrast to electrotaxis, the worms were trapped at the location of the maximum electric field intensity and were thus insensitive to the field's polarity. The DEP force acting on live and dead worms was modelled as multishell ellipsoids with and without a membrane, respectively. The live worm was modelled with a biological membrane between the cuticle and internal milieu of the worm, which was considered to filter low frequency electric fields from the worm's interior. In contrast, the dead worm's membrane was inactive in this way due to irreversible permeation. To trap a live worm by DEP, the electrostatic forces must overcome the worm's muscular force. As a result, dead worms were trapped much more readily than live ones, thus enabling DEP as a tool to remove dead worms from the suspension. Worms tethered by DEP exhibited behavioural responses to blue light, indicating that at least some of the nervous system functions were unimpaired by the field. Thus, Chuang *et al.* [186] have demonstrated that DEP is a useful tool to dynamically tether nematodes, sort them according to size and to separate dead from live ones.

11.6 Virions

A virus is a small infectious particle that replicates only inside a viable cell of an animal, plant or microorganism.

Box 11.3 DEP Acting against Randomizing Thermal (Brownian) Motion

Particles collected at an electrode by DEP diffuse away down their number concentration gradient ∇N when the applied field is removed. This diffusive pressure gradient $\nabla \Pi$ can be expressed using the derivative of the Van't Hoff equation:

$$\nabla \Pi = kT\nabla N$$

with k the Boltzmann constant (1.38×10^{-23} Nm/K) and T the absolute temperature. As pointed out by Pohl [22, p. 86] this may be regarded as the negative of the average osmotic force (F_{os}) per unit volume of particles, expressed per particle as:

$$F_{os} = \frac{-1}{N}\nabla \Pi = -kT\left(\frac{\nabla N}{N}\right)\cdot\left(\frac{1}{\Delta r}\right)\hat{r}_o \tag{11.5}$$

where $\nabla N/N$ is the fractional change in concentration along any direction r and ∇r is the change in ordinate along r. The maximum relative change of $\nabla N/N$ is unity, corresponding to the presence versus absence of a particle. The corresponding minimum change in radial distance ∇r is the diameter $2R$ of the particle. From Equation (11.5) the maximum diffusional force per particle is thus given by:

$$F_{os[\max]} = -\frac{kT}{2R} \tag{11.6}$$

For a particle of diameter 100 nm at $T = 300$ K, $F_{os[max]} = 4.14 \times 10^{-14}$ N per particle. For a particle to exhibit DEP it must oppose this diffusional force, which from Equation (10.9) implies that the following relationship must hold:

$$2\pi R^3 \varepsilon_o \varepsilon_m [CM] \nabla E^2 > 4.14 \times 10^{-14}\ \text{N} \tag{11.7}$$

For the case of a 100 nm diameter particle suspended in an aqueous medium and assuming a modest value of 0.5 for the Clausius–Mossotti factor $[CM]$, this gives the condition that ∇E^2 should exceed $1.5 \times 10^{17}\ V^2/m^3$. An alternative approach is to employ Equations (4.40) and (4.41), which give the time-averaged potential energy, per unit volume, of a polarized particle as $U = -(\alpha E^2)/2$. The rate of thermal-assisted escape from this potential energy well is proportional to $\exp(-U/kT)$. To be trapped by DEP the particle must compete against the thermal energy $(3kT)/2$ associated with Brownian motion, so we require $U \geq (3kT)/2$. The condition for this is:

$$(\alpha E^2)/2 = 2\pi R^3 \varepsilon_o \varepsilon_m [CM] E^2 \geq (3kT)/2 \tag{11.8}$$

For a 100 nm diameter particle suspended in an aqueous medium at room temperature, with $CM = 0.5$, this requires $E \geq 1.8 \times 10^5$ V/m. As an approximation for the local field gradient we take $\nabla E = E/R$, to give $\nabla E \geq 3.6 \times 10^{12}\ V/m^2$ and $(E\cdot\nabla)E \geq 6.5 \times 10^{17}\ V^2/m^3$. This condition is readily achieved using microelectrodes.

A complete virus particle is called a virion. The simplest virion consists of two basic components, namely nucleic acid, in the form of either DNA or RNA and a protective coat called the capsid, which protects these nucleic acids from digestion by enzymes (nucleases). The capsid also facilitates the process of infection by attaching to specific sites on the surface of the host cell. Once it has entered its host the virion can also envelope itself with a lipid bilayer, derived from the plasma membrane of the infected cell.

Pohl suggested that 'particles with sizes in the range of perhaps 10 nm to 100 nm might be expected to be the smallest ones readily controllable by DEP' [22, p. 89]. Virions range in size from a diameter of ~20 nm to 300 nm and so they just fit into this minimal size category. Pohl's estimate was based on the maximum value of the field gradient factor $(E\cdot\nabla)E$ that may reasonably be generated with the macroscopic electrode geometries he was familiar with at that time (see Figure 2.10). As shown in Box 11.3, if $(E\cdot\nabla)E$ is not large enough the DEP force acting on a particle is unable to compete against Brownian motion (a defeat over thermal scrambling, as Pohl would express it). On the other hand, the upper limit on the value for E is determined by the electrical breakdown strength of the medium. However, Pohl may well have had in mind the limitation presented by macroscopic electrodes, because he proceeds to write: 'The present estimate has been particularized by various choices of experimental parameters and hence should only be regarded as suggestive.'

To summarize the content of Box 11.3, a particle subjected to an imposed electric field and Brownian motion will have a total potential energy comprised a deterministic electrical polarization potential energy $(-\alpha E^2)/2$ and a randomizing (stochastic) thermal potential energy $(3kT)/2$. The threshold for DEP capture at an electrode corresponds to the situation where these two energies balance. For a spherical particle of radius R the particle will be trapped for a finite time period if the condition given by Equation (11.8) holds, namely:

$$(\alpha E^2)/2 = 2\pi R^3 \varepsilon_o \varepsilon_m [CM] E^2 \geq (3kT)/2$$

The total potential energy $U(r)$ of the particle at a location r is given by the sum of the Brownian thermal energy and the DEP electrical potential energy:

$$U(r) = U_{th} + U_{DEP} = \frac{3}{2}kT - \frac{1}{2}\alpha E(r)^2 \tag{11.9}$$

For the particle to be trapped, $U(r)$ must have a *negative* value – it should represent a deep potential energy well for the particle, rather like a small rubber ball at the bottom of a cup. The randomizing thermal energy has the same action as shaking the cup – every now and then the ball will bounce out of it. The probability that randomizing thermal fluctuations will dislodge the particle from its DEP trap is governed by the Boltzmann factor $\exp(-U_{DEP}/kT)$. Expressed another way, if we can view the particle for a sufficient period of time to witness it being trapped and released a large number of times, the average dwell time (τ) spent by the particle in the trap is given by the inverse of its probability to escape and so is proportional to $\exp(U_{DEP}/kT)$. Large values (compared to kT) for the factor $-(\alpha E^2)/2$ lead to the particle being strongly trapped. The particle will at some point be released by thermal agitation, but this may take a long time. When the external nonuniform field E is switched on, the polarized particles begin migrating towards the electrode at a velocity (u) determined by the balancing of the DEP force and the Stokes fluid frictional force ($6\pi\eta uR$). The concentration of particles will increase near to the electrode and an equilibrium distribution of the particles will be reached when their diffusion down their concentration gradient is equal to the rate at which the particles are brought to the electrode by the DEP force. The concentration $n(r)$ of the particles at a distance r from the electrode is given by:

$$n(r) = N_o \exp(U_{DEP}/kT) \quad (11.10)$$

where N_o is the initial uniform concentration of the particles in the medium. When the field is switched off, the particles return to their initial uniform distribution by diffusion, often at a rate that differs to that of their collection under the action of the DEP force. The interplay of the potential energy factors in Equation (11.9) is shown schematically in Figure 11.26.

Before proceeding to an overview of DEP studies involving viruses, it is of value to review and expand on the brief discussion given in section 10.3.6 of Chapter 10 regarding the influence of surface conductance K_s on the effective conductivity σ_p of a particle. A quantitative analysis of this can be approached using a modified form of Equation (10.64):

$$\sigma_p = \sigma_{bulk} + \frac{2(K_{Stern} + K_{diff})}{R} \quad (11.11)$$

where σ_p is the bulk conductivity of the particle and R is its radius, with K_{Stern} and K_{diff} the individual contributions to K_s arising from field-induced charge movements in the Stern layer and the diffuse part of the electrical double layer, respectively. The value of K_{Stern} is given by the product of the surface charge density (determined by the zeta potential) and the counterion mobility in the Stern layer. Equation (11.11) predicts that as a particle becomes smaller the influence of the surface conductance increases. It is therefore instructive to employ

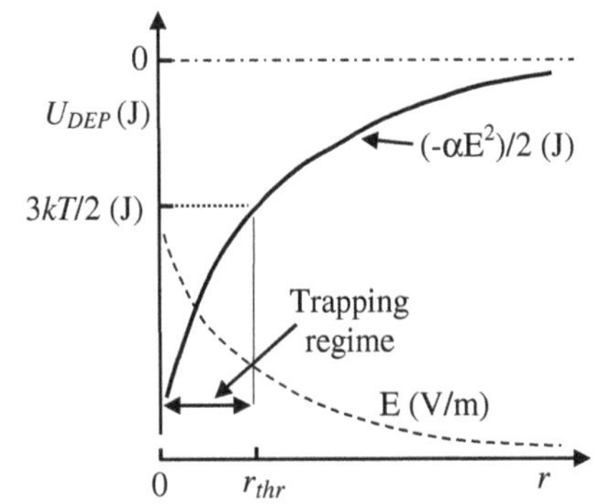

Figure 11.26 The solid line represents the potential energy (U_{DEP}) of a particle, polarized in an external nonuniform field, as a function of distance r from a surface. At a threshold distance r_{thr} the value of U_{DEP} is exactly balanced by the randomizing thermal energy ($3kT/2$). Nearer to the surface the DEP potential energy 'well' is deeper and the probability (proportional to $\exp[U_{DEP}/kT]$) that it will be trapped at the surface increases. The influence of a field associated with an electrical double layer at the surface is ignored in this description – but can be included in the model.

Equation (11.11) to model the DEP response of a particle of size comparable to that of a virus. The results of such modelling (using MATLAB) are shown in Figure 11.27. The DEP response of a particle of diameter 40 nm, bulk conductivity 1 mS/m and permittivity $10\varepsilon_o$ was determined by calculating the real component of the Clausius–Mossotti factor (CM^*), given by:

$$CM^* = \left(\frac{\sigma_p^* - \sigma_m^*}{\sigma_p^* + 2\sigma_m^*} \right)$$

as a function of frequency. The particle is assigned a surface conductance $K_s = 2$ nS and is taken to be suspended in an aqueous medium (permittivity $79\varepsilon_o$) whose conductivity can be increased from 1 mS/m up to 250 mS/m.

From Figure 11.27(a) it is clear that for a medium conductivity up to 100 mS/m the particle exhibits positive DEP, with a transition to negative DEP at ~34 MHz. For medium conductivities larger than 200 mS/m the particle exhibits negative DEP across the whole frequency range. This transition of DEP polarity can be depicted, as shown in Figure 11.27(b), in the form of a plot of the DEP crossover frequency (f_{xo}) as a function of medium conductivity. The results shown in Figure 11.27 are relatively insensitive to values below 1 mS/m chosen for the bulk conductivity of the particle.

The first DEP manipulation of a virus appears to be that reported for the tobacco mosaic virus (TMV) by Morgan and Green [188]. TMV is a cylindrical, nonenveloped virion of length 280 nm, diameter 18 nm, with a capsid that surrounds RNA. The TMV (strain U1) was raised in *Nicotiana tabacum*, purified by differential centrifugation and fluorescently labelled with Rhodamine

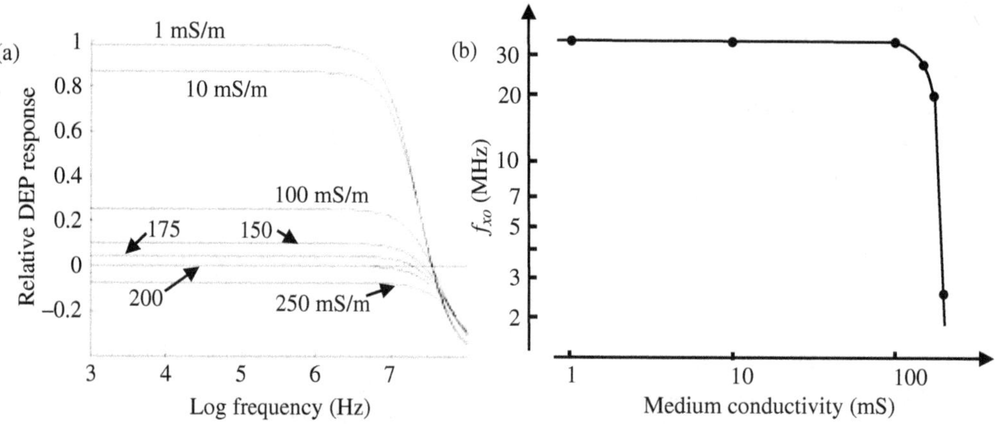

Figure 11.27 (a) The DEP-frequency response modelled for a particle of diameter 40 nm, surface conductance 2 nS, bulk conductivity 1 mS/m and bulk permittivity $10\varepsilon_o$, suspended in an aqueous medium of conductivity ranging from 1 mS/m to 250 mS/m. (b) The transition from positive to negative DEP is shown more dramatically as a plot of the DEP crossover frequency (f_{xo}) as a function of medium conductivity.

B [189]. For the DEP experiments, TMV samples were suspended in various strengths of potassium phosphate buffer to provide a range of medium conductivities. The virions were exposed to field gradients generated by an array of 'sawtooth' gold electrodes with 4 μm gaps, fabricated onto glass slides using standard photo- and electron-beam lithography. The value for (E·∇)E generated at the tips of the electrode teeth was calculated by finite element analysis to be ~5×10^{20} V^2/m^3 for an applied voltage of 2 V_{pk-pk} [188]. Positive DEP of TMV was observed over the frequency range 1 kHz to 1.5 MHz for medium conductivities in the range 170 ~mS/m to 1.4 S/m. It was not possible to observe single virions, only a faint fluorescent 'haze', which continually increased in brightness as more virions became trapped by positive DEP. The initial movement of TMV by DEP was seen to occur at a distance of ~1 μm from an electrode tip, corresponding to a value for (E·∇)E of ~ 5×10^{16} V^2/m^3. Above 1.5 MHz the virus did not collect at the electrode tips. At frequencies above approximately 20 MHz and for all conductivities examined the TMV were repelled from the electrodes under the action of negative DEP. The observation of the negative DEP was facilitated by using positive DEP to collect particles at the electrode edges before switching the voltage frequency to the negative DEP region. The experimentally determined values of the DEP crossover frequency (f_{xo1}), as a function of medium conductivity, are shown in Figure 11.28. On the logarithmic scales a linear increase in DEP crossover frequency (f_{xo1}) with medium conductivity was found, before an abrupt change at just below 0.1 S/m, beyond which conductivity value the DEP response of the virus was negative at all frequencies. To model this behaviour, TMV was considered to be a homogeneous prolate spheroid, with the appropriate depolarization factors along the major and minor axis, with a single value of permittivity and conductivity. The best fit curve shown in Figure 11.28 for TMV was obtained by assigning a permittivity value of $(55 \pm 2)\varepsilon_o$ and 85 ± 2 mS/m for the conductivity [188]. These values compare favourably to those ($60\varepsilon_o$ and 100 mS/m) obtained from dielectric studies of suspensions of the alfalfa mosaic virus [190].

The frequency-dependent DEP behaviour of the enveloped herpes simplex virus (HSV-1) was the next to be described [191]. Over the frequency range 10 kHz to 20 MHz and when suspended in an iso-osmotic EDTA/mannitol medium of conductivity 5 mS/m above quadrupole electrodes, both positive and negative DEP responses were observed. The transition between these two types of behaviour at a frequency ~4.5 MHz was in qualitative agreement with a simple model of the virus as a conducting particle surrounded by an insulating membrane. The positive DEP response, observed at low frequencies, was considered to arise from a surface

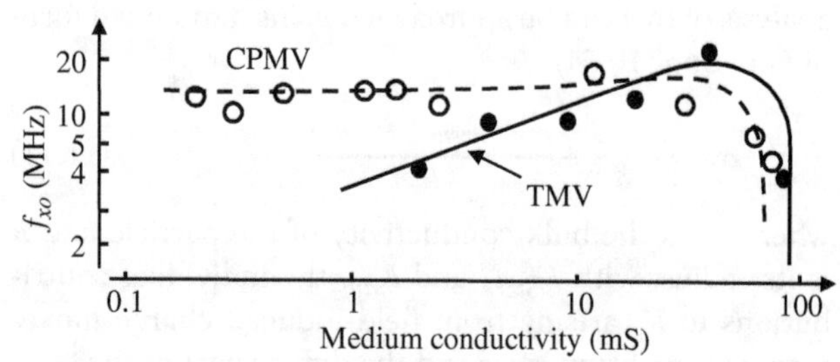

Figure 11.28 Values of the DEP crossover frequency (f_{xo}) determined for the tobacco mosaic virus (TMV) and the cow pea mosaic virus (CPMV) as a function of the conductivity of the suspending medium. The solid line represents the best fit for the TMV data, calculated by assigning to the virion an effective relative permittivity and conductivity of 55 and 85 mS, respectively (based on Morgan and Green [188] and Ermolina *et al.* [193]).

conductance arising from polarization of the diffuse double-layer surrounding the virus. The DEP characteristics of the HSV-1 capsid were also investigated [192]. In KCl solutions of low conductivity (below 0.5 mS/m) containing 280 mM mannitol, the capsids exhibited positive DEP at low frequencies and negative DEP at high frequencies. The DEP crossover occurred at ~3.5 MHz. As the medium conductivity was increased, the crossover frequency also increased. This behaviour could be modelled with the capsid comprising an outer protein shell, on average 15 nm thick (inner radius 47.5 nm), consisting of a structured network of protein molecules with channels of up to 5 nm in diameter. These channels connect the interior and exterior of the capsid, so that the internal space is equivalent to a chamber full of suspending fluid. At the centre of the capsid are the scaffolding proteins, treated as a solid protein sphere of diameter 60 nm. Using this multishell model the best fit to the experimental data was obtained by assigning: a particle permittivity of $30\varepsilon_o$; an internal conductivity of 30 mS/m; a shell permittivity of $30\varepsilon_o$; a Stern layer conductance (K_{Stern}) of 0.15 nS; and a zeta potential of 9.5 mV. The behaviour of capsids suspended in KCl combined with 280 mM mannitol exhibited very different DEP behaviour to those in KCl without mannitol. It was postulated that the effect of mannitol was to reduce the surface charge density of the capsid [192]. In later work, Ermolina *et al.* [193] investigated further the DEP characteristics of TMV and compared this to the DEP behaviour obtained for cow pea mosaic virus (CPMV), which is of similar size but different shape. Rather than being rod-shaped like TMV, CPMV is a spherical, isometric, unenveloped virus with a diameter of approximately 28 nm and an angular profile. In previous work [188, 191] the DEP crossover frequency (f_{xo1}) was determined by observation of the behaviour of single particles using fluorescence microscopy. In this later work [193] the value of f_{xo1} was determined by analysing the behaviour of an ensemble of particles using image processing. From an evaluation of the DEP data, the Stern layer conductance was determined to be 0.3 nS for CPMV, 0.38 nS for TMV and 0.52 nS for 27 nm diameter carboxy-latex beads that were also studied. It was concluded that the optimal condition for separation of TMV and CPMV is in a suspending medium of conductivity 1 mS/m or lower. The reason for this can be seen from Figure 11.28, where the f_{xo1} versus medium conductivity results obtained for CPMV have been added to those obtained for TMV. At a frequency of ~7 MHz, for example, CPMV should exhibit positive DEP and be trapped at the electrodes, whilst TMV should be repelled by negative DEP. It is also of interest to note that the spherical CPMV exhibited a similar variation of f_{xo1} versus medium conductivity to that of the spherical particle shown in Figure 11.27(b). The different behaviour exhibited by TMV could thus be related to its having a rodlike, rather than spherical, shape. It was assumed [188] that the TMV was always in an equilibrium orientation with its major axis along the field direction. However, this might not have been the case for medium conductivities less than ~3 mS/m. With a minor axes oriented along the field direction at low medium conductivities, the corresponding depolarization factor would have been larger (see Figure 7.11) and so alter its DEP-frequency response.

Other DEP studies involving viruses have largely been concerned with improving their capture and detection in microfluidic devices. Grom *et al.* [194] successfully accumulated and trapped hepatitis-A virus particles (27 nm diameter) in a microfluidic system by means of a combination of electrohydrodynamic (EHD) flow and DEP forces, within a field cage consisting of eight microelectrodes. The use of a suspending medium of high conductivity (0.3 S/m) resulted in sufficient Joule heating and the corresponding spatial variation of temperature, density and permittivity to induce EHD flow in the vicinity of this field cage. EHD-induced fluid flow vortices transported the viruses towards the center of the field cage, where they were then retained and aggregated by the DEP force. The results of numerical modeling of the spatial distribution of temperature, density and permittivity, as well as resulting EHD flow patterns, were in good agreement with the experimental observations. EHD fluid flow, rather than DEP, was also found by Docoslis *et al.* [195] to constitute the major mechanism for the transport of the vesicular stomatitis virus (VSV) when they are suspended in aqueous suspensions of physiological ionic strength and far from an electrode. The influence of DEP was calculated to be confined to within a few microns from the electrodes. The modeled collection patterns of both virus and fluorescently labelled particles near the electrodes were found to be in qualitative agreement with experiment observations. The simultaneous action of EHD and DEP forces led to the detection of 10^5 pfu/ml of VSV in two minutes. This can be compared to the finding that passive diffusion did not lead to detectable levels of virus captured on a surface when using titers of VSV as high as 10^7 pfu/ml [195]. Madiyar *et al.* [196] describe manipulation of bacteriophage virus particles using a nanostructured DEP device. The required nonuniform field for DEP was created using a 'point-plane' design where a nanoelectrode array of vertically aligned carbon nanofibres faced a macroscopic indium tin oxide electrode. At a concentration of 8.9×10^4 pfu/ml, the fluorescently stained virus particles were observed to collect one-by-one and to accumulate linearly with time over 30 s at the carbon fibre tips. Capture efficiencies of up to 60% were obtained. On switching off the voltage

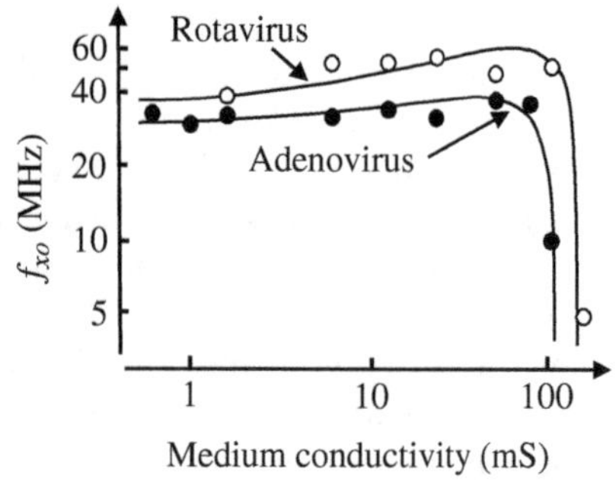

Figure 11.29 Modelled values of the DEP crossover frequency (f_{xo}) for the adenovirus and rotavirus as a function of the conductivity of the suspending medium (based on Nakano *et al.* [198].

signal the captured virus particles were released immediately. Ding *et al.* [197] have optimized a sawtooth-gradient, DC-insulator-based DEP device to concentrate the Sindbis virus. This enveloped virus is transmitted by mosquitoes and has an icosahedral structure with a diameter of 68 nm. The concentration of the Sindbis virus was increased by two to six times within seconds in an open fluidic channel (length 4 cm, width 500 μm, depth 20 μm) with a DC voltage as low as 70 V applied along the channel. The group led by Suehiro have advanced their DEP impedance measurement device [76,77], originally designed to detect bacteria in aqueous solutions, for the detection of pathogenic viruses [198]. Although the dielectric properties of the viruses were estimated to be different from those of bacteria, the impedance detection of their collection at the microelectrodes was similar to that obtained for bacteria. Nakano *et al.* [198] detected concentrations as low as 70 ng/ml for adenovirus and 50 ng/ml for rotavirus, within 60 s. The DEP-frequency responses obtained for these two inactivated, nonenveloped, virus types are shown in Figure 11.29, whilst their estimated dielectric properties are given in Table 11.6.

Table 11.6 Dielectric properties derived for: the adenovirus and rotavirus (Nakano *et al.* [198]; the herpes simplex virus-1 capsid (Hughes *et al.* [192]); the tobacco mosaic virus (Morgan and Green [188]); the cow pea mosaic virus (Ermolina *et al.* [193]).

Virus	Diameter (nm)	Surface conductance K_{Stern} (nS)	Zeta potential ζ (mV)	Relative permittivity
Adenovirus	90	1.74 ± 0.24	58.3	68.15 ± 1.90
Rotavirus	100	1.65 ± 0.73	83.5	73.21 ± 3.34
HSV-1	125	0.2	75	70
TMV	280×18	0.38	40	55 ± 2
CPMV	30	0.3	0–25	–

Finally, DEP studies of cells infected by viruses are of value because they can potentially aid the ability to detect and then separate them from noninfected cells. The Human Immunodeficiency Virus (HIV) is an example of a virus that can establish a long-term latent infection in B cells and T cells, for example. Even after effective antiretroviral therapy, these viral reservoirs persist in an HIV-infected person and are not detected by their immune system. Being able to selectively detect and isolate such latently infected cells could therefore result in a more effective therapy. The Kaposi's sarcoma-associated herpes virus (KSHV) is another example of a virus that can establish a persistent presence. Safavieh *et al.* [199] investigated the DEP responses and capture efficiency of latently KSHV-infected and uninfected B lymphoma cells in a microfluidic device. For the DEP studies the cells were suspended in buffers of conductivity 0.1, 0.16 and 10 mS/m. They observed sufficient differences in the DEP responses of the infected and uninfected cells, to the extent that with operating parameters of 1 V, 50 kHz and 0.02 μl/minute fluid flow ~37% of latently infected could be captured by positive DEP from a prepared mixture with uninfected cells. 55% of the captured cells remained viable after this 15 min procedure. The operating medium conductivity was not specified, but was probably 10 mS/m. An understanding of the factors dictating this cell separation is not at present possible, because specific differences in dielectric parameters (e.g., f_{xo1} and membrane capacitance) of the infected and uninfected B lymphoma cells were not obtained by Safavieh *et al.* [199]. This is a line of study worthy of further pursuit.

11.7 Nucleic Acids and Proteins

11.7.1 DNA

Most of the DEP studies described in this section were performed on double-stranded fragments or whole molecules of DNA (see Figure 8.22). With its two intertwined single strands of DNA running in opposing directions and the base pairs stacked with their molecular plane perpendicular to the helical axis, the dipole moments of paired nucleotides and dipoles in the main strands should all cancel out. In its double-stranded (ds) form the DNA molecule should therefore not possess a permanent dipole moment. The polarizability factor α in the term $(-\alpha E^2)/2$ describing the DEP potential energy of a polarized particle thus arises from electric field induced fluctuations of counterions surrounding the negatively charged phosphate groups of the DNA. This model of the dielectric behaviour of DNA in solution is described in section 8.5.1 of Chapter 8. Also, from Figure 8.26, the magnitude of the polarizability parameter α for DNA should exhibit a frequency dependence associated with the α- and β-dispersions. The DEP behaviour of DNA should therefore be expected to vary as a function of the frequency of an applied AC field.

The first submitted (July 1988) report of the studies of the DEP behaviour of DNA appears to be that of Washizu [200], who suspended DNA molecules in deionized water and introduced them into the gap between a planar pin-strip electrode arrangement (referred to as an edge-to-strip electrode geometry by Washizu). The DNA was obtained from the lambda phage virus (bacteriophage) that infects *E. coli* and visualization of individual molecules was achieved by fluorescently staining with the DNA-binding probe known as DAPI. To minimize heat dissipation and induced fluid flow, the λ-DNA was suspended in deionized water of conductivity 0.2 mS/m. With no voltage applied between the electrodes, uniform luminescence was observed in the entire electrode gap, indicating that the molecules were randomly dispersed. Upon application of a 1 MHz voltage signal the λ-DNA exhibited positive DEP, being attracted to the high-field region of the 'edge' electrode, with the long molecular axis directed along the field lines. This observation was confirmed in a follow-up paper by Washizu and Kurosawa [201], who supplied added information that orientation perpendicular to the field was observed at 40 kHz and that the DNA molecules appeared to have a random coil formation. With low DNA concentrations the stretching and alignment of single molecules was observed. Experiments with λ-DNA were also described [200, 201] using electrodes of the 'fluid intergrated circuit' design described by Masuda *et al.* [202]. This is the device described in Chapter 10 that employs insulating electrode structures in a microfluidic channel to distort an imposed electric field sufficiently enough to generate a DEP force on cells. This is now referred to as insulator-based DEP (iDEP).

From the comprehensive review by Hölzel [203] of the status in December 2008 of work on the dielectric and DEP properties of DNA, we find that for the preceding 10 years Washizu and his colleagues contributed significantly to our knowledge of electric field and DEP manipulation of DNA, with some proposed exciting applications [204–209]. Suzuki *et al.* [204] measured the field-intensity dependence of fluorescence anisotropy for λ-DNA and plasmid DNA, using interdigitated, sinusoidal shaped, microelectrodes. Comparing the results with an analytical model, the measured polarization factor was found to be several orders of magnitude larger than that of a conducting ellipsoid with the same dimension. This was explained by assuming a 'swelling' of the electrical equivalent diameter of DNA by 20 nm, comparable to the characteristic Debye length of the counterion cloud in the electrical double layer. The counterion concentration was varied by changing the pH of the medium, whilst keeping its conductivity constant. With increasing pH an increase in the anisotropy was observed, particularly between pH 5 and 6. This was attributed to an increase of the negative charge density on the DNA backbone, arising from dissociation of the phosphate groups. In a series of papers [205–207] that exploited the ability to stretch DNA by DEP, it was demonstrated that spatial information on the positions of specific base sequences could be achieved. Fluorescence-labeled restriction endonuclease (*Eco*RI) was observed at certain positions on λ-DNA, corresponding to the sequence (GAATTC) to which *Eco*RI binds. In a flowing fluid stream, *Eco*RI molecules were observed [207] to move along stretched DNA and were trapped at putative GAATTC sequences, providing evidence of sliding as a mechanism for relocation of *Eco* RI on DNA. It was suggested that this single-molecule-based, DEP-assisted method could take the form of a lab-on-a-chip device for mapping genomic DNA and analysing the motility of DNA-binding 'nanomachines' such as the *Eco*RI enzyme. Yamamoto *et al.* [207] created an array of immobilized, fully stretched DNA molecules between two electrode edges. Two enzymes that cut DNA strands were attached to latex beads and an optical tweezer was used to hold a bead and to press it against the strectched and immobilized DNA. The two enzymes chosen were DNaseI, which cuts DNA regardless of the base sequence, and the restriction enzyme HindIII, which cuts DNA at a specific base sequence. When a DNaseI-labeled bead was brought into contact with the immobilized DNA, its strand was cut instantaneously. On the other hand, when the restriction enzyme was used, the bead had to be moved along the DNA strand for a certain distance until it was finally cut. The interpretation of this enzyme dependence was that the restriction enzyme had to enter into the grooves of DNA to find the restriction sites, so the condition for the molecular contour fitting of the DNA and the enzyme were stricter compared to the case of the simple backbone-cutting enzyme DNaseI. It was proposed that this technique could be used for space-resolved molecular surgical operations, not just limited to dissections, but also for chemical modifications or insertion of genes.

Kawabata and Washizu [208, 209] describe a DEP-chromatography-based separation method for DNA and proteins, which they compared to the DEP-FFF concept introduced a few years before by others [210, 211]. The bioparticles are suspended in a fluid that passes at a constant rate over an array of planar electrodes, located at the bottom of a meandering channel (to minimize the substrate area). A 1 MHz voltage signal was applied to the electrodes, which was a compromise between the larger polarization of the molecules at lower frequencies and lower voltage drop at the insulator layer between electrodes at higher frequencies. The voltage level was adjusted so that the DEP force was not so strong as to perfectly trap the particles, as its function is just to increase the chance of them being trapped transiently. The particles experience a series of field gradients generated by the electrode array and so repeat their processes of trapping and release. The bioparticles of larger polarizability experience a stronger DEP force, so that the total time the particle is trapped becomes longer and as an average, the particle is retained on the electrode. We can view each electrode as a potential energy well of depth $U = (\alpha E^2)/2$ (see Box 11.3) with a particle attempting to escape from it at a rate proportional to $\exp(-U/kT)$. The deeper the energy well, in terms of multiples of kT, the less chance the particle has of a thermally assisted escape from an electrode. The mean escape, or trapped, time τ is given by the reciprocal of this attempt to escape rate, so that $\tau \propto \exp(U/kT)$. By increasing the number of electrodes in the array, one can trap every particle with equal probability as an average, which is not attained by having a single pair of electrodes. An increased number of electrodes also represents more theoretical separation 'plates' in the DEP-chromatography 'column', which enhances the resolution of particle separation. The DEP chromatograms were obtained for fluorescence-labelled DNA and proteins by monitoring the time course of the total fluorescence intensity at the outlet of the fluidic channel as the voltage signal was applied and then switched off. The percentage reduction of the fluorescence intensity was taken as a measure of the DEP trapping of the particles. The relative rates of DEP collection for various DNA and protein types are shown in Figure 11.30. For the same type of biomolecules (i.e., DNA or protein), it is seen that the larger their size the more readily they are trapped by DEP. This agrees with the standard theory for this ponderomotive (i.e., particle volume-dependent) effect and our understanding (see Chapter 8) of the factors that control the dielectric polarizability of DNA molecules. The fact that DNA appears to be more easily trapped by DEP than protein molecules was attributed [209] to the following reasons:

1. DNA has many polar groups (phosphates) that attract positive counterions and thus has larger polarizability due to the motion of the counter-ions [204].
2. As a result of the electrostatic repulsion between the negatively charged phosphate groups, DNA in solution is not compactly folded, but occupies a certain volume of larger equivalent diameter than closely packed molecules such as proteins.
3. DNA is stretched by the high-intensity field [201] so that it becomes more polarizable than that predicted for a closely packed model of a particle.

Figure 11.30 also reveals that separation of protein and DNA can readily be achieved by a proper choice of the field strength (e.g., 1.4 MV/m) and that separation of differently sized proteins or DNA should also be possible. Kawabata and Washizu also demonstrated the ability of DEP chromatography to detect molecular binding. This was accomplished by mixing a fluorescence-labelled probe with the sample. If the probe binds with a target molecule, its size becomes larger and so is more likely to be trapped by the DEP force. By monitoring the fluorescent intensity downstream of the channel, such molecular binding was detected by a decrease of the fluorescence signal. Separations of λ-DNA (48.5 thousand base pairs) and an oligonucleotide consisting of 22 bases were demonstrated in this manner [209].

Dalir *et al.* [212] have also investigated the electrical stretching of DNA molecules, finding that they were elongated from a compact random coil into an extended conformation orientated along the electric field lines. By trapping the DNA molecules onto an electrode using a 1 kHz field, the contributions of the applied voltage and frequency to the elongation of the DNA molecules were studied. Maximum elongation was found at ~100 kHz for two different-sized DNA fragments (25 kbp and 48 kbp) that increased with an increase of applied field strength, attaining full elongation at 1.9 MV/m. The DNA length

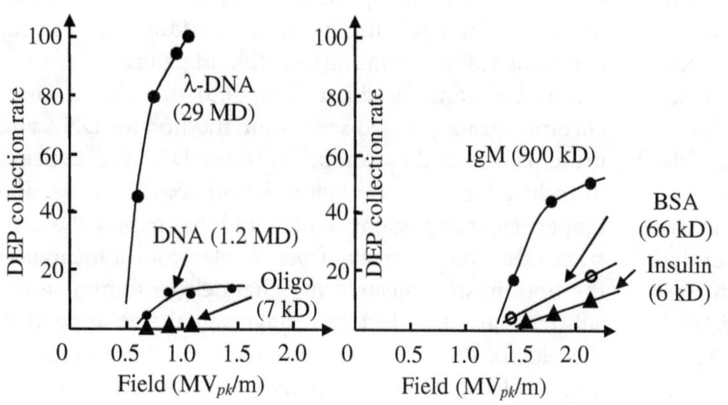

Figure 11.30 The DEP collection rate as a function of the applied 1 MHz field strength for various DNA and protein molecules. The relative sizes of the DNA, an oligonucleotide (Oligo), immunoglobulin M (IgM), bovine serum albumin (BSA) and insulin can be judged by their indicated molecular weights (based on Kawabata and Washizu [208,209]).

then decreased with increasing frequency up to 1 MHz. In an elastic model for double-stranded DNA, which quantitatively took into account the bending deformations of the backbone and the base-stacking interactions between adjacent DNA base pairs, an underlying scaling relation (δ) between the DNA extension and the applied voltage of the form $\delta \sim V^{1-1.5}$ was obtained.

Asbury and van den Engh [213] adopted a different approach to that of Washizu and co-workers regarding electrical manipulation of DNA. They did not stretch the DNA or attach it to an electrode, but instead aimed to manipulate DNA molecules by their inducible dipole moments (i.e., DEP) in conjunction with electrophoresis. A dilute sample of fluorescently stained DNA in deionized water of conductivity 0.18 mS/m was placed in a chamber formed of a quartz chip and a cover slip. One hundred parallel gold strips, spaced 30 μm apart, had been fabricated by lithography onto the top surface of the quartz chip. No electrical signals were applied directly to these gold strips – they acted in the opposite sense to the insulative (electrodeless) structures described by Washizu [200], in that they concentrated (rather than repelled) an electric field. The electric field was generated using two parallel platinum wires located on either side of the DNA sample in its chamber. Voltages were generated by a function generator and could be mixed with a DC voltage before amplification. Typically, a sinusoidal voltage of 200 V_{pk-pk} was applied between the platinum wires over the frequency range 1 Hz to 10 kHz. This applied voltage was divided among the 99 narrow (30 μm) gaps between adjacent gold strips and the 4 mm separation between the most peripheral strips and the platinum wires. For a 200 V_{pk-pk} sinusoid applied to the platinum wires, the voltage drop across each 30 μm gap was only ~0.5 V_{pk-pk}. Because the gold-film strips were very thin (6 nm), the electric field was highly concentrated near their edges, giving the large field gradient required for the desired DEP trapping of the DNA molecules at these metal edges. Movement of the DNA molecules was observed in an epifluorescence microscope and recorded with a sensitive video camera, while the voltage was turned on and then off. By integrating the digital images parallel to the gold-film strips, brightness profiles perpendicular to the strips were obtained. When the trapping field was switched on, narrow peaks of fluorescence (~5 μm in width) grew rapidly in the profiles, where the DNA molecules became concentrated at the high field gradient regions over the edges of the gold strips. When the field was switched off the fluorescence peaks broadened and shortened as the DNA molecules dispersed down their concentration gradient. The voltage required for DNA trapping increased with frequency in the range of 1 Hz to 10 kHz. At low frequencies (1–10 Hz) trapping could occur at voltages as low as 30 V_{pk-pk} applied to the platinum wires. Trapping at 1 kHz required 200 V_{pk-pk} across the platinum wires. For the applied voltage signal of 30 Hz, 200 V_{pk-pk}, used most often by Asbury and van den Engh [213], λ-DNA molecules (48.5 kb) were trapped rapidly and effectively. The field and field gradient intensity at the edges of the thin gold-film strips was estimated by solving the Laplace equation using a two-dimensional finite element model of a cross section through the fluid layer, near the edge of a gold-film strip [213]. This calculation indicated that field strengths as high as 190 kV/m and field gradients of 10 MV/m^2, occurred at the edges of the gold-film strips. This gives the value for the field factor ∇E^2 in the DEP force equation as $\nabla E^2 = 2(E{\cdot}\nabla)E = 3.8 \times 10^{12}\ V^2/m^3$. This is very small compared with the value for ∇E^2 of $\sim 10^{18}\ V^2/m^3$ at the tips of the DEP cell manipulator (tweezer) described by Menachery *et al.* [214] and especially so when compared with that of $\sim 10^{21}\ V^2/m^3$ generated by a 2 V_{pk-pk} signal to the 'sawtooth' gold electrodes described by Morgan and Green [188]. To place this into context regarding the putative DEP manipulation of a λ-DNA molecule, according to the treatment given in Box 11.3 the following relationship must hold for the DEP force (F_{DEP}) to be large enough to overcome the randomizing action of Brownian motion:

$$F_{DEP} = 2\pi R^3 \varepsilon o \varepsilon_m [CM] \nabla E^2 > 4.14 \times 10^{-14}\ N$$

where R is the radius of the particle and $[CM]$ is the Clausius–Mossotti factor. Assigning a reasonable value for $[CM]$ of 0.5, substituting the values for ε_o (8.85×10^{-12} F/m), ε_m (80), as well as the radius of gyration (~0.5 μm) determined for λ-DNA [213], then for $\nabla E^2 = 3.8 \times 10^{12}\ V^2/m^3$ the corresponding value determined for the DEP force is $\sim 2 \times 10^{-15}$ N. This is some 20-times smaller than that required to overcome the randomizing thermal energy $3kT/2$. Unless the value derived for the factor ∇E^2 was greatly underestimated, this implies that the DEP force acting on the single DNA molecules may not have been the major one causing their collection at the gold strip edges. One possibility is that the DNA molecules aggregated into larger, more polarizable, entities. Electroosmotic fluid flow induced by the applied DC field may also have been a contributory factor. Support for this suggestion is given by the observation [213] that trapped DNA molecules 'wiggled' near the edge of the gold strips, at a wiggle rate equal to the frequency of the applied voltage and giving the fluorescence images a streaked appearance. Asbury and van den Engh attributed this effect to electrophoretic motion caused by the net charge of the DNA, but the possible contribution of electroosmotic motion of the fluid was also suggested by them.

In other groundbreaking work, Chou *et al.* [215] demonstrated that 'electrodeless' DEP (also known as

Table 11.7 DNA polarizabilities (α) and diffusion coefficients (D) for various sizes of linear and supercoiled DNA fragments (derived from Regtmeier *et al.* [216,217]).

	Linear			Supercoiled	
DNA (kbp)	α (10^{-29} Fm^2)	D (μm^2/s)	DNA (kbp)	α (10^{-29} Fm^2)	D (μm^2/s)
6	1.5 ± 0.1	1.71 ± 0.11	7	0.5 ± 0.1	2.80 ± 0.15
12	2.5 ± 0.2	1.36 ± 0.10	10.3	1.7 ± 0.2	1.32 ± 0.05
12.2	2.8 ± 0.6	1.09 ± 0.07	12.2	2.3 ± 0.7	1.08 ± 0.04
48.5	3.1 ± 0.3	0.68 ± 0.09	15.5	2.8 ± 0.1	0.83 ± 0.03
164	5.8 ± 0.5	0.39 ± 0.05	21	2.9 ± 0.2	1.05 ± 0.06

insulator-based DEP (iDEP) and described in section 10.4.2 of Chapter 10) can be used for the concentration and patterning of both single-strand and double-strand DNA. They recognized that a combination of both a DEP and electrophoretic force would act on their DNA particles. By reversing the field direction and hence the electrophoretic force, they found evidence for the strong DEP response of the DNA in the frequency range 200 Hz to 1 kHz. On measuring the DEP force under different solvent viscosity conditions and relating this to the dielectric relaxation times obtained from the frequency responses, Chou *et al.* [215] determined that field-induced mobility of counterions in the electrical double layers around the DNA molecules were responsible for their observed DEP responses. Furthermore, for a given applied voltage the DEP force exhibited a significant size-dependent frequency response as the length of the DNA fragments was increased from 368 bp to 39.9 kbp. As a result of this finding it was proposed that by an appropriate choice of experimental parameters it would be possible to selectively trap one range of DNA molecules while removing others. Regtmeier *et al.* [216, 217] proceeded to demonstrate just this by manipulating DNA by electrophoresis and insulator-based DEP in an array of microinsulator posts. By maintaining a fixed DC component of an applied field (hence fixed electrophoretic force) and systematically increasing the magnitude of the AC field to increase the strength of the DEP traps, Regtmeier *et al.* [216] were able to perform efficient and fast DNA separation according to molecular length for two different DNA conformations, namely linear λ-DNA (48.5 kbp) and linear T2-DNA (164 kbp), together with supercoiled, covalently closed, circular plasmid DNA (7 and 14 kbp). With an applied AC voltage of 140 V, the DEP trapping force estimated for T2-DNA was 1.2×10^{-14} N and so of the same order as that specified in Box 11.3. The underlying migration mechanism of the molecules was a thermally induced escape process out of the DEP traps at the insulator posts in the direction of the electrophoretic force and was sensitive to different DNA fragments because of their length-dependent DNA polarizabilities (see Table 11.7). As long as the DEP trapping force was larger than the electrophoretic force, the DNA molecules were still trapped deterministically by the DEP potential and could escape only due to the thermal excitation. Having escaped from a trap, the DNA was electrophoretically driven along the fluidic channel until it was trapped at the next micropost along its way. For appropriate (not too large) applied AC field values, the escapes from the DEP traps were sufficiently fast to yield average migration velocities of the DNA in the μm/s range. If the DEP trapping force was smaller than the electrophoretic force (i.e., small applied AC voltage or large DC voltage) the traps completely disappeared and the migration velocity of the DNA was determined by its length-independent electrophoretic mobility.

Regtmeier *et al.* [217] extended their work on the DEP trapping and polarizability of linear DNA, to examine the roles of size, topology and spatial conformation. In this work all experiments were performed with a constant DC voltage and a fixed frequency of 60 Hz for the variable AC voltage magnitude. The reason for keeping the frequency fixed was based on the opinion that the escape times were the combined result of the molecules' polarizability and the details of the escape dynamics out of the traps, which both, in general, depend on the frequency of the applied AC voltage component. By keeping the frequency fixed these two contributions were considered to be disentangled. (The frequency dependence of polarizability is logical, but its influence on the thermally assisted escape dynamics is not so clear and deserves further investigation.) The DEP effects were quantified by deriving the DNA polarizability from their dwell times in the potential energy traps. The dwell times were obtained by recording the fluorescence intensities of the DNA molecules in a given region of interest (where the DNA had to pass four spatial periods of trapping to reach it) for a sequence of time instances. This polarizability data was combined with information about the spatial extension and configuration of DNA, which was obtained from measuring diffusion coefficients and from AFM images. As can be seen from Table 11.7, the polarizability values for linear and supercoiled DNA are of the same order of magnitude. For linear DNA fragments the results imply a direct scaling of the polarizability with the radius of gyration of the molecule (derived from their diffusion coefficients) consistent with the dipole moment models for DNA described in Chapter 8. For supercoiled DNA, the situation was found to be more complex. As determined with the help of AFM images, the polarizability as well as the coefficient of diffusion critically depend on whether the spatial conformation of the supercoiled DNA fragments take on that of a random coil, or have a branched, starlike (plectonemic) structure. From atomic force microscopy the 7, 10.3 and 15.5 kbp supercoiled DNA fragments listed in Table 11.7 exhibited branched plectonemic structures, whilst the 21 kbp DNA fragment was also supercoiled but had a random coil like configuration. Regtmeier *et al.* [217] speculate that the polarization process in this case involves a significant deformation of the plectonemic configuration (e.g., a dipole may be induced in every plectonemic arm such that they become aligned with the applied electric field) resulting in a prolate 'cigar'-like geometry. So, although the two 12.2 kbp DNA fragments with different conformation exhibit similar polarizabilities, they were distinguishable via DEP effects – as demonstrated by their DEP separation within 3.5 min [217]. One of the participating research groups (Bielefeld University) in this work later described a continuous-flow micronanofluidic device for the rapid and efficient detection or purification of mixtures of DNA fragments [218]. The separation takes place in the vicinity of a curved insulating ridge that spans across the channel, creating in effect a nanoslit. From a two-dimensional finite element numerical analysis it was found that the applied electrophoretic field lines do not cross this ridge at right angles – but at an angle of 15° or less. This results in a small tangential electrophoretic force component that drives the DNA, which is on top of the ridge, along the curved ridge towards the fluid channel wall. The time-average velocity of the molecules in the microchannel and along the ridge is thus determined by a combination of electrophoretic, electroosmotic and DEP forces, induced by the application of DC and AC electric fields. Depending on the relative magnitude of the DEP force, which will depend on the polarizability of a DNA molecule, a target molecule can be induced to travel along the curved ridge and be deflected into a flow stream further across the channel width. Molecules of insufficient polarizability pass over the ridge without being deflected. By this means Viefhues *et al.* [218] separated 2.686 kbp and 6.0 kbp linear DNA fragments, as well as circular parental and minicircle-DNA. The following DNA complexes were also detected: cancer drug-DNA complexes for three different ratios of drug per DNA base pair; protein–DNA complexes. All separations were with baseline separated resolution that validated the high sensitivity of the system. To verify and understand the electrokinetic separation mechanism in more detail the polarizabilities were also determined for the various DNA entities. At the AC frequencies used (300 to 550 Hz) the polarizability values scanned the range from 1.4 – 4.3×10^{-30} Fm^2.

Using metal nanoelectrode-based DEP and confocal microscopy, Tuukkanen *et al.* [219] determined the DEP trapping characteristics of DNA fragments of various base-pair lengths. A nonuniform field was created between two 'fingertip-type' electrodes that took the form of 100 nm wide gold 'wires' fabricated by vacuum deposition onto a SiO_2 coated silicon substrate. Double-stranded DNA fragments with varying lengths (27–8416 bp) were fabricated by the three different methods, namely by the annealing of synthetic oligonucleotides, the polymerase chain reaction and restriction enzyme digestion of the plasmids multiplied in bacteria. The DNA fragments were fluorescently labelled and diluted into Hepes/NaOH buffer (3 mM Hepes, 1 mM NaOH), which resulted in a suspending medium of pH 6.9 and conductivity 2 mS/m. The effect of the DNA length and the size of the gap between the gold nanoelectrodes on the DEP trapping efficiency of the DNA fragments were investigated from 0.2 to 10 MHz. To summarize the main experimental results, it was found that a larger voltage was required to trap the smaller DNA fragments, which reflected their smaller polarizability and higher Brownian motion compared to the larger molecules. It was also observed and explained by means of electric field simulations, that small changes in the gap between the nanoelectrodes (e.g., 80 nm compared to 130 nm) did not affect the DEP trapping efficiency as long as the gap was smaller than the physical dimension of the resolved 'trapping region'. With a fixed voltage across the electrode gap, although more DNA was trapped at the lower frequencies, the DNA was better localized between the electrodes at higher frequencies. This trade-off between efficiency and accuracy resulted in the optimum frequency for the measurements to be ~1 MHz. The three-dimensional electric field generated by the fingertip electrode structure inside a 8 μm^3 cubic region was obtained using finite element software to solve the 3D Poisson's equation. Based on this knowledge of the spatial variation of the field E, the DEP force was then estimated using the concepts described in Box 11.3 and Equation (11.9). The fluorescence intensity of the trapped DNA molecules was interpreted as being proportional to the dwell time τ given by $\tau \propto \exp(U_{DEP}/kT)$. Apart from the limited accuracy of the measurements performed by Tuukkanen *et al.* [219], there will also have been inherent approximations regarding the spherical shape (or otherwise) of the DNA particles and neglect of electrostatic

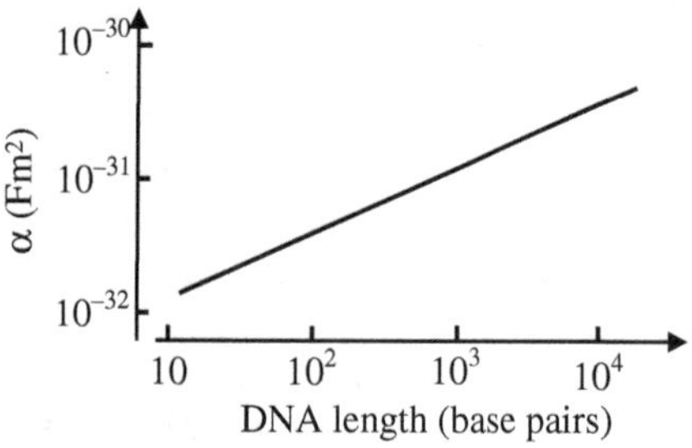

Figure 11.31 The trend in values of the polarizabilities α of different DNA fragments, as a function of their base pair length, determined by DEP in the frequency range 0.2 to 10 MHz (derived from Tukkanen *et al.* [219]).

particle-particle interactions that would have influenced their effective electrical potential energy. The results obtained for the polarizability of the DNA fragments as a function of the base pair length are shown in Figure 11.31. The frequency dependence of the DNA polarizability was found to be rather weak and so in this Figure the general trend (ignoring the spread of data points) across the whole range of frequencies is shown. Although the shorter DNA molecules exhibited smaller polarizability values than the longer ones, it is evident from the slope of the trend of values shown in Figure 11.31 that the polarizability *per base pair* was higher for small molecules than for large ones. This suggests that the longer DNA molecules did not behave as rod-shaped objects in the DEP capture process and that the polarization of DNA is related to counter-ion fluctuations along persistent lengths (as described in Chapter 8, section 8.5.1.1).

It is of interest to note that the polarizability values cited for DNA in Chapter 8, derived from other types of electrical measurements, are generally much smaller than the range shown in Figure 11.31. An example is the value of 3×10^{-33} Fm^2 obtained by Suzuki *et al.* (Chapter 8, reference [154]) from studies of field-induced orientation of λ-DNA and plasmid DNA. On the other hand, the value of 8×10^{-30} Fm^2 derived for plasmid DNA from time-domain dielectric spectroscopy by Bakewell *et al.* (Chapter 8, reference [130]) lies above the range shown in Figure 11.31. This wide range of values may reflect differences in the types and conductivities of the suspension buffers used and is worthy of further study, as well as which dielectric relaxation process of the DNA molecule was being investigated in the frequency range of measurements.

Other notable examples of where insulator-based DEP has been investigated for its potential to selectively sort DNA fragments include the description by Parikesit *et al.* [220] of how a DEP force was considered to induce size-dependent flow trajectories of DNA molecules. This was demonstrated for λ-DNA (48.5 kbp) and T4GT7 (165.6 kbp) DNA flowing continuously around a sharp corner of a glass post inside a fluidic channel of depth of 0.4 μm. The radius of gyration of the λ-DNA and the T4GT7-DNA was assumed to be 0.74 μm and 1.37 μm, respectively and so larger than the channel. Larger DNA particles were observed to deflect less strongly than smaller ones. This appears to contradict the fact that the DEP force should, with all other factors fixed, increase with particle size (as for example observed by Chou *et al.* [215] and Regtmeier *et al.* [216]). Numerical simulations of the electrokinetic force distribution inside the channels, which took into account the fact that the channel depth was smaller than the radius of gyration of the DNA, were in qualitative agreement with the experimentally observed trajectories. The electrokinetic response of the DNA particles was considered to have been influenced mainly by the electric fields in the electrical double layers of the channel wall and DNA particles. This was taken to explain why the observed DEP effects were not significantly influenced by changing the electric field applied between the inlet and outlet fluid ports. The apparent anomalous size-dependence was considered to result from the fact that the larger (T4GT7) DNA particle is exposed to a larger volume of the field distribution profile near the sharp corner and hence to regions of much lower $(E\cdot\nabla)E$ values than that sensed by the λ-DNA particles. Gallo-Villanueva *et al.* [221] describe the concentration of linear DNA particles (pET28b) using a combination of electroosmotic, electrophoretic and DEP forces, induced by direct current electric fields, in a glass microchannel containing an array of cylindrical insulating posts. DNA suspending media with conductivities between 100 and 120 mS/cm and pH values between 10.8 and 11.15 were employed, with applied fields between 50 kV/m and 200 kV/m. By varying the magnitude of the applied field it was possible to control the degree of negative DEP trapping of the DNA particles at the insulating posts, to a level that could achieve significant sample concentration. Concentration factors varying from eight to 24 times the feed concentration were measured at 200 kV/m after a processing time period of 20–40 s. Li *et al.* [222] describe a microchip device, fabricated on a silicon wafer, consisting of two microchannels connected by a series of wedge-shaped nanoslits. Electro-osmotic fluid flow was induced through the slits by applying a voltage between the microchannels, of polarity that depended on which channel was used for sample injection. As for other insulator-based DEP particle separators, the whole process of DEP trapping involved a combination of electroosmotic, electrophoretic and DEP forces. DNA fragments with length 2 kbp were fluorescently labelled and their trapping by DEP at the nanoslits was analysed by fluorescent intensity changes. This showed that the DNA fragments could be trapped at a nanoslit in both high

(≥ 1 S/m) and low (~1 mS/m) conductivity media using a lower applied electric field strength (10 V/cm) compared to the iDEP methods that had been described previously. This represented a significant improvement to suppress Joule heating effects that can occur in insulator-based DEP devices that operate at high electric field strengths.

As described in Chapter 10, carbon electrodes can serve as an alternative to metal electrodes and insulator structures for DEP applications. The first description of how carbon-electrode DEP can be used for the manipulation and trapping of bioparticles such as DNA was reported by Martinez-Duarte *et al.* [223]. An array of 3D carbon electrodes, contained in a microfluidic channel, was constructed, analysed theoretically and tested experimentally for the DEP manipulation of DNA. In particular, the DEP response of λ-DNA under various frequencies and fluid flow conditions necessary for its retention at the carbon electrodes was investigated. Negative DEP was observed at frequencies above 75 kHz and positive DEP the range below 75 kHz and down to 5 kHz. A theoretical model was also implemented to describe the experimental findings in sufficient detail. Theoretical considerations, based on reported scaling laws for linear and supercoiled DNA, implied that carbon electrode DEP devices can be employed in future analytical applications such as DNA preconcentration and fractionation.

The DEP properties of DNA have also been determined by measuring capacitance changes between planar microelectrodes, for DNA sizes ranging from 100 bp to 48 kbp and concentrations from below 0.1 to 70 μg/ml [224]. The DEP spectra exhibited maximum responses at ~3 kHz and 3 MHz. The strongest response was found for long DNA fragments above 10 kbp, as well as for short 100 bp fragments that correspond to the persistence length of DNA. Henning *et al.* [224] consider that the combination of impedance measurements with DEP collection serves at least two important purposes. First, the concentration of macromolecules can be determined without any chemical modifications of the analyte and detection is improved as compared to impedance measurements alone. Although electrical sensing only gives limited information about the type of molecules detected, specificity can readily be attained in microfluidic systems by functionalization of the electrode surface with binding molecules such as antibodies and oligonucleotides, or by combination with chromatographical methods, for example. In contrast to optical methods, there is in principle no lower limit (except signal-to-noise ratio) in geometrical resolution. Thus, detection of single nanoparticles and even molecules appears feasible. Second, the DEP response of a wide range of nanoparticles can be investigated rather easily, which should aid further insights into the mechanisms controlling this interaction. Li *et al.* [225] describe the combined use of DEP concentration and capacitive impedance measurement, using the same electrodes on a microchip, to characterize the DEP responses of two types of DNA of different and representative molecular conformations, namely pUC18-DNA in a supercoiled form and λ-DNA in a linear form. The microchip, a commercial surface acoustic wave resonator, contained an array of interdigitated aluminium electrodes (1.4 mm width, 1.1 mm gap) on a quartz substrate as the DEP concentrator of the target particle. Measurements were taken with a high precision impedance analyser, which also acted as the electrical excitation source, to induce a DEP response at various frequencies from 20 kHz to 5 MHz. To verify the DEP response, fluorescence microscope images were captured before and after the electric excitation of the interdigitated electrodes. A large change in impedance corresponded to positive DEP collection at the electrodes of the DNA, while little change corresponded to negative DEP. The strongest positive DEP effect and the maximum collection efficiency were observed around 300 kHz for supercoiled pUC18 and 100 kHz for linear λ-DNA. A significant advantage of the method was the DEP concentration of the DNA was achieved using physiological strength PBS solutions.

Finally, the ability to operate with high conductivity fluidic media is not just important for the development of insulator-based or impedance DEP devices, for example. A potential application, often suggested by authors for their DEP devices, is the detection and isolation of rare target cells or other bioparticles directly from clinical samples such as whole blood or plasma. The vast majority of DEP devices described in this chapter only operate with efficiency for suspending media of conductivity less than ~1 S/m. This means that the dilution of blood and other clinical samples is necessary, which has implications regarding the volume of sample that needs to be processed in a short time. One of the groups that have seriously addressed this issue is led by Heller [226–229]. Krishnan *et al.* [226] show how a commercially produced 100-microelectrode array (the NanoChip® 100) can be used for the separation of DNA particles in high conductivity media. The circular microelectrodes in the array are 80 mm in diameter and made of platinum, to which has been applied a 10 mm thick coating of porous polyacrylamide hydrogel. AC electric field conditions have been found that allow the separation of DNA nanoparticles to be achieved under high conductance (ionic strength) conditions. At frequencies in the 3 kHz–10 kHz range and 10 V_{pk-pk} applied voltages, the separation of 10 mm polystyrene particles into low electric field regions and 60 nm DNA-derivatized nanoparticles and 200 nm nanoparticles into high-field regions, was carried out in a buffer of conductivity 1.68 S/m (the conductivity of plasma). Sonnenberg *et al.* [227] extended this performance to enable the rapid isolation, concentration and detection of high molecular weight (hmw) DNA and nanoparticles directly from human and rat whole blood. At 20 V_{pk-pk} and 10 kHz a wide range of hmw-DNA and nanoparticles were concentrated into high-field regions by positive DEP, while the blood cells were concentrated into the low-field regions by negative DEP. A simple fluidic wash removes the blood cells while the DNA and nanoparticles remain concentrated in the DEP high-field regions where they can be detected by fluorescence. The DNA could be detected at 260 ng/ml, which is a detection level suitable for analysis of disease-related cell-free circulating DNA biomarkers. Fluorescent 40 nm nanoparticles could be detected at 9.5×10^9 particles/ml, which is a level suitable for monitoring drug delivery nanoparticles. Further development of the microelectrode array has enabled the isolation of hmw-DNA from serum and its detection at levels as low as 8–16 ng/ml, as well as fluorescently tagged T7 bacteriophage virus to be isolated directly from blood samples and fluorescently stained mitochondria to be isolated from biological buffer samples [228]. Heller's work has also been directed at designing DEP pipette-type formats for dipping into and recovering specific analytes from samples in microtiter plates. Micropipette tip devices were fabricated that contained a 2% agarose gel plug, a buffer chamber and platinum electrode as the DEP collection device [229]. Using this DEP pipette, operated at 10 kHz and 160 V_{pk-pk}, 200 nm fluorescent particles could be isolated into high-field regions of the microelectrode array and separated from 10 μm fluorescent microbeads in high conductance PBS buffer. The collected nanoparticles could then be transferred to a new buffer solution. The DEP isolation and separation of genomic DNA (>50 kbps) from the 10 μm microbeads in high conductance PBS buffer was also demonstrated, with transfer of collected DNA to another solution [229].

11.7.2 RNA

The RNA molecule is a linear polymer in which nucleotides are linked by phosphodiester units. As for DNA, RNA carries negative charges along its backbone associated with ionized phosphate groups and in aqueous solutions these charges are screened by counterions. Generally, RNA has regions that are self-complementary (A-U or G-C pairing) leading to the presence of loops and folded motifs. Ribosomes consist of RNA subunits, which translate mRNA into polypeptides during protein synthesis. All living organisms contain ribosomal RNA (rRNA). Consequently, rRNA has been used for evolutionary science and taxonomy and utilized to define and identify different organism species. The high copy numbers of rRNA molecules present in individual cells have also been used as a naturally amplified biomarker for the detection of bacteria in environmental and clinical studies. DEP offers a noninvasive method for efficient concentration, trapping and separation of RNA molecules.

The first (and apparently only one to date) investigation of the DEP behaviour of RNA is that reported by Giraud *et al.* [230] for the 16S and 23S subunits of *E. coli* rRNA, over the range 3 kHz to 50 MHz using interdigitated microelectrodes. Although the DEP properties of RNA had not been reported before this work, pertinent dielectric studies of the free RNA subunit of 70S rRNA had been performed and interpreted to indicate that counterion fluctuations dominate the dielectric dispersions centred around 9 MHz at 25 °C [231]. For the DEP measurements the RNA was fluorescently labelled and suspended in a medium of conductivity 12.8 mS/m. Quantitative measurement using total internal reflection fluorescence (TIRF) microscopy of the time dependent collection of the molecules at the electrodes indicated a positive DEP response, characterized by a plateau between 3 kHz and 1 MHz, followed by a decrease in response at higher frequencies. Negative DEP was observed above 9 MHz. From finite element analysis of the electric field generated by the electrodes, the value of ∇E^2 was determined as $\sim 10^{18}$ V^2/m^3 for an applied voltage signal of 2 V_{pk}. As a negative control, a solution of the fluorescently labelled rRNA was placed on the electrode array with no applied voltage. The background fluorescence increased slowly over a period of 30 min due to unspecific adsorption of the RNA molecules on the surface of the gold electrodes and its substrate. However, no preferential adsorption of the RNA at the electrodes was observed. This weak background fluorescence was subtracted from the total fluorescence signal during the DEP collection.

The voltage dependence of rRNA collection at the electrodes revealed three distinct regimes: (i) Between 0 and 2.2 V_{pk} the fluorescence signal intensity remained constant. It was considered that in this range the DEP force was too weak to overcome Brownian motion. (ii) From 2.2 to 3.8 V_{pk} the fluorescence signal was quadratic in applied electric field as expected from standard DEP theory. (iii) Above 3.8 V_{pk} the fluorescence intensity still increased – however, its voltage dependence was sub-quadratic.

The V^2-dependence of the fluorescence intensity ruled out the possible influence on the DEP measurements of fluid flow arising from electrothermal effects, because these were expected to exhibit a V^4-dependence. Also, AC electroosmosis, which does exhibit a V^2-dependence, peaks at a frequency several orders of magnitude smaller than the charge relaxation frequency ($\omega = \sigma/\varepsilon = 2.9$ MHz for the experiments) and falls off as a function of ω^2 thereafter. Electroosmosis was therefore considered to exhibit a negligible influence on the DEP collection above 100 kHz, so that the observed V^2-dependence of fluorescence intensity reflected the DEP collection of the rRNA samples at the electrode edges, with negligible interference from induced fluid flow effects. The sub-quadratic dependence on voltage was considered to arise from the number of collected RNA particles saturating the volume available to them in the space interrogated by the TIRF technique. The initial amount of RNA contained in the volume (1.1×10^{-11} dm^3) bounded by the TIRF depth of field (275 nm) and the electrode array (area 4×10^{-6} dm^2) was 4.2×10^{-19} moles. This is equivalent to 2.5×10^5 RNA particles, which occupy a total volume of $\sim 10^{-15}$ dm^3. From the increase in fluorescence intensity above the background observed for the DEP measurements between 1 kHz and 1 MHz, it was estimated that a 10^3-fold increase in RNA concentration occurred after 30 s of an applied 4 V_{pk} signal. At this concentration within the 'TIRF volume' mutual electrostatic repulsions between the charged rRNA molecules, as well as diffusion down their concentration gradient, began to counterbalance the DEP force and resulted in the observed saturation of fluorescence (see Equation 11.10).

Because RNA is single stranded, then unlike the case for double-stranded DNA there is a possibility for the rRNA molecule to exhibit a permanent dipole moment. The standard DEP force equation can be modified to the following form to accommodate both a permanent moment (m_p) and an induced dipole moment:

$$F_{DEP} = (m_p + pvE) \cdot \nabla E = m_p \cdot \nabla E + \frac{1}{2} pv\nabla E^2$$

where p is the polarizability per unit volume and v is the particle volume. The DEP voltage dependence measured at 1 MHz revealed that a quadratic relationship was dominant, which from the above equation indicates that any contribution of the permanent dipole moment m_p to the DEP of rRNA was minimal at 1 MHz. The standard, effective dipole moment method for modeling the DEP response was therefore adopted. In this model it was assumed that the rRNA molecules maintained their secondary structures, as well as some aspects of their tertiary structures, so that as a first approximation the molecules can be treated as spherical particles. An analysis, in the form of Equation (11.9) of the threshold voltage required for the positive DEP force to overcome Brownian motion provided an estimate for the induced dipole moment of 1.1×10^{-26} Cm ($\sim$3300 D) and corresponding molecular polarizability of 7.8×10^{-32} F.m^2. Analysis of the negative DEP response that occurred above 9 MHz indicated that the rRNA molecules exhibited a net negative moment of -8.3×10^{-28} Cm ($\sim$$-$250 D). This gave the molecule an effective permittivity value $78.5\varepsilon_0$, close to that of the aqueous suspending medium and a relatively small surface conductance value of $\sim$0.1 nS. This suggests that the rRNA samples had a fairly open structure accessible to the surrounding water molecules, with counterions strongly bound to the charged phosphate groups in the rRNA backbone. On the assumption that NaCl was the dominant ionic salt remaining in the extracted sample, the measured value of 12.8 mS/m for the conductivity of the solution leads to an estimate of $\sim$1 mol m^{-3} for the solutions ionic density. From Equation (12.30) the Debye screening length ($1/\kappa$) was estimated to be $\sim$10 nm, which is similar to the radius of the rRNA particle. This implies that the derived surface conductance value of K_s $\sim$0.1 nS in Equation (11.11) has contributions arising from charge movements in both the Stern layer and the diffuse part of the double layer.

11.7.3 Proteins

Nakano and Ros [232] provide an excellent and extensive review of the advances achieved (up to 2013) of the DEP manipulation and characterization of proteins. The interests of those working in protein DEP tend also to cover other types of nanoparticle. Therefore, some aspects of the review [232] have been covered in this chapter, an example being the pioneering steps by Kawabata and Washizu [208, 209] in devising a DEP-chromatography-based separation method for DNA, proteins and nanobeads. Nakano and Ros's particular contribution is the first demonstration of protein DEP streaming with insulator-based DEP [233]. They designed suitable post geometries (triangular and elliptical) of reduced size (smaller than 20 μm) to improve insulator-based DEP. Numerical simulations were also performed to calculate the electric field distribution, as well as the concentration of proteins according to a convection-diffusion model for both negative and positive DEP. In phosphate buffer solutions the DEP trapping of immunoglobulin G (IgG) and bovine serum albumin (BSA) occurred as protein aggregates, rather than single molecules. However, when a charged zwitterionic detergent was added, they observed DEP streamlining of immunoglobulin G and bovine serum albumin. Although the DEP force was not strong enough to trap the IgG and BSA molecules at the insulator posts, the proteins were concentrated along fluid streamlines under the action of positive DEP. In later work [234] they extended their study to include the influence on protein DEP of pH, surfactant concentration, aggregation, as well as electrophoretic and electroosmotic effects. Tuning of the DEP behaviour of proteins under DC conditions, from positive to negative DEP, was demonstrated through adequate choice of surfactants.

A striking feature of the review by Nakano and Ros [232] is the extent of the types of DEP devices that have been used to characterize, separate or focus DNA

Table 11.8 Examples of the DEP manipulation of Proteins (in chronological order).

Protein(s)	Investigators	Notes
Avidin; Concanavalin; Chymotrypsinogen; Ribonucease A.	Washizu *et al.* [235]	DEP occurs at much lower field than that predicted by theory
Actin	Asokan *et al.* [236]	Patterned using quadrupole electrodes
R-phycoerythrin	Hölzel *et al.* [237]	Trapping of single molecule
Kinesin-microtubules	Uppalapati *et al.* [238]	Microtubules collected and aligned
Bovine serum albumin	Lapizco-Encinas *et al.* [239]	First protein study using DC iDEP
Amyloid peptide nanotubes	Castillo *et al.* [240]	Single nanotubes immobilized
Streptavidin	Maruyama *et al.* [241]	Attachment to carbon nanotube
Immunoglobulin G, Bovine serum albumin	Nakano *et al.* [233]	DEP streamline concentration
Aβ amyloid	Staton *et al.* [242]	Used DC insulating gradient DEP
R-phycerythrin, IgG antibodies	Otto *et al.* [243]	Biological activity after DEP proven
Bovine serum albumin	Laux *et al.* [244]	Confirmed by atomic force microscopy

molecules, as well as the range of protein types studied. Examples of this are shown in Table 11.8, listed chronologically rather than in order of novelty or impact. Nearly a decade spans the time difference (1994–2003) between the demonstration by Washizu *et al.* [235] of their insulator-based DEP device for concentrating proteins by positive DEP and the work of Asokan *et al.* [236] in orienting and collecting actin filaments. From 2008 the field of protein DEP really takes off and only some of this action is encapsulated in Table 11.8. Read the review by Nakano and Ros [232] to fill in the gaps.

A particularly interesting entry in Table 11.8 is the reference to the work of Hölzel *et al.* [237], which describes the trapping of freely diffusing protein between two sharp gold nanoelectrodes spaced 500 nm apart. As a suitable molecule for this demonstration, *R*-phycoerythrin from red algae was chosen because of its intense autofluorescence. It is a 240 kDa protein of disklike shape with a diameter of 11 nm and thickness of 6 nm. Based on finite element modelling, the highest value for the field gradient factor at the tip of the electrodes was determined as $\nabla E^2 \geq 10^{21}\,V^2/m^3$, which translated to a DEP holding force of more than 0.1 pN per protein molecule. The DEP response of the protein was found to be maximal at 0.1 MHz, of similar strength at 1 MHz and much lower, though still positive, at 5 MHz. This accomplishment was significant enough – but what raises the interest level of this work further is that it was challenged in the literature by Ying *et al.* [245]. Their first comments (reproduced with permission) were as follows:

> The authors used *R*-phycoerythrin (RPE) as the sample protein and carried out fluorescence correlation spectroscopy measurements in a free solution and single molecule fluorescence measurements on a glass surface where RPE molecules were adsorbed. These two experiments showed that the RPE was not aggregated before the trapping experiment. However, these experiments alone cannot rule out the possibility of protein adsorption onto the gold electrodes, to which the authors applied alternating voltage to generate a dielectrophoretic trap and therefore having the risk of forming aggregates on the electrodes when a voltage is applied.

Hölzel *et al.* respond [246] as follows (reproduced with permission):

> Although we were able to detect and quantify single molecules adsorbed to a cleaned glass surface, as represented by the fluorescence intensity histogram, we never found fluorescing spots on the electrodes without field application. Fluorescing spots having appeared during field application immediately vanished after switching off the field. Thus, under the present experimental conditions, the proteins clearly did not stick to the gold electrodes or to the silicon surface.

To the following comment made by Ying *et al.* [245]:

> Surprisingly, they attributed the two bright fluorescent spots in Fig. 2 after 10 s field application to *just two* single RPE molecules, stating the reason that the fluorescence intensity of the spot is 'more than fourfold of the *detection limit*'; no absolute values were given . . . We suspect that the

> bright spots may be originated from the fluorescence of many molecules concentrated in the trapping region rather than from a single molecule.
>
> In conclusion, the authors need to improve their experiment and provide conclusive evidence to support the claim of single-molecule trapping. [245]

Hölzel *et al.* respond:

> Following their recommendation to compare the fluorescence intensity values of the intensity histogram with those of Fig. 2 actually reveals that each bright spot in Fig. 2 corresponds to not more than one or two *R*-phycoerythrin molecules. Hence, this rough estimation even further supports our view of single-molecule trapping. Without additional changes in field conditions or protein concentration, dielectrophoretic attraction continues increasing the local protein concentration.
>
> In conclusion, we are very thankful for the opportunity to discuss our original work in more detail, resulting in further evidence for the single-molecule character of the experiments. [246]

Here we have pertinent and great questions, followed by thoughtful and informative responses. More exchanges like this in the peer-reviewed literature are to be welcomed and would also be of considerable benefit to the DEP community.

Two other citations to the work from the group of Bier and Hölzel appear in Table 11.8, namely that of Otto *et al.* [243] and Laux *et al.* [244]. Otto *et al.* describe a silicon-based chip device with a regular array of more than 100 000 cylindrical submicroelectrodes for the DEP manipulation of nanoparticles and molecules in solution. The device was fabricated by a standard complementary metal oxide semiconductor (CMOS) process. The distribution of the electric field gradient was analysed using finite element software and the electrically induced heating was determined microscopically using a temperature sensitive fluorescent dye. Depending on voltage and frequency, the determined temperature increase was found to be compatible with protein function. Successful DEP controlled immobilization from solution was demonstrated with the autofluorescent protein R-phycoerythrin (RPE) and with fluorescently labelled IgG antibodies. Biological activity after this DEP procedure was proven by immobilization of an anti-RPE antibody and subsequent binding of RPE. These results demonstrate that DEP-directed immobilization of proteins onto microelectrodes can be achieved without the need for any chemical modification and that protein function is preserved. Being based on standard lithographical methods, further miniaturization and on-chip integration of electronics towards a multiparameter single cell analysis system is a real possibility. Laux *et al.* [244] introduced atomic force microscopy to verify the permanent immobilization of proteins by DEP and to compare this with results obtained using fluorescence microscopy. Experimental parameters such as the magnitude and duration of the applied voltage were varied systematically and the influence on the amount of immobilized proteins was investigated. A linear correlation to the duration of field application was found by atomic force microscopy and, as validation of DEP being the principal effect responsible, both methods yielded a square-law dependence of the amount of immobilized proteins on the applied voltage. While fluorescence microscopy allows real-time imaging, atomic force microscopy revealed immobilized proteins obscured in fluorescence images due to low signal-to-noise ratios. In this way the patterning of the protein molecules was found to agree with the calculated field gradient distribution. Furthermore, the higher spatial resolution of the atomic force microscope permitted the visualization of the protein distribution on single nanoelectrodes.

Other work from the group of Bier and Hölzel should also be described. Stanke *et al.* [247] have developed a system that allows for the simultaneous observation of fluid flow above and around energized microelectrodes in all three directions in space. As well as conventional microscopic inspection from above, lateral observation through the same objective was made possible using two small mirrors placed next to the electrodes. Fluid flow and movement of fluorescent nanoparticles above interdigitated electrodes were monitored by fluorescence microscopy and digital imaging and was further analysed by image processing. In contrast to 3D laser scanning microscopy the method allows much higher frame rates and can also be combined with phase contrast or interference techniques, for example. The electrical conductivity of the fluid was monitored *in situ* in the actual measuring volume. The aim in developing this system was to investigate AC electrokinetic phenomena over a very wide frequency range, covering eight frequency decades from 10 Hz to 1 GHz at up to 30 V_{pk-pk}. Over this frequency range, several mechanisms of field–particle interaction occur and can be investigated as a function of applied field and the conductivity of the fluid. To gain a better insight into the underlying mechanisms of the voltage dependence on particle velocity, Stanke *et al.* [247] tested the system at four frequencies, namely: 1 kHz (often associated with AC electroosmosis for insulator-based DEP), 10 kHz (the transition region to high frequency DEP); 1 MHz (used in electrode-based DEP of cells and macromolecules); and 500 MHz (not previously accessible). For all four frequencies, the particle velocity (u_p) varied with voltage as a power law of the type $u_p \propto V^n$. At 1 kHz this relationship took the form $u_p \propto V^{1.48}$, when in fact an exponent $n = 2$ due to AC electro-osmosis or $n = 4$ for electrothermal flow is expected from theory [248]. For the other frequencies the relationship $u_p \propto V^{1.88}$ was obtained at 10 kHz; at 1 MHz $u_p \propto V^{2.25}$; whilst at 500 MHz the result was $u_p \propto V^{2.23}$. The general trend found that the exponent n increased with increasing frequency is consistent with theoretical prediction, since at lower frequencies electroosmosis should be dominant (i.e., $n = 2$), whilst at higher frequencies electrothermal effects should be the main driving force of fluid motion (i.e., $n = 4$) [248]. In the theoretical models [248] the fluid flow was examined above the electrode plane, within a layer of thickness of the order of the electrode width or gap, whereas Stanke *et al.* [247] measured the particle velocities at a height of 100 μm height above electrodes of less than 2 μm width. This is probably the cause for the lower than expected values found for the exponent n. The geometry of a particular microelectrode array should also be a significant factor in this respect. It is clear that much work of fundamental importance to understanding and developing the performance of DEP devices can be accomplished using this innovative system.

Table 11.8 includes an example of where DEP has been used to pattern a protein on a substrate [236]. The importance of patterning proteins in this way is that it offers the advantage of permanent immobilization suitable for binding and recognition events in biosensing. A critical property for this is that the immobilized proteins retain their structural integrity as an antibody or enzyme over a long period. Bier and Hölzel, with their co-workers, have demonstrated that DEP is able to immobilize horseradish peroxidase (HRP) molecules while retaining their activity [249]. This enzyme was immobilized by DEP on a square array of tungsten nanopins, at a field frequency of 10 kHz to circumvent electrolysis of water at lower frequencies and to minimize fluid flow at high frequencies. As observed for the case of rRNA by Giraud *et al.* [230], it was found for HRP that the expected square dependence on voltage for DEP collection commenced at a threshold voltage. At 0.35 V no immobilization of HRP was observed, whilst at 1.8 V about 25% of the electrodes were covered with the enzyme, giving a threshold value of somewhat below 1.8 V. Effective immobilization at nearly all the electrodes was achieved at 3.5 and 7.1 V. However, at 7.1 V increased streaming of the fluid and enzyme aggregation occurred. According to these results, immobilization of HRP was considered to be optimal at a frequency of 10 kHz and a voltage of 3.5 V. A negative control was conducted to ensure that the HRP enzymes were immobilized as a consequence of DEP action and not by adherence of HRP preferentially on the tungsten electrode or tungsten oxide substrate. Preservation of the enzymatic function of HRP after its DEP immobilization was demonstrated by oxidizing dihydrorhodamine 123 with hydrogen peroxide as co-oxidant to create its fluorescent form, rhodamine 123.

Laux *et al.* [250] also describe the immobilization and alignment by DEP at planar nanoelectrodes of the enhanced green fluorescent protein (eGFP). This protein is commonly used as a fluorescent label in molecular biology. Applying fluorescence polarization microscopy, Laux *et al.* demonstrated for the first time a purely field-induced alignment and immobilization of protein molecules. According to X-ray data of green fluorescent protein crystals, the protein structure can be approximated by a cylinder with a diameter of 2.4 nm and a length of 4.2 nm. A single chromophore is covalently bound and situated in the center of the so-called β-barrel structure with the chromophore plane at an angle of ~60° to the symmetry axis of the cylinder. The chromophore's conformation is stabilized by additional hydrogen bonds and can be assumed as fixed in relation to the barrel. As a result, the transition dipole moment, which is nearly identical for absorption and emission, could be used to determine the orientation of the protein. Alignment was found to follow the molecule's geometrical shape with its longitudinal axes parallel to the electric field. Simultaneous DEP attraction and AC electroosmotic flow was identified as the dominant forces causing protein movement and alignment. Molecular orientation was determined by fluorescence microscopy based on polarized excitation of the proteins' chromophores. The orientation found with respect to the whole molecule was in agreement with the X-ray crystal data.

Finally, to bring us back full circle to the early days of DEP, we refer to the studies of the synthetic polyamino acids, poly-γ-benzyl L-glutamate and poly-n-butyl isocyanate, by Eisenstadt and Scheinberg in 1972 [251, 252]. These biopolymers exhibit a permanent dipole moment and it was the interaction of this moment with the nonuniform field, rather than an induced moment, which was considered responsible for their observed DEP migration to a 10 μm diameter platinum wire within a 0.3 cm diameter, 2 cm long, platinum cylinder. The DEP collection was monitored as a change in capacitance of this arrangement, the time course of which led to a determination of the diffusion coefficients of the macromolecules to within ± 5% precision and in good agreement with the values obtained by standard methods. This capacitance method was adopted from earlier work by Peter Debye (the father of dielectrics and, as described in Chapter 6, the formulator of the concept of dipole moments) in his investigations of the affect of inhomogeneous electric fields on polymer solutions [253]. Debye and his co-workers also used razor blades as electrodes to create the large field gradients required to collect polymer particles and to detect this collection by measuring the optical diffraction pattern just above the edge of the razor blade. Debye also instigated and guided the work of Prock and McConkey [254], who used the capacitance method of detecting DEP collection to determine the molecular weight distribution of polymers in a polydisperse polymer solution. These early studies [251, 252, 254] of DEP have largely been ignored, but they contain valuable insights into the dynamics of the DEP collection of small particles.

11.8 Summary

Fifty years have passed (at the time of writing) since the first reported use of DEP to collect living cells by Herb Pohl and Ira Hawk [1]. This was also the first demonstration of the separation of living from dead cells ((*S. cerevisiae*) by purely physical means. Twelve years later (1978), Pohl was in a position to describe in some detail the DEP characterization of yeast cells and several types of bacteria [22]. He could also provide preliminary data for blood platelets, chloroplasts, erythrocytes, green algae and mitochondria, as well as the continuous DEP separation of some cell types from cell mixtures and the use of DEP to form masses of living cells. The preliminary results were either described in the theses of Pohl's MSc students (e.g., Chen [66]) or formed part of work in progress with Kaler for later publication [2]. This work with Kaler describes the continuous DEP separation of yeast from green algae (*Netrium digitus*), as well as different types of green algae from their mixtures (*Chlorella vulgaris* with *Netrium digitus*; *Ankistrodesmus falcatus* with *Staurastrum gracile*), using a continuous separator of the form shown in Figure 11.4 (but with an isomotive rather than cylindrical geometry). To maximize the DEP force the cells were suspended in low conductivity media. However, with an insightful view of where the subject would go, it was pointed out that although yeast and algae could tolerate the low conductivity media used in their DEP experiments, cellular organisms that require higher osmolarity should be suspended in nonionic (isotonic) solutes such as sucrose and mannose, for example.

The electrodes used to create large field gradients in these early studies of DEP were of macroscopic form, such as razor blade edges or fine metal wires as used in earlier studies of polymer solutions in inhomogeneous fields [253, 254] and later by Eisenstadt and Scheinberg in their determination of the diffusion constant of synthetic polyamino acids by DEP [251, 252]. The introduction of microfabricated electrodes to the field of DEP occurred independently around the globe and reported initially at international conferences between 1987 and 1991 [255–257]. This, together with the development of microfluidics, led to 'an explosion of DEP publications' in the 1990s [258].

The fallout from this 'explosion' has led to a greater understanding of the physico-chemical and physiological cell-state parameters that are involved in the ability of DEP to distinguish between live and dying / dead mammalian cells, yeast and bacteria. Whereas Pohl was able only to provide preliminary data for the DEP characteristics of red blood cells, chloroplasts, green algae and mitochondria, for example, the contents of sections 11.3 to 11.6 describe in some detail the DEP properties of a much wider range of cell types (white blood cells, cancer and stem cells, neurons, oocytes, plant cells) as well as spermatozoa, oocysts, virions and even worms. Herb Pohl would be dually amazed and delighted to learn of the progress that has been made in the DEP manipulation of nucleic acids and proteins, and of the promise this brings to the realization of novel diagnostic and assay techniques, as well as biosensors, for environmental and biomedical applications. At the time (June 2016) of completing this chapter, there are at least six commercial products that incorporate DEP technology. The Panasonic bacteria counter determines the concentration of bacteria through measurement of the change of electrical impedance between microelectrodes to which bacteria are captured by DEP. This device is based on the earlier research work of Suehiro *et al.* [77]. The Shimadzu IG-1000 nanoparticle analyser exploits the fact that particles collected by DEP can form an optical grating. By monitoring the change in intensity of diffracted laser light as the particles disassemble on removal of the DEP force, particle size ranging from 0.5 to 200 nm can be determined (small particles disperse more rapidly than large ones). The Silicon Biosystems DEPArray^{TM} uses a system of electrodes to create DEP 'field cages' around cells, so that target cells identified by image analysis can be spatially manipulated and then selectively isolated for further characterization. This technology grew out of the development by Manaresi *et al.* [259] of a CMOS chip for manipulating cells by DEP. With the DEPtech-3DEP system the dielectric properties (e.g., mean membrane capacitance and cytoplasm conductivity) of cells in suspension is determined, based on changes in the suspension's optical density as the cells respond to DEP forces over a range of applied voltage frequencies [90, 91]. Two DEP technologies have also been developed for separating target cells and bioparticles from blood, namely: ApoStream^{TM} technology for isolating metastatic cancer cells [123, 124]; and microarray devices developed by Biological Dynamics for isolating nanoparticulate biomarkers (e.g., cell-free circulating DNA, RNA, exosomes) [229]. These six commercial devices share a common aspect in using metal-based electrodes to generate the field factor (E·∇)E required to collect bioparticles. To this author's knowledge other forms of DEP (i.e., insulator-, liquid-, photoconductive- or carbon-electrode based) devices have yet to be commercialized and exploit some of the other aspects of DEP described in Chapter 10 and this one. Hopefully, some of this work will soon translate to the development of new tools and sensors to address unmet needs in clinical therapy, stem cell research and tissue engineering, for example.

11.9 References

1 Pohl, H. A. and Hawk, I. (1966) Separation of living and dead cells by dielectrophoresis. *Science* **152**(3722), 647–649.
2 Pohl, H. A. and Kaler, K. (1979) Continuous dielectrophoretic separation of cell mixtures. *Cell Biophys.* **1**, 15–28.
3 Wyllie, A. H. (1992) Apoptosis and the regulation of cell numbers in normal and neoplastic tissues – an overview. *Cancer Metastasis Rev.* **11**, 95–103.
4 Fadok, V. A., Voelker, D. R., Campbell, P. A. *et al.* (1992) Exposure of phosphatidylserine on the surface of apoptotic lymphocytes triggers specific recognition and removal by macrophages. *J. Immunol.* **148**, 2207–2216.
5 Majno, G. and Joris, I. (1995) Apoptosis, oncosis, and necrosis: An overview of cell death. *Am. J. Pathol.* **146**, 3–15.
6 Savill, J., Dransfield, I., Gregory C. and Haslett, C. (2002) A blast from the past: Clearance of apoptotic cells regulates immune responses. *Nat. Rev. Immunol.* **2**, 965–975.
7 Savill, J., Gregory C. and Haslett, C. (2003) Eat me or die. *Science* **302**(5650), 1516–1517.
8 Krysko, D. V., VandenBerghe, T., Parthoens, E. *et al.* (2008) Methods for distinguishing apoptotic from necrotic cells and measuring their clearance. *Methods Enzymol.* **442**, 307–341.
9 Feinstein-Rotkopf, Y. and Arama, E. (2009) Can't live without them, can live with them: roles of caspases during vital cellular processes. *Apoptosis* **14**, 980–995.
10 Gregory, C. D. and Pound, J. D. (2011) Cell death in the neighbourhood: direct micro-environmental effects of apoptosis in normal and neoplastic tissues. *J. Pathol.* **223**, 177–194.
11 Green, D. R. (2011) *Means to an End: Apoptosis and Other Cell Death Mechanisms*, Cold Spring Harbor Laboratory Press, New York, NY.
12 Ouyang, L., Shi, Z., Zhao, S. *et al.* (2012) Programmed cell death pathways in cancer: a review of apoptosis, autophagy and programmed necrosis. *Cell Proliferation* **45**, 487–498.
13 VandenBerghe, T., Grootjans, S., Goossens, V. *et al.* (2013) Determination of apoptotic and necrotic cell death *in vitro* and *in vivo*. *Methods* **61**(2), 117–129.
14 Patel, P. and Markx, G. H. (2008) Dielectric measurement of cell death. *Enzyme Microbiol. Technol.* **43**, 463–470.
15 Carnochan, P. (1982) Dielectric properties of biological cell suspensions. Ph.D. thesis, University of Wales, Bangor.
16 Feldmann, H. (2010) *Yeast: Molecular and Cell Biology*, Wiley-Blackwell, Weinheim, Ch. 4.
17 Crane, J. S. and Pohl, H. A. (1968) A study of living and dead yeast cells using dielectrophoresis. *J. Electrochem. Soc.* **115**, 584–586.
18 Pohl, H. A. and Crane, J. S. (1971) Dielectrophoresis of Cells. *Biophys. J.* **11**, 711–719.
19 Mason, B. D. and Townsley, P. M. (1971) Dielectrophoretic separation of living cells. *Can. J. Microbiol.* **17**, 879–888.
20 Pohl, H. A. (1951) The motion and precipitation of suspensoids in divergent electric fields. *J. Appl. Phys.* **22**, 869–871.
21 Pohl, H. A. (1958) Some effects of nonuniform fields on dielectrics. *J. Appl. Phys.* **29**, 1182–1188.
22 Pohl, H. A. (1978) *Dielectrophoresis: The Behaviour of Neutral Matter in Nonuniform Fields*, Cambridge University Press, Cambridge.
23 Asami, K. and Yonezawa, T. (1996) Dielectric behavior of wild-type yeast and vacuole-deficient mutant over a frequency range of 10 kHz to 10 GHz. *Biophys. J.* **71**(4), 2192–2200.
24 Raicu, V., Raicu, G. and Turcu, G. (1996) Dielectric properties of yeast cells as simulated by the two-shell model. *Biochim. Biophys. Acta* **1274**, 143–148.
25 Asami, K., Hanai, T. and Koizumi, N. 1976. Dielectric properties of yeast cells, *J. Membrane Biol.* **28**, 169–180.
26 Asami, K. (1977) Dielectric properties of protoplasm, plasma membrane and cell wall in yeast cells. *Bull. Inst. Chem. Res. Kyoto Univ.* **55**, 394–414.
27 Carstensen, E. L. and Marquis, R. E. (1968) Passive electrical properties of microorganisms. III. Conductivity of isolated bacterial cell walls. *Biophys. J.* **8**, 536–548.
28 Halfmann, H. J., Emeis, C. C. and Zimmermann, U. (1983) Electrofusion of haploid *Saccaromyces* yeast cells of identical mating type. *Arch. Microbiol.* **134**, 1–4.

29 Förster, E. and Emeis, C. C. (1985) Quantitative studies on the viability of yeast protoplasts following dielectrophoresis. *FEMS Microbiol. Lett.* **26**, 65–69.

30 Huang, Y., Hölzel, R., Pethig, R. and Wang, X.-B. (1992) Differences in the AC electrodynamics of viable and non-viable yeast cells determined through combined dielectrophoresis and electrorotation studies. *Phys. Med. Biol.* **37**, 1499–1517.

31 Wang, X.-B., Pethig, R. and Jones, T. B. (1992) Relationship between dielectrophoresis and electrorotation of polarised particles. *J. Phys. D: Appl. Phys.* **25**, 905–912.

32 Markx, G. H., Talary, M. S. and Pethig, R. (1994) Separation of viable and nonviable yeast using dielectrophoresis. *J. Biotechnology* **32**, 29–37.

33 Markx, G. H. and Pethig, R. (1995) Dielectrophoretic separation of cells – continuous separation. *Biotechnol. Bioeng.* **45**, 337–343.

34 Patel, P. M., Bhat, A. and Markx, G. H. (2008) A comparative study of cell death using electrical capacitance measurements and dielectrophoresis. *Enzyme Microbiol. Technol.* **43**, 523–530.

35 Wang, X-B., Huang, Y., Burt, J. P. H. *et al.* (1993) Selective dielectrophoretic confinements of bioparticles in potential energy wells. *J. Phys. D: Appl. Phys.* **26**, 1278–1285.

36 Huang, Y., Wang, X-B., Tame, J. A. and Pethig, R. (1993) Electrokinetic behaviour of colloidal particles in traveling electric fields – studies using yeast cells. *J. Phys. D: Appl. Phys.* **26**, 1528–1535.

37 Talary, M. S., Burt, J. P. H., Tame, J. A. and Pethig, R. (1996) Electromanipulation and separation of cells using travelling electric fields. *J. Phys. D: Appl. Phys.* **29**, 2198–2203.

38 Nudurupati, S., Aubry, N. and Singh, P. (2006) Electrohydrodynamics of yeast cells in microchannels subjected to travelling electric fields. *J. Phys. D: Appl. Phys.* **39**, 3425–3439.

39 Mernier, G., Piacentini, N., Tornay, R. *et al.* (2011) Cell viability assessment by flow cytometry using yeast as cell model, Sensor. *Actuator. B-Chemical* **154**, 160–163.

40 Patel, S., Showers, D., Vedantam, P. *et al.* (2012) Microfluidic separation of live and dead yeast cells using reservoir-based dielectrophoresis. *Biomicrofluidics* **6**, 034102.

41 Tang, S. Y., Yi, P., Soffe, R. *et al.* (2015) Using dielectrophoresis to study the dynamic response of single budding yeast cells to lyticase. *Anal. Bioanal. Chem.* **407**, 3437–3448.

42 Vahey, M. D., Pesudo, L. Q., Svensson, J. P. *et al.* (2013) Microfluidic genome-wide profiling of intrinsic electrical properties in *Saccharomyces cerevisiae*. *Lab Chip* **13**, 2754–2763.

43 Smith, A. M., Heisler, L. E., Mellor, J. *et al.* (2009) Quantitative phenotyping via deep barcode sequencing. *Genome Res.* **19**, 1836–1842.

44 Giaever G. and Nislow, C. (2014) The yeast deletion collection: A decade of functional genomics. *Genetics* **197**, 451–465.

45 Vahey, M. D. and Voldman, J. (2009) High-throughput cell and particle characterisation using isodielectric separation. *Anal. Chem.* **81**, 2446–2455.

46 Schmidt, K. P., Glaser, R., Pescheck, C. and Krause, G. (1979) Kinetics of pearl-chain formation, a parameter for measuring dielectrohoresis of cells. *Studia Biophysics* **75**, 81–89.

47 Venkatesh, A. G. and Markx, G. H. (2007) On the height of cell aggregates formed with positive dielectrophoresis. *J. Phys. D: Appl. Phys.* **40**, 106–113.

48 Kadaksham, J., Singh, P. and Aubry, N. (2005) Dielectrophoresis induced clustering of viable yeast cells. *Electrophoresis* **26**, 3738–3744.

49 Talary, M. and Pethig, R. (1994) Optical technique for measuring the positive and negative dielectrophoretic behaviour of cells and colloidal suspensions. *IEE Proc. Sci. Meas. Technol.* **141**, 395–399.

50 Hölzel, R. (2002) Single particle characterization and manipulation by opposite field dielectrophoresis, *J. Electrostat.* **56**, 435–447.

51 Prasad, B., Du, S., Badawy, W. and Kaler, K. V. I. S. (2005) A real-time multiple-cell tracking platform for dielectrophoresis (DEP)-based cellular analysis. *Meas. Sci. Technol.* **16**, 909–924.

52 Abdallat, R. G., Tajuddin, A. S. A., Gould, D. H. *et al.* (2013) Process development for cell aggregate arrays encapsulated in a synthetic hydrogel using negative dielectrophoresis. *Electrophoresis* **34**, 1059–1067.

53 Zhu, Z., Frey, O., Ottoz, D. S. *et al.* (2012) Microfluidic single-cell cultivation chip with controllable immobilization and selective release of yeast cells. *Lab Chip* **12**, 906–915.

54 Jaeger, M. S., Uhlig, K., Schnelle, T. and Mueller, T. (2008) Contact-free single-cell cultivation by negative dielectrophoresis. *J. Phys. D: Appl. Phys.* **41**, 175502.

55 Perch-Nielsen, I. R., Bang, D. D., Poulsen, C. R. *et al.* (2003) Removal of PCR inhibitors using dielectrophoresis as a selective filter in a microsystem. *Lab Chip* **3**, 212–216.

56 Jaramillo, M. C., Martinez-Duarte, R., Huttener, M. *et al.* (2013) Increasing PCR sensitivity by removal of polymerase inhibitors in environmental samples by using dielectrophoresis. *Biosensor. Bioelectron.* **43**, 297–303.

57 Jaramillo, M. C., Huttener, M., Alvarez, J. M. *et al.* (2015) Dielectrophoresis chips improve PCR detection of the food-spoiling yeast *Zygosaccharomyces rouxi* in apple juice. *Electrophoresis* **36**, 1471–1478.

58 Hunt, P., Issadore, D. and Westervelt, R. M. (2008) Integrated circuit / microfluidic chip to programmably trap and move cells and droplets with dielectrophoresis. *Lab Chip* **8**, 81–87.

59 Li, Y. L., Dalton, C., Crabtree, H. J. *et al.* (2007) Continuous dielectrophoretic cell separation microfluidic device. *Lab Chip* **7**, 239–248.

60 Abidin, Z. Z., Downes, L. and Markx, G. H. (2007) Novel electrode structures for large scale dielectrophoretic separations based on textile technology. *J. Biotechnol.* **130**, 183–187.

61 Higginbotham, S. N. and Sweatman, D. R. (2008) A combined travelling wave dielectrophoresis and impedance sensing device for sensing biological cell suspensions. *J. Phys. D: Appl. Phys.* **41**, 175503.

62 Suehiro, J., Zhou, G. B., Imamura, M. and Hara, M. (2003) Dielectrophoretic filter for separation and recovery of biological cells in water. *IEEE Trans. Ind. Appl.* **39**, 1514–1521.

63 Terpitz, U., Letschert, S., Bonda, U. *et al.* (2012) Dielectric analysis and multi-cell electrofusion of the yeast *Pichia pastoris* for electrophysiological studies. *J. Membr. Biol.* **245**, 815–826.

64 Tang, S. Y., Zhang, W., Soffe, R. *et al.* (2015) High resolution scanning electron microscopy of cells using dielectrophoresis. *PloS One* **9**, e104109.

65 Gilbert, P. and Moore, L. E. (2005) Cationic antiseptics: diversity of action under a common epithet. *J. Appl. Microbiol.* **99**, 703–715.

66 Chen, C. S. (1972) On the nature and origins of biological dielectrophoresis. M.Sc thesis, Oklahoma State University.

67 Reigada, R. (2011) Influence of chloroform in liquid-ordered and liquid-disordered phases in lipid membranes. *J. Phys. Chem.* **115**, 2527–2535.

68 Zhou, X-F., Markx, G. H., Pethig, R. and Eastwood, I. M. (1995) Differentiation of viable and non-viable bacterial biofilms using electrorotation. *Biochim. Biophys. Acta*, **1245**, 85–93.

69 Hoettges, K. F., Dale, J. W. and Hughes, M. P. (2007) Rapid determination of antibiotic resistance in *E. coli* using dielectrophoresis. *Phys. Med. Biol.* **52**, 6001–6009.

70 Chung, C. C., Cheng, I. F., Yang, W. H. and Chang, H. C. (2011) Antibiotic susceptibility test based on the dielectrophoretic behavior of elongated *Escherichia coli* with cephalexin treatment. *Biomicrofluidics* **5**, 021102.

71 Cabeen, M. T. and Jacobs-Wagner, C. (2005) Bacterial cell shape. *Nat. Rev. Microbiol.* **3**, 601–610.

72 Chung, C. C., Cheng, I. F., Chen, H. M. *et al.* (2012) Screening of antibiotic susceptibility to beta-lactam-induced elongation of Gram-negative bacteria based on dielectrophoresis. *Anal. Chem.* **84**, 3347–3354.

73 Jones, P. V., Hilton, S. H., Davis, P. E. *et al.* (2015) Biophysical separation of *Staphylococcus epidemidis* strains based on antibiotic resistance. *Analyst* **140**, 5152–5161.

74 Elitas, M., Martinez-Duarte, R., Dhar, N. *et al.* (2014) Dielectrophoresis-based purification of antibiotic-treated bacterial subpopulations. *Lab Chip* **14**, 1850–1857.

75 Donato, S. S., Chu, V., Prazeres, D. M. F. Conde, J. P. (2013) Metabolic viability of *Escherichia coli* trapped by dielectrophoresis in microfluidics. *Electrophoresis* **34**, 575–582.

76 Suehiro, J., Yatsunami, R., Hamada, R. and Hara, M. (1999) Quantitative estimation of biological cell concentration suspended in aqueous medium by using dielectrophoretic impedance measurement method. *J. Phys. D: Appl. Phys.* **32**, 2814–2820.

77 Suehiro, J., Hamada, R., Noutomi, D. *et al.* (2003) Selective detection of viable bacteria using dielectrophoretic impedance measurement method, *J. Electrostat.* **57**, 157–168.

78 Zhou, R., Wang, P. and Chang, H-C. (2006) Bacteria capture, concentration and detection by alternating current dielectrophoresis and self-assembly of dispersed single-wall carbon nanotubes. *Electrophoresis* **27**, 1376–1385.

79 Amako, E., Enjoji, T., Uchida, S. and Tochikubo, F. (2011) Estimation of metabolic characteristics of heat-injured bacteria by dielectrophoretic impedance measurement. *Electron. Comm. Japan* **94**, 9–16.

80 Jones, P. G., Mitta, M., Kim, Y. *et al.* (1996) Cold shock induces a major ribosomal-associated protein that unwinds double-stranded RNA in *Escherichia coli. Proc. Natl. Acad. Sci. USA* **93**, 76–80.

81 Blair, D. F. (1991) The bacterial rotary motor. *Nanotechnology* **2**, 123–133.

82 Washizu, M., Kurahashi, Y., Iochi, H. *et al.* (1993) Dielectrophoretic measurement of bacterial motor characteristics. *IEEE Trans. Ind. Appl.* **29**, 286–294.

83 Berry, R. M. and Berg, H. C. (1999) Torque generated by the flagellar motor of *Escherichia coli* while driven backward. *Biophys. J.* **76**, 580–587.

84 Hughes, M. P. and Morgan, H. (1999) Measurement of bacterial flagellar thrust by negative dielectrophoresis. *Biotechnol. Prog.* **15**, 245–249.

85 Pethig, R. (2005) Biodetection using micro-physiometry tools based on electrokinetic phenomena, in *Defence against Bioterror: Detection Technologies, Implementation Strategies and Commercial Opportunities* (eds D. Morrison, F. Milanovich, D. Ivniski and T. R. Austin). Springer Verlag, Dordrecht, pp. 129–142.

86 Ratanachoo, K., Gascoyne, P. R. C. and Ruchirawat, M. (2002) Detection of cellular responses to toxicants by dielectrophoresis. *Biochim. Biophys. Acta* **1564**, 449–458.

87 Wang, X., Becker, F. F. and Gascoyne, P. R. C. (2002) Membrane dielectric changes indicate induced apoptosis in HL-60 cells more sensitively than surface phosphatidylserine expression or DNA fragmentation. *Biochim. Biophys. Acta* **1564**, 412–420.

88 Pethig, R. and Talary, M. S. (2007) Dielectrophoretic detection of membrane morphology changes in Jurkat T-cells undergoing etoposide-induced apoptosis. *IET Nanobiotechnol.* **1**, 2–9.

89 Mulhall, H. J., Cardnell, A., Hoettges, K. F. *et al.* (2015) Apoptosis progression studied using parallel dielectrophoresis electrophysiological analysis and flow cytometry. *Intergr. Biol.* **7**, 1396–1401.

90 Hoettges, K. F., Hübner, Y., Broche, L. M. *et al.* (2008) Dielectrophoresis-activated multiwell plate for label-free high-throughput drug assessment. *Anal. Chem.* **80**, 2063–2068.

91 Broche, L. M., Hoettges, K. F., Ogin, S. L. *et al.* (2011) Rapid, automated measurement of dielectrophoretic forces using DEP-activated microwells. *Electrophoresis* **32**, 2393–2399.

92 Lv, Y., Zeng, L., Zhang, G. *et al.* (2013) Systematic dielectrophoretic analysis of the Ara-C-induced NB4 cell apoptosis combined with gene expression profiling. *Int. J. Nanomed.* **8**, 2333–2350.

93 Chung, C., Waterfall, M., Pells, S. *et al.* (2011) Dielectrophoretic characterisation of mammalian cells above 100 MHz. *J. Electr. Bioimp.* **2**, 64–71.

94 Broche, L. M., Labeed, F. H. and Hughes, M. P. (2005) Extraction of dielectric properties of multiple populations from dielectrophoretic collection spectrum data. *Phys. Med. Biol.* **50**, 2267–2274.

95 Nikolic-Jaric, M., Cabel, T., Salimi, E. *et al.* (2013) Differential electronic detector to monitor apoptosis using dielecrophoresis-induced translation of flowing cells (dielectrophoresis cytometry). *Biomicrofluidics* **7**, 024101.

96 Butler M. (2005) Animal cell cultures: Recent achievements and perspectives in the production of biopharmaceuticals. *Appl. Microbiol. Biotechnol.* **68**, 283–291.

97 Walsh, G. (2010) Biopharmaceutical benchmarks 2010. *Nat. Biotechnol.* **28**, 917–924.

98 Braasch, K., Nikolic-Jaric, M., Cabel, T. *et al.* (2013) The changing dielectric properties of CHO cells can be used to determine early apoptotic events in a bioprocess. *Biotechnol. Bioeng.* **110**, 2902–2914.

99 Chin, S., Hughes, M. P., Coley, H. M. and Labeed, F. H. (2006) Rapid assessment of early biophysical changes in K562 cels during apoptosis determined using dielectrophoresis. *Int. J. Nanomed.* **3**, 333–337.

100 Labeed, F. H., Coley, H. M. and Hughes, M. P. (2006) Differences in the biophysical properties of membrane and cytoplasm of apoptotic cells revealed using dielectrophoresis. *Biochim. Biophys. Acta* **1760**, 922–929.

101 Tsoneva, I. Ch., Zhelev, D. V. and Dimitrov, D. S. (1986) Red blood cell dielectrophoresis in axisymmetric fields, *Cell Biophysics* **8**, 89–101.

102 Srivastava, S. K., Daggolu, P. R., Burgess, S. C. and Minerick, A. R. (2008) Dielectrophoretic characterisation of erythrocytes: Positive ABO blood types. *Electrophoresis* **29**, 5033–5046.

103 Leonard, K. M. and Minerick, A. R. (2011) Explorations of ABO-Rh antigen expressions on erythrocyte dielectrophoresis: Changes in cross-over frequency. *Electrophoresis* **32**, 2512–2522.

104 Srivastava, S. K., Aremiou, A. and Minerick, A. R. (2011) Direct current insulator-based dielectrophoretic characterization of erythrocytes: ABO-Rh human blood typing. *Electrophoresis* **32**, 2530–2540.

105 Xu, C., Wang, Y., Cao, M. and Lu, Z. (1999) Dielectrophoresis of human red cells in microchips. *Electrophoresis* **20**, 1829–1831.

106 Aceti, A., Bonincontro, A., Cametti, C. *et al.* (1990) Electrical conductivity of human erythrocytes infected with *Plasmodium falciparum* and its modification following quinine therapy. *Trans. Roy. Soc. Trop. Med. Hygiene* **84**, 671–672.

107 Gascoyne, P., Pethig, R., Satayavivad, J. *et al.* (1997) Dielectrophoretic detection of changes in erythrocyte membranes following malarial infection. *Biochim. Biophys. Acta* **1323**, 240–252.

108 Gascoyne, P., Mahidol, C., Ruchirawat, M. *et al.* (2002) Microsample preparation by dielectrophoresis: isolation of malaria. *Lab Chip* **2**, 70–75.

109 Gascoyne, P., Satayavivad, J. and Ruchirawat, M. (2004) Microfluidic approaches to malaria detection. *Acta Tropica* **89**, 357–369.

110 Piacentini, N., Mernier, G., Tornay, R. and Renaud, P. (2011) Separation of platelets from other blood cells in continuous-flow by dielectrophoresis field-flow fractionation. *Biomicrofluidics* **5**, 034122.

111 Pethig, R., Bressler, V., Carswell-Crumpton, C. *et al.* (2002) Dielectrophoretic studies of the activation of human T lymphocytes using a newly developed cell profiling system. *Electrophoresis* **23**, 2057–2063.

112 Huang, Y., Wang, X.-B., Gascoyne, P. R. C. and Becker, F. F. (1999) Membrane dielectric responses of human T-lymphocytes following mitogenic stimulation. *Biochim. Biophys. Acta* **1417**, 51–62.

113 Hu, X., Arnold, W. M. and Zimmermann, U. (1990) Alterations in the electrical-properties of lymphocyte-T and lymphocyte-B membranes induced by mitogenic stimulation-activation monitored by electrorotation of single cells. *Biochim. Biophys. Acta* **1021**, 191–200.

114 Burt, J. P. H., Pethig, R., Gascoyne, P. R. C. and Becker, F. F. (1990) Dielectrophoretic characterisation of Friend murine erythroleukaemic cells as a measure of induced differentiation. *Biochim. Biophys. Acta* **1034**, 93–101.

115 Burt, J. P. H., Al-Ameen, T. A. K. and Pethig, R. (1989) An optical dielectrophoresis spectrometer for low-frequency measurements on colloidal suspensions. *J. Phys. E: Sci. Instrum.* **22**, 952–957.

116 Gascoyne, P. R. C., Huang, Y., Pethig, R. *et al.* (1992) Dielectrophoretic separation of mammalian cells studied by computerized image analysis. *Meas. Sci. Technol.* **3**, 4394–4345.

117 Pethig, R., Huang, Y., Wang, X.-B. and Burt, J. P. H. (1992) Positive and negative dielectrophoretic collection of colloidal particles using interdigitated castellated electrodes. *J. Phys. D: Appl. Phys.* **24**, 881–888.

118 Gascoyne, P. R. C., Noshari, J., Becker, F. F. and Pethig, R. (1994) Use of dielectrophoretic collection spectra for characterizing differences between normal and cancerous cells. *IEEE Trans. Ind. Appl.* **30**, 829–834.

119 Gascoyne, P. R. C. and Becker, F. F. (1990) Alterations in electrophoretic mobility, diaphorase activity, and terminal differentiation induced in murine erythroleukemia cells by differentiating agents. *Cell. Physiol.* **142**, 309–315.

120 Becker, F. F., Wang, X.-B., Huang, Y. *et al.* (1994) The removal of human leukaemia cells from blood using interdigitated microelectrodes. *J. Phys. D: Appl. Phys.* **27**, 2659–2662.

121 Hayes, D. F., Cristofanilli, M., Budd, G. T. *et al.* (2006) Circulating tumor cells at each follow-up time point during therapy of metatastic breast cancer patients predict progression-free and overall survival. *Clin. Cancer Res.* **12**, 4218–4224.

122 Gascoyne, P. R. C., Noshari, J., Anderson, T. J. and Becker, F. F. (2009) Isolation of rare cells from cell mixtures by dielectrophoresis. *Electrophoresis* **30**, 1388–1398.

123 Shim, S., Stemke-Hale, K., Tsimberidou, A. M. *et al.* (2013) Antibody-independent isolation of circulating tumor cells by continuous-flow dielectrophoresis. *Biomicrofluidics* **7**, 011807.

124 Gupta, V., Jafferji, I., Garza, M. *et al.* (2012) ApoStream^{TM}, a new dielectrophoretic device for antibody independent isolation and recovery of viable cancer cells from blood. *Biomicrofluidics* **6**, 024133.

125 Shim, S., Stemke-Hale, K., Noshari, J. *et al.* (2013) Dielectrophoresis has broad applicability to marker-free isolation of tumor cells from blood by microfluidic systems. *Biomicrofluidics* **7**, 011808.

126 An, J., Lee, J., Lee, S. H. *et al.* (2009) Separation of malignant human breast cancer epithelial cells from healthy epithelial cells using an advanced dielectrophoresis-activated cell sorter (DACS). *Anal. Bioanal. Chem.* **394**, 801–809.

127 Yang, F., Yang, X. M., Jiang, H. *et al.* (2010) Dielectrophoretic separation of colorectal cancer cells. *Biomicrofluidics*, **4**, 013204.

128 Moon, H. S., Kwon, K., Kim, S. I. *et al.* (2011) Continuous separation of breast cancer cells from blood samples using multi-orifice flow fractionation (MOFF) and dielectrophoresis (DEP). *Lab Chip*, **11**, 1118–1125.

129 Salmanzadeh, A., Romera, L., Shafiee, H. *et al.* (2012) Isolation of prostate tumor initiating cells (TICs) through their dielectrophoretic signature. *Lab Chip* **12**, 182–189.

130 Salmanzadeh, A., Kittur, H., Sano, M. B. *et al.* (2012) Dielectrophoretic differentiation of mouse ovarian surface epithelial cells, macrophages, and fibroblasts using contactless dielectrophoresis. *Biomicrofluidics*, **6**, 024104.

131 Salmanzadeh, A., Sano, M. B., Gallo-Villanueva, R. C. *et al.* (2013) Investigating dielectric properties of different stages of syngeneic murine ovarian cancer cells. *Biomicrofluidics* **7**, 011809.

132 Mulhall, H. J., Labeed, F. H., Kazmi, B. *et al.* (2011) Cancer, pre-cancer and normal oral cells distinguished by dielectrophoresis. *Anal. Bioanal. Chem.* **401**, 2455–2463.

133 Pethig, R., Menachery, A., Pells, S. and De Sousa, P. (2010) Dielectrophoresis: A review of applications for stem cell research. *J. Biomed. Biotechnol.* 182581.

134 Talary, M. S., Mills, K. I., Hoy, T. *et al.* (1995) Dielectrophoretic separation and enrichment of $CD34^+$ cell subpopulation from bone marrow and peripheral blood stem cells. *Med. Biol. Eng. Comput.* **33**, 235–237.

135 Stephens, M., Talary, M. S., Pethig, R. *et al.* (1996) The dielectrophoretic enrichment of $CD34^+$ cells from bone peripheral blood stem cell harvests. *Bone Marrow Transplantation* **18**, 777–782.

136 Vykoukal, J., Vykoukal, D. M., Freyberg, S. *et al.* (2008) Enrichment of putative stem cells from adipose tissue using dielectrophoretic field-flow fractionation. *Lab Chip*, **8**, 1386–1393.

137 Wu, A. Y. and Morrow, D. (2012) Clinical use of dielectrophoresis separation for live adipose derived stem cells. *J. Translational Med.* 10.99.

138 Flanagan, L. A., Jente, L., Marchenko, S. A. *et al.* (2008) Unique dielectric properties distinguish stem cells and their differentiated progeny. *Stem Cells* **26**, 656–665.
139 Labeed, F. H., Lu, J., Mulhall, H. J. *et al.* (2011) Biophysical characteristics reveal neural stem cell differentiation potential. *PloS One* **6**, e25458.
140 Nourse, J. L., Prieto, J. L., Dickson, A. R. *et al.* (2014) Membrane biophysics define neuron and astrocyte progenitors in the neural lineage. *Stem Cells* **32**, 706–716.
141 Velugotla, S., Pells, S., Mjoseng, H. K. *et al.* (2012) Dielectrophoresis based discrimination of human embryonic stem cells from differentiating derivatives. *Biomicrofluidics* **6**, 044113.
142 Adams, T. N. G., Turner, P. A., Janorkar, A. V. *et al.* (2014) Characterizing the dielectric properties of human mesenchymal stem cells and the effects of charged elastin-like polypeptide copolymer treatment. *Biomicrofluidics* **8**, 054109.
143 Pethig, R. and Kell, D. B. (1987) The passive electrical properties of biological systems: their significance in physiology, biophysics and biotechnology. *Phys. Med. Biol.* **32**, 933–970.
144 Muratore, M., Srsen, V., Waterfall, M. *et al.* (2012) Biomarker-free dielectrophoretic sorting of differentiating myoblast multipotent progenitor cells and their membrane analysis by Raman spectroscopy. *Biomicrofluidics* **6**, 034113.
145 Muratore, M., Mitchell, S. and Waterfall, M. (2013) Plasma membrane characterization, by scanning electron microscopy, of multipotent myoblasts-derived populations sorted using dielectrophoresis. *Biochem. Biophys. Res. Comm.* **438**, 666–672.
146 Heida, T., Rutten, W. L. C. and Marani, E. (1998) Characterization of the behaviour of dissociated neurons exposed to dielectrophoretic forces. *Proc. Ann. Int. Conf. IEEE Eng. Med. Biol. Soc.* **20**, 2935–2938.
147 Heida, T., Rutten, W. L. C. and Marani, E. (1998) Dielectrophoretic trapping of dissociated fetal cortical rat neurons. *IEEE Trans Biomed. Eng.* **48**, 921–930.
148 Heida, T., Vutto, P., Rutten, W. L. C. and Marani, E. (2001) Viability of dielectrophoretically trapped neural cortical cells in culture. *J.Neurosci. Meth.* **110**, 37–44.
149 Prasad, S., Zhang, X., Yang, M. *et al.* (2004) Separation of individual neurons using dielectrophoretic alternative fields. *J. Neurosci. Meth.* **135**, 79–88.
150 Zhou, T., Perry, S. F., Ming, Y. *et al.* (2015) Separation and assisted patterning of hippocampal neurons from glial cells using positive dielectrophoresis. *Biomed. Microdev.* **17**, 62.
151 Gentet, L. J., Stuart, G J. and Clements, J. D. (2000) Direct measurement of specific membrane capacitance in neurons. *Biophys. J.* **79**, 314–320.
152 Fuhr, G., Müller, T., Baukloh, V. and Lucas, K. (1998) High-frequency electric field trapping of individual human spematozoa. *Human Reproduction* **13**, 136–141.
153 Boon, J. B. and Marcos, Y. K. (2014) Effect of dielectrophoresis on spermatozoa. *Microfluid. Nanofluid.* **17**, 613–622.
154 MacQueen, L. A., Buschmann, M. D. and Wertheimer, M. R. (2010) Mechanical properties of mammalian cells in suspension measured by electro-deformation. *J. Micromech. Microeng.* **20**, 065007.
155 Chen, J., Abdelgawad, M., Yu, L. *et al.* (2011) Electrodeformation for single cell mechanical characterization. *J. Micromech. Microeng.* **21**, 054012.
156 Guido, I., Jaeger, M. S. and Duschl, C. (2011) Dielectrophoretic stretching of cells allows for characterization of their mechanical properties. *Eur. Biophys. J.* **40**, 281–288.
157 Lin, Y-W. and Wang, G-J. (2013) Dielectrophoresis measurement of bovine endothelial cells adhesion on different materials. *Microsyst. Technol.* **19**, 945–955.
158 Haque, Md. M., Moisescu, M. G., Valkai, S. *et al.* (2014) Stretching of red blood cells using an electro-optics trap. *Biomed. Optic. Express* **6**, doi: 10.1364.
159 Zhang, X., Chu, H. K., Zhang, Y. *et al.* (2015) Rapid characterizaion of the biomechanical properties of drug-treated cells in a microfluidic device. *J. Micromech. Microeng.* **25**, 105004.
160 Schmid-Schonbein, G. W., Sung, K. L., Tozeren, H. *et al.* (1981) Passive mechanical properties of human leukocytes. *Biophys. J.* **36**, 243–256.
161 Zheng, Y., Wen, J., Nguyen, J. *et al.* (2014) Decreased deformability of lymphoctyes in chronic lymphocytic leukemia. *Sci. Rep.* **5**, 7613.
162 Markx, G. H., Huang, Y., Zhou, X-F. and Pethig, R. (1994) Dielectrophoretic characterization and separation of micro-organisms. *Microbiol.* **140**, 585–591.
163 Brown, A. P., Betts, W. B., Harrison, A. B. and O'Neill, J. G. (1999) Evaluation of a dielectrophoretic bacterial counting technique. *Biosens. Bioelectron.* **14**, 341–351.
164 Castellarnau, M., Errachid, A., Madrid, C. *et al.* (2006) Dielectrophoresis as a tool to characterize and differentiate isogenic mutants of *Escherichia coli*. *Biophys. J.* **91**, 3937–3945.
165 Suehiro, J., Ohtsubo, A., Hatano, T. and Hara, M. (2006 Selective detection of bacteria by a dielectrophoretic impedance measurement method using an antibody-immobilized electrode chip. *Sens. Actuator.* **B. 119**, 319–326.

166 Sanchis, A., Brown, A. P., Sancho, M. *et al.* (2007) Dielectric characterization of bacterial cells using dielectrophoresis. *Bioelectromag.* **28**, 393–401.

167 Johari, J., Hübner, Y., Hull, J. C. *et al.* (2003) Dielectrophoretic assay of bacterial resistance to antibiotics. *Phys. Med. Biol.* **48**, 193–198.

168 Kaler, K. V. I. S. and Jones, T. B. (1990) Dielectrophoretic spectra of single cells determined by feedback-controlled levitation. *Biophys. J.* **57**, 173–182.

169 Kaler, K. V. I. S., Xie, J.-P., Jones, T. B. and Paul, R. (1992) Dual-frequency dielectrophoretic levitation of canola protoplasts. *Biophys. J.* **63**, 58–69.

170 Kaler, K. V. I. S., Jones, T. B., and Paul, R. J. (1995) Low-frequency micromotions of DEP-levitated plant protoplasts, I. *J. Colloid Interface Sci.* **175**, 108–117.

171 Inoue, T., Pethig, R., Al-Ameen, T. A. K. *et al.* (1988) Dielectrophoretic behaviour of *Micrococcus lysodeikticus* and its protoplast. *J. Electrostatics* **21**, 215–223.

172 Wang, X.-B., Pethig, R. and Jones, T. B. (1992) Relationship of dielectrophoretic and electrorotational behaviour exhibited by polarized particles. *J. Phys. D: Appl. Phys.* **25**, 905–912.

173 Barrie, G., Paul, R., Kaler, K. V. I. S. and Jones, T. B. (1995) Low-frequency micromotion of DEP-levitated bioparticles, II: Spectral and stability analysis. *J. Colloid Interface Sci.* **175**, 97–107.

174 Paddison, S. J., Paul, R., and Kaler, K. V. I. S. (1995) Hysteresis loops in the low frequency region of the Clausius–Mossotti polarization factor: the result of non-linear boundary conditions at the particle interface. *Bioelectrochem. Bioenerg.* **38**, 321–331.

175 Gimsa, J., Marszalek, P., Loewe, U. and Tsong, T. Y. (1991) Dielectrophoresis and electrorotation of neurospora slime and murine myeloma cells. *Biophys. J.* **60**, 749–760.

176 Chisti, Y. (2007) Biodiesel from microalgae. *Biotechnol. Adv.* **25**, 294–306.

177 Singh, A. and Olsen, S. I. (2011) A critical review of biochemical conversion, sustainability and life cycle assessment of algal biofuels. *Appl. Energy* **88**, 3548–3555.

178 Deng, Y.-L., Chang, J.-S. and Juang, Y.-J. (2013) Separation of microalgae with different lipid contens by dielectrophoresis. *Bioresource Technol.* **135**, 137–141.

179 Hadady, H., Wong, J. J., Hiibel, S. R. *et al.* (2014) High frequency dielectrophoretic response of microalgae over time. *Electrophoresis* **35**, 3533–3540.

180 Fiorina, P., Shapiro, A. M. J., Ricordi, C. and Secchi, A. (2008) The clinical impact of islet transplantation. *Am. J. Transplant.* **8**, 1990–1997.

181 Pethig, R., Jakubek, L. M., Sanger, R. H. *et al.* (2005) Electrokinetic measurements of membrane capacitance and conductance for pancreatic β-cells. *IEE Proc. Nanobiotechnol.* **152**, 189–193.

182 Pethig, R., Menachery, A., Heart, E. *et al.* (2008) Dielectrophoretic assembly of insulinoma cels and fluorescent nanosensors into three-dimensional pseudo-islet constructs. *IET Nanobiotechnol.* **2**, 31–38.

183 Burgarella, S., Merlo, S., Figliuzzi, M. and Remuzzi, A. (2013) Isolation of Langerhans islets by dielectrophoresis. *Electrophoresis* **34**, 1068–1075.

184 Dessie, S.-W., Rings, F., Hölker, M. *et al.* (2007) Dielectrophoretic behaviour of *in vitro*-derived metaphase II oocytes and zygotes and its relation to *in vitro* embryonic developmental competence and mRNA expression pattern. *Reproduction* **133**, 931–946.

185 Goater, A. D., Burt, J. P. H. and Pethig, R. (1997) A combined travelling wave dielectrophoresis and electrorotation device: applied to the concentration and viability determination of *Cryptosporidium. J. Phys. D: Appl. Phys.* **30**, L65–L69.

186 Chuang, H.-S., Raizen, D. M., Lamb, A. *et al.* (2011) Dielectrophoresis of *Caenorhabditis elegans. Lab Chip* **11**, 599–604.

187 Rezai, P., Siddiqui, A., Selvaganapathy, P. R. and Gupta, B. P. (2010) Electrotaxis of *Caenorhabditis elegans* in a microfluidic environmemt. *Lab Chip*, **10**, 220–226.

188 Morgan, H. and Green, N. G. (1997) Dielectrophoretic manipulation of rod-shaped viral particles. *J. Electrostatics* **42**, 279–293.

189 Green, N. G., Morgan, H. and Milner, J. J. (1997) Manipulation and trapping of nanoscale bioparticles using dielectrophoresis. *J. Biochem. Biophys. Methods* **35**, 89–102.

190 Van der Touw, F., Briede, J. W. H. and Mandel, M. (1973) Electric permittivity of alfalfa mosaic virus in aqueous solutions. *Biopolymers* **12**, 111–119.

191 Hughes, M. P., Morgan, H., Rixon, F. J. *et al.* (1998) Manipulation of herpes simplex virus type 1 by dielectrophoresis. *Biochim. Biophys. Acta.* **1425**, 119–126.

192 Hughes, M. P., Morgan, H. and Rixon, F. J. (2001) Dielectrophoretic manipulation and characterization of herpes simplex virus-1 capsids. *Eur. Biophys. J.* **30**, 268–272.

193 Ermolina, I., Milner, J. and Morgan, H. (2006) Dielectrophoretic investigation of plant virus particles: Cow pea mosaic virus and tobacco mosaic virus. *Electrophoresis* **27**, 3939–3948.

194 Grom, F., Kentsch, J., Müller, T. *et al.* (2006) Accumulation and trapping of hepatitis-A particles by

electrohydrodynamic flow and dielectrophoresis. *Electrophoresis* **27**, 1386–1393.

195 Docoslis, A., Espinoza, L. A. T., Zhang, B. *et al.* (2007) Using nonumifom electric fields to accelerate the transport of viruses to surfaces from media of physiological ionic strength. *Langmuir* **23**, 3840–3848.

196 Madiyar, F. R., Syed, L. U., Culbertson, C. T. and Li, J. (2013) Manipulation of bacteriophages with dielectrophoresis on carbon nanofiber nanoelectrode arrays. *Electrophoresis* **34**, 1123–1130.

197 Ding, J., Lawrence, R. M., Jones, P. V. *et al.* (2016) Concentration of Sindbis virus with optimized gradient insulator-based dielectrophoresis. *Analyst* **141**, 1997–2008.

198 Nakano, M., Ding, Z. and Suehiro, J. (2016) Dielectrophoresis and dielectrophoretic impedance detection of adenovirus and rotavirus. *Jpn. J. Appl. Phys.* **55**, 017001.

199 Safavieh, M., Khetani, S., Juillard, F. *et al.* (2016) Electrical response of a B lymphoma cell line latently infected with Kaposi's sarcoma herpesvirus. *Biosens. Bioelectron.* **80**, 230–236.

200 Washizu, M. (1990) Electrostatic manipulation of biological objects, *J. Electrostat.* **25**, 109–123.

201 Washizu, M. and Kurosawa, O. (1990) Electrostatic manipulation of DNA in microfabricated structures. *IEEE Trans. Ind. Appl.* **26**, 1165–1172.

202 Washizu, M., Nanba, T. and Masuda, S. (1990) Handling biological cells using a fluid integrated circuit. *IEEE Trans. Ind. Appl.* **26**, 352–358.

203 Hölzel, R. (2009) Dielectric and dielectrophoretic properties of DNA. *IET Nanobiotechnol.* **3**, 22–45.

204 Suzuki, S., Yamanashi, T., Tazawa, S. *et al.* (1998) Quantitative analysis of DNA orientation in stationary AC electric fields using fluorescence anisotropy. *IEEE Trans. Ind. Appl.* **34**, 75–83.

205 Kabata, H., Kurasawa, O., Arai, I. *et al.* (1993) Visualization of single molecules of RNA polymerase sliding along DNA. *Science* **262**, 1561–1563.

206 Kabata, H., Okada W. and Washizu, M. (2000) Single-molecule dynamics of the Eco RI enzyme using stretched DNA: its application to *in situ* sliding assay and optical DNA mapping. *Jpn. J. Appl. Phys.* **39**, 7164–7171.

207 Yamamoto, T., Kurosawa, O., Kawabata, H. *et al.* (2000) Molecular surgery of DNA based on electrostatic micromanipulation. *IEEE Trans. Ind. Appl.* **36**, 1010–1017.

208 Kawabata, T. and Washizu, M. (2001) Separation of biomolecules using dielectrophoretic chromatography. *Trans. Inst. Elect. Eng. Jpn.* **119**, 454–459.

209 Kawabata, T. and Washizu, M. (2001) Dielectrophoretic detection of molecular bindings. *IEEE Trans. Ind. Appl.* **37**, 1625–1633.

210 Huang, Y., Wang, X.-B., Becker, F. F. and Gascoyne, P. R. C. (1997) Introducing dielectrophoresis as a new field for field-flow-fractionation. *Biophys. J.* **73**, 1118–1129.

211 Markx, G. H., Pethig, R. and Rousselet, J. (1997) The levitation of latex beads, with reference to field-flow-fractionation. *J. Phys. D: Appl. Phys.* **30**, 2470–2477.

212 Dalir, H., Yanagida, Y. and Hatsuzawa. T. (2009) Probing DNA mechanical characteristics by dielectrophoresis. *Sensor. Actuator.* **B, 136**, 472–478.

213 Asbury, C. L. and van den Engh, G. (1998) Trapping of DNA in nonuniform oscillating electric fields. *Biophys. J.* **74**, 1024–1030.

214 Menachery, A., Graham, D., Messerli, S. M. *et al.* (2010) Dielectrophoretic tweezer for isolating and manipulating target cells. *IET Nanobiotechnol.* **5**, 1–7.

215 Chou, C-F., Tegenfeldt, J. O., Bakajin, O. *et al.* (2002) Electrodeless dielectrophoresis of single- and double-stranded DNA. *Biophys. J.* **83**, 2170–2179.

216 Regtmeier, J., Duong, T. T., Eichhorn, R. *et al.* (2007) Dielectrophoretic manipulation of DNA: Separation and polarizability. *Anal. Chem.* **79**, 3925–3932.

217 Regtmeier, J., Eichhorn, R., Bogunovic, L. *et al.* (2010) Dielectrophoretic trapping and polarisability of DNA: The role of spatial conformation. *Anal. Chem.* **82**, 7141–7149.

218 Viefhues, M., Regtmeier, J. and Anselmetti, D. (2013) Fast and continuous-flow separation of DNA-complexes and topological DNA variants in microfluidic chip format. *Analyst* **138**, 186–196.

219 Tuukkanen, S., Kuzyk, A., Toppari, J. J. *et al.* (2007) Trapping of 27 bp-8 kbp DNA and immolization of thiol-modified DNA using dielectrophoresis. *Nanotechnology* **2007**, **18**, 295204.

220 Parikesit, G. O. F., Markesteijn, A. P., Piciu, O. M. *et al.* (2008) Size-dependent trajectories of DNA macromolecules due to insulative DEP in submicrometer-deep fluidic channels. *Biomicrofluidics* **2**, 024103.

221 Gallo-Villanueva, R. C., Rodriguez-López, C. E., Díaz-de-la-Garza, R. I. *et al.* (2009) DNA manipulation by means of insulator-based DEP employing direct current electric fields. *Electrophoresis* **30**, 4195–4205.

222 Li, S., Hui, Y. S., Gao, Y. *et al.* (2015) On-chip DNA preconcentration in different media conductivities by electrodeless dielectrophoresis. *Biomicrofluidics* **9**, 054115.

223 Martinez-Duarte, R., Camacho-Alanis, F., Renaud, P. and Ros, A. (2013) Dielectrophoresis of lambda-DNA

using 3D carbon electrodes. *Electrophoresis* **34**, 1113–1122.

224 Henning, A., Bier, F. A. and Hölzel, R. (2010) Dielectrophoresis of DNA: Quantification by impedance measurements. *Biomicrofluidics* **4**, 022803.

225 Li, S., Yuan, Q., Morshed, B. I. *et al.* (2013) Dielectrophoretic responses of DNA and fluorophore in physiological solution by impedimetric characterization. *Biosensor. Bioelectron.* **41**, 649–655.

226 Krishnan, R., Sullivan, B. D., Mifflin, R. L. *et al.* (2008) Alternating current electrokinetic separation and detection of DNA nanoparticles in high-conductance solutions. *Electrophoresis* 2008, **29**, 1765–1774.

227 Sonnenberg, A., Marciniak, J. Y., Krishnan, R. and Heller, M. J. (2012) Dielectrophoretic isolation of DNA and nanoparticles from blood. *Electrophoresis*, **33**, 2482–2490.

228 Sonnenberg, A., Marciniak, J. Y., McCanna, J. *et al.* (2013) Dielectrophoretic isolation and identification of cfc-DNA nanoparticulate biomarkers and virus from blood. *Electrophoresis*, **34**, 1076–1084.

229 Song, Y., Sonnenberg, A., Heaney, Y. and Heller, M. J. (2015) Device for dielectrophoretic separation and collection of nanoparticles and DNA under high conductance conditions. *Electrophoresis* **36**, 1107–1114.

230 Giraud, G., Pethig, R., Schulze, H. *et al.* (2011) Dielectrophoretic manipulation of ribosomal RNA. *Biomicrofluidics* **5**, 024116.

231 Bonincontro, A., Mari, C., Mengoni, M. and Risuleo, G. (1997) A study of the dielectric properties of *E-coli* ribosomal RNA and proteins in solution. *Biophys. Chem.* **67**, 43–50.

232 Nakano, A. and Ros, A. (2013) Protein dielectrophoresis: Advances, challenge, and applications. *Electrophoresis* **34**, 1085–1096.

233 Nakano, A., Chao, T.-C., Camacho-Alanis, F. and Ros, A. (2011) Immunoglobulin G and bovine serum albumin streaming dielectrophoresis in a microfluidic device. *Electrophoresis*, **32**, 2314–2322.

234 Nakano, A., Camacho-Alanis, F., Chao, T.-C. and Ros, A. (2012) Tuning direct current streaming dielectrophoresis of proteins. *Biomicrofluidics* **6**, 034108.

235 Washizu, M., Suzuki, S., Kurosawa, O. *et al.* (1994) Molecular dielectrophoresis of biopolymers. *IEEE Trans. Ind. Appl.* **30**, 835–843.

236 Asokan, S. B., Jawerth, L., Carroll, R. L. *et al.* (2003) Two-dimensional manipulation and orientation of actin-myosin systems with dielectrophoresis. *Nano Lett.* **3**, 431–437.

237 Hölzel, R., Calander, N., Chiragwandi, Z. *et al.* (2005), Trapping single molecules by dielectrophoresis. *Phys. Rev. Lett.* **95**, 128102.

238 Uppalapati, M., Huang, Y.-M., Jackson, T. N. and Hancock, W. O. (2008) Microtubule alignment and manipulation using AC electrokinetics. *Small* **4**, 1371–1381.

239 Lapizco-Encinas, B. H., Ozuna-Chacón, S. and Rito-Palomares, M. (2008) Protein manipulation with insulator-based dielectrophoresis and direct current electric fields. *J. Chromatogr. A* **1206**, 45–51.

240 Castillo, J., Tanzi, S., Dimaki, M. and Svendsen, W. (2008) Manipulation of self-assembly peptide nanotubes by dielectrophoresis. *Electrophoresis* **29**, 5026–5032.

241 Maruyama, H. and Nakayama, Y. (2008) Trapping protein molecules at a carbon nanotube tip using dielectrophoresis. *Appl. Phys. Express* **1**, 124001.

242 Staton, S. J. R., Jones, P. V., Ku, G. *et al.* (2012) Manipulation and apture of Aβ amyloid fibrils and monomers by DC insulator gradient dielectrophoresis (DC-iGDEP). *Analyst*, **137**, 3227–3229.

243 Otto, S., Kaletta, U., Bier, F. F. *et al.* (2014) Dielectrophoretic immobilisation of antibodies on microelectrode arrays. *Lab Chip* **14**, 998–1004.

244 Laux, E.-V., Knigge, X., Bier, F. F. *et al.* (2015) Dielectrophoretic immobilization of proteins: Quantification by atomic force microscopy. *Electrophoresis*, **36**, 2094–2101.

245 Ying, L., Zhou, D. and Bruckbauer, A. (2006) Comment on 'Trapping single molecules by dielectrophoresis'. *Phys. Rev. Lett.* **96**, 199801.

246 Hölzel, R., Calander, N., Chiragwandi, Z. *et al.* (2006) Reply. *Phys. Rev. Lett.* **96**, 199802.

247 Stanke, S., Bier, F. F. and Hölzel, R. (2011) Fluid streaming above interdigitated electrodes in dielectrophoresis experiments. *Electrophoresis* **32**, 2448–2455.

248 Castellanos, A., Ramos, A., Gonzalez, A. *et al.* (2003) Electrohydrodynamics and dielectrophoreis in microsystems: scaling laws. *J. Phys. D: Appl. Phys.* **36**, 2584–2597.

249 Laux, E.-V., Kaletta, U. C., Bier, F. F. *et al.* (2014) Functionality of dielectrophoretically immobilized enzyme molecules. *Electrophoresis*, **35**, 459–466.

250 Laux, E.-V., Knigge, X., Bier, F. F. *et al.* (2016) Aligned immobilization of proteins using AC electric fields. *Small* **11**, 1514–1520.

251 Eisenstadt, M. and Scheinberg, I. H. (1972) Dielectrophoresis of macromolecules: Determination of the diffusion constant of poly- γ benzyl-L-gluatamate. *Science* **176**, 1335–1337.

252 Eisenstadt, M. and Scheinberg, I. H. (1973) Dielectrophoresis of macromolecules. *Biopolymers* **12**, 2491–2512.

253 Debye, P., Debye, P. P., Eckstein, B. H. *et al.* (1954) Experiments on polymer solution in inhomogeneous electrical fields. *J. Chem. Phys.* **22**, 152–153.
254 Prock, A. and McConkey, G. (1960) Inhomogeneous field method for the study of large polarizable particles. *J. Chem. Phys.* **32**, 224–236.
255 Price, J. A. R., Burt, J. P. H. and Pethig, R. (1987) An optical technique for measurement of cell dielectrophoresis. Institute of Physics Conference 'Electrostatics '87' (Ser. No. 85), Oxford, April, pp. 75–79.
256 Washizu, M. and Kurosawa, O. (1989) Electrostatic manipulation of DNA in microfabricated structures. IEEE Industry Applications Society, Annual Meeting, San Diego, Oct.
257 Fuhr, G., Hagedorn, R. and Müller, T. (1991) Linear motion of dielectric particles and living cells in microfabricated structures induced by travelling electric fields, 4th IEEE Workshop on Micro Mechanical Systems, 259–264 (MEMS 91, Nara, Jpn.).
258 Martinez-Duarte, R. (2012) Microfabrication technologies in dielectrophoresis applications – a review. *Electrophoresis* **33**, 3110–3132.
259 Manaresi, N., Romani, A., Medoro, G. *et al.* (2003) A CMOS chip for individual cell manipulation and detection. *IEEE J. Solid-State Circ.* **38**, 2297–2305.

12

Microfluidic Concepts of Relevance to Dielectrophoresis

12.1 Introduction

The devices described in Chapter 10 incorporate microfluidic channels along which particles are directed through and then away from the influence of nonuniform electric fields. To qualify as a microfluidic device it should incorporate fluid flow in structures that have at least one dimension in the scale between 1 μm and 1 mm. Electrokinetic and microfluidic phenomena are therefore co-actors in most DEP devices.

This chapter describes some of the basic physical properties and hydrodynamic behaviour of fluids in microchannels. This includes pressure-driven as well as electrokinetic-driven fluid flow; laminar fluid flow; fluid flow resistance of channels and DEP chambers; electrical double layers at channel surfaces and electro-osmosis. An important concept is the *characteristic* dimension of a fluidic channel. This length is incorporated into dimensionless parameters such as the Knudson, Reynolds and Peclet number, which can be used to predict how a fluid should behave in a particular microchannel. The electrical analogue of fluid flow in channels is the flow of current through resistors. Kirchhoff's rules, as applied to the analysis of electrical networks, can thus be employed to analyse fluid flow in simple channel networks.

12.2 Gases and Liquids

Although they respond very differently to changes in pressure and temperature, the term 'fluid' includes both liquids and gases. They share the common feature that, unlike solids, they do not resist shearing forces such as those acting at a solid surface and continue to deform as long as the force is applied. Gases and liquids assume the shape of the solid boundary, whereas solids can resist such shear and maintain an unsupported shape. Their motion, as shown in Figure 12.1, is controlled by the interaction and internal shear between fluid layers.

Gases can be expanded and compressed more easily than liquids due to the lower density and larger spacing between molecules. At the molecular scale ($\sim 10^{-9}$ m) the interaction between fluid layers involves collisions of many molecules, whilst at the macroscale scale ($>10^{-4}$ m) the physical properties of a fluid result from the statistical averages of such molecular interactions. In this case, the effects of individual molecular collisions can be ignored and we can deal with the liquid's bulk, or continuum, properties.

12.2.1 Gases

The molecules in a gas are widely spaced and interactions between them (apart from collisions) are weak, especially at low pressures. An increase of temperature increases the kinetic energy of the molecules, mass transfer between gas layers increases and viscosity increases. In gases, except for extremely high pressures, viscosity is independent of pressure. At a sufficiently low pressure, where intermolecular interactions are negligible, all gases obey the *Ideal Gas Law:*

$$PV = nRT \tag{12.1}$$

where P is the pressure, V the volume, n the amount (moles) of substance of gas molecules and T the absolute temperature. In this equation R is the gas constant given by $R = kN_A$, where k is Boltzmann's constant $= 1.38 \times 10^{-23}$ J K^{-1} and N_A is Avogadro's constant $= 6.022 \times 10^{23}$ mol^{-1}. The Ideal Gas Law follows from experiment (e.g., Boyle's Law) and Avogadro's Hypothesis (formulated in 1810) that: *equal volumes of gases at the same temperature and pressure contain the same number of molecules.*

Thus, a mole of hydrogen (2 g) and a mole of oxygen (32 g) at the same temperature and pressure occupy the same volume. At standard temperature and pressure (STP: 273.15 K, 100 kPa) this volume is 22.414 l.

Although on average the molecules in a gas are widely spaced apart, they are in constant motion and often

Dielectrophoresis: Theory, Methodology and Biological Applications, First Edition. Ronald Pethig.

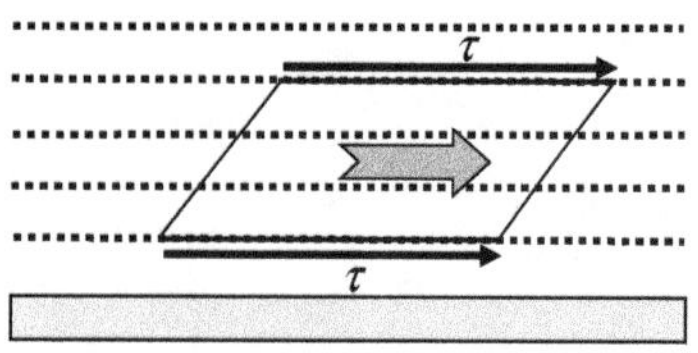

Figure 12.1 Fluids can be considered as a series of parallel laminas. Under the action of a shearing force τ, a fluid laminar transmits this shear force to its neighbour and is, in turn, sheared by those it touches. This is the process by which a fluid is deformed under the action of a shearing force.

collide with each other. As shown in Box 12.1 the mean free path length L_{mfp} between such collisions is given by:

$$L_{mfp} = \frac{kT}{\sqrt{2}\pi P d^2} \tag{12.2}$$

Example 12.1 Average Separation Distance and Mean Free Path Length of Gas Molecules
Estimate 1. the average separation distance and 2. the mean free path length between collisions of the molecules in a gas at STP.

Solution 12.1

1. The volume occupied by a mol of any gas at STP (273.15 K, 100 kPa) is 22.414 l. Based on Avogadro's constant there are thus $6.022 \times 10^{23}/22.4 = 2.7 \times 10^{22}$ gas molecules per dm^3 at STP. Spheres cannot be packed together to fill all space and so we will assume that each molecule occupies a cube of side l. The average spacing between molecules can thus be estimated from the relationship:

$$l^3\,(2.7 \times 10^{22}) = 1, \text{ to give } l = 3.34\text{ nm}.$$

2. Typical molecular diameters d fall in the range 0.2~0.3 nm. Assuming $d = 0.25$ nm, from Equation (12.2) the mean free path length between collisions of the molecules in a gas at STP is given by:

$$L_{mfp} = \frac{kT}{\sqrt{2}\pi P d^2} = \frac{1.38 \times 10^{-23} \times 273.15}{\sqrt{2}\pi 10^5 (2.5 \times 10^{-10})^2}$$
$$= 136\text{ nm}$$

Thus, the average distance between collisions of a gas molecule is more than 500 times their molecular diameter (~0.25 nm) and some 40 times larger than their average molecular separation distance of 3.34 nm. This demonstrates that molecules in a gas typically travel in straight paths over significant distances at the molecular scale before they collide with another molecule.

12.2.2 Liquids

The average centre-to-centre distance between molecules in a liquid is just a little more than its molecular diameter d. Cohesive forces such as those arising from induced dipole-dipole interactions give rise to viscous effects. Glass and molten polymers are highly viscous because their large molecules become entangled. Water has a higher viscosity than liquids such as benzene

Box 12.1 Mean Free Path between Molecular Collisions in a Gas

The frequency of collisions between sets of two molecules depends on their relative velocity v_{rel} of approach. For two molecules this is given by the vector difference of their velocities v_1 and v_2, so that:

$$v_{rel} = \sqrt{v_{rel} \cdot v_{rel}} = \sqrt{(v_1 - v_2) \cdot (v_1 - v_2)}$$
$$= \sqrt{v_1 \cdot v_2 - 2v_1 \cdot v_2 + v_2 \cdot v_1}$$

Velocities v_1 and v_2 are random and uncorrelated and, because the same average velocity $\langle v_{rel} \rangle$ is associated with each molecule, then:

$$\langle v_{rel} \rangle = \sqrt{v_1^2 + v_2^2} = \sqrt{2}\langle v \rangle$$

Over time t a collision cross section associated with one molecule moving with an average velocity $\langle v \rangle$ will travel a path length $\langle v \rangle t$. Treating each molecule as a hard sphere of diameter d the effective collision area for two colliding molecules can be taken as a circle of diameter $2d$. The effective cross-sectional collision area A_c for a molecule is thus given by $A_c = \pi d^2$. The volume V_c of collision space swept through during time t is:

$$V_c = A_c \langle v_{rel} \rangle t = \pi d^2 \sqrt{2} \langle v \rangle t$$

The mean free path length L_{mfp} can then be taken as the path length $\langle v \rangle t$ divided by the number of molecular collisions:

$$L_{mfp} = \frac{\langle v \rangle t}{\pi d^2 \sqrt{2} \langle v \rangle t N_v} = \frac{1}{\pi d^2 \sqrt{2} N_v}$$

In this equation N_v is the number of molecules per unit volume, calculated from Avogadro's number and the Ideal Gas Law given by Equation (12.1) as:

$$N_v = \frac{nN_A}{V} = \frac{PN_A}{RT} = \frac{P}{kT}$$

so that

$$L_{mfp} = \frac{kT}{\sqrt{2}\pi P d^2}$$

or alcohols because of its network of cohesive hydrogen bonds. With increasing pressure, the energy required for relative movement of molecules is increased, so that the viscosity increases. As the temperature increases there is an increase of molecular kinetic energy, which reduces the cohesive forces and hence also the viscosity. An increase of molecular kinetic energy also facilitates an increased molecular interchange between the fluid layers, which will increase viscosity. However, this produces a relatively small effect compared to the reduction of cohesive forces, so the net result is that liquids show a reduction in viscosity for an increase in temperature.

We can appreciate the difference at the molecular level between a liquid and a gas by noting that 1 dm^3 of liquid nitrogen weighs ~800 g, whilst at STP 1 dm^3 of gaseous nitrogen weighs ~1.2 g. This informs us that at the molecular level a nitrogen molecule on average occupies ~670 times more space than it does as a liquid. To what extent can we treat a gas in a microfluidic device to have the properties of a continuum? A dimensionless parameter, known as the *Knudsen number Kn*, provides an important test of the validity of the continuum approximation. *Kn* compares the characteristic dimensions of a microfluidic device to the mean free path between molecular collisions and is defined as:

$$Kn = \frac{L_{mfp}}{L}$$

with L being the characteristic length of the flow field. For a microfluidic device, this characteristic length can be taken as either the hydraulic diameter of a channel, or the gradient of a bulk fluid property such as density ($\rho/\mathrm{d}\rho/\mathrm{dx}$). The hydraulic diameter, also known as the wetted diameter, D_H is defined as:

$$D_H = \frac{4 \times Area}{Wetted\ Perimeter}$$

For the case of a channel with circular cross section, D_H is equal to the channel's diameter. For a channel of rectangular cross section, width w and height h:

$$D_H = \frac{2wh}{(w+h)}$$

The following ranges of the Knudson number provide a rough guide as to whether a gaseous fluid may be treated as a continuum or as an assembly of discrete molecular particles:

- $Kn < 0.001$: the mean free path between molecular collisions is very small compared to the distance between the fluidic boundaries. The continuum model and zero slip of the fluid at boundaries may be assumed.
- $0.001 < Kn < 0.1$: the continuum model may be assumed, but finite fluid slip occurs at boundaries.
- $0.1 < Kn < 10$: this defines the *mesoscale* region between the continuum approximation and a model that involves discontinuous dynamics.
- $Kn > 10$: the molecules bounce off the boundary walls more often than they do each other. The continuum approximation is not valid. A particle-based method, such as a Monte Carlo simulation, should be used.

Therefore, as a rough guide, we can adopt the continuum model if the characteristic scale of our microfluidic system is more than ~500 times larger than the molecular mean free path length. From Example 12.1 we can estimate that the characteristic size of a microfluidic system *below which* the properties of a gas at STP should *not* be modelled as a continuum is ~500 × 136 nm ≈ 68 μm. At the other end of the scale, where the molecular mean free path is ten times larger than the characteristic length of our system, the molecular particles collide with the physical boundaries of the system more often than they do with each other. We cannot treat the gas as a homogeneous medium, instead, the dynamic behaviour of each molecule must be considered. The mesoscale region covers the change in physics between the continuum approximation and discontinuous molecular models. The lower limit of the mesoscale can be taken to be around 100 molecular diameters (i.e., ~25 nm). The upper limit, corresponding to where the continuum approximating laws are violated, is not so well defined. For example, a rarefied gas might invalidate the use of continuum physics up to scales of ~10 μm, whereas for a dense liquid the continuum laws could be valid down to scales below 1 μm.

The transition between the continuum and molecular regions for liquids goes through the same stages as for gases, but there is no parameter to act as a guide throughout the transition. The Knudson number cannot be defined, as there is no concept of mean free path for liquid flows – the molecules are in a constant state of collision and move over much shorter distances (comparable to the molecular diameter). Travis *et al.* [1] modelled the velocity profile and heat flux profile of an atomic liquid in a narrow channel, using molecular dynamics and the Navier–Stokes equations (to be described later in this chapter). For a channel width of 5.1 molecular diameters the two simulations of the velocity profiles differed significantly. The heat flux profile did not agree with that predicted by Navier–Stokes hydrodynamics, but exhibited significant oscillations located about one molecular diameter from the walls. However, classical Navier–Stokes behaviour was approached for a channel width greater than 10 molecular diameters. As an example of experimental investigations at the boundary between the continuum and mesoscale, Pfahler *et al.* [2] constructed three channels of rectangular cross-section ranging in area from 80 to 7200 square microns. In the relatively large flow channels the experimental observations were in rough agreement with predictions from the Navier–Stokes equations but significant deviations were found for the smallest of the channels. Mala and Li [3] studied water flow through microtubes with diameters ranging from 50 to 254 μm. Results in rough agreement with conventional continuum theory were obtained for the large diameters, but not for the smaller diameters.

12.3 Fluids Treated as a Continuum

When treated as a continuum, the properties of a fluid such as density, pressure and velocity remain constant at any defined point and changes in these properties due to molecular motions are taken to be negligible. The physical properties of fluids can be defined as continuous functions of time and space.

12.3.1 Density

This is defined as the mass contained within a unit volume and is computed as the product of molecular mass m and the number of molecules N per unit volume V:

$$\rho = \frac{Nm}{V}$$

Molecular mass m is the mass of the molecule given by $m = M_w m_u$, where M_w is the molecular weight (molecular mass relative to $^{12}C = 12$) and m_u is the atomic mass unit (1.6606×10^{-27} kg). The Ideal Gas Law of Equation (12.1) can be interpreted as stating that pressure is linearly proportional to the product of temperature and density.

12.3.2 Temperature

Temperature relates to the translational kinetic energy E of N molecules in a particular volume domain, with each molecule having velocity v_j and mass m:

$$E = \sum_{j=1}^{N} \frac{1}{2} m v_j^2$$

The kinetic theory of gases uses statistical mechanics to relate the average kinetic energy of the atoms to the temperature of the system. In one dimension we have:

$$\langle E \rangle = \frac{1}{2} m \langle v_x^2 \rangle = \frac{kT}{2}$$

For a three-dimensional domain:

$$\langle E \rangle = \frac{1}{2} m \langle v_x^2 + v_y^2 + v_z^2 \rangle = \frac{3}{2} kT$$

This relationship is important to understanding the concept of gas pressure, which is the force imparted by collisions of gas molecules against a unit area of surface. This can be evaluated in terms of Newton's Law as the momentum lost per molecular collision with the surface, averaged over a large number of molecules. The macroscopic pressure of a gas therefore relates directly to the average kinetic energy per molecule and to the gas density. This application of the laws of classical physics can be applied to the particles of a gas, but the molecules or atoms in the liquid or solid state have to obey the laws of quantum statistics. In which case only certain energy values, rather than a continuous distribution of particle velocities, are permitted.

12.3.3 Viscosity

Viscosity is the measure of the effort required to deform a fluid. This can be described in terms of *Couette* flow, corresponding to the situation shown in Figure 12.2 where a fluid is contained between a moving plate and a parallel stationary plate. The fluid velocity immediately next to a surface of a plate will equal the velocity of that plate. This is referred to as *zero slip*. If the fluid is a *Newtonian* fluid, such as water, the fluid velocity changes smoothly from zero at the stationary surface to the velocity of the moving surface. Thus, as shown in Figure 12.2, the spatial gradient of the fluid velocity dv/dy is constant.

The *dynamic* viscosity η is defined as the proportionality constant between the shear stress τ applied to the fluid and the resulting rate of shear strain. The shear strain is defined as the ratio dy/dx of the lateral deformation dy to the thickness dx of the layer being displaced. The rate of shear strain is given by $(dy/dx)/dt = dv/dx$, corresponding to the induced velocity gradient. For a Newtonian fluid the following relationship exists between stress, rate of shear strain and the dynamic viscosity:

$$\tau = \eta \, dv/dx \tag{12.3}$$

We can interpret this relationship as indicating that, for $\eta = 1$ Pa.s and $\tau = 1$ Pa, the mobile plate moves in one second a distance equal to the thickness of the fluid layer

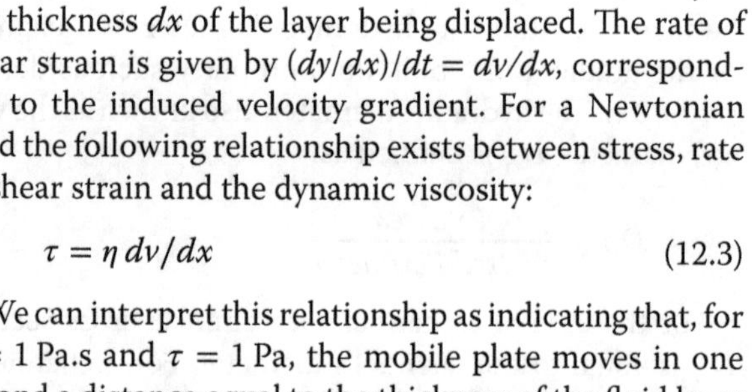

Figure 12.2 Couette flow is shown induced in a fluid bounded by a stationary surface and a moving plate. For the case of a Newtonian fluid, the fluid velocity changes smoothly from zero at the stationary surface to the velocity of the moving plate. There is no slip of the fluid at each surface and the spatial velocity gradient dv/dx has a constant value.

Table 12.1 Viscosity values of some liquids (293 K, unless specified).

Liquid	Viscosity (Pa s)
Water (liquid)	1.0×10^{-3}
Water (vapour, 373 K)	1.3×10^{-5}
Blood (whole, 310 K)	$3{\sim}4 \times 10^{-3}$
Blood (plasma, 310 K)	1.5×10^{-3}
Ethyl alcohol	1.2×10^{-3}
Glycerine	1.49
Oil (light)	0.11
Oil (heavy)	0.66

between the plates. Values of the dynamic viscosity for some liquids are given in Table 12.1.

The viscosity of a Newtonian fluid depends only on temperature and concentration (if diluted with another miscible fluid). For some fluids, particularly molten polymers or biological fluids such as blood, their viscosity depends also on the internal stress. These are classed as non-Newtonian fluids. Their viscosity decreases with an increase of the rate of the applied shear stress $d\tau/dt$ applied to a fluid flowing between two parallel surfaces, one moving at a constant velocity and the other one stationary and is defined by:

$$d\tau/dt = v/h$$

where v is the velocity of the moving surface and h is the distance between the two parallel surfaces. Non-Newtonian fluids exhibit viscoelastic behaviour (shear thinning) and some, such as whole blood, require an initial shear stress that must be applied before they begin to flow. Viscoelastic fluids exhibit a relaxation time, typically ranging from milliseconds to seconds, given by the reciprocal of the critical shear rate. The critical shear rate corresponds to the shear threshold at which the viscosity begins to change or, for the case of molten polymers, where the polymer chains make the transition from a coiled to a stretched configuration.

12.4 Basic Fluid Statics and Fluid Dynamics

12.4.1 Static Fluid Pressure and Pascal's Law

The pressure exerted by a static fluid arises from the weight of that fluid and so depends only upon the fluid depth h, its density ρ and the acceleration of gravity g:

$$P_{static\ fluid} = \frac{weight}{area} = \frac{mg}{A} = \frac{\rho V g}{A} = \rho g h \qquad (12.4)$$

Because the fluid pressure at a given depth h does not depend upon the total mass or total volume of the liquid, the shape of the fluid container is also not relevant. Examples of this are given in Figure 12.3.

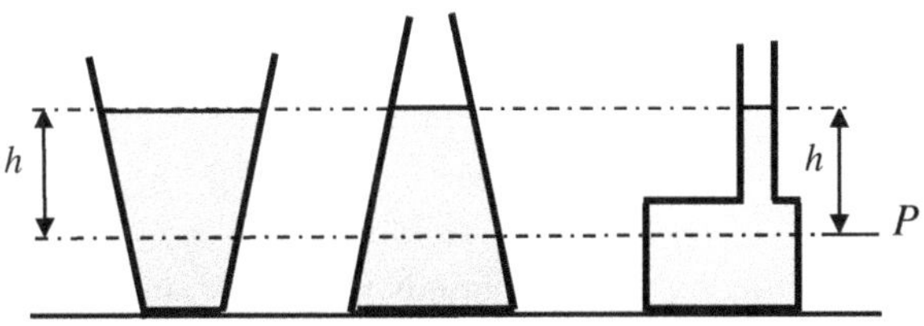

Figure 12.3 The static fluid pressure P at a given depth h in a liquid does not depend on the total volume or shape of the vessel.

It follows from Equation (12.4) that the change in pressure between two elevations is due to the weight of the fluid between the elevations regardless of the geometry of the container. This leads directly to Pascal's Law, which states that: *the pressure exerted anywhere in an enclosed, incompressible, static fluid is transmitted equally in all directions throughout the fluid.*

Pascal's Law can be interpreted to indicate that any *change* in pressure applied at any given point of the fluid is transmitted *undiminished* throughout the fluid.

12.4.2 Conservation of Mass Principle (Continuity Equation)

Fluid flowing steadily through a cylindrical channel of reducing cross-sectional area is shown in Figure 12.4. No fluid can exit or enter the channel between areas A_1 and A_2. Let ρ_2 be the density of the fluid flowing through A_2. The rate of fluid flow Q through the channel can be determined as either volumetric flow (dm^3/s) or mass flow (g/s). In terms of mass flow $Q = A_2\rho_2 v_2$ g/s, where v_2 is the mean velocity of the fluid flow through A_2. Because no fluid exits or enters the channel between areas A_1 and A_2, then from the conservation of mass principle the mass of fluid crossing each section of the pipe per unit time must be the same:

fluid flow rate through A_1
= *fluid flow rate through* A_2

$$A_1\rho_1 v_1 = A_2\rho_2 v_2$$

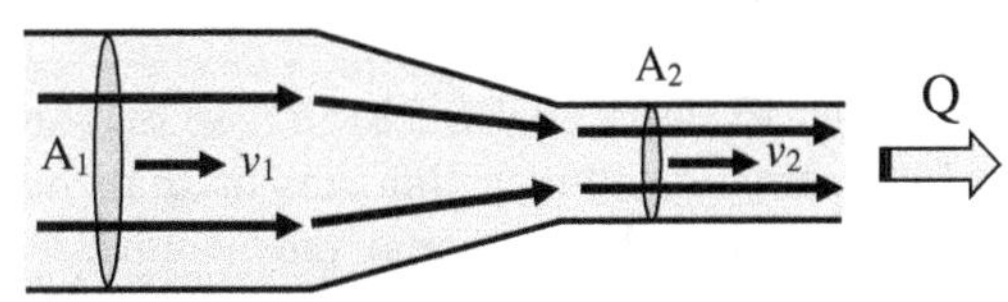

Figure 12.4 Flow of a liquid through a constriction in a cylindrical channel, where its cross-sectional area reduces from A_1 to A_2. As detailed in the main text, the principle of conservation of mass dictates that $A_1v_1 = A_2v_2$, where v_1 and v_2 are the average fluid velocities through A_1 and A_2, respectively. Because $A_1 > A_2$, then $v_1 < v_2$. This is also known as the Venturi effect and is the principle of operation of a bath showerhead.

This relationship, which expresses the *equation of continuity*, also expresses the *Law of Conservation of Mass* in fluid dynamics, namely:

$$A\rho v = \text{constant}$$

We will consider only liquids (i.e., not gases), so that because they are incompressible the density of the fluid will be constant (i.e., $\rho_1 = \rho_2$). The conservation of mass principle can thus be written as:

$$A_1 v_1 = A_2 v_2$$

or

$$\text{A}v = \text{constant}$$

The reduction in channel area shown in Figure 12.4 ($A_1 > A_2$) indicates that $v_2 > v_1$.

Example 12.2 Fluid Flow through a Channel of Reducing Cross Section

Water flows through a channel of circular cross-section at a rate of 0.1 mL per minute. The channel has the same geometrical profile as that depicted in Figure 12.4, with the radius constricting down from 100 μm to 60 μm. Calculate the effective velocity of fluid flow through areas A_1 and A_2.

Solution 12.2 The volumetric flow is given by:

$$Q_v = 0.1/60 = 1.67 \times 10^{-3}\text{mL s}^{-1}$$
$$= 1.67 \times 10^{-9}\text{m}^3\text{s}^{-1}$$
$$\text{Area: } A_1 = \pi(10^{-4})^2 = 3.14 \times 10^{-8}\text{m}^2$$
$$\text{Velocity: } v_1 = Q/A_1$$
$$= 1.67 \times 10^{-9}\text{m}^3\text{ s}^{-1}/(3.14 \times 10^{-8}\text{m}^2)$$
$$= 5.3 \times 10^{-2}\text{m s}^{-1}$$

From the principle of conservation of mass:

$$A_1 v_1 = A_2 v_2$$
$$\therefore v_2 = (3.14 \times 10^{-8}\text{m}^2)(5.3 \times 10^{-2}\text{ m s}^{-1})/$$
$$(3.14 \times (6 \times 10^{-5})^2) = 0.15\text{ m s}^{-1}$$

12.4.3 Bernoulli's Equation (Conservation of Energy)

With reference to Figure 12.4 and Example 12.2, the principle of conservation of mass informs us that the fluid velocity is greatest in that section of the cylindrical channel having the smaller cross-sectional diameter. On flowing through a constriction to a smaller cross-sectional area the fluid velocity increases. This corresponds to an acceleration of the fluid mass, which in turn requires an unbalanced force in the form of a pressure gradient exerted on the fluid by the walls of the channel. As shown in Figure 12.5 the pressure P_1 in the large area of the pipe

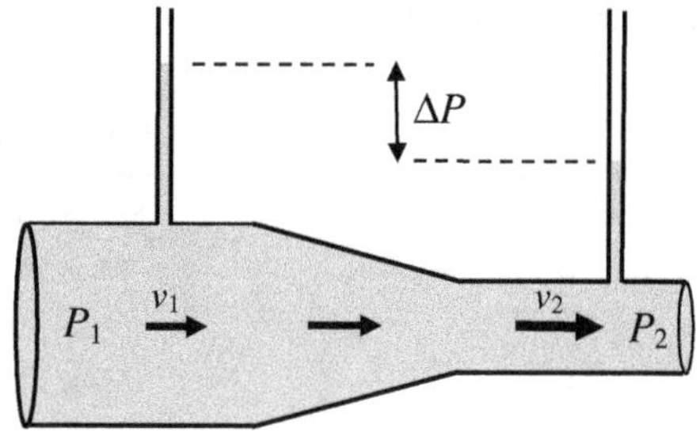

Figure 12.5 A pressure drop ($P_1 - P_2 = \Delta P$) occurs in a fluid flowing through a constriction in a channel, driving the acceleration of the fluid mass through the smaller cross-sectional area of the channel.

must be greater than P_2 in order to accelerate the fluid. Likewise, if the fluid flow is reversed, P_1 must exceed P_2 in order to decelerate the fluid mass.

Bernoulli's principle states that for *viscous free* fluid flow an increase of the fluid velocity occurs simultaneously with a decrease in fluid pressure, or with a decrease in the fluid's potential energy. This principle can be applied to various types of fluid flow and quantified using various forms of what is known as Bernoulli's equation. A simple form of this equation is valid for incompressible fluids and for compressible gases moving at speeds well below the velocity of sound in a particular gas. This equation can be derived from the principle of conservation of energy, which states that along a steady fluid flow path the sum of all forms of mechanical energy remains constant. The fluid possesses kinetic energy due to its motion and, because of its location in the earth's gravitational field, it also possesses potential energy. Work is also being done on the fluid due to the static pressure acting on it. If there are no frictional losses we can apply the Law of Conservation of Energy and write Bernoulli's equation as:

$$P + \rho gh + {}^1\!/_2\rho v^2 = \text{constant} \tag{12.5}$$

where P is the static pressure, h the height above some reference level, v the mean fluid velocity, ρ the fluid density and g the acceleration due to gravity at any chosen elemental volume in the fluid flow line. The term $(\frac{1}{2}\rho v^2)$ is known as the dynamic pressure and the total pressure is the sum of the static pressure P and this dynamic pressure. The sum of the elevation h and static pressure head $(P/\rho g)$ is known as the hydraulic head.

A consequence of Bernoulli's principle is demonstrated in Figure 12.6, which shows an aerofoil-shaped object in flowing fluid. Because of the differences in path length, the fluid flows more rapidly over the top surface than over the lower surface of the aerofoil. Faster fluid flow implies a lower pressure, so that the pressure will be greater on the bottom surface of the aerofoil and produce an upward lift force. This is the principle used in the design of aircraft wings and propeller blades, for example.

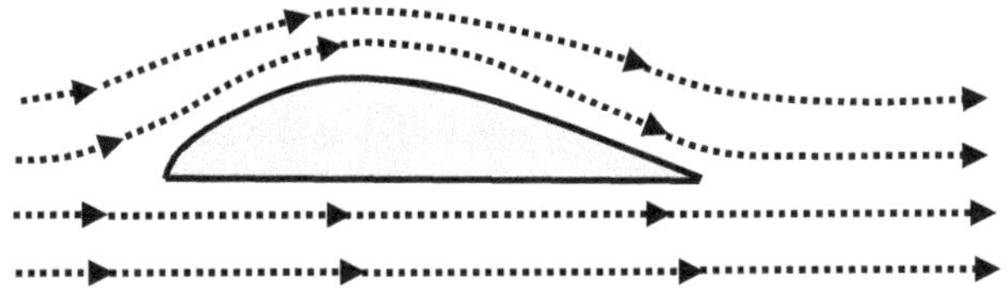

Figure 12.6 The fluid velocity above the upper surface of this object is greater than that in the shorter path lengths below the object. According to Bernouilli's principle the fluid pressure below the object is greater than that above it, which results in a lift force.

An inverted version of the shape shown in Figure 12.6 will result in a downward force, which is an effect used in racing car spoilers.

Example 12.3 Pressure Drop across a Fluid Constriction
Water flows through a horizontal channel of geometrical profile similar to that shown in Figure 12.5, with the exit port open to the atmosphere. The mean fluid velocity v_1 before the constriction is 0.1 ms^{-1} and the fluid exit velocity v_2 is 15 ms^{-1}. Neglecting frictional losses, calculate the pressure P_1 in the main channel and the pressure drop across the constriction. (The density of water is 1000 kg m^{-3} and atmospheric pressure is 100 kPa.)

Solution 12.3 Ignoring frictional losses we can employ Bernoulli's equation (12.5). The channel is horizontal and so potential energy differences arising from changes of fluid height can also be ignored. Equation (12.5) therefore takes the form:

$$P_1 + {}^1\!/_2\rho v_1^2 = P_2 + {}^1\!/_2\rho v_2^2$$

We are given ρ = 1000 kg m^{-3}, v_1 = 0.1 ms^{-1}, v_2 = 15 ms^{-1}, P_2 = 100 kPa (10^5 N m^{-2}).

Pressure P_1 is given by:

$$\begin{aligned} P_1 &= P_2 + {}^1\!/_2\rho(v_2^2 - v_1^2) \\ &= 10^5 + 500(225 - 10^{-2}) \\ &= 2.125 \times 10^5 \text{ N m}^{-2} (\sim 2.1 \text{ atm}) \end{aligned}$$

The value for ΔP in Figure 12.5 is given by:

$$\Delta P = P_1 - P_2 = 1.125 \times 10^5 \text{ N m}^{-2} (\sim 1.1 \text{ atm})$$

Example 12.4 Lift Force Acting on an Object in a Flowing Fluid
Measurements of the flow of fluid around the object shown in Figure 12.6 give the upper and lower flow velocities as 0.3 m s^{-1} and 0.25 m s^{-1}, respectively. The bottom surface is flat, with a width of 1 cm and length 10 cm. Calculate the lift force acting on the object.

Solution 12.4 Ignoring any differences in fluid height, from Equation (12.5):

$$\begin{aligned} \Delta P = P_1 - P_2 &= {}^1\!/_2\rho(v_2^2 - v_1^2) \\ &= 500(0.09 - 0.0625) = 13.75 \text{ N m}^{-2} \end{aligned}$$

The lift (upward) force F acting on this object is $A.\Delta P$, where A is the lower surface area of the object:

$$F = (10^{-2} \times 10^{-1} \text{ m}^2) \times 13.75 \text{ N m}^{-2} = 0.014 \text{ N}$$

12.4.4 Poiseuille's Law (Flow Resistance)

Bernoulli's principle assumes the fluid flow is not influenced by viscous forces. In fact, for the case of smooth, turbulence free, fluid flow the viscous shearing forces shown in Figure 12.1 determines the fluid velocity profile across a channel. There is no relative motion of the fluid at the surfaces of the channel walls (i.e., zero fluid slip) and the flow velocity increases towards the centre line of the channel. The consequence of this is that in order to pump a viscous fluid along a channel a pressure difference ΔP must be applied between its inlet and outlet, irrespective of any changes of the channel diameter. This is equivalent to the channel having the properties of flow resistance. In the 1840s Poiseuille experimentally and then theoretically derived the following relationship for fluid flow in pipes of circular cross section:

$$\Delta P = \frac{8\eta LQ}{\pi r^4} \tag{12.6}$$

where L is the length of the pipe, r its internal radius and η is the dynamic viscosity of the fluid.

The flow resistance R_f of a channel is defined from the relationship

$$R_f = \Delta P / Q \tag{12.7}$$

where Q is the volumetric flow rate. From Equations (12.6) and (12.7) the flow resistance of a channel of circular cross section is given as:

$$R_f = \frac{8\eta L}{\pi r^4} \tag{12.8}$$

In practice, fluidic channels of either a rectangular or semicircular cross section are easier to fabricate than those of circular cross section (e.g., by placing a flat plate on top of a rectangular or rounded trench). The fluidic resistance of a rectangular channel with a high aspect ratio (i.e., width $w \gg$ height h) can be calculated using the formula:

$$R_f = \frac{12\eta L}{wh^3} \tag{12.9}$$

For a channel of semicircular cross section defined by a radius of curvature r:

$$R_f = \frac{64\eta L}{3r^4} \quad (12.10)$$

Thus, for any specified channel geometry, the flow resistance is directly proportional to the viscosity of the fluid. The viscosity values for gases are considerably smaller than those for liquids. For example, the viscosity of air is 1.8×10^{-5} Pa.s (cf. 1×10^{-3} Pa.s for water). Bernoulli's approximation that fluid flow is not influenced by viscous forces may thus be adequate for the flow of gases, but should not be adopted when considering the flow of liquids. The flow resistance of water is about 50 times greater than for the flow of a gas along the same channel and this difference increases to a factor of $\sim 10^5$ for the flow of a highly viscous fluid such as glycerine.

Example 12.5 Fluid Flow Resistance of a DEP Chamber

A chamber of rectangular cross section, width 5 mm, height 200 μm and length 2 cm contains a planar electrode array on its floor for the purpose of trapping target cells by positive DEP. A pump is to be used to flow a relatively dilute, aqueous, suspension of cells ($\sim 10^5$ cells per ml) through this chamber at a volumetric flow rate of 5 μl/s into a collection vessel open to atmospheric pressure (100 kPa).

1. Calculate the fluid flow resistance of the DEP chamber.
2. What pressure must the pump produce to achieve the desired flow rate?
3. What is the mean flow velocity of the cell suspension through the DEP chamber?
4. Calculate the pressure required if the chamber height is reduced to 25μm.

Solution 12.5

1. The DEP chamber has a relatively high aspect ratio (width $w \gg$ height h) so that Equation (12.9) can be used to estimate its fluid flow resistance. Assuming that the viscosity of the aqueous cell suspension is close to that of water (1×10^{-3} Pa.s) the flow resistance is:

$$R_f = \frac{12\eta L}{wh^3} = \frac{12(1 \times 10^{-3})(0.02)}{(5 \times 10^{-3})(200 \times 10^{-6})^3}$$
$$= 6 \times 10^9 \text{ Pa m}^{-3}\text{s}$$

2. The pressure drop ΔP across the DEP chamber for $Q = 5\ \mu\text{l s}^{-1}$ ($5 \times 10^{-9}\ \text{m}^3\ \text{s}^{-1}$):

$$\Delta P = QR_f = (5 \times 10^{-9}\ \text{m}^3\text{s}^{-1})(6 \times 10^9 \text{Pa m}^{-3}\text{s})$$
$$= 30 \text{ Pa}$$

Total pressure output required of pump = ΔP + atmospheric pressure = 100.03 kPa.

3. The mean flow velocity v is given by $v = Q/A$:

$$v = \frac{Q}{A} = \frac{5 \times 10^{-9}}{(5 \times 10^{-3})(100 \times 10^{-6})} = 10^{-2}\ m/s$$

4. If the DEP chamber height is reduced (by one-quarter) from 200 to 25 μm, then from Equation (12.9):

$$R_f \propto h^{-3}. \therefore \text{ new value for}$$
$$R_f = (6 \times 10^9)/(1/4)^3 = 3.84 \times 10^{11} \text{Pa m}^{-3}\ \text{s}$$
$$\Delta P = QR_f = (5 \times 10^{-9}\ \text{m}^3\ \text{s}^{-1})$$
$$\times (3.84 \times 10^{11}\ \text{Pa m}^{-3}\ \text{s}) = 1.92 \text{ kPa}$$

Total pressure output required of pump = ΔP + atmospheric pressure = 101.92 kPa

12.4.5 Laminar Flow

Figure 12.1 depicts a fluid moving in laminas with successively higher velocity. The flow velocity is zero in the vicinity of a stationary wall and increases away from that wall. The fluid flow is a function of the x-coordinate and not of the y- and z-directions. This is termed as laminar flow. The laminar flow in a channel of circular cross-section is depicted in Figure 12.7 to take the form of concentric, thin-walled, tubes of fluid whose velocities increase from zero at the channel wall to a maximum at the centre line of the channel. The flow is directed along the channel's axis and there are no pressure gradients across the channel diameter. A shear stress τ exists between each tube and increases by $d\tau$ for each tube. A pressure drop between the ends of the fluid tube is required to overcome the shear stress. It is normally assumed for a channel of constant cross-section that the pressure declines uniformly with distance down the fluid stream. The pressure gradient $\Delta P/\Delta L$ is thus assumed to be constant.

Consider the elemental fluid tube shown in Figure 12.7, of length ΔL, at radius r and thickness dr. If τ is the shear

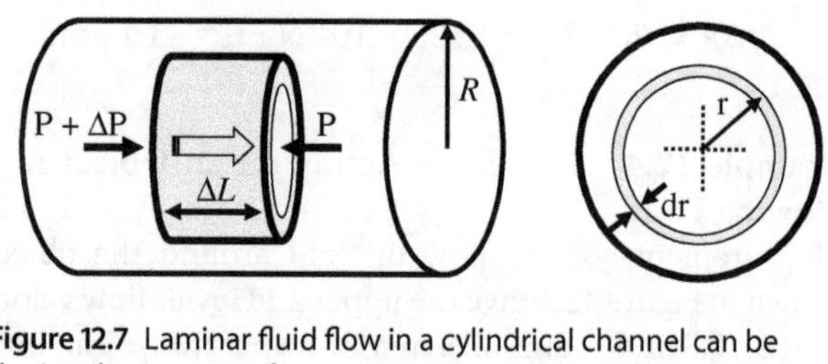

Figure 12.7 Laminar fluid flow in a cylindrical channel can be depicted as a series of concentric 'stream tubes' of length ΔL whose velocities increase towards the centre of the channel as a function of the distance $(R - r)$ from the channel wall.

stress per unit area acting on the surface of this tube, the shear force F_s is given by

$$F_s = 2\pi r \Delta L \tau$$

From Equation (12.3):

$$\tau = \eta dv/dx = -\eta dv/dr (x = R - r)$$

to give

$$F_s = -2\pi r \Delta L \eta \, dv/dr$$

At equilibrium this shear force balances the force acting on the ends of the fluid tube as a result of the pressure difference ΔP, so that:

$$\Delta P \pi r^2 = -2\pi r \Delta L \eta \, dv/dr$$

to give

$$dv = -\frac{\Delta P}{2\eta \Delta L} r dr$$

The velocity v of a fluid tube at any radius r is found by integrating between the limits $v = 0$ ($r = R$) and $u = u$ for $r = r$:

$$\int_0^v dv = -\frac{\Delta P}{2\eta \Delta L} \int_R^r r dr$$

from which

$$v(r) = -\frac{\Delta P}{2\eta \Delta L}(r^2 - R^2) = \frac{\Delta P}{2\eta \Delta L}(R^2 - r^2) \quad (12.11)$$

This equation describes a parabolic fluid velocity profile across the channel, as shown in Figure 12.8, with zero velocity at the channel walls and a maximum velocity along the central axis (at $r = 0$). The maximum velocity is given as

$$\hat{v} = \frac{\Delta p R^2}{4\eta \Delta L}$$

The mean velocity <v> is the averaged velocity in the cross section

$$\langle v \rangle = \frac{1}{\pi R^2} \int_0^R v(r) 2\pi r dr = \frac{\Delta p R^2}{8\eta \Delta L}$$

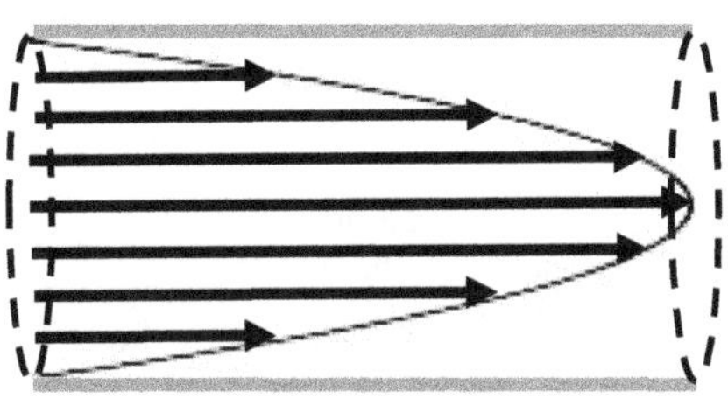

Figure 12.8 In laminar flow the fluid velocity is zero at the channel wall and reaches a maximum value at the centre axis. The velocity profile of the fluid layers takes the form of the parabola defined by Equation (12.11).

which corresponds to half the maximum value. The volumetric flow rate Q is given by the product of the mean velocity and the cross-sectional area:

$$Q = \frac{\Delta p R^2 \pi R^2}{8\eta \Delta L} = \frac{\pi R^4 \Delta p}{8\eta \Delta L}$$

which corresponds to the Poiseuille relationship of Equation (12.6).

We should note that the Poiseuille relationship is derived assuming that the walls of the channel are perfectly smooth, so that the fluid flow has a unique axial component and no transverse components. If the walls are sufficiently rough to induce three-dimensional components of fluid flow near the wall surfaces, the pressure drop will tend to be greater than that predicted by Equation (12.6) and the fluid flow resistance will also be larger.

12.4.5.1 Reynolds Number: Laminar or Turbulent Flow?

All fluid flow, whether around an object, in channels or in a river, can be broadly classified as either laminar or turbulent. These two flow regimes behave markedly differently, with significant implications for mass and heat transport. Whether fluid flow is laminar or turbulent depends on the relative importance of the inertial forces ($\rho_m v^2/L$) versus viscous forces ($\eta v/L^2$) in the flow (i.e., ratio of the momentum of the fluid and the friction force imparted by the channel walls). This ratio is defined as the *Reynolds number* (Re):

$$\text{Re} = \frac{(\rho_m v^2/L)}{(\eta v/L^2)} = \frac{\rho_m v L}{\eta} \quad (12.12)$$

This dimensionless number was originally proposed by Osborne Reynolds in 1883, where v is the bulk velocity of the flow, ρ_m is fluid mass density and η is the fluid's *dynamic* viscosity. (An alternative version of Re expresses it in the form vL/υ, where υ is the *kinematic* viscosity ($\upsilon = \eta/\rho_m$) with units of m^2/s.) The characteristic length L can be taken as the diameter or wetted perimeter of a fluid channel, or the diameter of a spherical object in a fluid stream, for example.

A low Reynolds-number flow is a laminar, or layered, flow in which fluid streams flow parallel to each other and mix only through advective and molecular diffusion. Laminar flow is dominated by viscous forces and has fluid velocity at all locations invariant with time when boundary conditions are constant. There is advective mass transport only in the direction of fluid flow. An excellent example of laminar flow is shown in Figure 12.9 for a certain brand of tooth paste. Brands such as the one shown have two or more components, typically varying in both colour and composition. When such toothpaste is squeezed out of its tube, the colours do not mix because the paste's high viscosity ensures a low Reynolds number and thus laminar flow.

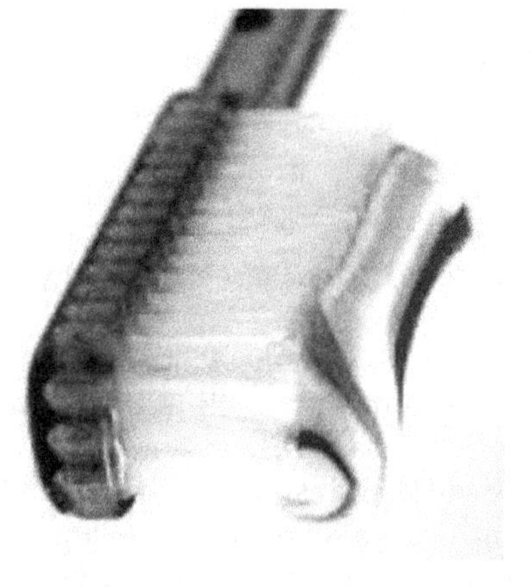

Figure 12.9 Some brands of toothpaste contain two or more components. The high viscosity of the past ensures that when it is squeezed from its tube it exhibits laminar flow. Mixing of the components takes place very slowly through molecular diffusion.

In contrast, a high Reynolds-number flow is a turbulent flow in which inertial forces dominate and various parts of the fluid exhibit motions that are simultaneously random in both space and time. Significant advective mass transport occurs in all directions. This is the kind of flow we can see in rapidly flowing streams, or when we vigorously stir cream into coffee, for example. This difference in the behaviour of laminar and turbulent flow is illustrated schematically in Figure 12.10. The transition between laminar and turbulent flow typically occurs above Re ≈ 2000, although some experiments (e.g., [4]) suggest transition in gas flows in microchannels may occur at Re as low as 400. From Equation (12.12) the mean flow velocity is given by:

$$v = \frac{\eta \mathrm{Re}}{\rho_m L}$$

The DEP fluidic device described in Example 12.5 has a hydraulic diameter given by:

$$D_H = \frac{2wh}{(w+h)} = \frac{2(5\times10^{-3})(2\times10^{-4})}{5.2\times10^{-3}} = 3.85\times10^{-4}\ m$$

The transition from laminar flow to turbulent flow in this DEP device (taking $L = 385\ \mu m$, $\eta = 10^{-3}$ Pa s, $\rho_m = 10^3\ \mathrm{kg\ m^{-3}}$) would therefore occur at a mean fluid velocity of ~5.2 m/s. From a practical perspective this is a difficult flow velocity to achieve in a typical DEP or lab-on-chip fluidic system. For the device of Example 12.5, a pressure drop of 30 Pa was required to achieve a flow velocity of 10^{-2} m/s. To achieve a flow of ~5.2 m/s would require a pressure drop of ~15 600 Pa. Apart from the fact that DEP forces are not able to compete with viscous drag forces for flow rates much above 10^{-2} m/s, a typical DEP device, constructed using polymer or glass parts, would develop fluid leaks or fall apart well below sustaining such physical stress. In most situations it is therefore safe to assume that laminar, rather than turbulent, flow takes place in a DEP device. Laminar flow is not established immediately after a fluid is injected into a channel or chamber, but over a distance known as the *entry length* [5]. This distance is of the order of the characteristic length L, which can be taken as the hydraulic diameter D_H.

Figure 12.10 Schematic representations of laminar flow (Re <~2000) and turbulent flow (Re >~2000) along a channel.

From experimental data laminar flow is identified by a linear proportionality between the log of the pressure drop in the channel and the volume flow rate, i.e., a straight line on a log-log plot of pressure loss versus flow rate. If the flow transitions to turbulence at higher flow rates, the same linear proportionality no longer holds and the slope of the line changes at that flow rate, as depicted in Figure 12.11. The balance of the inertial forces and viscous forces in electrolyte fluid flow can be disturbed by applying an electric field across the channel. If the inertial and electric forces are balanced and both are larger than the viscous force, the convective time scale can be much smaller than the related viscous diffusion time and lead to chaotic flow [6]. Although the difference between turbulent flow and chaotic flow is not clear cut [7], Wang *et al.* [8] have demonstrated that *micro electrokinetic turbulence*, driven by electrokinetic forcing rather than pressure driven channel macroflows, can be induced in microchannels at low Reynolds numbers. This is considered to have the potential to control flow and particle mixing in lab-on-chip devices.

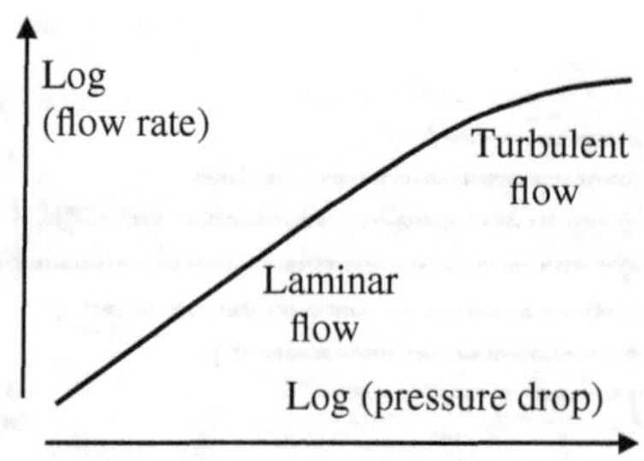

Figure 12.11 The transition to turbulent fluid flow from laminar flow is identified as a deviation from linearity of a logarithmic plot of the pressure drop in a channel and the corresponding volumetric flow rate.

Fluid flow with high Reynolds number is characterized by thin boundary layers. This can be understood by considering what happens when a flat plate at rest receives a step-function impulse of force, causing it to move in its own plane at velocity v. A fluid boundary layer will develop at the plate surface due to the non-slip of fluid at this boundary. If L is the characteristic length scale, then the characteristic time τ_c for transport of material by convection down the resulting fluid flow is $\tau_c = L/v$. The boundary layer will widen at a rate proportional to the fluid viscosity. The *kinematic* viscosity υ ($\upsilon = \eta/\rho$) has units of m^2s^{-1} so that the characteristic time τ_c for a viscosity controlled effect to be transmitted normal to the fluid flow can be given as $\tau_{visc} = L^2/\upsilon$. The ratio of viscous to convective time scales is

$$\frac{\tau_{visc}}{\tau_c} = \frac{(L^2/\upsilon)}{(L/v)} = \frac{\rho v L}{\eta} = \text{Re}$$

Thus the Reynolds number is a measure of the viscous and convective time scales. A large Reynolds number means that viscous effects propagate slowly into the fluid. This is the reason why boundary layers are thin in high Reynolds number flows because the fluid is convected along the flow direction at a much faster rate than the spreading of the boundary layer, which is normal to the flow direction.

12.4.6 Application of Kirchhoff's Laws (Electrical Analogue of Fluid Flow)

Equation (12.8) can be rearranged to give:

$$\Delta P = Q\,R_f$$

This takes the same form as Ohm's Law, which relates the current I generated along an electrical conductor of resistance R_e as a result of the application of a voltage potential difference ΔV between the two ends of the conductor:

$$\Delta V = I\,R_e$$

This implies that Kirchhoff's rules, as applied to the analysis of electrical networks, can be employed to analyse fluid flow in channel networks. An example now follows.

Example 12.6 Applying Kirchhoff's Laws to the Fluidics of a DEP Device

A proposed DEP device for capturing dead cells by positive DEP is shown in Figure 12.12. The requirement is that enriched viable cells flow at a volumetric rate J_3 into a collection vessel open to atmospheric pressure (100 kPa).

1. Derive an equation for calculating the fluid flow J_3 in terms of the channel fluidic resistances R_1, R_2 and R_3 and the pressures applied by syringe pumps P1 and P2.

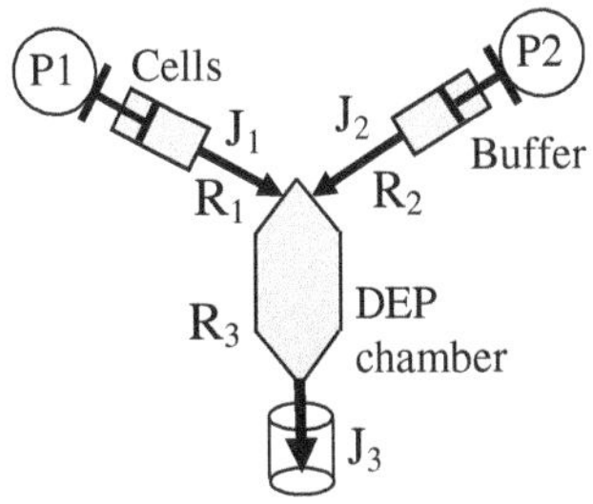

Figure 12.12 A suspension of cells is injected at a volumetric rate J_1 into a DEP chamber, where it mixes with a buffer solution injected at a rate J_2. The cells and buffer are pressure driven (P_1 and P_2) through channels of fluidic resistance R_1 and R_2, respectively. Fluid flow J_3 through the DEP chamber (fluidic resistance R_3,) is collected into a vessel open to atmospheric pressure.

2. Use this equation to calculate the fluid flow J_3 that would exit into a chamber at atmospheric pressure for $R_1 = R_2 = 2 \times 10^{11}$ Pa m^{-3} s; $R_3 = 2 \times 10^{10}$ Pa m^{-3} s and where pumps P1 and P2 exert pressures of 200 kPa and 250 kPa, respectively.

Solution 12.6

1. We will apply Kirchhoff's laws to the electrical analogue of the fluidic T-network shown in Figure 12.13.
   (a) Current Law (algebraic sum of currents at a junction is zero)

$$J_3 = J_1 + J_2 \qquad \text{(i)}$$

   (b) Voltage Law (Algebraic sum of voltage drops around a closed circuit is zero)

$$\Delta P1 = J_1R_1 + J_3R_3 \qquad (\Delta P1 = P1 - P_{atm})$$

to give

$$J_1 = (\Delta P1 - R_3J_3)/R_1 \qquad \text{(ii)}$$

$$\Delta P2 = J_2R_2 + J_3R_3 \qquad (\Delta P2 = P2 - P_{atm})$$

to give

$$J_2 = (\Delta P2 - R_3J_3)/R_2 \qquad \text{(iii)}$$

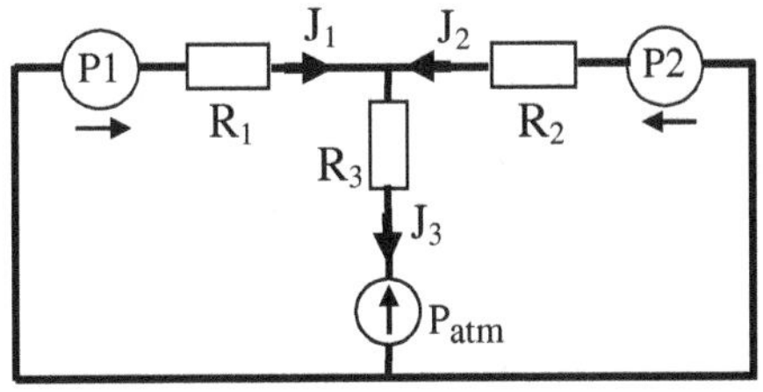

Figure 12.13 The electrical circuit analogue for the fluidic system of the DEP device shown in Figure 12.12.

Substitute J_1 obtained from (ii) and J_2 obtained from (iii) into (i):

$$J_3 = (\Delta P1 - R_3J_3)/R_1 + (\Delta P2 - R_3J_3)/R_2$$

Rearranging:

$$J_3 = (\Delta P1.R_2 + \Delta P2.R_1)/(R_1R_2 + R_2R_3 + R_1R_3) \quad \text{(iv)}$$

2. In equation (iv) we define $\Delta P1 = P1 - P_{atm}$ and $\Delta P2 = P2 - P_{atm}$. Therefore

$$\Delta P1 = 200 - 100 = 100 \text{ kPa};$$
$$\Delta P2 = 250 - 100 = 150 \text{ kPa}$$

Substituting the given values for R_1, R_2 and R_3 into (iv):

$$J_3 = (5 \times 10^{16} \text{ Pa}^2 \text{ m}^{-3} \text{ s})/(4.8 \times 10^{22} \text{ Pa}^2 \text{ m}^{-6} \text{ s}^2) = 1.04 \times 10^{-6} \text{ m}^3 \text{ s}^{-1} = 10.4 \text{ µl/s}$$

This example demonstrates how Kirchhoff's laws, which are used to analyse current flows in electrical circuits, can also be used to control and design for liquid flow in DEP devices.

12.5 Navier–Stokes Equations

The Navier–Stokes equations are widely used to describe the behaviour of fluids in terms of continuous functions of space and time. They encapsulate the three conservation laws of mass, energy and momentum and are considered in terms of flux rather than changes of their instantaneous values. In mathematical terms this is represented as partial derivatives of the dependent variables.

The calculation of fluid velocities and pressures at the macroscopic scale is based on the assumption that the fluid can be treated as a continuum. Apart from fluid velocity v and pressure P, for the most general situation that includes compressible and incompressible fluids we also require knowledge of the mass density ρ_m, viscosity η, specific heat C_p and temperature T of the fluid. Pressure and temperature characterize the energy state and number of molecules present in a given volume of fluid. If the pressure and temperature do not vary too greatly within this volume element, analytical functions can be derived that relate the density, viscosity and specific heat to the pressure and temperature. In a three-dimensional system we are therefore left with five unknowns, namely P, T, v_x, v_y and v_z. These five unknowns are related by a system of equations that describe:

- the conservation of mass;
- the conservation of momentum;
- the conservation of energy.

The equations describing these three conservation laws are often referred to as the Navier–Stokes equations, but

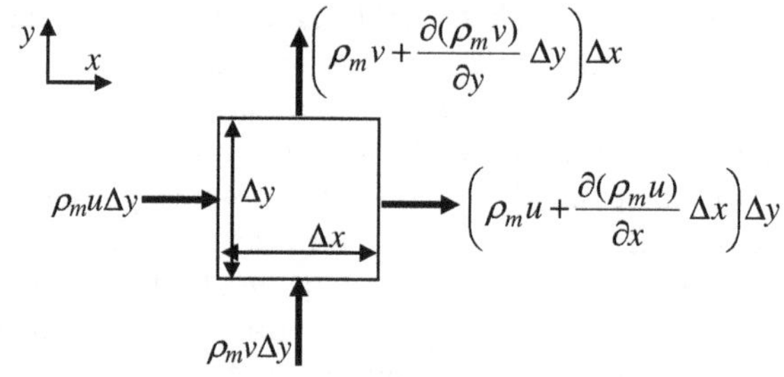

Figure 12.14 Relationships describing the conservation of fluid mass for flow through a two-dimensional element $\Delta x \Delta y$.

it is more correct to reserve this description to the equations that describe conservation of momentum. Conservation of energy usually concerns heat flow in fluid systems in which a temperature gradient is created by an energy source or sink, associated with chemical reactions or heating and cooling devices. For most microfluidic flows in DEP devices the temperature is constant, in which case the conservation of energy equation is redundant. We will thus focus on the derivations of the conservation of mass and conservation of momentum equations.

12.5.1 Conservation of Mass Equation

In Chapter 3, Equation (3.57) describes the condition for the *conservation of electrical charge*, also known as the *continuity equation of electrical current flow*. In simple terms this states that the sum of all the sources of charge minus the sinks of charge within a defined region gives the net charge flow (current) out or into that region. This is coupled to the fact that charge can neither be instantaneously created nor destroyed. This fact also applies to mass. In terms of a linear, one-dimensional fluid flow, the corresponding conservation or continuity of mass equation is:

$$\frac{d\rho_m}{dt} + \rho_m \nabla \cdot \mathrm{u} = 0 \tag{12.13}$$

The rate of change of mass density along a flow path is equal to $\rho_m u$, where u is the fluid velocity. A two-dimensional element (Δx, Δy) is shown in Figure 12.14, with fluid velocities u and v in the x- and y-directions, respectively. In due course this is generalized to the three-dimensional case.

For the system of fluid flow shown in Figure 12.14 the conservation of mass is given by:

$$\frac{\partial(\rho_m \Delta x \Delta y)}{\partial t} = \rho_m u \Delta y + \rho_m v \Delta x - \left[\rho_m u + \frac{\partial(\rho_m u)\Delta x}{\partial x}\right]\Delta y - \left[\rho_m v + \frac{\partial(\rho_m v)\Delta y}{\partial y}\right]\Delta x$$

Dividing by ΔxΔy we obtain

$$\frac{\partial \rho_m}{\partial t} + \frac{\partial(\rho_m u)}{\partial x} + \frac{\partial(\rho_m v)}{\partial y} = 0$$

which can be written as

$$\frac{\partial \rho_m}{\partial t} + \frac{u\partial \rho_m}{\partial x} + \frac{\partial \rho_m}{\partial y} + \rho_m \left[\frac{\partial u}{\partial x} + \frac{\partial v}{\partial y}\right] = 0 \tag{12.14}$$

Defining the operator D/Dt in three-dimensional Cartesian coordinates as

$$\frac{D}{Dt} = \frac{\partial}{\partial t} + u\frac{\partial}{\partial x} + v\frac{\partial}{\partial y} + w\frac{\partial}{\partial z}$$

we can write Equation (12.14) in the vector form of Equation (12.13):

$$\frac{D\rho_m}{Dt} + \rho_m \nabla \cdot \mathrm{V} = 0 \tag{12.15}$$

where V is the velocity vector (u, v, w). We are dealing with an incompressible liquid, so terms such as $\partial\rho_m/\partial t$, $\partial\rho_m/\partial x$ and $D\rho/Dt$ are zero, with density ρ_m remaining constant. Equations (12.14) and (12.15) thus reduce, for the three-dimensional case, to

$$\frac{\partial u}{\partial x} + \frac{\partial v}{\partial y} + \frac{\partial w}{\partial z} = 0 \tag{12.16a}$$

and

$$\nabla . \vec{V} = 0 \tag{12.16b}$$

12.5.2 Conservation of Momentum Equation (Navier–Stokes Equation)

The change of momentum in a fluid element is given by the balance between the inlet and outlet fluid momentum and the tangential and normal stresses acting on that element. These are considered separately in Figures 12.15 and 12.16 for the two-dimensional case.

For Newtonian fluids the tangential stress τ and normal stress σ are given as

$$\tau_{xy} = \eta\left(\frac{\partial v}{\partial x} + \frac{\partial u}{\partial y}\right) \tag{12.17a}$$

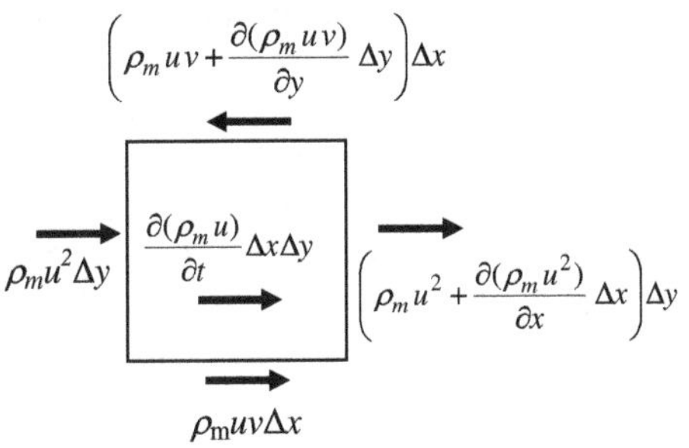

Figure 12.15 Inlet and outlet fluid momentum in the x-direction for a fluid element ΔxΔy.

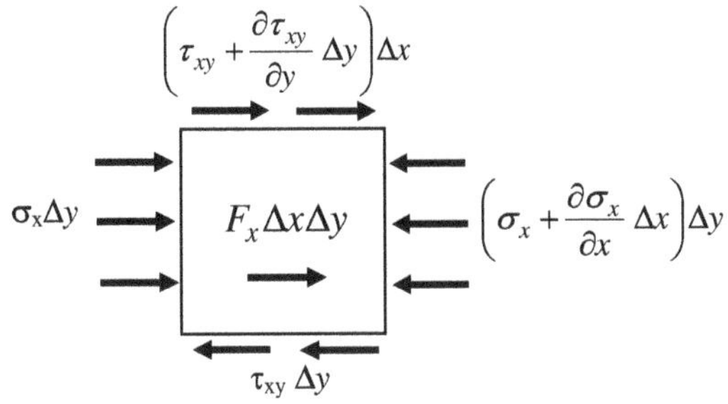

Figure 12.16 The normal and tangential stresses acting on the fluid element shown in Figure 12.15.

and

$$\sigma_x = P - 2\eta\frac{\partial u}{\partial x}\left(\frac{\partial u}{\partial x} + \frac{\partial v}{\partial y}\right) \tag{12.17b}$$

Summing the forces shown in Figure 12.16 in the x-direction and using the mass conservation Equations (12.16) we obtain

$$\rho_m \frac{Du}{Dt} = -\frac{\partial \sigma_x}{\partial x} + \frac{\partial \tau_{xy}}{\partial y} + \mathrm{F}_x$$

Combining this result with Equation (12.17) gives the Navier–Stokes equation:

$$\rho_m\left(\frac{\partial u}{\partial t} + u\frac{\partial u}{\partial x} + v\frac{\partial u}{\partial y} + w\frac{\partial u}{\partial z}\right) = -\frac{\partial P}{\partial x} + \eta\left(\frac{\partial^2 u}{\partial x^2} + \frac{\partial^2 u}{\partial y^2} + \frac{\partial^2 u}{\partial z^2}\right) + \mathrm{F}_x \tag{12.18a}$$

Extending this to three-dimensions

$$\rho_m\left(\frac{\partial v}{\partial t} + u\frac{\partial v}{\partial x} + v\frac{\partial v}{\partial y} + w\frac{\partial v}{\partial z}\right) = -\frac{\partial P}{\partial y} + \eta\left(\frac{\partial^2 v}{\partial x^2} + \frac{\partial^2 v}{\partial y^2} + \frac{\partial^2 v}{\partial z^2}\right) + \mathrm{F}_y \tag{12.18b}$$

$$\rho_m\left(\frac{\partial w}{\partial t} + u\frac{\partial w}{\partial x} + v\frac{\partial w}{\partial y} + w\frac{\partial w}{\partial z}\right) = -\frac{\partial P}{\partial z} + \eta\left(\frac{\partial^2 w}{\partial x^2} + \frac{\partial^2 w}{\partial y^2} + \frac{\partial^2 w}{\partial z^2}\right) + \mathrm{F}_z \tag{12.18c}$$

and in vector form:

$$\rho_m \frac{D\mathrm{V}}{Dt} = -\nabla P + \eta\nabla^2 \mathrm{V} + \mathrm{F} \tag{12.18d}$$

where V is the velocity vector (u, v, w) and F is the force per unit volume acting on the element (Δx, Δy, Δz). If the fluid is an electrolyte containing ionic solutes and an external electric field is applied to the fluid element, an electrical force contributes to the total force F [5, 6]. The electrical force (F = qE) acting on each ionic particle in the electrolyte is transferred to the solvent liquid through collisions at the molecular level. The free ions individually transfer their momentum to the liquid. This gives rise to fluid motion induced by electro-osmosis, as described in section 12.8.

12.5.3 Conservation of Energy Equation

To derive this equation we identify either a source or sink of heat S_H and specify the specific heat C_p and heat conductivity κ of the liquid. The specific heat is defined as the amount of heat Q per unit mass required to raise the temperature of a substance by one degree Celsius:

$$Q = C_p\, m\Delta T$$

The thermal conductivity of a substance is defined in terms of the quantity of heat Q conducted per unit time Δt down a unit temperature gradient ΔT in a direction normal to a surface of unit area ΔA:

$$k = Q\Delta T/(\Delta t\Delta A) = -Q/(\partial T\partial n)$$

The heat conduction must arise only from the temperature gradient and not from a secondary heat source or chemical reaction, for example. The specific heat of water is 4.186 J g^{-1}K^{-1} and its thermal conductivity is ~0.6 W m^{-1}K^{-1}.

In three-dimensional Cartesian coordinates the conservation of energy equation is:

$$\rho_m C_p\left(\frac{\partial T}{\partial t} + u\frac{\partial T}{\partial x} + v\frac{\partial T}{\partial y} + w\frac{\partial T}{\partial z}\right)$$
$$= \frac{\partial}{\partial x}\left(\kappa\frac{\partial T}{\partial x}\right) + \frac{\partial}{\partial y}\left(\kappa\frac{\partial T}{\partial y}\right) + \frac{\partial}{\partial z}\left(\kappa\frac{\partial T}{\partial z}\right) + S_H$$

12.6 Diffusion

From section 12.3 we learn that a molecule in thermal equilibrium with a surrounding fluid of absolute temperature T has an average kinetic energy of $3kT/2$, with an average velocity $(3kT/m)^{1/2}$ associated with motion along each of the three axes in a three-dimensional volume. Diffusion is the random migration of molecules or small particles from multiple collisions arising from the kinetic motion of neighbouring molecules. A schematic of this process is shown in Figure 12.17, where a cluster of gas molecules is shown occupying the corner of an otherwise empty container. As a simplification we may assume that the time τ and average mean free path length between collisions remains constant. The rate of randomizing collisions is thus $1/\tau$. After a sufficiently large number n of such collisions the molecules are evenly distributed in the container (after time $n\tau$).

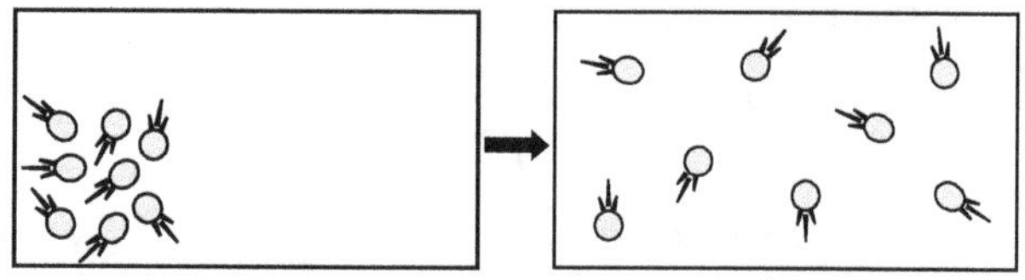

$t = t(0)$ $\qquad$ $t = t(0) + n\tau$

Figure 12.17 Through a process called diffusion, randomizing collisions at a rate of $1/\tau$ per second result in the cluster of molecules shown at time $t = 0$ being evenly distributed in a container after n such collisions.

From their independent analyses of Brownian motion (the buffeting of macroscopic particles through collisions with fluid molecules) Einstein and Smoluchowski derived the following expression for the diffusion coefficient D

$$D = \frac{L_{mfp}^2}{2\tau} \tag{12.19}$$

where L_{mfp} is the mean free path length given in Equation (12.2). An excellent discussion of the origins and validity of this so-called Einstein–Smoluchowski equation is given by Isla [9]. For nitrogen gas at room temperature and atmospheric pressure we can estimate that $L_{mfp} = 14.4 \times 10^{-10}$ m and $\tau = 3.1 \times 10^{-10}$ s. From Equation (12.19) this provides an estimate for D of 3.3 × 10^{-9} m^{2}s^{-1}. Einstein also demonstrated that for macroscopic particles exhibiting Brownian motion in a fluid, the particle's diffusion coefficient is given as:

$$D = \frac{kT}{6\pi\eta a} \tag{12.20}$$

where a is the particle's effective hydrodynamic radius and η is the fluid viscosity. This is called the Stokes–Einstein equation, whose origin and validity has also been discussed by Isla [9]. For particles suspended in water, the effective hydrodynamic radius is defined as the radius of a rigid uncharged sphere, which exhibits the same hydrodynamic behaviour as the solvated molecule in solution. This should therefore include water of hydration, which is too firmly bound to the particle's surface to participate in the viscous shearing process as it moves through the aqueous medium. Equation (12.20) was derived on the assumption that the solute molecule is large compared to the solvent. Nevertheless the equation has been experimentally confirmed for suspended particles with radii as small as 5 nm and for large colloidal particles with suspension volume fractions up to 3%. It can also provide good approximations for the diffusion of molecular species in water. Thus sucrose ($a \approx 0.5$ nm) can be estimated to have a diffusion coefficient in water ($\eta = 1 \times 10^{-3}$ Pa) at 298 K of ~3.9 × 10^{-10} m^{2}s^{-1}, which can be favourably compared to the value of 5.2 × 10^{-10} m^{2}s^{-1} cited in Table 12.2. Approximate values of diffusion

Table 12.2 Diffusion coefficients for various molecules and ions in water at 298 K.

Molecule	D (10^{-9} m^2 s^{-1})
Water	2.26
Sucrose	0.52
Methanol (CH_3OH)	1.6
Glycine	1.06
NaCl	1.7
H^+	9.3
OH^-	5.3
Na^+	1.33
K^+	1.96
Cl^-	2.03

coefficients for some biologically relevant particles are given in Table 12.3.

A description of particle diffusion can be made in terms of a one-dimensional random walk, often described in terms of the 'drunken sailor' problem outlined in Figures 12.18 and 12.19. At each new step forward, the drunken sailor is equally likely to stagger one step to the left as he is to the right. We can use this analogy to describe the resulting random direction that a molecule follows after colliding with another molecule.

After a number of random steps the spatial distribution of particles along a one-dimensional axis takes the form of a probability distribution described by the factorials of the binomial coefficients. Applying Stirling's approximation for these factorials, then for a sufficiently large number of thermal collisions we can represent the probability distribution as a Gaussian or normal distribution. In one-dimension the probability $P(x)dx$ of finding a particle between x and $x + dx$ at time t is given by Isla [9] as:

$$P(x)dx = \frac{1}{(4\pi Dt)^{1/2}} e^{-x^2/4Dt} dx$$

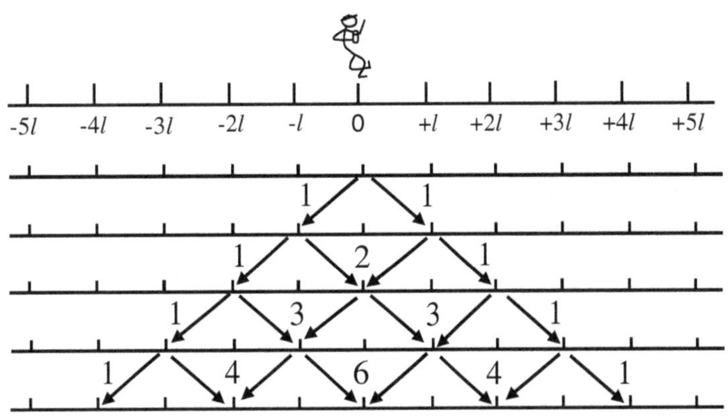

Figure 12.18 The probability distribution for a one-dimensional random (drunken sailor) walk is given by the factorials of the binomial coefficients as given by Pascal's triangle. The three-dimensional case is given by Pascal's pyramid.

Table 12.3 Diffusion coefficients for various macromolecules and particles in water at 293 K, derived using the Stokes–Einstein relation given by Equation (12.20).

Macromolecule	D (m^2 s^{-1})
Ribonuclease	1.2×10^{-10}
Lysozyme	1.0×10^{-10}
Serum albumin	5.9×10^{-11}
Haemoglobin	6.9×10^{-11}
Urease	3.5×10^{-11}
Collagen	6.9×10^{-12}
Viruses, bacteria, cells	$10^{-13} \sim 10^{-16}$

The mean displacement $\langle x^2 \rangle$of the particle is thus given by

$$\langle x^2 \rangle = \frac{\int_0^\infty x^2 dP}{\int_0^\infty dP} = 2Dt$$

An alternative way to derive this relationship is to employ Equation (12.19) as follows:

$$\langle x^2 \rangle = nL_{mfp}^2 = \frac{t}{\tau} L_{mfp}^2 = 2Dt$$

to give

$$\langle x \rangle = \sqrt{2Dt} \tag{12.21}$$

We can consider $\langle x \rangle$ as the mean diffusion length for a molecule interacting through collisions with neighbouring molecules. Values for this diffusion length are given in Table 12.4 for times of 1 ms and 10 s. We can see that small sugar molecules like glucose will diffuse a distance of around 1 μm after 1 ms and 0.1 mm after 10 s. These can be significant distances in microfluidic systems. For particles of the size of bacteria, however, the corresponding diffusion lengths are much less at 20 nm and 2 μm, respectively.

Diffusion of molecules and particles tends to occur down their concentration gradient – also referred to as diffusion gradients (see Figure 12.20).

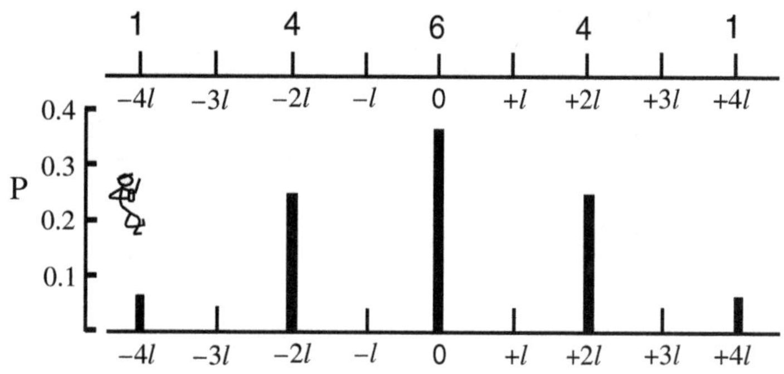

Figure 12.19 The probability distribution for a one-dimensional random walk after time $t = 4\tau$. After a period of 4τ the probability of the sailor standing straight ahead of his original location is 6/16 = 0.375.

This diffusion process can be described by Fick's First Equation of Diffusion:

$$J_x = -D\frac{\partial C}{\partial x}$$

which states that the net flux J_x (moles $m^{-2}s^{-1}$) of diffusing molecules or particles is proportional to the concentration gradient and diffusion constant of the molecule/particle (the negative sign indicates that the molecules diffuse *down* the concentration gradient). Unless the concentration gradient is artificially maintained (e.g., with a continuous source and sink of the molecules or particles) the factor $\partial C/\partial x$ will change as a function of time. This leads to Fick's Second Equation:

$$\frac{\partial C}{\partial t} = D\frac{\partial^2 C}{dx^2}$$

This equation can be used (with the appropriate boundary conditions) to determine how a nonuniform distribution of molecules or particles will redistribute itself as a function of time. Diffusion along a microfluidic channel is effectively a one-dimensional problem. In this case the solutions of Fick's Second Equation are:

$$\frac{\partial C}{\partial x} = \frac{C_0}{(4\pi Dt)^{1/2}}e^{-x^2/4Dt} \quad \text{and} \quad \frac{\partial C}{\partial t} = -\frac{x}{2t}\frac{\partial C}{\partial x}$$

12.6.1 The Peclet Number: Transport by Advection or Diffusion?

The constant motion of molecules in fluids ensures that, when one fluid is placed adjacent to a second fluid, its molecules proceed to enter into the bulk of that second fluid by a process called diffusion. When we employ the continuum concept, instead of calculating each individual motion, we calculate the average motion of a statistically significant number of molecules. It then becomes convenient to separate the actual diffusion process into two conceptual transport mechanisms: a molecular process modelled as a statistical random walk, which is proportional to the degree of kinetic energy in the system; and an advective process in which molecules are carried along by the average velocity of the flow. The common practice is to restrict the word *diffusion* to describe the first process and label the second process *advection* (*convection* if heat is being transferred). The relative importance of these two conceptual transport mechanisms is given by the Peclet Number, the ratio of advection and diffusion:

$$\text{Pe} = vL/D$$

in which v is the fluid velocity, D is the diffusion coefficient of the solute in the solvent and L is the characteristic

Table 12.4 Approximate diffusion coefficients for some biologically relevant particles in water at 293 K. Values for the mean diffusion distance (diffusion layer thickness), defined by Equation (12.21), are given for time intervals of 1 ms and 10 s.

Particle	Diffusion coefficient m^2s^{-1}	Diffusion layer thickness (m) 10^{-3} s	10 s
Small ions	2×10^{-9}	2×10^{-6}	2×10^{-4}
Sugar molecules	5×10^{-10}	1×10^{-6}	1×10^{-4}
Small proteins (e.g., lysozyme)	1×10^{-10}	4.5×10^{-7}	4.5×10^{-5}
50-base pair DNA	2.5×10^{-11}	2.2×10^{-7}	2.2×10^{-5}
Large proteins (e.g., collagen)	7×10^{-12}	1.2×10^{-7}	1.2×10^{-5}
Virus	4×10^{-12}	9×10^{-8}	9×10^{-6}
5000-base pair DNA	1×10^{-12}	4.5×10^{-8}	4.5×10^{-6}
Bacteria	2×10^{-13}	2×10^{-8}	2×10^{-6}

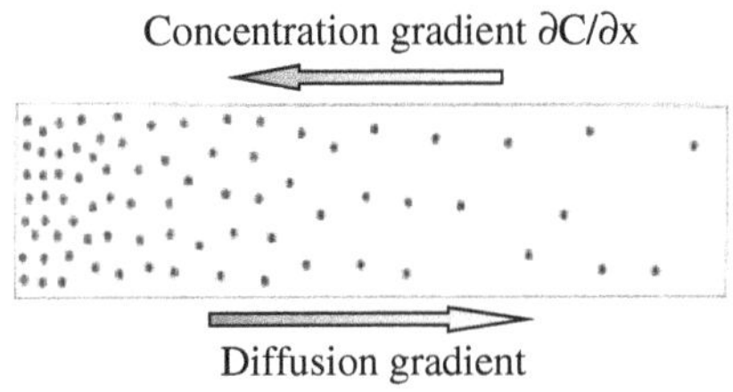

Figure 12.20 Molecules tend to diffuse down a concentration gradient – also termed as a diffusion gradient.

dimension of the fluid channel. When L is so small that the Peclet number is less than 1000, molecular diffusion becomes an important mechanism for mixing. Stirring may be appropriate for mixing in a macroscale device, but a diffusion-based approach should be used in a low Pe device.

The isodielectric cell separator described in Chapter 10 (section 10.4.6) must operate in an advection-dominated transport regime to ensure that the conductivity gradient is preserved along the length of the fluidic channel. The two length scales of relevance are the channel width (w) from which the time scale for diffusion is obtained ($\sim w^2/D$) and the channel length (l), which enters through the advective time constant ($\sim l/v$). The transport in the device is best governed in terms of a modified Peclet number (w^2v/Dl), which defines for a given separation a critical Pe value below which cell separation is no longer possible. For example, if the cells in the sample exhibit effective conductivities varying in value by a factor of 5, it was concluded that the Pe number must be maintained above ~18 to accommodate this range.

12.7 Ionic (Electrical) Double Layer

The distribution of ions around a charged particle is determined by the balance between electrostatic forces and thermal agitation. A quantitative description of the distribution of ions in thermal equilibrium in an electrostatic field can be obtained by combining Poisson's equation (Equation 3.30) with the Boltzmann distribution. Poisson's equation generalizes the electrostatic Coulomb potential $V(r)$ to a volume distribution of charge density $\rho(r)$:

$$\nabla^2 V(r) = -\frac{\rho(r)}{\varepsilon_o \varepsilon_m} \tag{12.22}$$

Equation (12.22) reduces to the Coulomb potential described by Equation (4.23) when $\rho(r)$ is a point charge in a uniform dielectric medium. For an arbitrary collection of ions of number densities (m^{-3}) c_i and valences z_i the charge distribution is given by

$$\rho(r) = q \sum_i z_i c_i(r)$$

where q is the charge on an electron. The earliest model, known as the Helmholtz model, describes the distribution of ions at the interface between a charged surface and an electrolyte as a parallel-plate capacitor. One plate of this capacitor contains the charge on the electrified surface and the other plate (known as the Helmholtz plane) contains the ions, of opposite charge polarity to that on the surface, electrostatically attracted to it from the electrolyte. The thickness of this electrical double layer is taken to be the diameter of the ions attracted to the charged surface. If we define σ to be the surface charge density, it is balanced by an equivalent amount of ionic charge of opposite polarity in the solution. This balance can be expressed by the relationship

$$\sigma = -\int_0^\infty \rho(r)dr \tag{12.23}$$

In the Helmholtz model the counter ionic charge density $\rho(r)$ takes the form of a layer of charges at the charged surface. An obvious oversimplification of the Helmholtz model is that thermal motions of the ions in the electrolyte are ignored. These thermal motions will cause $\rho(r)$ to form a diffused distribution rather than a layer.

The concentration c_i of ions in thermodynamic equilibrium with the electrolyte solution, as a function of their distance r from a charged surface, is related to the electrostatic potential $\phi(r)$ using the Boltzmann distribution as follows:

$$c_i(r) = c_{i\infty} \exp\left(\frac{-qz_i V(r)}{kT}\right) \tag{12.24}$$

The parameter $c_{i\infty}$ is the concentration of ions in the bulk solution, far enough away from the charged object that the value of $V(r)$ is zero. If the electrolyte is an aqueous sodium chloride solution, for example, this corresponds to equal concentrations [Na^+] and [Cl^-] of the sodium and chloride ions. As we move from the bulk solution towards a negatively charged surface, for example, we would find that [Na^+] increases and [Cl^-] decreases. Substituting Equation (12.24) into Equation (3.22) gives

$$\rho(r) = q \sum_i z_i c_{i\infty} \exp\left(\frac{-qz_i V(r)}{kT}\right) \tag{12.25}$$

Using this result to eliminate ρ(r) from Equation (12.21) we obtain the Poisson–Boltzmann equation:

$$\nabla^2 V(r) = -\frac{q}{\varepsilon_o \varepsilon_r} \sum_i z_i c_{i\infty} \exp\left(\frac{-qz_i V(r)}{kT}\right) \tag{12.26}$$

This equation describes the electrical potential $V(r)$ at the interface between a charged object and an electrolyte solution, taking into account the screening of this potential by counterions. For example, it can, in principle, describe the spatial composition of the ionic 'atmosphere' around an ion or a charged particle. For such situations, where r is the only relevant coordinate, the appropriate form of the vector operator ∇^2 involves spherical coordinates. However, Equation (12.26) cannot be solved analytically using spherical coordinates. Depending on the geometry of the system and the boundary conditions, solving Equation (12.26) may require the use of approximations. One such approximation is to assume that the electrostatic interactions of the ions in the solution are weak ones, so that $qz_i V(r)/kT \ll 1$. This allows the linear form ($e^x = 1 + x + x^2/2! + \ldots$) of the exponential function to be used in Equation (12.26) to give:

$$\nabla^2 V(r) = -\frac{q}{\varepsilon_o \varepsilon_r} \sum_i z_i c_{i\infty} \left(1 - \frac{qz_i V(r)}{kT}\right) \tag{12.27}$$

A further simplification can be made by noting that for a sufficiently large distance r the electrical potential $V(r)$ is zero and $dV(r)/dr$ is also zero. This corresponds to electrical neutrality of the solution, so that

$$q \sum_i z_i c_{i\infty} = 0$$

Adopting this boundary condition as r tends to infinity, Equation (12.27) can then be written as:

$$\nabla^2 V(r) = \frac{q^2}{\varepsilon_o \varepsilon_m kT} \sum_i z_i^2 c_{i\infty} V(r) \tag{12.28}$$

Equation (12.28) is referred to as the linear form of the Poisson–Boltzmann equation and can be written as

$$\nabla^2 V(r) = \kappa^2 V(r) \tag{12.29}$$

12.7.1 The Debye Screening Length

If the variable $V(r)$ in Equation (12.29) is transformed to a variable with units of $1/\kappa$, this equation will contain no parameters. This means that $1/\kappa$ must represent a fundamental unit when considering electrostatic interactions in ionic solutions. From Equations (12.28) and (12.29)

$$\kappa^2 = \frac{q^2}{\varepsilon_o \varepsilon_m kT} \sum_i z_i^2 c_{i\infty}$$

From this we can determine that $1/\kappa$ has units of length. It is an important parameter known as the Debye length. We can interpret its significance by stating that, for distances r shorter than the Debye length, the electrostatic interactions will be strong, but for much larger distances the interactions will be weak because of ionic screening. The factor $1/\kappa$ can thus be taken to be the ionic screening distance. Its value is given by

$$1/\kappa = \sqrt{\frac{\varepsilon_o \varepsilon_m kT}{2q^2 I N_A 10^3}} \tag{12.30}$$

In this equation N_A is the Avogadro constant (6.022×10^{23} mol^{-1}) and we have converted the ionic density c (m^{-3}) to the ionic strength I (mol l^{-1}) of the solution by assuming a simple monovalent salt ($z_i = 1$) such as NaCl, so that

$$\sum_i z_i^2 c_{i\infty} = 2IN_A 10^3$$

For an aqueous 10 mM solution of NaCl at 298 K we can calculate $1/\kappa$ to be 3.07 nm (assuming $\varepsilon_r = 80$). From Equation (12.30) we note that the Debye length is inversely proportional to the square root of the solution's ionic strength, so that for a 1 M solution it decreases to 0.31 nm. We would expect the ionic screening to increase as the number of ions per unit volume increases. The ionic strength also increases as z_i^2, so that solutions containing multivalent salts (e.g., $CaCl_2$) will be more effective at screening electrostatic interactions.

12.7.2 The Gouy–Chapman Equation

If we wish to consider the case of a charged membrane surface, or to approximate the curved surface of a particle as a planar surface, the only important dimension is the distance normal to the surface, which we will take to be the x-direction. In one dimension Equation (12.22) is written as

$$\frac{d^2 V(x)}{dx^2} = -\frac{\rho(x)}{\varepsilon_o \varepsilon_m} \tag{12.31}$$

so that Equation (12.26) takes the form:

$$\frac{d^2 V(x)}{dx^2} = -\frac{q}{\varepsilon_o \varepsilon_r} \sum_i z_i c_{i\infty} \exp\left(\frac{-qz_i V(x)}{kT}\right) \tag{12.32}$$

This form of the Poisson–Boltzmann equation can be solved analytically without converting it to the linear form. For a monovalent salt solution such as NaCl (i.e., $z = \pm 1$) of number density c and noting that, as the distance x tends to infinity, $d\phi/dx$ tends to zero, integration of Equation (12.32) gives:

$$\frac{dV(x)}{dx} = \sqrt{\frac{2kTc}{\varepsilon_o \varepsilon_r}} \cdot (\exp(-qV(x)/2kT) - \exp(qV(x)/2kT)) \tag{12.33}$$

Using the relationship between the surface charge density σ and the counter charge density $\rho(r)$ given by Equation (12.23), then from Equation (12.31):

$$\sigma = \varepsilon_o \varepsilon_m \int_0^{\infty} \frac{d^2V(x)}{dx^2} dx = -\varepsilon_o \varepsilon_m \frac{dV(0)}{dx} \qquad (12.34)$$

This equation relates the electric field (i.e., the gradient of the electric potential) at the surface to the surface charge density, where we can define $V(0)$ to be the surface potential. Substituting Equation (12.33) into Equation (12.34) we obtain the important Gouy–Chapman equation [10]:

$$\sigma = -\sqrt{2\varepsilon_o \varepsilon_m kTc} \cdot [\exp(-qV(0)/2kT) - \exp(qV(0)/2kT] \qquad (12.35)$$

Adopting the linear approximation for the exponential terms and inserting the Debye screening length $1/\kappa$ introduced for Equation (12.30), the Gouy–Chapman equation reduces to the simple form:

$$\sigma = \varepsilon_o \varepsilon_m \kappa V(0) \qquad (12.36)$$

Equation (12.36) describes a proportionality between the surface charge σ and the surface potential $V(0)$ analogous to the relationship $q = VC$ between charge and applied voltage for a capacitance C. From Equation (12.36) the term $\varepsilon_o \varepsilon_r \kappa$ can be considered to be an effective capacitance per unit area, with the Debye length $1/\kappa$ representing the distance between the two charge-carrying plates. This further supports the concept of the charge distribution of ions at the interface between a charged surface and an electrolyte taking the form of an electrical double layer.

By integrating Equation (12.33) and adopting the assumption that $qz_i\phi(r)/kT \ll 1$, together with the linear form of the exponential, we obtain the relationship:

$$V(x) = V(0)\exp(-xk) \qquad (12.37)$$

Equation (12.37) indicates that the electrostatic potential falls exponentially with distance into the electrolyte, reaching a value of $0.37V(0)$ at a distance equal to the Debye screening length $1/\kappa$.

12.7.3 Stern's Modification of the Gouy–Chapman Equation

The assumptions used in deriving Equation (12.35) do not hold for high surface charge densities and high potential gradients (fields). The predicted concentrations of ions attracted to the charged surface can be unrealistically high, sometimes above the saturation level for a salt. Multivalent salt ions can also be attracted so strongly to

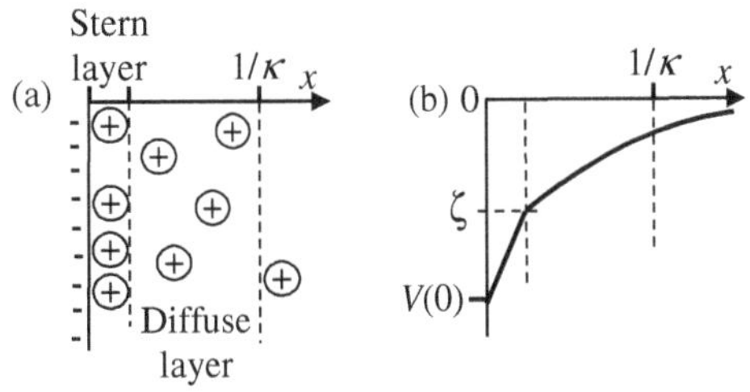

Figure 12.21 The Gouy–Chapman–Stern treatment of the ionic (electrical) double layer at a charged surface. (a) Cations are shown strongly bound in the Stern layer at a negatively charged surface. Other cations form a diffuse layer within a distance corresponding to the Debye screening length $1/\kappa$. (b) The electrostatic potential $V(r)$ falls linearly within the Stern layer before following an exponential fall to zero in the bulk fluid. Beyond the Debye length the charged surface is effectively screened from the bulk electrolyte. The zeta potential ζ is shown located at the hydrodynamic plane of shear.

the surface as to bind to it. Otto Stern introduced two modifications to the Gouy–Chapman theory, described in detail in the book by Aveyard and Haydon [11]. The first modification simply takes account of the fact that an ion cannot get closer to the charged surface than its own radius and the second modification is to allow for specific binding of ions to the charged surface in what is called the Stern layer. Within a distance of the Debye length other ions form a diffuse layer and the electrostatic potential falls exponentially as described by Equation (12.37), with the potential $V(0)$ being replaced with the value $V(a)$ at the interface of the Stern and diffuse layers. These two ion distributions and the corresponding profile of the electrostatic potential are shown in Figure 12.21. Bedzyk *et al.* [12] determined the ion distribution in an electrolyte solution in contact with a charged polymerized phospholipid membrane using x-ray standing waves and found it to qualitatively agree with the Gouy–Chapman–Stern model.

The ions shown in Figure 12.21 appear as bare charges, but in fact they interact and attract neighbouring water molecules. This interaction involves the torques induced on the water molecules by the electrostatic interactions of their dipole moments with the electric fields around the ions. This restricts rotational motions of the water molecules and creates a 'hydration shell' of oriented water molecules around an ion, as schematically depicted in Figure 12.22. This hydration shell is not a solid structure, but does increase the effective diameter of an ion and reduces ion-ion electrostatic interactions. It also contributes to some extent to the effective width of the Stern layer and more so to the effective width of the diffuse ionic layer.

The charge densities σ_{St} and σ_{Dl} in the Stern layer and the diffuse layer add up to the total charge σ on

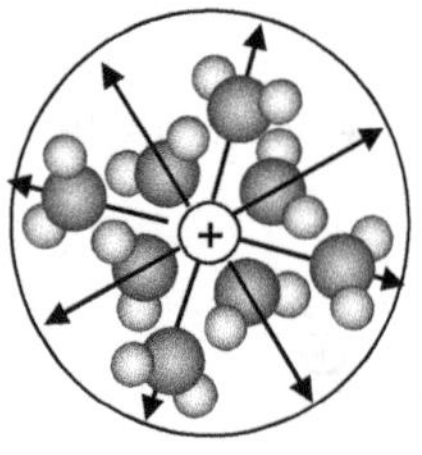

Figure 12.22 Electrostatic interactions between a solvated ion result in the attraction and alignment of the dipole moments of surrounding water molecules along the ion's field lines. This restricts the rotational motion of the water molecules and creates a so-called hydration shell.

the surface and together these charged layers act as two capacitors in series. The total capacitance C is given by:

$$C = \frac{C_{St}C_{Dl}}{C_{St} + C_{Dl}} \tag{12.38}$$

The contribution to the total capacitance of the Stern layer tends to be unaffected by changes in the ionic strength of the solution. At low ionic strengths the Debye screening length $1/\kappa$ is relatively large so that the effective capacitance $\varepsilon_o\varepsilon_r\kappa$ of the diffuse layer is low and the total capacitance C given by Equation (12.38) tends to be dominated by the diffuse layer. At high ionic strengths the Debye length is small, C_{Dl} is large and now the total capacitance tends to be dominated by C_{St}.

12.7.4 Hydrodynamic Plane of Shear and the Zeta Potential

As described in the discussion of electrophoretic mobility in Chapter 2, the effective charge interacting with the external electric field is less than the intrinsic surface charge of the particle. A proportion of the counterions and their hydration shells remain attached to the particle, with the region close to surface of the Stern layer forming the hydrodynamic plane of shear (the slip plane) with the surrounding bulk fluid. As given by Equation (2.8) in Box 2.3, the electrophoretic mobility is defined as:

$$\mu_e = \frac{\varepsilon_o\varepsilon_m\varsigma}{\eta}$$

where η is the dynamic viscosity of the bulk fluid and ζ is the zeta potential. This potential is less than the potential V(o) at the surface of the particle and its approximate location is shown in Figure 12.21.

12.8 Electro-osmosis

Chapter 10 describes how fluid flow in insulator-based DEP devices is driven by electro-osmosis. This effect is the bulk motion of an aqueous electrolyte fluid next to a charged surface as a result of applying an electric field. If the field is applied along the axis of the channel, a volume coulombic force ($\rho.E$) is exerted on the fluid. This scheme is shown in Figure 12.23. The net charge density ρ in the fluid is significantly different from zero only in a thin annular region, within the Debye length region close to the channel wall, as shown in Figure 12.21. In the bulk fluid away from the channel walls the anions balance the cations in number, so that $\rho = 0$. Therefore, only the counterions in the fluid close to the wall experience a coulombic accelerating force and induce fluid movement, in a direction that depends on the field direction and polarity of the counterions. Due to the radial velocity gradient that is formed, the adjacent fluid annuli is accelerated by the momentum transfer caused by viscous forces until the velocity gradient approaches zero across the whole radius of the capillary. A charged fluid layer effectively 'drags' the adjacent fluid layer along, until finally the entire channel moves at a uniform velocity. The 'stationary surface / moving-plate' scheme outlined in Figure 12.2 has in effect been created. Numerical simulations by Dose and Guiochon [13] demonstrated that this process develops on a time scale between 100 μs and 1 ms. After that time, the whole fluid inside the channel moves at a constant velocity, with the resulting flow profile across the capillary being of a rectangular 'plug' shape as shown in Figure 12.23 [14]. This uniform velocity profile occurs if the channel's characteristic length (e.g., diameter) is at least seven times that of the electric double layer thickness (Debye length) and if other sources of fluid acceleration such as convection due to Joule heating are absent. For a 10 mM solution of NaCl, for example,

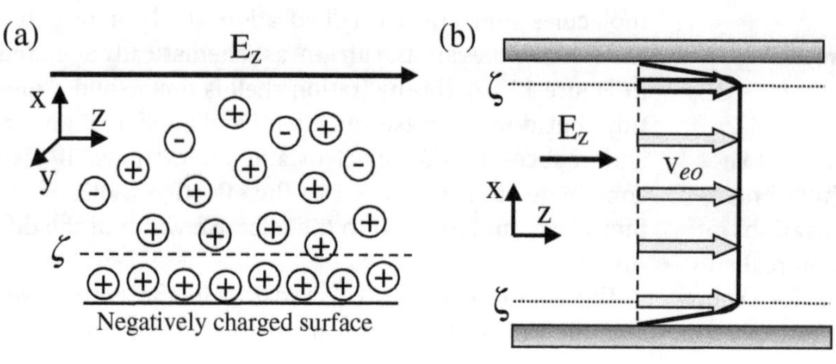

Figure 12.23 (a) An external electric field Ez is shown applied tangentially to the charged surface of a fluidic channel with its electrical double layer. (b) The resulting electroosmotic flow velocity profile across the channel.

this represents a minimum diameter of ~22 nm and is therefore not a practical limitation for the effectiveness of electroosmotic flow in insulator-based DEP devices. The pluglike velocity profile of electroosmotic induced fluid flow is very different from that of pressure-driven flow, which has the parabolic profile shown in Figure 12.8. As a special characteristic of electroosmotic pumped systems, fluid zones can be transported without significant hydrodynamic dispersion. This is of particular relevance in capillary electrophoresis and other aspects of microfluidic devices.

The Navier–Stokes Equation (12.18d) can be used to derive the relationship between the electroosmotic velocity $\mathbf{v}_{eo}$ and the applied electric field:

$$\rho_m \frac{D\mathrm{V}}{Dt} = -\nabla P + \eta \nabla^2 \mathrm{V} + \mathrm{F}$$

where V is the velocity vector and F is the coulombic force ($\rho \mathrm{E}_z$) per unit volume acting on a fluid element. We are dealing with a microfluidic channel, so that (as for laminar flow) the fluid flow is dominated by viscous effects and not inertial ones. We also do not have pressure driven flow. The inertial and pressure gradient terms can therefore be ignored, so that the Navier–Stokes equation reduces to the form:

$$\eta \nabla^2 \mathrm{V} = -\mathrm{F} \tag{12.39}$$

With the aid of Equation (12.31) the coulombic force is given by:

$$\mathrm{F} = \rho(x)\mathrm{E}_z = -\varepsilon_o \varepsilon_m \frac{d^2 V(x)}{dx^2} \mathrm{E}_z \tag{12.40}$$

where E_z is the applied uniform field acting along the channel and $V(x)$ is the potential across the double layer, normal to the charged surface. The potential does not vary along the y- and z-axes. The electro-osmotic fluid flow shown in Figure 12.23(b) is also essentially one-dimensional, along the z-axis. Taking these considerations into account, from Equations (12.39) and (12.40), the Navier–Stokes equation takes the form:

$$\eta \frac{d^2 \mathrm{v}_{eo}}{dx^2} = \varepsilon_o \varepsilon_m \frac{d^2 V(x)}{dx^2} \mathrm{E}_z$$

The channel width will greatly exceed the width of the fluid channel. We will also make the approximation that the slip plane and hence zeta potential ζ, is located at the channel surface. Integrating this equation twice, from $x = 0$ at the slip plane to $x = \infty$ in the bulk fluid, we obtain the relationship:

$$\mathrm{v}_z = -\frac{\varepsilon_o \varepsilon_m \varsigma \mathrm{E}_z}{\eta} \left[1 - \frac{V(x)}{\varsigma} \right]$$

The reference value for the potential, $V(x) = 0$, occurs in the bulk fluid at $x = \infty$. Therefore in the bulk fluid (as x tends to ∞) we obtain the following expression for the electroosmotic velocity:

$$\mathrm{v}_{\mathrm{eo}} = -\frac{\varepsilon_o \varepsilon_m \varsigma \mathrm{E}_z}{\eta} \tag{12.41}$$

We can define the electroosmotic mobility μ_{eo} as:

$$\mu_{\mathrm{eo}} = \frac{\mathrm{v}_{\mathrm{eo}}}{\mathrm{E}_z} = -\frac{\varepsilon_o \varepsilon_m \varsigma}{\eta} \tag{12.42}$$

This is the same expression as that obtained in Box 2.3 for electrophoretic mobility (not a surprising result, since we have used the same concepts and mathematical procedure). The velocity v_p of a charged particle located in a channel in which the fluid is driven by electro-osmosis will be determined by a combination of its own electrophoretic mobility μ_{ep} and the electroosmotic mobility μ_{eo}. The apparent electrokinetic mobility μ_{ek} of the particle will be v_p/E, so that:

$$\mu_{ek} = \mu_{ep} + \mu_{eo}$$

If the particle carries the same polarity of charge as the counterions in the electrical double layer of the channel wall, the particle will exhibit a mobility that is *greater* than its true electrophoretic mobility. The opposite will result if the particle carries the same polarity of charge as the co-ions. The charge densities (hence zeta potential) of the channel wall and the particle will be controlled by the ionization of surface groups, which can be altered by changing the pH of the fluid in the channel (e.g., see Example 2.1).

For electro-osmosis to occur, immobilized electric charges must be present on the inner surface of a channel wall in contact with the fluid. This surface charge can arise from ionizable groups that form part of the wall's chemical structure, or from the adsorption of charged species present in the fluid. A way to ensure that the walls are charged is to fabricate them from glass or fused silica, where deprotonation of Si-OH silanol groups occurs above pH 3 to form negatively charged silanoate (Si-O^-) groups. This surface charge induces the formation of an electric double layer by attracting cations from the bulk (electrolyte) fluid, as shown in Figures 12.21 and 12.23. Microfluidic devices made from glass tend to have a well characterized surface charge that varies predictably as a function of fluid pH and composition. Under physiological conditions, glass and silica have a negative zeta potential. With reduction of the local pH, the silanol groups on the glass surface become protonated and the zeta potential falls in magnitude. Many different surface modification techniques have been developed for glass, which allow the user to change the surface charge or to alter its biocompatibility (e.g., cell adhesion or non-fouling coatings). However, with an increasing use of

Table 12.5 Adjustable parameters that can be used to influence electro-osmosis in a fluidic channel, as well as the value of the electroosmotic mobility (μ_{eo}).

Parameter	Effect	Comments
Electric field	μ_{eo} changes proportionately	Joule heating may result if too high. Efficiency and resolution may decrease if lowered too much
Solution pH	μ_{eo} decreased at low pH and increased at high pH	Simple and practical. May change charge or structure of a solute such as a protein
Ionic strength of solution	If increased, the zeta potential and μ_{eo} decrease	High ionic strength can generate high current and Joule heating
Temperature	Changes fluid viscosity (2–3% per °C)	Can be controlled automatically
Surfactant	Adsorbs to channel wall via hydrophobic and/or ionic interactions	Anionic surfactants can increase μ_{eo}, whilst cationic ones can decrease or reverse μ_{eo}
Organic modifier	Usually decreases μ_{eo} by changing zeta potential and viscosity	Often requires experimentation to determine complex changes. Can significantly alter selectivity
Covalent coating	Bonding of chemicals to channel wall.	Can alter hydrophilicity and surface charge of wall May not be stable
Neutral hydrophilic polymer	Adsorbs to channel wall via hydrophobic interactions	Decreases μ_{eo} by shielding surface charge and increasing viscous drag

polymeric components (silicone, Mylar, Teflon) electro-osmotic behaviour is less predictable. Biological fluids, in particular, can lead to problems such as protein adsorption on polymeric surfaces.

Some of the advantages and disadvantages of employing electro-osmosis to drive fluid flow in microchannels and DEP devices can be summarized as follows:

Advantages:

- Uniform flow profile. This results in uniform retention times for all particles in a given section of a fluidic system, which can greatly simplify calculations and analysis. Because fluids move as a bolus, the leading and trailing edges of materials are minimized. This reduces the time and material required to change solutions in a device.
- No moving-part pumps are required.
- A simple fluidic interface. The interface between the source of pumping (electrodes) can be as simple as two wires placed into holes at the end of the channel or chamber. Unlike pressure-driven flow, a leak-tight interface between the source of the hydraulic force and the fluid being driven is not required.

Disadvantages:

- Strong dependence on the electrochemical properties of channel wall and fluid. If a device is expected to process a variety of fluids or a fluid of unknown pH and ionic strength, the electroosmotic mobility value will be unpredictable.
- Often requires high voltages (typically in the kV range). This requires isolation of the electrodes from the sample fluid to avoid the products of electrolysis (bubbles, acid or base production) from entering the sample fluid, whilst at the same time retaining electrical connectivity.
- Heat produced by the electric field may have to be dissipated.

A summary of the various ways to control electroosmotic induced fluid flow in microfluidic devices is given in Table 12.5.

Finally, the electroosmotic effect described so far here is a DC phenomenon. A reversal of the applied field reverses the direction of induced fluid flow in the channel. Application of an AC field should result in no net fluid flow along the channel. However, localized fluid motion can be generated when AC voltage signals are applied to microelectrode structures, such as the interdigitated design shown in Figure 10.27. Consider the situation shown in Figure 12.24, where negative counterions occupy the Stern layer and diffuse layer above the positively charged electrode and positive counterions are located above the negatively charged electrode. The tangential component E_T of the field E between the electrodes exerts a coulombic force on these counterions, causing them and the fluid to flow along the paths

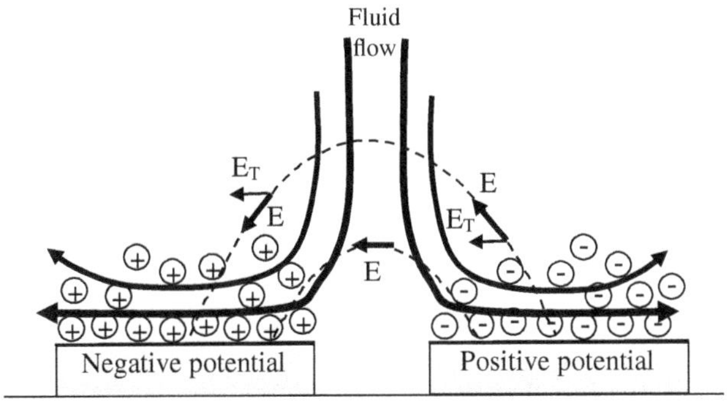

Figure 12.24 Electroosmotic fluid flow above a pair of interdigitated planar electrodes. The tangential component E_T of the field E between the electrodes exerts a coulombic force on the counterions, causing them and the fluid to flow along the paths indicated.

shown in Figure 12.24. This situation has been analysed experimentally and analytically by González *et al.* [15] to show that this electro-osmotically induced fluid flow has a strong dependency on the frequency of the applied voltage signal. The time-averaged fluid flow is small at low frequencies and above around 1 kHz, typically peaking at ~100 Hz [15]. This effect, coupled with electrode polarization effects, limits the ability to perform electrode-based DEP experiments below ~1 kHz.

12.9 Summary

In the design and performance analysis of a DEP device, consideration should be made of the nature of fluid flow through its microchannels and chambers. The term 'fluid' includes both liquids and gases. The fluid used in most bio-DEP devices is a liquid, usually an aqueous electrolyte. The characteristic length of its flow field, which for a channel of circular cross-section is its diameter, is normally greater than 1 μm. This greatly simplifies our task, because the Knudson number is very low (< 0.001) and we can use the continuum model when dealing with the fluid. Its physical properties, such as density, pressure, viscosity and velocity can be defined as continuous functions of time and space. The maximum value of the flow field's characteristic length is also usually below 1 mm. Unless the fluidic system can sustain a very high internal pressure, the volumetric flow velocity will be well below that required to achieve turbulent (chaotic) flow. We can usually assume that we have conditions of laminar flow in a DEP device, characterized by a low Reynolds number (Re). In such flow we can assume that there is no fluid slip at the wall of channel. The fluid velocity profile across a channel will be of the parabolic form shown in Figure 12.8. The transition from laminar to turbulent flow for water occurs at Re ≈ 2000. So that the DEP force acting on a particle can compete with viscous fluid friction, the fluid velocity should not be too high. This leads to a Reynolds number below 10 – well into the laminar flow regime.

The continuum model also allows the behaviour of the fluid in a DEP device to be treated using the three conservation laws of mass, energy and momentum, where properties such as density, pressure, viscosity and flow velocity can be considered in terms of flux rather than changes of their instantaneous values. In mathematical terms this is represented as partial derivatives of the dependent variables. The most important of these is the Navier–Stokes equation, describing the conservation of fluid momentum, given by Equation (12.18d). This equation was used to derive the expressions for the electroosmotic velocity and mobility given by Equations (12.41) and (12.42). The velocity profile for electro-osmosis driven fluid flow in a channel is shown in Figure 12.23. This profile takes the form of a rectangular 'plug', rather than the parabolic profile for pressure driven laminar flow.

To understand electrode polarization and electro-osmosis requires an appreciation of the origin and properties of ionic (electrical) double layers. An electrical double layer is formed by the electrostatic attraction of counterions to a charged surface. Some of these counterions are strongly bound to the surface in what is called the Stern layer, whilst the others form a diffuse layer, which merges into the electrically neutral bulk electrolyte. This in effect resembles the two charged plates of a very thin capacitor. An important parameter defining the width of the electrical double layer is the Debye screening length, which defines the distance away from the charged surface where the potential decays to 1/e of its value at the charged surface. For distances beyond the Debye length the ions in the bulk fluid are effectively screened from the strong field produced by the charges on the channel surface. From Equation (12.30) this screening distance can be calculated as 3.1 nm for the case of an aqueous 10 mM solution of NaCl at 298 K. This distance varies as the inverse of the square root of the solution's ionic strength, so that for a 1 M solution it decreases to 0.31 nm.

Electrode-based DEP devices are susceptible to electrode polarization effects at frequencies below ~1 kHz. At low frequencies, localized fluid motion can also be generated near electrode surfaces by what we can term as AC electro-osmosis, the basics of which are shown in Figure 12.24. This situation offers niche applications for insulator-based DEP devices, especially for frequencies below ~1 kHz where electrode-based DEP devices are compromised by effects associated with the formation of electrical double layers at the electrodes.

12.10 References

1 Travis, K. P., Todd, B. D. and Evans, D. J. (1997) Departure from Navier–Stokes hydrodynamics in confined liquids. *Phys. Rev. E* **55**, 4288–4295.

2 Pfahler, J., Harley, J. and Bau, B. (1989) Liquid transport in micron and submicron channels. *Sensor. Actuator.* **22**, 431–434.

3 Mala Gh. M. and Li, D. (1999) Flow characteristics of water in microtubes. *Int. J. Heat Fluid Flow* **20**, 142–148.

4 Wu, P. and Little, W. A. (1983) Measurement of friction factors for the flow of gases in very fine channels used for micro-miniature Joule–Thompson refrigerators. *Cryogenics* **23**, 272–277.

5 Tritton, D. J. (1977) *Physical Fluid Dynamics*, Van Nostrand Reinhold, New York, NY, p. 12.

6 Baygents, J. C. and Baldessari, F. (1998) Electrohydrodynamic instability in a thin fluid layer with an electrical conductivity gradient. *Phys. Fluid.* **10**, 301–311.

7 Ottino, J. M. (1990) Mixing, chaotic advection and turbulence. *Ann. Rev. Fluid Mechanics* **22**, 207–253.

8 Wang, G. R., Yang, F. and Zhao, W. (2014) There can be turbulence in microfluidics at low Reynolds number. *Lab Chip* **14**, 1452–1458.

9 Isla, M. A. (2004) Einstein–Smoluchowski diffusion equation: a discussion. *Physica Scripta* **70**, 120–125.

10 Grahame, D. C. (1947) The electrical double layer and the theory of electrocapillarity. *Chem. Rev.* **41**, 441–501.

11 Aveyard, R. and Haydon, D. A. (1973) *An Introduction to the Principles of Surface Chemistry*, Cambridge University Press, Cambridge, p. 231.

12 Bedzyk, M. J., Bommarito, G. M., Caffrey, M. and Penner, T. L. (1990) Diffuse-double layer at a membrane-aqueous interface measured with X-ray standing waves. *Science* **248** (4951), 52–56.

13 Dose, E. V. and Guiochon, G. (1993) Timescales of transient processes in capillary electrophoresis. *J. Chromatog. A* **652**(1), 263–275.

14 Rice, C. L. and Whitehead, R. (1993) Electrokinetic flow in a narrow cylindrical capillary. *J. Phys. Chem.* **69**(11), 4017–4024.

15 González, A., Ramos, A., Green, N. G. *et al.* (2000) Fluid flow induced by non-uniform fields in electrolytes on microelectrodes. II. A linear double-layer analysis. *Phys. Rev. E.* **61**(4), 4019–4028.

Appendices

Appendix A: Values of Fundamental Physical Constants

Quantity	Symbol	Value
Elementary charge	e	$1.60217(6) \times 10^{-19}$ C
Electric constant	ε_o	$8.85418(8) \times 10^{-12}$ F m^{-1}
Magnetic constant	μ_o	$4\pi \times 10^{-7}$ N A^{-2}
Speed of light in vacuum	c	$2.99792(4) \times 10^{8}$ m s^{-1}
Planck constant	h	$6.62606(9) \times 10^{-34}$ J s
Rest mass of electron	m_e	$9.10938(2) \times 10^{-31}$ kg
Rest mass of proton	m_p	$1.67262(2) \times 10^{-27}$ kg
Boltzmann constant	k	$1.60217(6) \times 10^{-23}$ J K^{-1}
Avogadro constant	N_A	$6.02214(2) \times 10^{23}$ mol^{-1}
Faraday constant	F	$9.64853(4) \times 10^{4}$ C mol^{-1}
Molar gas constant	R	$8.31447(2)$ J K^{-1} mol^{-1}

Appendix B: SI Prefixes

Prefix	Symbol	Factor	Prefix	Symbol	Factor
yotta	Y	10^{24}	deci	d	10^{-1}
zetta	Z	10^{21}	centi	c	10^{-2}
exa	E	10^{18}	milli	m	10^{-3}
peta	P	10^{15}	micro	μ	10^{-6}
tera	T	10^{12}	nano	n	10^{-9}
giga	G	10^{9}	pico	p	10^{-12}
mega	M	10^{6}	femto	f	10^{-15}
kilo	k	10^{3}	atto	a	10^{-18}
hecto	h	10^{2}	zepto	z	10^{-21}
deca	da	10^{1}	yacto	y	10^{-24}

Appendix C: The Base Quantities in the SI System of Units

Quantity	SI name	SI symbol
Length (L)	metre	m
Time (t)	second	s
Mass (M)	kilogram	kg
Temperature (T)	kelvin	K
Electric current (I)	ampere	A
Amount of substance	mole	mol
Luminous intensity	candela	cd

Appendix D: Derived Physical Quantities, their Defining Equation or Law and Dimensions

Quantity	Defining equation/ law	Dimension	Dimension (SI units)	Name
Area	$A = \int dxdy$	L^2	m^2	
Volume	$V = \int dxdydz$	L^3	m^3	
Velocity	$v = dx/dt$	Lt^{-1}	ms^{-1}	
Acceleration	$a = d^2x/dt^2$	Lt^{-2}	ms^{-2}	
Mass Density	$\rho = M/V$	ML^{-3}	kg m^{-3}	
Concentration	mole/V	mol L^{-3}	mol m^{-3}	
Force	$F = Ma$	MLt^{-2}	kg m s^{-2}	newton (N)
Stress/Pressure	$p = F/A$	$ML^{-1} t^{-2}$	kg m^{-1} s^{-2} (N m^{-2})	pascal (Pa)
Dynamic Viscosity	$\eta = p/(dv/dy)$	$ML^{-1} t^{-1}$	kg m^{-1} s^{-1}	poiseuille
Work/Energy	$W = \int Fdx$	$ML^2 t^{-2}$	kg m^2 s^{-2} (N m)	joule (J)
Surface Tension	$T = W/A$	Mt^{-2}	kg s^{-2} (N m^{-1})	
Power	$P = dW/dt$	$ML^2 t^{-3}$	kg m^2 s^{-3} (J s^{-1})	watt (W)
Frequency	$f = 1/t$	t^{-1}	s^{-1}	hertz (Hz)
Charge	$Q = \int I\,dt$	It	A s	coulomb (C)
Electromotive Force (Voltage)	$E = P/I$	$ML^2 t^{-3} I^{-1}$	kg m^2 s^{-3} A^{-1}	volt (V)
Capacitance	Q/E	$M^{-1}L^{-2}t^4 I^2$	kg^{-1} m^{-2} s^4 A^2	farad (F)
Resistance	E/I	$ML^2t^{-3} I^{-2}$	kg m^2 s^{-3} A^{-2}	ohm (Ω)
Conductance	I/E	$M^{-1}L^{-2}t^3 I^2$	kg^{-1} m^{-2} s^3 A^2	siemens (S)

Dielectrophoresis: Theory, Methodology and Biological Applications, First Edition. Ronald Pethig.

Appendix E: Diffusion Coefficients for Molecules and Ions in Water at 298 K

Molecule	D (10^{-9} m^2 s^{-1})
Water	2.26
Sucrose	0.52
Methanol (CH_3OH)	1.6
Glycine	1.06
NaCl	1.7
H^+	9.3
OH^-	5.3
Na^+	1.33
K^+	1.96
Cl^-	2.03

Appendix F: Diffusion Coefficients for Bio-Particles in Water at 293 K

Macromolecule	D (m^2 s^{-1})
Ribonuclease	1.2×10^{-10}
Lysozyme	1.0×10^{-10}
Serum albumin	5.9×10^{-11}
Haemoglobin	6.9×10^{-11}
Urease	3.5×10^{-11}
Collagen	6.9×10^{-12}
Viruses, bacteria, cells	$10^{-13} \sim 10^{-16}$

Source: Derived using the Stokes–Einstein relation: $D = kT/(6\pi\eta a)$ where 'a' is the hydrodynamic radius of a spherical particle.

Appendix G: Viscosity and Surface Tension Values for Liquids at 293 K

Liquid	η (Pa s)	T_s (N m^{-1})
Water	1.002×10^{-3}	7.275×10^{-2}
Blood (37 °C)	$3 \sim 4 \times 10^{-3}$	5.5×10^{-2}
Ethanol	1.074×10^{-3}	2.21×10^{-2}
Methanol	5.94×10^{-4}	2.27×10^{-2}
Mercury	1.55×10^{-3}	47.2×10^{-2}
Benzene	6.04×10^{-4}	2.89×10^{-2}
Chloroform	6.96×10^{-4}	2.75×10^{-2}
Glycerol (100%)	1.41	6.4×10^{-2}

Appendix H: Activity Coefficients for Common Compounds that Dissociate into Ions in Solution

Substance	0.01 M	0.05 M	0.1 M	0.5 M	1 M
KCl	0.901	0.816	0.768	0.649	0.604
NaCl	0.903	0.822	0.779	0.681	0.657
$MgCl_2$	0.734	0.590	0.535	0.485	0.577
$CaCl_2$	0.727	0.577	0.528	0.444	0.495
HCl	0.905	0.832	0.797	0.759	0.811
H_2SO_4	0.542	0.325	0.251	0.146	0.125

Source: Derived from the *CRC Handbook of Chemistry and Physics*, 87th edn. CRC Press, Boca Raton, FL, 2006–2007.

Appendix I: Electrical Mobility of Ions at 25 °C in Dilute Aqueous Solution

Cation	Mobility (10^{-8} m^2/V.s)	Anion	Mobility (10^{-8} m^2/V.s)
H^+, H_3O^+	36.2	OH^-	20.6
K^+	7.6	Cl^-	7.9
Na^+	5.2	F^-	5.7

Source: Atkins, P. W. and De Paula, J. (2002) *Physical Chemistry*, 7th edn. W. H. Freeman, Ch. 27.)

Appendix J: Buffering Systems and their pH Buffering Range

Buffering system	Useful buffering range (pH)
Hydrochloric acid and potassium chloride	1.0–2.2
Hydrochloric acid and glycine	2.2–3.6
Citric acid and sodium citrate	3.0–6.2
Acetic acid and sodium acetate	3.7–5.6
Sodium hydroxide and potassium dihydrogen phosphate	5.8–8.0
Sodium tetraborate and hydrochloric acid	8.1–9.2
Sodium hydroxide and glycine	8.6–10.6
Sodium hydroxide and sodium bicarbonate	9.6–11.0
Sodium hydroxide and sodium hydrogen phosphate	11.0–11.9
Sodium hydroxide and potassium chloride	12.0–13.0

Appendix K: Composition of 1 μL of Human Blood

(Equivalent to a volume of ~1/40th of a droplet of blood.)

Red blood cells (Erythrocytes)			5 ~ 6 million
White blood cells (Leukocytes)			~7000
Comprising:	4400 granulocytes		
	400 monocytes		
	2200 lymphocytes		
	Comprising:	1500 T-cells	
		400 NK-cells	
		300 B-cells	
Platelets			~250 000

Appendix L: Blood Cells, Platelets and Some Pathogenic Bioparticles

L.1 Blood Fractionation

When a sample of whole blood is centrifuged, the cells sediment to the bottom of the tube to leave ~55% of the sample at the top in the form of a slightly alkaline (pH 7.4) and pale yellow fluid, known as the blood plasma fraction. This comprises 90% water and 10% solid matter (nine parts organic and one part mineral). The organics include amino acids, glucose, hormones, lipids, proteins and vitamins. The minerals take the form of ions such as Na^+, K^+, Ca^{2+}, Mg^{2+} and Cl^-.

About 55% of the sample appears as a deep red fluid at the bottom half of the tube and contains the erythrocytes (red blood cells). A thin band of pale fluid, comprising less than 1% of the sample, appears above the erythrocyte fraction and below the plasma. This is known as the buffy coat and consists of the leukocytes (white blood cells) mixed with platelets.

L.1.1 Erythrocytes

These cells are rich in haemoglobin (~250 million per cell), which is a protein able to bind oxygen and thus responsible for providing oxygen to tissues. Where there is a high concentration of oxygen in the body, such as in the alveoli of the lungs, each haemoglobin molecule binds four oxygen molecules to form oxyhaemoglobin. When an erythrocyte reaches tissue with low oxygen concentration the haemoglobin releases these oxygens. Erythrocytes are also partly responsible for recovering carbon dioxide produced as waste, but most CO_2 is carried by plasma in the form of soluble carbonates. The mean lifetime of erythrocytes is about 120 days, at which time they are retained by the spleen and then phagocyted (eaten) by macrophages.

In man and in all mammals, erythrocytes are devoid of a nucleus and have the shape of a biconcave lens, which allows more room for haemoglobin and raises the cell surface and cytoplasmic volume ratio. These characteristics maximize the efficiency of oxygen diffusion by these cells. In fishes, amphibians, reptilians and birds, erythrocytes do have a nucleus.

L.1.2 Leukocytes

These cells are responsible for the defence of the organism, and are of two types, namely *granulocytes* and *lymphoid cells*. The ratio of white to red blood cells in normal human blood is ~ 1 : 700.

Granulocytes contain granules in their cytoplasm, which have different properties including a different affinity towards neutral, acid or basic stains. Granulocytes can thus be distinguished as neutrophils, eosinophil (or acidophils) and basophils;

- *Neutrophils*: act to phagocyte bacteria and are present in large numbers in the pus of wounds. They are unable to renew the lysosomes used in digesting the bacteria and die after having phagocyted a few of them.
- *Eosinophils*: attack parasites and phagocyte antigen-antibody complexes.
- *Basophils*: possess a phagocytory capability, but also secrete anticoagulant and vasodilatory substances such as histamines and serotonin.

Lymphoid cells consist of two types, namely lymphocytes and monocytes:

- Lymphocytes are cells, which besides being present in blood populate the lymphoid tissues and organs (e.g., thymus, bone marrow, spleen). They are slightly larger than erythrocytes, and have a nucleus that occupies nearly all of the internal cellular volume. They are also the main constituents of the immune system, which is the defence against the attack of pathogenic microorganisms such as viruses, bacteria, fungi and protista (e.g., unicellular organisms). Lymphocytes produce *antibodies*, which appear on their outer plasma membrane. An antibody is a molecule able to 'recognize' and bind itself to molecules called *antigens*. As for all proteins, these antibodies are coded by genes. Based on a recombination mechanism of some of these genes, every lymphocyte produces antibodies of a specific molecular shape. The number of lymphocytes circulating in the blood is so large that they are able to recognize practically all the chemicals existing in the organism, both its own natural and foreign ones. They recognize hundreds of millions of different molecules!

The cells of the immune system, chiefly lymphocytes, cooperate amongst themselves to activate, boost or make more precise the immune response. To attain this scope, there exist different types of lymphocytes, with different functions, namely B and T lymphocytes. When the B cells are activated, they quickly multiply and secrete hosts of antibodies, which on meeting microorganisms with complementary shape (*epitopes*) bind to them and form complexes to immobilize the microorganisms. Other cells, which are not specific but able to recognize antibodies, phagocyte these complexes. In their turn, the T cells are divided into three categories: Tc (cytotoxic) cells kill infected cells directly by inducing them to undergo apoptosis (programmed cell death); Th (helpers) assist in activating B cells to make antibody responses; Ts (suppressors) suppress the activity of other T cells and are crucial for self-tolerance. The immune system also produces *memory cells*, which are deactivated lymphocytes ready to be reactivated on further encounters with the same antigen.

Another population of lymphocytes in the peripheral blood and lymphoid organs does not have receptors for antigens. These lymphocytes have a nonspecific defence function that is not activated by Th lymphocytes. These cells represent the more ancient component of the immune system and they are characterized by their cytotoxic activity. They are called *Natural Killer (NK)* cells. Apart from killing viruses, bacteria, infected and neoplastic (abnormal) cells, these lymphocytes also regulate the production of other haematic cells such as erythrocytes and granulocytes.

- Monocytes are the precursors of *macrophages*. They are larger blood cells, which after attaining maturity in the bone marrow, enter the blood circulation where they stay for 24–36 hours. Then they migrate into the connective tissue, where they become macrophages and move within the tissues. In the presence of an inflammation site, monocytes quickly migrate from blood vessels and start an intense phagocytory activity. Macrophages also cooperate in the immune defence by exposing molecules of digested bodies on their membrane, presenting them to more specialized cells such as B and T lymphocytes.

L.1.3 Platelets (Thrombocytes)

The main function of platelets is to prevent the loss of blood in injured tissues, by aggregating and releasing chemicals to promote blood coagulation. Released substances include serotonin, which reduces the diameter of damaged blood vessels, and fibrin to trap cells and form a clot. They have a diameter of 2∼3 microns and in mammalian blood are not considered to be real cells. They do not have a nucleus and are derived from the cytoplasm of cells (megakaryocytes) in the bone marrow. In birds and amphibians, platelets circulate in the blood as mononuclear cells.

L.2 Bacteria

Bacteria are prokaryotic cells and are about one-tenth the size of eukaryotic cells, typically 0.5∼5.0 μm in length. They display a wide range of morphologies. Most are either spherical (cocci), rod-shaped (bacilli) or spiral shaped (spirilla). Many bacterial species exist simply as single cells, but others associate in characteristic patterns. For example, *Streptococcus* form chains and *Staphylococcus* group together into clusters. Only a small number of bacterial species cause disease in humans. Some of those that do so can only replicate inside the body and are called *obligate pathogens*. Other bacteria, called *facultative pathogens*, replicate in environments such as water or soil and only cause disease on encountering a susceptible host. *Opportunistic pathogens* are bacteria that are normally harmless but have a latent ability to cause disease in an injured or immunocompromised host. Basic details of their structure, division into Gram-positive and Gram-negative types, and pathogenecity are given in Chapter 9, section 9.6.

Bacteria are often grown in solid media, such as agar plates, to isolate and identify pure cultures of a bacterial strain. However, liquid growth media are used when measurement of growth or large volumes of cells are required. Growth in stirred liquid media occurs as an even cell suspension, making the cultures easy to divide and transfer, although isolating single bacteria from liquid media is difficult. The use of selective media (media with specific nutrients or antibiotics added) can help identify specific organisms. Bacterial growth follows three phases. When bacteria first enter a high-nutrient environment that allows growth, the cells need to adapt to their new environment. The first phase of growth is the *lag phase*, a period of slow growth when the cells are adapting to the high-nutrient environment and preparing for fast growth. The lag phase has high biosynthesis rates, as proteins necessary for rapid growth are produced. The second phase of growth is the logarithmic phase, marked by rapid exponential growth. The rate at which cells grow during this phase is known as the *growth rate*, and the time it takes the cells to double is known as the *doubling time* or *generation time*. During log phase, nutrients are metabolised at maximum speed until one of the nutrients is depleted and starts limiting growth. The final phases are the *stationary phase*, followed by the *death phase*, caused by depleted nutrients. The cells reduce their metabolic activity, consume nonessential cellular proteins, and then die.

A practical way to monitor bacterial growth and to determine the doubling time is to periodically measure the optical absorbance A of a sample of a bacterial suspension. The size of a typical bacterium is such that it will scatter light of wavelengths around 450~600 nm. During the *log phase* of growth the number of bacteria n with time t, commencing with an initial number n_0, can be written as:

$$n(t) = n_0 \exp(kt) = n_0 2^{t/T}$$

where k is the growth constant and T is the time it takes for the number of bacteria to double. During the log growth phase, a plot of $\log_2(A)$ against time should be a straight line, the slope of which gives the time for the culture to double in number density. The doubling times for bacteria such as *M. luteus* and *E. coli* are often given in texts as 30 and 20 minutes, respectively. However, these values are valid only for optimum conditions of temperature, nutrient concentration and cell density and when no growth suppressing substances are present. Many bacteria produce such substances if their cell density becomes too high. Examples of other typical doubling times are 2 hours and 24 hours, exhibited by *B. subtilis* and *M. tuberculosis*, respectively.

L.3 Fungal and Protozoal Cells

Fungi include both unicellular yeast cells, such as *Saccharomyces cerevisiae*, and filamentous, multicellular moulds, such as those found on mouldy fruit or bread. Most pathogenic fungi exhibit *dimorphism*, which is the ability to grow in either yeast or mould form. The yeast-to-mould or mould-to-yeast transition is frequently associated with infection. For example, some fungi grow as a mould at low temperatures in soil but then change to a harmful yeast form when inhaled into the lungs.

Protozoan parasites exist as single cells and frequently require more than one host in a complex life cycle. The most common protozoal disease is malaria, transmitted to humans by the bite of the female of any of 60 species of *Anopheles* mosquito. The most intensively studied of the malaria-causing parasites, *Plasmodium falciparum*, exists in eight distinct forms, and requires both human and mosquito hosts to complete its life cycle. Because fungi and protozoan parasites are eukaryotes, their pathogenic varieties are difficult to kill with drugs without harming the host. The tendency of fungal and parasitic infecting organisms to switch among several different forms during their life cycles also makes them more difficult to treat. A drug that is effective at killing one form is often ineffective at killing another form, which therefore survives the treatment. As a result, antifungal and antiparasitic drugs are often less effective and more toxic than antibiotics.

L.4 Viruses

Viruses are not cells. They are bioparticles that can vary from simple helical or icosahedral shapes, to more complex structures. They can reproduce only inside a host cell. They are about 1/100th the size of bacteria, with diameters ranging from around 10 to 300 nm. Unlike cells and bacteria which can be viewed using a conventional light microscope, most viruses can therefore only be seen using scanning and transmission electron microscopes. A complete virus particle, known as a virion, consists of nucleic acid (DNA or RNA) surrounded by a protective coat of protein called a capsid, made from proteins encoded by the viral genome. The capsid shape serves as the basis for morphological distinction (helical, icosahedral, envelope, complex).

Some species of virus surround themselves with a modified form of one of the host cell membranes, either the outer membrane of the infected host cell, or internal membranes such as nuclear membrane or endoplasmic reticulum. The virus thus gains an outer lipid bilayer, known as a viral envelope. This membrane is studded with proteins coded for by the viral genome and host genome; the lipid membrane itself and any carbohydrates present originate entirely from the host cell. The influenza virus and HIV use this strategy. Most membrane enveloped viruses are dependent on the envelope for their infectivity. The *complex* viruses possess a capsid that is neither purely helical, nor purely icosahedral, and may possess extra structures such as protein tails or a complex outer wall. For example, the T4 bacteriophage has a complex structure consisting of an icosahedral head bound to a helical tail with protruding protein tail fibres. This tail structure acts like a molecular syringe, attaching to the bacterial host and then injecting the viral genome into the cell.

L.5 Prions

Prions are infectious agents in the form of misfolded proteins that replicate and propagate in the host cell. They cause various neurodegenerative diseases in mammals, a well known example being *bovine spongiform encephalopathy* (BSE), otherwise known as mad cow disease. This fatal disease (Creutzfeld–Jacob disease) can be transmitted to humans who eat infected beef, and can also be transmitted from human to human via blood transfusions. The brain tissue develops holes and takes on a spongelike appearance. The DNA code for making prion protein is in a gene that all mammals possess and is mainly active in nerve cells.

The prion has the identical amino acid sequence to the normal form of the protein. The only difference between them is in their folded three-dimensional structure. The misfolded protein can cause a normal folded form to unfold and to aggregate with other prions to produce regular helical structures called *amyloid* fibres. The prion is thus able to cause the normal protein form to adopt its misfolded prion conformation, causing it to become infectious. This is equivalent to prions being able to replicate themselves in the host cell. If an amyloid fibre is broken into smaller pieces, each piece can initiate the prion polymerization process in a new cell. The prion can therefore propagate as well as replicate. Furthermore, if consumed by another host organism, the newly formed misfolded prions may transmit the infection to that organism. How polypeptide chains explore their conformation-energy space and fall into a global free energy minimum state corresponding to their correct three-dimensional folded form is poorly understood. The linear order of the peptides is reliably given by the pertinent gene's DNA sequence but, referring to Figure 8.14 (p 184), we can appreciate that the number of possible polypeptide folding possibilities must be enormous (as overheard by the author, Sydney Brenner once remarked that there are possibly as many folding possibilities as there are proteins!). The chances of misfolded protein production must be high, so what special circumstances are required to generate the relatively uncommon prion disease? Finding the answers to such questions is currently the objective of active research.

Author Index

The chapter (bold number) and reference number is given.

Dielectrophoresis: Theory, Methodology and Biological Applications, First Edition. Ronald Pethig.

Subject Index

www.ingramcontent.com/pod-product-compliance
Lightning Source LLC
LaVergne TN
LVHW062138200926
848484LV00007B/1830